THE HANDBOOK OF
AD HOC WIRELESS NETWORKS

The Electrical Engineering Handbook Series

Series Editor
Richard C. Dorf
University of California, Davis

Titles Included in the Series

The Handbook of Ad Hoc Wireless Networks, Mohammad Ilyas
The Avionics Handbook, Cary R. Spitzer
The Biomedical Engineering Handbook, 2nd Edition, Joseph D. Bronzino
The Circuits and Filters Handbook, Second Edition, Wai-Kai Chen
The Communications Handbook, 2nd Edition, Jerry Gibson
The Computer Engineering Handbook, Vojin G. Oklobdzija
The Control Handbook, William S. Levine
The Digital Signal Processing Handbook, Vijay K. Madisetti & Douglas Williams
The Electrical Engineering Handbook, 2nd Edition, Richard C. Dorf
The Electric Power Engineering Handbook, Leo L. Grigsby
The Electronics Handbook, Jerry C. Whitaker
The Engineering Handbook, Richard C. Dorf
The Handbook of Formulas and Tables for Signal Processing, Alexander D. Poularikas
The Handbook of Nanoscience, Engineering, and Technology, William A. Goddard, III,
   Donald W. Brenner, Sergey E. Lyshevski, and Gerald J. Iafrate
The Industrial Electronics Handbook, J. David Irwin
The Measurement, Instrumentation, and Sensors Handbook, John G. Webster
The Mechanical Systems Design Handbook, Osita D.I. Nwokah and Yidirim Hurmuzlu
The Mechatronics Handbook, Robert H. Bishop
The Mobile Communications Handbook, 2nd Edition, Jerry D. Gibson
The Ocean Engineering Handbook, Ferial El-Hawary
The RF and Microwave Handbook, Mike Golio
The Technology Management Handbook, Richard C. Dorf
The Transforms and Applications Handbook, 2nd Edition, Alexander D. Poularikas
The VLSI Handbook, Wai-Kai Chen

Forthcoming Titles

The CRC Handbook of Engineering Tables, Richard C. Dorf
The Engineering Handbook, Second Edition, Richard C. Dorf
The Handbook of Optical Communication Networks, Mohammad Ilyas and
   Hussein T. Mouftah

THE HANDBOOK OF
AD HOC WIRELESS NETWORKS

Edited by
Mohammad Ilyas
Florida Atlantic University
Boca Raton, Florida

CRC PRESS

Boca Raton   London   New York   Washington, D.C.

Library of Congress Cataloging-in-Publication Data

The handbook of ad hoc wireless networks / edited by Mohammad Ilyas.
    p. cm. -- (The electrical engineering handbook series)
    Includes bibliographical references and index.
    ISBN 0-8493-1332-5 (alk. paper)
    1. Wireless LANs. I. Ilyas, Mohammad, 1953- II. Series

TK5105.78 .H36 2002
621.382--dc21
    2002031316

Visit the CRC Press Web site at www.crcpress.com

© 2003 by CRC Press LLC

No claim to original U.S. Government works
International Standard Book Number 0-8493-1332-5
Library of Congress Card Number 2002031316
Printed in the United States of America 1 2 3 4 5 6 7 8 9 0
Printed on acid-free paper

Preface

To meet the need for fast and reliable information exchange, communication networks have become an integral part of our society. The success of any corporation largely depends upon its ability to communicate. Ad hoc wireless networks will enhance communication capability significantly by providing connectivity from anywhere at any time. This handbook deals with wireless communication networks that are mobile and do not need any infrastructure. Users can establish an ad hoc wireless network on a temporary basis. When the need disappears, so will the network.

As the field of communications networks continues to evolve, a need for wireless connectivity and mobile communication is rapidly emerging. In general, wireless communication networks provide wireless (and hence) mobile access to an existing communication network with a well-defined infrastructure. Ad hoc wireless networks provide mobile communication capability to satisfy a need of a temporary nature and without the existence of any well-defined infrastructure. In ad hoc wireless networks, communication devices establish a network on demand for a specific duration of time. Such networks have many potential applications including the following:

- Disaster recovery situations
- Defense applications (army, navy, air force)
- Healthcare
- Academic institutions
- Corporate conventions/meetings

This handbook has been prepared to fill the need for comprehensive reference material on ad hoc wireless networks. The material presented in this handbook is intended for professionals who are designers and/or planners for emerging telecommunication networks, researchers (faculty members and graduate students), and those who would like to learn about this field.

The handbook is expected to serve as a source of comprehensive reference material on ad hoc wireless networks. It is organized in the following nine parts:

- Introduction
- Wireless transmission techniques
- Wireless communication systems and protocols
- Routing techniques in ad hoc wireless networks — part I
- Routing techniques in ad hoc wireless networks — part II
- Applications of ad hoc wireless networks
- Power management in ad hoc wireless networks
- Connection and traffic management in ad hoc wireless networks
- Security and privacy aspects of ad hoc wireless networks

The handbook has the following specific salient features:

- It serves as a single comprehensive source of information and as reference material on ad hoc wireless networks.
- It deals with an important and timely topic of emerging communication technology of tomorrow.
- It presents accurate, up-to-date information on a broad range of topics related to ad hoc wireless networks.
- It presents material authored by experts in the field.
- It presents the information in an organized and well-structured manner.

Although the handbook is not precisely a textbook, it can certainly be used as a textbook for graduate courses and research-oriented courses that deal with ad hoc wireless networks. Any comments from readers will be highly appreciated.

Many people have contributed to this handbook in their unique ways. The first and foremost group that deserves immense gratitude is the group of highly talented and skilled researchers who have contributed 32 chapters. All of them have been extremely cooperative and professional. It has also been a pleasure to work with Nora Konopka, Helena Redshaw, and Susan Fox of CRC Press, and I am extremely grateful for their support and professionalism. My wife Parveen and my four children Safia, Omar, Zakia, and Maha have extended their unconditional love and strong support throughout this project, and they all deserve very special thanks.

Mohammad Ilyas
Boca Raton, Florida

The Editor

Mohammad Ilyas is a professor of computer science and engineering at Florida Atlantic University, Boca Raton, Florida. He received his B.Sc. degree in electrical engineering from the University of Engineering and Technology, Lahore, Pakistan, in 1976. In 1978, he was awarded a scholarship for his graduate studies, and he completed his M.S. degree in electrical and electronic engineering in June 1980 at Shiraz University, Shiraz, Iran. In September 1980, he joined the doctoral program at Queen's University in Kingston, Ontario. He completed his Ph.D. degree in 1983. His doctoral research was about switching and flow control techniques in computer communication networks. Since September 1983, he has been with the College of Engineering at Florida Atlantic University. From 1994 to 2000, he was chair of the Department of Computer Science and Engineering. During the 1993–94 academic year, he spent a sabbatical leave with the Department of Computer Engineering, King Saud University, Riyadh, Saudi Arabia.

Dr. Ilyas has conducted successful research in various areas including traffic management and congestion control in broadband/high-speed communication networks, traffic characterization, wireless communication networks, performance modeling, and simulation. He has published one book and more than 120 research articles. He has supervised several Ph.D. dissertations and M.S. theses to completion. He has been a consultant to several national and international organizations. Dr. Ilyas is an active participant in several IEEE technical committees and activities and is a senior member of IEEE.

List of Contributors

George N. Aggélou
Institute of Technology
Athens, Greece

Roberto Baldoni
Universita' di Roma, "La Sapienza"
Roma, Italy

Roberto Beraldi
Universita' di Roma, "La Sapienza"
Roma, Italy

Ezio Biglieri
Politecnico di Torino
Torino, Italy

Satyabrata Chakrabarti
Sylvaine Algorithmics
Aurora, Illinois

Chaou-Tang Chang
National Chiao Tung University
Hsinchu, Taiwan

Chih Min Chao
National Central University
Chung-Li, Taiwan

Xiao Chen
Southwest Texas State University
San Marcos, Texas

Chua Kee Chaing
National University of Singapore
Singapore, Singapore

Marco Conti
Consiglio Nazionale delle Ricerche
Pisa, Italy

José Ferreira de Rezende
Federal University of Rio de Janeiro
Rio de Janeiro, Brazil

Nelson Fonseca
State University of Campinas
Campinas, Brazil

Holger Füßler
University of Mannheim
Mannheim, Germany

Silvia Giordano
LCA-IC-EPFL
Lausanne, Switzerland

Zygmunt J. Haas
Cornell University
Ithaca, New York

Hannes Hartenstein
NEC Europe Ltd.
Heidelberg, Germany

Xiao Hannan
National University of Singapore
Singapore, Singapore

Hossam S. Hassanein
Queen's University
Kingston, Ontario, Canada

Chih-Shun Hsu
National Central University
Chung-Li, Taiwan

Cheng-Ta Hu
National Central University
Chung-Li, Taiwan

Pei-Kai Hung
National Central University
Chung-Li, Taiwan

Aditya Karnik
Indian Institute of Science
Bangalore, India

Won-Ik Kim
ETRI
Taejon, South Korea

Anurag Kumar
Indian Institute of Science
Bangalore, India

Dong-Hee Kwon
POSTECH
Pohang, South Korea

Chiew-Tong Lau
Nanyang Technological University
Singapore, Singapore

Ben Lee
Oregon State University
Corvallis, Oregon

Bu-Sung Lee
Nanyang Technological University
Singapore, Singapore

Bo Li
Hong Kong University of Science
  and Technology
Kowloon, Hong Kong

Michele Lima
State University of Parana West
Cascavel, Brazil

Ting-Yu Lin
National Chiao-Tung University
Hsinchu, Taiwan

Jiang Chuan Liu
Hong Kong University of Science
and Technology
Kowloon, Hong Kong

Pascal Lorenz
Universtiy of Haute Alsace
Colmar, France

Martin Mauve
University of Mannheim
Mannheim, Germany

Amitabh Mishra
Virginia Polytechnic Institute and
State University
Blacksburg, Virginia

Sangman Moh
ETRI
Taejon, South Korea

Hussein T. Mouftah
Queen's University
Kingston, Ontario, Canada

Ketan M. Nadkarni
Virginia Polytechnic Institute and
State University
Blacksburg, Virginia

**Panagiotis
Papadimitratos**
Cornell University
Ithaca, New York

Marc R. Pearlman
Cornell University
Ithaca, New York

Matthew Sadiku
Prairie View A&M University
Prairie View, Texas

Ahmed M. Safwat
Queen's University
Kingston, Ontario, Canada

Prince Samar
Cornell University
Ithaca, New York

Boon-Chong Seet
Nanyang Technological University
Singapore, Singapore

Jang-Ping Sheu
National Central University
Chung-Li, Taiwan

Yantai Shu
Tianjin University
Tianjin, People's Republic of China

Kazem Sohraby
Lucent Technologies
Lincroft, New Jersey

Ivan Stojmenovic
University of Ottawa
Ottawa, Ontario, Canada

Young-Joo Suh
POSTECH
Pohang, South Korea

Yu-Chee Tseng
National Chiao-Tung University
Hsinchu, Taiwan

Kuochen Wang
National Chiao Tung University
Hsinchu, Taiwan

Lei Wang
Tianjin University
Tianjin, People's Republic of China

Jörg Widmer
University of Mannheim
Mannheim, Germany

**Seah Khoon Guan
Winston**
National University of Singapore
Singapore, Singapore

Jie Wu
Florida Atlantic University
Boca Raton, Florida

Oliver Yang
University of Ottawa
Ottawa, Ontario, Canada

Sal Yazbeck
Barry University
Palm Beach Gardens, Florida

Hee Yong Youn
Sungkyunkwan University
Jangangu Chunchundong, South
Korea

Chansu Yu
Cleveland State University
Cleveland, Ohio

Qian Zhang
Microsoft Research
Beijing, People's Republic of China

Dan Zhou
Florida Atlantic University
Boca Raton, Florida

Wenwu Zhu
Microsoft Research
Beijing, People's Republic of China

Table of Contents

I

Introduction

Body, Personal, and Local Ad Hoc Wireless Networks

Marco Conti
Consiglio Nazionale delle Ricerche

Abstract

A mobile ad hoc network (MANET) represents a system of wireless mobile nodes that can freely and dynamically self-organize into arbitrary and temporary network topologies, allowing people and devices to seamlessly internetwork in areas without any preexisting communication infrastructure. While many challenges remain to be resolved before large scale MANETs can be widely deployed, small-scale mobile ad hoc networks will soon appear. Network cards for single-hop ad hoc wireless networks are already on the market, and these technologies constitute the building blocks to construct small-scale ad hoc networks that extend the range of single-hop wireless technologies to few kilometers. It is therefore important to understand the qualitative and quantitative behavior of single-hop ad hoc wireless networks. The first part of this chapter presents the taxonomy of single-hop wireless technologies. Specifically, we introduce the concept of Body, Personal, and Local wireless networks, and we discuss their applicative scenarios. The second part of the chapter focuses on the emerging networking standards for constructing small-scale ad hoc networks: IEEE 802.11 and Bluetooth. The IEEE 802.11 standard is a good platform to implement a single-hop local ad hoc network because of its extreme simplicity. Furthermore, multi-hop networks covering areas of several square kilometers could be built by exploiting the IEEE 802.11 technology. On smaller scales, the Bluetooth technologies can be exploited to build ad hoc wireless Personal and Body Area Networks, i.e., networks that connect devices placed on a person's body or inside a small circle around it. The chapter presents the architectures and protocols of IEEE 802.11 and Bluetooth. In addition, the performance of these two technologies is discussed.

1.1 Introduction

In recent years, the proliferation of mobile computing devices (e.g., laptops, handheld digital devices, personal digital assistants [PDAs], and wearable computers) has driven a revolutionary change in the computing world. As shown in Fig. 1.1, we are moving from the Personal Computer (PC) age (i.e., one computing device per person) to the Ubiquitous Computing age in which individual users utilize, at the same time, several electronic platforms through which they can access all the required information whenever and wherever they may be [47]. The nature of ubiquitous devices makes wireless networks the easiest solution for their interconnection. This has led to rapid growth in the use of wireless technologies for the Local Area Network (LAN) environment. Beyond supporting wireless connectivity for fixed, portable, and moving stations within a local area, wireless LAN (WLAN) technologies can provide a mobile and ubiquitous connection to Internet information services [10]. It is foreseeable that in the not-so-distant future, WLAN technologies will be utilized largely as means to access the Internet.

WLAN products consume too much power and have excessive range for many personal consumer electronic and computer devices [40]. A new class of networks is therefore emerging: *Personal Area Networks*. A Personal Area Network (PAN) allows the proximal devices to dynamically share information with minimum power consumption [49].

LANs and PANs do not meet all the networking requirements of ubiquitous computing. Situations exist where carrying and holding a computer are not practical (e.g., assembly line work). A wearable computer solves these problems by distributing computer components (e.g., head-mounted displays, microphones, earphones, processors, and mass storage) on the body [21,49]. Users can thus receive job-critical information and maintain control of their devices while their hands remain free for other work. A network with a transmission range of a human body, i.e., a *Body Area Network* (BAN), constitutes the best solution for connecting wearable devices. Wireless connectivity is envisaged as a natural solution for BANs.

One target of the ubiquitous computing revolution is the ability of the technology to adapt itself to the user without requiring that users modify their behavior and knowledge. PCs provided to their users a large set of new services that revolutionized their lives. However, to exploit the PC's benefits, users have had to adapt themselves to PC standards. The new trend is to help users in everyday life by exploiting technology and infrastructures that are hidden in the environment and do not require any major change in the users' behavior. This new philosophy is the basis of the *ambient intelligence* concept [1]. The objective of ambient intelligence is the integration of digital devices and networks into the everyday environment. This will render accessible, through easy and "natural" interactions, a multitude of services and applications. Ambient intelligence places the user at the center of the information society. This view heavily relies on wireless and mobile communications [36,37]. Specifically, advances in wireless communication will

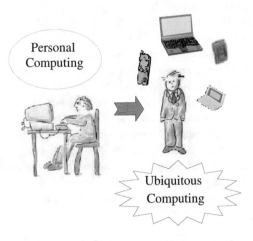

FIGURE 1.1 From PC age (one-to-one) to ubiquitous computing (one-to-many).

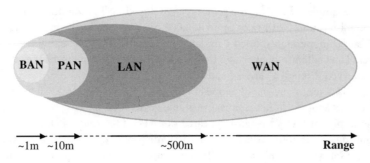

FIGURE 1.2 Ad hoc networks taxonomy.

enable a radical new communication paradigm: self-organized information and communication systems [17]. In this new networking paradigm, the users' mobile devices are the network, and they must cooperatively provide the functionality that is usually provided by the network infrastructure (e.g., routers, switches, and servers). Such systems are sometimes referred to as mobile ad hoc networks (MANETs) [33] or as infrastructure-less wireless networks.

1.2   Mobile Ad Hoc Networks

A Mobile Ad hoc NETwork (MANET) is a system of wireless mobile nodes that dynamically self-organize in arbitrary and temporary network topologies. People and vehicles can thus be internetworked in areas without a preexisting communication infrastructure or when the use of such infrastructure requires wireless extension [17].

As shown in Fig. 1.2, we can classify ad hoc networks, depending on their coverage area, into four main classes: Body, Personal, Local, and Wide Area Networks.

Wide area ad hoc networks are mobile multi-hop wireless networks. They present many challenges that are still to be solved (e.g., addressing, routing, location management, security, etc.), and they are not likely to become available for some time. On smaller scales, mobile ad hoc networks will soon appear [6]. Specifically, ad hoc, single-hop BAN, PAN, and LAN wireless networks are beginning to appear on the market. These technologies constitute the building blocks to construct small multi-hop ad hoc networks that extend the range of the ad hoc networks' technologies over a few radio hops [16,17].

1.2.1   Body Area Network

A Body Area Network is strongly correlated with wearable computers. The components of a wearable computer are distributed on the body (e.g., head-mounted displays, microphones, earphones, etc.), and a BAN provides the connectivity among these devices. Therefore, the main requirements of a BAN are [18,19]:

1. The ability to interconnect heterogeneous devices, ranging from complete devices (e.g., a mobile phone) to parts of a device (microphone, display, etc.)
2. Autoconfiguration capability (Adding or removing a device from a BAN should be transparent to the user.)
3. Services integration (Isochronous data transfer of audio and video must coexist with non–real time data, e.g., Internet data traffic.)
4. The ability to interconnect with the other BANs (to exchange data with other people) or PANs (e.g., to access the Internet)

The communicating range of a BAN corresponds to the human body range, i.e., 1–2 meters. As wiring a body is generally cumbersome, wireless technologies constitute the best solution for interconnecting wearable devices.

One of the first examples of BAN is the prototype developed by T.G. Zimmerman [48], which could provide data communication (with rates up to 400,000 bits per second) by exploiting the body as the channel. Specifically, Zimmerman showed that data can be transferred through the skin by exploiting a very small current (one billionth of an amp). Data transfer between two persons (i.e., BAN interconnection) could be achieved through a simple handshake.

A less visionary BAN prototype was developed in the framework of the MIThril project where a wired Ethernet network was adopted to interconnect wearable devices [35].

Marketable examples of BANs have just appeared (see [19], [32], and [38]). These examples consist of a few electronic devices (phone, MP3 player, headset, microphone, and controller), which are directly connected by wires integrated within a jacket. In the future, it might be expected that more devices (or parts of devices) will be connected using a mixture of wireless and wired technologies.

1.2.2 Personal Area Network

Personal area networks connect mobile devices carried by users to other mobile and stationary devices. While a BAN is devoted to the interconnection of one-person wearable devices (see part a of Fig. 1.3), a PAN is a network in the environment around the person. A PAN communicating range is typically up to 10 meters, thus enabling (see part b of Fig. 1.3):

1. The interconnection of the BANs of people close to each other
2. The interconnection of a BAN with the environment around it

Wireless PAN (WPAN) technologies in the 2.4 GHz ISM band are the most promising technologies for widespread PAN deployment. Spread spectrum is typically employed to reduce interference and utilize the bandwidth.

The technologies for WPANs offer a wide space for innovative solutions and applications that could create radical changes in everyday life. We can foresee that a WPAN interface will be embedded not only in such devices as cellular phones, mobile computers, PDAs, and so on, but in every digital device. Obvious applications arise from the possibility of forming ad hoc networks with our workspace electronic devices, e.g., a PDA that automatically synchronizes with the desktop computer to transfer e-mail, files, and schedule information. Moreover, PANs make possible the design of innovative pervasive applications. For example, let us imagine a PDA (with a PAN interface) that upon your arrival at a location (e.g., home, office, airport, etc.) automatically synchronizes (via the PAN network interface) with all the electronic devices within its 10-meter range. For example, when you arrive at home, your PDA can automatically unlock the door, turn on the house lights while you are getting in, and adjust the heat or air conditioning to your preset preferences. Similarly, when you arrive at the airport you can avoid the line at the check-in desk by using your handheld device to present an electronic ticket and automatically select your seat.

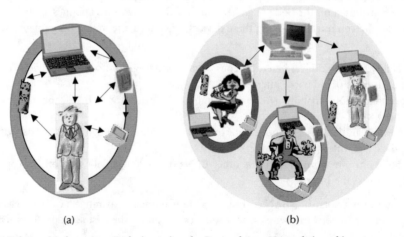

(a)   (b)

FIGURE 1.3 Relationship between a Body (part a) and a Personal Area Network (part b).

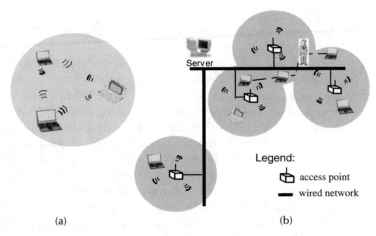

FIGURE 1.4 WLAN configurations: (a) ad hoc networking; (b) infrastructure-based.

1.2.3 Wireless Local Area Network

In the last few years, the use of wireless technologies in the LAN environment has become more and more important, and it is easy to foresee that the wireless LANs (WLANs) — as they offer greater flexibility than wired LANs — will be the solution for home and office automation.

Like wired LANs, a WLAN has a communication range typical of a single building or a cluster of buildings, i.e., 100–500 meters.

A WLAN should satisfy the same requirements typical of any LAN, including high capacity, full connectivity among attached stations, and broadcast capability. However, to meet these objectives, WLANs should be designed to face some issues specific to the wireless environment, such as security on the air, power consumption, mobility, and bandwidth limitation of the air interface [39].

Two different approaches can be followed in the implementation of a WLAN (see Fig. 1.4): an *infra-structure-based* approach or an *ad hoc networking* approach [39]. An infrastructure-based architecture imposes the existence of a centralized controller for each cell, often referred to as the *Access Point* (see Fig. 1.4b). The Access Point is normally connected to the wired network, thus providing Internet access to mobile devices. In contrast, an ad hoc network is a peer-to-peer network formed by a set of stations within the range of each other that dynamically configure themselves to set up a temporary network (see Fig. 1.4a). In the ad hoc configuration, no fixed controller is required, but a controller is dynamically elected among all the stations participating in the communication.

1.3 Technologies for Ad Hoc Networks

The success of a network technology is connected to the development of networking products at a competitive price. A major factor in achieving this goal is the availability of appropriate networking standards. Currently, two main standards are emerging for ad hoc wireless networks: the IEEE 802.11 standard for WLANs [25] and the Bluetooth specifications[1] for short-range wireless communications [3,4,34].

The IEEE 802.11 standard is a good platform for implementing a single-hop WLAN ad hoc network because of its extreme simplicity. Multi-hop networks covering areas of several square kilometers could also be built by exploiting the IEEE 802.11 technology. On smaller scales, technologies such as Bluetooth can be used to build ad hoc wireless Body and Personal Area Networks, i.e., networks that connect devices on the person, or placed around the person inside a circle with a radius of 10 meters.

---

[1] The Bluetooth specifications are released by the Bluetooth Special Interest Group [12].

Here we present the architecture and protocols of IEEE 802.11 and Bluetooth. In addition, the performances of these technologies are analyzed. Two main performance indices will be considered: the throughput and the delay.

As far as throughput is concerned, special attention will be paid to the Medium Access Control (MAC) protocol capacity [15,30], defined as the maximum fraction of channel bandwidth used by successfully transmitted messages. This performance index is important because the bandwidth delivered by wireless networks is much lower than that of wired networks, e.g., 1–11 Mb/sec vs. 100–1000 Mb/sec [39]. Since a WLAN relies on a common transmission medium, the transmissions of the network stations must be coordinated by the MAC protocol. This coordination can be achieved by means of control information that is carried explicitly by control messages traveling along the medium (e.g., ACK messages) or can be provided implicitly by the medium itself using the carrier sensing to identify the channel as either active or idle. Control messages or message retransmissions due to collision remove channel bandwidth from that available for successful message transmission. Therefore, the capacity gives a good indication of the overheads required by the MAC protocol to perform its coordination task among stations or, in other words, of the effective bandwidth that can be used on a wireless link for data transmission.

The delay can be defined in several forms (access delay, queuing delay, propagation delay, etc.) depending on the time instants considered during its measurement (see [15]). In computer networks, the response time (i.e., the time between the generation of a message at the sending station and its reception at the destination station) is the best value to measure the Quality of Service (QoS) perceived by the users. However, the response time depends on the amount of buffering inside the network, and it is not always meaningful for the evaluation of a LAN technology. For example, during congested periods, the buffers fill up, and thus the response time does not depend on the LAN technology but it is mainly a function of the buffer length. For this reason, hereafter, the MAC delay index is used. The MAC delay of a station in a LAN is defined as the time between the instant at which a packet comes to the head of the station transmission queue and the end of the packet transmission [15].

1.4   IEEE 802.11 Architecture and Protocols

In 1997, the IEEE adopted the first wireless local area network standard, named IEEE 802.11, with data rates up to 2 Mb/sec [27]. Since then, several task groups (designated by letters) have been created to extend the IEEE 802.11 standard. Task groups 802.11b and 802.11a have completed their work by providing two relevant extensions to the original standard [25]. The 802.11b task group produced a standard for WLAN operations in the 2.4 GHz band, with data rates up to 11 Mb/sec. This standard, published in 1999, has been very successful. Currently, there are several IEEE 802.11b products available on the market. The 802.11a task group created a standard for WLAN operations in the 5 GHz band, with data rates up to 54 Mb/sec. Among the other task groups, it is worth mentioning task group 802.11e (which attempts to enhance the MAC with QoS features to support voice and video over 802.11 networks) and task group 802.11g (which is working to develop a higher-speed extension to 802.11b).

The IEEE 802.11 standard specifies a MAC layer and a physical layer for WLANs (see Fig. 1.5). The MAC layer provides to its users both contention-based and contention-free access control on a variety of physical layers. Specifically, three different technologies can be used at the physical layer: infrared, frequency hopping spread spectrum, and direct sequence spread spectrum [27].

The basic access method in the IEEE 802.11 MAC protocol is the *Distributed Coordination Function* (DCF), which is a *Carrier Sense Multiple Access with Collision Avoidance* (CSMA/CA) MAC protocol. Besides the DCF, the IEEE 802.11 also incorporates an alternative access method known as the *Point Coordination Function* (PCF). The PCF operates similarly to a polling system [15]; a point coordinator provides (through a polling mechanism) the transmission rights at a single station at a time. As the PCF access method cannot be adopted in ad hoc networks, in the following we will concentrate on the DCF access method only.

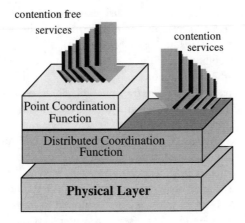

FIGURE 1.5   IEEE 802.11 architecture.

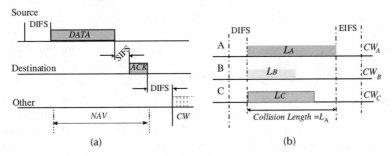

FIGURE 1.6   IEEE 802.11 DCF: (a) a successful transmission; (b) a collision.

1.4.1   IEEE 802.11 DCF

The DCF access method, hereafter referred to as *Basic Access*, is summarized in Fig. 1.6. When using the DCF, before a station initiates a transmission, it senses the channel to determine whether another station is transmitting. If the medium is found to be idle for an interval that exceeds the *Distributed InterFrame Space* (DIFS), the station continues with its transmission.[2] The transmitted packet contains the projected length of the transmission. Each active station stores this information in a local variable named *Network Allocation Vector* (NAV). Therefore, the NAV contains the period of time the channel will remain busy (see Fig. 1.6a).[3]

The CSMA/CA protocol does not rely on the capability of the stations to detect a collision by hearing their own transmissions. Hence, immediate positive acknowledgments are employed to ascertain the successful reception of each packet transmission. Specifically, the receiver after the reception of the data frame (1) waits for a time interval, called the *Short InterFrame Space* (SIFS), which is less than the DIFS, and then (2) initiates the transmission of an acknowledgment (ACK) frame. The ACK is not transmitted if the packet is corrupted or lost due to collisions. A Cyclic Redundancy Check (CRC) algorithm is adopted to discover transmission errors. Collisions among stations occur when two or more stations start transmitting at the same time (see Fig. 1.6b). If an acknowledgment is not received, the data frame is presumed to have been lost, and a retransmission is scheduled.

---

[2] To guarantee fair access to the shared medium, a station that has just transmitted a packet and has another packet ready for transmission must perform the backoff procedure before initiating the second transmission.

[3] This prevents a station from listening to the channel during transmissions. This feature is useful to implement (among others) power-saving policies.

After an erroneous frame is detected (due to collisions or transmission errors), the channel must remain idle for at least an *Extended InterFrame Space* (EIFS) interval before the stations reactivate the backoff algorithm to schedule their transmissions (see Fig. 1.6b).

To reduce the collision probability, the IEEE 802.11 uses a mechanism (*backoff mechanism*) that guarantees a time spreading of the transmissions.

When a station *S*, with a packet ready for transmission, observes a busy channel, it defers the transmission until the end of the ongoing transmission. At the end of the channel busy period, the station *S* initializes a counter (called the *backoff timer*) by selecting a random interval (*backoff interval*) for scheduling its transmission attempt. The backoff timer is decreased for as long as the channel is sensed as idle, stopped when a transmission is detected on the channel, and reactivated when the channel is sensed as idle again for more than a DIFS. The station transmits when the backoff timer reaches zero. Specifically, the DCF adopts a slotted binary exponential backoff technique. The time immediately following an idle DIFS or EIFS is slotted, and a station is allowed to transmit only at the beginning of each *Slot Time*.[4] The backoff time is uniformly chosen in the interval $(0, CW - 1)$, defined as the Backoff Window, also referred to as the *Contention Window*. At the first transmission attempt $CW = CW_{min}$, and then *CW* is doubled at each retransmission up to CW_{max}. The CW_{min} and CW_{max} values depend on the physical layer adopted. For example, for the frequency hopping, CW_{min} and CW_{max} are 16 and 1024, respectively [27].

An IEEE 802.11 WLAN can be implemented with the access points (i.e., infrastructure based) or with the ad hoc paradigm. In the IEEE 802.11 standard, an ad hoc network is called an *Independent Basic Service Set* (IBSS). An IBSS enables two or more IEEE 802.11 stations to communicate directly without requiring the intervention of a centralized access point or an infrastructure network. Due to the flexibility of the CSMA/CA algorithm, synchronization (to a common clock) of the stations belonging to an IBSS is sufficient for correct receipt or transmission of data. The IEEE 802.11 uses two main functions for the synchronization of the stations in an IBSS: (1) synchronization acquisition and (2) synchronization maintenance.

Synchronization Acquisition — This functionality is necessary for joining an existing IBSS. The discovery of existing IBSSs is the result of a scanning procedure of the wireless medium. During the scanning, the station receiver is tuned on different radio frequencies, searching for particular control frames. Only if the scanning procedure does not result in finding any IBSS may the station initialize a new IBSS.

Synchronization Maintenance — Because of the lack of a centralized station that provides its own clock as common clock, the synchronization function is implemented via a distributed algorithm that shall be performed by all of the members of the IBSS. This algorithm is based on the transmission of beacon frames at a known nominal rate. The station that initialized the IBSS decides the beacon interval.

1.4.1.1   IEEE 802.11 DCF Performance

In this section we present a performance analysis of the IEEE 802.11 basic access method by analyzing the two main performance indices: the capacity and the MAC delay. The physical layer technology determines some network parameter values relevant for the performance study, e.g., SIFS, DIFS, backoff, and slot time. Whenever necessary, we choose the values of these technology-dependent parameters by referring to the frequency hopping spread spectrum technology at a transmission rate of 2 Mb/sec. Specifically, Table 1.1 reports the configuration parameter values of the IEEE 802.11 WLAN analyzed in this chapter [27].

1.4.1.1.1   Protocol Capacity

The IEEE 802.11 protocol capacity was extensively investigated in [14]. The main results of that analysis are summarized here. Specifically, in [14] the theoretical throughput limit for IEEE 802.11 networks was

---

[4] A slot time is equal to the time needed at any station to detect the transmission of a packet from any other station.

TABLE 1.1   IEEE 802.11 Parameter Values

Parameter Value	t_{slot}	τ	DIFS	EIFS	SIFS	ACK	CW_{min}	CW_{max}	Bit Rate
	50 μsec	≤1 μsec	2.56 t_{slot}	340 μsec	0.56 t_{slot}	240 bits	8 t_{slot}	256 t_{slot}	2 Mb/sec

analytically derived,[5] and this limit was compared with the simulated estimates of the real protocol capacity. The results showed that, depending on the network configuration, the standard protocol can operate very far from the theoretical throughput limit. These results, summarized in Fig. 1.7a, indicate that the distance between the IEEE 802.11 and the analytical bound increases with the number of active stations, M. In the IEEE 802.11 protocol, due to its backoff algorithm, the average number of stations that transmit in a slot increases with M, and this causes an increase in the collision probability. A significant improvement of the IEEE 802.11 performance can thus be obtained by controlling the number of stations that transmit in the same slot.

Several works have shown that an appropriate tuning of the IEEE 802.11 backoff algorithm can significantly increase the protocol capacity [2,13,46]. In particular, in [13], a distributed algorithm to tune the size of the backoff window at run time, called *Dynamic IEEE 802.11 Protocol*, was presented and evaluated. Specifically, by observing the status of the channel, each station gets an estimate of both the number of active stations and the characteristics of the network traffic. By exploiting these estimates, each station then applies a distributed algorithm to tune its backoff window size in order to achieve the *theoretical throughput limit* for the IEEE 802.11 network.

The Dynamic IEEE 802.11 Protocol is complex due to the interdependencies among the estimated quantities [13]. To avoid this complexity, in [7] a *Simple Dynamic IEEE 802.11 Protocol* is proposed and evaluated. It requires only simple load estimates for tuning the backoff algorithm. An alternative and interesting approach for tuning the backoff algorithm, without requiring complex estimates of the network status, has been proposed in [5]. In this work a distributed mechanism is defined, called *Asymptotically Optimal Backoff* (AOB), which dynamically adapts the backoff window size to the current load. AOB guarantees that an IEEE 802.11 WLAN asymptotically (i.e., for a large number of active stations) achieves its optimal channel utilization. The AOB mechanism adapts the backoff window to the network contention level by using two load estimates: the slot utilization and the average size of transmitted frames. These estimates are simple and can be obtained with no additional costs or overheads.

It is worth noting that the above mechanisms that tune the IEEE 802.11 protocol to optimize the protocol capacity also guarantee quasi-optimal behavior from the energy consumption standpoint (i.e., minimum energy consumption). Indeed, in [11] it is shown that the optimal capacity state and the optimal energy consumption state almost coincide.

1.4.1.1.2 MAC Delay

The IEEE 802.11 capacity analysis presented in the previous section is performed by assuming that the network operates in asymptotic conditions (i.e., each LAN station always has a packet ready for transmission). However, LANs normally operate in normal conditions, i.e., the network stations generate an aggregate traffic that is lower·(or slightly higher) than the maximum traffic the network can support. In these load conditions, the most meaningful performance figure is the MAC delay (see Section 1.3 and [15]). Two sets of MAC delay results are presented here, corresponding to traffic generated by 50 stations, made up of short (two slots) and long (100 slots) messages, respectively. Stations alternate between idle and busy periods. In the simulative experiments, the channel utilization level is controlled by varying the idle periods' lengths.

Figure 1.7b (which plots the average MAC delay vs. the channel utilization) highlights that, for light load conditions, the IEEE 802.11 exhibits very low MAC delays. However, as the offered load approaches

---

[5] That is, the maximum throughput that can be achieved by adopting the IEEE 802.11 MAC protocol and using the optimal tuning of the backoff algorithm.

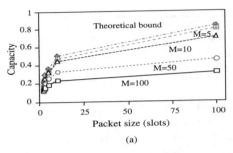
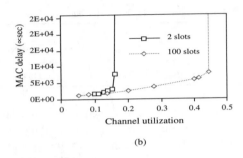

(a)   (b)

FIGURE 1.7   IEEE 802.11 performance: (a) protocol capacity; (b) average MAC delay

the capacity of the protocol (see Fig. 1.7a), the MAC delay sharply increases. This behavior is due to the CSMA/CA protocol. Under light-load conditions, the protocol introduces almost no overhead (a station can immediately transmit as soon as it has a packet ready for transmission). On the other hand, when the load increases, the collision probability increases as well, and most of the time a transmission results in a collision. Several transmission attempts are necessary before a station is able to transmit a packet, and hence the MAC delay increases. It is worth noting that the algorithms discussed in the previous section (i.e., SDP, AOB, etc.) for optimizing the protocol capacity also help prevent MAC delays from becoming unbounded when the channel utilization approaches the protocol capacity (see [5] and [7]).

1.4.2   IEEE 802.11 RTS/CTS

The design of a WLAN that adopts a carrier-sensing random access protocol [24], such as the IEEE 802.11, is complicated by the presence of hidden terminals [42]. A pair of stations is referred to as being *hidden* from each other if a station cannot hear the transmission from the other station. This event makes the carrier sensing unreliable, as a station wrongly senses that the wireless medium has been idle while the other station (which is hidden from its standpoint) is transmitting. For example, as shown in Fig. 1.8, let us assume that two stations, say S_1 and S_2, are hidden from each other, and both wish to transmit to a third station, named *Receiver*. When S_1 is transmitting to *Receiver*, the carrier sensing of S_2 does not trigger any transmission, and thus S_2 can immediately start a transmission to *Receiver* as well. Obviously, this event causes a collision that never occurs if the carrier sensing works properly.

The hidden stations phenomenon may occur in both infrastructure-based and ad hoc networks. However, it may be more relevant in ad hoc networks where almost no coordination exists among the stations. In this case, all stations may be transmitting on a single frequency, as occurs in the WaveLAN IEEE 802.11 technology [45].

To avoid the hidden terminal problem, the IEEE 802.11 basic access mechanism was extended with a virtual carrier sensing mechanism, called *Request To Send (RTS)/Clear To Send (CTS)*.

In the RTS/CTS mechanism, after access to the medium is gained and before transmission of a data packet begins, a short control packet, called RTS, is sent to the receiving station announcing the upcoming transmission. The receiver replies to this with a CTS packet to indicate readiness to receive the data. RTS and CTS packets contain the projected length of the transmission. This information is stored by each active station in its NAV, the value of which becomes equal to the end of the channel busy period. Therefore, all stations within the range of at least one of the two stations (receiver and transmitter) know how long the channel will be used for this data transmission (see Fig. 1.9).

The RTS/CTS mechanism solves the hidden station problem during the transmission of user data. In addition, this mechanism can be used to capture the channel control before the transmission of long packets, thus avoiding "long collisions." Collisions may occur only during the transmissions of the small RTS and CTS packets. Unfortunately, as shown in the next section, other phenomena occur at the physical layer making the effectiveness of the RTS/CTS mechanism quite arguable.

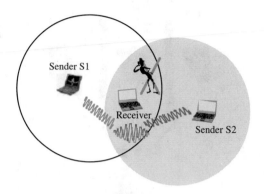

FIGURE 1.8 The hidden stations phenomenon.

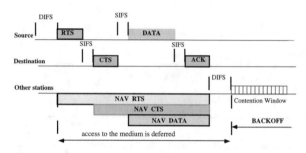

FIGURE 1.9 The RTS/CTS mechanism.

1.4.2.1 RTS/CTS Effectiveness in Ad Hoc Networks

The effectiveness of the RTS/CTS mechanism was studied in [44] in a real field trial. The main results of that study are summarized here. The testbed analyzed the performance of the TCP protocol over an IEEE 802.11 ad hoc network. To reduce the complexity of the study, static ad hoc networks were considered, i.e., the network nodes did not change their positions during an experiment. Both indoor and outdoor scenarios were investigated.

1.4.2.1.1 Indoor Experiments

In this case the experiments were performed in a scenario characterized by hidden stations. The scenario is shown in Fig. 1.10. Nodes 1, 2, and 3 are transferring data, via ftp, toward node 4. As these data transfers are supported by the TCP protocol, in the following the data flows will be denoted as TCP #*i*, where *i* is the index of the transmitting station.

In the analyzed scenario, a reinforced concrete wall (the black rectangle in the figure) is located between node 1 and node 2 and between node 2 and node 3. As a consequence, the three transmitting nodes are hidden from each other, e.g., nodes 2 and 3 are outside the transmission range of node 1.[6] Node 4 is in the transmission range of all the other nodes.

Two sets of experiments were performed using the DCF mechanism with or without the RTS/CTS mechanism. In Table 1.2, the results of the experiments are summarized. Two main conclusions can be reached from these experiments:

1. No significant performance differences exist between adopting the RTS/CTS mechanism vs. the basic access mechanism only.
2. Due to the additional overheads of the RTS and CTS packets, the aggregate network throughput with the RTS/CTS mechanism is a bit lower with respect to the basic access mechanism.

---

[6] Specifically, the ping application indicated no delivered packet.

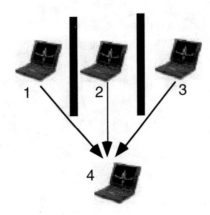

FIGURE 1.10   Indoor scenario.

TABLE 1.2   Indoor Results — Throughput (Kbytes/sec)

	TCP #1	TCP #2	TCP #3	Aggregate
No RTS/CTS	42	29.5	57	128.5
RTS/CTS	34	27	48	109

These results seem to indicate that the carrier sensing mechanism is still effective even if transmitting stations are "apparently" hidden from each other. Indeed, a distinction must be made between transmission range, interference range, and carrier sensing range, as follows:

- The Transmission Range (TX_Range) represents the range (with respect to the transmitting station) within which a transmitted packet can be successfully received. The transmission range is mainly determined by the transmission power and the radio propagation properties.
- The Physical Carrier Sensing Range (PCS_Range) is the range (with respect to the transmitting station) within which the other stations detect a transmission.
- The Interference Range (IF_Range) is the range within which stations in receive mode will be "interfered with" by a transmitter and thus suffer a loss. The interference range is usually larger than the transmission range, and it is a function of the distance between the sender and receiver and of the path loss model.

Normally, the following relationship exists between the transmission, carrier sensing, and interference ranges: TX_Range ≤ IF_Range ≤ PCS_Range.[7] The relationship among TX_Range, IF_Range, and PCS_Range helps in explaining the results obtained in the indoor experiments: even though transmitting nodes are outside the transmission range of each other, they are inside the same carrier sensing range. Therefore, the physical carrier sensing is effective, and hence adding a virtual carrier sensing (i.e., RTS/CTS) is useless.

1.4.2.1.2   Outdoor Experiments
The reference scenario for this case is shown in Fig. 1.11. The nodes represent four portable computers, each with an IEEE 802.11 network interface. Two ftp sessions are contemporary active. The arrows represent the direction of the ftp sessions.

Several experiments were performed by varying the transmission, the carrier sensing, and the interference ranges. This was achieved by modifying the distance, d, between nodes 2 and 3. In all the experiments, the receiving node was always within the transmission range of its transmitting node —

---

[7] For example, in NS2 the following values are used: TX_Range = 250 m, IF_Range = PCS_Range = 550 m.

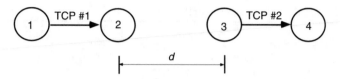

FIGURE 1.11 Outdoor reference scenario.

TABLE 1.3 Outdoor Results — Throughput (Kbytes/sec)

	Exp #1		Exp #2		Exp #3	
	TCP #1	TCP #2	TCP #1	TCP #2	TCP #1	TCP #2
No RTS/CTS	61	54	123	0.5	122.5	122
RTS/CTS	59.5	49.5	81	6.5	96	100

i.e., node 2 (4) was within the transmitting range of node 1 (3) — while, by varying the distance *d*, the other two nodes[8] could be:

1. In the same transmitting range (Exp #1)
2. Out of the transmitting range but inside the same carrier sensing range (Exp #2)
3. Out of the same carrier sensing range (Exp #3)

The achieved results, summarized in Table 1.3, show the following:

- Exp #1. In this case (all stations are inside the same TX_Range), a fair bandwidth sharing is almost obtained: the two ftp sessions achieve (almost) the same throughput. The RTS/CTS mechanism is useless as (due to its overheads) it only reduces the throughput.

- Exp #3. In this case the two sessions are independent (i.e., outside their respective carrier sensing ranges), and both achieve the maximum throughput. The RTS/CTS mechanism is useless as (due to its overheads) it only reduces the throughput.

- Exp #2. In the intermediate situation, a "capture" of the channel by one of the two TCP connections is observed. In this case, the RTS/CTS mechanism provides a little help in solving the problem.

The experimental results confirm the results on TCP unfairness in ad hoc IEEE 802.11 obtained, via simulation, by several researchers, e.g., see [43]. As discussed in previous works, the TCP protocol, due to flow control and congestion mechanisms, introduces correlations in the transmitted traffic that emphasize/generate the capture phenomena. This effect is clearly pointed out by experimental results presented in Table 1.4. Specifically, the table reports results obtained in the Exp #2 configuration when the traffic flows are either TCP or UDP based. As shown in the table, the capture effect disappears when the UDP protocol is used.

To summarize, measurement experiments have shown that, in some scenarios, TCP connections may suffer significant throughput unfairness, even capture. The causes of this behavior are the hidden terminal problem, the 802.11 backoff scheme, and large interference ranges. We expect that the methods discussed in the section "IEEE 802.11 DCF Performance" for optimizing the IEEE 802.11 protocol capacity area are moving in a promising direction to solve the TCP unfairness in IEEE 802.11 ad hoc networks. Research activities are ongoing to explore this direction.

---

[8] That is, the couple (3,4) with respect to the couple (1,2) and vice versa.

TABLE 1.4   UDP vs. TCP performance (Exp #2) — Throughput (Kbytes/sec)

	TCP Traffic		UDP Traffic	
	Flow #1	Flow #2	Flow #1	Flow #2
No RTS/CTS	123	0.5	83	84
RTS/CTS	81	6.5	77.5	68

1.5   A Technology for WBAN and WPAN: Bluetooth

The Bluetooth technology is a de facto standard for low-cost, short-range radio links between mobile PCs, mobile phones, and other portable devices [3,34]. The Bluetooth Special Interest Group (SIG) releases the Bluetooth specifications. Bluetooth SIG is a group consisting of industrial leaders in telecommunications, computing, and networking [12]. In addition, the IEEE 802.15 Working Group for Wireless Personal Area Networks has just approved its first WPAN standard derived from the Bluetooth specification [26]. The IEEE 802.15 standard is based on the lower portions of the Bluetooth specification.

The Bluetooth system is operating in the 2.4 GHz industrial, scientific, and medicine band. A Bluetooth unit, integrated into a microchip, enables wireless ad hoc communications of voice and data in stationary and mobile environments. Because the cost target is low, it can be envisaged that Bluetooth microchips will be embedded in all consumer electronic devices.

1.5.1   A Bluetooth Network

From a logical standpoint, Bluetooth belongs to the contention-free token-based multi-access networks [24]. In a Bluetooth network, one station has the role of master, and all other Bluetooth stations are slaves. The master decides which slave has access to the channel. The units that share the same channel (i.e., are synchronized to the same master) form a *piconet*, the fundamental building block of a Bluetooth network. A piconet has a gross bit rate of 1 Mb/sec.[9] A piconet contains a master station and up to seven *active* (i.e., participating in data exchange) slaves, contemporarily. Figure 1.12 shows an example of two partially overlapping piconets. In the figure, we denote with M and S a master and a slave, respectively. Stations marked with P (*Parking* state) are stations that are synchronized with the master but are not participating in any data exchange.

Independent piconets that have overlapping coverage areas may form a *scatternet*. A scatternet exists when a unit is active in more than one piconet[10] at the same time. A slave may communicate with the different piconets it belongs to only in a time-multiplexing mode. This means that, for any time instant, a station can only transmit on the single piconet to which (at that time) its clock is synchronized. To transmit on another piconet it has to change the synchronization parameters.

The complete Bluetooth protocol stack contains several protocols: Bluetooth radio, Baseband, Link Manager Protocol (LMP), Logical Link Control and Adaptation Protocol (L2CAP), and Service Discovery Protocol (SDP). For the purpose of this chapter, we will focus only on the Bluetooth radio, Baseband, and (partially) L2CAP protocols. A description of the Bluetooth architecture can be found in [9].

Bluetooth radio provides the physical links among Bluetooth devices, while the Baseband layer provides a transport service of packets on the physical links. In the next subsections, these layers will be presented in detail. The L2CAP services are used only for data transmission. The main features supported by L2CAP are protocol multiplexing (the L2CAP uses a protocol type field to distinguish between upper layer protocols) and segmentation and reassembly. The latter feature is required because the Baseband packet size is smaller than the usual size of packets used by higher layer protocols.

---

[9] 1 Mb/sec represents the channel capacity before considering the overhead introduced by the Bluetooth protocols and polling scheme.

[10] A unit can be master into only one piconet.

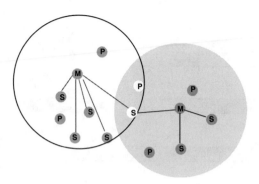

FIGURE 1.12 Two partially overlapping piconets.

A Bluetooth unit consists of a radio unit operating in the 2.4 GHz band. In this band are defined 79 different radio frequency (RF) channels spaced 1 MHz apart. The radio layer utilizes as transmission technique the *Frequency Hopping Spread Spectrum* (FHSS). The hopping sequence is a pseudo-random sequence of 79-hop length, and it is unique for each piconet (it depends on the master local parameters). The FHSS system has been chosen to reduce the interference of nearby systems operating in the same range of frequency (for example, IEEE 802.11 WLAN) and to make the link robust [22,23]. The nominal rate of hopping between two consecutive RF is 1600 hop/sec.

A Time Division Duplex (TDD) scheme of transmission is adopted. The channel is divided into time slots, each 625 μsec in length, and each slot corresponds to a different RF hop frequency. The time slots are numbered according to the Bluetooth clock of the master. The master has to begin its transmissions in even-numbered time slots. Odd-numbered time slots are reserved for the beginning of slaves' transmissions. The first row of Fig. 1.13 shows a snapshot of the master transmissions.

The transmission of a packet nominally covers a single slot, but it may also last for three or five consecutive time slots (see the second and third rows of Fig. 1.13, respectively). For multi-slot packets, the RF hop frequency to be used for the entire packet is the RF hopping frequency assigned to the time slot in which the transmission began.

1.5.1.1 Bluetooth Piconet Formation

The Bluetooth technology has been devised to provide a flexible wireless connectivity among digital devices. Before starting a data transmission, a Bluetooth unit needs to discover if any other Bluetooth unit is in its operating space. To do this, the unit enters the *inquiry state*. In this state, it continuously sends an inquiry message, i.e., a packet with only the access code.[11] During the inquiry message transmission, the inquiring unit uses a frequency hopping sequence of 32 frequencies derived from the access code. These 32 frequencies are split into two trains, each containing 16 frequencies. A single train must be repeated at least 256 times before a new train is used. Several (up to three) train switches must take place to guarantee a sufficient number of responses. As a result of this inquiring policy, the inquiry state lasts at most 10.24 seconds. A unit can respond to an inquiry message only if it is listening to the channel to find an inquiry message, and its receiver is tuned to the same frequency used by the inquiring unit. To increase the probability of this event, a unit scans the inquiry access code (on a given frequency) for a time long enough to completely scan for 16 inquiry frequencies. Obviously, a unit is not obliged to answer an inquiring message, but if it responds it has to send a special control packet, the FHS packet, which contains its Bluetooth device address and its native clock.

After the inquiry, a Bluetooth unit has discovered the Bluetooth device address of the units around it and has collected an estimation of their clocks. If it wants to activate a new connection, it has to

---

[11] The inquiring unit can adopt a General Inquiry Access Code (GIAC) that enables any Bluetooth device to answer the inquiry message or a dedicated inquiry access code (DIAC) that enables only Bluetooth devices belonging to certain classes to answer the inquiry message.

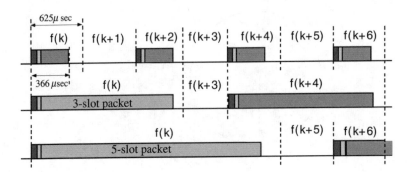

FIGURE 1.13 One-slot and multi-slot packet transmission.

distribute its own Bluetooth device address and clock. This is the aim of paging routines. The unit that starts the paging is (automatically) elected the master of the new connection, and the paged unit is the slave. The paging unit sends a page message, i.e., a packet with only the device access code (DAC). The DAC is derived directly from the Bluetooth device address of the paged unit that, therefore, is the only one that can recognize the page message. After the paging procedure, the slave has an exact knowledge of the master clock and of the channel access code. Hence, the master and that slave can enter the connection state. However, a real transmission will begin only after a polling message from the master to the slave.

When a connection is established, the active slaves maintain the synchronization with the master by listening to the channel at every master-to-slave slot. Obviously, if an active slave is not addressed, after it has read the type of packet it can return to sleep for a time equal to the number of slots the master has taken for its transmission.

Most devices that will adopt the Bluetooth technology are mobile and handheld devices for which power consumption optimization is a critical matter. To avoid power consumption (caused by the synchronization), the Bluetooth specification has defined some power saving states for connected slaves: Sniff, Hold, and Park Modes. We redirect the interested reader to [4,34].

1.5.1.2 Bluetooth Scatternet

The Bluetooth specification defines a method for the interconnection of piconets: the scatternet. A scatternet can be dynamically constructed in an ad hoc fashion when some nodes belong, at the same time, to more than one piconet (inter-piconet units). For example, the two piconets in Fig. 1.12 share a slave, and hence they can form a scatternet. The traffic between the two piconets is delivered through the common slave. Scatternets can be useful in several scenarios. For example, we can have a piconet that contains a laptop and a cellular phone. The cellular phone provides access to the Internet. A second piconet contains the laptop itself and several PDAs. In this case, a scatternet can be formed with the laptop as the inter-piconet unit. By exploiting the scatternet, the PDAs can exploit the cellular phone services to access the Internet.

The current Bluetooth specification only defines the notion of a scatternet but does not provide the mechanisms to construct the scatternet.

A node can be synchronized with only a single piconet at a time, and hence it can be active in more piconets only in a time-multiplexed mode. As the inter-piconet traffic must go through the inter-piconet units, the presence of the inter-piconet units in all the piconets to which they belong must be scheduled in an efficient way.

The scatternet formation algorithms and the algorithm for scheduling the traffic among the various piconets are hot research issues, see [31,50].

1.5.2 Bluetooth Data Transmission

Two types of physical links can be established between Bluetooth devices: a *Synchronous Connection-Oriented* (SCO) link, and an *Asynchronous Connection-Less* (ACL) link. The first type of physical link is a point-to-point, symmetric connection between the master and a specific slave. It is used to deliver delay-sensitive traffic, mainly voice. The SCO link rate is 64 kb/sec, and it is settled by reserving two consecutive slots for master-to-slave transmission and immediate slave-to-master response. The SCO link can be considered as a circuit-switched connection between the master and the slave. The second kind of physical link, ACL, is a connection between the master and all slaves participating in the piconet. It can be considered as a packet-switched connection between the Bluetooth devices. It can support the reliable delivery of data by exploiting a fast *Automatic Repeat Request* (ARQ) scheme. An ACL channel supports point-to-multipoint transmissions from the master to the slaves.

As stated before, the channel access is managed according to a polling scheme. The master decides which slave is the only one to have access to the channel by sending it a packet. The master packet may contain data or can simply be a polling packet (NULL packet). When the slave receives a packet from the master, it is authorized to transmit in the next time slot. For SCO links, the master periodically polls the corresponding slave. Polling is asynchronous for ACL links. Figure 1.14 presents a possible pattern of transmissions in a piconet with a master and two slaves. Slave 1 has both a SCO and an ACL link with the master, while Slave 2 has an ACL link only. In this example, the SCO link is periodically polled by the master every six slots, while ACL links are polled asynchronously. Furthermore, the size of the packets on an ACL link is constrained by the presence of SCO links. For example, in the figure the master sends a multi-slot packet to Slave 2, which replies with a single-slot packet only because the successive slots are reserved for the SCO link.

A piconet has a gross bit rate of 1 Mb/sec. The polling scheme and the protocol control information obviously reduce the amount of user data that can be delivered by a piconet. The limiting throughput performances of a piconet were discussed in [9] by analyzing a single master-slave link in which both stations operate in *asymptotic conditions,* i.e., the stations always have a packet ready for transmission. Here, the Bluetooth performances are analyzed under realistic traffic conditions where several slaves are active inside a piconet. In this case, the master must implement a scheduling algorithm to decide the slaves' polling order. The Bluetooth specification indicates as a possible solution the Round Robin (RR) polling algorithm: slaves are polled in a cyclic order. However, it has been shown (e.g., see [9]) that, under unbalanced traffic conditions, the RR algorithm may cause (due to a large number of NULL packets) severe bandwidth wastage. Several authors have proposed new schedulers suitable for Bluetooth [8,9,20,28,29]. An effective scheduling algorithm, called *Efficient Double-Cycle* (EDC), was proposed in [8,9]. EDC tunes the polling order to the network traffic conditions to limit the channel bandwidth wastage caused by the polling of empty stations. A detailed EDC specification through pseudo-code can be found in [9]. Due to space constraints, only a high-level description of EDC is provided here.

The EDC algorithm is based upon two main ideas. First, it is necessary to avoid NULL transmissions towards and from the slaves; furthermore, the fairness typical of a Round Robin scheme should be preserved. These targets can be accomplished if the selection of the slave to be polled takes into consideration the master's knowledge of the traffic from and to the slaves. Hereafter, we indicate as *uplink* the link direction from the slaves to the master, and as *downlink* the link direction from the master towards the slaves.

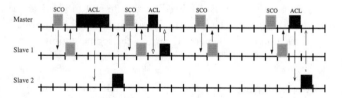

FIGURE 1.14 Transmissions in a piconet.

For the downlink (i.e., master-to-slaves traffic), the master has a deterministic knowledge of the packets it has to send to each slave. In the other direction (uplink), the master does not have any knowledge; at most it can only estimate the probability that a slave will send a NULL packet. This probability can be estimated by exploiting the knowledge of each slave's behavior in the previous polling cycles.

An additional problem in guaranteeing fair and efficient scheduling in Bluetooth is caused by the coupling between the transmissions in uplink and downlink, i.e., a master-to-slave transmission implies also a polling of the slave and hence a possibly NULL transmission from the slave to the master. Therefore, it is not possible to remove a slave from the polling cycle without blocking, at the same time, the master's transmissions towards this slave (and vice versa). To introduce a (partial) decoupling in the scheduling of the transmissions in uplink and downlink, EDC introduces the idea of a double polling cycle: an *uplink polling sub cycle*, $Cycle_{UP}$, and a *downlink polling sub cycle*, $Cycle_{DW}$. The main task of the scheduler is to identify the slaves eligible for the polling in $Cycle_{UP}$ and $Cycle_{DW}$, hereafter denoted as $E(UP)$ and $E(DW)$, respectively. $E(DW)$ is computed by considering only the traffic from the master to the slaves, whereas $E(UP)$ is computed by considering only the estimated slaves' activity, i.e., the traffic from the slaves to the master. Slaves that have no traffic to transmit (to the master) are removed from the eligible slaves during the $Cycle_{UP}$, while $E(DW)$ contains only those slaves for which the master has packets to transmit.[12]

The scheduler defines the eligible slaves at the beginning of each polling cycle, and then it polls the slaves contained in $E(DW)$ or in $E(UP)$. During a cycle, a slave is polled at most once.

The scheduler has no problem defining the $E(DW)$ set: it has a deterministic knowledge of the downlink traffic. On the other hand, for the uplink, it can only exploit the knowledge of the slaves' behavior in the previous polling cycles. To this end, EDC uses the rate of null packets returned by a slave as an indication of that slave's transmission activity. Specifically, the basic behavior of EDC is derived from the backoff algorithms used in random access protocols. These backoff algorithms increase the time between transmission attempts when the number of consecutive collisions increases. In the EDC case, the number of consecutive NULL packets returned by a slave, say x, indicates its transmission requirements: the larger x is, the longer can be the polling interval for that slave. To implement this idea, EDC adopts a *truncated binary exponential backoff* algorithm. Specifically, a polling interval c_i and a polling window w_i are associated to each slave S_i. The values of these variables are updated as follows:

For each polling to S_i (in $Cycle_{UP}$ or in $Cycle_{DW}$), if S_i returns a NULL packet, c_i is increased by 1, otherwise it is set to 0.

After each polling to S_i, the polling window of S_i is set equal to $w_i = \min\{w_{max}, 2^{c_i}\}$, where w_{max} is the maximum length (measured in number of polling cycles) of a slave polling interval.

After each polling cycle, $w_i = \max[0, w_i - 1]$.

In a polling cycle, a slave S_i is eligible only if $w_i = 0$.

1.5.2.1   Internet Access via Bluetooth: A Performance Evaluation Study

Ubiquitous Internet access is expected to be one of the most interesting Bluetooth applications. For this reason, we evaluate here the scheduler impact on the performance experienced by Bluetooth slaves when they access remote Internet servers. Specifically, via simulation, we analyze a scenario made up of a Bluetooth piconet with seven slaves. Bluetooth slaves (through the master) download/upload data from/to remote Internet servers. In each slave of the piconet, the traffic (generated by either an ftp application or a Constant Bit Rate (CBR) source) is encapsulated into the TCP/IP protocol stack, the L2CAP protocol, and the baseband protocol, and finally it is delivered on the Bluetooth physical channel. Large L2CAP packets are segmented into smaller packets before their transmission. The transmission of a new L2CAP packet cannot start until all fragments (generated during the.segmentation at the MAC layer) of the

---

[12] The distinction between the downlink and the uplink polling introduces a "fairness separation": in the downlink (uplink) subcycle fairness is guaranteed only in the downlink (uplink) direction, i.e., only the slaves with traffic in the downlink (uplink) are eligible for polling.

TABLE 1.5 Simulative Scenario

	Data Flow Direction	Traffic Type	Activity Interval (sec)	Rate (kb/sec)
Slave 1	Downloading	ftp - TCP	[0–90]	—
Slave 2	Downloading	ftp - TCP	[15–75]	—
Slave 3	Downloading	ftp - TCP	[30–75]	—
Slave 4	Downloading	CBR - UDP	[40–70]	30
Slave 5	Uploading	CBR - UDP	[45–70]	10
Slave 6	Uploading	CBR - UDP	[60–90]	5
Slave 7	Downloading	CBR - UDP	[65–90]	15

previous L2CAP packet have been successfully transmitted. The segmentation procedure is accomplished just before the transmission, in such a way as to maximize the amount of data conveyed by each baseband packet (see [8])

Table 1.5 summarizes the details of the simulated scenario. For each slave, the direction of the data flow is indicated (downloading if data are retrieved from a remote Internet server or uploading if data are sent from the slave towards the Internet), along with the application and the transport protocol adopted.[13] In addition, by denoting with 0 the time instant at which each simulative experiment starts, the table reports the time interval in which each data flow is active (activity interval). The different activity intervals highlight the dynamic behavior of the scheduling algorithm. Finally, only for UDP flows, the table reports the source transmission rate.

Results reported here have been derived by assuming an ideal channel with no errors and using constant size packets — a TCP packet of 1024 bytes, a UDP packet of 500 bytes, and TCP ACKs of 20 bytes.

Figure 1.15 shows the throughput for the TCP connection of slave 1 when the scheduler adopts either the EDC or the Round Robin (RR) algorithms. First, we can observe that EDC guarantees a throughput that is always (significantly) higher than that achieved with a RR scheduler.

Figure 1.15 clearly shows the dynamic behavior of the EDC algorithm. In the first time interval, [0, 15 sec], only slave 1 is active, and hence the throughput obtained with EDC is more than twice that achieved with RR. This is expected since EDC adapts the polling rate to the sources' activity level. As the number of active sources increases, the difference between the RR and EDC performance decreases. The minimum distance between EDC and RR is achieved (as expected) when all sources are active. However, also in this case by adopting EDC, the slave 1 performances are always better than those it achieves with RR. EDC exploits the inactivity periods of the CBR sources to increase the polling frequency of the TCP slaves. One may argue that the slave 1 performance improvements are achieved by decreasing the

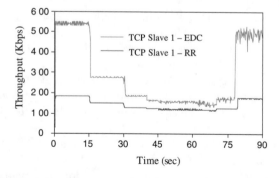

FIGURE 1.15 TCP throughput of Slave 1 connection.

---

[13] The TCP version considered is the TCP-Reno [41].

performance of the other flows inside the piconet. The results presented in [8] indicate that this is not true. Indeed, those results show that EDC is fair as:

1. The three TCP flows, when active, achieve exactly the same throughput.
2. The throughput of each CBR flow is equal to the rate of the corresponding CBR source.

However, it must be pointed out that a small degree of unfairness may exist when EDC is adopted. Unfairness may exist among TCP flows depending on their direction (i.e., master-to-slave vs. slave-to-master). Specifically, experimental results (see [8]) show that the TCP throughput slightly increases when the data packet flow is from the slave towards the master. This is due to the different polling rate (during $Cycle_{UP}$) to the slaves in the two cases. When the TCP data flow is from the master to the slave, the slave queue contains the acknowledgment traffic. When the master sends a fragment of a TCP packet to the slave, it often receives a NULL packet from the slave (the ACK cannot be generated by the TCP receiver until the TCP packet is completely received); therefore, the polling interval for that slave increases, and the scheduler will avoid polling it for some successive uplink polling subcycles. This slows down the delivery of the acknowledgment traffic and as a consequence (due to TCP congestion and flow control mechanisms), also reduces the TCP data delivery rate. On the other hand, in the slave-to-master scenario the slave queue contains the data traffic, and hence it is always highly probable to find a queued TCP packet when the master polls that slave (the TCP source is asymptotic). Therefore, in this scenario, the TCP connection is always eligible for polling in $Cycle_{UP}$. Furthermore, as soon as the ACK for the slave is generated, the master will serve it in the first available $Cycle_{DW}$ without introducing any additional delay.

To summarize, the results presented so far demonstrate that EDC significantly improves the throughput performance of TCP flows in a piconet, when compared to a RR scheduler. However, the decoupling of scheduler decisions between uplink and downlink can introduce some unfairness among data flows when the traffic in the two directions is correlated, as happens in a TCP connection.

Acknowledgment

This work was partially supported by NATO Collaborative Linkage Grant PST.CLG.977405 "Wireless access to Internet exploiting the IEEE 802.11 technology." The author thanks Giuseppe Anastasi, Raffaele Bruno, Enrico Gregori, and Veronica Vanni for fruitful discussions and their help in producing the results presented in this chapter.

References

[1] J. Ahola, Ambient Intelligence, ERCIM NEWS (European Research Consortium for Information and Mathematics), No. 47, October 2001.

[2] G. Bianchi, L. Fratta, and M. Oliveri, Performance Evaluation and Enhancement of the CSMA/CA MAC Protocol for 802.11 Wireless LANs, *Proceedings of PIMRC*, Taipei, October 1996, pp. 392–396.

[3] C. Bisdikian, An Overview of the Bluetooth Wireless Technology, *IEEE Communication Magazine*, December 2001.

[4] Specification of the Bluetooth System, Version 1.0B, December 1999.

[5] L. Bononi, M. Conti, and E. Gregori, Design and Performance Evaluation of an Asymptotically Optimal Backoff Algorithm for IEEE 802.11 Wireless LANs, Proc. HICSS-33, Maui, January 4–7, 2000.

[6] R. Bruno, M. Conti, and E. Gregori, WLAN Technologies for Mobile Ad Hoc Networks, Proc. HICSS-34, Maui, January 3–6, 2001.

[7] R. Bruno, M. Conti, and E. Gregori, A simple protocol for the dynamic tuning of the backoff mechanism in IEEE 802.11 networks, *Computer Networks*, 37, 33–44,

[8] R. Bruno, M. Conti, and E. Gregori, Wireless Access to Internet via Bluetooth: Performance Evaluation of the EDC Scheduling Algorithm, 1st ACM Workshop on Wireless Mobile Internet, Rome, 2001, pp. 43–49.

[9]   R. Bruno, M. Conti, and E. Gregori, Bluetooth: architecture, protocols and scheduling algorithms, *Cluster Computing Journal,* 5,(2) April 2002.

[10] R. Bruno, M. Conti, and E. Gregori, Traffic integration in personal, local and geographical wireless networks, in *Handbook of Wireless Networks and Mobile Computing,* I. Stojmenovic, Ed., John Wiley & Sons, New York, 2002.

[11] R. Bruno, M. Conti, and E. Gregori, Optimization of efficiency and energy consumption in p-persistent CSMA-based wireless LANs, *IEEE Transactions on Mobile Computing,* 1(1), 2002.

[12] Web site of the Bluetooth Special Interest Group: http://www.bluetooth.com/

[13] F. Calì, M. Conti, and E. Gregori, Dynamic IEEE 802.11: design, modeling and performance evaluation, *IEEE Journal on Selected Areas in Communications,* 18, 1774–1786, 2000.

[14] F. Calì, M. Conti, and E. Gregori, Dynamic tuning of the IEEE 802.11 protocol to achieve a theoretical throughput limit, *IEEE/ACM Transactions on Networking,* 8, 785–799, 2000.

[15] M. Conti, E. Gregori, and L. Lenzini, Metropolitan Area Networks, Springer Limited Series on Telecommunication Networks and Computer Systems, November 1997.

[16] M. Conti and S. Giordano, special issue on "Mobile Ad Hoc Networking," *Cluster Computing Journal,* 5(2), April 2002.

[17] M.S. Corson, J.P. Maker, and J.H. Cernicione, Internet-based Mobile Ad Hoc Networking, *IEEE Internet Computing,* July-August 1999, pp. 63–70.

[18] K. Van Dam, S. Pitchers, and M. Barnard, Body Area Networks: Towards a Wearable Future, Proc. Wireless World Research Forum, Munich, March 2001.

[19] K. Van Dam, S. Pitchers, and M. Barnard, From PAN to BAN: Why Body Area Networks, Proc. Wireless World Research Forum, Munich, March 2001.

[20] A. Das, A. Ghose, A. Razdan, H. Saran, and R. Shorey, Efficient Performance of Asynchronous Data Traffic over Bluetooth Wireless Ad-hoc Network, Proc. IEEE INFOCOM 2001, Anchorage, April 2001.

[21] S. Ditlea, The PC Goes Ready to Wear, IEEE Spectrum, October 2000.

[22] S. Galli, K.D. Wong, B.J. Koshy, and M. Barton, Bluetooth Technology: Link Performance and Networking Issues, Proc. European Wireless 2000, Dresden, Germany, September 2000.

[23] J.C. Haartsen and S. Zurbes, Bluetooth Voice and Data Performance in 802.11 DS WLAN Environment, Technical Report Ericsson, May 1999.

[24] J.L. Hammond and P.J.P. O'Reilly, *Performance Analysis of Local Computer Networks,* Addison-Wesley Publishing Company, Reading, MA, 1988.

[25] Web site of the IEEE 802.11 WLAN: http://grouper.ieee.org/grups/802/11/main.html

[26] Web site of the IEEE 802.15 WPA.N Task Group 1: http://www.ieee802.org/15/pub/TG1.html

[27] IEEE standard for Wireless LAN: Medium Access Control and Physical Layer Specification, P802.11, November 1997. See also IEEE P802.11/D10, 14 January 1999.

[28] N. Johansson, U. Korner, and P. Johansson, Performance Evaluation of Scheduling Algorithm for Bluetooth, Proc. IFIP Broadband Communications, Hong Kong, November 1999.

[29] M. Kalia, D. Bansal, and R. Shorey, Data Scheduling and SAR for Bluetooth MAC, Proc. IEEE VTC 2000, Tokyo, May 2000.

[30] J.F. Kurose, M. Schwartz, and Y. Yemini, Multiple access protocols and time constraint communications, *ACM Computing Surveys,* 16, 43–70,

[31] C. Law, A.K. Mehta, and K.Y. Siu, A New Bluetooth Scatternet Formation Protocol, *ACM/Kluver Mobile Networks and Applications Journal,* Special Issue on Ad Hoc Networks, A.T. Campbell, M. Conti, and S. Giordano, Eds., 8(5), Oct. 2003.

[32] Levis® ICD+™, http://www.levis-icd.com

[33] The IETF Mobile Ad-Hoc Networking Page (MANET): http://www.ietf.org/html.charters/Manet-charter.html

[34] B.A. Miller and C. Bisdikian, *Bluetooth Revealed,* Prentice Hall, New York, 2000.

[35] The MIThril Home Page, http://www.media.mit.edu/wearables//mithril/index.html

[36] NFS Tetherless T3 and Beyond Workshop, interim report, November 1998.

[37] NFS Wireless Information Technology and Networks Program Announcement NSF 99–68, 1999.

[38] http://www.research.philips.com/password/pw3/pw3_4.html

[39] W. Stallings, *Local & Metropolitan Area Networks*, Prentice Hall, New York, 1996.

[40] M. Stemm and R.H. Katz, Measuring and Reducing Energy Consumption of Network Interfaces in Hand-Held Devices, Proc. 3rd International Workshop on Mobile Multimedia Communications (MoMuC-3), Princeton, NJ, September 1996.

[41] W.R. Stevens, *TCP/IP Illustrated*, Vol. 1, Addison Wesley, Reading, MA, 1994.

[42] F.A. Tobagi and L. Kleinrock, Packet Switching in Radio Channels: Part II, *IEEE Trans on Comm*, 23, 1417–1433, 1975.

[43] S. Xu and T. Saadawi, Does the IEEE 802.11 MAC Protocol Work Well in Multihop Wireless Ad Hoc Networks? *IEEE Communications Magazine*, June 2001, pp. 130–137.

[44] V. Vanni, Misure di prestazioni del protocollo TCP in reti locali Ad Hoc, Computer Engineering Laurea Thesis, Pisa, February 2002 (in Italian).

[45] WaveLAN IEEE 802.11, PC Cards User's Guide, Lucent Technology, 1999.

[46] J. Weinmiller, M. Schläger, A. Festag, and A. Wolisz, Performance study of access control in wireless LANs—IEEE 802.11 DFWMAC and ETSI RES 10 HIPERLAN, *Mobile Networks and Applications*, 2, 55–67, 1997.

[47] M. Weiser, The Computer for the Twenty-First Century, *Scientific American*, September 1991.

[48] T.G. Zimmerman, Personal Area Networks: near-field intrabody communication, *IBM Systems Journal*, 35(3&4), 1996.

[49] T.G. Zimmerman, Wireless networked devices: a new paradigm for computing and communication, *IBM Systems Journal*, 38(4), 1999.

[50] G. Zussman and A. Segall, Capacity Assignment in Bluetooth Scatternets — Analysis and Algorithms, Proc. Networking 2002, LNCS 2345.

2

Multicasting Techniques in Mobile Ad Hoc Networks

Xiao Chen
Southwest Texas State University

Jie Wu
Florida Atlantic University

Abstract

This chapter gives a general survey of multicast protocols in mobile ad hoc networks (MANETs). After giving a brief summary of two multicast protocols in wired networks — *shortest path multicast tree protocol* and *core-based tree multicast protocol* — we point out limitations of these protocols when they are applied in the highly dynamic environment of MANETs. Four multicast protocols — On-Demand Multicast Routing Protocol (ODMRP), Multicast Ad Hoc On-Demand Distance Vector Routing Protocol (Multcast AODV), Forwarding Group Multicast Protocol (FGMP), and Core-Assisted Mesh Protocol — are discussed in detail with a focus on how the limitations of multicast protocols in wired networks are overcome. A brief overview of other multicast protocols in MANETs is provided. The chapter ends with two important related issues: QoS multicast and reliable multicast in MANETs.

2.1   Introduction

Multicasting is the transmission of packets to a group of zero or more hosts identified by a single destination address. Multicasting is intended for group-oriented computing. Typically, the membership of a host group is dynamic: that is, hosts may join and leave groups at any time. There is no restriction on the location or number of members in a host group. A host may be a member of more than one group at a time. A host does not have to be a member of a group to send packets to it.

In the wired environment, there are two popular network multicast schemes: the shortest path multicast tree and core-based tree. The shortest path multicast tree guarantees the shortest path to each destination, but each source needs to build a tree. Therefore, too many trees exist in the network. The core-based tree cannot guarantee the shortest path from a source to a destination, but only one tree is constructed for each group. Therefore, the number of trees is greatly reduced.

Currently, one particularly challenging environment for multicast is a *mobile ad hoc network* (MANET). A MANET consists of a dynamic collection of nodes with sometimes rapidly changing multihop topologies that are composed of relatively low-bandwidth wireless links. There is no assumption of an underlying fixed infrastructure. Nodes are free to move arbitrarily. Since each node has a limited *transmission range*, not all packets may reach all the intended hosts. To provide communication through the whole network, a source-to-destination path could pass through several intermediate neighbor nodes. For example, two nodes can communicate directly with each other only if they are within each other's transmission range. Otherwise, the communication between them has to rely on other nodes. In the mobile ad hoc network shown in Fig. 2.1, nodes A and B are within each other's transmission range (indicated by the circles around A and B, respectively). If A needs to send a packet to B, it can send it directly. A and C are not within each other's range. If A wants to send a packet to C, it has to first forward the packet to B and then use B to route the packet to C.

Unlike typical wired routing protocols, routing protocols for mobile ad hoc networks must address a diverse range of issues. In general, the main characteristics of mobile computing are low bandwidth, mobility, and low power. Wireless networks deliver lower bandwidth than wired networks do, and hence, information collection during the formation of a routing table is expensive. Mobility of hosts, which causes topological changes of the underlying network, also increases the volatility of network information. In addition, the limitation of power leads users to disconnect mobile units frequently in order to limit power consumption.

The goal of MANETs is to extend mobility into the realm of autonomous, mobile, wireless domains, where a set of nodes forms the network routing infrastructure in an ad hoc fashion. The majority of applications for the MANET technology are in areas where rapid deployment and dynamic reconfiguration are necessary and the wired network is not available. These include military battlefields, emergency search and rescue sites, classrooms, and conventions where participants share information dynamically using their mobile devices. These applications lend themselves well to multicast operations. In addition, within a wireless medium, it is even more crucial to reduce the transmission overhead and power consumption. Multicasting can improve the efficiency of the wireless link when sending multiple copies of messages by exploiting the inherent broadcast property of wireless transmission. However, besides the issues for any ad hoc routing protocol listed above, wireless mobile multicasting faces several key challenges. Multicast group members move, thus precluding the use of a fixed multicast topology. Transient loops may form during multicast tree reconfiguration, so tree reconfiguration schemes should be simple to keep channel overhead low.

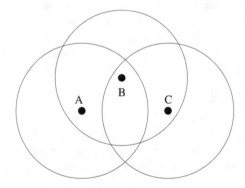

FIGURE 2.1 An example of a mobile ad hoc network.

In mobile ad hoc networks, there are three basic categories of multicast algorithms. A naive approach is to simply flood the network. Every node receiving a message floods it to a list of neighbors. Flooding a network acts like a chain reaction that can result in exponential growth. The proactive approach precomputes paths to all possible destinations and stores this information in routing tables. To maintain an up-to-date database, routing information is periodically distributed throughout the network. The final approach is to create paths to other hosts on demand. The idea is based on a query-response mechanism or reactive multicast. In the query phase, a node explores the environment. Once the query reaches the destination, the response phase starts and establishes the path.

The rest of this chapter is organized as follows: in the next section, we review two multicast routing protocols, shortest path multicast tree and core-based tree, that are widely used in wired networks. In Section 2.3, we describe four extensions in mobile ad hoc networks: two distinct on-demand multicast protocols, forwarding group multicast protocol (FGMP), and core-assisted mesh protocol. Other multicast protocols used in mobile ad hoc networks are briefly summarized in Section 2.4. Section 2.5 discusses two related issues: QoS multicast and reliable multicast. The chapter concludes in Section 2.6.

2.2 Multicast Protocols in Wired Networks

In this section, we review two multicast protocols in wired networks, namely, the shortest path multicast tree protocol and the core-based tree multicast protocol. To facilitate the discussion, in the figures in the chapter, we use black nodes to represent group members, sources, and destinations; gray nodes for forwarding nodes; and white for non-group members.

2.2.1 Shortest Path Multicast Tree

The single shortest path multicast tree can be constructed by applying Dijkstra's spanning tree algorithm [Cormen et al., 1997]. Each path from the root of the tree to a destination is a shortest path.

In this protocol, to do multicast routing, each node computes a spanning tree covering all other nodes in the network. For example, in Fig. 2.2a, we have a network with two groups, 1 and 2. Some nodes are attached to hosts that belong to one or both of these groups, as indicated in the figure. A spanning tree for node S is shown in Fig. 2.2b.

When a process sends a multicast packet to a group, the first node examines its spanning tree and prunes it, removing all lines that do not lead to hosts that are members of the group. In our example, Fig. 2.2c shows the pruned spanning tree for group 1. Similarly, Fig. 2.2d shows the pruned spanning tree for group 2. Multicast packets are forwarded only along the appropriate spanning tree.

One potential disadvantage of this algorithm is that it scales poorly to large networks. Suppose that a network has n groups, each with an average of m members. For each group, m pruned spanning trees

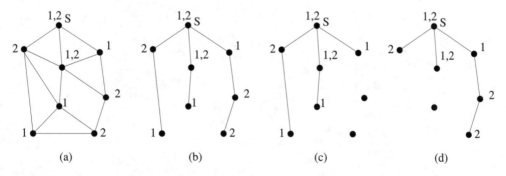

FIGURE 2.2 (a) A network. (b) A spanning tree for node S. (c) A multicast tree for group 1. (d) A multicast tree for group 2.

must be stored, for a total of *mn* trees. When many large groups exist, considerable storage is needed to store all the trees.

An alternative design uses *core-based trees* [Ballardie et al., 1993] (discussed in the following subsection). Here, a single spanning tree per group is computed, with the root (the core) near the middle of the group. To send a multicast message, a host sends it to the core, which then does the multicast along the spanning tree. Although this tree will not be optimal for all sources, the reduction in storage costs from *m* trees to one tree per group is a major saving.

2.2.2 Core-Based Trees Multicast Protocol

A core-based tree (CBT) involves having a single node, known as the *core* of the tree, from which branches emanate. These branches are made up of other nodes, so-called *noncore* nodes, which form a shortest path between a member-host's directly attached node and the core. A node at the end of a branch shall be known as a *leaf* node on the tree. The core need not be topologically centered between the nodes on the tree, since multicasts vary in nature, and so can the form of a core-based tree.

CBT involves having a single core tree per group, with additional cores to add an element of robustness to the model. Since there exists no polynomial time algorithm that can find the center of a dynamic multicast spanning tree, a core should be "hand-picked," i.e., selected by external agreement based on a judgment of what is known about the network topology among the current members.

A node can join the group by sending a JOIN_REQUEST. This message is then forwarded to the next-hop node on the path to the core. The join continues its journey until it either reaches the core or reaches a CBT-capable node that is already part of the tree. At this point, the join's journey is terminated by the receiving node, which normally sends back an acknowledgment by means of a JOIN_ACK. It is the JOIN_ACK that actually creates a tree branch. Figure 2.3 shows the procedure of a node joining a group.

A noncore node can leave the group by sending a QUIT_REQUEST. A QUIT_REQUEST may be sent by a node to detach itself from a tree if and only if it has no members for that group on any directly attached subnets, and it has received a QUIT_REQUEST on each of its children for that group. The QUIT_REQUEST is sent to the parent node. The parent immediately acknowledges the QUIT_REQUEST with a QUIT_ACK and removes that child from the tree. Any noncore node that sends a QUIT_ACK in response to receiving a QUIT_REQUEST should itself send a QUIT_REQUEST upstream if the criteria described above are satisfied.

For any noncore node, if its parent node or path to the parent fails, that noncore node has one of two options for failure recovery: it can either attempt to rejoin the tree by sending a JOIN_REQUEST to the highest-priority reachable core or alternatively, the node subordinate to the failure can send a

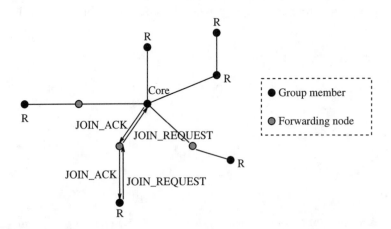

FIGURE 2.3 The member join procedure in CBT.

FLUSH_TREE message downstream, thus allowing each node to independently attempt to reattach itself to the tree.

For reasons of robustness, we need to consider what happens when a primary core fails. There are two approaches we can take:

- *Single-core CBT trees.* If paths or cores fail, a single tree can itself become partitioned. To cater for tree partitions, we have multiple "backup" cores to increase the probability that every network node can reach at least one of the cores of a CBT tree. At any one time, a noncore node is part of a single-core CBT tree.

- *Multi-core CBT trees.* Multi-core CBT trees are most useful for groups that are topologically widespread. Each core is then strategically placed where the largest "pockets" of members are located so as to optimize the routes between those members. Each core must be joined to at least one other, and a reachability/maintenance protocol must operate between them. No ordering between the multiple cores exists, and senders send multicasts preferably to the nearest core.

2.3   Multicast Protocols in Mobile Ad Hoc Networks

In the highly dynamic environment of mobile ad hoc networks, the traditional multicast approaches used in wired networks are no longer suitable. Because nodes in these networks move arbitrarily, network topology changes frequently and unpredictably. Moreover, bandwidth and battery power are limited. These constraints, in combination with the dynamic network topology, make multicasting in mobile ad hoc networks extremely challenging. The general solutions used in the protocols to solve these problems are: avoid global flooding and advertising, dynamically build routes and maintain memberships, etc. In this section, we introduce four extensions of multicast protocols in mobile ad hoc networks.

2.3.1   On-Demand Multicast Routing Protocol (ODMRP)

ODMRP (on-demand multicast routing protocol) [Bae et al., 2000] is mesh-based and uses a forwarding group concept (only a subset of nodes forwards the multicast packets). A soft-state approach is taken in ODMRP to maintain multicast group members. No explicit control message is required to leave the group.

In ODMRP, group membership and multicast routes are established and updated by the source on demand. Consider the example in Fig. 2.4a. The source S, desiring to send packets to a multicast group but having no route to the multicast group, will broadcast a JOIN_DATA control packet to the entire network. This JOIN_DATA packet is periodically broadcast to refresh the membership information and update routes.

When an intermediate node receives the JOIN_DATA packet, it stores the source ID and the sequence number in its message cache to detect any potential duplicates. The routing table is updated with the appropriate node ID (i.e., backward learning) from which the message was received for the reverse path back to the source node. If the message is not a duplicate and the time-to-live (TTL) is greater than zero, it is rebroadcast.

When the JOIN_DATA packet reaches a multicast receiver, it creates and broadcasts a JOIN_TABLE to its neighbors. When a node receives a JOIN_TABLE, it checks to see if the next hop node ID of one of the entries matches its own ID. If it does, the node realizes that it is on the path to the source and thus is part of the forwarding group and sets the FG_FLAG (forwarding group flag). It then broadcasts its own join table built on matched entries. The next hop node ID field is filled by extracting information from its routing table. In this way, each forward group member propagates the JOIN_TABLE until it reaches the multicast source S via the selected path (shortest). Figure 2.4b shows how these packets are forwarded to S. On receiving JOIN_TABLES, a node also has to build its multicast table for forwarding future multicast packets. For example, when B receives R_2's JOIN_TABLE, it will add R_2 as its next hop. The final multicast table for each host is shown in Fig. 2.4c. This whole process constructs (or updates) the routes from sources to receivers and builds a mesh of nodes called the forwarding group.

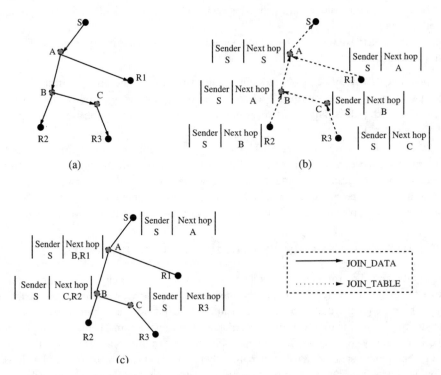

FIGURE 2.4 An example of ODMRP. (a) Propagation of JOIN_DATA packets. (b) Propagation of JOIN_TABLE packets. (c) The final multicast tables.

After the forwarding group establishment and route construction process, sources can multicast packets to receivers via selected routes and forwarding groups. While it has data to send, the source periodically sends JOIN_DATA packets to refresh the forwarding group and routes. When receiving the multicast data packet, a node forwards it only when it is not a duplicate and the setting of the FG_FLAG for the multicast group has not expired. This procedure minimizes the traffic overhead and prevents sending packets through stale routes.

In ODMRP, no explicit control packets need to be sent to join or leave the group. If a multicast source wants to leave the group, it simply stops sending JOIN_DATA packets, since it does not have any multicast data to send to the group. If a receiver no longer wants to receive from a particular multicast group, it does not send the join reply for that group. Nodes in the forwarding group are demoted to nonforwarding nodes if not refreshed (no join tables received) before they time out.

2.3.2 Multicast Ad Hoc On-demand Distance Vector Routing Protocol (Multicast AODV)

The MAODV routing protocol [Royer and Perkins, 1999] discovers multicast routes on demand using a broadcast route-discovery mechanism. A mobile node originates a route request (*RREQ*) message when it wishes to join a multicast group or when it has data to send to a multicast group but does not have a route to that group. Figure 2.5a illustrates the propagation of RREQ (represented in the graph by solid arrow) from a host *S*. Only a member of the desired multicast group may respond to a join RREQ. If the RREQ is not a join request, any node with a fresh enough route (based on group sequence number) to the multicast group may respond. If an intermediate node receives a join RREQ for a multicast group of which it is not a member, or if it receives a RREQ and it does not have a route to that group, it rebroadcasts the RREQ to its neighbors.

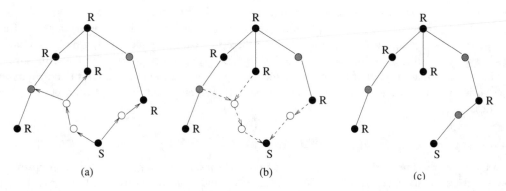

FIGURE 2.5　An example of MAODV protocol. (a) The propagation of RREQ packets. (b) The propagation of RREP packets. (c) The final multicast tree.

As the RREQ is broadcast across the network, nodes set up pointers to establish the reverse route in their route tables. A node receiving an RREQ first updates its route table to reverse route; entry may later be used to relay a response back to S. For join RREQs, an additional entry is added to the multicast route table. This entry is not activated unless the route is selected to be part of the multicast tree.

If a node receives a join RREQ for a multicast group, it may reply if it is a member of the multicast group's tree and its recorded sequence number for the multicast group is at least as great as that contained in the RREQ. The responding node updates its route and multicast route tables by placing the requesting node's next-hop information in the tables, and then unicasts a request response *RREP* (represented in the graph by dashed arrow) back to S (shown in Fig. 2.5b). As nodes along the path to the source node S receive the RREP, they add both a route table and a multicast route table entry for the node from which they received the RREP, thereby creating the forward path.

When S broadcasts a RREQ for a multicast group, it often receives more than one reply. The source node S keeps the received route with the greatest sequence number and shortest hop count to the nearest member of the multicast tree for a specified period of time and disregards other routes. At the end of this period, it enables the selected next hop in its multicast route table, and unicasts an activation message to this selected next hop. The next hop, on receiving this message, enables the entry for S in its multicast route table. If this node is a member of the multicast tree, it does not propagate the message any further. However, if this node is not a member of the multicast tree, it will have received one or more RREPs from its neighbors. It keeps the best next hop for its route to the multicast group, unicasts an activation message to that next hop, and enables the corresponding entry in its multicast route table. This process continues until the node that originated the RREP (member of tree) is reached. The activation message ensures that the multicast tree does not have multiple paths to any tree node. Nodes only forward data packets along activated routes in their multicast route tables. Figure 2.5c illustrates the final multicast tree that is created.

The first member of the multicast group becomes the leader for that group. The multicast group leader is responsible for maintaining the multicast group sequence number and broadcasting this number to the multicast group. This is done through a group hello message. The group hello contains extensions that indicate the multicast group Internet Protocol (IP) address and sequence numbers (incremented every group hello) of all multicast groups for which the node is the group leader. Nodes use the group hello information to update their request tables.

Since AODV maintains hard state in its routing table, the protocol has to actively track and react to changes in this tree. If a member terminates its membership with the group, the multicast tree requires pruning. Links in the tree are monitored to detect link breakages. When a link breakage is detected, the node that is furthest from the multicast group leader (downstream of the break) is responsible for repairing the broken link. If the tree cannot be reconnected, a new leader for the disconnected downstream node is chosen as follows. If the node that initiated the route rebuilding is a multicast group member, it becomes the new multicast group leader. On the other hand, if it was not a group member and has only one next hop for the tree, it prunes itself from the tree by sending its next hop a prune message. This continues until a group member is reached. Once these two partitions reconnect, a node eventually receives a group hello for the multicast group that contains group leader information that differs from the information it already has. If this node is a member of the multicast group, and if it is a member of the partition in which the group leader has the lower IP address, it can initiate reconnection of the multicast tree.

2.3.3 Forwarding Group Multicast Protocol (FGMP)

In a highly dynamic network such as a mobile ad hoc network, multicast protocols based on upstream and downstream links (such as CBT [Ballardie et al., 1993] and DVMRP [Deering and Cheriton, 1990]) are not efficient because creating and maintaining upstream and downstream link status in a wireless network cause a lot of overheads.

In [Chiang et al., 1998], the authors put forward *forwarding group multicast protocol*. The protocol keeps track not of links but of groups of nodes that participate in multicast packets forwarding. To each multicast group G is associated a forwarding group, FG. Any node in FG is in charge of forwarding (broadcast) multicast packets of G. That is, when a forwarding node (a node in FG) receives a multicast packet, it will broadcast this packet if it is not a duplicate. All neighbors can hear it, but only neighbors that are in FG will first determine if it is a duplicate and then broadcast it in turn. Figure 2.6 shows an example of a multicast group containing two senders and two receivers. Four forwarding nodes take the responsibility to forward multicast packets. This scheme can be viewed as "limited scope" flooding. That is, flooding is contained within a properly selected forwarding set. It is interesting to note that with proper selection of the forwarding group, the FG scheme can emulate any of the existing schemes. For example, to produce global flooding, the FG must include all nodes in the network. For CBT, the FG is restricted to the nodes on the shared tree except the leaf nodes. In DVMRP, FG includes all the nonleaf nodes on the source trees.

Only one flag and a timer are needed for each forwarding node. When the forwarding flag is set, each node in FG forwards data packets belonging to G until the timer expires. Only the forwarding flag and timer are stored, thus reducing the storage overhead and increasing the flexibility and performance. In FGMP, only small size control messages are flooded and with less frequency.

The major problem of FGMP is how to elect and maintain the set FG of forwarding nodes. The size of FG should be as small as possible to reduce wireless channel overhead, and the forwarding path from senders to receivers should be as short as possible to get high throughput. Three schemes are discussed in the following three subsections.

One way to advertise the membership is to let each receiver periodically and globally flood its member information. When a sender receives the join request from receiver members, it updates the

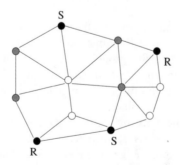

FIGURE 2.6 An example of FGMP.

member table. Expired receiver entries will be deleted from the member table. The sender will broadcast multicast data packets only if the member table is not empty. After updating the member table, the sender creates from it the forwarding table *FW*. Next hop information is obtained from preexisting routing tables. The forwarding table *FW* is broadcast by the sender to all neighbors; only neighbors listed in the next hop list (next hop neighbors) accept this forwarding table (although all neighbors can hear it). Each neighbor in the next hop list creates its forwarding table by extracting the entries where it is the next hop neighbor and again using the preexisting routing table to find the next hops, etc. After the *FW* table is built, it is then broadcast again to neighbors and so on, until all receivers are reached. The forwarding group is created and maintained via the forwarding table *FW* exchanges. At each step, nodes on the next hop neighbor list after receiving the forwarding table enable the forwarding flag and refresh the forwarding timer.

Another way to advertise the membership is to let senders flood sender information. Sender advertising is more efficient than receiver advertising if the number of senders is less than the number of receivers. Most multicast applications belong to this category. As with receiver advertising, senders periodically flood the sender information. Receivers will collect senders' status, then periodically broadcast "joining tables" to create and maintain the forwarding group *FG*. The "joining table" has the same format as the "forwarding table" except that the joining table contains the sender IDs while the forwarding table contains receiver IDs. The forwarding flag and timer are set when a node receives the joining table. The forwarding group is maintained by the senders in the receiver advertising scheme and by the receivers in sender advertising scheme.

Wu [2002] proposed a method of *FW* maintenance via connected dominating set. A set is dominating if all the nodes in the network are either in the set or neighbors of nodes in the set. Wu and Li [2001] proposed a simple localized algorithm to define a connected dominating set: a host is selected as a dominating set if it has two unconnected neighbors. The size of the dominating set is reduced by two distributed pruning rules; a dominating node can be removed if its neighbor set is covered either by the neighbor set of another node with higher ID or by the union of neighbor sets of two connected nodes both with higher IDs. With connected dominating sets as the basic "spine" of the network, FGMP is built and maintained on top of the set.

2.3.4   Core-Assisted Mesh Protocol

Many multicasting protocols today involve routing trees. Multicast trees can achieve efficiency and simplicity by forcing a single path between any pair of nodes. If multiple sources must transmit information to the same set of destinations, using routing trees requires that either a shared multicast tree be used for all sources or that a separate multicast tree be established for each source. Using a shared multicast tree has the disadvantage that packets are distributed to the multicast group along paths that can be much longer than the shortest paths from sources to receivers. Using a separate multicast tree for each source of each multicast group forces the nodes that participate in multiple multicast groups to maintain an entry for each source in each multicast group, which does not scale as the number of groups and sources per group increases. In addition, because trees provide minimal connectivity among the members of a multicast group, the failure of any link in the tree partitions the group and requires the nodes involved to reconfigure the tree.

For these reasons, Garcia and Madruga [1999] put forward *multicast meshes* and the corresponding *Core-Assisted Mesh Protocol* (CAMP). CAMP builds and maintains a multicast mesh for information distribution within each multicast group. A multicast mesh is a subset of the network topology that provides at least one path from each source to each receiver in the multicast group. CAMP ensures that the shortest paths from receivers to sources (called reverse shortest paths) are part of a group's mesh. Packets are forwarded through the mesh along the paths that first reach the nodes from the sources, i.e., the shortest paths from sources to receivers that can be defined within the mesh. CAMP does not predefine such paths along the mesh. A node keeps a cache of the identifiers of those packets it has forwarded recently, and forwards a multicast packet received from a neighbor if the packet identifier is not in its

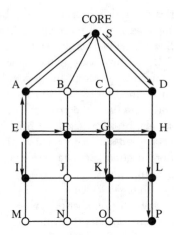

FIGURE 2.7 Traffic flow from node *E* in a multicast mesh.

cache and the node has been told by at least one neighbor that the node is the successor in a reverse shortest path to any group source.

Because a member node of a multicast mesh has redundant paths to any other node in the same mesh, topology changes are less likely to disrupt the flow of multicast data and to require the reconstruction of the routing structures that support packet forwarding. Figure 2.7 illustrates how packets are forwarded from node *E* to the rest of the group in CAMP. In the figure, solid arrows indicate the flow of traffic along the reverse shortest path in CAMP. Note that CAMP delivers data along shorter paths than a core-based tree protocol does: e.g., *E* to *F* takes only one hop in this example but four hops in core-based tree protocol. The length of paths incurred in multicasting over mobile ad hoc networks is very important because longer paths require more nodes forwarding packets.

In CAMP, cores are used to limit the control traffic needed for receivers to join multicast groups. One or multiple cores can be defined for each mesh. Cores need not be part of the mesh of their group. The use of cores in CAMP eliminates the need for flooding, unless all cores are unreachable from a connected component.

2.4   Other Multicast Protocols in Mobile Ad Hoc Networks

In recent years, many new multicast protocols in mobile ad hoc networks have been proposed by considering Global Positioning System (GPS), IP multicast, prefix routing, zone routing, differential destination, real-time, bandwidth, and ID numbers. These protocols are explained as follows.

In [Basagni et al., 2000], an on-demand location aware multicast protocol (OLAM) is introduced. The protocol assumes that, through the use of positioning system devices such as GPS devices, each node knows its own position and the current (global) time, and it is able to efficiently distribute these measures, including its current transmission radius, to all other nodes. As the measures are received, each node updates its local snapshot of the complete network topology. When a packet is to be multicast to a group, a heuristic is then used to locally compute the Steiner (i.e., multicast) tree for the addressed multicast group based on the snapshot rather than maintaining the tree in a distributed manner. The resulting Steiner tree is then optimally encoded by using its unique Prufer sequence and included along with the packet, extending the length of the header by no more than the header of packets in source routing (unicast) techniques. All local computations are executed using efficient (i.e., polynomial time) algorithms. The protocol has been simulated in mobile ad hoc networks with 30 and 60 nodes and with different multicast group sizes. The results show that OLAM delivers packets to all the nodes in a destination group in more than 85% of the cases. Furthermore, compared to flooding, OLAM achieves improvements of up to 50% on multicast completion delay.

In [Bommaiah et al.], an approach for robust IP multicast in mobile ad hoc networks by exploiting user-multicast tree and dynamic logical cores, called Adhoc Multicast Routing Protocol (AMRoute) is presented. It creates a bidirectional shared tree for data distribution using only group senders and receivers as tree nodes. Unicast tunnels are used as tree links to connect neighbors on the user-multicast tree. Thus, group state cost is incurred only by group senders and receivers, and tree structure does not need to change even in case of a dynamic network topology. Certain tree nodes are designated by AMRoute as logical cores and are responsible for initiating and managing the signaling component of AMRoute, such as detection of group members and tree setup. Simulation results demonstrate that AMRoute signaling traffic and join latency remain at relatively low levels for typical group sizes. The results also indicate that group members receive a high proportion of data multicast by a sender, even in the case of a dynamic network.

In [Chen and Jia, 2001], an efficient routing algorithm for MANETs is discussed. The fundamental idea is to build a spanning tree generation by generation with a generation table associated with each mobile station. Once the tree and generation tables are built, routing can proceed along the tree branches. This routing algorithm has a small transmission delay and needs only a small and easy-to-update generation table for each station. In addition, the communication channels in the routing algorithm are dynamically assigned, which allows a large MANET to use a limited number of communication channels. This routing algorithm is combined with the *prefix routing* algorithm in [Chen et al., 2002]. Prefix routing is a special type of routing with a compact routing table (called generation table in the paper) associated with each station. Basically, each station is assigned a special label, and it is selected as a forwarding node if its label is a prefix of the label of the destination node. The routing process follows a two-phase process of first going up and then going down the spanning tree, with a possible cross transmission (shortcut) between two branches of the tree between two phases.

In [Devarapalli and Sidhu, 2001], a multicast routing protocol based on zone routing (MZR) is proposed. MZR is a source-initiated on-demand protocol, in which a multicast delivery tree is created using a concept called the zone routing mechanism. It is a source tree based protocol and does not depend on any underlying unicast protocol. The protocol's reaction to topological changes can be restricted to a node's neighborhood instead of propagating throughout the network.

In [Ji and Corson, 2001], a protocol termed differential destination multicast (DDM) is proposed. It differs from other approaches in two ways. First, instead of distributing membership control throughout the network, DDM concentrates this authority at the data sources (i.e., senders), thereby giving sources knowledge of group membership. Second, differentially encoded, variable-length destination headers are inserted in data packets, which are used in combination with unicast routing tables to forward multicast packets towards multicast receivers. Instead of requiring the multicast forwarding state to be stored in all participating nodes, this approach also provides the option of stateless multicasting. Each node independently has the choice of caching the forwarding state, having its upstream neighbor insert this state into self-routed data packets, or some combination thereof. The protocol is best suited for use with small multicast groups operating in dynamic networks of any size.

In [Kondylis et al., 2000], a protocol for multicasting real-time data called the Wireless Ad Hoc Real-Time Multicast (WARM) protocol is proposed. The protocol is distributed, highly adaptive, and flexible. Multicast affiliation is receiver initiated. The messaging is localized to the neighborhood of the receiving multicast member, and thus the overhead consumed is low. The protocol enables spatial bandwidth reuse along a multicast mesh (a connected structure of multicast group members). The real time connection is guaranteed quality of service (QoS) in terms of bandwidth. For VBR traffic, a combination of reserved and random access mechanisms is used. The protocol is self-healing in the sense that the mesh structure has the ability to repair itself when members move or relays fail. The simulation results show that the throughput is above 90% for pedestrian environments.

In [Ozaki et al., 2001], a bandwidth-efficient multicast routing protocol for mobile ad hoc networks is proposed and investigated. The proposed protocol achieves low communication overhead; it requires a small number of control packet transmissions for route setup and maintenance. The proposed protocol also achieves high multicast efficiency; it delivers multicast packets to receivers with a small number of transmissions. In order to achieve low communication overhead and high multicast efficiency, the proposed protocol employs the following mechanisms:

1. On-demand invocation of the route setup and route recovery processes to avoid periodic transmissions of control packets
2. A new route setup process that allows a newly joining node to find the nearest forwarding node to minimize the number of forwarding nodes
3. A route optimization process that detects and removes unnecessary forwarding nodes to eliminate redundant and inefficient routes

The simulation results show that the proposed protocol achieves high multicast efficiency with low communication overhead compared with other existing multicast routing protocols, especially in the case where the number of receivers in a multicast group is large.

In [Wu et al., 1998], a multicast protocol called AMRIS, short for Ad hoc Multicast Routing protocol utilizing Increasing ID numbers, is introduced. The conceptual idea behind AMRIS is to assign every node in a multicast session an ID number. A delivery tree rooted at a particular node called Sid joins up the nodes participating in the multicast session. The relationship between the ID numbers (and the node that owns them) and Sid is that the ID numbers increase in numerical value as they radiate from the root of the delivery tree. The significance of the Sid is that it has the smallest ID number within that multicast session. Utilizing the ID numbers, nodes are able to adapt rapidly to changes in link connectivity. Recovery messages due to link breakages are confined to the region where the breakage occurred.

2.5 Related Issues

The main characteristic of a mobile ad hoc network is its ability to start and maintain the communication setup without the support of any existing wired or wireless infrastructure. However, scarce bandwidth, highly dynamic network topology, and an unreliable communication medium pose special challenges on the design of such a network. The traditional best effort traffic scenarios are not suitable for the mobile ad hoc network. In the following, we will discuss two aspects of design related to multicast: QoS multicast and reliable multicast.

2.5.1 QoS Multicast

The notion of QoS (quality of service) was proposed to capture the qualitatively or quantitatively defined performance contract between the server and client. Specifically, QoS is a guarantee by the network to satisfy a set of predetermined service performance constraints for the client in terms of the end-to-end delay, available bandwidth, probability of packet loss, and so on. QoS in multicasting (routing) typically deals with multiple constraints on the selected routing tree (path). Assume $m(u,v)$ is the performance metric for the link (u,v) connecting host u to host v, and a path $(u, u_1, u_2,..., u_k, v)$ is a sequence of links in the multicast tree. Three types of constraints on the path are given in [26]:

1. *Additive constraints:* A constraint is additive if

$$m(u, v) = m(u, u_1) + m(u_1, u_2) + ... + m(u_k, v)$$

For example, the end-to-end *delay*(u,v) is an additive constraint that is equivalent to the summation of delays at each link.
2. *Multiplicative constraints:* A constraint is multiplicative if

$$m(u, v) = m(u, u_1) \times m(u_1, u_2) \times ... \times m(u_k, v)$$

The probability *prob*(u,v) for a packet to reach v from u is the product of individual link probabilities.
3. *Concave constraints:* A constraint is concave if

$$m(u, v) = min\{m(u, u_1), m(u_1, u_2), \ldots, m(u_k, v)\}$$

The bandwidth *band(u,v)* available along the path from u to v is the minimum bandwidth among the links on the path.

Based on the above classification of constraints, Wang and Hou [2000] gave a list of twelve combinations with multiple constraints. It has been proven in [Wang and Crowcroft, 1996] that any multiple constraints with two or more type 1 and/or 2 constraints are NP-complete; otherwise, they are tractable. Various approximation methods exist for QoS constraints that are NP-complete. One commonly used approach is *sequential filtering*, where paths based on a single primary metric (say bandwidth) are selected first, and a subset of them is eliminated by optimizing over the secondary metric (say delay), and so on.

The mobility of mobile ad hoc networks adds another dimension of difficulty. Highly mobile hosts in the ad hoc network will make any QoS constraints unobtainable. Therefore, it is assumed that the mobile ad hoc network under consideration is *combinatorially stable* [Chakrabarti and Mishra, 2001]: under a specific time window, the topology changes occur sufficiently slowly to allow successful propagation of all topology updates as necessary.

QoS routing (multicast) depends on the accurate availability of the current network state, which is expensive to maintain because of network dynamics and aggregation in large networks. The *imprecise network state model* [Chen, 1999] is a promising approach and provides a cost-effective solution for QoS routing (multicast) based on imprecise network information. Most QoS routing (multicast) is reservation-based; probe signals are sent out to find QoS route(s) to the destination (one or more connecting hosts on the multicast or core tree). Because of network dynamics and imprecise state information, reserved QoS route(s) need to be reaffirmed periodically by sending special control packets, called *refreshers*, from the destination (connecting host on the multicast or core tree) back to the source. Another approach is the use of *soft state* to tree/state maintenance: the state and reservation kept at each node periodically time out.

The main issues in providing QoS multicasting in mobile ad hoc networks are (1) locating a QoS route and (2) maintaining desired QoS on a multicast or core tree. We use the core-based tree as an example. In this case, the shortest path from a requesting host to the core may not be the best QoS route. In [Banerjea et al., 2000], a QoS route is found through a TTL-based bid-order broadcast. On-tree routes that receive the broadcast message become candidates and return bid messages. To maintain desired QoS on a multicast or core tree, each join request message carries relevant QoS parameters. The core/on-tree node conducts a set of eligibility tests to decide whether or not a new member can join.

Normally, routing and resource reservation are treated separately. It is an open problem whether to consider these two related issues at one stage, rather than two separate stages. The tradeoff between the design complexity of QoS multicast protocols and the resulting performance improvement, especially in large-scale networks, still remains as a challenging issue. The models of imprecise information to support QoS routing/multicasting still need to be developed.

2.5.2 Reliable Multicast

The design of reliable multicast depends on the following three decisions [Petitt, 1997]: (1) by whom errors are detected, (2) how error messages are signaled, and (3) how missing packets are retransmitted. The first two of these decisions are normally handled jointly.

In the sender-initiated approach, the sender is responsible for the error detection. Error messages are signaled using ACK signals sent from each receiver. A missing piece of data at a receiver is detected if the sender does not receive an ACK from the receiver. In this case, the need to retransmit a missing packet is handled by retransmitting the missing data from the source through a unicast. When several receivers have missing packets, the sender may decide to remulticast the missing packets to all receivers in the multicast group.

In the receiver-initiated approach, each receiver is responsible for error detection. Instead of acknowledging each multicast packet, each receiver sends a NACK once it detects a missing packet. If multicast packets are timestamped using a sequence number, a missing packet can be detected by a gap between sequence numbers of the receiving packets.

When the sender-initiated approach is applied, only the sender (which keeps the history of multicast packets) is responsible for retransmitting the missing packet, and the corresponding retransmitting method is called sender-oriented. Note that when the sender receives ACK signals from all the receivers, the corresponding packet can be removed from the history.

There are three ways to retransmit the missing packet when the receiver-initiated approach is used: (1) sender-oriented, (2) neighborhood-oriented, and (3) fixed-neighborhood-oriented. These methods differ by the locations of the copies of missing packets. These locations are also called copy sites, which include the sender. Note that when several receivers have the same missing packet, multicast NACK signals will be sent to the copy site(s). To ensure that at most one NACK is returned to the sender per packet transmission, when a receiver detects a missing error, it waits a random period of time before broadcasting a NACK to the sender and all other receivers. This process is called *NACK suppression* since a receiver will cancel its broadcast if it receives a NACK that corresponds to a packet it has missed.

In the sender-oriented approach, senders can either unicast to a receiver (that needs the missing packet) or multicast to all the receivers in the multicast group. In the neighborhood-oriented approach, the receiver that needs the missing packet searches its neighborhood for a group member that has kept a copy of the missing packet. The search process uses a TTL-based unicast process or TTL-based broadcast process. The search space is either limited to the multicast tree (but now it is rooted at the receiver) or without limitation. In the fixed-neighborhood-oriented approach, the copy sites are fixed to a subgroup or each receiver has a "buddy" in the multicast group; buddies back up each other.

Mobility of mobile ad hoc networks adds complexity in achieving reliability. When a host moves from one neighborhood to another, proper handoff protocols are needed. For example, when host u has just completed its forwarding process to its neighbor v, host w, a neighbor of v, moves away from the neighborhood of v and enters the neighborhood of u. To ensure that host w gets a copy of the packet, u needs to keep the copy for a while and will reforward the packet (with a proper tag indicating this is a reforwarding packet) whenever a change of its neighborhood is detected.

Reliability can be achieved through other means. For example, *forward error correction* [Lucas et al., 1995] adds redundant information, which allows lost packets to be reconstructed from correctly received packets received from either a single path or multicast paths [Tsirigos and Haas, 2001]. Two other requirements, ordering and delivery semantics that are commonly used in the traditional distributed system, are still uncharted territory. Also, the way to integrate reliable multicast and QoS multicast still remains an open issue.

2.6 Conclusions

In this chapter, we first reviewed two wired multicast protocols, shortest path multicast tree and core-based tree methods, that are used widely in wired networks. Then, we described four extensions of these wired methods to mobile ad hoc networks: two distinct on-demand multicast protocols, forwarding group multicast protocol (FGMP), and core-assisted mesh protocol. Other multicast protocols used in mobile ad hoc networks have also been briefly summarized. Finally, two related issues, QoS multicast and reliable multicast, were discussed. There is a lot of work to be done in this field in the future. Simulations need to be conducted to examine the various tradeoffs and alternatives of multicast routing algorithms suitable for mobile ad hoc networks. Also, work needs to be done on the development of the multicast gateway for interconnecting wired network multicast with ad hoc based multicast.

Acknowledgment

This work was supported in part by NSF grant CCR 9900646 and grant ANI 0073736.

References

[1] citeseer.nj.nec.com/lucas95distributed.html.

[2] S.H. Bae, S.J. Lee, W. Su, and M. Gerla, The Design, Implementation, and Performance Evaluation of the On-demand Multicast Routing Protocol in Multihop Wireless Networks, *IEEE Network*, Jan./Feb. 2000, pp. 70–77.

[3] T. Ballardie, P. Francis, and J. Crowcroft, Core Based Trees (CBT): An Architecture for Scalable Inter-Domain Multicast Routing, *Proc. of ACM SIGCOMM '93*, 1993, p. 85.

[4] A. Banerjea, M. Faloutsos, and R. Pankaj, Designing QoSMIC: a Quality of Service Sensitive Multicast Internet Protocol, submitted as Internet Draft IETF in the IDMR working group, 2000.

[5] S. Basagni, I. Chlamtac, V.R. Syrotiuk, and R. Talebi, On-demand Location Aware Multicast (OLAM) for Ad Hoc Networks, *Proc. of Wireless Communications and Networking Conference*, Sep. 2000, Vol. 3, pp. 1323–1328.

[6] E. Bommaiah, M. Liu, A. McAuley, and R. Talpade, AMRoute: Ad Hoc Multicast Routing Protocol, Internet Draft, http://www.ietf.org/internet-drafts/draft-talpade-manet-amroute-00.txt.

[7] S. Chakrabarti and A. Mishra, QoS Issues in Ad Hoc Networks, *IEEE Communications Magazine*, Feb. 2001, pp. 142–148.

[8] S. Chen, Routing Support for Providing Guaranteed End-to-End Quality-of-Service, http://www.cs.uiuc.edu/Dienst/UI/2.0/Describe/ncstrl.uiuc_cs/UIUCDCS-R-99–2090, UIUCDCS-R-99–2090, University of Illinois at Urbana–Champaign, July 1999.

[9] X. Chen and X.D. Jia, Package Routing Algorithms in Mobile Ad Hoc Networks, *Proc. of the Workshop on Wireless Networks and Mobile Computing* held in conjunction with the 2001 International Conference on Parallel Processing, Sep. 2001, pp. 485–490.

[10] X. Chen, J. Wu, and X.D. Jia, Prefix Routing in Wireless Ad Hoc Networks of Mobile Stations, Technical Report, FAU-CSE-02–13, Florida Atlantic University, Boca Raton, FL, May 2002.

[11] C.-C. Chiang, M. Gerla, and L. Zhang, Forwarding Group Multicast Protocol (FGMP) for Multihop, Mobile Wireless Networks, *Cluster Computing*, 1998, p. 187.

[12] T.H. Cormen, C.E. Leiserson, and R.L. Rivest, *Introduction to Algorithms*, MIT Press, Cambridge, MA, 1997.

[13] S.E. Deering and D.R. Cheriton, Multicast Routing in Datagram Internetworks and Extended LANs, *ACM Transactions on Computer Systems*, 1990, pp. 85–111.

[14] V. Devarapalli and D. Sidhu, MZR: A Multicast Protocol for Mobile Ad Hoc Networks, *Proc. of IEEE International Conference on Communications*, June 2001, Vol. 3, pp. 886–891.

[15] M. Faloutsos, A. Banerjea, and R. Pankaj, QoSMIC: Quality of Service Sensitive Multicast Internet Protocol, SIGCOMM citeseer.nj.nec.com/faloutsos98qosmic.html, 1998, pp. 144–153.

[16] J.J. Garcia-Luna-Aceves and E.L. Madruga, A Multicast Routing Protocol for Ad Hoc Networks, *Proc. of IEEE INFOCOM '99*, Mar. 1999, pp. 784–792.

[17] L.S. Ji and M.S. Corson, Differential Destination Multicast-a MANET Multicast Routing for Multihop, Ad Hoc Network, *Proc. of IEEE INFOCOM*, Vol. 2, Apr. 2001, p. 1192–1201.

[18] G.D. Kondylis, S.V. Krishnamurthy, S.K. Dao, and G.J. Pottie, Multicasting Sustained CBR and VBR Traffic in Wireless Ad Hoc Networks, *Proc. of IEEE International Conference on Communications*, June 2000, Vol. 1, pp. 543–549.

[19] S.J. Lee, M. Gerla, and C.C. Chiang, On-demand Multicast Routing Protocol (ODMRP) for Ad Hoc Networks, internet draft, draft-retf-manet-odmrp-01.txt, June 1999, work in progress.

[20] M.T. Lucas, B.J. Dempsey, and A.C. Weaver, Distributed Error Recovery for Continuous Media Data in Wide-Area Multicast, CS-95–52, University of Virginia, Charlottesville, July 1995.

[21] T. Ozaki, J.B. Kim, and T. Suda, Bandwidth-efficient Multicast Routing for Multihop, Ad Hoc Networks, *Proc. of IEEE INFOCOM*, Vol. 2, Apr. 2001, pp. 1182–1191.

[22] C. Perkins and E.M. Royer, Ad Hoc on Demand Distance Vector (AODV) Routing (internet draft), Aug. 1998.

[23] D.G. Petitt, Reliable Multicast Protocol Design Choices, *MILCOM 97 Proceedings*, Nov. 1997, Vol. 1, pp. 242–246.

[24] E.M. Royer and C.E. Perkins, Multicast Operation of the Ad Hoc On-Demand Distance Vector Routing Protocol, *Proc. of MobiCom*, Seattle, WA, 1999, pp. 207–218.

[25] A. Tsirigos and Z.J. Haas, Multipath Routing in the Presence of Frequent Topological Changes, *IEEE Communications Magazine*, Nov. 2001, pp. 132–138.

[26] B. Wang and C.-J. Hou, A Survey on Multicast Routing and its QoS Extension: Problems, Algorithms, and Protocols, *IEEE Network Magazine*, Jan./Feb. 2000, pp. 22–36.

 27] Z. Wang and J. Crowcroft, QoS routing for supporting resource reservation, *IEEE Journal on Selected Areas in Communications*, 14, 1228–1234, 1996.

[28] C.W. Wu, Y.C. Tay, and C.-K. Toh, Ad Hoc Multicast Routing Protocol Utilizing Increasing ID-numbers (AMRIS) Functional Specification, internet-draft, draft-ietf-manet-amris-spec00.txt, Nov. 1998, work in progress.

[29] J. Wu, Dominating-Set-Based Routing in Ad Hoc Networks, in *Wireless Networks and Mobile Computing*, I. Stojmenovic, Ed., 2002, pp. 425–460.

[30] J. Wu and H. Li, A Dominating-Set-Based Routing Scheme in Ad Hoc Wireless Networks, *Telecommunication Systems Journal*, 3, 63–84, 2001.

3

Quality of Service in Mobile Ad Hoc Networks

Satyabrata Chakrabarti
Sylvaine Algorithmics

Amitabh Mishra
Virginia Polytechnic Institute and State University

Abstract

Wireless mobile ad hoc networks consist of mobile nodes interconnected by wireless multi-hop communication paths. Unlike conventional wireless networks, ad hoc networks have no fixed network infrastructure or administrative support. The topology of such networks changes dynamically as mobile nodes join or depart the network or radio links between nodes become unusable. Supporting appropriate quality of service for mobile ad hoc networks is a complex and difficult task because of the dynamic nature of the network topology and generally imprecise network state information, and has become an intensely active area of research in the last few years. This chapter presents the basic concepts of quality of service support in ad hoc networks for *unicast* communication, reviews the major areas of current research and results, and addresses some new issues. The principal focus is on routing and security issues associated with quality of service support. The chapter concludes with some observations on the open areas for further investigation.

3.1   Introduction

Conventional wireless networks require as prerequisites a fixed network infrastructure with centralized administration for their operation. In contrast, the so-called (wireless) mobile ad hoc networks, consisting of a collection of wireless nodes, all of which may be mobile, dynamically create a wireless network among themselves without using any such infrastructure or administrative support [1,2]. Ad hoc wireless networks are *self-creating, self-organizing, and self-administering*. They come into being solely by interactions among their constituent wireless mobile nodes, and it is only such interactions that are used to provide the necessary control and administration functions supporting such networks.

Mobile ad hoc networks offer unique benefits and versatility for certain environments and certain applications. First, since they have no fixed infrastructure including base stations as prerequisites, they can be created and used any time, anywhere. Second, such networks could be intrinsically fault resilient, for they do not operate under the limitations of a fixed topology. Indeed, since all nodes are allowed to be mobile, the composition of such networks is necessarily time varying. Addition and deletion of nodes occur only by interactions with other nodes; no other agency is involved. Such perceived advantages elicited immediate interest in the early days among military, police, and rescue agencies in the use of such networks, especially under disorganized or hostile conditions, including isolated scenes of natural disaster or armed conflict. See Fig. 3.1 for a conceptual representation. In recent days, home or small office networking and collaborative computing with laptop computers in a small area (e.g., a conference or classroom, single building, convention center, etc.) have emerged as other major areas of application. These include commercial applications based on progressively developing standards such as Bluetooth [3], as well as other frameworks such as Piconet [4], HomeRF Shared Wireless Access Protocol [5], etc. In addition, people have recognized from the beginning that ad hoc networking has obvious potential use in all the traditional areas of interest for mobile computing.

Mobile ad hoc networks are increasingly being considered for complex multimedia applications, where various *quality of service* (QoS)[1] attributes for these applications must be satisfied as a set of predetermined service requirements. At a minimum, the QoS issues pertaining to delay and bandwidth management are of paramount interest. In addition, because of the use of ad hoc networks by the military and police and of increasingly common commercial applications, various security issues need to be addressed as well. Cost-effective resolution of these issues at appropriate levels is essential for widespread general use of ad hoc networking.

Mobile ad hoc networking emerged from studies on extending traditional Internet services to the wireless mobile environment. All current works, as well as our presentation here, consider ad hoc networks as a wireless extension to the Internet based on the ubiquitous Internet Protocol (IP) networking mechanisms and protocols. Today's Internet possesses an essentially static infrastructure where network elements are interconnected over traditional wire-line technology, and these elements, especially the elements providing the routing or switching functions, do not move. In a mobile ad hoc network, by definition, all the network elements move. As a result, numerous more stringent challenges must be overcome to realize the practical benefits of ad hoc networking. These include effective routing, medium (or channel) access, mobility management, power management, and security issues, all of which have effects on quality of the service experienced by the user.

The absence of a fixed infrastructure for ad hoc networks means that the nodes communicate directly with one another in a peer-to-peer fashion. The mobility of these nodes imposes limitations on their power capacity, and hence, on their transmission range; indeed, these nodes often must satisfy stringent weight limitations for portability. Mobile hosts are no longer just end systems; to relay packets generated by other nodes, each node must be able to function as a router as well. As the nodes move in and out

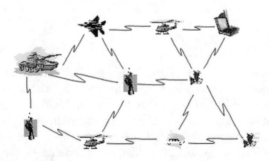

FIGURE 3.1 Conceptual representation of a mobile ad hoc network.

---

[1]We have added an acronym guide as an appendix to this chapter.

of range with respect to other nodes, including those that are operating as routers, the resulting topology changes must somehow be communicated to all other nodes as appropriate. In accommodating the communication needs of the user applications, the limited bandwidth of wireless channels and their generally hostile transmission characteristics impose additional constraints on how much administrative and control information may be exchanged, and how often. Ensuring effective routing is one of the great challenges for ad hoc networking.

The lack of fixed base stations in ad hoc networks means that there is no dedicated agency for managing the channel resources for the network nodes. Instead, carefully designed distributed medium access techniques must be used for channel resources, and hence, mechanisms must be available to recover efficiently from the inevitable packet collisions. Traditional carrier sensing techniques cannot be used, and the hidden terminal problem [6,7] may significantly diminish the transmission efficiency [8]. An effectively designed protocol for medium access control (MAC) is essential to the quest for QoS.

All the challenges enumerated above are potential sources of service impairment in ad hoc networks and hence may degrade the quality of service seen by the users. As of now, the Internet has only supported "best effort" service — best effort in the sense that it will do its best to transport the user packets to their intended destination, although without any guarantee. QoS support is recognized as a challenging issue for the Internet, and a vast amount of research on this issue has appeared in the literature during the last decade or so [9]. With the Internet as the basic model, ad hoc networks have been initially considered only for "best effort" services as well, especially given their peculiar challenges when compared against traditional wire-line or even conventional wireless networks. Indeed, just as the QoS accomplishments for a wired network such as the Internet cannot be directly extended to the wireless environment, the QoS issues become even more formidable for mobile ad hoc networks. Happily, during the last few years, QoS for ad hoc networks has emerged as an active and fertile research topic of a growing number of researchers [e.g., 8–39], and many major advances are expected in the next few years. See [12,16] for comprehensive references on QoS routing in ad hoc networks, with [16] presenting an exhaustive review of the state of the art, *circa* 1999. The URLs of [35] are good sources of more up-to-date information in this area.

Performance of these various protocols under field conditions is, of course, the final determinant of their efficacy and applicability. Relative comparisons of computational and communication complexities of various routing protocols for ad hoc networks have appeared in the past [12,16,18,23,40,41], providing the foundation for more application-oriented assessment of their effectiveness. On the other hand, performance studies have started to appear only recently [28,42–49; also see 10]. The mathematical analysis of ad hoc networks even under the simplest assumptions about the dynamics of the topology changes as well as the traffic processes poses formidable challenges, and even their simulation is considerably more difficult than that of their static counterpart. Performance studies of ad hoc networks with QoS constraints remain an open area of research.

None of the QoS studies cited so far addresses the critical issue of ensuring security, i.e., robustness against maliciously, and also inadvertently or accidentally, created service-impairing conditions in ad hoc networks. RFC 1636 [50] identifies four essential ingredients for Internet security: end-system security, end-to-end security, secure QoS, and secure network infrastructure. As an extension of the Internet, ad hoc networks inherit the same general security considerations. Several aspects of these security considerations are profoundly important for fulfilling the QoS requirements for an ad hoc network [51]. For example, in the absence of any protective mechanism, a host in an ad hoc network may create a denial-of-service condition by overwhelming the network resources with superfluous messages; such messages may be generated deliberately by a malicious attacker or inadvertently by a malfunctioning host. In another service-threatening scenario, a rogue node may launch false routing messages into the network. Yet another scenario is that of selective relaying, where a node refuses to relay packets intended for another destination. Such threats are difficult to eliminate owing to the very nature of ad hoc networking. Indeed, the traditional Internet approaches for mitigating such threats may be ineffective for ad hoc networks. Recently, however, a number of innovative mechanisms have begun to be developed for minimizing various security threats against ad hoc networks [52–61,99] with particular emphasis on secure routing.

Although [63] specifically addresses the QoS issues for the Internet, secure QoS routing remains essentially an open area of research for ad hoc networks. See [54] for a survey of the security issues for mobile ad hoc networks *circa* 2001; an even more recent article [60] also includes a thorough review of the work on secure routing for such networks.

Our discussion is limited to unicast communication only; multicasting adds additional layers of complexity to the problems of unicast communication and requires its own separate survey. See [16] and [35] for additional information on the QoS issues associated with multicast routing.

The organization of the rest of the chapter is as follows: Section 3.2 presents a brief review of the operating principles of an ad hoc network and introduces some networking concepts pertinent to routing and QoS. The general issue of routing in mobile ad hoc networks is reviewed in Section 3.3. QoS routing as it is done in today's Internet is the topic of Section 3.4. Section 3.5 addresses the QoS routing issues for ad hoc networks and the current state of research in this area. The more specialized area of secure QoS routing is reviewed in Section 3.6. Finally, Section 3.7 presents concluding remarks and directions for future research. A separate appendix includes an acronym guide.

3.2   The Ad Hoc Wireless Network: Operating Principles

We start with a description of the basic operating principles of a mobile ad hoc network. Figure 3.2 depicts the peer-level multi-hop representation of such a network. Mobile node A communicates with another such node B directly (single-hop) whenever a radio channel with adequate propagation characteristics is available between them. Otherwise, multi-hop communication, in which one or more intermediate nodes must act as a relay (router) between the communicating nodes, is necessary. For example, there is no direct radio channel (shown by the lines) between A and C or A and E in Fig. 3.2. Nodes B and D must serve as intermediate routers for communication between A and C, and A and E, respectively. Indeed, a distinguishing feature of ad hoc networks is that all nodes must be able to function as routers on demand. To prevent packets from traversing infinitely long paths, an obvious essential requirement for choosing a path is that the path must be loop free. A loop-free path between a pair of nodes is called a *route*.

An ad hoc network begins with at least two nodes broadcasting their presence (*beaconing*) with their respective address information. As discussed later, they may also include their location information, obtained for example by using a system such as the Global Positioning System (GPS), for more effective QoS routing. If node A is able to establish direct communication with node B in Fig. 3.2, verified by exchanging suitable control messages between them, they both update their routing tables. When a third node C joins the network with its beacon signal, two scenarios are possible. In the first, both A and B determine that single-hop communication with C is feasible. In the second, only one of the nodes, say B, recognizes the beacon signal from C and establishes the availability of direct communication with C. The distinct *topology updates*, consisting of both *address and route updates*, are made in all three nodes immediately afterwards. In the first case, all routes are direct. For the other, shown in Fig. 3.3, the route update first happens between B and C, then between B and A, and then again between B and C, confirming the mutual reachability between A and C via B.

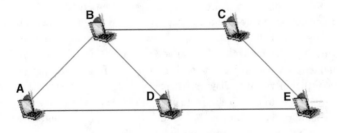

FIGURE 3.2   Example of an ad hoc network.

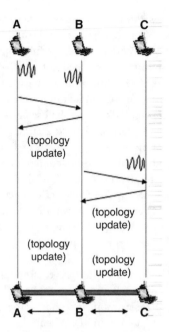

FIGURE 3.3 Bringing up an ad hoc network.

The mobility of nodes may cause the reachability relations to change in time, requiring route updates. Assume that for some reason, the link between B and C is no longer available as shown in Fig. 3.4. Nodes A and C are still reachable from each other, although this time only via nodes D and E. Equivalently, the original loop-free route $[A \leftrightarrow B \leftrightarrow C]$ is now replaced by the new loop-free route $[A \leftrightarrow D \leftrightarrow E \leftrightarrow C]$. All five nodes in the network are required to update their routing tables appropriately to reflect this topology change, which will be detected first by nodes B and C, then communicated to A and E, and then to D.

The reachability relation among the nodes may also change for other reasons. For example, a node may wander too far out of range, its battery may be depleted, or it may suffer a software or hardware failure. As more nodes join the network or some of the existing nodes leave, the topology updates become more numerous, complex, and usually, more frequent, thus diminishing the network resources available for exchanging user information.

Finding a loop-free path as a legitimate route between a source–destination pair may become impossible if the changes in network topology occur too frequently. Here, "too frequently" means that there was not enough time to propagate to all the pertinent nodes all the topology updates arising from the last network topology changes, or worse, before the completion of determining all loop-free paths accommodating the last topology changes. The ability to communicate degrades with accelerating

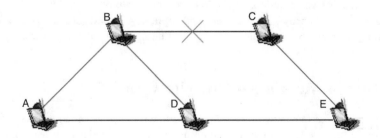

FIGURE 3.4 Topology update due to a link failure.

rapidity, as the knowledge of the network topology becomes increasingly inconsistent. Given a specific time window, we call (the behavior of) an ad hoc network *combinatorially stable* if and only if the topology changes occur sufficiently slowly to allow successful propagation of all topology updates as necessary. Clearly, combinatorial stability is determined not only by the connectivity properties of the networks, but also by the complexity of the routing protocol in use and the instantaneous computational capacity of the nodes, among other factors. Combinatorial stability is an essential consideration for attaining QoS objectives in an ad hoc network, as we shall see below. We address the general issue of routing in mobile ad hoc networks separately in the next section.

The shared wireless environment of mobile ad hoc networks requires the use of appropriate medium access control (MAC) protocols to mitigate the medium contention issues, allow efficient use of limited bandwidth, and resolve the so-called hidden/exposed terminal problems. These are basic issues, independent of the support of QoS; the QoS requirements add extra complexities for the MAC protocols, mentioned later in Section 3.5. The efficient use of bandwidth and the hidden/exposed terminal problem have been studied exhaustively and are well understood in the context of accessing and using any shared medium. We briefly discuss the "hidden terminal" problem [6] as an issue especially pertinent for wireless networks.

Consider the scenario of Fig. 3.5, where a barrier prevents node B from receiving the transmission from D, and vice versa, or as usually stated, B and D cannot "hear" each other. The "barrier" does not have to be physical; large enough distance separating two nodes is the most commonly occurring "barrier" in ad hoc networks. Node C can "hear" both B and D. When B is transmitting to C, D, which is unable to "hear" B, may transmit to C as well, thus causing a collision and exposing the *hidden terminal* problem. In this case, B and D are "hidden" from each other. Now consider the case when C is transmitting to D. Since B can "hear" C, B cannot risk initiating a transmission to A for fear of causing a collision at C. Figure 3.5 shows an example of the *exposed terminal* problem, where B is "exposed" to C.

A simple message exchange protocol solves both problems. When D wishes to transmit to C, it first sends a Request-to-Send (RTS) message to C. In response, C broadcasts a Clear-to-Send (CTS) message that is received by both B and D. Since B has received the CTS message unsolicited, B knows that C is granting permission to send to a hidden terminal and hence refrains from transmitting. Upon receiving the CTS message from C in response to its RTS message, D transmits its own message.

Not only does the above (simplified outline of the) dialogue solve the hidden terminal problem, it also solves the exposed terminal problem, for after receiving an unsolicited CTS message, B refrains from transmitting and cannot cause a collision at C. After an appropriate interval, determined by the attributes of the channel (i.e., duration of a time slot, etc.), B can send its own RTS message to C as the prelude to a message transmission.

Limitation on the battery power of the mobile nodes is another basic issue for ad hoc networking. Limited battery power restricts the transmission range (hence the need for each node to act as a router) as well as the duration of the active period for the nodes. Below some critical threshold for battery power, a node will not be able to function as a router, thus immediately affecting the network connectivity, possibly isolating one or more segments of the network. Fewer routers almost always mean fewer routes and therefore, increased likelihood of degraded performance in the network. Indeed, QoS obviously becomes meaningless if a node is not able to communicate owing to low battery power. Since exchange of messages necessarily means power consumption, many ad hoc networking mechanisms, especially routing and security protocols, explicitly include minimal battery power consumption as a design objective [63-70].

3.3 Routing in Mobile Ad Hoc Networks

All routing protocols for ad hoc networks need to perform a set of basic functions in the form of route identification and route reconfiguration. For communication to be possible, at least one route (i.e., a loop-free path) must exist between any pair of nodes. Route identification functions, as the name suggests, identify a route between a pair of nodes as a prerequisite to communication. Route reconfiguration

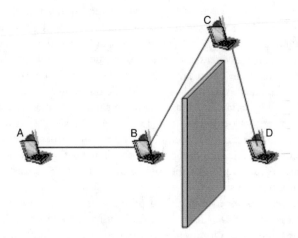

FIGURE 3.5 Example of the hidden/exposed terminal problem.

functions are invoked to recover from the effects of undesirable events, such as host or link failures of various kinds, and traffic congestions appearing within a subnetwork. Evidently, recognition of changes in the network topology and the topology update functions constitute an indispensable subset of the route reconfiguration functions. A separate category of resource management functions is also considered to ensure that all the network resources are available, to the extent possible, in support of special objectives such as those associated with QoS or security. Different authors use different classification schemes for these basic routing functions.

Routing in ad hoc networks, as in its wired counterparts, has traditionally used the knowledge of the instantaneous connectivity of the network with emphasis on the state of the links. This is the so-called *topology-based* approach [41]; the associated routing protocols can be generally classified into three categories, *periodic* (also called *proactive* or *table-driven*), *on-demand* (also called *reactive*), and *hybrid* protocols.

Networks using periodic protocols attempt to maintain the knowledge of every current route to every other node by periodically exchanging routing information, regardless of whether the routes are being used for carrying packets. Each node maintains the necessary routing information, and each node is responsible for propagating topology updates in response to instantaneous connectivity changes in the network. Examples of such protocols include those based on Destination-Sequenced Distance-Vector (DSDV) routing [71] and its derivatives such as Clusterhead Gateway Switch Routing (CSGR) [72], or various link-state routing approaches, such as the Wireless Routing Protocol (WRP) [73], STAR [74], Fisheye State Routing (FSR) [75], and others [76,77]. As a class, these protocols tend to suffer from wasted bandwidth due to the large control overhead in maintaining unused routes, especially during frequent changes in network topology, although some of the newer link-state routing protocols [75,76] present approaches for reducing the overhead.

The on-demand protocols, in contrast to the above, create routes only when necessary for carrying traffic. As a result, a route discovery process is a prerequisite to establishing communication between any two nodes, and a route is maintained as long as the communication continues. Examples of on-demand protocols include Dynamic Source Routing (DSR) [78], Temporally Ordered Routing Algorithm (TORA) [79], Ad Hoc On-demand Distance Vector (AODV) routing [80], Associativity-Based Routing (ABR) [81], and others [82,83,95,96].

The on-demand protocols also tend to generate large overheads and suffer loss of packets in transit as topology changes become more frequent. However, in general, these protocols perform better than their periodic counterparts, especially when topology changes are infrequent [42–49]. A survey of periodic and on-demand routing protocols including their relative time and communication complexities, *circa* 1999, is presented in [40].

The hybrid approach combines aspects of both periodic and on-demand routing. For example, the Zone Routing Protocol (ZRP) [84,85] allows the use of a periodic routing protocol within a local zone, while an on-demand routing scheme is used globally. Thus, at least at the level of interzone routing, if the topology changes are not too frequent, the benefits of on-demand routing are available. The performance of ZRP clearly depends on the organization of the zones within the network and the traffic patterns within the zones, neither of which is particularly predictable under all circumstances.

A different approach, called *location-based* (or *position-based*) routing, may have the potential to reduce some of the drawbacks of topology-based routing [15,17,39,41,62,70,86–92]. A recent survey, including comparative information on the time and communication complexities of various protocols of this category, is presented in [41]. In addition to topology-based information, protocols of this type also use information about the physical location of the mobile hosts. The GPS is used often by the nodes to determine their respective positions.[2]

A distinguishing characteristic of location-based protocols is that to forward packets, a node only requires its own position, that of the destination (obviously), and those of its adjacent (one-hop) neighbors. A transmitting node uses a location service to determine the location of the destination, and includes this location information as part of the destination address in its messages. Routes do not need to be established or maintained explicitly; thus there is no need to store routing tables at the nodes and no need for routing table updates. Adjacent nodes are typically identified by broadcasting limited range beaconing messages and various time-stamping mechanisms. The beaconing message includes distance limits; a receiving node discards the message if its location lies beyond the distance limit.

Availability of accurate location information at each node is essential for location-based routing to work. This, in turn, requires timely and reliable location updates as nodes change their locations. One or more nodes, designated to act as *location servers*, coordinate these *location service* functions, which are necessarily decentralized because of the mobility of the nodes. A large part of the ongoing research, as the references cited above show, is focused on designing efficient location services.

Performance studies on location-based routing, similar to those reported above for topology-based routing, have yet to appear in the literature. Little has appeared on QoS support for ad hoc networks using location-based routing beyond the work being done [34] under the iMAQ Project by the MONET group at the University of Illinois at Urbana–Champaign (http://cairo.cs.uiuc.edu), and by the Termin-odes group at the Swiss Federal Institute of Technology at Lausanne, Switzerland (http://www.termin-odes.org/publications.html). At first glance, it appears reasonable to expect that QoS objectives would be easier to meet by avoiding routing updates. More work is needed to confirm this expectation. Finally, we are not aware of any reported work on security issues for such protocols.

A high-level overview of routing with QoS constraints follows next as a prologue to the more detailed discussion of these issues for ad hoc networks.

3.4   Routing with Quality of Service Constraints

RFC 2386 [93] characterizes QoS as a set of service requirements to be met by the network while transporting a packet stream from the source to the destination. Intrinsic to the notion of QoS is an agreement or a guarantee by the network to provide a set of measurable prespecified service performance constraints for the user in terms of end-to-end delay, delay variance (jitter), available bandwidth, probability of packet loss, etc. The cost of transport and total network throughput may also be included as parameters. Obviously, enough network resources must be available during the service invocation to honor the guarantee. The first essential task is to find a suitable loop-free path through the network, or route, between the source and destination that will have the necessary resources available to meet the QoS constraints for the desired service. The task of resource (request, identification) and reservation is the other indispensable ingredient of QoS. By QoS routing, we mean both these tasks together.

---

[2] See [62] for a GPS-free approach.

FIGURE 3.6 A flow: QoS is meaningful only for a flow between a specific source–destination pair.

The Internet of today operates in a connectionless and stateless mode. The network of routers is not aware of any association between the source and destination except on a per-packet basis. Each packet is routed individually without any information about the state of the flow of packets between the source and destination. On the other hand, QoS is meaningful only for a *flow* of packets between the source and destination, and thus depends on the notion of a logical association, or logical connection, between them for the *duration of the flow*, as represented in Fig. 3.6. Also, the network must guarantee the availability of a set of resources associated with the flow. Consequently, appropriate routers must remain aware of the logical connection and state of the flow to ensure that adequate network resources such as link bandwidth, nodal buffers, processing power, etc., are available for the duration of the logical connection, and their underlying routes.[3] QoS guarantees can be attained only with appropriate resource reservation techniques. The most important element among them is *QoS routing*, i.e., the process of choosing the routes to be used by the flow of packets of a logical connection in attaining the preestablished QoS guarantee.

Consider Fig. 3.7, where the numbers next to the links represent their respective bandwidths, say in Mb/sec. To minimize delay and for better use of network resources, minimizing the number of intermediate hops is one of the principal objectives in determining suitable routes. However, suppose that the packet flow from A to E requires a bandwidth guarantee of 3 Mb/sec. QoS routing will then select the route [A →B →C →E] over the route [A →D →E] because the latter is unable to meet the bandwidth need although it has fewer hops. The only other alternative, [A →B →D →E], will also be rejected for failing to meet the bandwidth need.

QoS routing offers serious challenges even for the static environment of today's Internet. Different service types, e.g., voice, live video, and document transfer, have significantly different objectives for delay, bandwidth, and packet loss. Determining the QoS capability of candidate links is not simple for such scenarios (for multicast services, the difficulties are even greater). We have already noted that the route computation cannot take "too long." Consequently, the computational and communication complexities of route selection criteria must also be taken into account. The presence of more than one QoS constraint often makes the QoS routing problem NP-complete [94]. Suboptimal algorithms such as sequential filtering are often used, especially for large networks, where an optimal path based on a single primary metric (e.g., bandwidth) is selected first, and a subset is eliminated by optimizing over the secondary metric (e.g., delay), and so on, until all the metrics have been taken into account (a random selection is made if more than one choice remains after considering network throughput as the last metric). All else remaining the same, the same route is used for all the packets in the flow as long as the QoS constraints are satisfied.

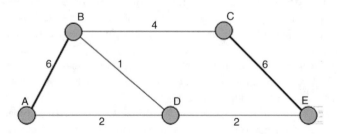

FIGURE 3.7 Links with fixed bandwidth in a network.

---

[3] For an interesting approach to providing guaranteed services without per flow management for the Internet, see [98].

Candidate routes for a flow with specific QoS objectives are determined by using various QoS metrics associated with its constituent nodes and links. These metrics collectively characterize the *state* of the nodes and links. A typical *link-state* is an ordered tuple of its specific QoS metrics of interest, and is usually represented as follows:

$$link\text{-}state \equiv\ <\ bandwidth,\ propagation\ delay,\ cost\ >$$

where bandwidth is the maximum residual bandwidth that the link can support, and "cost" here is used as a generic catch-all for other parameters such as packet loss statistics, service class (if multiple service classes are to be supported, each with its own QoS requirements), etc. "Cost" often does represent a single number. Observe also that each link is assigned a unique direction in terms of the source and sink nodes. The *state* of a node is likewise characterized as an ordered tuple of typical QoS metrics as follows:

$$node\text{-}state \equiv\ <\ CPU\ bandwidth\ capacity,\ delay\ distribution,\ cost\ >$$

where the CPU bandwidth is the minimum rate at which the node can place data into the link, delay distribution at a minimum includes the mean and the variance of the queueing delay, and "cost" is again a generic term for many other parameters that need be considered for different service classes for different traffic types, including service classes with multiple priorities. Frequently, the node-state is incorporated into the state of each of the links incident on it, as we do in this chapter. In such cases, we only have (augmented) link-states, where the link bandwidth is now the minimum of the residual link bandwidth and the CPU bandwidth of its source node, and the delay is the (random) sum of the link propagation and node queueing delays. Finally, the cost is determined appropriately by considering its component metrics.

Accurate location information has to be included as part of the local state information, if location-based routing is to be considered.

The state of a route such as $[A \rightarrow B \rightarrow C \rightarrow E]$ of Fig. 3.7, then, follows immediately as the appropriate numerical operations of the various components of the (augmented) states of its constituent links. The bandwidth of a route is the minimum of that of its components, the delay is the (random) sum of all link delays, and cost is either the sum (if it is an additive quantity) or another appropriate deterministic or stochastic numerical operation of all the component costs. For a given flow, a *feasible* route is one with sufficient available resources to satisfy the QoS requirements. It is evident then that the QoS routing problem is a constrained combinatorial (graph) optimization problem,[4] and it is solved as such.

Once a route has been selected for a specific flow, the necessary resources, e.g., bandwidth, buffer space in routers, etc., must be reserved for the flow. These resources will not be available to other flows until the end of this particular flow. Consequently, the amount of remaining network resources available to accommodate the QoS requests of other flows will have to be recalculated and propagated to all other pertinent nodes as part of the topology update information.

Since QoS routing is dependent on the accurate availability of the current network state, we briefly consider the nature of such information. The first is the *local state* information maintained at each node, which includes every pertinent component (for a given flow) of the node-state, as well as the link-state for each of its outgoing links. The totality of the local state information for all nodes constitutes the *global state* of the network, which is also maintained at each node. The instantaneous network connectivity is part of the global state information. While the local state information may be assumed to be always available at any particular node, the global state information is constructed by exchanging the local state information for every node among all the network nodes at appropriate moments. The process of updating the global state information, also loosely called topology updates, may significantly affect the QoS performance of the network, as we have mentioned before.

---

[4] Most categories of the general constrained combinatorial optimization problems for graphs are known to be NP-complete.

The global state update may be done by broadcasting the local state of each node to every other node (*link-state protocol*), or by periodically exchanging suitable "distance vector" information among the adjacent nodes only (*distance-vector protocol*). A distance vector is usually maintained as a table in each node with an entry for each QoS metric. Given a node K, for each QoS metric, the corresponding entry is a triple consisting of the following:

- Address of a destination node for every possible destination
- Best attainable value of the metric over the best route from K to the destination
- Address of the node immediately adjacent (next hop) to K on the best route with respect to the value of the metric

Consider node A in the example of Fig. 3.7. The bandwidth entry for the distance-vector at the node A will have a representation as follows:

Bandwidth Entry at Node A

Destination	B	C	D	E
Bandwidth	6	4	2	4
Adjacent Node	B	B	D	B

For destination B, the available routes are [A →B], [A →D →E], and [A →D →E →C →B]. Likewise for E, the available routes are [A →D →E], [A →B →D →E], [A →B →C →E], and [A →D →B → C →E], and so on.

Since the topology updates throughout the network cannot happen instantaneously, the global state information may only be an approximation of the true current network state. For ad hoc networks with highly mobile nodes, the global state information may *never* be accurate.

Three distinct route-finding techniques are used for determining an optimal path satisfying the QoS constraints. These are *source routing, distributed routing,* and *hierarchical routing*.

In source routing, a feasible route is locally computed at the source node using the locally stored global state information, and then all other nodes along this feasible route are notified by the source of their adjacent preceding and successor nodes. A link-state protocol is almost always used to update the global state at every node. State update is done by using either a distance-vector (most common) or a link-state protocol.

In distributed or hop-by-hop routing, the source and other nodes are involved in the path computation by identifying the adjacent router to which a node must forward the packet associated with the flow. Practical considerations for large networks with many nodes and high connectivity sometimes compel the use of the so-called *aggregated* global state information, by first partitioning the network into a hierarchical cluster of some form, and then only considering the suitable state information associated with these clusters. Such information is necessarily a partial representation of the true global state.

Hierarchical routing, as the name suggests, uses the aggregated partial global state information for determining a feasible path using source routing where the intermediate nodes are actually logical nodes representing a cluster; for more details see [14].

Flooding is not an option for QoS routing, except for broadcasting control packets under appropriate circumstances, e.g., for *beaconing*, or at the start of a route discovery process. See [40] for a comparative discussion of the advantages and disadvantages of various algorithms associated with each of these three approaches to routing.

One may reasonably expect that all packet exchanges will not be treated with equal priority in a QoS network. The exchange of control packets should receive higher priority than user data packets in a network designed for QoS. Indeed, except for instances of "thin" low-traffic (relative to the network capacity) networks, control packets should receive *preemptive* priority over user data packets. Second, the QoS policy may allow different priorities to exist even among different flows of user packets. Clearly, in accommodating packets with preemptive priorities, the network may not be able to preserve the QoS guarantee for ordinary flows. Appropriate admission control policies could also offer additional benefits. Indeed, QoS routing allowing preemption and admission control policies is an open area for further research.

Handling of user data with multiple priorities presents potential security threats as well. When a user requests QoS with a certain priority, the network first needs to authenticate such a request by exchanging appropriate control packets. Too many authentication requests, by themselves, may degrade the operational performance of a large QoS network. Next, the network must find a route with the requested QoS for a higher priority against other all other flows with lesser priority, even if they are allocated identical QoS parameters in all other respects. In heavy traffic situations, guaranteeing QoS for lesser priority traffic may be extremely difficult or impossible. The development of QoS routing policies, algorithms, and protocols for handling user data with multiple priorities is also an open area.

Similar challenges exist in designing QoS routing schemes supporting multiple service classes. For more discussion, see [93]; for additional details, see [16; Chap. 3].

Our discussion, up to this point, has been limited to *unicast* routing. The essential problem here is to find a feasible path from a source node to a single destination node that satisfies a set of QoS constraints and possibly some other additional optimization criteria such as minimum cost and maximum network throughput. The multicast routing problem, on the other hand, is distinguished by more than one destination node, where the objective is to find not a single path but a feasible *tree* rooted at the source. Each path from the source to one of the destination nodes in the tree is required to satisfy the specified set of QoS constraints, together with additional optimization criteria, if any, simultaneously. As observed in [16], many of the associated optimization problems are NP-complete. We do not address the topic of multicast routing in this chapter.

We have presented only a broad-brush overview of QoS routing. Many issues, such as the effect of imperfect knowledge of network state information on routing and hierarchical aggregation of routing information for scalability, have not been mentioned. All these issues profoundly affect the QoS in ad hoc wireless networks and are considered in the next section.

3.5   QoS Routing in Ad Hoc Networks

The basic concepts of QoS routing discussed in the previous section constitute the foundation for QoS routing for ad hoc networks [12,16]. The core of our discussion is based on topology-based routing; observations pertinent to location-based routing are added as appropriate. We assume that each node carries a unique identity recognizable within the network. Following [16], we assume the existence of all necessary basic capabilities, such as suitable protocols for medium access control and resource reservation, resource tracking, state updates, etc.

The mobile nodes, as mentioned earlier, use some form of multiple access technique with suitable collision avoidance and "hidden terminal" mitigation for accessing radio resources. The larger the number of nodes contending for radio resources, the larger is the delay (random variable) in accessing the radio channel for transmitting a packet. Enough reserve of radio channel capacity must be available to ensure an upper bound on end-to-end delay as part of QoS. The MAC protocols such as [7,100] do provide for QoS support with bandwidth reservation.

Each node periodically broadcasts a *beacon* packet identifying it and its pertinent QoS characteristics, thus allowing each node to learn of its adjacent neighbors (i.e., those with which it can communicate directly). The beaconing mechanism, regardless of whether it is based on topology or location, lies at the heart of ad hoc networking, for otherwise, a node will not even know its adjacent neighbors, which change dynamically in such networks. The knowledge of adjacent neighbors, of course, is indispensable for routing.

A principal objective of network engineering, as emphasized earlier, is the minimization of routing updates, for such updates consume network bandwidth and router CPU capacity. Second, frequently changing routes could increase the delay jitter experienced by the users. This objective is extremely difficult to attain in wireless networks because of the involuntary network state changes as nodes join in or depart, traffic loads vary, and link quality swings dramatically. To accommodate real time traffic needs such as those of voice or live video, both the overall delay and the delay variance must be kept under a certain bound. This is accomplished primarily by minimizing, as far as possible, the number of hops, or the intermediate routers, in the path. With potentially unpredictable topology changes in an ad hoc network, this last objective is difficult to attain.

Combinatorial stability, therefore, is a critical consideration for QoS in an ad hoc network. Combinatorial stability follows directly when the geographical distribution of the mobile nodes does not change much relative to one another during the time interval of interest. Such is the case, for example, for the Internet and in a classroom setting for communication among laptop computers as ad hoc nodes. The routes among network nodes in such cases will change little or not at all. There are other cases, for example, in rescue operations, refugee migrations, etc., where route updates do occur during the intervals of interest, but not sufficiently frequently to violate the limits of combinatorial stability. In such cases, it is possible that the topology updating takes long enough so that by following the now unacceptable characteristics of the last used route, the QoS guarantees cannot be met. Indeed, the old route may even cease to exist during the topology update. This is entirely possible for geographically dispersed networks with a large number of nodes and sparse connectivity, where each route consists of many intermediate nodes like a string of beads.

The topology of an ad hoc network may be combinatorially just right so that the QoS guarantees are maintained during any topology updating. Observe that it is not just the connectivity that affects the QoS. Equally essential is the availability of enough resources along the previous and the new routes during and after the transition. We call an ad hoc network *QoS-robust* with respect to a specific set of QoS guarantees, only if such guarantees are maintained *regardless* of the topology updates that may occur within the network; guaranteeing QoS robustness under all circumstances is possible only with unlimited resources. More narrowly, we call such a network *QoS-preserving* if it can continue to maintain the QoS guarantees *during* the interval from the end of a successful topology update until the occurrence of the next topology change event. A QoS-robust ad hoc network is, by definition, QoS-preserving; the converse is obviously false.

A mobile node may lose connectivity with the rest of the network simply because it has wandered off too far or its power reserve has dropped below a critical threshold. Since the network cannot control the occurrence of such events, *we* must exclude them in considering the QoS guarantees.[5] Topology updates occur when a new node joins the network or an existing node is detected to have become unavailable with respect to a particular flow. One naturally expects that such topology updates should not affect the QoS for the rest of the nodes as long as the topology of the rest of the network (as a subnetwork) remains unchanged. So far, with the exception of [16], little has appeared on the preservation of QoS guarantees under various failure conditions in ad hoc networks as a specific area of study.

Two routing techniques are considered in [16], both limited to combinatorially stable, QoS-preserving networks. One is based on the availability of only local state information, and the other assuming possibly inaccurate knowledge of global states. When an existing feasible route becomes unavailable, a new feasible path is determined, and the flow is rerouted to the new feasible path. During the interval immediately following the disappearance of the existing path and the establishment of the new route, data packets are sent as *best-effort* traffic.

For QoS routing using only the local state information, [16] introduces two different distributed routing algorithms, called *source initiated routing* and *destination-initiated routing*. Both use only the local state information stored at each node. Both rely on the use of *probe* packets with appropriate nodal identity and QoS information in identifying a feasible route with the desired QoS characteristics. The

---

[5] Recall that in developing various routing and other algorithms for ad hoc networks, minimizing power consumption has been explicitly investigated by many researchers. See [70] for results on a location-aided power-aware routing protocol for such networks. Minimization of power consumption and QoS support do not appear to be mutually consistent objectives at the current "state of the art."

source and the intermediate routers use a form of flooding to send the probe packets. Various mechanisms are considered in [16] to mitigate the penalties of flooding and to minimize the number of probe packets to be used, and the advantages of destination-initiated routing over the other methods established under certain conditions.

Preestablished network policies should determine the steps to be taken in case no feasible route could be found during the route establishment phase. The service request may be rejected and the node blocked, or the network may negotiate for a service with lower QoS by exchanging control packets using best-effort routing, assuming that such alternative QoS is available. Such considerations offer opportunities for further research.

Efficient source-initiated routing results from a number of innovative techniques introduced in [16]. Avoiding unnecessary probes by noting their respective sources is one. The second is the novel concept of *local multicast*, which limits the broadcast of probes to only an appropriate subset of the adjacent nodes. The third relies on caching the distance information by counting the number of hops traversed by the probe up to that point. By maintaining at each node the relevant state information of all its *n-hop* neighbors, a route to any other node can be determined by using only the local information. Although [16] cites the work of [95,96] in this connection, no explicit indication appears on the possible use of any location-based routing approach. It is evident that the location-based routing techniques mentioned in [41] perform similar functions without the need for route updates and offer opportunities for potential improvement in efficiency.

The destination-initiated routing approach of [16] actually relies on the best available estimate of the distance between the source and destination. Here the destination node identifies a feasible route by sending probe packets toward the source on the basis of restricted flooding. Of course, it is the source that initiates a flow by sending a control message to the destination with the necessary QoS information by using one of the many best effort routing algorithms mentioned in Section 3.3. The control message counts the number of hops it traverses while following the best effort route as an estimate (upper bound) of the distance between the source and the destination. This hop count is used at the destination node to limit the flooding range for its probe packets back toward the source. More precise location information used in location-based routing should result in more accurate restriction on the flooding range, thus offering opportunities for greater efficiency.

The initial control packet launched by the source in the destination-initiated routing also checks for the values of the pertinent QoS metrics, e.g., bandwidth, for each link in its route. If the QoS specifications can be met by this initial route, no flooding of probe packets needs to follow, and one immediately recognizes this route as a feasible route. In such a situation, the destination node returns an acknowledgment message along the same route back to the source causing all necessary resources to be reserved along the route.

The techniques based on imprecise knowledge of global states in [16] uses the notion of *ticket-based* probing for identifying a feasible route. Each probe from the source toward the destination carries at least one ticket to control the number of alternate paths to be searched, thus minimizing the routing overhead. The lower the likelihood of finding a route with the desired QoS requirements, the larger the number of tickets carried by the probe. Attempts are made to send the probes along links the QoS characteristics of which are relatively constant (or slowly varying) in time. The basic routing mechanism is distributed or hop-by-hop; in [16], the information for multiple feasible routes is stored in the probes, instead of the within the intermediate routers.

Multiple mechanisms are considered in [16] for QoS-preserving QoS routing by detecting broken routes and then either repairing the broken route or rerouting the flow on an alternate route with the desired QoS. The use of redundant routes of various kinds further reduces the likelihood of QoS violation. A broken route is detected by a node on the route using a mechanism similar to the beaconing protocols for detecting adjacent neighbors. When a node detects a broken route, it sends a "route failure" message back to the source. After receiving the "route failure" message, the source switches the flow over to an alternate route, as discussed below, and sends a "resource release" message along the original route so that all nodes on the route receiving this message can release all QoS resources previously reserved for the flow. Obviously, the "resource release" message will not reach those nodes on the now broken route that are no longer reachable from the source. Even then, their resources will not remain associated with the now-rerouted flow indefinitely by using the following "timeout" mechanism. The existence of the QoS route between a source–destination pair needs to be reaffirmed periodically when routing with imprecise information by sending suitably constructed control packets, called *refresher* packets in [16], from the destination back to the source. When an intermediate node receives the refresher packet, it resets the "refresher timer" and sends the refresher packet to the adjacent node upstream. The receiving node always sends an acknowledgment back to the sending node. If such a refresher packet fails to arrive within a predetermined timeout interval, the QoS route is declared unavailable and the associated resources released. Likewise, if a node expecting to receive an acknowledgment to the refresher packet does not receive it before the "refresher acknowledgment timeout" expires, it releases all of its own resources associated with the particular QoS route. This also accommodates the failure to deliver various unavailability notifications to their intended recipients using additional timeout mechanisms, such as timeouts on timeout messages.

A simple example is shown in Fig. 3.8, where Fig. 3.8a depicts the normal operation of a QoS route for a certain flow. Suppose that the route fails because the link between B and C is now broken. The route failure event is detected at both B and C either by beaconing, or as a result of some timeout(s). B is able to send the "route failure" message back to the source over the still-intact segment; C cannot. After receiving the route failure message, the source sends back the "resource release" message for the flow, as shown in Fig. 3.8b, which when received by nodes A and B will cause them to release all QoS resources associated with the flow.

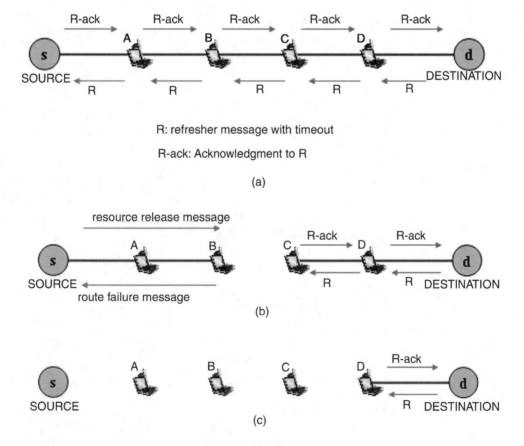

FIGURE 3.8 Detection of a route failure and release of resources.

We assume that no communication is possible between B and C. Nodes C and D will not receive the resource release message from the source; their resources will be released by the refresher message mechanism. In Fig. 3.8b, C can receive a refresher message from D and forward it to B. Node B, however, does not receive the message, and therefore, C receives no acknowledgment to the refresher message. The refresher acknowledgment timeout eventually expires at C, and C releases all its resources associated with the QoS flow. As a result. the link between C and D also becomes unavailable, as shown in Fig. 3.8c, and hence D will no longer receive an acknowledgment to its refresher message to C, causing D to release all pertinent QoS resources at the expiration of the refresher acknowledgment timeout, and so on. A node will release all its resources because either it did not receive a refresher message within the prescribed timeout period or it did not receive a refresher acknowledgment before the expiration of the refresher acknowledgment timeout, or the third and the only remaining possibility, that it did receive a "resource release" message from the source. These multiple timeout mechanisms ensure that resources will not be locked in indefinitely because of any communication failure.

When the source receives the notification of route unavailability, it seeks an alternate route with the same QoS characteristics, as shown in Fig. 3.9. The unusable route is shown using thin lines; thick lines show the new alternate route. If such a route can be found, the flow is rerouted to it after the necessary route updates among the pertinent nodes.

Multiple redundant routing mechanisms are also considered in [16] for minimizing the likelihood of QoS violation owing to route failures. Consider Fig. 3.10, representing the highest level of redundancy. At the highest level of redundancy, multiple alternate routes with the same QoS guarantee are established for the flow and are used simultaneously. The alternate routes should be preferably disjoint,[6] although this may not always be possible, as shown in Fig. 3.11. At the next lower level of redundancy, the routes

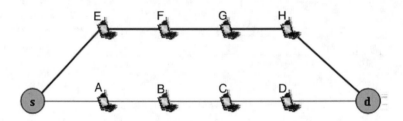

FIGURE 3.9 Alternate routing.

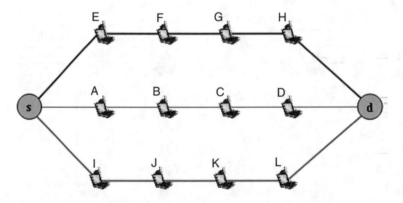

FIGURE 3.10 Redundant routing: all routes are disjoint.

---

[6] Two routes are disjoint if and only if the source and destination nodes are only common nodes in the routes.

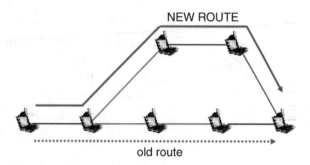

FIGURE 3.11 Redundant routing: all routes are not disjoint.

and the associated resources are reserved and rank ordered but not used unless the primary route fails or the first choice for the alternate route fails while the primary is unavailable and so forth. When not in use for the QoS-guaranteed flow, the alternate route is used to carry best-effort packets. At the lowest level of redundancy, only the route is identified; no resource is reserved. When the primary path fails, the alternate paths are checked to determine whether the necessary resources are still available. An explicit discovery process for rerouting is initiated if none of the alternate routes is found to be able to support the desired QoS. In all cases, the duplicate packets are discarded at the destination.

Variations of the above approach are possible; one is where an attempt is made to repair the route solely on the basis of local adjacency information, instead of switching over to a new route. If node B in Fig. 3.12 determines that the link to C is broken, it does not send a route failure message back to the source. Instead, B attempts to repair the path as follows. Using the beaconing mechanism, B sends a "repair request" message to its adjacent neighbors with the query whether any of these other nodes may be able to offer at least the same QoS support as was done by C. An adjacent neighbor E will send an affirmative response only if it is also an adjacent neighbor to C with a link [E → C], and it has adequate residual resources. If node E sends an affirmative response, B will add the link [E → C] as an element to the QoS route from **s** to **d** and send a "path repair" message to E. After receiving the path repair message, E will dedicate the necessary resource to the QoS flow and update its own route information. The new information will become the part of regular topology update as required.

None of these scenarios explicitly considers location-based routing. The recently introduced notion of *predictive location-based QoS routing* [34] is an attempt to exploit explicitly the potential advantages of location-based routing in connection with alternate routing. In this approach, the route failures are predicted beforehand, and new routes are determined using these predictions before the current route becomes unavailable. In principle, the QoS flow can be transferred onto one of these predicted alternate routes without any packet loss. The nodes in the network use flooding for state updates consisting of their respective locations and resource availability. The location updates are used to predict the future locations of the nodes, and alternate routes are determined using the predicted locations such that the connectivity between the source and destination on the new route will be preserved. However, this approach does not reserve resources for the alternate routes, and as would be expected, does not promise

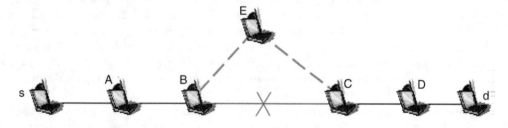

FIGURE 3.12 Route repair.

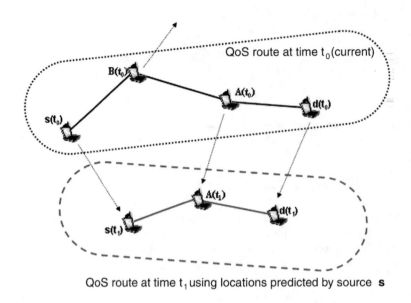

FIGURE 3.13 Example of predictive location-based routing.

any hard QoS guarantees. The availability of resources on all active routes is appropriately monitored against the QoS objectives to ensure the occurrence of the necessary route switching when the available resources drop below acceptable thresholds.

The following discussion paraphrasing [34] and Fig. 3.13, adapted from the same reference, presents an example of finding an alternate route using this predictive approach. The upper part depicts the current QoS route, say at time t_0, between the source **s** and destination **d**; $s(t_0)$ and $d(t_0)$ denote their respective current locations. Likewise, $A(t_0)$ and $B(t_0)$ denote the respective current positions of the nodes A and B. Using the current and past location information, the source **s** predicts the future location of these nodes at time t_1, $t_1 > t_0$, and determines a new route from **s** to **d** based on the predicted location. Observe that node B is no longer a part of the predicted route at time t_1. In other cases, new routes with more intermediate nodes may result from the location prediction to preserve connectivity and maintain the QoS as close to the original specification as possible.

A bandwidth-constrained QoS routing algorithm using a distance vector protocol is proposed in [10], but without accommodating the effects of imprecise network state information. Using concepts from multi-layer adaptive control, [14] presents a sophisticated approach for controlling QoS in large ad hoc networks by using hierarchically structured multi-clustered organizations. Roles of cluster dynamics and mobility management, as well as the effects of resource reservation, route repair, and router movement on QoS, are addressed in detail. Two new QoS routing schemes, both based on link-state protocols as the underlying mechanism, appear in [19]. Both attempt to reduce the routing update overhead, one by selectively adjusting the frequencies of routing table updates, and the other by reducing the size of the update messages using a hierarchical addressing approach. Another novel approach for QoS routing [21] uses the notion of a *core* as a self-organizing set of nodes for routing. The overhead of routing updates is reduced by decreasing the number of nodes doing route computation and limiting the propagation of link-state information for highly transient nodes. A distance vector based routing mechanism with focus on bandwidth control and with explicit consideration of broken routes is proposed in [20].

Admission control policies, common in wire-line networks, offer opportunities for preserving existing QoS in an ad hoc network. In wire-line networks, these are most frequently used with multiple QoS classes with different QoS attributes and priorities. For mobile ad hoc networks, the "best effort" traffic is the most natural QoS class, say service class 0. Multimedia traffic, with and without live video, may be assigned its own QoS classes. When a new node attempts to find a QoS route for service with a QoS class q, the route discovery will fail if there is no feasible QoS route available at that moment to accommodate q. An admission control policy for ad hoc networks should answer questions such as whether the requesting node could negotiate with the destination for a "lower QoS," q'; if such negotiations are allowed, then the policy will also specify how many, how often, and for how many different "values" for q'. The default option, as considered in [16], could be to switch to "best effort" service only. More complex options will require addressing issues such as whether to preengineer for alternate QoS options or use adaptive negotiations. A robust admission control policy will take into account the effect of additional control traffic on the QoS capacity of the network.

We have mentioned earlier the option of assigning control packets *preemptive* priorities over other "data" packets as part of strengthening the QoS support. One may reasonably expect that all packet exchanges will not be treated with equal priority in a QoS network. Likewise, different levels of QoS may include different priorities for different flows. The routing protocol must find a route with the requested QoS for a higher priority against all other flows with lesser priority, even if they are allocated identical QoS parameters in all other respects. In heavy traffic situations, guaranteeing QoS for lesser priority traffic may be difficult or even impossible. In such cases, the admission control policy needs to address whether the QoS guarantee for flows could be preserved and what actions to take in case they are not. The development of QoS routing with admission control policies, algorithms, and protocols, with or without control packet priorities or multiple levels of priorities for user data, is an open area for further research.

We have repeatedly noted that all the policies, protocols, and algorithms in an ad hoc network with QoS support must be QoS-preserving. How badly do the rapid topology changes militate against the QoS guarantees? Let τ_c and T_u denote the interval between two consecutive topology change events and the time it takes to detect the change, complete the calculation, and propagate the topology updates resulting from the last topology change, respectively, to all pertinent nodes. Recall that an ad hoc network is combinatorially stable only if $T_u < \tau_c$.[7] If the just-computed feasible route ceases to exist during the corresponding topology update, the QoS guarantee becomes meaningless. Maintaining bounds on delay jitter may also become impractical even in a combinatorially stable network if τ_c remains "close" to T_u. It may be necessary to investigate more rigorous criteria for different *degrees* of combinatorial stability for different QoS constraints. Since combinatorial stability is governed by random processes arising from random changes in the topology and link traffic intensity, making the network QoS-robust for a particular flow and its associated QoS constraints is clearly impossible as a deterministic objective for an arbitrary ad hoc network. This is why no QoS routing algorithm offers hard QoS guarantees now, and why it may not be possible to do so in the future. Any such guarantee at best could only be statistical in nature, where QoS robustness is specified as a probability bound for QoS violation during a topology update the duration of which does not exceed a fixed upper bound. This is why any performance study of ad hoc networks with QoS support is meaningful only for combinatorially stable networks. This is why one also assumes that the connectivity between a node and the rest of the network is never lost because of low battery power or because a mobile node has wandered far enough away. The smaller the value of T_u, the smaller is the probability of QoS violation, assuming that resources remain available for use whenever necessary. Redundant routing as in Figs. 3.9 through 3.11 clearly could help accomplish both. It is obvious that QoS is a realistic goal to pursue for static ad hoc networks, e.g., a classroom setting. It is equally obvious that considerable additional work is necessary to understand better both the specific conditions and extents under which various QoS objectives could be satisfied for dynamic ad hoc networks in the real world. It is in this latter context that the use of admission control, multiple classes of service with possibly different priorities, preemptive priority of control messages, and segregation of dedicated resources for QoS-robust ad hoc networking offer promising areas of investigation.

---

[7] In practice, these are random variables.

3.6   QoS Routing with Security Constraints

Robustness of network operations is an obvious prerequisite for QoS support in an ad hoc network, and indeed, for any network. The objective of a robust network security policy is to maintain the operational integrity of the network against unintended degradation or malicious attacks. Ad hoc network operations are intrinsically vulnerable because of their intermittent connectivity, dynamic topology, open admission policy, and reliance on cooperation among the nodes, as well as the absence of dedicated agents for ensuring reliable operation of the essential functions. In spite of the importance, as observed in a recent survey [54], most of the work on the security issues for ad hoc networks is fairly recent.

The threats to the operational integrity of any network, and hence to its ability to support QoS, can be classified into one of two categories. The first includes the actions or events that impair the basic functions such as routing, resource scheduling, or medium access control; the other includes those that compromise the security mechanisms themselves (e.g., management of the encryption keys). Observe that all threats do not arise from deliberate, malicious actions; often they are results of poor design or just accidents. Threats to basic operations have the most conspicuously immediate effects on the QoS; this is our focus here. We do not consider issues such as jamming or eavesdropping, effective solutions to which are already in use, nor do we consider issues such as the physical security of equipment; for other considerations, see [55,61], as well as [54] and the references cited there.

Naïve operation of ad hoc networks is entirely vulnerable to rogue nodes because of the networks' dependence on perfect cooperation among the nodes, where all nodes automatically enjoy perfect mutual trust. As an example [54], consider the MAC layer, the successful operation of which is dependent on cooperation among the nodes for proper allocation of the channel resources. If a rogue node violates the rules, fair allocation of the shared bandwidth will be unworkable and the QoS will be ruined.

The discussion of the previous sections repeatedly noted the security vulnerabilities of the routing protocols in ad hoc networks. Indeed, since all nodes are required to act as routers, any node can utterly spoil normal operations by inserting bogus routing information or by just refusing to relay packets to their intended destinations. Consequently, secure routing has received the most attention so far in the published literature [52–54,56–60,97], and we do the same here.

Routing misbehavior [56] may occur in many forms. A selfish[8] or rogue node that refuses to forward packets destined for other nodes, i.e., drops the packets, causes a denial-of-service attack. If the control packets are *selectively* dropped, various timeouts will be repeatedly invoked at assorted other nodes, causing wasted CPU resources and bandwidth. Indeed, depending on the instantaneous role of the node as a router, it may cause repeated route discovery and rerouting attempts, causing the network to become combinatorially unstable and fail to support some or all of its QoS routes. We mention this first, for as observed in [52], no effective and general protection against deletion of routing information exists. While an overloaded node will drop *all* packets for the duration of its overload conditions and may not be able to respond to control packets from other nodes, there may be no detectable pattern for use by other nodes to bypass a rogue node that only drops some of the control packets. It is even conceivable that a node suffering from some software fault may selectively drop some of the control packets (as well as other packets).

It is evident that any sufficiently sustained denial-of-service attack, depending on the amount of resources it will cause the network to waste, will destroy QoS routing, and even worse, cause an ad hoc network to become combinatorially unstable.

The second category of misbehavior involves the use of bogus information by a rogue node;[9] we call these *type II routing attacks* in the following. In [58], the specific behaviors are classified into three categories: *modification, impersonation,* and *fabrication,* which are explained as follows:

---

[8] These are called *selfish* in [56] because of their unwillingness to forward packets for reasons such as saving their own battery power or conserving their available CPU or link bandwidth. However, it may not be so easy to be selfish in a *really* dynamic ad hoc network environment, for a node spends battery energy even while listening.

[9] The probability of such events occurring spontaneously because of a software fault in a node is small indeed.

In attacks using *modification*, a rogue node alters the appropriate fields of the control packets of a routing protocol, thereby causing misdirected network traffic and hence a denial-of-service attack. Such attacks also include occasions when the rogue node simply launches routing messages with inaccurate but real values in these fields (e.g., lowers or increases a sequence number or hop count) or forwards the traffic to its intended destination after corrupting the user data contained in it. *Coordinated* attacks by more than one node colluding in repeatedly sending superfluous but legitimate control packets (e.g., beaconing, route discovery, probe packets, etc.), with or without deliberately caused delays, are difficult to prevent, and it may be impossible to do so in general.

Attacks using *impersonation*, or *spoofing*, are those in which a rogue node masquerades as another node by replacing its own MAC or IP address with the legitimate address of another node; [58] shows how with such attacks even a single node may cause a routing loop to form. A single node can easily launch both modification and spoofing attacks simultaneously.

Finally, in attacks using *fabrication*, a rogue node generates routing messages with false information that another node is waiting to receive. As shown in [58], such attacks may result in deletion of routing table entries or corruption of information in route caches, etc.

We now address the reported approaches to deal with these security attacks. The most obvious is that of making the nodes immune to tampering. Unfortunately, this is not a simple issue for arbitrary ad hoc networks allowing unrestricted admission. An interesting approach towards tamper resistance for essentially static networks is considered in [61]. Most of the current research is more appropriately focused on developing secure routing protocols and policies that are intrinsically robust against the attacks associated with routing misbehavior of the second category, or type II routing attacks.

All countermeasures against type II routing attacks attempt to protect the routing information (and the routing protocol) by incorporating various forms of strong authentication with cryptography[10] mechanisms as their foundation. These include the authentication of message sources (i.e., they are indeed who they say they are), data integrity, and protection of message sequencing and timeliness, which in turn, provide protection for the QoS as well. Authentication of sources involves the use of cryptographic checksums and public-key–based signatures. Proper sequencing and timeliness of messages are protected by using sequence numbers and timestamps.

Additional other measures also become necessary to protect against certain types of fabrication attacks, two examples of which are mentioned in [52]. In the first, in using link-state routing, a rogue node may send incorrect information about its links. Since the rogue node will pass the origin authentication test, protection against such an attack will require a mechanism to compare and verify the actual link-states at the two nodes connected by the link. A second example is where a rogue node announces the presence of fictitious routes or links. The proposed protection against such an attack requires validating all routing information against an "infrequently changing configuration map of the network" [52]. Although any assumption of "infrequent configuration change" will manifestly fail to be true for an arbitrary dynamic ad hoc network, we also note that the emphasis of [52] was not on such networks.

One of the first papers on Type II routing attacks [56][11] systematically addresses the problem in which a few of the nodes renege on their promise to forward packets to all destinations, and presents efficient solutions under certain assumptions. However, as observed in [53] and [60], the efficiency is attainable only under rather rigid conditions, and more important, the approach does not succeed in many commonly encountered practical situations. In a similar vein, [60] also presents the potential benefits and problems for the proposed approaches in [54] and [59]. Other sources, such as [52] and [58], also propose solutions that are effective for various limited classes of problems.

One of the most recent contributions is the Secure Routing Protocol (SRP) of [60], which offers secure routing against a wide class of attack scenarios with a flexible QoS-friendly solution. As is the case for other secure routing protocols, the SRP uses cryptographic tools for authentication of data integrity. In

---

[10] The cryptography mechanisms are used for authentication and data integrity, not for data secrecy.

[11] The authors coined the phrase "routing misbehavior."

addition, it also relies on the notion of "trust," where two "trusted" nodes can authenticate each other by using a cryptographic shared secret-key. For noncolluding rogue nodes, this protocol guarantees that modified, fictitious, or just repeated responses to route discovery requests are rejected either by the originating node or by some other benign node somewhere along the return path. In addition, the SRP also provides guaranteed timeliness of the response messages.

The use of shared cryptographic keys by SRP implicitly requires tamper immunity for the benign nodes. This, as we know, is a challenging problem for ad hoc networks. Reliance on the binding of the MAC address of a node to its IP address for immunity against IP spoofing is a potential weakness mentioned by the authors themselves.

We find the SRP particularly attractive, because it can be readily combined with some of the existing protocols mentioned earlier, including DSR [78], ZRP [85], and ABR [81]. Consequently, the SRP with its potential performance benefits can be used as the secure routing adjunct to QoS routing protocols discussed in the previous section. Indeed, the SRP should be highly effective in environments such as the military where "trust" and noncollusion, as well as tamper resistance, are often encountered as intrinsic operational attributes.

The current research on secure routing protocols generally relies on the assumption of noncollusion among rogue nodes and different degrees of trust among the benign nodes. Truly Byzantine robustness [101] will likely remain elusive for dynamic ad hoc networks, and owing to their generally "short-lived" nature for many communication applications, such robustness is most likely unnecessary. Even with less stringent security goals, the indispensable issues of performance and scalability remain open for all secure routing protocols. Comprehensive performance studies are a prerequisite to making secure QoS useful for mobile ad hoc networks. Since none exists at all, the goals of simultaneously meeting security and QoS objectives for mobile ad hoc networks offer exceptionally challenging research opportunities.

Current research on security in ad hoc networks concentrates on developing countermeasures against malicious behavior. This is understandable, for ad hoc networks are intrinsically resilient against broken or congested links. Yet, the effects of transient overloads have to be taken into account to ensure QoS support. We consider the control of such overloads an element of QoS security for the simple reason that all denial-of-service attacks result in overload conditions of various durations, and the presence of overload could violate QoS objectives. An overload control policy should determine whether to impose limits on the total traffic a node is allowed to generate for different QoS classes, and if so, of what kinds and under what conditions, as far as the relevant parts of the rest of the network are concerned. After all, a faulty node may generate too many legitimate control packets with the likelihood of degrading the operational performance of a large QoS network. The development of suitable overload control policies, algorithms, and protocols for preserving QoS in a mobile ad hoc network is also an open area.

3.7 Conclusion and Areas of Future Research

We have attempted a brief introduction to the new but rapidly growing area of research on guaranteeing QoS in ad hoc mobile wireless networks. The issues are challenging, many of the underlying algorithmic problems are currently perceived as generally intractable (NP-complete), and opportunities exist for creating more effective heuristics. The issues are complicated by the lack of sufficiently accurate knowledge, both instantaneous and predictive, of the states of the network, e.g., the quality of the radio links and availability of routers and their resources. Successful QoS routing includes the necessary knowledge of the network state and algorithms for feasible route selection and resource reservation. Location-based routing, including predictive QoS routing, is now an active area of research.

Clearly, QoS is a realistic goal to pursue for static ad hoc networks, e.g., a classroom setting. It is equally clear that considerable additional work is necessary to better understand both the specific conditions and extents under which various QoS objectives could be satisfied for dynamic ad hoc networks in the real world. Indeed, guaranteeing QoS in such a network may be impossible if the nodes are too mobile. Even the size of the ad hoc network becomes an issue beyond a certain level because of the increased computational load and difficulties in propagating network updates within the given time bounds.

Minimization of power consumption and QoS support do not appear to be mutually consistent objectives at the current state of the art. Will the network have to be treated, as some have already suggested earlier [14], as some form of a hierarchically ordered collection of subnetworks where at each level the pertinent size is not an issue? Is such an ordering always possible? The challenges increase even more for those ad hoc networks which, like their conventional wireless counterparts, support both best-effort services and those with QoS guarantees, allow different classes of service, and are required to interwork with other wireless and wire-line networks, both connection oriented and connectionless. Algorithms, policies, and protocols for coordinated admission control, resource reservation, and routing for QoS under such models are only beginning to receive attention. In the latter context, the use of preemptive priority of control messages, class of service mechanisms, and segregation of dedicated resources for QoS-robust ad hoc networking offer promising areas of investigation. The general issue of QoS robustness is as yet uncharted territory. The same is also true for accommodating traffic with multiple priorities, including preemptive priorities.

Secure QoS routing introduces an entirely new dimension to the existing challenges. It is obvious that any sufficiently sustained denial-of-service attack, depending on the amount of resources it will cause the network to waste, will destroy QoS routing and even worse, cause an ad hoc network to become combinatorially unstable. While secure routing is now an active area of research, QoS issues are yet to be addressed explicitly in this connection. So far, with the exception of [16], little has appeared on the preservation of QoS guarantees under various failure conditions in ad hoc networks as a specific area of study. The development of suitable overload handling policies, algorithms, and protocols for preserving QoS in a mobile ad hoc network is also an open area.

Performance and scalability studies of ad hoc networks with QoS constraints remain an open area of research, although important results are now appearing in this area for general routing issues associated with such networks. The indispensable issues of performance and scalability remain open for all secure routing protocols. Comprehensive performance studies are a prerequisite to making secure QoS useful for mobile ad hoc networks. Since none exist at all, the goals for simultaneously meeting security and QoS objectives for mobile ad hoc networks offer exceptionally challenging research opportunities.

Support of multicast services such as video conferencing is one of the principal attractions of ad hoc networks. We have not even mentioned the manifold complex issues of adding QoS support to multi-casting in mobile ad hoc networks.

Much work remains to be done on cost-effective implementation issues to bring the promise of ad hoc networks within the reach of the public.

References

1. Z.J. Haas, M. Gerla, D.B. Johnson, C.E. Perkins, M.B. Pursley, M. Steenstrup, and C.-K. Toh, Guest editorial, *IEEE J. Select. Areas Commun.*, 17, 1329–1332, 1999.
2. D.B. Johnson and D.A. Maltz, Protocols for Adaptive Wireless and Mobile Networking, *IEEE Personal Commun.*, Feb. 1996, pp. 34–42.
3. C. Bisdikian, An Overview of the Bluetooth Wireless Technology, *IEEE Commun. Mag.*, Dec. 2001, pp. 86–94. (For additional sources of comprehensive information on Bluetooth, see the official Web sites, http://www.bluetooth.com/ and http://www.bluetooth.org/; an excellent compendium of tutorials and references is available at http://kjhole.com/Bluetooth/.)
4. F. Bennett, D. Clarke, J. B. Evans, A. Hopper, A. Jones, and D. Leask, Piconet: Embedded Mobile Networking, *IEEE Personal Commun.*, Oct. 1997, pp. 8–15.
5. K.J. Negus, J. Waters, J. Tourrilhes, C. Romans, J. Lansford, and S. Hui, HomeRF and SWAP: Wireless Networking for the Connected Home, *ACM SIGMOBILE Mobile Computing Commun. Rev.*, Oct. 1998, pp. 28–37. (Also see HomeRF Working Group, http://www.homerf.org/.)

6. F.A. Tobagi and L. Kleinrock, Packet switching in radio channels–part 2: the hidden terminal problem in carrier sense multiple-access and the busy tone solution, *IEEE Trans. Commun.*, COM-23, 1417–1433, 1985.

7. C.R. Lin and M. Gerla, MACA/PR: An Asynchronous Multimedia Multihop Wireless Network, *Proc. 16th Annual Joint Conf. of IEEE Computer and Communications Societies (INFOCOM 1997)*, vol. 1, 1997, pp. 118–125.

8. J. L. Sobrinho and A.S. Krishnakumar, Quality-of-service in ad hoc carrier sense multiple access wireless networks, *IEEE J. Select. Areas Commun.*, 17, 1353–1414, 1999.

9. S. Chen and K. Nahrstedt, An Overview of Quality-of-Service Routing for the Next Generation High-Speed Networks: Problems and Solutions, *IEEE Network*, Nov./Dec. 1998, pp. 64–79.

10. J.T.-C. Tsai, T. Chen, and M. Gerla, QoS Routing Performance in Multihop, Multimedia Wireless Network, *Proc. IEEE ICUPC '97*, 1997.

11. S. Chen and K. Nahrstedt, Distributed QoS Routing with Imprecise State Information, *Proc. IEEE 7th Intl. Conf. on Computer Commun. and Networks (I3CN)*, Lafayette, LA, Oct. 1998.

12. T.-W. Chen, Efficient Routing and Quality of Service Support for Ad Hoc Wireless Networks, dissertation, University of California at Los Angeles, 1998.

13. S.-B. Lee and A.T. Campbell, INSIGNIA: IN-BAND Signaling Support for QoS in Mobile Ad Hoc Networks, *Proc. 5th Intl. Workshop on Mobile Multimedia Commun. (MoMuC '98)*, Berlin, Oct. 1998.

14. R. Ramanathan and M. Steenstrup, Hierarchically-organized, multihop mobile wireless networks for quality-of-service support, *ACM/Baltzer Mobile Networks and Applications*, 3, 101–119, 1998.

15. S. Basagni, I. Chlamtac, V.R. Syrotiuk, and B.A. Woodward, A Distance Routing Effect Algorithm for Mobility (DREAM), *Proc. 4th Annual ACM/IEEE Intl. Conf. on Mobile Computing and Networking (MobiCom '98)*, Dallas, Oct. 25–30, 1998, pp. 76–84.

16. S. Chen, Routing Support For Providing Guaranteed End-To-End Quality-Of-Service, thesis, University of Illinois at Urbana–Champaign, 1999.

17. P. Bose, P. Morin, I. Stojmenovic, and J. Urrutia, Routing with Guaranteed Delivery in Ad Hoc Wireless Networks, *Proc. 3rd ACM Intl. Workshop on Discrete Algorithms and Methods for Mobile Computing and Communications (DIAL M '99)*, Seattle, Aug. 20, 1999, pp. 48–55.

18. S. Chen and K. Nahrstedt, Distributed quality-of-service routing in ad hoc networks, *IEEE J. Select. Areas Commun.*, 17, 1488–1505, 1999.

19. A. Iwata, C.-C. Chiang, G. Pei, M. Gerla, and T.-W. Chen, Scalable routing strategies for ad hoc wireless networks, *IEEE J. Select. Areas Commun.*, 17, 1369–1379, 1999.

20. C.R. Lin and J.-S. Liu, QoS Routing in ad hoc wireless networks, *IEEE J. Select. Areas Commun.*, 17, 1426–1438, 1999.

21. R. Sivakumar, P. Sinha, and V. Bharghavan, CEDAR: a core extraction distributed ad hoc routing algorithm, *IEEE J. Select. Areas Commun.*, 17, 1454–1465, 1999.

22. D.H. Cansever, A.M. Michelson, and A.H. Levesque, Quality of Service Support in Mobile Ad-Hoc IP Networks, *Proc. MILCOM '99*, Oct. 1999.

23. S.-J. Lee, Routing and Multicasting Strategies in Wireless Mobile Ad hoc Networks, thesis, University of California at Los Angeles, 2000 (available from http://citeseer.nj.nec.com/lee00routing.html).

24. Y.-C. Hu and D.B. Johnson, Caching Strategies in On-Demand Routing Protocols for Wireless Ad Hoc Networks, *Proc. 6th Annual ACM/IEEE Intl. Conf. Mobile Computing and Networking (MobiCom '00)*, Boston, MA, Aug. 2000, pp. 231–242.

25. C.R. Lin, On-Demand QoS Routing in Multihop Mobile Networks, *Proc. 20th Annual Joint Conference of IEEE Computer and Communications Societies (INFOCOM 2001)*, Apr. 2001.

26. P. Sinha, R. Sivakumar, and V. Bharghavan, Enhancing Ad Hoc Routing with Dynamic Virtual Infrastructures, *Proc. 20th Annual Joint Conf. of IEEE Computer and Communications Societies (INFOCOM 2001)*, Apr. 2001.

27. Baochun Li, QoS-aware Adaptive Services in Mobile Ad-hoc Networks, *Proc. 9th IEEE Intl. Workshop on Quality of Service (IWQoS '01)*; also *Lecture Notes in Computer Science*, Vol. 2092, Springer-Verlag, Karlsruhe, Germany, June 6–8, 2001, pp. 251–268.

28. S.-B. Lee, G.-S. Ahn, and A.T. Campbell, Improving UDP and TCP Performance in Mobile Ad Hoc Networks with INSIGNIA, *IEEE Commun. Mag.*, Jun. 2001, pp. 156–165.

29. Y.-C. Hu and D.B. Johnson, Implicit Source Routes for On-Demand Ad Hoc Network Routing, *Proc. 2001 ACM Intl. Symp. on Mobile Ad Hoc Networking and Computing (MobiHoc '01)*, Aug. 2001, pp. 1–10.

30. K. Wu and J. Harms, QoS support in mobile ad hoc networks, *Crossing Boundaries*, 1, 92–106, 2001.

31. R. Leung, J. Liu, E. Poon, A.-L.C. Chan, and B. Li, MP-DSR: A QoS-aware Multi-path Dynamic Source Routing Protocol for Wireless Ad-hoc Networks, *Proc. 26th IEEE Annual Conf. Local Computer Networks (LCN 2001)*, Tampa, Nov. 15–16, 2001

32. K. Nahrstedt, D. Xu, D. Wichadakul, and B. Li, QoS-Aware Middleware for Ubiquitous and Heterogeneous Environments, *IEEE Commun. Mag.*, Nov. 2001.

33. C. Zhu and M.S. Corson, QoS Routing for Mobile Ad Hoc Networks, *Tech. Rept. CSHCN TR 2001–18 (ISR TR 2001–28)*, The Center for Satellite and Hybrid Communication Networks, University of Maryland (www.isr.umd.edu/CSHCN/), 2001.

34. S.H. Shah and K. Nahrstedt, Predictive Location-Based QoS Routing in Mobile Ad Hoc Networks, *Tech. Rept. UIUCDCS-R-2001–2242 - UILU-ENG-2001–1749*, Dept. of Computer Science, University of Illinois at Urbana-Champaign, Sep. 2001; to appear in *Proc. IEEE Intl. Conf. Commun. (ICC '02)*, New York, NY, Apr. 28–May 2, 2002.

35. For extensive additional information on Routing in Mobile Ad Hoc Networks, including QoS Routing, see *inter alia* the web pages of the MONET Group (including K. Nahrstedt: http://cairo.cs.uiuc.edu/cgi-bin/getpaper.pl?action = list&area = routing&type = all and http://cairo.cs.uiuc.edu/cgi-bin/getpaper.pl?action = list&area = all&type = all, Shigang Chen: http://www-sal.cs.uiuc.edu/~s-chen5/publications.html, Baochun Li, now at U. of Toronto: http://www.eecg.toronto.edu/~bli/research.html, et al.) and the CEDAR group (including V. Bharghavan: http://www.timely.crhc.uiuc.edu/Projects/cedar/papers.html, *et al*) of University of Illinois at Urbana-Champaign, of the Terminodes project at the Swiss Federal Institute of Technology (including: J.-P. Hubaux: http://www.terminodes.org/publications.html, et al.), and the INSIGNIA Team at Columbia University (including A.T. Campbell: http://comet.ctr.columbia.edu/insignia/publications.html, et al.). This is only a representative list; there are many others.

36. Y.-C. Hsu and T.-C. Tsai, Bandwidth Routing in Multihop Packet Radio Environment, *Proc. 3rd Intl. Mobile Computing Workshop*, 1997.

37. A. Michail and A. Ephremides, A Distributed Routing Algorithm for Supporting Connection-Oriented Service in Wireless Networks with Time-Varying Connectivity, *Proc. 3rd IEEE Symposium on Computers and Communications (ISCC '98)*, Athens, June 1998, pp. 587–591.

38. A. Michail, W. Chen, and A. Ephremides, Distributed Routing and Resource Allocation for Connection-Oriented Traffic in Ad Hoc Wireless Networks, *Proc. Conf. on Information Sciences and Systems (CISS '98)*, 1998.

39. S. Basagni, I. Chlamtac, and V.R. Syrotiuk, Geographic Messaging in Wireless Ad Hoc Networks, *Proc. 4th Annual ACM/IEEE Intl. Conf. on Mobile Computing and Networking (MobiCom '98)*, Dallas, Oct. 25–30, 1998, pp. 76–84.

40. E.M. Royer and C.-K. Toh, A Review of Current Routing Protocols for Ad Hoc Mobile Wireless Networks, *IEEE Personal Commun.*, Apr. 1999, pp. 46–55.

41. M. Mauve, J. Widner, and H. Hartenstein, A Survey on Position-Based Routing in Mobile Ad-Hoc Networks, *IEEE Network*, Nov./Dec. 2001, pp. 30–39.

42. J. Broch, D.A. Maltz, D.B. Johnson, Y.-C. Hu, and J. Jetcheva, A Performance Comparison of Multi-Hop Wireless Ad Hoc Network Routing Protocols, *Proc. 4th Annual ACM/IEEE Intl. Conf. on Mobile Computing and Networking (MobiCom '98)*, Dallas, Oct. 25–30, 1998, pp. 85–97.

43. S.R. Das, R. Castaneda, J. Yan, and R. Sengupta, Comparative Performance Evaluation of Routing Protocols for Mobile, Ad Hoc Networks, *Proc. IEEE 7th Intl. Conf. on Computer Communication and Networks (I3CN)*, Lafayette, LA, Oct. 1998, pp. 153–161.

44. S.R. Das, R. Castaneda, and J. Yan, Simulation Based Performance Evaluation of Mobile, Ad Hoc Network Routing Protocols, *ACM/Baltzer Mobile Networks and Applications Journal*, July 2000, pp. 179–189.

45. S.R. Das, C.E. Perkins, and E. Royer, Performance Comparison of Two On-demand Routing Protocols for Ad Hoc Networks, *Proc. 19th Annual Joint Conf. of IEEE Computer and Communications Societies (INFOCOM 2000)*, Tel Aviv, Mar. 2000, pp. 3–12.

46. P. Jacquet and L. Viennot, Overhead in Mobile Ad-hoc Network Protocols, *INRIA Res. Rept. RR-3965*, INRIA, Rocquencourt, France, 2000.

47. P. Johansson, T. Larsson, N. Hedman, B. Mielczarek, and M. Degermark, Scenario-Based Performance Analysis of Routing Protocols for Mobile Ad-hoc Networks, *Proc. 5th Annual ACM/IEEE International Conference on Mobile Computing and Networking (MobiCom '99)*, Seattle, WA, Aug. 15–19, 1999, pp. 195–206.

48. D.A. Maltz, J. Broch, J. Jetcheva, and D.B. Johnson, The effects of on-demand behavior in routing protocols for multi-hop wireless ad hoc networks, *IEEE J. Select. Areas Commun.*, 17, 1439–1453, 1999.

49. M.K. Marina and S.R. Das, Performance of Route Caching Strategies in Dynamic Source Routing, *Proceedings of the 2nd Wireless Networking and Mobile Computing (WNMC)*, (in conjunction with the Intl. Conf. on Distributed Computing Systems [ICDCS]), Phoenix, Apr. 2001.

50. R. Braden, D. Clark, S. Croker, and C. Huitema, Report of IAB Workshop on Security in the Internet Architecture, IETF, RFC 1636.

51. S. Chakrabarti and A. Mishra, QoS Issues in Ad Hoc Wireless Networks, *IEEE Commun. Mag.*, Feb. 2001, pp. 142–148.

52. R. Hauser, T. Przygienda, and G. Tsudik, Lowering Security Overhead in Link State Routing, *Computer Networks and ISDN Systems*, 31, 885–894, 1999.

53. L. Zhou and Z.J. Haas, Securing Ad Hoc Networks, *IEEE Network*, Nov./Dec. 1999, pp. 24–30.

54. J.-P. Hubaux, L. Buttyán, and S.Čapkun, The quest for security in mobile ad hoc networks, *Proc. 2001 ACM Intl. Symp. on Mobile Ad Hoc Networking and Computing (MobiHoc '01)*, Long Beach, CA, 2001, pp. 146–155.

55. N. Asokan and P. Ginzboorg, Key agreement in ad hoc networks, *Computer Commun.*, 23, 1627–1637, 2000.

56. S. Marti, T.J. Giuli, K. Lai, and M. Baker, Mitigating Routing Misbehavior in Mobile Ad Hoc Networks, *Proc. 6th ACM Annual Intl. Conf. on Mobile Computing and Networking (MobiCom '00)*, Boston, Aug. 2000, pp. 255–265.

57. Y. Zhang and W. Lee, Intrusion Detection in Wireless Ad-Hoc Networks, *Proc. 6th ACM Annual Intl. Conf. on Mobile Computing and Networking (MobiCom '00)*, Boston, Aug. 2000, pp. 275–283.

58. B. Dahill, B.N. Levine, E. Royer, and C. Shields, A Secure Routing Protocol for Ad Hoc Networks, *Tech. Rept. UM-CS-2001–037*, University of Massachusetts at Amherst, Aug. 28, 2001.

59. S. Yi, P. Naldurg, and R. Kravets, Security-Aware Ad Hoc Routing for Wireless Networks, *Tech. Rept. UIUCDCS-R-2001–2241, UILU-ENG-2001–1748*, University of Illinois at Urbana–Champaign, Aug. 2001.

60. P. Papadimitratos and Z.J. Haas, Secure Routing for Mobile Ad hoc Networks, *SCS Communication Networks and Distributed Systems Modeling and Simulation Conf. (CNDS 2002)*, San Antonio, TX, Jan. 27–31, 2002.

61. F. Stajano and R. Anderson, The Resurrecting Duckling: Security Issues for Ad-Hoc Wireless Networks, *Proc. 7th International Workshop on Security Protocols, Lecture Notes in Computer Science*, Springler-Verlag, Berlin, Apr. 1999. Available from http://www.cl.ac.uk/~fms27/duckling/duckling.htm.

62. S. Čapkun, M. Hamdi, and J.P. Hubaux, GPS-free Positioning in Mobile Ad-Hoc Networks, *Proc. 34th Annual Hawaii Intl. Conf. on System Sciences,* Jan. 2001.

63. F. Wang, B. Vetter, and S.F. Wu, Secure Routing Protocols: Theory and Practice, *Tech. Rept.,* University of California at Davis, May 1997. Available from http://shang.csc.ncsu.edu/papers.htm.

64. C.-K. Toh, Maximum Battery Life Routing to Support Ubiquitous Mobile Computing in Wireless Ad Hoc Networks, *IEEE Commun. Mag.,* Jun. 2001, pp. 138–147.

65. S. Singh, M. Woo, and C.S. Raghavendra, Power-Aware Routing in Mobile Ad Hoc Networks, *Proc. 4th Annual ACM/IEEE Intl. Conf. on Mobile Computing and Networking (MobiCom '98),* Dallas, Oct. 25–30, 1998, pp. 181–190.

66. V. Rodoplu and T.H.-Y. Meng, Minimum energy mobile wireless networks, *IEEE J. Select. Areas Commun.,* 17, 1333–1344, 1999.

67. J. Chang and L. Tassiulas, Energy Conserving Routing Wireless Ad Hoc Networks, *Proc. 19th Annual Joint Conf. of IEEE Computer and Communications Societies (INFOCOM 2000),* vol. 1, 2000, pp. 22–31.

68. P. Bahl, R. Wattenhofer, L. Li, and Y. Wang, Distributed Topology Control for Power Efficient Operation in Multihop Wireless Ad Hoc Networks, *Proc. 20th Annual Joint Conf. of IEEE Computer and Communications Societies (INFOCOM 2001),* 2001.

69. L. Li and J. Halpern, Minimum-Energy Mobile Wireless Networks Revisited, *Proc. IEEE Intl. Conf. Commun. (ICC '01),* 2001.

70. Y. Xue and B. Li, A Location-aided Power-aware Routing Protocol in Mobile Ad Hoc Networks, *Proc. IEEE Symp. on Ad Hoc Mobile Wireless Networks/IEEE GLOBECOM 2001,* San Antonio, TX, Nov. 25–29, 2001.

71. C.E. Perkins and P. Bhagwat, Highly Dynamic Destination-Sequenced Distance Vector Routing, *ACM SIGCOMM Computer Commun. Rev.,* Oct. 1994, pp. 234–244.

72. C.-C. Chiang, Routing in Clustered Multihop Mobile Wireless Networks with Fading Channel, *Proc. IEEE Singapore Intl. Conf. on Networks (SICON '97),* Singapore, Apr. 1997, pp. 197–211.

73. S. Murthy and J.J. Garcia-Luna-Aceves, An efficient routing protocol for wireless networks, *ACM Mobile Networks and Applications Journal,* 1, 183–197, 1996.

74. J.J. Garcia-Luna-Aceves and M. Spohn, Source-Tree Routing in Wireless Networks, *Proc. IEEE 7th Intl. Conf. on Network Protocols (ICNP '99),* Toronto, Oct. 31–Nov. 3, 1999.

75. G. Pei, M. Gerla, and T.-W. Chen, Fisheye State Routing: A Routing Scheme for Ad Hoc Wireless Networks, *Proc. 2000 ICDCS Workshop on Wireless Networks and Mobile Computing,* Taipei, Apr. 2000, pp. D71-D78.

76. J.J. Garcia-Luna-Aceves and M. Spohn, Bandwidth-efficient link-state routing in wireless networks, in *Ad Hoc Networking,* C. E. Perkins, Ed., Addison-Wesley Longman, Reading, MA, 2001.

77. C.A. Santiváñez, R. Ramanathan, and I. Stavrakakis, Making Link-State Routing Scale for Ad Hoc Networks, *Proc. 2001 ACM Intl. Symp. on Mobile Ad Hoc Networking and Computing (MobiHoc '01),* Aug. 2001, pp. 11–21.

78. D.B. Johnson and D.A. Maltz, Dynamic Source Routing in Ad Hoc Wireless Networks, in *Mobile Computing,* T. Imielinski and H. Korth, Eds., Kluwer Academic Publishers, Dordrecht, The Netherlands, 1996, pp. 153–181.

79. V.D. Park and M.S. Corson, A Highly Adaptive Distributed Routing Algorithm for Mobile Wireless Networks, *Proc. 16th Annual Joint Conf. of IEEE Computer and Communications Societies (INFOCOM 1997),* 1997.

80. C.E. Perkins and E.M. Royer, Ad-Hoc On-Demand Distance Vector Routing, *Proc. 2nd IEEE Workshop on Mobile Computing Systems and Applications (WMCSA '99),* Feb. 1999, pp. 90–100.

81. C.-K. Toh, Associativity based routing for ad hoc mobile networks, *Wireless Personal Communication Journal,* 4, 1–36, 1997.

82. R. Dube, C.D. Rais, K.-Y. Wang, and S.K. Tripathi, Signal Stability Based Adaptive Routing (SSA) for Ad Hoc Mobile Networks, *IEEE Personal Commun.,* Feb. 1997, pp. 36–45.

83. M. Spohn and J.J. Garcia-Luna-Aceves, Neighborhood Aware Source Routing, *Proc. 2001 ACM Intl. Symp. on Mobile Ad Hoc Networking and Computing (MobiHoc '01)*, Long Beach, CA, Aug. 2001, pp. 11–21.

84. Z.J. Haas and M.R. Pearlman, Determining the optimal configuration for the zone routing protocol, *IEEE J. Select. Areas Commun.*, 17, 1395–1414, 1999.

85. Z.J. Haas and M.R. Pearlman, The performance of query control schemes for the zone routing protocol, *ACM/IEEE Trans. on Networking*, 9, 427–438, 2001.

86. D.S.J. De Couto and R. Morris, Location Proxies and Intermediate Node Forwarding for Practical Geographic Forwarding, *Proc. 4th Annual ACM/IEEE Intl. Conf. on Mobile Computing and Networking (MobiCom '98)*, Dallas, Oct. 25–30, 1998.

87. Z.J. Haas and B. Liang, Ad-hoc mobility management with uniform quorum systems, *IEEE/ACM Trans. on Networking*, 7, 228–240, 1999.

88. B.N. Karp, Geographic Routing for Wireless Networks, thesis, Harvard University, Cambridge, MA, 2001.

89. B.N. Karp and H.T. Kung, GPSR: Greedy Perimeter Stateless Routing for Wireless Networks, *Proc. 6th Annual ACM/IEEE Intl. Conf. on Mobile Computing and Networking (MobiCom '00)*, Boston, Aug. 2000, pp. 243–254.

90. Y.-B. Ko and N.H. Vaidya, Location-Aided Routing (LAR) in Mobile Ad Hoc Networks, *Proc. 4th Annual ACM/IEEE Intl. Conf. on Mobile Computing and Networking (MobiCom '98)*, Dallas, Oct. 1998, pp. 66–75. Also in *ACM/Baltzer Wireless Networks Journal*, 6, 307–321, 2000.

91. Y.-B. Ko and N.H. Vaidya, Using Location Information in Wireless Ad Hoc Networks, *IEEE Vehicular Technology Conf. (VTC '99)*, May 1999.

92. J. Li, J. Jannotti, D.S.J. De Couto, D.R. Karger, and R. Morris, A Scalable Location Service for Geographic Ad Hoc Routing, *Proc. 6th Annual ACM/IEEE Intl. Conf. on Mobile Computing and Networking (MobiCom '00)*, Boston, Aug. 2000, pp. 120–130.

93. E. Crawley, R. Nair, B. Rajagopalan, and H. Sandick, A Framework for QoS-based Routing in the Internet, RFC 2386, http://www.ietf.org/rfc/rfc.2386.txt, Aug. 1998.

94. Z. Wang and J. Crowcroft, Quality of service routing for supporting multimedia applications, *IEEE J. Select. Areas Commun.*, 14, 1228–1234, 1996.

95. B. Das, R. Sivakumar, and V. Bharghavan, Routing in Ad-Hoc Networks Using a Spine, *IEEE Intl. Conf. on Communications (ICC '97)*, Las Vegas, Sep. 1997.

96. R. Sivakumar, B. Das, and V. Bharghavan, An Improved Spine-based Infrastructure for Routing in Ad Hoc Networks, *IEEE Symp. on Computers and Communications '98*, Athens, June 1998.

97. S. Cheung and K.N. Levitt, Protecting Routing Infrastructures from Denial of Service Using Cooperating Intrusion Detection, *Proc. New Security Paradigms Workshop*, Cumbria, UK, Sep. 1997.

98. I. Stoica and H. Zhang, Providing Guaranteed Services Without Per Flow Management, *Proc. ACM Conf. Applications, Technologies, Architectures, and Protocols for Computer Commun. (SIGCOMM '99)*, Cambridge, MA, August 1999, pp. 81–94.

99. S. Buchegger and J.-Y. Le Boudec, Nodes Bearing Grudges: Towards Routing Security, Fairness, and Robustness in Mobile Ad Hoc Networks, *Proc. 10th Euromicro Workshop on Parallel, Distributed and Network-based Processing*, Canary Islands, Jan. 2002.

100. M. Gerla and J.T.-C. Tsai, Multicluster, mobile, multimedia radio network, *ACM/Baltzer Wireless Networks Journal*, 1, 255–265, 1995.

101. R. Perlman, Network Layer Protocols with Byzantine Robustness, dissertation, Massachusetts Institute of Technology, Cambridge, MA, MIT Tech. Rept. LCS TR-429, October 1988.

Appendix

Acronym Guide

Acronym	Expansion
AODV	Ad-hoc On-demand Distance Vector (routing)
CSGR	Clusterhead Gateway Switch Routing
DSDV	Destination-Sequenced Distance-Vector (routing)
DSR	Dynamic Source Routing
FSR	Fisheye State Routing
GPS	Global Positioning System
MAC	Medium Access Control
QoS	Quality of Service
SRP	Secure Routing Protocol
STAR	Source-Tree Adaptive Routing
TORA	Temporally Ordered Routing Algorithm
WRP	Wireless Routing Protocol
ZRP	Zone Routing Protocol

4

Power-Conservative Designs in Ad Hoc Wireless Networks

Yu-Chee Tseng
National Chiao-Tung University

Ting-Yu Lin
National Chiao-Tung University

Abstract

Since the basic components of ad hoc wireless networks are mostly battery-operated portable devices, power conservation is one of the central issues of such networks. Power-conservative designs for ad hoc networks pose many challenges due to the lack of central coordination facilities. The purpose of this chapter is to offer a comprehensive discussion of the power-conservative issues for mobile ad hoc networks. We generally divide power-conservative protocols into two categories: transmitter power control mechanisms and power management algorithms. The latter can be further classified into MAC-layer, network-layer, and higher-layer implementations. This chapter will present a variety of existing power-conservative designs for ad hoc networks in order, based on the above classification.

4.1 Background

In recent years, more and more communication environments have become wireless oriented. Unlike traditional wired networks, in which end hosts are fixed in location, wireless networks include a variety of mobile terminals, such as notebooks, personal digital assistants (PDAs), cellular phones, and microsensors. Mobile/portable devices are inevitably battery powered, and thus battery lifetime becomes crucial for wireless communications and mobile computing.

Battery technology has lagged compared to the advancements in communication and computing technology in the past decade. Now that batteries' capacity cannot be significantly improved, efforts should be put into designing energy-efficient software and hardware. A portable device typically has several main hardware components that consume battery power: display monitor, disk, CPU, memory,

TABLE 4.1   Doze/Receive/Transmit Powers of Lucent ORiNOCO Wireless LAN PC Card

	Doze Mode	Receive Mode	Transmit Mode
Lucent ORiNOCO WLAN PC card	0.05 watt	0.9 watt	1.4 watt

Source: Data from http://www.agere.com, dated September 13, 2000.

and wireless network interface card (WNIC). According to [22], the WNIC component can consume 10–50% of overall system energy, which explains why notebooks' lifetimes reduce significantly when inserted with WNICs. Thus, it is essential to support some low-power modes in WNICs. Table 4.1 lists doze/receive/transmit powers of a manufactured IEEE 802.11b WLAN card [1] for readers' reference. Note that doze mode consumes nonzero battery power, and transmission operation consumes more energy than reception does. One implication from these statistics is that the upper layer software should cooperate with WNICs to tune to the power mode at the proper time so as to reduce the communication energy expenditure. The other implication is that application software should reduce unnecessary transmissions as much as possible.

Several works have addressed *non–network-related* power-saving software techniques for CPU [26–28] and disk [29–31]. This chapter will focus on *network-related* power issues, specifically targeting wireless communication environments. Reference [2] provides an overview towards power-sensitive designs in wireless networks. The main emphasis is on networking issues, taking into account both *computation* and *communication* costs. Efforts should be made to balance between these two costs to achieve overall power conservation [34].

Wireless architectures can be classified as *infrastructured* and *ad hoc.* With infrastructure (base stations), energy-efficient techniques are usually easier to develop due to the availability of central coordination. Without such central coordination, power-conservative designs pose more challenges for ad hoc networks. The purpose of this chapter is to offer a comprehensive discussion of the related power-conservative issues for mobile ad hoc networks. We generally divide power-conservative protocols into two categories:

1. Transmitter power control mechanisms
2. Power management algorithms

The latter can be further classified into MAC-layer (where MAC stands for Medium Access Control), network-layer, and higher-layer implementations.

In Section 4.2, we present power control techniques. Section 4.3 reviews several MAC-layer power-conservative designs. Network-layer power-aware unicasting/multicasting/broadcasting protocols are introduced in Section 4.4. We describe several higher-layer energy-efficient approaches in Section 4.5. Finally, Section 4.6 summarizes this chapter and draws out future research directions.

4.2   Transmitter Power Control Mechanisms

In wireline networks, the network topologies are fixed and can be easily determined by wireline deployments. In contrast, the topology of a wireless network is usually changeable, even if no host mobility is present. Both adjusting antenna directions and tuning transmission powers can change the network topology. *Power control* refers to the technique of tuning hosts' transmission powers to the proper range. Power control has two advantages:

1. It can conserve battery energy.
2. It can reduce radio interference and thus increase spatial reuse of wireless bandwidth.

For example, consider an ad hoc network based on a single shared medium as shown in Fig. 4.1. Figure 4.1a shows the case where hosts A and B are communicating, but A uses an improper (too large) transmission power. The circle centered at A shows its interference region. This results in interference at hosts D and F, thus destroying their receiving activity. With proper power control, as shown in Fig. 4.1b, if we intelligently control the transmission powers of A, C, and E, then three pairs of communications

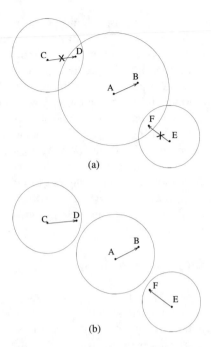

FIGURE 4.1 Advantage of power control — channel spatial reuse: (a) without power control and (b) with power control.

(AB, CD, and EF) can coexist simultaneously. This will greatly increase the utilization of the precious wireless bandwidth.

Although power control contributes to network throughput by enabling simultaneous transmissions, the drawbacks include lower reliability, higher error rate, and weaker network connectivity. Balancing these factors is an important design issue. Below, we review several power control techniques.

Assuming that all hosts' physical locations are known, Reference [4] proposes a distributed algorithm for topology control. Each host conducts local computation involving a planar triangulation process. This is dealt with as the starting topology, which is then transformed into the so-called *Delaunay triangulation* by switching links based on certain criteria. The resultant topology is optimized in terms of host degrees. However, this work did not address adaptive power control for mobile networks.

In [5], given a set of hosts with their transmission powers undetermined, two centralized algorithms are proposed to construct a *1-connected* network and a *biconnected* network. The output is the transmission power of each host. The algorithms are designed for static networks with the assumption that all hosts' locations are known. The construction is similar to the greedy, minimum-cost spanning tree approach. On top of these, two distributed heuristics are presented, where hosts are allowed to constantly adjust their transmission powers to maintain their degrees. Whenever its degree is below a certain threshold (say, three), a host is allowed to increase its transmission power. Whenever its degree is beyond a threshold, a host is allowed to decrease its transmission power. However, without knowing global connectivity, these heuristics may lead to a disconnected network, which is undesirable.

Reference [6] further proposes a distributed *cone-based* topology control mechanism, which guarantees global network connectivity. In this work, a receiver is assumed to be able to determine the direction of the sender when receiving a message. Two phases are performed. In the first phase, each host should increase its transmission power until there exists at least one neighbor in each of its cones, where a *cone* is an angle with the host itself situated as the center (say, from 90 to 120° on the Euclidean plane). The second phase is to remove redundant links without affecting the global network connectivity. Simulation results indicated that this approach can result in longer network lifetime and reduce radio interference due to decreased transmission powers and low host degrees.

Different from the above pure power controls, [7] and [8] incorporate power control into MAC protocols. In [7], the authors study the IEEE 802.11 MAC protocol and suggest modifying header formats of CTS and DATA packets to support power control. Ten transmission power levels are defined for selection. The receiver informs the sender of the appropriate power level through a CTS packet, while the sender informs the receiver via a DATA packet. Hence, during one RTS–CTS–DATA–ACK exchange, both sender and receiver can determine proper transmission powers to use. The performance results show a 10–20% reduction in power consumption and a 15% improvement in throughput.

On the other hand, [8] proposes a new MAC protocol that combines RTS/CTS dialogue and two busy tones. The two busy tones are *transmission busy tone* and *receive busy tone,* the purposes of which are to warn the neighborhood of a host's sending and receiving activities, respectively. Since transmissions with lower powers may suffer higher collision probability, the busy tones can help relieve this problem. Through theoretical analyses and experiments, the protocol is verified to be able to significantly increase channel utilization due to the reduced signal overlapping. Energy conservation and interference reduction are achieved simultaneously.

4.3   MAC-Layer Power Management

Based on the OSI model, the MAC layer is a sublayer of the data link layer right above the physical layer. The MAC layer is mainly responsible for access scheduling of shared medium. According to the studies in [9], a device's power consumption strongly depends on the MAC policy adopted. Power-efficient designs are even more challenging in ad hoc networks, which lack central coordination. Below, we present several existing MAC-layer power-conservative designs for ad hoc networks.

In [10,11], the authors propose a Distributed Contention Control (DCC) protocol. Based on the slot utilization and number of transmission attempts, DCC estimates the probability of successful transmission before each frame is actually transmitted. If the probability of success is too low, the transmission is deferred to reduce the potential retransmission overhead. Otherwise, the frame is transmitted immediately. Battery energy is saved in the sense that, by observing the channel congestion level that may increase the packet loss probability, many unnecessary retransmissions, which consume significant power, are avoided. The proposed DCC is compatible with the IEEE 802.11 DCF (Distributed Coordination Function) mode. The authors suggest implementing DCC between the physical and the MAC layers. The DCC protocol can temporarily suspend the transmission of frames with low success probability. In addition, a Power Save DCC (PS-DCC) is further proposed to incorporate the energy concerns into transmission decision making by deriving another evaluation formula. This mechanism is reported to yield a quasi-optimal power consumption. Interested readers can refer to [11] for more details.

In [12], a Power-Aware Multi-Access protocol with Signaling (PAMAS) is introduced for power conservation in ad hoc networks. PAMAS is a RTS/CTS-based MAC protocol with separate data and signaling channels for data and control packets, respectively. The signaling channel is used for exchanging RTS/CTS packets, busy tones, and probe control packets. The basic observation is that, due to the broadcast nature of radio transmission, a host's energy is often wasted on overhearing packets that are not destined for it. As Fig. 4.2 shows, during the transmission from A to B, it would be beneficial for C to turn off its

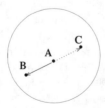

FIGURE 4.2   A scenario that shows that host A is sending to host B, and host C can turn off its receiver since no receiving activity is possible for it for a while.

radio to conserve power because any frame destined to it is doomed to fail. In PAMAS, battery power is conserved by judiciously turning off hosts' radios when no transmission/reception is possible.

The separate signaling channel in [12] enables a host to determine when and for how long to power its antenna off. It is easy for a host to decide its power-off duration if RTS/CTS packets can be correctly received. However, if a host misses some RTS/CTS packets during its power-off period, PAMAS provides a probe protocol to estimate the duration of others' remaining transmissions. The ultimate goal is to reduce power consumption without increasing delay and decreasing throughput. Simulation results indicate 10 to 60% power savings in most cases.

Rather than proposing new MAC protocols as previous works do, [13] presents three power-saving mechanisms to effectively manage the IEEE 802.11 Power-Saving (PS) mode for multi-hop ad hoc networks. IEEE 802.11 supports two power modes: *active* and *power-saving* (PS). When staying in the PS mode, PS hosts wake up periodically. In the ad hoc mode, the short interval that PS hosts wake up is called the *ATIM window*. Figure 4.3 plots an example, where host A wants to transmit a packet to host B. During the ATIM window, an ATIM frame is sent from A to B. In response, B replies with an ACK. After the ATIM window finishes, A tries to send out its data packet based on the regular DCF contention mechanism.

Since ad hoc networks are characterized by no clock synchronization mechanism, predicting when PS hosts will wake up to receive packets becomes a nontrivial task. In [13], the authors propose three power management mechanisms based on the IEEE 802.11 PS mode. They define their *MTIM* frames, which serve the similar purpose as ATIM frames in IEEE 802.11. Several notations are used to facilitate the presentation: *BI* (length of a beacon interval), *AW* (length of an active window), *BW* (length of a beacon window), and *MW* (length of an MTIM window). The proposed protocols are described in the following:

1. *Dominating-awake-interval protocol* — The basic idea is to impose a PS host to stay awake sufficiently long so as to ensure that neighboring hosts can discover each other. "Dominating-awake" means that a PS host should stay awake for at least half of *BI* in each beacon interval. This guarantees a PS host's active window to overlap with any neighboring PS host's active window. Furthermore, the sequence of beacon intervals is alternatively labeled as odd and even intervals. Each odd beacon interval starts with an active window, which is led by a beacon window followed by an MTIM window. On the other hand, each even beacon interval also starts with an active window, which is terminated by an MTIM window followed by a beacon window. The structures

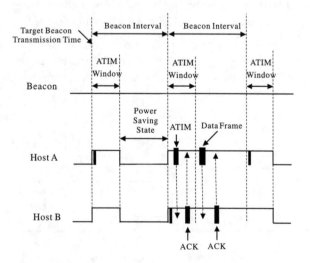

FIGURE 4.3 Operations of PS mode in IEEE 802.11 under the ad hoc mode.

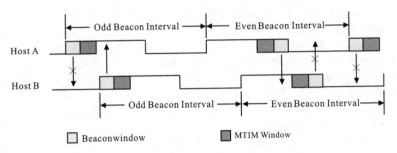

FIGURE 4.4   Dominating-awake-interval protocol.

of odd and even intervals are illustrated in Fig. 4.4. The different structures for odd and even intervals guarantee that a PS host's active window always overlaps with any neighboring PS host's active window in every other beacon interval.

2. *Periodically-fully-awake-interval protocol* — The previous protocol requires PS hosts to keep active more than half of the time, which is not energy conserving. To reduce the active time, this protocol designs two types of beacon intervals: *low-power intervals* and *fully-awake intervals*. In a low-power interval, the length of the active window is reduced to the minimum, while in a fully-awake interval, the length of the active window is extended to the maximum. Since fully-awake intervals are more energy-consuming, they only appear periodically and interleaved with low-power intervals. Specifically, the fully-awake intervals arrive every T intervals, and the rest of the intervals are low-power. Figure 4.5 shows an example with $T = 3$. This protocol guarantees that a PS host's beacon windows overlap with any neighbor's fully-awake intervals in every T beacon intervals.

3. *Quorum-based protocol* — In the previous two protocols, a PS host had to send a beacon in *every* beacon interval. This protocol is based on the concept of a *quorum*, where a PS host only needs to send beacons in *some* beacon intervals. Thus this protocol can be more power conservative when the transmission cost is high. In the literature of distributed systems, a quorum is a set of identities from which one has to obtain permission to perform some action. Typically, two quorum sets always have nonempty intersection so as to guarantee the atomicity of a transaction. By adopting this concept to design PS hosts' wake-up patterns, this protocol can guarantee that a PS host always has at least two entire beacon windows that are fully covered by another PS host's active windows in every certain period. Figure 4.6 shows an example.

Simulation results indicate that the periodically-fully-awake-interval protocol can balance both power consumption and neighbor discovery time and thus may be practical in most cases. The proposed protocols are directly applicable to the current IEEE 802.11 standard with little modification. Interested readers can refer to [13] for more details.

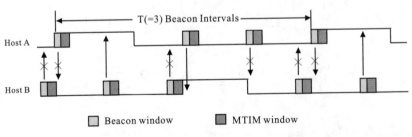

FIGURE 4.5   Periodically-fully-awake-interval protocol with fully awake intervals arrive every $T = 3$ beacon intervals.

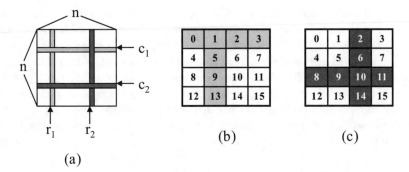

(a)   (b)   (c)

FIGURE 4.6 Quorum-based protocol: (a) intersections of two PS hosts' quorum intervals, (b) host A's quorum intervals, and (c) host B's quorum intervals.

4.4 Network-Layer Power Management

In this section, we review several power-saving techniques implemented in the network layer for ad hoc networks. Residing above the data link layer, the network layer is responsible for routing packets toward destinations. For wired environments, routing paths are usually fixed since end terminals are static. However, in wireless networks, mobile hosts are free to move, resulting in frequently changed routing paths. The wireless network layer hence needs to take care of mobility management, such as location tracking/update, which is not considered in static wired networks.

In an ad hoc network, a packet typically needs to be relayed by several hosts before reaching its final destination. Network-layer power-aware routing is thus a unique issue for ad hoc networks. Traditional ad hoc routing protocols adopt metrics, such as hop-count, link quality, and location stability, for route selection. These metrics, however, do not take battery power conditions into account, which may lead to improper energy expenditure and thus reduced network lifetime. Figure 4.7 illustrates one routing

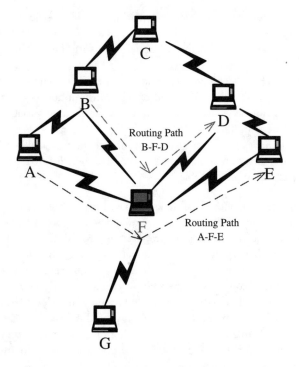

FIGURE 4.7 A scenario that may quickly drain off the energy of host F, an articulation point in the network.

scenario. Based on the hop-count metric, host A selects the shortest route A-F-E to reach E. Similarly, B chooses route B-F-D to reach D. Both communication pairs go through F, which may quickly drain off F's battery energy. This may make F die earlier, forcing G to become isolated from the network. Consequently, energy cost should be taken into account to lengthen the network lifetime. The design principle of power-aware routing is to equally balance energy expenditure among mobile hosts to prolong network lifetime, while at the same time conserving overall power consumption as much as possible.

Below, we present power-conservative ad hoc routing protocols for unicasting and multicasting/broadcasting, separately. Unicasting is one-to-one communication with a single receiver. Multicasting is one-to-many communication with a set of receivers, while broadcasting is one-to-all communication with all nodes as receivers.

4.4.1   Unicast Data Communications

Several power-aware routing protocols have been proposed for unicast communication [14–17]. In [14], the authors propose five metrics for power-aware routing. The underlying MAC protocol is PAMAS, which was reviewed earlier in Section 4.3. This work can further improve power consumption by 5–30% over traditional shortest-path routing protocols. The five metrics are:

1. *Minimize energy consumed per packet* — Obviously, by minimizing energy consumption for each packet, the total power consumption will be reduced. However, to conserve power, this metric tends to select routes around congested areas. How to share energy cost among hosts is left unaddressed. So the network lifetime may not be improved.
2. *Maximize time to network partition* — Given an ad hoc network, there exists a minimal set of hosts whose removal will partition the network. The set is called the *cut-set*. Energies of the hosts in the cut-set are critical and decisive for network lifetime. This metric attempts to divide network load among the cut-set hosts to prolong the time before the network becomes partitioned. However, this optimization problem is similar to a load-balancing problem, which is known to be NP-complete.
3. *Minimize the variance of hosts' power levels* — The goal of this metric is to make all hosts process about the same number of packets to achieve an equal power draining rate at each host. The basic idea is that each host sends packets through a neighbor with the least amount of data waiting to be transmitted.
4. *Minimize cost per packet* — This metric avoids selecting hosts with depleted battery energy on routing paths. It is possible that delay and energy consumption per packet will be increased. Nevertheless, this is not necessarily a drawback because lifetime variation of hosts is reduced, resulting in longer network lifetime.
5. *Minimize maximum node cost* — The goal is to minimize the cost paid by a host relaying a packet. Using the metric, the time of first host failure is expected to be delayed. Unfortunately, this metric is an ideal case and not implementable.

In short, power-aware routing should be designed based on shortest-cost instead of shortest-hop metrics, where "cost" can be defined differently but should be energy related.

Two more power-aware routing protocols are proposed in [15,16]. Reference [15] proposes a distributed algorithm to maximize battery drained-out time at each host. The idea is to balance the energy draining rates among all hosts proportional to their energy resources. Different from [14], this work incorporates transmitter power control into the routing protocol where several transmission levels are available. In [16], the authors assume static networks and propose three localized routing protocols with energy concerns. Each host is equipped with a GPS receiver and knows the locations of itself, one-hop neighbors, and destinations. Three energy-efficient routing mechanisms are proposed, namely, *power-aware*, *cost-aware*, and *power–cost* routing protocols. The power-aware routing tries to minimize total power consumption by choosing well-positioned neighbors for relaying packets. The cost-aware routing attempts to increase hosts' lifetimes by favoring hosts with more remaining battery power. The power–cost routing is a combination of the above two, which aims to minimize overall power needed and avoid hosts with low power levels.

An energy consumption model is provided in [17] to compare the performance of two existing ad hoc routing protocols: Dynamic Source Routing (DSR) and Ad hoc On-demand Distance-Vector (AODV). Mobility is also modeled in the analysis. In DSR, a receiver's continuous listening for caching and route responses may consume significant energy. On the other hand, AODV's frequent broadcasting is also very energy consuming.

4.4.2. Multicast/Broadcast Data Communications

What makes multicasting/broadcasting different from unicasting is the number of destined receivers. The idea for power-aware multicast/broadcast routing is to construct multicast/broadcast trees that are energy efficient. This is usually accomplished by transmitter power control [18–21]. Using a higher power level, a transmitter can reach more receivers within one hop. On the other hand, with a lower power level, fewer receivers can be reached at a time, but energy expense is conserved. The construction of power-aware multicast/broadcast trees should balance between these two tradeoffs, with the ultimate goal of reducing total power consumption and increasing host/network lifetime, while maintaining acceptable throughput and delays.

4.5 Higher-Layer Power Management

4.5.1 Transport Layer

Residing above the network layer, the transport layer provides end-to-end communication functionality. The most widely adopted transport protocol is TCP (Transmission Control Protocol). Unfortunately, TCP was originally designed for reliable wired environments, where packet error/loss due to interference/fading/handoff is scarce. When packet losses do happen, TCP will interpret this as channel congestion and invoke congestion control to reduce the transmission rate. This misinterpretation results in reduced throughput and unacceptable delays. Furthermore, because of the bursty-error characteristic of wireless links, transmissions under bad conditions are probably continuing to be lost. This introduces many unnecessary retransmissions and energy waste, but it can be avoided by queuing data for later delivery. Energy-efficient transport protocols attempt to reduce delays and the number of unnecessary retransmissions, both of which are energy consuming. References [22] and [23] address power-efficient transport protocols based on TCP to reduce unnecessary battery energy expenditure. In [24], an energy consumption model is proposed to analyze the performance of TCP variants, namely, Tahoe, Reno, and NewReno. The results show that energy efficiency is very sensitive to the TCP version adopted and protocol parameters chosen.

4.5.2 Operating System (OS)/Application Layer

Power management strategies are suggested to be implemented at the operating system (OS) level by [25]. The authors classify the power-saving software issues into three categories:

1. *Transmission problem* — When should a hardware component switch from one mode to another?
2. *Load-change problem* — How can the functionality needed by a hardware component be modified so that it can be put into low-power modes more often?
3. *Adaptation problem* — How can software be modified to permit power-saving usage of hardware components?

In [25], the authors first describe particular characteristics of each hardware component and then discuss solutions to the above three problems. Four units are addressed: secondary storage, processing unit, wireless communication unit, and display unit.

Going up to the application layer, several techniques [33,34] have been proposed to include energy efficiency in the software design considerations. Reference [33] focuses on organizing the invalidation report for cache invalidation to conserve energy of mobile hosts. The idea is to allow mobile hosts to selectively tune to the channel for the portions of reports that are of interest to them. This kind of selective tuning saves unnecessary energy spent on useless listening. In [34], the authors address a power-sensitive video processing mechanism. The basic idea is to reduce the number of transmitted bits. Video frames should be discarded intelligently before transmission without degrading video quality too severely. Other application-level energy-efficient strategies are provided in [3] and [32].

4.6 Conclusions and Future Research Directions

In this chapter, we have presented a variety of power-conservative designs for ad hoc networks. From MAC layer up to application layer, we have tried to provide a complete overview addressing the important design goal of power conservation. Considering today's portable/wireless devices, batteries are essential components. If battery power is not used efficiently, these devices may soon become burdens to us once their energy is drained out, no matter how fantastic the applications/services they can provide.

In the near future, ad hoc networks are very likely to become popular due to their flexibility. With similar advantages, sensor networks, also in the form of ad hoc networks, are gaining more and more attention as well. Since small sensors are also battery operated, power conservation is bound to be a major issue for sensor networks [35,36].

References

1. Lucent Technologies http://www.lucent.com, User's Guide for ORiNOCO PC Card, Sep. 2000.
2. N. Bambos, Toward Power-Sensitive Network Architectures in Wireless Communications: Concepts, Issues, and Design Aspects, *IEEE Personal Communications*, 1998.
3. J. Flinn and M. Satyanarayanan, PowerScope: A Tool for Profiling the Energy Usage of Mobile Applications, *IEEE WMCSA*, 1999.
4. L. Hu, Topology Control for Multihop Packet Radio Networks, *IEEE Transactions on Communications*, 1993.
5. R. Ramanathan and R. Rosales-Hain, Topology Control of Multihop Wireless Networks Using Transmit Power Adjustment, *IEEE INFOCOM*, 2000.
6. R. Wattenhofer, L. Li, P. Bahl, and Y.-M. Wang, Distributed Topology Control for Power Efficient Operation in Multihop Wireless Ad Hoc Networks, *IEEE INFOCOM*, 2001.
7. S. Agarwal, S.V. Krishnamurthy, R.H. Katz, and S.K. Dao, Distributed Power Control in Ad Hoc Wireless Networks, *IEEE International Symposium on Personal, Indoor and Mobile Radio Communications (PIMRC)*, 2001.
8. S.-L. Wu, Y.-C. Tseng, and J.-P. Sheu, Intelligent Medium Access for Mobile Ad Hoc Networks with Busy Tones and Power Control, *IEEE Journal on Selected Areas in Communications*, 2000.
9. H. Woesner, J.-P. Ebert, M. Schlager, and A. Wolisz, Power-Saving Mechanisms in Emerging Standards for Wireless LANs: The MAC Level Perspective, *IEEE Personal Communications*, 1998.
10. L. Bononi, M. Conti, and L. Donatiello, A Distributed Contention Control Mechanism for Power Saving in Random-Access Ad-Hoc Wireless Local Area Networks, *IEEE International Workshop on Mobile Multimedia Communications (MoMuC)*, 1999.
11. L. Bononi, M. Conti, and L. Donatiello, A Distributed Mechanism for Power Saving in IEEE 802.11 Wireless LANs, *ACM/Kluwer Mobile Networks and Applications (MONET)*, 2001.
12. C.S. Raghavendra and S. Singh, PAMAS — Power-Aware Multi-Access Protocol with Signaling for Ad Hoc Networks, *ACM Computer Communication Review*, 1998.
13. Y.-C. Tseng, C.-S. Hsu, and T.-Y. Hsieh, Power-Saving Protocols for IEEE 802.11-Based Multi-Hop Ad Hoc Networks, *IEEE INFOCOM*, 2002.

14. S. Singh, M. Woo, and C.S. Raghavendra, Power-Aware Routing in Mobile Ad Hoc Networks, *ACM International Conference on Mobile Computing and Networking (MobiCom)*, Dallas, TX, 1998.

15. J.-H. Chang and L. Tassiulas, Energy Conserving Routing in Wireless Ad Hoc Networks, *IEEE INFOCOM*, 2000.

16. I. Stojmenovic and X. Lin, Power-Aware Localized Routing in Wireless Networks, *IEEE Transactions on Parallel and Distributed Systems*, 2001.

17. L.M. Feeney, An Energy Consumption Model for Performance Analysis of Routing Protocols for Mobile Ad Hoc Networks, *ACM/Kluwer Mobile Networks and Applications (MONET)*, 2001.

18. J.E. Wieselthier, G.D. Nguyen, and A. Ephremides, Multicasting in Energy-Limited Ad-Hoc Wireless Networks, *IEEE MILCOM*, 1998.

19. J.E. Wieselthier, G.D. Nguyen, and A. Ephremides, On the Construction of Energy-Efficient Broadcast and Multicast Trees in Wireless Networks, *IEEE INFOCOM*, 2000.

20. J.E. Wieselthier, G.D. Nguyen, and A. Ephremides, Algorithms for Energy-Efficient Multicasting in Static Wireless Networks, *ACM/Kluwer Mobile Networks and Applications (MONET)*, 2001.

21. S. Singh, C.S. Raghavendra, and J. Stepanek, Power-Aware Broadcasting in Mobile Ad Hoc Networks, Technical Report, Oregon State University, Corvallis, 1999.

22. R. Kravets and P. Krishnan, Power Management Techniques for Mobile Communication, *ACM International Conference on Mobile Computing and Networking (MobiCom)*, Dallas, TX, 1998.

23. R. Kravets, K. Schwan, and K. Calvert, Power-Aware Communication for Mobile Computers, *IEEE International Workshop on Mobile Multimedia Communications (MoMuC)*, 1999.

24. M. Zorzi and R. Rao, Energy Efficiency of TCP in a Local Wireless Environment, *ACM/Kluwer Mobile Networks and Applications (MONET)*, 2001.

25. J.R. Lorch and A.J. Smith, Software Strategies for Portable Computer Energy Management, *IEEE Personal Communications*, 1998.

26. K. Govil, E. Chan, and H. Wasserman, Comparing Algorithms for Dynamic Speed-Setting of a Low-Power CPU, *ACM International Conference on Mobile Computing and Networking (MobiCom)*, Berkeley, CA, 1995.

27. J.R. Lorch and A.J. Smith, Reducing Processor Power Consumption by Improving Processor Time Management in a Single-User Operating System, *ACM International Conference on Mobile Computing and Networking (MobiCom)*, Rye, NY, 1996.

28. M. Weiser, B. Welch, A. Demers, and S. Shenker, Scheduling for Reduced CPU Energy, *Proceedings of the First Symposium on Operating System Design and Implementation (OSDI)*, Nov. 1994.

29. F. Douglis, P. Krishnan, and B. Marsh, Thwarting the Power Hungry Disk, *Proceedings of the 1994 Winter USENIX Conference*, Jan. 1994.

30. D. Helmbold, D.D.E. Long, and B. Sherrod, A Dynamic Disk Spin-Down Technique for Mobile Computing, *ACM International Conference on Mobile Computing and Networking (MobiCom)*, Rye, NY, 1996.

31. K. Li, R. Kumpf, P. Horton, and T. Anderson, A Quantitative Analysis of Disk Drive Power Management in Portable Computers, *Proceedings of the 1994 Winter USENIX Conference*, Jan. 1994.

32. K. Naik and D.S.L. Wei, Software Implementation Strategies for Power-Conscious Systems, *ACM/Kluwer Mobile Networks and Applications (MONET)*, 2001.

33. K.-L. Tan, Organization of Invalidation Reports for Energy-Efficient Cache Invalidation in Mobile Environments, *ACM/Kluwer Mobile Networks and Applications (MONET)*, 2001.

34. P. Agrawal, J.-C. Chen, S. Kishore, P. Ramanathan, and K.M. Sivalingam, Battery Power Sensitive Video Processing in Wireless Networking, *IEEE International Symposium on Personal, Indoor and Mobile Radio Communications (PIMRC)*, 1998.

35. W.R. Heinzelman, A. Chandrakasan, and H. Balakrishnan, Energy-Efficient Communication Protocol for Wireless Microsensor Networks, *IEEE HICSS*, Jan. 2000.

36. W.R. Heinzelman, A. Sinha, A. Wang, and A.P. Chandrakasan, Energy-Scalable Algorithms and Protocols for Wireless Microsensor Networks, *ICASSP*, 2000.

5

Performance Analysis of Wireless Ad Hoc Networks

Anurag Kumar
Indian Institute of Science

Aditya Karnik
Indian Institute of Science

Abstract

With the growing importance of wireless ad hoc networks (particularly in applications such as wireless local area networks and ad hoc sensor networks), it is important to develop an understanding of their performance. Performance analysis of wireless ad hoc networks is a challenging task because such analysis must take into account the interactions between the wireless physical layer, radio propagation, multiple access, random topology, routing, and the characteristics of the application that generates the traffic carried by the network. In this paper, we discuss the major issues in the performance of wireless ad hoc networks in light of recent research in this area. Among the topics surveyed are capacity scaling results, stochastic capacity, Bluetooth performance, and the performance of the transmission control protocol (TCP).

5.1 Introduction

For the purpose of performance analysis, one needs to abstract out the essential (performance governing) aspects of a system so as to build a mathematical model. Wireless ad hoc networks (henceforth abbreviated as WANETs) are communication networks in which even the network topology depends on the way the network is operated. Evidently, unlike the situation with wired and fixed topology networks, understanding and optimizing the performance of WANETs is a difficult undertaking, owing to the complex interaction between the various "layers" of the network. Nevertheless, in order to talk about the issues

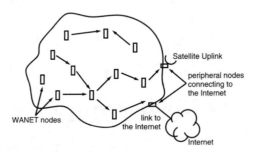

FIGURE 5.1 A schematic of a wireless ad hoc network, showing packet flow between the nodes and attachments to the wide-area communication infrastructure via satellite or terrestrial links.

systematically, we can begin by developing a layered view of a WANET, akin to the one utilized so effectively by the OSI (Open System Interconnection) model for communication network protocols. This will also set down our view of a WANET, for our subsequent discussions.

Key features of WANETs include the following:

1. *Wireless physical communication:* At the physical level, a WANET comprises several nodes (e.g., handheld, laptop, or personal computers), each equipped with a digital radio unit (see Fig. 5.1 for a schematic depiction). The antennas are (typically) omnidirectional, and the radios use a portion of the spectrum that does not require complicated spectrum licensing, coordination, and planning. Thus, for example, the popular IEEE 802.11b standard and the emerging Bluetooth system use the unlicensed 2.4 GHz ISM band. To combat multi-path and interference, spread spectrum modulation is usually employed. Beginning with 1 or 2 Mb/sec, with modulation schemes such as OFDM and better coding techniques, data rates greater than 50 Mb/sec are now achievable. We note that the digital communication system employed, the transmit power used, and the radio propagation characteristics of the environment determine the "links" in the network, and hence the topology of a WANET (see Section 5.2). Thus the most basic performance aspects of a WANET (which nodes are neighbors, which nodes have a path between them, and at what speeds neighbor nodes can communicate) are determined by the physical communication layer.

2. *Multiple access mechanism:* Since all the radios share the same band and the antennas are omnidirectional, clearly nodes cannot arbitrarily communicate with each other. Hence a WANET employs a multiple access mechanism which permits the nodes to coordinate their transmissions in a decentralized manner. A simple random access mechanism can be used, in which the nodes just attempt, and then resend if there is a collision. Such extreme lack of coordination obviously leads to poor performance. Such a simple mechanism, however, is amenable to analysis, at least to yield performance bounds (such an analysis is presented in Section 5.3.2). The IEEE 802.11 standard employs the CSMA/CA multiple access algorithm. A large class of multiple access schemes is based on the "RTS-CTS" dialogue. A node wanting to transmit first requests the destination node (Request To Send [RTS]) and transmits only when it receives a grant (Clear To Send [CTS]) from it. This ensures that the destination node is in the receiving mode during transmission. Other nodes that hear this dialogue refrain from transmitting for some duration, thus reducing interference. The Bluetooth system, essentially designed for ad hoc wireless interconnection of devices (laptop, cordless phone, printer, computer) in an office environment, uses a polling-based multiple access mechanism. In each so-called *piconet* there is a master device that polls the other devices; all data transfer takes place through the piconet master. We will present an analysis of the Bluetooth system in Section 5.4.

3. *Multihop packet communication:* Nodes in an ad hoc network are usually battery-powered handheld devices. This constrains their transmission power. Even though a node may not be able to directly send a packet to its destination, it can forward that packet to a node to which it can, hoping that the latter node has a path to the final destination; hence we have a multihop packet radio network.

It is not necessarily good for a node to send as far as it possibly can, as this may increase the interference it causes to other transmissions. For effective communication (low probability of error) and also to conserve battery power, nodes need to choose appropriate transmission ranges that determine their neighboring nodes.

4. *Adaptive routing protocols:* A routing protocol, typically derived from one of the Internet routing protocols, permits nodes to forward packets in such a way that if there is a path between a pair of nodes, then packets can be sent between those nodes. Since such a routing protocol is adaptive, the nodes can actually move, and new routes are learned as the topology changes. If nodes fail (owing to damage or loss of power), the network will automatically learn new paths between nodes that can still communicate. Routing protocols vary in their handling of topology updates. Some protocols are proactive; they constantly keep learning new routes and updating routing tables. Other protocols are "on-demand"; they search for routes when such a request is made. We will not touch on this topic any further in this chapter. A recent book [1] is devoted mainly to ad hoc network routing protocols. Several simulation studies comparing various ad hoc routing protocols are available (see [2] and [3]).

5. *Ad hoc network applications:* We can broadly classify WANETs into wireless ad hoc internets and special purpose ad hoc networks. Nodes in an ad hoc internet would like to run the same applications as any other node attached to the Internet. Hence the TCP/IP protocol suite has to be extended across such a WANET. Ad hoc internet nodes encapsulate their data into IP (Internet Protocol) packets. The transmission control protocol (TCP), in the applications in the end-systems, takes care of lost packets and prevents network congestion. By adopting the TCP/IP networking protocols, applications on ad hoc network nodes can communicate seamlessly with applications in the Internet. The performance of TCP's end-to-end adaptive window based packet transmission mechanism has been a subject of much concern over point-to-point wireless links (we will summarize an analytical approach to such problems in Section 5.5). The problems with TCP over multihop packet radio links are only recently beginning to be understood (for simulation based performance analysis, see [4] and [5], and for some scaling results, see [6]). We will also touch upon the performance of TCP over a multihop radio path in Section 5.5.

An important emerging class of applications of WANETs arises from the availability of inexpensive miniature multifunction devices. Such a device would have embedded sensors (for light, chemicals, temperature, for example) and would have embedded computing, a digital radio, and a long idle-life battery. These devices could be independent units to be randomly strewn into an area or could be deliberately embedded in common systems such as watches, appliances, building walls, etc. Functions of such ad hoc sensor networks could be:

- Monitoring contamination levels after radioactive or chemical leaks
- Forest fire detection, timber management, and wildlife tracking
- Materials and people tracking and security management in factories, airports, and hospitals
- Distributed instrumentation in large machines or vehicles (such as ships)

Ad hoc sensor networks would not need to use TCP. In fact, the challenge would be for them to achieve their function by only explicitly communicating with their neighbors. See [7] for a survey of this emerging area.

Ad hoc networks are hard to analyze, and most analytical models are intractable because of the various interdependencies that need to be accounted for. The earliest papers aimed at determining the network capacity for a given topology and a combination of channel access and capture schemes [8–10] (see [11] for a survey). These were essentially extensions of the analyses for single hop networks; hence the effort was made to formulate Markovian models. The general models assumed exogenous traffic as independent Poisson processes, exponentially distributed packet lengths, and transmission scheduling process for packets from any node i to any node j as independent Poisson processes which included new and rescheduled packets. The network topology was specified as a graph where two nodes were connected by a link if they could hear each other's transmissions. Throughputs of ALOHA, CSMA, and CDMA schemes with various collision management and capture assumptions were analyzed. Results, however, could be obtained only for simple network topologies since the state description became formidable for large networks.

Since the devices will, in general, be randomly located, the graph of a WANET is actually a *random graph*. Hence, the "specification" of the topology in the above analyses turned out to be a major limitation of these approaches. Recent papers, therefore, have focused on obtaining bounds on the network capacity for random topologies [12–15]. These analyses have led to the so-called "scaling laws," i.e., scaling of the network capacity with the number of nodes in the network. These results will be discussed in Sections 5.2 and 5.3.

The following is a summary of the results surveyed in this chapter. In Section 5.2, we discuss the basic tradeoff between spatial reuse and WANET connectivity. We discuss how rapidly the transmission range can decrease when the number of network nodes increases in a given area so that the network stays connected. In Section 5.3, we first review how this leads to a scaling law that shows that the per node capacity of a WANET scales poorly with the number of nodes. Then, turning to stochastic capacity, we provide a packet flow model of WANET and show how the performance of the network depends on the physical layer and the way nodes organize themselves and operate. In Section 5.4, we analyze a specific WANET, namely Bluetooth, and seek its scaling law, i.e., how the performance of a cluster of piconets degrades as the number of piconets increases in a given area. We answer this question in terms of the stationary outage probability and the temporal correlation in the outage process. In Section 5.5, we first review the performance analysis of TCP over a single-hop wireless link. Then we discuss the performance of multihop TCP connections over WANETs. Using a simple example, we illustrate the complexity involved in this task. We conclude in Section 5.6.

5.2 WANET Topology, Spatial Reuse, and Connectivity

One of the most basic questions about a network is whether any two nodes in the network are able to communicate, perhaps by routing their data through other nodes. Such a question is formally answered by representing the topology of the network by a graph on the set of nodes of the network. In the context of a WANET, where a radio spectrum is shared among several contending nodes, there is the additional, and conflicting, issue of spatial reuse. To promote spatial reuse, the transmission ranges of the nodes need to be kept small, while the opposite is required for ensuring that the network is connected. We will begin by examining this issue.

5.2.1 The WANET Graph

Let $\mathcal{N}$ ($= \{1, 2, ..., n\}$) denote the WANET nodes. In the WANET topology graph, there is a directed link (i,j) (from node i to node j) if node i can send data to node j with the desired level of reliability (which may, for example, be specified as a bit error rate of 10^{-4}, or perhaps a packet error rate of 0.01). This definition of a link is not as straightforward as it may at first seem. If several nodes are simultaneously transmitting, and if node i is attempting to transmit to node j, then this transmission may not succeed owing to excessive radio interference at j. On the other hand, if i is the only transmitting node, then the signal power to receiver noise ratio at j may be sufficient to provide reliable communication. For the purpose of defining the WANET topology, we will define a link in this more "optimistic" sense. The point is that if two nodes are not connected in this graph, then they cannot be connected when arbitrary sets of nodes are allowed to transmit. Further, we will assume in our discussions that if link (i,j) exists then so does link (j,i), and hence we can take the WANET topology graph to be undirected. (This assumption basically requires that the propagation channel between the two nodes is reciprocal.) Let $\mathcal{L}$ denote these links, and let $\mathcal{G} = (\mathcal{N}, \mathcal{L})$ denote the WANET graph.

Thus, we say that the WANET is connected if the graph $\mathcal{G}$ is connected. (There is a path between every pair of nodes of the graph.)

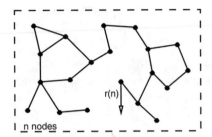

FIGURE 5.2 The graph of a wireless ad hoc network. There are *n* nodes in a given area. The radio communication range is *r(n)*.

5.2.2 Spatial Reuse and Connectivity

Given a particular transmitter and receiver design and a transmitter power, there is a range (say, *r*) over which two nodes can communicate. Now let us consider *n* nodes distributed uniformly in a given area, and the effect of increasing *n*. If *r* is kept fixed as we increase *n*, then transmissions from any node interfere with a large number of nodes, and the number of simultaneous transmissions possible in the network (or the spatial reuse) is just $O(1/r^2)$, which does not increase with increasing *n*. Evidently, in order to increase the spatial reuse, the transmitter powers must be reduced so that the range (say, *r(n)*) is a decreasing function of *n*. Figure 5.2 shows a WANET in which the nodes have transmission range *r(n)*; any two nodes that are within a distance *r(n)* of each other are shown as being neighbors in the WANET graph. Observe that, if *r(n)* is too small, then the WANET may not be connected. An important question that follows from this discussion is, "How fast can *r(n)* decrease so that the network remains connected?"

Consider *n* nodes uniformly distributed over a square field of unit area, and a graph $\mathcal{G}$ on these nodes obtained by putting a link between any two nodes that are separated by a distance no more than *r(n)*. What we obtain is a probabilistic model, called a *random graph*. Let I_i denote the event that node *i* is isolated (i.e., has no links to any other node), and consider the probability that at least one node is isolated, i.e., $P(\cup_{i=1}^{n} I_i)$. We proceed in the spirit of the arguments in [16]. Since the area within the range of a node, and also within the square field, is at least $1/4 \, \pi \, r^2 (n)$, we have for all *i*, $1 \leq i \leq n$,

$$P(I_i) \leq (1 - \frac{1}{4}\pi r^2(n))^{(n-1)}$$

Using the union bound we then have

$$P(\cup_{i=1}^{n} I_i) \leq \sum_{i=1}^{n} P(I_i)$$

$$\leq n(1 - \frac{1}{4}\pi r^2(n))^{(n-1)}$$

$$= e^{(\ln n + (n-1)\ln(1 - \frac{1}{4}\pi r^2(n)))}$$

$$= e^{(\ln n - (n-1)O(\frac{1}{4}\pi r^2(n)))}$$

$$= e^{\ln n(1 - \frac{(n-1)}{\ln n}O(\frac{1}{4}\pi r^2(n)))}$$

where we have used the fact that *r(n)* decreases to zero as $n \to \infty$. It follows that if *r(n)* shrinks more slowly than

$$\sqrt{\frac{\ln\ n}{n}}$$

then the probability of some node being isolated goes to zero as $n \to \infty$. In fact, using some results from random graph theory, it has been shown in [16] that in order for the WANET graph to stay *connected* with probability 1, $r(n)$ must shrink more slowly than

$$\sqrt{\frac{\ln\ n}{n}}$$

More precisely, it has been shown in [17] that if the range $r(n)$ scales as

$$O\left(\sqrt{\frac{\ln\ n + c(n)}{n}}\right)$$

then the graph stays connected, as $n \to \infty$, if and only if $c(n) \to \infty$. Note that $c(n)$ can go to ∞ arbitrarily slowly.

5.3   The Capacity of a WANET

Consider again n nodes constituting a WANET in a given area Figure 5.3 is a depiction of such a network. Each node is running some applications (e.g., packet telephony, web browsing, or e-mail) that generate packets that need to be transported to other nodes in the network. We can view these application-generated packets as arrivals into the network. These arrivals are shown by the straight arrows pointing into the nodes in Fig. 5.3. Packets are transmitted to neighboring nodes, from which they may need to be relayed to other nodes; i.e., some packets are multihopped. Finally, each packet reaches its destination node, and this can be viewed as a departure, shown by the straight arrows pointing away from some of the nodes in Fig. 5.3.

With the above view of the traffic flow, it is natural to think of the WANET as a service system into which customers (i.e., packets) arrive, get serviced (i.e., transported over the wireless medium), and finally depart. Figure 5.4 shows a queuing model for the entire WANET. The model shows one queue at each node, and this queue holds all the packets (new or transit) waiting to be transmitted out of the node. The digital radio communication and multiple access mechanism of the network can be viewed as a complex service mechanism that serves packets from these queues. Multihopped packets are fed back to other queues.

For this queuing model, the first question that can be asked is: "How fast can packets arrive into the nodes so that the nodal queues remain stable?" Stability could be defined in the usual sense of the joint queue length random process converging to a proper distribution.

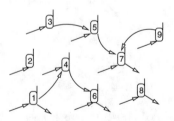

FIGURE 5.3 Packet flow in a multihop wireless ad hoc network.

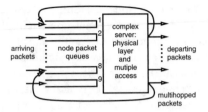

FIGURE 5.4 A queuing model representing the entire wireless ad hoc network.

5.3.1 A Scaling Law from Spatial Reuse

Let the rate of arrival of new packets at each node be $\lambda(n)$. The transmission range $r(n)$ scales with n, as discussed above. Let $h(n)$ be the average number of hops over which a packet is relayed. Hence, the total rate of packets that need to be transmitted by the WANET is $n\lambda(n)h(n)$. Assuming that the nodes are uniformly distributed, and all destinations are equally likely, it follows that $h(n) = O(r(n)^{-1})$.

Let us denote the service capacity of the network by $S(n)$; i.e., $S(n)$ is the total rate at which the network can serve packets from the queues. In general, $S(n)$ will depend on the amount of spatial reuse, on the physical layer and the multiple access, on the routing protocol, and on the prevailing propagation conditions. In any case, the rate of packets to be transported can be no more than $S(n)$; i.e.,

$$n\lambda(n)h(n) \leq S(n)$$

If we consider only spatial reuse, then $S(n) \leq O(r(n)^{-2})$. Hence if $r(n)$ scales down no faster than

$$\sqrt{\frac{\ln n}{n}}$$

(in order to maintain connectivity), we can put together the above expressions to yield

$$\lambda(n) \leq O\left(\frac{1}{\sqrt{n \ln n}}\right)$$

Thus the per node capacity of a WANET scales very poorly with n, and this can be seen to be primarily because of multihopping.

It follows that dense WANETs would serve the best for applications in which there is primarily local communication between nodes. The basic wireless ad hoc networking paradigm is extremely well suited to ad hoc sensor networks. When strewn over an area, these devices sense their immediate environment and can be viewed as constituting a *data space*. The sensor network can then be queried ("What is the maximum level of chemical contamination?"), and by using a distributed algorithm, which only requires the devices to communicate with their neighbors, the query can be answered and the results returned.

5.3.2 Stochastic Capacity

After discussing the above fairly general scaling results, we now turn to a packet flow model of a WANET. Packet flow models are of interest because various performance measures can be directly calculated from them. The question of particular interest is the following: If the network consists of n randomly distributed nodes in a given area A, then for given input traffic rates $\{\gamma_1, \gamma_2, \ldots \gamma_n\}$, does there exist a channel access, routing, and packet scheduling scheme so that the network is stable? In this section, however, we will instead discuss the following simpler questions: For a given multiple access, routing, and packet scheduling scheme, what is the throughput capacity of the network, and what is the maximum input traffic rate for which the network is stable?

We consider the following network model. Nodes are randomly distributed in the plane according to a Poisson point process of intensity λ per m^2. In the simulation results presented, we consider a Poisson field of 5000 network nodes. Each radio node transmits on the same frequency f with power P using an omnidirectional antenna. Also, it cannot transmit and receive simultaneously. For the propagation model, we assume only path loss with exponent $\eta = 4$. We say that a transmission can be decoded if its signal to interference ratio (SIR) exceeds a given threshold β.[1] Typical values of β are in the range of 10–20 dB. Channel access is random, i.e., nodes make decisions to transmit or receive independently, and time is slotted. In each slot, a node decides to transmit with probability α and decides to receive with probability $(1-\alpha)$ independently; α is called the attempt probability. If the transmission is successful, the packet is removed from the queue (i.e., this is an assumption of instantaneous acknowledgments). A successfully received packet at a node joins the queue for next hop transmission with probability v, otherwise the packet departs the network. A node addresses[2] a transmission only to one of its neighbors; the neighbors are the nodes within a given distance R from a node. Thus, a transmission is successful when the node to which it is addressed is in receive mode, and the SIR of the transmission exceeds β at the addressed node. Note that R need not be the radio range of a node, in fact, we show later that a smaller R improves the performance. In each slot, the one-hop destination for the head-of-the-line (HOL) packet is chosen randomly from among the neighbors, i.e., given the number of neighbors K, a neighbor is chosen as the destination of the HOL packet with probability $1/K$ (for a typical node, the number of its neighbors is Poisson distributed with mean $\lambda \pi R^2$). Then the random selection of nodes to which to transmit is basically the assumption that the routing and traffic pattern are uniform and homogeneous. Thus, each packet will traverse a geometrically distributed number of hops before departing the network.

5.3.2.1 Interference Analysis

With the channel attempt model described above, in each slot, the nodes that are in the transmit mode form an independent Poisson process[3] (see [18] for some early work in this framework). At a typical node, the average power received in a slot, from transmitters further than a distance d_0, is proportional to

$$\lambda \alpha \int_{d_0}^{\infty} \left(\frac{r}{d_0} \right)^{-\eta} 2\pi r\, dr$$

where d_0 denotes the near field cut-off distance. The integral converges for path loss exponents $\eta > 2$, and hence even with infinite number of transmitters the average power received is finite. We now assume that $\eta = 4$. It can be further shown that the characteristic function of the distribution of total received power at a node in a slot is given by

$$\Psi(w) = \exp^{\lambda \pi d_0^2 jw \int_0^1 t^{-1/2} \exp^{jwt} dt}$$

It can be shown from this that the mean interference is $2\lambda \alpha \pi d_0^2$, and the variance is $\frac{4}{3} \lambda \alpha \pi d_0^2$

We approximate the interference as Gaussian

$$\mathcal{N} \left(2\lambda \alpha \pi d_0^2, \frac{4}{3} \lambda \alpha \pi d_0^2 \right)$$

---

[1] Given a modulation and coding scheme, β actually specifies the maximum BER, assuming interference to be Gaussian (which is a fair assumption with a large number of interferers). In this way, details of the actual modulation and coding scheme can be abstracted.

[2] By actually inserting a physical or IP address in the packet header.

[3] In general, the spatial point processes of transmitters and receivers depend upon the channel access scheme.

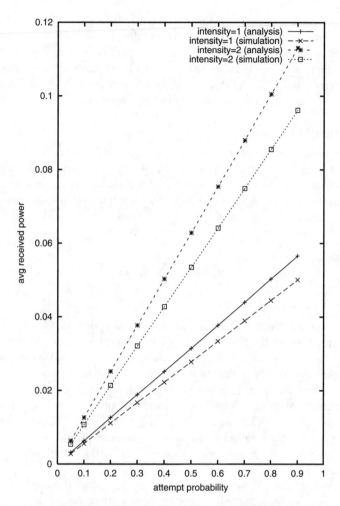

FIGURE 5.5 Variation of mean interference at a node with attempt probability, for $\lambda = 1$ and 2 per square meter.

where $\mathcal{N}(\mu,\sigma^2)$ denotes a Gaussian distribution with mean μ and variance σ^2. Note that the mean is linear in both λ and α; this is shown in Fig. 5.5, where simulation results are plotted along with the analytical values. The value of d_0 is taken to be 0.1 m, and $\lambda = 1$ nodes per square meter. For the simulation, 5000 nodes were generated, and the simulation was run for 20,000 slots. An average was taken over several realizations of the node positions and for several target nodes in the middle of the network. We observe that the analysis and simulation results are qualitatively the same, and the analysis is also a good approximation.

Consider a transmitter–receiver pair at a distance r in this Poisson field of points. Their communication is successful with probability

$$p_s(r,\alpha) := P\left(\frac{P_s}{N+I} \geq \beta\right)$$

where $P_s = (r/d_0)^{-\eta}$ denotes the intended signal power (normalized to the power at the reference distance d_0), I the interference [distributed as $\mathcal{N}\left(2\lambda\alpha\pi d_0^2, \frac{4}{3}\lambda\alpha\pi d_0^2\right)$], and N the thermal noise at the receiver.

Figure 5.6 shows the variation of $p_s(r,\alpha)$ with r for various values of α, for $\lambda = 1$ per m². For these calculations, we take the transmit power to be 1 mW, and the thermal noise $N = 5 \times 10^{-7}$ mW. The curves are decreasing with α; here $\alpha = 1$ is just indicative of the performance if all the nodes in the network were transmitting simultaneously and thus interfering with reception at our reference node. Note the sharp decay of $p_s(r,\alpha)$ with r. The probability of success is very poor beyond a few centimeters from a node; clearly random access is not a good way to operate a WANET.

5.3.2.2 Saturation Throughput

The saturation throughput of a network is defined as the rate of transmission in the network when a transmitted packet from the HOL of a queue at a node is immediately replaced by another packet. Saturation throughput is akin to a *service rate* provided by the complex channel *server* (see Fig. 5.4). In some models, the saturation throughput is shown to yield a sufficient condition for network stability. Let γ_i denote the saturation throughput of the ith node.[4] Then,

$$\gamma_i := \lim_{n \to \infty} \frac{1}{n} \sum_{k=1}^{n} Z_k = p_t$$

where $Z_k = 1$ if the transmission is successful in slot k, and p_t denotes the probability of successful transmission. Figure 5.7 shows the probability of successful transmission as α varies for various values of R. Recall that we use a Poisson field of 5000 nodes. Here $\lambda = 1$ per m^2. We have again used $\eta = 4$. R specifies the "neighborhood" of a node; the mean number of neighbors is $\lambda \pi R^2$. Observe that, as R increases, the probability decreases. This is to be expected. A node attempts transmissions uniformly randomly to its neighbors. With a smaller R, a node's neighbors are closer, and the probability of successful transmission to these nodes is higher. We also see from Fig. 5.7 that, for a given R, there is an *optimum* attempt probability. Thus, for a range of 0.5 m, an attempt probability $\alpha = 0.3$ provides the best throughput. This analysis shows that the performance of the network can be optimized based on how nodes organize themselves.

The maximum rate of transmission is obtained when, instead of addressing its transmission to a particular neighbor,[5] a node broadcasts it. Its transmission is, thus, successful *if at least one node in its radio range* decodes it. Let C denote the maximum rate of transmission of a node in saturation (normalized to average number of nodes in its radio range). Then, it can be shown that

$$C < \int_0^{\frac{1}{\beta}-N} \frac{\alpha(1-\alpha)}{\pi \sqrt{2.6\lambda\alpha d_0^2}} e^{-(y-2\lambda\alpha\pi d_0^2)^2 \left(2.6\lambda\alpha\pi d_0^2\right)} dy \qquad (5.1)$$

This bound, in a sense, is a measure of network capacity of our model. Figure 5.8 shows its variation with α for $\lambda = 1$.

It appears from Figs. 5.7 and 5.8 that when $R = 0.5$, the throughput exceeds the capacity bound. The answer to this apparent paradox lies in the fact that when R is small, the network is not connected and consists of many isolated nodes. Since isolated nodes do not transmit at all (since these nodes do not have neighbors to address their transmissions to), the overall interference to transmitting nodes is reduced, thereby greatly increasing the probability of success. The capacity bound, on the other hand, is

---

[4] The network is assumed to be homogenous; therefore, the subscript i can be omitted. If the network consists of N nodes, then the network throughput is $\sum_{j=1}^{N} \gamma_j$.

[5] Recall that the success of a transmission depends on the addressed node being in receive mode.

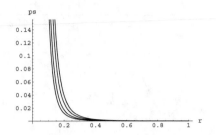

FIGURE 5.6 Variation of probability of successful transmission with distance for $\alpha = 0.1, 0.2, 0.5, 1$, and $\lambda = 1$ per m².

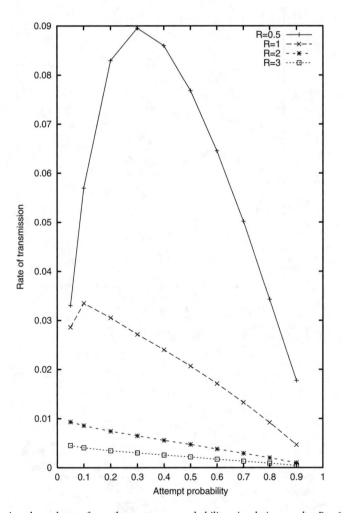

FIGURE 5.7 Saturation throughput of a node vs. attempt probability; simulation results. $R = 0.5, 1, 2, 3$ m.

calculated assuming every node is able to transmit. This leads to important questions regarding the performance of the network and the organization of nodes, which are beyond the scope of this chapter.

The above analysis was all for WANETs in which nodes use random access. There appears to be little literature available on the analytical performance evaluation of RTS-CTS type protocols in dense WANETs with random node placement.

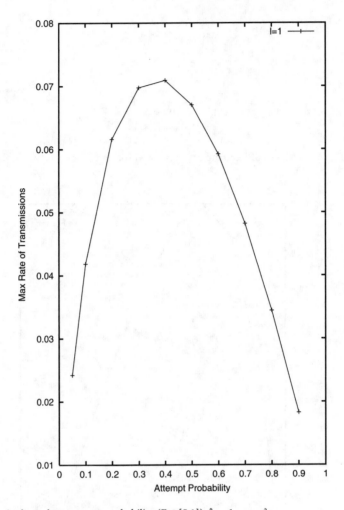

FIGURE 5.8   Capacity bound vs. attempt probability (Eq. [5.1]). $\lambda = 1$ per m^2.

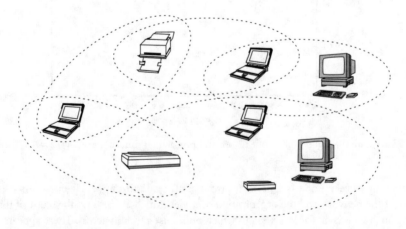

FIGURE 5.9   A system of overlapping Bluetooth piconets.

5.4 Performance of Bluetooth

5.4.1 The Model and Its Analysis

We consider a single circular room (radius R) office environment (LOS propagation), with a number of Bluetooth devices distributed randomly in it to form M piconets. Because of the ad hoc nature of the formation of piconets, the devices in a piconet are spatially randomly distributed in the room leading to overlap among the piconets. Note that in our results M will be a parameter, i.e., we are interested in analyzing the performance of the Bluetooth system as the number of piconets increases in a given area.

The propagation of radio waves inside a building is a complicated process and depends significantly on the indoor environment (e.g., office, factory), and topography (LOS, obstructed). The statistics of the indoor channel vary with time due to motion of people and equipment. The indoor channel is characterized by high path losses (path loss exponent η is 1.5–1.8 for LOS; see [19]) and large variations in losses. Motion causes signal fading which compares well with the Rician distribution with K equal to 6.8 dB for the office environment. If the degree of motion is small (as in an office environment), the fading is extremely slow and the channel is said to be quasi-static. For practical data rates, fading is quasi–wide-sense stationary. In our analysis, since we consider LOS propagation in an open office environment, we take $\eta = 2$ and a Rician signal distribution with $K = 6$ dB.

Since Bluetooth operates in the ISM 2.4 GHz band, there are many sources of interference other than Bluetooth devices, e.g., microwave ovens and IEEE 802.11 Wavelans. We, however, focus on the interference generated within the Bluetooth system only. A piconet experiences interference when a frequency hit occurs, i.e., the transmission frequency of a packet in a piconet matches with that being used in one or more other piconets for some overlapping duration. We assume an ideal frequency hopping pattern (with N_f frequency hops), and model it as a discrete time Markov chain; a state denotes a frequency. Hopping patterns of piconets are independent and identically distributed. T denotes the hop time. Note that the piconets are not synchronized in time. We assume that the time offset of the ith piconet, denoted by t_i, with respect to the reference piconet are uniformly distributed in $[0,T]$.

Since the devices in the piconets are spatially distributed, the *interfering piconet* is not a fixed location or device but a random device in that piconet transmitting at a given time. We assume a physically stationary environment (little or infrequent movement) so that the channel is quasi-static and quasi–wide-sense stationary. Hence, for the *duration of a frequency hit* on the single frequency, constant interference power can be assumed to be received from the individual interferers.

We consider a reference piconet and denote by $\{\Gamma(t), t \geq 0\}$ the SIR process:

$$\Gamma(t) = \frac{P_S(t)}{\sum_{j=1}^{N(t)} P_I^j(t) + \frac{N_0 W}{2}}$$

where, at time t, $P_S(t)$ is the desired signal power, $P_I^j(t)$ is the interference power due to the jth interfering piconet, and $N(t)$ is the number of interferers. Recall that in Bluetooth a device can transmit in alternate slots. In addition, a slave can transmit only when polled. Hence, the receiver is not a fixed device, and the SIR process we characterize is of the reference piconet as a whole.

Outage occurs when the bit SIR falls below the resistance ratio (denoted by β) specified by the standard (11–14 dB). We denote by Γ_b the bit SIR and seek the stationary outage probability P_{out}.

$$P_{out} := P\left(\frac{P_s}{\sum_{j=1}^{N_b} P_I^j + \frac{N_0 W}{2}} < \beta \right)$$

P_S is the desired signal power and P^j_I denotes the power received from the jth interferer. N_b is a random variable denoting the number of interferers in a bit duration. We assume that the devices are distributed uniformly in the room with respect to the center of the room.[6]

In order to characterize the temporal correlation in the outage process, we consider a bit b and obtain the one step joint distribution of bit b and $b + 1$ being in outage. This outage process is then approximated by a Markov process. This gives us a two-state Markov model as follows. We say that the channel is in *good* state, denoted by "g," if $\Gamma_b \geq \beta$ or BER on the channel is below the specified value, and that it is in *bad* state, denoted by "b," when $\Gamma_b < \beta$, i.e., when the outage occurs. Let P_{ij} denote the transition probability of going from state i to state j. Then, the outage duration (in number of bits) is geometrically distributed with mean $1/(1 - P_{bb})$. Also the outage probability calculated from the Markov model, denoted by $P_{out}{}^M$, is given by $P_{gb}/(P_{gb} + P_{bg})$.

Details of the modeling assumptions and the analysis can be found in [20].

5.4.2   Results and Discussion

We take $R = 5$ m. Since the calculation of exact outage probability is complicated, even numerically, we resort to finding bounds. Table 5.1 shows lower bounds on P_{out} with signal fading

Using the two-state Markov chain approximation and and lower bounds on the outage probability, we calculate approximate values of P_{gb} and P_{bg} for $\beta = 14$ dB and 400 bits packet size. We also calculate the outage probability from this Markov model (P_{out}^M). Results are shown in Table 5.2. Table 5.2 also shows in bits the mean outage duration (EB) and the mean duration in good state (EG). Note that for Bluetooth, $N_f = 79$.

It is very complicated to calculate even the approximate values of P_{out} as the number of piconets increases. We, therefore, have results only for $M \leq 3$ piconets. However, the insight gained with these results allows us to predict the performance for larger values of M. The results indicate that when a frequency hit occurs it is difficult to maintain the resistance ratio of 14 dB as specified by the standard. Hence, co-channel interference is avoided mainly through the use of a large number of frequencies. For larger values of M, we expect that P_{out} will increase almost linearly, in the form of $(M - 1)/N_f$. Though the Markov model is only an approximation (as seen from Table 5.2), we expect that when the number of piconets is small, outages are infrequent (once in 20,000 bits). This is because piconets are not time-synchronized, and frequency hits occur with probability $1/N_f$. However, outages persist for approximately 200 bits, which is a considerable duration considering 1/3 and 2/3 FEC used in Bluetooth. Also in certain packet types FEC is not mandatory. As M increases, the duration of the good state will reduce and there will be longer outage durations.

TABLE 5.1   Lower Bounds on Outage Probability for $M = 2,3$ Piconets

$M - 1$	Lower Bound on P_{out}	
	$\beta = 14$ dB	$\beta = 11$dB
1	0.0114	0.0103
2	0.0227	0.0204

TABLE 5.2   Parameters of the Two-State Markov Model

$M - 1$	P_{gb}	P_{bg}	P_{out}^M	EB	EG
1	0.00005	0.005	0.010	200	20000
2	0.000055	0.0035	0.016	285	18018

---

[6] A Bluetooth receiver has –70 dBm sensitivity level to meet BER < 0.001. With 1 Mb/sec bit rate, the noise power is approximately 5×10^{-7} mW.

5.5 Performance of TCP Controlled Transfers over a WANET

One of the applications of WANETs is to create ad hoc internets. Each WANET node in such a network is an endpoint, as well as a packet forwarding device (even a fully fledged router, participating in a distributed routing protocol). In an internet, most store-and-forward applications (e-mail, file transfers, web browsing) operate over the end-to-end TCP (Transmission Control Protocol). The TCP protocol serves three functions in the Internet:

1. TCP converts the nonsequential and unreliable packet transport service provided at the IP layer into a sequential and reliable service.
2. TCP implements end-to-end flow control, thereby preventing a fast sender (e.g., a high performance computer) from swamping out a slow receiver (e.g., a slow mechanical device such as a printer).
3. The applications that use TCP's services are *elastic*, in the sense that they can work satisfactorily even if the network offers them a time-varying packet transfer rate. TCP determines the way the network shares bandwidth among the competing flows.

Since TCP sessions comprise over 90% of the Internet traffic, it can be said that to a large extent TCP governs the dynamics and the performance of the Internet.

Since elastic applications will continue to be the most popular ones to be deployed on ad hoc Internets, it is important to understand the impact of TCP on the performance of these networks.

We divide our discussion into two parts. First we consider the situation in which the TCP session is over a single hop between neighboring nodes. This will be by far the most common situation within offices and homes, as one can expect that the WANET node will be near a wired access point. On the other hand, ad hoc internets can be expected to be deployed in situations where there is no wired infrastructure (e.g., in disaster zones). The performance of multihop TCP sessions over such a network will, therefore, also be of interest.

5.5.1 Single Hop Performance

The motivating scenario is shown in Fig. 5.10. A WANET node is downloading a large file from a server on a high-speed LAN. The propagation delay over this wireless link is much smaller than the packet transmission times. We identify random epochs $(T_1, T_2, ..., T_k, ...)$ as follows. Let $T_0 = 0$, and the connection starts off with the congestion window $W = 1$. As packets are transmitted, the window grows (slow-start, if necessary, and then congestion avoidance). Eventually a loss occurs on the lossy link at the epoch l_1; we denote the window achieved at this epoch by X_1. Owing to the local area and high-speed LAN assumptions, this window includes all increments due to any successful transmissions before the lost packet. Some random time later (depending on the version of the protocol and the recovery method used), at the epoch T_1, the normal congestion window algorithm resumes with a slow-start threshold of $\left\lceil \dfrac{X_1}{2} \right\rceil$. In this way we identify the sequence of random vectors $((X_0, T_0), (X_1, T_1), ..., (X_k, T_k), ...)$; see Fig. 5.11. With Bernoulli packet loss [21], or Markov modulated packet loss [22] and appropriate assump-

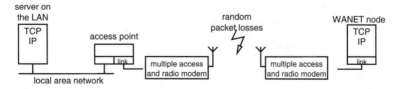

FIGURE 5.10 A WANET node transferring data over a TCP connection from a server on a LAN.

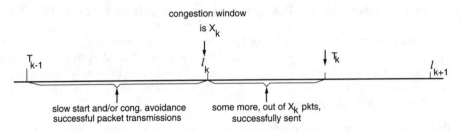

FIGURE 5.11 Analysis of TCP performance over a lossy link. The epochs $\{T_k\}$ and the Markov chain $\{X_k\}$.

tions (the window increase during congestion avoidance is taken as a probabilistic increase), we can show that:

1. $\{X_k\}$ is a Markov chain.
2. $\{(X_k, T_k)\}$ is a Markov renewal process.

Denoting the number of successful packets in the *k*th cycle by a "reward" V_k, we find that the V_k depends only on $X_{(k-1)}$, and we have a Markov renewal reward process. Letting U denote the random variable for the cycle length, and taking expectations under the stationary distribution $\pi(\cdot)$ of $\{X_k\}$, the throughput in packets per second is given by

$$\gamma = \frac{E_\pi V}{E_\pi U}$$

The various versions of TCP are distinguished by the structure of the Markov chain $\{X_k\}$, and the cycle times $\{T_k\}$; e.g., for the original Van Jacobson algorithm (TCP "Old" Tahoe) $T_k = l_k + \text{timeout}$.

In spite of many approximations, the analysis is quite accurate as demonstrated in Fig. 5.12 (taken from [21]), where analytically obtained results are compared with those obtained from actual TCP code running over an emulated lossy link.

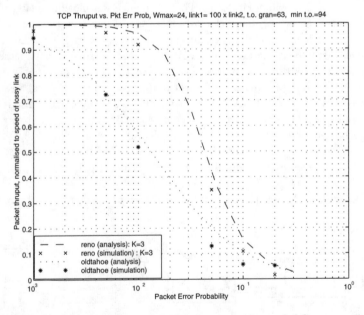

FIGURE 5.12 Throughput of TCP Old Tahoe and TCP Reno vs. packet loss probability; comparison of simulation and analysis. (Maximum congestion window is 24 packets; the LAN is 100 times faster than the lossy link; K is the duplicate ack threshold for fast retransmit.)

5.5.2 Performance of Ad Hoc Internets with Multihop TCP Connections

A detailed model, such as the one above, for multihop TCP connections in a WANET is not yet available. In fact, several basic issues that have been well studied for TCP over wired networks are not yet clear in the context of TCP over multihop wireless links. For example, if there is a single TCP connection over an *h* hop wired connection, with each link's propagation delay being very small compared to the packet transmission time, then the TCP window should optimally be *h* (ignoring the acknowledgment transmission times). In a multihop wireless connection, however, the links are half-duplex, and furthermore arbitrary links cannot be simultaneously used owing to interference.

In Figs. 5.13 and 5.14, we show the evolution of end-to-end window controlled packet transmission, over a three-hop connection, with the nodes labeled *a,b,c,d*. This would be the case if TCP with a fixed window is used. It is assumed that each transmission takes one slot, whether it is a TCP packet or a TCP ack, and there is perfect centralized scheduling of transmissions (i.e., which nodes should transmit and to whom). We further assume that nodes more than two hops away (e.g., *a* and *c*, or *b* and *d*) are out of range; this implies that if *a* transmits to *b*, and *d* to *c*, the interference at *b*, caused by the transmission from *d* to *c*, is acceptable. The numbers to the left of Fig. 5.14 are the slot numbers. The open triangles

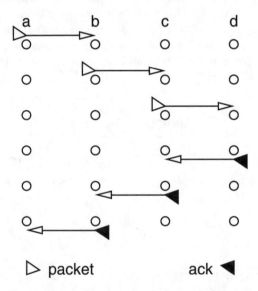

FIGURE 5.13 Window controlled transmission over a three-hop wireless route; window = 1.

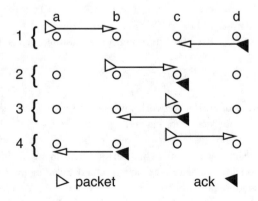

FIGURE 5.14 Window controlled transmission over a three-hop wireless route; window = 2.

are packets, and the shaded ones are acknowledgments. A triangle with an arrow indicates that the packet or ack is being transmitted. A triangle without an arrow means that the packet or ack is queued. The triangles "point" in the direction in which the packet is traveling.

In Fig. 5.13, the window is 1, and it takes six slots to transmit one packet and get back its acknowledgment, thus yielding a connection throughput of 1/6 packets per slot. Can we use a window of 2, and thus improve the throughput? We note that not all hops can be active at the same time. For example, node *a* cannot send to node *b* while node *c* sends to node *d*, as the transmission from node *c* will interfere at node *b* with the transmission from node *a*. Figure 5.14 shows a transmission schedule with a window of 2. The schedule is assumed to have "started" some time back and will continue even after slot 4. At the end of schedule, the acknowledgment reaching the TCP transmitter and the packet reaching the TCP receiver would generate a new packet and an acknowledgment, thus bringing the schedule back to the state that we had in slot 1. Notice that two "half" packets were transmitted in four slots, thus yielding a throughput of 1/4.

The above simple example illustrates the complexity of the question "What is the optimal TCP window, and what is the corresponding throughput?" The answer will depend on the topology of the path, the way the links interfere, and the multiple access protocol (i.e., the link scheduling mechanism).

We notice from the above discussion that even if a link can be scheduled, and if TCP is being used, this link may not be used; see the example in Fig. 5.13. Consider a WANET confined to a fixed (say, unit) area. There are *n* nodes. Suppose that TCP is being used, and each node, when given a turn to transmit, is only able to use a fraction of turns $\Phi(n)$. It follows, by continuing the arguments in Section 5.3.1, that the per node throughput will scale as

$$\lambda(n) \leq O\left(\frac{1}{\sqrt{n \ln n}}\right)\Phi(n)$$

Some directions to determining $\Phi(n)$ have been provided in [6]. In [14], the authors have reported empirical results, where it has been found that the per node throughput in an ad hoc internet (of UDP controlled sessions, i.e., no error recovery) scales as

$$O\left(\frac{1}{n^{1.68}}\right)$$

Since the transfers do not take place over TCP, this scaling probably captures the inefficiencies of the multiaccess protocol. One would expect that the performance would be even worse with TCP.

Recent research (see [4] and [5]) has shown that TCP over multihop wireless connections has serious performance limitations. A single TCP connection provides good throughput only if the window is properly tuned; several TCP connections sharing a WANET tend to lock each other out (almost completely) over long periods of time, and these lockouts occur at random times. We may wish to conclude that dense WANETs should not be used for building traditional Internet-like packet networks.

5.6 Conclusion

In the performance evaluation of ad hoc networks, we need to work with models that incorporate the properties of all the layers. Models that are general and address these interdependencies are hard to analyze. We have surveyed some of the recent progress in the area. Much recent work has emphasized network scaling laws, which show how the network performance changes with increasing node density. We have also provided a model for stochastic capacity and a detailed analysis of a Bluetooth scatternet, taking into account propagation phenomena. The stochastic model for a general WANET assumed network synchronization along slots; although this is impractical, the model provides useful insights. This analysis shows that in order for the network to operate efficiently, optimization of performance needs to be done simultaneously at various levels. The performance of the network depends on the physical layer and the way nodes organize themselves and operate. Similar insights were obtained from our discussion of TCP over multihop wireless routes. The most challenging task is to build upon such results and insights and develop distributed algorithms for efficient resource management of WANETs.

References

1. C.E. Perkins, Ed., *Ad Hoc Networking*, Addison Wesley, Reading, MA, 2001.
2. J. Broch, D.A. Maltz, D.B. Johnson, Y.-C. Hu, and J. Jetcheva, A Performance Comparison of Multi-Hop Ad Hoc Network Routing Protocols, *ACM/IEEE Int. Conf. on Mobile Computing and Networking*, Oct. 1998.
3. C.E. Perkins, E.M. Royer, S.R. Das, and M.K. Marina, Performance Comparison of Two On-Demand Routing Protocols for Ad Hoc Networks, *IEEE Communications Magazine*, Feb. 2001.
4. M. Gerla, R. Bagrodia, L. Zhang, K. Tang, and L. Wang, TCP over Wireless Multi-hop Protocols: Simulations and Experiments, *IEEE ICC*, 1999.
5. S. Xu and T. Saadawi, Does the IEEE 802.11 MAC Protocol Work Well in Multihop Wireless Ad Hoc Networks? *IEEE Communications Magazine*, June 2001, pp. 130–137.
6. S. Bansal, R. Gupta, R. Shorey, I. Ali, A. Razdan, and A. Misra, Energy Efficiency and Throughput for TCP Traffic in Multihop Wireless Networks, *IEEE Infocom*, June 2002.
7. I.F. Akyildiz, W. Su, Y. Sankarasubramaniam, and E. Cayirci, Wireless sensor networks: a survey, *Computer Networks*, 38, 393–422, 2002.
8. M.-S. Chen and R. Boorstyn, Throughput Analysis of Code Division Multiple Access (CDMA) Multihop Packet Radio Networks in the Presence of Noise, *IEEE Infocom*, 1985.
9. R.R. Boorstyn and A. Kershenbaum, Throughput Analysis of Multihop Packet Radio, *IEEE Infocom*, 1980.
10. O. deSouza, P. Sen, and R.R. Boorstyn, Performance Analysis of Spread Spectrum Packet Radio Networks, *IEEE Milcom*, 1988.
11. F.A. Tobagi, Modeling and performance analysis of multihop packet radio networks, *Proceedings of the IEEE*, 75, 135–154, 1987.
12. P. Gupta and P.R. Kumar, The capacity of wireless networks, *IEEE Transactions on Information Theory*, IT46, 388–404, 2000.
13. M. Grossglauser and D. Tse, Mobility Increases the Capacity of Ad-hoc Wireless Networks, *IEEE Infocom*, 2000.
14. P. Gupta, R. Gray, and P.R. Kumar, An Experimental Scaling Law for Ad Hoc Networks, Preprints.
15. J. Li, C. Blake, D.S.J. De Couto, H.I. Lee, and R. Morris, Capacity of Ad Hoc Wireless Networks, *ACM MobiCom*, Rome, 2001.
16. P. Panchapakesan and D. Manjunath, On the Transmission Range in Dense Ad Hoc Radio Networks, *Conference on Signal Processing and Communications SPCOM*, IISc, Bangalore, July 2001.
17. P. Gupta and P.R. Kumar, Critical Power for Asymptotic Connectivity in Wireless Networks, A Volume in Honour of W.H. Fleming in Stochastic Analysis, Control, Optimisation and Applications, 1998.
18. E.S. Sousa and J.A. Silvester, Optimum transmission ranges in a direct-sequence spread-spectrum multihop packet radio network, *IEEE Journal on Selected Areas in Communications*, 8, 762–771, June 1990.
19. H. Hashemi, The indoor radio propagation channel, *Proceedings of the IEEE*, 81, 943–968, 1993.
20. A. Karnik and A. Kumar, Performance of the Bluetooth Physical Layer, *IEEE ICPWC*, Dec. 2000.
21. A. Kumar, Comparative performance analysis of versions of TCP in a local area network with a lossy link, *IEEE/ACM Transactions on Networking*, 6, 485–498, 1998.
22. A. Kumar and J.M. Holtzman, Performance analysis of versions of TCP in a local network with a mobile radio link, *SADHANA: Indian Academy of Sciences Proceedings in Engineering Sciences*, 23, 113–129, 1998, also a WINLAB Technical Report, Rutgers University, New Brunswick, NJ, 1996.

II

Wireless Transmission Techniques

6

Coding for the Wireless Channel

Ezio Biglieri
Politecnico di Torino

Abstract

We consider the design and the performance of coding schemes for a channel affected by fading and additive noise. Optimum coding schemes for this channel lead to the development of new criteria for code design, differing markedly from the Euclidean-distance criterion which is commonplace over the additive white Gaussian noise (AWGN) channel. In fact, for frequency flat, slow fading channels the code performance depends strongly, rather than on the minimum Euclidean distance of the code, on its minimum Hamming distance (the "code diversity"). If the channel model is not stationary, as happens for example in a mobile radio communication system, where it may fluctuate in time between the extremes of Rayleigh and AWGN, then a code designed to be optimal for a fixed channel model might perform poorly when the channel varies. Therefore, a code optimal for the Rayleigh fading channel may be actually suboptimal for a substantial fraction of the time, and rather than for an optimum solution one should look for a robust solution, i.e., one with performance that is not critically dependent on the environment. In these conditions, antenna diversity with maximum-gain combining may prove useful: in fact, under fairly general conditions, a channel affected by fading can be turned into an AWGN channel by increasing the number of diversity branches. Another robust solution is based on bit interleaving, which yields a large code diversity thanks to the choice of powerful error-control codes coupled with a bit interleaver and the use of a suitable bit metric. An important feature of bit-interleaved coded modulation is that it lends itself quite naturally to "pragmatic" designs, i.e., to coding schemes that keep as their basic engine an off-the-shelf decoder. Yet another solution is based on controlling the transmitted power so as to compensate for the attenuations due to fading. Adaptive techniques, allowing the coding/modulation scheme to be modified according to the channel conditions, are also examined. Finally, the effect of multiple transmit and receive antennas on the performance of a radio system is analyzed.

6.1   Introduction and General Considerations

The increasing practical relevance of digital mobile radio transmission systems has led of late to a great deal of interest in coding for fading channels. For these channels, the textbook wisdom developed for the additive white Gaussian noise (AWGN) channel may not be valid anymore, and a fresh look at code design philosophies is called for. Specifically, system designers' choices are often driven by their knowledge of coding over the AWGN channel: consequently, coding solutions might be applied that are far from optimum on channels where nonlinearities, Doppler shifts, fading, shadowing, and interference from other users make the channel far from Gaussian.

A considerable body of work has been reversing this "Gaussian" perspective, and it is now widely accepted that coding solutions for the wireless channel should be selected by taking into account the distinctive features of fading. Here we examine in particular the effects of three features that make the fading channel differ from AWGN: namely, the fading channel is generally not memoryless (unless infinite-depth interleaving is assumed, an assumption that may not be realistic in several instances); it has a signal-to-noise ratio that is a random variable rather than a constant; and the propagation vagaries may make the channel model vary with time, so that any chosen model may be able to represent the channel only for a fraction of the time.

6.1.1   Speech vs. Data: The Delay Issue

A relevant factor in the choice of a coding scheme is the decoding delay that one may allow: for example, some extremely powerful codes (the "turbo codes") suffer from a considerable decoding delay, and hence their applicability is restricted. Consider for example real-time speech transmission: here a strict decoding delay is imposed (e.g., 100 msec, at most). In this case, the transmission of a code word may span only a few TDMA channel bursts, over which the channel fading is strongly correlated. Thus, a code word experiences only a few significant fading values, which makes the assumption of a memoryless channel, normally achieved through ideal or very long interleaving, no longer valid. On the contrary, with data traffic a large interleaving delay is tolerable, so that very effective coding techniques become available. For example, convolutional codes, bit interleaving, and high-level modulation (such as 8PSK or 16QAM) can be used. These techniques are generally referred to as Bit-Interleaved Coded Modulation (BICM) [1]. Capacity calculations show that with large interleaving, BICM performs as well as optimal coding over more complicated alphabets, and its complexity is much lower, so that the performance–complexity trade-off of BICM is very attractive. Moreover, capacity calculations show that constant-power constant-rate transmission performs very close to optimal transmission schemes where power and rate are adapted dynamically to the channel conditions via a perfect feedback link. Then, with large interleaving and powerful coding, there is no need for implementing such complicated adaptive techniques and feedback links.

The delay constraints can be easily taken into account when designing a coding scheme if a "block-fading" channel model is used. In this model, the fading process is about constant for a number of symbol intervals. On such a channel, a single code word may be transmitted after being split into several blocks, each suffering from a different attenuation, thus realizing an effective way of achieving diversity.

The "block-fading" channel model, introduced in [2], is motivated by the fact that, in many mobile radio situations, the channel coherence time is much longer than one symbol interval, and hence several transmitted symbols are affected by the same fading value. Use of this channel model allows one to introduce a delay constraint for transmission, which is realistic whenever infinite-depth interleaving is not a reasonable assumption.

This model assumes that a code word of length $n = MN$ spans M blocks of length N (a group of M blocks is referred to as a *frame*). The value of the fading in each block is constant. M turns out to be a measure of the interleaving delay of the system: in fact, $M = 1$ corresponds to $N = n$, i.e., to no interleaving, while $M = n$ corresponds to $N = 1$, and hence to ideal interleaving. Thus, the results for different values of M illustrate the downside of nonideal interleaving.

With no delay constraint, a code word can span an arbitrarily large number M of fading blocks. If this is the case, then capacity is a good performance indicator. This applies for example to variable-rate systems (e.g., wireless data networks). On the other hand, most of today's mobile radio systems carry real-time speech (cellular telephony), for which constant-rate, constrained-delay transmission should be considered. In the latter case, that is, when each code word must be transmitted and decoded within a frame of M < ∞ blocks, *information outage rate*, rather than capacity, is the appropriate performance limit indicator [3]. An additional important definition related to the probability of outage is the *zero-outage*, or *delay-limited* capacity. This is the maximum rate for which the minimum outage probability is zero.

6.1.2 Diversity

Receiver diversity techniques have been known for a long time to improve the fading channel quality. Their synergy with coding has been extensively investigated [4]. The standard approach to antenna diversity is based on the fact that, with several diversity branches, the probability that the signal will be simultaneously faded on all branches can be made small. A philosophically different approach [4] is based upon the observation that, under fairly general conditions, a channel affected by fading can be turned into an additive white Gaussian noise (AWGN) channel by increasing the number of diversity branches. Consequently, a coded modulation scheme designed to be optimal for the AWGN channel will perform asymptotically well also on a fading channel with diversity, at only the cost of an increased receiver complexity. An advantage of this solution is its robustness, since changes in the physical channel affect the reception very little. This allows us to argue that the use of "Gaussian" codes along with diversity reception provides indeed a solution to the problem of designing robust coding schemes for the mobile radio channel.

6.1.3 Multi-User Detection: The Challenge

The design of coding schemes is further complicated when a multi-user environment is accounted for. The main problem here, and in general in communication systems that share channel resources, is the presence of multiple-access interference (MAI). This is generated by the fact that every user receives, besides the signal that is specifically directed to it, also some power from transmission to other users. This is true not only when CDMA is used, but also with space division multiple access, in which intelligent antennas are directed towards the intended user. The earlier studies devoted to multi-user transmission simply neglected the presence of MAI. Typically, they were based on the naive assumption that, due to some version of the ubiquitous "Central Limit Theorem," signals adding up from a variety of users would coalesce to a process resembling Gaussian noise. Thus, the effect of MAI would be an increase of thermal noise, and any coding scheme designed to cope with the latter would still be optimal, or at least near-optimal, for multi-user systems.

Of late, it was recognized that this assumption was groundless, and consequently several of the conclusions that it prompted were wrong. The central development of multi-user theory was the introduction of the optimum multi-user detector; rather than demodulating each user separately and independently, it demodulates all of them simultaneously. A simple example should suffice to appreciate the extent of the improvement that can be achieved by optimum detection; in the presence of vanishingly small thermal noise, optimum detection would provide error-free transmission, while standard ("single-user") detection is affected by an error probability floor which increases with the number of users.

Multi-user detection was born in the context of terrestrial cellular communication, and hence implicitly assumed a MAI-limited environment where thermal noise is negligible with respect to MAI (high signal-to-noise ratio [high-SNR] condition). For this reason coding was seldom considered, and hence most multi-user detection schemes known from the literature are concerned with symbol-by-symbol decisions.

6.1.4 Unequal Error Protection

In some analog source coding applications, such as speech or video compression, the sensitivity of the source decoder to errors in the coded symbols is typically not uniform; the quality of the reconstructed analog signal is rather insensitive to errors affecting certain classes of bits, while it degrades sharply when errors affect other classes. This happens, for example, when analog source coding is based on some form of hierarchical coding, where a relatively small number of bits carries the "fundamental information" and a larger number of bits carries the "details," as in the case of the MPEG2 standard.

Assuming that the source encoder produces frames of binary coded symbols, each frame can be partitioned into classes of symbols of different "importance" (i.e., of different sensitivity). Then, it is apparent that the best coding strategy aims at achieving lower bit error probability (BER) levels for the important classes while admitting higher BER levels for the unimportant ones. This feature is referred to as *unequal error protection* (UEP). On the contrary, codes for which the BER is (almost) independent of the position of the information symbols are referred to as *equal error protection* (EEP) codes.

6.1.5   Where We Are (The Gaussian Channel)

While the discipline of coding for the wireless channel is still in a relatively early stage, the problem of finding capacity-achieving codes for the (nonfaded) additive white Gaussian noise channel has been essentially solved in the last few years.

With uncoded transmission at a BER of 10^{-5}, we are about 9.4 dB away from the theoretical (Shannon's) limit. With a powerful convolutional code, we may obtain an improvement close to 5.7 dB over uncoded transmission. Binary convolutional codes with sequential decoding were shown in the 1960s to be an implementable solution for operating about 3 dB away from Shannon's limit, and in the last decade this 3-dB barrier was eventually broken. Up until the last few years, a code obtained by concatenating a Reed–Solomon code with a convolutional code was considered to be the state of the art; at a BER of 10^{-5}, this system was roughly 2.3 dB from Shannon's limit.

"Turbo codes" with properly designed interleavers can now achieve an error performance extremely close to the limit [13]. The first turbo code, introduced in 1993, was roughly 0.5 dB from the limit at a BER of 10^{-5}. Turbo codes perform extremely well for BERs above 10^{-4}–10^{-5} ("waterfall" region); however, they have a significantly weakened performance at lower BER, due to the fact that their component codes have a relatively poor minimum Euclidean distance, which manifests its effects at these BERs. The fact that these codes do not have large minimum distances causes the BER curve to decrease its slope at BERs below 10^{-5}, a phenomenon known as *error floor*. It has been argued that the presence of this error floor makes turbo codes not suitable for applications requiring extremely low BERs. Their poor minimum distance and their lack of error-detection capability (due to the fact that in turbo decoding only information bits are decoded) make these codes perform badly in terms of block error probability. In turn, poor block error performance also makes these codes not suitable for certain communication applications. Another relevant factor that may guide in the choice of a coding scheme is the decoding delay that one should allow: in fact, turbo codes suffer from a considerable decoding delay, and hence their application might be useful for data transmission more than for real-time speech.

6.2   The Frequency-Flat, Slow Rayleigh Fading Channel

We now focus attention on the Rayleigh fading channel. This channel model assumes that the duration of a modulated symbol is much greater than the delay spread caused by the multipath propagation. If this occurs, then all frequency components in the transmitted signal are affected by the same random attenuation and phase shift, and the channel is frequency flat. This implies that the fading affects the transmitted signal multiplicatively. If in addition the channel varies very slowly with respect to the symbol duration, then the fading process remains approximately constant during the transmission of one symbol (if this does not occur, the fading process is called *fast*).

While the assumption of nonselectivity allows one to model the fading as a process affecting the transmitted signal in a multiplicative form, the assumption of a slow fading allows us to model this process as a constant random variable during each symbol interval. In summary, if $x(t)$ denotes the complex envelope of the modulated signal transmitted during a symbol interval, then the complex envelope $y(t)$ of the signal received at the output of a channel affected by slow, flat fading and additive white Gaussian noise can be expressed in the form

$$y(t) = Re^{j\Theta}x(t) + n(t)$$

where $n(t)$ is a complex Gaussian noise, and $Re^{j\Theta}$ is a Gaussian random variable, with R having a Rice or Rayleigh probability density function and unit second moment. If we can further assume that the fading is so slow that we can estimate the phase shift Θ with sufficient accuracy and hence compensate for it, then coherent detection is feasible. The previous channel model can be further simplified to

$$y(t) = Rx(t) + n(t)$$

With this simple model of the fading channel, the only difference with respect to an AWGN channel resides in the fact that R, instead of being a constant attenuation, is now a random variable, the value of which affects the amplitude and hence the power of the received signal. Assume finally that the value taken by R is known at the receiver; we describe this situation by saying that we have perfect *channel state information* (CSI). Channel state information can be obtained for example by inserting a pilot tone in a notch of the spectrum of the transmitted signal, and by assuming that the signal is faded exactly in the same way as this tone.

Figure 6.1 compares error probabilities over the Gaussian channel with those over the Rayleigh fading channel with perfect CSI. The modulation is binary PSK, and detection is coherent [5]. It is seen that the loss in error probability is considerable. On the other hand, capacity calculations show that the loss in capacity due to fading is much smaller than the loss in terms of error probability. This observation suggests that coding for the fading channel can compensate for a substantial amount of the latter loss and hence be highly beneficial.

6.2.1 The Robustness Issue

If the channel model is not stationary, as happens for example in a mobile radio communication system, then a code designed to be optimum for a fixed channel model might perform poorly when the channel

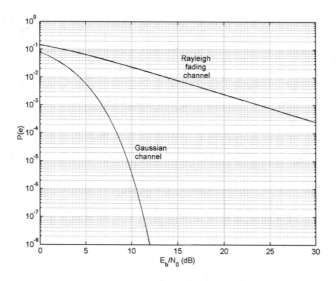

FIGURE 6.1 Bit error rate of binary antipodal modulation with coherent demodulation on the AWGN and Rayleigh-fading channel.

varies. Therefore, a code optimal for the AWGN channel may actually be suboptimal for a substantial fraction of time. Under these conditions, antenna diversity with maximum ratio combining may prove useful; in fact, under fairly general conditions, a channel affected by fading can be turned into an AWGN channel by increasing the number of diversity branches [4]. Another robust solution is based on bit interleaving, which yields a large diversity gain thanks to the choice of powerful convolutional codes coupled with a bit interleaver and the use of a suitable bit metric [1]. Yet another solution is adapting the transmitted power so as to compensate for the attenuations due to fading (see below).

6.2.2 Code Design Criteria

A standard code design criterion, when soft decoding is used, is to choose coding schemes that maximize the minimum Euclidean distance between code words. This is correct on the Gaussian channel with high SNR (although not when the SNR is very low) and is often accepted, *faute de mieux*, on channels that deviate little from the Gaussian model (e.g., channels with a moderate amount of intersymbol interference). However, the Euclidean distance criterion should be outright rejected over the Rayleigh fading channel, unless antenna diversity is used. In fact, analysis of coding for the Rayleigh fading channel proves that Hamming distance (also called "code diversity" in this context) plays the central role here.

Assume transmission of a coded sequence $\mathbf{x} = (x_1, x_2, \ldots, x_n)$, where the components of $\mathbf{x}$ are signal vectors selected from a constellation S. We do not distinguish here among block or convolutional codes (with soft decoding), or block- or trellis-coded modulation.

6.2.2.1 No Delay Constraint: Infinite-Depth Interleaving

We also assume for the moment infinite-depth interleaving, which makes the fading random variables affecting the various symbols x_k independent. Hence we write, for the components of the received sequence $\mathbf{y} = (y_1, y_2, \ldots, yn)$:

$$y_k = R_k x_k + n_k$$

where the R_k are independent random variables, and, under the assumption that the noise is white, the Gaussian noise random variables are also independent.

Coherent detection of the coded sequence, with the assumption of perfect channel-state information, is based upon the search for the code word $\mathbf{x} = (x_1, x_2, \ldots, x_n)$ that minimizes the Euclidean distance

$$\sum_{k=1}^{n} \left| y_k - R_k x_k \right|^2$$

To evaluate the error probability of this scheme, the standard procedure consists of using the upper "union" bound obtained by summing, over all possible pairs of code words $\mathbf{x} \neq \hat{\mathbf{x}}$, the pairwise error probabilities $P(\mathbf{x} \to \hat{\mathbf{x}})$. These are defined as the probability that, when x is the transmitted code word, the decoder prefers to it the competing word **x**. The pairwise error probability can be upper bounded in this case as

$$P(x \to \hat{x}) \leq \frac{1}{\left[\delta^2 / 4N_0 \right]^d}$$

where d denotes the Hamming distance between the code words $\mathbf{x}, \hat{\mathbf{x}}$, that is, the number of components in which the two differ, and the "squared product distance"

$$\delta^2 = \left[\prod_{k \in K} \left| x_k - \hat{x}_k \right|^2 \right]^d$$

(where K is the set of indices k such that $x_k \neq \hat{x}_k$) is the geometric mean of the nonzero squared Euclidean distances between the components of the two code words $\mathbf{x}, \hat{\mathbf{x}}$.

The latter result shows the important fact that the error probability is (approximately) inversely proportional to the product of the squared Euclidean distances between the components of the two code words that differ, and, to a more relevant extent, to a negative power of the signal-to-noise ratio, the exponent of which is the Hamming distance between the code words. This exponent reflects the slope of the error-probability curve as plotted versus the signal-to-noise ratio.

If the pairwise error probabilities are summed up to obtain an upper union bound, it can be seen that for high signal-to-noise ratios a few equal terms will dominate the bound [5]. These correspond to pairs of code words at the smallest Hamming distance. This observation may be used to design coding schemes for the Rayleigh fading channel, which should be chosen so as to maximize the minimum Hamming distance between code words. Here no central role is played by the Euclidean distance, which is the basic parameter used in the design of coding schemes for the AWGN channel.

For uncoded systems ($n = 1$), the results above hold with the position $d = 1$, which shows that the error probability decreases as N_0. A similar result could be obtained for maximal-ratio combining in a system with diversity d. In this context, the various diversity schemes may be seen as implementations of the simplest among the coding schemes, the repetition code, which provides a diversity equal to the number of diversity branches.

6.2.2.2   The Block-Fading Channel

The above analysis holds, *mutatis mutandis*, for the block-fading channel. In this situation, it can be shown that the relevant criterion becomes the Hamming block distance, i.e., the number of blocks (rather than the symbols) in which two code words differ.

6.3   Adaptive C/M Techniques

Since wireless channels exhibit a time-varying response, adaptive transmission strategies look attractive to prevent insufficient utilization of the channel capacity. The basic idea behind adaptivity consists of allocating transmitted power and code rate to take advantage of favorable channel conditions by transmitting at high speeds, while at the same time counteracting bad conditions by reducing the throughput. For an assigned quality of service (QoS), the goal is to increase the average spectral efficiency by taking advantage of the transmitter having knowledge of the CSI. The amount of performance improvement provided by such knowledge can be evaluated in principle by computing the Shannon capacity of a given channel with and without it. However, it should be kept in mind that capacity results refer to a situation in which complexity and delay are not constrained. Thus, for example, for a Rayleigh fading channel the capacity with CSI at the transmitter and the receiver is only marginally larger than for a situation in which only the receiver has CSI. This implies that if very powerful and complex codes are used, then CSI at the transmitter can buy little. However, in a delay- and complexity-constrained environment, a considerable gain can be achieved.

Adaptive techniques are based on two steps:

1. Measurement of the parameters of the transmission channel
2. Selection of one or more transmission parameters based on the optimization of a preassigned cost function

A basic assumption here is that the channel does not vary too rapidly, otherwise the parameters selected might be badly matched to the channel. Thus, adaptive techniques can only be beneficial in a situation where the Doppler spread is not too wide. This makes adaptive techniques especially attractive in an indoor environment, where propagation delays are small and the relative speed between transmitter and receiver is typically low. In these conditions, adaptive techniques can work on a frame-by-frame basis. Some adaptive solutions are categorized as follows.

Adapting power level — Through power control, the transmission level is varied according to the channel fluctuations. This strategy increases the transmitter peak-power requirements, and, in a multi-user environment, the level of cochannel interference, which may reduce channel capacity if coordination among users is not allowed.

Adapting constellation size — Among adaptive transmission techniques, adaptive modulation plays a central role because it increases the data transmission efficiency without increasing the multi-access interference power. In essence, adaptive modulation consists of transmitting at the highest possible rate compatible with an assigned QoS, as specified by higher-layer requirements such as packet error rate, packet delay, etc. This is obtained by using a hierarchy of different constellations of increasing size. Adaptive constellation size may be implemented so as to maintain a constant transmit power while providing a target QoS. The number of signals in the modulator constellation can be varied in such a way that the short-term error rate is approximately constant while the short-term bit rate varies, or vice versa. In a single-user environment, adaptive modulation can provide a 5–10 dB gain over a fixed-rate system having only power control.

Adapting code rate — The coding scheme can be changed so as to respond to the channel state by selecting the optimum code rate. Punctured convolutional codes are especially useful for this purpose because they enable adaptive encoding and decoding without modifying the basic structure of the encoder and the decoder.

Adapting power level and constellation size — Both modulation scheme and transmit level can be adapted in a single-user environment or for a multi-user channel. This combination leads to a significant throughput increase as compared to no power control.

Adapting constellation size and symbol rate — Both constellation size and symbol transmission rate can be adapted. The system selects the optimum modulation parameters so as to maximize the bit rate while satisfying the required error rate. Here a lower symbol rate is achieved by consecutively transmitting identical modulation symbols at the maximum symbol rate (this is equivalent to repetition coding).

Adapting power and transmission rate — Both the transmission rate and the power can be selected so as to maximize the spectral efficiency while satisfying average power and BER constraints.

Adapting modulation size and coding scheme — This adaptation strategy refers to trellis-coded modulation. By holding the number of information bits entering the encoder fixed, and adapting the number of bits left uncoded according to the estimate of the CSI, the trellis structure is changed. This adaptation scheme might not be robust to estimation errors, especially on Rayleigh fading channels. If this is the case, bit-interleaved coded modulation [1] may be a more attractive proposition because it increases the time diversity. With this scheme, the code is left fixed, while the signal constellation is adapted to the channel conditions (hence, it could be categorized under the rubric "adapting the constellation size"). Analyses suggest that the code structure of BICM is more suitable for adaptive systems that must support highly mobile users.

Adapting code rate, symbol rate, and constellation size — Code rate, symbol rate, and constellation size can be adapted simultaneously. Code rate adaptation is obtained by puncturing a convolutional code, while the constellation size is selected by setting SNR thresholds. If the target error rate cannot be achieved under any combination of parameters, the system transmits no data.

6.4   Transmission with Multiple Antennas

It has long been known that multiple receive antennas can be used as an alternative to coding, or in conjunction with it, to provide diversity. Recent work has explored the performance of systems in which multiple antennas are used at both the transmitter and receiver sides; these systems have the potential of increasing the capacity and the error performance of wireless transmission without any additional expenditure of bandwidth or power. It has been shown that, in a system with t transmit and r receive antennas (Fig. 6.2) and a slow fading channel modeled by an $r \times t$ matrix with random independent identically distributed complex Gaussian entries (the "independent Rayleigh fading" assumption), the average channel capacity with perfect CSI at the receiver is about $m = \min (t, r)$ times larger than that of a single-antenna

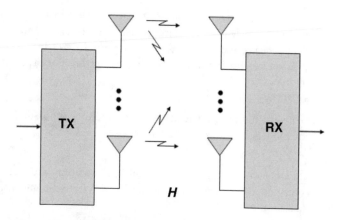

FIGURE 6.2 A radio communication system with multiple transmit and receive antennas.

system for the same transmitted power and bandwidth. The capacity increases by about m bit/sec/Hz for every 3-dB increase in SNR and reaches levels that cannot be achieved in any other way with present-day technology. A further performance improvement can be obtained under the assumption that CSI is available at the transmitter as well. Obtaining transmitter CSI from multiple transmitting antennas is particularly challenging because the transmitter should achieve instantaneous information about the fading channel. On the other hand, if transmit CSI is missing, the coding/modulation scheme employed should guarantee good performance with the majority of possible channel realizations.

Codes specifically designed for a multiple transmit antenna system use degrees of freedom in both space and time and are usually called space-time codes [6]. The symbols in code words are distributed across transmit antennas and time; by introducing correlation among signals transmitted by different antennas as well as those transmitted at different times, a coding gain is obtained without any sacrifice in bandwidth and with a relatively simple receiver structure. Space-time codes are already finding their way to modern wireless system standards.

6.4.1 Preliminaries

Consider channel transmission with t trasmit and r receive antennas. Assuming two-dimensional modulations (e.g., PSK or QAM) throughout, the channel model is

$$y = Hx + z$$

where x is a complex t-vector, y is a complex r-vector, H is a complex $r \times t$ matrix with entries h_{ij} that describe the gains of each transmission path to a receive from a transmit antenna, and z is a complex noise vector, the components of which are independent identically distributed Gaussian random variables with independent real and imaginary components. We also assume that $E[zz'] = I$, that is, the noises affecting the different receivers are independent, and the signal power coincides with the signal-to-noise ratio (the prime denotes conjugate, or Hermitian, transpose).

The ith component of vector x is the signal transmitted from antenna i; the jth component of vector y is the signal received by antenna j. Explicitly, we have

$$y_j = \sum_{i=1}^{t} h_{ij} x_i + z_j$$

which shows how the received signal includes a linear combination of the signals emitted by each antenna. We say that y is affected by *spatial interference* from the signals transmitted by the various antennas. This interference has to be removed, or controlled in some way, in order to separate the single transmitted signals. In the following, we shall see how this can be done. For the moment, we may just observe that the problem of removing interference makes the tools for the analysis of multiple-antenna transmission have much in common with those used in the study of digital equalization (where interference is generated by symbols adjacent in time) and of multi-user detection (where interference is generated by users sharing the same channel).

6.4.2 Channel Capacity

Here we evaluate the capacity of a multiple-antenna, or MIMO (multiple-input, multiple-output) transmission system [7]. Several models for the matrix H can be considered:

1. H is deterministic.
2. H is a random matrix, chosen according to a probability distribution, and each channel use (viz., the simultaneous transmission of a symbol from each of the t transmit antennas) corresponds to an independent realization of H.
3. H is a random matrix, but once it is chosen it remains fixed for the transmission of an entire code word.

When H is random, as in cases (2) and (3) above, we assume that its entries form an independent identically distributed Gaussian collection with zero-mean, independent real and imaginary parts, each with variance $1/2$. Equivalently, each entry of H has uniform phase and Rayleigh magnitude. This choice models Rayleigh fading with enough separation within antennas such that the fades for each transmit/receive antenna pair are independent. We also assume that the CSI (that is, the realization of H) is known at the receiver, while the distribution of H is known at the transmitter (the latter assumption is necessary for capacity computations, since the transmitter must choose an optimum code for that specific channel).

Model (3) is applicable to an indoor wireless data network or a personal communication system with mobile terminals moving at walking speed, so that the channel gain, albeit random, varies so slowly with time that it can be assumed as constant along transmission of a long block of data. More generally, fading blocks can be thought of as separated in time (e.g., in a time-division system), as separated in frequency (e.g., in a multicarrier system), or as separated both in time and in frequency (e.g., with slow time-frequency hopping). This model, the block-fading (BF) channel introduced in Section 6.1.1, allows the incorporation of delay constraints. These imply that, even though very long code words are transmitted, perfect (i.e., infinite-depth) interleaving cannot be achieved. With this fading model, the channel may or may not be ergodic (ergodicity is lost when there is a delay constraint). Without ergodicity, the average mutual information over the ensemble of channel realizations cannot characterize the achievable transmission rates; since the instantaneous mutual information of the channel turns out to be a random variable, the quantity that characterizes the quality of a channel when coding is used is not channel capacity, but rather the *information outage probability*, i.e., the probability that this mutual information is lower than the rate of the code used for transmission [2]. This outage probability is closely related to the code word error probability, as averaged over the random coding ensemble and over all channel realizations; hence, it provides useful insight on the performance of a delay-limited coded system.

6.4.2.1 Deterministic Channel

We derive the capacity by maximizing the average mutual information $I(x;y)$ between input and output of the channel over the choice of the distribution of x. Singular-value decomposition of the matrix H yields

$$H = UDV'$$

where U is an $r \times t$ complex matrix, V a $t \times t$ complex matrix with the properties $U'U = I_r$ and $V'V = I_t$ (that is, U and V are unitary), and D is a real $r \times t$ matrix with the form $D = [I_r : 0]$ if $r < t$, or the form

$$D = \begin{bmatrix} I_t \\ \cdots \\ 0 \end{bmatrix}$$

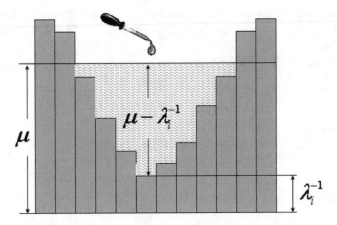

FIGURE 6.3 Illustrating the water-filling principle.

if $r > t$. The diagonal entries of $\mathbf{D}$ are the square roots of the eigenvalues of $\mathbf{HH}'$; the columns of $\mathbf{U}$ are the eigenvectors of $\mathbf{HH}'$, and the columns of $\mathbf{V}$ are the eigenvectors of $\mathbf{H}'\mathbf{H}$. We can write

$$\mathbf{y} = \mathbf{UDV}'\mathbf{x} + \mathbf{z}$$

which shows that the original channel is equivalent to the channel described by the input-output relationship

$$\tilde{y} = \mathbf{D}\tilde{\mathbf{x}} + \tilde{\mathbf{z}}$$

where $\tilde{\mathbf{y}} = \mathbf{U}'\mathbf{y}$, $\tilde{\mathbf{x}} = \mathbf{U}'\mathbf{x}$, and $\tilde{\mathbf{z}} = \mathbf{U}'\mathbf{z}$ is zero-mean Gaussian, with independent, identically distributed real and imaginary parts and $\mathrm{E}[\tilde{\mathbf{z}}\tilde{\mathbf{z}}'] = \mathbf{I}_t$. Now, the rank of $\mathbf{H}$ is at most $m = \min\{t, r\}$, and hence at most m of its singular values are nonzero. Denote these by $\sqrt{\lambda_i}$, $i=1,\dots,m$, and rewrite the previous equation componentwise in the form:

$$\tilde{y} = \sqrt{\lambda_i}\tilde{x}_i + \tilde{z}_i, \quad i = 1,\dots,m$$

The remaining components of $\tilde{\mathbf{y}}_i$ (if any) are equal to the corresponding components of the noise vector $\mathbf{z}_i$: we see that, for $i > m$, $\tilde{y}_i$ is independent of the transmitted signal, and $\tilde{x}_i$ does not play any role.

To maximize the mutual information we need to choose $\tilde{x}_i$, $i = 1,\dots,m$, to be independent, with each $\tilde{x}_i$ having independent Gaussian, zero-mean real and imaginary parts. Their variances

$\mathrm{E}[\Re\tilde{x}_i]^2 = \mathrm{E}[\Im\tilde{x}_i]^2$ should be chosen via "water-filling" (Fig. 6.3):

$$\mathrm{E}[\Re\tilde{x}_i]^2 = \mathrm{E}[\Im\tilde{x}_i]^2 = \frac{1}{2}\left(\mu - \lambda_i^{-1}\right)^+$$

where $a^+ = \max(0,a)$. This comes from a result of Information Theory concerning parallel channels [8]. With μ chosen so as to meet the power constraint, we have

$$P(\mu)\sum_i\left(\mu - \lambda_i^{-1}\right)^+$$

and the capacity takes on the value

$$C(\mu) = \sum_i\left(\log(\mu\lambda_i)\right)^+$$

(Here and in the following, log denotes base 2 logarithm.) Since the nonzero eigenvalues of $\mathbf{H'H}$ are the same as those of $\mathbf{HH'}$, the capacities of the channels corresponding to $\mathbf{H}$ and to $\mathbf{H'}$ are the same. A sort of "reciprocity" holds in this case.

Consider for example the all-1 channel matrix: we have the SVD

$$
\mathbf{H} = \begin{bmatrix} \sqrt{\dfrac{1}{r}} \\ \sqrt{\dfrac{1}{r}} \\ \vdots \\ \sqrt{\dfrac{1}{r}} \end{bmatrix} \sqrt{rt} \begin{bmatrix} \sqrt{\dfrac{1}{t}} & \sqrt{\dfrac{1}{t}} & \cdots & \sqrt{\dfrac{1}{t}} \end{bmatrix}
$$

In this case we have $m = 1$, $\sqrt{\lambda_1} = \sqrt{rt}$, and hence $\lambda_1 = rt$. Hence, for $P > 0$,

$$
P = \left(\mu - \frac{1}{rt} \right)^+ = \mu - \frac{1}{rt}
$$

and hence the capacity is

$$
C = \log\left[\left(P + \frac{1}{rt} \right) rt \right] = \log(1 + rtP)
$$

The signals achieving this capacity can be described as follows. We have that $\tilde{\mathbf{x}}$ has only one component, and

$$
\mathbf{V} = \sqrt{\frac{1}{t}} \begin{bmatrix} 1 \\ 1 \\ \vdots \\ 1 \end{bmatrix}
$$

Thus, the components of $\mathbf{x} = \mathbf{V}\tilde{\mathbf{x}}$ are all equal, i.e., the TX antennas all send the same signal. Each TX antenna sends a power P/t; because of the structure of $\mathbf{H}$ the signals *add coherently* at the receiver, so that at each receiver we have the voltage $t\sqrt{P/t}$, and hence the power Pt. Since each receiver sees the same signal, and the noises are uncorrelated, the overall SNR is rtP, as shown by the capacity formula $C = \log(1 + rtP)$.

For another example, take $r = t = m$, and $\mathbf{H} = \mathbf{I}_m$. Here we have m parallel independent AWGN channels, and consequently the capacity increases linearly with m:

$$
C = m\log(1 + P/m)
$$

where we recognize the term P/m as the power transmitted by each TX antenna, and the factor m as being generated by having m parallel independent channels, each with capacity $\log(1 + P/m)$.

6.4.2.2   Ergodic Rayleigh Fading Channel

We assume here that the entries of $\mathbf{H}$ are independent and zero-mean Gaussian with independent real and imaginary parts, each with the same variance $1/2$. Equivalently, each entry of $\mathbf{H}$ has uniformly distributed phase and Rayleigh-distributed magnitude, with expected squared magnitude equal to one. We also assume that for each channel use an independent realization of $\mathbf{H}$ is drawn, so that the channel is memoryless. The following holds [7]: The channel capacity is achieved when **x** is circularly symmetric complex Gaussian with mean zero and covariance $(P/t)\mathbf{I}_r$. The capacity is

$$C = E\left[\log \det\left(\mathbf{I}_r + \frac{P}{t}\mathbf{HH}'\right)\right]$$

Note that for fixed r and as $t \to \infty$, by the strong law of large numbers

$$\frac{1}{t}\mathbf{HH}' \to \mathbf{I}_r$$

almost surely. Thus, as $t \to \infty$ the capacity equals

$$\log \det (\mathbf{I}_r + P\mathbf{I}_r) = \log(1 + P)^r = r\log(1 + P)$$

so that the capacity increases *linearly* with r.

The reciprocity observed for deterministic channels does not hold in this case. If $C(r,t,P)$ denotes the capacity of a channel with r receive antennas, t transmit antennas, and total transmit power P, we have

$$C(a,b,Pb) = C(b,a,Pa)$$

Thus, for example, $C(r,1,P) = C(1,r,rP)$, which shows that with TX rather than RX diversity we need r times as much transmit power to achieve the same capacity.

Choose $t = r = 1$ as the baseline; this channel yields one additional bit/sec/Hz for every 3 dB of SNR increase. In fact, for large SNR,

$$C = \log(1 + \text{SNR}) \approx \log \text{SNR}$$

and hence, if $\text{SNR} \to 2\text{SNR}$ we have $C \to \log \text{SNR} + 1$. For multiple antennas with $t = r$, for every 3 dB of SNR increase we have t more bit/sec/Hz.

6.4.2.3   Nonergodic Rayleigh Fading Channel

When **H** is chosen randomly at the beginning of the transmission and held fixed for all channel uses, average capacity has no meaning. In this case the quantity to be evaluated is, rather than capacity, outage probability. The conclusions that can be drawn in this situation are similar to those related to ergodic channels [3,7].

6.4.3   Influence of Channel-State Information

A crucial factor in determining the performance of a multi-antenna system is the availability of the channel-state information (CSI), that is, the knowledge of the values of the fading gains in each one of the transmission paths. In a fixed wireless environment, the fading gains can be expected to vary slowly, so their estimate can be obtained by the receiver with a reasonable accuracy, even in a system with a large number of antennas. One way of obtaining this estimate is by periodically sending pilot signals on the same channel used for data signals; if the channel is assumed to remain constant for I symbol periods, then we may write, $I = I_r + I_d$, where I_r is the number of pilot symbols, while data transmission occupies I_d symbols. Since the transmission of pilot symbols lowers the information rate, there is a tradeoff between system performance and transmission rate.

6.4.3.1 Perfect CSI at the Receiver

The most commonly studied situation is that of perfect CSI available at the receiver, which is the assumption under which we developed our study of multiple-antenna systems above.

6.4.3.2 No CSI

The fundamental limits of noncoherent communication, i.e., one taking place in an environment where estimates of the fading coefficients are not available, are derived in [9,10]. In the channel model assumed in [9,10], the fading gains are Rayleigh distributed and remain constant for I symbol periods before changing to new independent realizations. Under these assumptions, further increasing the number of transmit antennas beyond I cannot increase capacity.

6.4.3.3 Imperfect CSI at the Receiver

In the real world, the receiver has an imperfect knowledge of the CSI. In [11], answers are provided to two related questions, referred to as transmission architecture to be described below and called BLAST:

1. How long should the training interval be for satisfactory operation?
2. What effects do estimation errors have on performance?

It is shown in [11] that we must have $I_r \geq t$, i.e., the duration of the training interval must be at least as great as the number of transmit antennas. Moreover, the optimum training signals are orthogonal with respect to time among the transmit antennas, and each transmit antenna is fed equal energy. As for the effects of channel estimation errors, the training interval required to control the probability of outages induced by these errors is approximately proportional to the number of receive antennas. Thus, if the total interval I for training and for transmitting data is limited, the situation is the following. If we assume $t = r$ for simplicity, the capacity in units of bits per symbol is approximately proportional to t: thus, $I_r = \alpha t$ for some constant α, and this leaves $I - I_r = I - \alpha t$ symbols for sending the message. The capacity is $C = \beta t$ for some constant β, so the total number of bits that can be sent is $CI_d = \beta t(I - \alpha t)$. The optimum number of transmit antennas from the standpoint of maximizing the number of message bits is $t = I/2\alpha$, implying that, whatever the values of α and β, the training interval is $I_r = I/2$: that is, half of the available interval should be used for training.

6.4.3.4 CSI at the Transmitter and at the Receiver

It is also possible to envisage a situation in which channel state information is known to the receiver and to the transmitter; the latter can take the appropriate measures to counteract the effect of channel attenuations by suitably modulating its power. To ensure causality, the assumption of CSI available at the transmitter is valid if it is applied to a multicarrier transmission scheme in which the available frequency band (over which the fading is selective) is split into a number of subbands, as with OFDM. The subbands are so narrow that fading is frequency-flat in each of them, and they are transmitted simultaneously, via orthogonal subcarriers. From a practical point of view, the transmitter can obtain the CSI either from a dedicated feedback channel (some existing systems already implement a fast power-control feedback channel) or by time-division duplex, where the uplink and the downlink time-share the same subchannels and the fading gains can be estimated from the incoming signal.

Reference [12] derives the performance limits of a channel with additive white Gaussian noise, delay and transmit-power constraints, and perfect channel-state information available at both transmitter and receiver. Because of a delay constraint, the transmission of a code word is assumed to span a finite (and typically small) number M of independent channel realizations; therefore, the channel is nonergodic, and the relevant performance limits are the information outage probability and the delay-limited capacity. The coding scheme that minimizes the information outage probability is also derived in [12]. This scheme can be interpreted as the concatenation of an optimal code for the AWGN channel without fading to an optimal linear beamformer, the coefficients of which change whenever the fading changes. For this optimal scheme, minimum-outage probability and delay-limited capacity are derived in [12]. For a fairly large class of fading channels, the asymptotic delay-limited capacity slope, expressed in bit/sec/Hz per dB of transmit SNR, turns out to be proportional to $\min(t,r)$ and independent of the number of fading blocks M. Since M is a measure of the time diversity (induced by interleaving) or of the frequency diversity of the system, this result shows that, if channel-state information is available also to the transmitter, very high rates with asymptotically small error probabilities are achievable without need of deep interleaving or high frequency diversity. Moreover, for a large number of antennas, the delay-limited capacity approaches the ergodic capacity. Finally, the availability of CSI at the transmitter makes transmit-antenna diversity equivalent, in terms of capacity improvement, to receive-antenna diversity, in the sense that reciprocity holds.

6.4.4 Simple Practical Schemes for Multiple-Antenna Systems

6.4.4.1 Delay Diversity

One of the first schemes proposed is called *delay diversity*. The problem with multiple transmit antennas is to *separate* the signals received at each receive antenna; in fact, what the latter receives is a *linear combination* of the transmitted signals, from which it must extract the signals sent by each transmit antenna. We have a kind of orthogonalization problem here.

The idea with delay diversity is to separate the TX signals via a delay. Assume for simplicity $t = 2$. A signal is transmitted from the first antenna, then delayed by one time slot and transmitted from the second antenna. The received signal turns out to be a linear combination of delayed replicas of the transmitted signals; thus, the resulting channel, as viewed by a single receive antenna, is reminiscent of an intersymbol-interference channel. By doing equalization the original signals can be recovered.

6.4.4.2 Alamouti Scheme

We describe first this orthogonalization scheme by considering the simple case $t = 2$, $r = 1$. The scheme is illustrated in Fig. 6.4. During the first symbol interval, the signal s_0 is transmitted from antenna 0, while signal s_1 is transmitted from antenna 1. During the next symbol period, antenna 0 transmits signal $-s_1^*$, and antenna 1 transmits signal s_0^*. Thus, the signals received in two adjacent time slots are

$$r_1 = h_0 s_0 + h_1 s_1 + n_0$$

and

$$r_2 = -h_0 s_1^* + h_1 s_0^* + n_1$$

where h_0, h_1 denote the path gains from the two transmit antennas to the receive antenna. The combiner of Fig. 6.4, which has perfect CSI and hence knows the values of the path gains, generates the signals

$$\tilde{s}_0 = h_0^* r_1 + h_1 r_2^*$$

and

$$\tilde{s}_1 = h_1^* r_1 - h_0 r_2^*$$

so that

$$\tilde{s}_1 = \left(\left|h_0\right|^2 + \left|h_1\right|^2\right)s_0 + \left(h_0^* n_0 + h_1 n_1^*\right)$$

and similarly

$$\tilde{s}_1 = \left(\left|h_0\right|^2 + \left|h_1\right|^2\right)s_1 + \left(h_1^* n_0 - h_0 n_1^*\right)$$

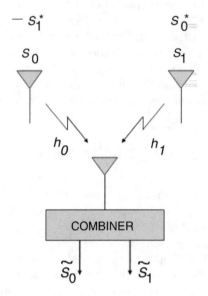

FIGURE 6.4 Alamouti scheme with two transmit and one receive antennas.

Thus, s_0 and s_1 are separated. This scheme has the same performance as a scheme with $t = 1$, $r = 2$, and maximal-ratio combining (provided that each TX antenna transmits the same power as the single antenna for $t = 1$). In fact, observe that if the signal s_0 is transmitted, the two receive antennas observe $h_0 s_0 + n_0$ and $h_1, s_0 + n_1$, respectively, and after maximal-ratio combining the decision variable is

$$h_0^*\left(h_0 s_0 + n_0\right) + h_1^*\left(h_1 s_0 + n_1\right) = \left(\left|h_0\right|^2 + \left|h_1\right|^2\right)s_0 + \left(h_0^* n_0 + h_1 n_1^*\right) = \tilde{s}_0$$

This scheme can be generalized to larger values of r; for example, with $t = r = 2$ and the same transmission scheme as before, if r_1, r_2, r_3, r_4 denote the signals received by antenna 0 at time 1, by antenna 0 at time 2, by antenna 1 at time 1, and by antenna 1 at time 2, respectively, we have (see Fig. 6.5):

$$r_1 = h_0 s_0 + h_1 s_1 + n_1$$

$$r_2 = -h_0 s_1^* + h_1 s_0^* + n$$

$$r_3 = h_2 s_0 + h_3 s_1 + n_3$$

$$r_4 = -h_2 s_1^* + h_3 s_0^* + r_i$$

The combiner generates

$$\tilde{s}_0 = h_0^* r_1 + h_1 r_2^* + h_2^* r_3 + h_3 r_4^*$$

$$\tilde{s}_1 = h_1^* r_1 - h_0 r_2^* + h_3^* r_3 - h_2 r_4^*$$

which yields

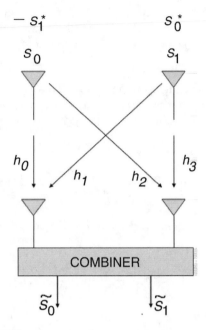

FIGURE 6.5 Alamouti scheme with two transmit and two receive antennas.

$$\tilde{s}_0 = \left(|h_0|^2 + |h_1|^2 + |h_2|^2 + |h_3|^2\right)s_0 + \text{noise}$$

$$\tilde{s}_1 = \left(|h_0|^2 + |h_1|^2 + |h_2|^2 + |h_3|^2\right)s_1 + \text{noise}$$

As above, it can be seen that the performance of this $t = 2$, $r = 2$ scheme is equivalent to that of a $t = 1$, $r = 4$ scheme with maximal-ratio combining (again, provided that each transmit antenna transmits the same power as with $t = 1$).

A general scheme with $t = 2$ and a general value of r can be exhibited; it has the same performance of a single-transmit-antenna scheme with $2r$ receive antennas and maximal-ratio combining.

6.4.5 Coding for Multiple-Antenna Systems

Given that, as capacity analyses show, considerable gains are potentially achievable by a multi-antenna system, the challenge is to design coding schemes that perform close to capacity; a code whose words are distributed in both space and time is called a *space-time code*. Assume that a space-time code $\mathcal{X}$ is used with block length N. Each code word of x is represented by the $t \times N$ matrix $\mathbf{X} = (\mathbf{x}_1, \mathbf{x}_2, \ldots, \mathbf{x}_N)$. The row index of $\mathbf{X}$ indicates space, while the column index indicates time; to wit, the ith component of the t-vector $\mathbf{x}_n$ is a complex number describing the signal transmitted by antenna i at discrete time n, $n = 1, \ldots, N$. The received signal is the $r \times N$ matrix

$$\mathbf{Y} = \mathbf{HX} + \mathbf{Z}$$

where $\mathbf{Z}$ is a matrix of complex Gaussian random variables with zero mean and independent real and imaginary parts with the same variance $N_0/2$ (i.e., they are circularly distributed). Thus, the noise affecting the received signal is spatially and temporally independent, with $\text{E}[\mathbf{ZZ}'] = NN_0\mathbf{I}_r$ where $\mathbf{I}_r$ denotes the $r \times r$ identity matrix. The channel is described by the $r \times t$ matrix $\mathbf{H}$, the entries of which are independent complex Gaussian random variables, circularly distributed with variance of their real and imaginary parts equal to 1/2. Equivalently, each entry of $\mathbf{H}$ has uniformly distributed phase and Rayleigh-distributed magnitude, with expected squared magnitude equal to 1. $\mathbf{H}$ is independent of both $\mathbf{X}$ and $\mathbf{Z}$; it remains constant during the transmission of an entire code word (the "quasi-static," or "block-fading," assumption), and its realization (the channel state information) is known at the receiver.

6.4.5.1 Maximum Likelihood Detection

Under the assumption of channel-state information perfectly known by the receiver, and of additive white Gaussian noise, maximum likelihood detection and decoding[1] corresponds to choosing the code word $\mathbf{X}$ which minimizes the squared "Frobenius" norm $\|\mathbf{Y} - \mathbf{HX}\|^2$, where we define, for an $m \times n$ matrix $\mathbf{A}$ with elements a_{ij}:

$$\|\mathbf{A}\|^2 = \sum_{i=1}^{m} \sum_{j=1}^{n} |a_{ij}|^2$$

Explicitly, maximum likelihood detection and decoding corresponds to the minimization of the quantity

$$\|\mathbf{Y} - \mathbf{HX}\|^2 = \sum_{i=1}^{m} \sum_{j=1}^{n} \left| y_{in} - \sum_{j=1}^{t} h_{ij} x_{jn} \right|^2$$

6.4.5.2 Pairwise Error Probability

As mentioned before, for practical computations of error probabilities it is customary to resort to the union bound to error probability

$$P(e) \leq \frac{1}{|X|} \sum_{\mathbf{X} \in X} \sum_{\hat{\mathbf{X}} \in X, \hat{\mathbf{X}} \neq \mathbf{X}} P(\mathbf{X} \to \hat{\mathbf{X}})$$

where in this case the pairwise error probability $P(\mathbf{X} \to \hat{\mathbf{X}})$ is given by

$$P(\mathbf{X} \to \hat{\mathbf{X}}) = E\left[Q\left(\frac{\|\mathbf{H}(\mathbf{X} - \hat{\mathbf{X}})\|}{\sqrt{2N_0}} \right) \right] \leq E\left[\exp(-\|\mathbf{H}(\mathbf{X} - \hat{\mathbf{X}})\|^2 /4N_0) \right]$$

The expectation has to be taken with respect to the random entries of matrix H, and can be computed in closed form to yield

$$P(\mathbf{X} \to \hat{\mathbf{X}}) = \det\left[\mathbf{I}_t + (\mathbf{X} - \hat{\mathbf{X}})(\mathbf{X} - \hat{\mathbf{X}})' /4N_0 \right]^{-r}$$

Since the determinant of a matrix is equal to the product of its eigenvalues, we have

$$P(\mathbf{X} \to \hat{\mathbf{X}}) = \prod_{j=1}^{t} \left(1 + \lambda_j / 4N_0 \right)^{-r}$$

where λ_j denotes the jth eigenvalue of $(\mathbf{X} - \hat{\mathbf{X}})(\mathbf{X} - \hat{\mathbf{X}})'$. We also have

---

[1] We distinguish here between *detection*, i.e., processing of the signals observed by the r receiving antennas, and *decoding*, i.e., deciding on the transmitted code word $\mathbf{X}$. Decoding is soft, i.e., maximum likelihood, as performed by a Viterbi decoder in which the branch metrics are generated by the interface used in the detection process.

$$P(\mathbf{X} \to \hat{\mathbf{X}}) \le \prod_{j \in J} \left(1 + \lambda_j / 4N_0\right)^{-r}$$

where J is the index set of the nonzero eigenvalues of $(\mathbf{X} - \hat{\mathbf{X}})(\mathbf{X} - \hat{\mathbf{X}})'$. Denoting by r the number of elements in J, and rearranging the eigenvalues so that $\lambda_1, \ldots, \lambda_p$ are the nonzero eigenvalues, we finally obtain

$$P(\mathbf{X} \to \hat{\mathbf{X}}) \le \left(\prod_{j=1}^{p} \lambda_j\right)^{-r} \left(\frac{\text{SNR}}{4}\right)^{-rp}$$

From this expression, we see that the total diversity order of the coded system is $r\rho_{\min}$, where $\rho_{\min}$ is the minimum rank of $(\mathbf{X} - \hat{\mathbf{X}})(\mathbf{X} - \hat{\mathbf{X}})'$ across all possible pairs of code words ("diversity gain"). Moreover, the pairwise error probability depends on the power r of the inverse of the product of eigenvalues of $(\mathbf{X} - \hat{\mathbf{X}})(\mathbf{X} - \hat{\mathbf{X}})'$. This does not depend on the SNR, and displaces the error probability curve rather than changing its slope. This is called the "coding gain." A good space-time code for high SNR should consequently maximize its diversity gain and its coding gain [14].

For an example of space-time trellis code designed according to the above criteria, see Fig. 6.6, which shows a section of its trellis. This code has $t = 2$, four states, and transmits 2 bit/channel use by using 4PSK, the signals of which are denoted 0, 1, 2, and 3. With $r = 1$, its diversity is 2, and with $r = 2$, its diversity is 4. Label xy means that signal x is transmitted by antenna 0, while signal y is simultaneously transmitted by antenna 1. Decoding of trellis space-time codes is performed by using a Viterbi algorithm.

6.4.6 BLAST Architecture

Extremely large spectral efficiencies can be achieved on a wireless link if the number of transmit and receive antennas is large. However, as t and r increase, the complexity of space-time coding with maximum-likelihood detection may become too large. This motivates the design of suboptimal receiver schemes with low complexity and still achieving a good portion of the spectral efficiency predicted by the theory.

A layered architecture that meets these requirements was proposed by Foschini [3]. This architecture is called BLAST (Bell Laboratories Layered Space Time) and is capable of realizing a significant fraction of the theoretical capacity using only ordinary modulation and coding techniques. The theory was confirmed in experimental demonstrations.

With BLAST, the information stream is demultiplexed into t substreams, which are independently encoded by t encoders. The encoders are connected to the t transmit antennas through a diagonal interleaving scheme schematically represented in Fig. 6.7. The interleaver is designed so that the symbols of a given code word are cyclically sent over all the t antennas, in order to guarantee the necessary diversity order. Diagonals are written from top to bottom, and the letters in each rectangle of Fig. 6.7 denote the corresponding code symbol index, i.e., indicate the sequence in which diagonals are filled. We assume for simplicity that each rectangle contains only one symbol (generalizations are straightforward). Each column of t symbols of the diagonal interleaver array is transmitted in parallel, from the t antennas. At the receiver side, the output of antennas is processed by a reduced-complexity suboptimal detector that includes the removal of a part of spatial interference (that due to symbols already detected) through the subtraction of the previous decisions. For example, if layer c is detected and we assume that layers a and b were already successfully detected, the spatial interference generated by these layers can be subtracted out. The interference due to layers not yet detected can be reduced by using linear processing techniques similar to equalization.

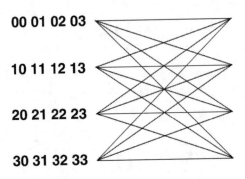

FIGURE 6.6   A simple trellis space-time code with four states and four symbols.

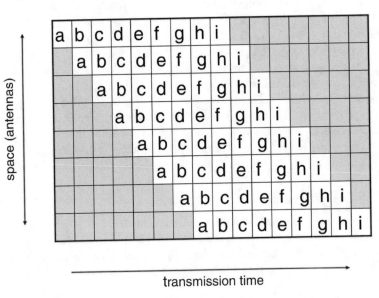

FIGURE 6.7   Structure of diagonal BLAST.

References

1. G. Caire, G. Taricco, and E. Biglieri, Bit-interleaved coded modulation, *IEEE Trans. Inform. Theory*, 44, 927–946, 1998.
2. L. Ozarow, S. Shamai, and A.D. Wyner, Information theoretic considerations for cellular mobile radio, *IEEE Trans. Vehic. Tech.*, 43(2), May 1994.
3. G.J. Foschini, Layered space-time architecture for wireless communication in a fading environment when using multi-element antennas, *Bell Labs Tech. J.*, 1, 41–59, 1996.
4. J. Ventura-Traveset, G. Caire, E. Biglieri, and G. Taricco, Impact of diversity reception on fading channels with coded modulation — part I: coherent detection, *IEEE Trans. Commun.*, 45(5), May 1997.
5. S. Benedetto and E. Biglieri, *Principles of Digital Transmission with Wireless Applications*, Kluwer/Plenum, New York, 1999.
6. A.F. Naguib, N. Seshadri, and A.R. Calderbank, Increasing data rate over wireless channels, *IEEE Signal Processing Magazine*, May 2000, pp. 49–61.
7. E. Telatar, Capacity of multi-antenna Gaussian channels, *Eur. Trans. Telecomm.*, 10, 585–595, 1999.
8. R.G. Gallager, *Information Theory and Reliable Communication*, John Wiley & Sons, New York, 1968.

9. L. Marzetta and B.M. Hochwald, Capacity of a mobile multiple-antenna communication link in Rayleigh flat fading, *IEEE Trans. Inform. Theory*, 45, 139–157, 1999.

10. B. Hochwald and T. Marzetta, Space-time modulation scheme for unknown rayleigh fading environment, *Proc. 36th Annual Allerton Conference on Communication, Control and Computing*, Monticello, IL, Sep. 1998.

11. T.L. Marzetta, BLAST Training: Estimating channel characteristics for high capacity space-time sireless, *Proc. 37th Annual Allerton Conference on Communication, Control and Computing*, Monticello, IL, Sep. 22–24, 1999, pp. 958–966.

12. E. Biglieri, G. Caire, and G. Taricco, Limiting performance of block-fading channels with multiple antennas, *IEEE Trans. Inform. Theory*, 47, 1273–1289, 2001.

13. C. Berrou and A. Glavieux, Near optimum error correcting coding and decoding: Turbo-codes, *IEEE Trans. Commun.*, Oct. 1996.

14. V. Tarokh, N. Seshadri, and A.R. Calderbank, Space-time codes for high data rate wireless communication: Performance criterion and code construction, *IEEE Trans. Inform. Theory*, 44, 744–765, 1998.

7

Unicast Routing
Techniques for Mobile
Ad Hoc Networks

Roberto Beraldi

Universita' di Roma, "La Sapienza"

Roberto Baldoni

Universita' di Roma, "La Sapienza"

Abstract

A Mobile Ad Hoc Network (MANET) is an autonomous system of functionally equivalent mobile nodes, which must be able to communicate while moving, without any kind of wired infrastructure. To this end, mobile nodes must cooperate to provide the routing service. Routing in mobile environments is challenging due to the constraints existing on the resources (transmission bandwidth, CPU time, and battery power) and the required ability of the protocol to effectively track topological changes.

This chapter discusses the issue of routing in mobile ad hoc networks by focusing on the main solutions proposed in the literature to cope with mobility. The chapter surveys the techniques adopted for the case of unicast routing, i.e., when it is required to send a packet from a source node to a destination node, by illustrating how they are introduced into the main representative protocols.

7.1   Introduction

The term MANET (Mobile Ad hoc NETwork) refers to a set of wireless mobile nodes that can communicate and move at the same time. No fixed infrastructures are required to allow such communications; instead, all nodes cooperate in the task of routing packets to destination nodes. This is required, since each node of the network is able to communicate only with those nodes located within its transmission radius R, while a source node S and a destination node D of the MANET can be located at a distance much greater than R. When S wants to send a packet to D, the packet may have to cross many intermediate nodes. For this reason, MANETs belong to the class of multi-hop wireless networks.

This chapter considers the problem of *unicast routing*, responsible for routing packets from a single source node to a single destination node, and illustrates the main techniques adopted in designing routing

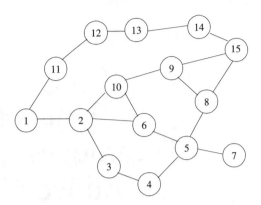

FIGURE 7.1 A graph representation of a network.

protocols for such a wireless multi-hop environment. If the destination is composed of a set of nodes, multicast routing is required. This issue is beyond the scope of this chapter and will not be considered further. Let us first briefly review how the routing problem can be formulated in an abstract setting. A network is usually modeled as a graph $G = <V,E>$, where $V = \{1,\ldots,n\}$ is the set of nodes (the routers) of the graph and $E \subseteq V \times V$ is the set of edges (the links) [1]. See Fig. 7.1.

When used to model MANETs, elements belonging to the vertex set V of G correspond to mobile stations (assumed fixed in number), and edges belonging to set of edge E correspond to wireless links in the network. A wireless link among two nodes i and j is established when the physical distance $PD(i,j)$ is less than or equal to the transmission radius R (a typical value for R is 250 m). Due to the random node mobility, new wireless links are continuously created and existing ones deleted, leading to a so-called random unit graph [15].

A change in the set E of the graph is called a topological change and is driven by node mobility. Topological changes also occur in wired networks, but in that case they are a consequence of a link fault rather than the normal operation mode of the network.

A path P(S,D) from S to D is specified as a sequence of nodes $P(S,D) \equiv <N_0,N_1,\ldots, N_K>$, where $N_0 = S, N_K = D, N_i \neq N_j, (N_i,N_{i+1}) \in E$ (for $i \neq j, 0 \leq i \leq k-1$). The length of the path, $|P(S,D)|$, is the number of links (hops) a packet performs before reaching the destination node, thus $|<N_0,N_1,\ldots,N_K>| = K$. The term route is also used as a synonym of path. The task of routing can be divided into two sub-tasks:

1. The computation of P(S,D)
2. Data packet forwarding along the route

Paths are calculated by a distributed algorithm running at network level, which includes some form of path maintenance to deal with topological changes. Maintenance can be provided by "sampling" the current network topology with periodic route updates and, optionally, sending updates upon a change in network topology (event-triggered update). In this way, the algorithm should track topological changes and consistently update paths. Also, in the very likely case that many paths exist between S and D, the routing protocol has to calculate the best path P*, with respect to some metric. For example, a widely used metric is the path length; in this case P* is the shortest-path, i.e., $P^*(S,D) = \min_P \{|P(S,D)|\}$.

A node $N_i \neq D$ that receives a packet addressed to D retransmits the packet on the link connected to the next-hop node of P(S,D), N_{i+1}. There are two main techniques to realize this forwarding function. In the first one, each node has access to a local *routing table* (RT), with entries for all nodes in the network, indicating the next-hop node along the path. The next-hop node does not depend on the source node S. As an alternative, each packet stores the full path into its header, so that the next-hop node — N_{i+1} — is learned from the packet itself. Since the path is stored in the packet by the source, this technique is called *source routing*.

Several important structural differences exist between wired and wireless networks that make routing very different in the two environments [4–7]:

1. In a MANET, the rate of topological changes is very high compared to that of wired networks. A conventional routing protocol should continuously be forced to send and receive topology updates to maintain up-to-date routing tables. In an event-triggered Link State protocol, for example, any topological change would trigger a flooding, resulting in a flooding rate equal to the topological change rate. In this scenario, nodes cannot easily preserve their own battery power by switching to "stand-by" or "sleep" operation mode. Moreover, the protocol could not react fast enough to changes. On the other hand, consistent and fast route table updates are mandatory. The possible effects of delay or time skew in table updates are:
   - A packet being routed through a nonoptimal path (an increase in the end-to-end packet delay)
   - A packet loss, due to a temporary inconsistency in routing tables (loops or a broken link along the path)

2. Links may be unidirectional: a node i can receive the signal (and thus a packet) from a node j, but the contrary can be not true (in this case G is a direct graph). As a consequence, a transmission that requires a handshake between i and j fails, while best effort transmissions — such as broadcast — succeed.

3. Several technological limitations on the use of resources exist, namely battery power, transmission bandwidth, and CPU time. The two main consequences of such limitations are:
   A. A blind route update mechanism, either periodic or event-triggered, can waste resources, since updates are sent even when no data transmission at all occurs in the network.
   B. Some suitable metrics that capture these costs need to be defined to compare different paths. For example, selection criteria based only on the path length could not be suitable: the shortest path can have a shorter "live" compared to a longer one and thus trigger more repair activities. The remaining battery lifetime of the nodes of a path can also be an important factor to include in the metric: paths composed of nodes with a high value for the remaining battery lifetime should be preferred to those with a lower value. On the other hand, fairness is also a concern. All nodes should equally contribute to the packet routing effort.

4. Security is challenging in the MANET environment [8–10]. In the absence of any authentication mechanism, due to the nature of radio transmissions, a malicious node can easily corrupt route tables, caches, and other similar information, for example advertising false route information.

When the topological changes are extremely high, little can be done to ensure that routing algorithms converge fast enough to track topological changes, and routing can only be achieved by flooding [2,3]. In the other cases, the design of routing protocols is challenging.

This chapter is organized as follows. Section 7.2 provides a brief review of conventional routing protocols. Section 7.3 describes the main techniques adopted in routing protocols for MANETs. Section 7.4 discusses the issue of protocol performance. Finally, conclusions are given in Section 7.5.

7.2 A Brief Review of Traditional Routing Protocols for Wired Networks

Conventional routing protocols are based on routing tables, which store paths to all possible destinations. To guarantee that routing tables are up to date and reflect the actual network topology, nodes continuously exchange route updates and recalculate the paths.

Such protocols are divided into two "complementary" classes [1]:

- Distance Vector (DV) algorithms, in which a node sends to its neighbors the whole routing table (its distance vector)
- Link State (LS) algorithms, in which a node sends to all the other ones the state of the link with the current neighbors via a reliable flooding

Route updates are sent periodically or when a topological change is detected. Path calculation in DV algorithms is based on a distributed version of the classical Bellman–Ford algorithm (DBF). Every node i maintains a set of distances for each destination j and for each neighbor node k. Let the distance be denoted as H^k_j. The next-hop node of the path from i to j is the node k^* such that $H^{k^*}_j = \min_k(H^k_j)$. Ties are broken arbitrarily. The succession of the next-hop nodes chosen in this manner leads to the shortest path from i to j.

A node running a DV protocol does not know the network topology, i.e., it does not store the graph G. A well-known example of a DV algorithm is RIP (Routing Information Protocol) [16].

In LS algorithms, each node stores the whole network topology and calculates the best path autonomously from the others. Information about the topology is advertised in the form of link-state. The link-state indicates whenever the links with the neighbors are up or down and the cost associated with it.

Due to flooding, each node eventually receives all the link-states from all the other nodes of the network. A node can then build the graph G of the network and can calculate the path from itself to any other node, usually according to the Dijkstra's shortest path algorithm. A well-known example of an LS algorithm is OSPF (Open Shortest Path First) [17]. The following simple result guarantees that routes calculated independently by nodes are consistent.

Lemma: Subpaths of a shortest path are also shortest path [1].

Routing information in the routing table of the nodes is stored in a distributed manner according to a "next-hop" fashion, so that globally they provide a shortest path spanning tree routed at the destination. See Fig 7.2.

DV protocols may react badly to a topology change, since it suffers from very low convergence (count-to-infinity problem) and may create temporary loops. On the contrary, LS converges faster then DV, but it requires a higher overhead [1].

7.3 Unicast Routing Protocols for MANETs

A first attempt to cope with mobility is to use specific techniques aimed at tailoring the conventional protocols to the mobile environment while preserving their nature. For this reason the protocols designed around such techniques are referred to as *table-driven* or *proactive* protocols. The term proactive refers

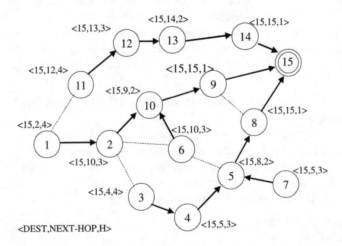

FIGURE 7.2 Route table entries for the destination node 15 showing the Minimum Spanning Tree (MST) obtained from Fig. 7.1

to the ability of the protocol to calculate all possible paths independently of their effective use. Proposals for proactive protocols include pure DV protocols, pure LS protocols, and mixed LS and DV protocols.

A different approach in designing a protocol is to calculate a path when strictly needed for data transmission. If data traffic is not generated by nodes, then the routing activity is totally absent. Protocols of this family are dubbed *reactive* protocols or *on-demand* routing protocols. Reactive protocols are characterized by the elimination of the conventional routing tables at nodes and consequently the need of their updates to track changes in the network topology. The computation of a path boils down to a discovery procedure based on flooding, used as a last resort when previously discovered routes are no longer valid. Proactive and reactive approaches are merged in *hybrid* protocols [11]. Also, reactive features can be added in the proactive framework [12].

Hierarchical routing protocols, which superimpose a logical topology on top of the physical one, are another proposed approach [2]. The hierarchy is a new dimension in the protocol design space, with solutions both in the proactive and reactive realms [13,14]. Finally, the knowledge of the location can help the task of routing. In the location-based routing protocol, the node location is the only criterion adopted in routing packets. For a survey of position-based routing techniques the reader may refer to [15].

7.3.1 Proactive Protocols

The advantage of these protocols is that a path to the destination is immediately available, so no delay is experienced when an application needs to send packets. In some cases this can be useful, as for interactive applications. The main mechanisms adopted in proactive protocols are the following:

- Increasing the amount of topology information stored at each node (to avoid loops and speed up protocol convergence)
- Varying dynamically the size of route updates and/or the update frequency
- Optimizing flooding
- Combining DV and LS features

7.3.1.1 Tailoring Distance Vector Protocols: DSDV

The Destination Sequenced Distance Vector (DSDV) protocol is one of the first attempts to customize the classical DBF protocol [18]. The key elements of DSDV are:

1. An aging mechanism based on monotonically increasing sequence numbers, which indicates the freshness of the route and which is used to avoid routing loops and the count-to-infinity problem
2. The use of full route updates, sent periodically every update interval, or incremental route updates sent on topological changes
3. The delay of route updates for routes that are likely to be unstable, i.e., those for which a new update is on the way towards a node

Entries in the classical routing table are enriched with a sequence number. An odd number indicates a distance equal to infinity and is used for those destinations that become unreachable, while even numbers are used by the destination to stamp route updates.

For each route update for a destination j, the corresponding entry in the routing table is updated with the new received information if the route has a more recent sequence number or if the route has the same sequence number but a better metric. A route entry is deleted from the table if no update has been received for a given number of update intervals.

In DSDV, routes with a metric of ∞ are advertised without a delay, while the others can be delayed according to an average *settling time*. By doing so, it is possible to prevent the advertisement of unstable routes, i.e., those routes that will very likely be replaced by a better route update received in the near future.

This can happen since updates for the same destination can arrive at a node in any order, for example, because they were received from different neighbors.

To calculate the average settling time, a node uses a table that stores, keyed by the first field:

- The destination address
- The last settling time
- The average settling time

Route updates are sent periodically (at an interval of ΔT sec) and incrementally as topological changes.

7.3.1.2 Tailoring Link-State Algorithm: OLSR

The Optimized Link State Routing (OLSR) protocol is an optimization over the classical link state protocol, tailored for mobile ad hoc networks [19,20]. The key idea of OLSR is the use of *multipoint relay* (MPR) nodes to flood the network in an efficient way by reducing duplicate packets in the same region. The protocol also selects bidirectional links for the purpose of routing, so that the problem of packet transfer over unidirectional links is avoided. Each node i selects, independently from the other nodes, a minimal (or near minimal) set of multipoint relay nodes, denoted as MPR(i), from among its one-hop neighbors. The nodes in MPR(i) have the following property: every node in the symmetric two-hops neighborhood of i must have a symmetric link towards MPR(i). In other words, the union of the one-hop neighbor set of MPR(i) contains the whole two-hops neighbor set. See Fig. 7.3. The MPR sets permit flooding to be realized efficiently: when a node i wants to flood a message, it sends the message only to the nodes in MPR(i), which in turn send the message to their MPR nodes and so on.

The *Multipoint Relay Selector* set (MPR selector set) of a node j is composed of the set of neighbors that have selected it as MPR. Each node periodically floods its MPR selector set, using the flooding technique described above, and a special type of control message called a Topology Control (TC) message. Using TC messages, a node announces to the network that it has reachability to the nodes of its MPR selector set (it is its last-hop node). A TC message is stamped with a sequence number, incremented when the MPR selector set changes.

To increase the reaction to topology changes while limiting the protocol overhead, the time interval between two consecutive TC message transmissions can be decreased to a minimum, if a change in the MPR selector is detected. Also, if a node has an empty MPR selection set, it may not generate any TC message. However, when its MPR selection set becomes empty, it should still send empty TC messages for a period of time in order to invalidate the previous TC messages. Information gained from TC messages is used to build the network topology and then the routing tables.

A node running OLSR thus adopts the following three data structures:

- Neighbor sensing information base
- Topology information base
- Routing table

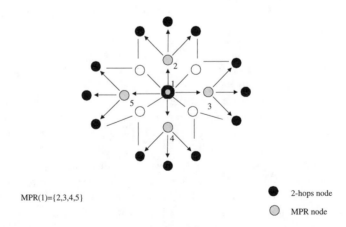

MPR(1)={2,3,4,5}

● 2-hops node
○ MPR node

FIGURE 7.3 An example of flooding using MPR nodes.

7.3.1.3   Merging Distance Vector and Link-State Behaviors: FSR

The Fisheye State Protocol (FSR) is a proactive protocol based on the so-called "fisheye technique," proposed to reduce the size of information required to represent graphical data [21].

The novelties in FSR are:

1. The transmission of link-state packets to neighbors instead of by flooding (a method borrowed from the Global State Routing protocol, GSR [22])
2. The introduction of the notion of scope to define regions of networks with different accuracy in routing information

FSR is similar to link-state protocols as each node maintains a topology table of the network. A node also maintains a route table and a neighbor list.

Unlike in link-state protocols, which flood route updates to the network, in FSR link-state packets are exchanged with neighbors only, while sequence numbers are used to indicate the freshness of the information, as in DSDV. A route update message includes of a sequence of <destination address, neighbors list> pairs. To realize the fisheye technique, FSR introduces the notion of scope. The scope at a node i is defined as the set of nodes that can be reached within h hops from i. See Fig. 7.4.

Routing updates are generated at different rates, with the higher frequency for nodes with the smaller scope. This produces a reduction in the number and size of update messages and thus in the protocol overhead. Basically, FRS is based on a route optimality–route cost tradeoff. It maintains distance and quality information about the neighborhood of a node with progressively less detail as the distance increases. Such an imprecise knowledge of the best path to the destination can lead to the use of some nonoptimal routes; however, the protocol has good potential to scale to large networks, since topological changes occurring in regions located far away from a node do not produce the same amount of traffic as they would in a single-scope proactive protocol. Moreover, when the packet gets closer to the destination, it reaches scopes with increasingly precise knowledge about the destination, and this mitigates the lack of precise route information. It is evident how the scope is a new dimension in the protocol design space.

7.3.1.4   Path-Finding Algorithms: WRP

The Wireless Routing Protocol (WRP) belongs to the general class of Path-Finding Algorithms (PFA) [23–25], defined as the set of distributed shortest-path algorithms that calculate the paths using information regarding the length and second-to-last hop (predecessor) of the shortest path to each destination.

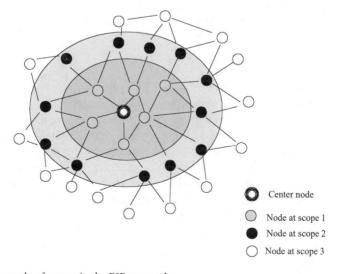

FIGURE 7.4   An example of scopes in the FSR protocol.

PFA algorithms eliminate the count-to-infinity problem of DBF. WRP also addresses the problem of avoiding short-term loops that can still be present in such algorithms for the paths specified by the predecessor node [24]. The protocol uses the following four data structures:

- Distance table
- Routing table
- Link-cost table
- Message Retransmission List (MRL)

The distance table of node i is a matrix containing, for each destination j and each neighbor of i (say k), the distance to j and the predecessor node reported by k.

The routing table of a node i is enriched with the predecessor and the successor of the path to the destination node, as well as a tag that specifies whether the entry corresponds to a simple path, a loop, or a destination that has not been tagged. With this information at hand, WRP is able to avoid the count-to-infinity problem and to highly reduce routing loops by forcing each node to perform consistency checks.

MRL is used to store the identification of those nodes that have not acknowledged a route update. After a timeout, route updates are retransmitted to them. Also, WRP relies on the transmission of HELLO messages to detect the connectivity with neighbors in case no route update is received within a suitable time interval.

7.3.2   Reactive Protocols

A different approach for routing in mobile environments is routing on demand. This approach is characterized by the elimination of the conventional routing tables at nodes and consequently the need of their updates to track changes in the network topology.

On-demand routing protocols calculate a path before data transmission. If data traffic is not generated by nodes, then the routing activity is totally absent. For this reason, they are also called *reactive protocols*.

A reactive protocol is characterized by the following procedures, used to manage paths:

- Path discovery
- Path maintenance
- Path deletion (optional)

Data forwarding is accomplished according to two main techniques:

- Source routing
- Hop-by-hop

The discovery procedure is based upon a query–reply cycle that adopts flooding of queries. The destination is eventually reached by the query and at least a reply is generated. Path discovery is triggered asynchronously on demand when there is a need for the transmission of a data packet and no path to the destination is known. Route discovery is not required for the transmission of every single data packet, since the discovered path is likely to be valid for a period of time that allows many successive transmissions to the same destination.

As a result of a path discovery, the network nodes acquire a new "routing state," which stores the paths learned during the discovery. Routing information is maintained by a maintenance procedure until it is either no longer used or explicitly deleted.

In the following, some representative routing protocols characterized by different maintenance procedures and the use of different techniques to record the routing state are described: route caches, temporary routing tables, and logical structures.

7.3.2.1 DSR

The distinguishing features of the Dynamic Source Routing protocol (DSR) are:

1. Packet forwarding via source routing
2. Aggressive use of route caches that store full paths to destinations [26]

Source routing presents the following advantages:

- It allows packet routing to be trivially loop-free.
- It avoids the need for up-to-date routing information in the intermediate nodes through which packets are forwarded.
- It allows nodes to cache route information by overhearing data packets.

The DSR protocol is composed of two main mechanisms, Route Discovery and Route Maintenance. Route Discovery adopts route request (RREQ)–route reply (RREP) control packets and is triggered by a node S, which attempts to send a packet to a destination node D and does not have a path into its cache. Discovery is based on flooding the network with a RREQ packet, which includes the following fields: the sender address, the target address, a unique number to identify the request, and a route record [27].

On receiving a RREQ control packet, an intermediate node can:

- Reply to S with a RREP if a path to the destination is stored in its cache (in this case the returned path is the concatenation of the path accumulated from S to the node and the path in the cache from the node to D)
- Discard the packet, if already received
- Append its own ID into the route record and broadcast the packet to its neighbors, in the other cases

On receiving a RREQ packet, the destination replies to S with a RREQ packet. An RREQ packet is routed through the path obtained by reversing the route stored in the RREQ's route record (the links are assumed bidirectional) and containing the accumulated path. See Figs. 7.5 and 7.6. An extension of DSR, called RODA, that supports asymmetric links has been also proposed. In this case the destination triggers another route discovery that piggybacks the accumulated route in the RREQ packet [28].

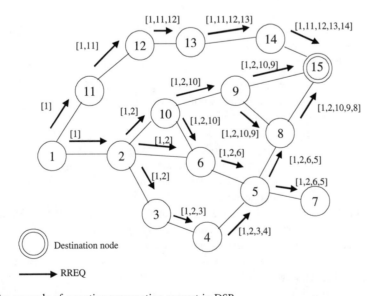

FIGURE 7.5 An example of a routing propagation request in DSR.

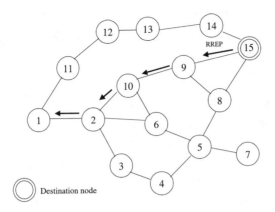

FIGURE 7.6 An example of route reply in DSR.

The base mechanism for route maintenance is as follows. When an intermediate node detects that the link to its next-hop node toward the destination is broken, it removes this link from its route cache and returns a Route Error control message to S. The source S then triggers a new route discovery.

Several improvements to this base mechanism have been proposed; they include:

1. *Promiscuous mode operation* — A node can run its radio interface in promiscuous mode to listen to control packets not addressed to it to gain information useful for cache management.
2. *Salvaging* — Instead of sending an error route message back to the source, an intermediate node can use a path to the destination from its own cache when the link to the next hop is broken.
3. *Random delays* — These can be used when sending RREP to avoid a reply storm [29].

Cache management is quite critical in DSR to ensure good performance. Especially under high load conditions, if stale routes are not immediately removed from the cache, they can be used in the reply to other discovery requests and quickly be cached by other nodes. Several schemes have been proposed [30,31]:

- Fixed lifetime
- Adaptive lifetime
- Negative cache
- Wider error notification

The first case is straightforward: each entry receives the same constant value for its lifetime. However, care must be taken when selecting the lifetime value, since a wrong value can result in performance even worse than that of a DSR with no cache at all.

Adaptive lifetime estimates the value of lifetime for each new entry through a heuristic based on significant events observed (for example, error-route notification, link breakage, etc.).

In the negative cache scheme, a node caches the broken links seen recently via link layer feedback or error route packets. This information disables the writing of any new cache entry that includes reference to those broken links for a given time interval.

The wider notification scheme broadcasts error routes notification to all the source nodes (i.e., the ones that forwarded packets along the broken route) in a tree fashion starting from the point of failure. The mechanism is able to efficiently remove cached routes also from neighbors of a source node.

Simulation results showed that the adaptive lifetime scheme can achieve the same performance as a well-tuned fixed lifetime. However, performance results with a combination of adaptive lifetime and active deletion with wider notification are superior to the ones obtained using deletion schemes separately.

7.3.2.2 AODV

The Ad Hoc On Demand Distance Vector routing (AODV) borrows the use of the sequence number from DSDV to supersede stale cached routes and to prevent loops, while the discovery procedure is derived from the one adopted in DSR [32,33]. The main difference from DSR is that a discovered route is stored locally at nodes rather than included in the packet's header.

The route discovery process is triggered by a node S when it needs to send a packet to a node D for which it has no routing information in its routing table. Route discovery is based on flooding a RREQ packet similarly to DSR.

As a node forwards the packet request, it sets up a reverse path from itself to S by recording the address of the neighbor from which it received the first copy of the RREQ. Similarly, when a RREQ control packet is forwarded toward the destination, a node automatically sets up the reverse path from all nodes back to the source. The other reverse paths are deleted after a timeout period. See Figs. 7.7 and 7.8.

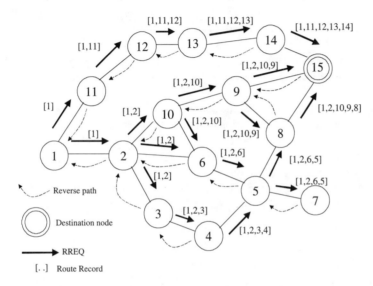

FIGURE 7.7 An example of propagation of RREQ in AODV and reserve path setup.

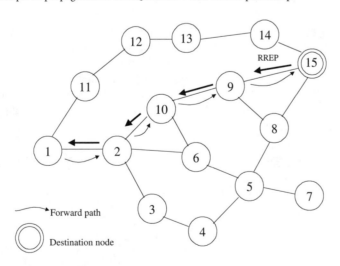

FIGURE 7.8 An example of forward path setup in AODV.

At the end of the discovery phase, as a result of the request–reply packet transmissions, a new routing state is created at nodes. The state is composed of the routing table entries that record the next-hop node to the destination of the forward path. A forward path is deleted if not used within a given route expiration time interval. Each time the route is used, the expiration time interval is reset.

Route maintenance is based on the periodic transmission of HELLO messages. Upon detecting a link-failure, a node sends an Unsolicited Route Reply packet to all its active upstream neighbors invalidating all the routes using the broken link. Those nodes, in turn, relay the packets to their respective upstream nodes so that eventually all active sources are notified. After receiving the Unsolicited Route Reply, the source issues another route request.

7.3.2.3   ADV

The Adaptive Distance Vector (ADV) is a distance vector protocol that maintains routes for *active receivers*, advertising route updates just as with any distance vector algorithm [12]. In other words, the distance vector algorithm is dormant unless activated for particular nodes. Authors classify ADV as a distance vector protocol that exhibits some on-demand characteristics.

In ADV, a node is tagged as an active receiver if it is the destination of any currently active connection. Before any data transmission can take place, the source node notifies, through an *init-connection* control packet flooded to all other nodes in the network, that it needs to open a connection with the destination node.

Similarly, when a connection is closed, the source node broadcasts network-wide an *end-connection* control packet. The destination, in turn, if it has no further active connection, broadcasts network-wide a *non-receive-alert control packet*. As a result of this mechanism, all nodes in the network are able to tag routing table entries with a receiver flag to indicate whenever a destination is an active receiver, i.e., the destination node becomes an active receiver.

ADV also reduces the routing overhead by varying the frequency and the size of routing updates in response to traffic and node mobility.

7.3.2.4   TORA

The Temporally Ordered Routing Algorithm (TORA) belongs to a general family of "link reversal" algorithms [3,34]. TORA is designed to react efficiently to topological changes and to deal with network partitions. The name of the protocol is derived from the assumption of having synchronized clocks (for example via GPS), required in ordering events occurring in the network.

TORA provides routing by exploiting a completely different approach from those described so far. Route optimality is a secondary concern in TORA. The main goal is to find stable routes that can be quickly and locally repaired. The protocol builds a direct acyclic graph (DAG) routed at the desired destination for this purpose. The DAG is obtained by assigning a logical direction to the links, on the basis of a "height" or reference level assigned to nodes. If (i,j) is a direct link of the DAG, i is called the upstream node and j the downstream node. The DAG has the following property: there is only one sink node (the destination), while all other nodes have at least one, but usually many, outgoing links. See Fig. 7.9. The destination node can be reached from a node by following any of its outgoing links. Loops are trivially avoided due to the property of the DAG.

Protocol functioning can be divided into three separated phases: route discovery, route maintenance, and route deletion.

The first phase relies on an exchange of short query–reply control packets. See Fig. 7.10. During the transmission of the reply messages, which is also done by flooding, links receive a logical direction (upstream or downstream) based on their logical relative height, so that a DAG, routed at the destination, is created at the end of the phase. This routing state can be viewed as network of tubes, with water flowing downhill toward the destination node, which has the lowest height in the network.

Route maintenance is activated to maintain the DAG and is based on a finite sequence of "link reversal" operations. A key feature of TORA is that many topological changes may trigger no reaction at all. In fact, if one of the outgoing links of a node breaks, but the node has at least one other downstream node,

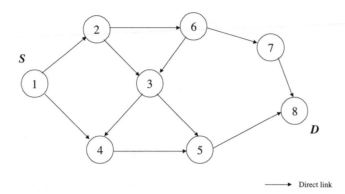

FIGURE 7.9 An example of DAG rooted at node 8.

the destination can still be reachable through another path, and thus no repairing activities are required. See Fig. 7.11.

On the contrary, when a node detects that it has no downstream nodes, it generates a new reference level in order to become a global maximum. The new reference level is propagated into the network, causing partial link reversal for those nodes which, as a result of the new reference level, have lost all routes to the destination. At the end of the repairing activities, localized near the node, the DAG is reestablished. See Fig. 7.12.

TORA is able to detect network partitions. On detecting a network partition, a node floods a clear packet that resets the routing state.

7.3.2.5 ABR

The Associativity Based Routing protocol (ABR) is an on-demand protocol carefully designed to work in mobile environments [35]. The key idea is the use of the longevity of routes, instead of the route length, as the main selection criterion. In ABR, a longer-lived route is preferred to a shorter-lived one, even if the length of the former is less than that of the latter. By doing this, the protocol uses the most stable routes, i.e., the ones that are likely to require the fewest maintenance activities. A similar approach has also been proposed in the Signal Stability Adaptive protocol (SSA) [36]. The estimation of a route lifetime includes the combination of several straightforward measurements (remaining power lifetime, signal strength, etc.) and a new metric, the so-called "associativity" between nodes, which captures the degree of stability of one node with respect to another node over time and space.

The technique proposed to capture associativity is as follows. Each node generates a periodic beacon and counts the beacons received from its neighbors to update their "associativity ticks." The associativity ticks are reset if the beacon signal is not received for a suitable period of time.

The value of the associativity ticks allows classification of a node as having a high or a low mobility state with its neighbors. A high value of a node i with respect to a node j indicates, in fact, that the node i was able to receive many consecutive beacons from j. As such, the protocol assumes that it is very likely for the two nodes to remain close each other. The node then exhibits a low mobility state with respect to j, and the link from i to j is classified as long-lived. In contrast, a low value for the associativity ticks indicates a transiting neighbor and thus a high mobility state.

The protocol consists of three phases: route discovery phase, route reconstruction phase, and route deletion phase. As with other on-demand protocols, route discovery is based on flooding. Intermediate nodes are not allowed to reply to a request. The path is selected by the destination node and is stored as the active path at intermediate nodes in a hop-by-hop fashion.

A source node acquires a new route to a destination node by broadcasting a Broadcast Query (BQ) control packet. As a packet is propagated, intermediate nodes append their own ID and route quality value, which includes the associativity ticks.

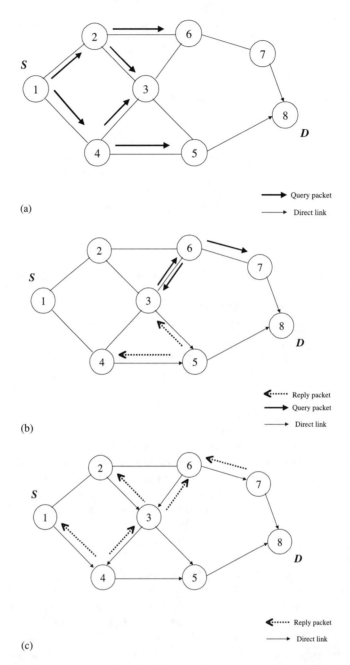

FIGURE 7.10 An example of a query–reply cycle: (a) generation of route request and its first propagation; (b) request propagation and initial reply; (c) reply propagation.

The destination node waits a suitable period of time after receiving the first BQ packet so that it can receive other request packets forwarded along other paths. In this way, the node can select the best path according to the following criteria. If a route is composed of nodes with a high value for the associativity ticks, then the route is preferred to a path with a shorter number of hops. The number of hops is considered only to select among two routes with the same overall associativity degree.

Once a route has been selected, the destination sends a BQ reply control packet (BQ-reply) to the source along the selected route. As the BQ-reply is propagated backwards to the source, intermediate

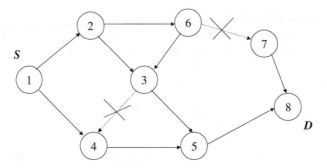

FIGURE 7.11 An example of topological changes that do not require maintenance.

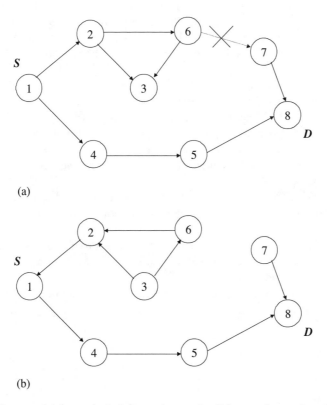

FIGURE 7.12 (a) An example of a topological change that requires link reversal operations; (b) DAG obtained after the link reversal operations.

nodes involved in its retransmission are able to set up a route entry, thus activating a forward path, as already described for AODV.

The route reconstruction phase (RRC) is invoked when the association stability relationship is violated. In some cases, the procedure can try to repair the corrupted subpaths, triggering a new subpath discovery, without notifying the break to the source. In the worst case, a Route Notification (RN) control message is sent to the source and a new discovery phase is initiated. The final phase of ABR is the route deletion phase. This phase consists of flooding the network with a Route Deletion (RD) control packet. As an alternative to such a costly deletion procedure, ABR also proposes a less expensive one, called the soft state approach, in which the route entries are deleted upon time out.

7.3.3 Hybrid Protocols: ZRP

ZRP is a hybrid routing protocol that aims to combine the advantages of both the proactive and reactive approaches, i.e., mainly to reduce the latency necessary to acquire a new route and the protocol overhead [37]. Central to such a protocol is the notion of *zone*. A zone $Z(k,n)$ for a node n with radius k, is defined as the set of nodes at a distance not greater than k hops:

$$Z(k,n) = \{i \mid H(n,i) \le k\},$$

where $H(i,j)$ is the distance in number of hops between node i and node j. The node n is called the *central* node of the routing zone, while node b such that $H(n,b) = k$ is called the *peripheral* node of n. See Fig. 7.13.

The value for k is usually small compared to the network diameter and can be optimized under different scenarios characterized by various mobility and traffic degrees.

The protocol's architecture is organized into four main components: the IntrAzone Routing Protocol (IARP), the IntErzone Routing Protocol (IERP), the Bordercast Protocol (BRP), and a layer-2 Neighbor Discovery/Maintenance Protocol (NDP) [38,39]. The IARP provides routes proactively to those nodes located inside the source's routing zone. It can be based upon any proactive protocol with the difference that route updates are propagated to a distance no greater than k hops. IARP uses NDP to learn about a node's neighbors.

For those nodes located at a distance $k' > k$ from the source, ZRP relies on the IERP to calculate on demand an interzone path. The IERP uses a form of selective flooding to exploit the underlying zone structure generated by the IARP. Specifically, flooding is based on sending query packets only to the peripheral nodes (also called border nodes), using a special kind of multicast transmission, dubbed *bordercast* (see Fig. 7.14). When a node receives the query packet, it can either reply to the source — if D is a member of its routing zone — or bordercast the query packet to its peripheral nodes. Eventually the query packet reaches a node having D as a member of its zone, so that a reply control packet is generated and sent back to the source. A route to D can be accumulated in the query packet during forwarding (as for DSR) or — to reduce the query packet length — in the reply control packet during the reply phase.

Packet forwarding along an interzone path adopts a modified source routing. A routing path contains only the border nodes that have to be traversed. Forwarding along border nodes is table driven, since the distance between border nodes is k.

Since there is no coordination among nodes, zones heavily overlap. A node can be a member as well as a border node of many zones. In this way, the basic search mechanism can perform even worse than

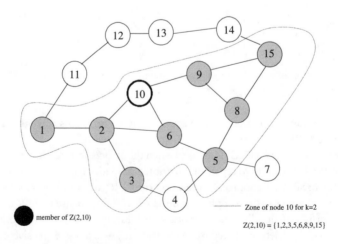

FIGURE 7.13 An example of a zone in ZRP.

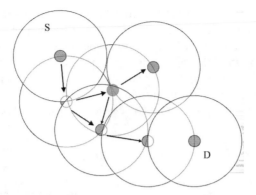

FIGURE 7.14 A sketch of query propagation using bordercast.

a standard flooding. The literature provides several solutions to deal with this problem by stopping and controlling redundant query threads.

Route maintenance is responsible for maintaining interzone paths. The use of a local repair procedure aiming at repairing the broken link by a mini-path search is also suggested to reduce the need of global route discovery. An improvement over ZRP — the Distributed Dynamic Routing algorithm (DDR) — has also been proposed [40]. The protocol is based on the construction of a forest of nonoverlapping dynamic zones.

7.3.4 Position Aided Protocols: LAR

The novelty in the Location Aided Routing protocol (LAR) is the estimation of the position of the destination node, used to increase the efficiency of the discovery procedure [41,42]. Only a subset of nodes is queried during the discovery phase, specifically those in the so-called "request zone" near the estimate. LAR uses standard flooding as a last resort when no estimations are available. The position can be obtained, for example, from the GPS.

The request zone is calculated from the "expected zone," which is defined as the zone where the destination node D should be located from the viewpoint of the source node S at the current time t_1, given that S had known its position at some time t_0 in the past and its average velocity.

The request zone is defined as the smallest rectangle that includes the current location of S and the expected zone. See Fig. 7.15. In the LAR scheme 1, the request zone (i.e., its four corners) is included in the route discovery packet. A node forwards a route discovery packet only if it belongs to the request zone and it is receiving the packet for the first time.

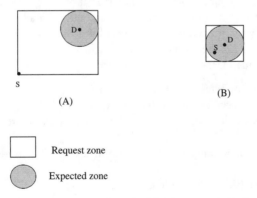

FIGURE 7.15 Examples of request and routing zones. (A) S outside the D's expected zone; (B) S inside the D's expected zone.

The destination D replies with a packet that contains its current position, the local time and — optionally — its current speed or its average value over some time interval.

In the LAR scheme 2, the discovery packet contains (1) the distance *Dist* from the node that is sending the packet to the destination and (2) the position (x,y) of the destination. Let D_i be the distance from a node *i* to the destination. When the source node sends the packet it sets $Dist = D_s$. When a node *i* forwards the packet it sets $Dist = D_i$ before the retransmission.

A node *j* that receives a request packet from a node *i* forwards the packet only if it was received for the first time and

$$Dist + \delta > D_j$$

i.e., if *j* is "at most δ farther" from (x,y) than node *i*. The value δ can be set to a value greater than zero to take into account location errors or to trade off the probability of finding a route on the first attempt with the cost of finding the route. Simulation results have shown that using location information results in significant lower routing overhead, as compared to algorithms that do not use location information.

7.4   Protocol Performance Analysis

A protocol performance analysis aims to measure two complementary aspects. The first is how efficiently the routing service is provided wrt a given set of cost indexes, e.g., bandwidth, battery energy consumption, CPU cycles; the second aspect concerns application-oriented metrics, i.e., the performance seen externally by an application in using the protocol.

To evaluate these two aspects, protocol complexity analysis and protocol simulation have been used. The aim of the complexity analysis is to measure — in a simplified and abstract setting — the resources required by the protocol in performing a single "protocol operation," for example, the reaction to a link breakage. More specifically:

1. Computation complexity is the number of computation steps executed by a node to perform a protocol operation.
2. Space complexity is the storage space required by a node to store routing information.
3. Communication complexity is the number of packets exchanged among nodes in performing a single protocol operation. The values are obtained as function of some network parameter (i.e., the number of nodes and/or the diameter of the network, etc.) and usually assumes the worst case protocol's behavior and synchronous execution (i.e., all nodes are able to execute single computation steps at fixed points in time) [3,24].

Complexity analysis is not able to capture all protocol performance issues, such as the measurement of other relevant application-oriented metrics and their relationships with many important factors — such as the mobility pattern, communication efficiency, and offered load.

Such a study is normally carried out through protocol simulations, as realistic analytical models are not feasible due to the large number of factors to be included. Protocol simulations aim to estimate several metrics, which include [43]:

1. Throughput (number of bytes delivered per second)
2. Packet delivery ratio (packets delivered/packets generated)
3. End-to-end data packet delay (time interval from when a packet is ready for the transmission to its delivery)
4. Routing overhead (number of control bits transmitted/data bit delivered; or control packets/data packets delivered)
5. Route discovery time (time required to compute a new route)
6. Route optimality (cost of the path used/cost of the optimal one)
7. Number of out-of-order packets
8. Power consumption (energy consumed per delivered bit or packet)

The first three metrics are the ones relevant for an application. The others provide insights on the efficiency of the routing service. It is important to note that when measured at TCP level, the throughput is usually very poor — despite the excellent values measured for connectionless traffic. This is due to the frequent temporary link failures and route changes that trigger many false congestion detections. The design of suitable TCP protocols for MANETs is an active research topic [44–47].

Two main factors influence the above metrics: the rate of topological changes (also called *mobility*) and the rate of packet transmission (*offered load*). By scanning the protocol performance in the mobility–offered load space, it is possible to determinate the domain where the protocol is suited to work.

A fair comparison of routing protocols through simulation, under realistic conditions, is not a trivial task. Simulations have to be carried out under a wide set of different scenarios and must use realistic and validated models for many physical phenomena, such as radio propagation and interference, 802.11 medium access control, etc. Widely used simulation tools include ns-2, developed at the University of California at Berkeley, with wireless extensions provided by the CMU Monarch Project [48], and GloMoSim developed at the University of California at Los Angeles [49]. Detailed and homogeneous simulation studies have been carried out for the following protocols: DSDV, DSR, ADV, AODV, and TORA. The main conclusions are given below; for an in-depth discussion of the results, the reader may refer to the original papers [50–52]. To summarize briefly:

The DSDV protocol is suitable only for low mobility. For a high value of mobility, the protocol fails to converge and thus tables contain stale routes. However, for low values of offered load, the periodic route update results in a high value for the normalized routing overhead.

The TORA protocol is suitable for low mobility with low-to-medium offered load. Performance is negatively affected by the existence of short-lived routing loops, raising during link reversal, and by the proactive nature of route maintenance (paths are monitored and repaired even if not used).

Both AODV and DSR are suitable under high mobility and offered load. However, AODV is suitable also when traffic diversity (number of active connections) increases, a condition with which DSR is not able to cope.

Simulation results also showed the benefit of combining both the proactive and reactive techniques. ADV outperforms AODV and DSR in terms of higher throughput (up to 50% in high mobility), lower packet delays, less routing and control overhead packets [12].

7.5 Conclusions

Many protocols have been proposed in the literature to provide routing in a mobile environment. In this chapter, we have illustrated the main techniques proposed to deal with mobility in the case of unicast flat routing protocols. Tables 7.1 and 7.2 summarize the main characteristics of each protocol described in this paper.

The experience gained in wired networks has influenced the design of mobile ad hoc routing protocols in many ways. However, wireless mobile networks introduce several concerns that make the routing issue completely different from the one provided for the fixed counterpart. For example, the limitations on the battery lifetime and transmission bandwidth, as well as the notion of route stability, lead to a new notion of "best" route. Also, fairness is an important property of a protocol, since always using the same route results in battery drain-out for the samemobiles.

Besides the above considerations, other more evident desirable properties include:

1. Loop avoidance
2. Fast converge upon link changes
3. Localized reaction to topology changes
4. Multiple routes information
5. Unidirectional link support
6. Quality of Service (QoS) support
7. Network partition support

TABLE 7.1   Comparison of On-Demand Protocols

Protocol		ABR	AODV	DSR	LAR	TORA	ZRP
Route discovery	Query	Flooding	Flooding	Flooding	Selective flooding	Flooding	Selective flooding
	Storing mechanism	Hop-by-hop	Hop-by-hop	Accumulation	No	Hop-by-hop	Hop-by-hop/accumulation
	Reply transmission	Unicast	Unicast	Unicast	Flooding	Flooding	Unicast
	Intermediate node reply	No	Yes	Yes	No	No	Yes
	Multiple routes	No	No	Yes	No	Yes	Yes
Route maintenance	Proactive/reactive	Proactive (beacon)	Proactive (hello)	Reactive	Reactive	Proactive (hello)	Proactive (IARP)
	Local partial repair	Yes	No	Yes	No	Yes	Yes
	Route deletion	Hard/soft	Soft	Soft	Soft	Hard	Soft
Route state		Route table	Route table	Route cache	Cache at source	DAG	Route tables

TABLE 7.2   Comparison of Reactive Protocols

		ADV	DSDV	FISHEYE	OLSR	WRP
Classification		DV	DV	Hybrid DV and LS	LS	PFA
Route update	Main content	<D,NH,H,SN>	<D,H,NH,SN>	<D,NL,SN>*	MPR set	<D,H,NH,PR,SN>
	Target nodes	Neighbors	Neighbors	Neighbors	All	Neighbors
	Transmission	Periodic/on event	Periodic/on event	Periodic *	Periodic	Periodic/on event
Maintain information			Forwarding table	Topology table	Topology table	Distance table
		Route Table	Setting table time	Neighbor list	Neighbor list	Routing table
				Routing table	Routing table	Link cost table
						Mess. retransm. List

* = Advertised destinations and frequency depend on the scope

D = destination; NH = next hop node; H = number of hops; NL = neighbor list; SN = sequence number; PR = predecessor node; DV = distance vector; LS = link state; PFA = path finding algorithm.

As a final remark, it is evident that while many interesting ideas have been proposed, still other efforts are required to fully understand how all the needs can be satisfied in a cost-effective way.

References

1. A.S. Tanenbaum, *Computer Networks*, 3rd ed., Prentice Hall, Englewood Cliffs, NJ, 1996.
2. C.-C. Chiang, G. Pei, M. Gerla, and T.-W. Chen, Scalable routing strategies for ad hoc wireless networks, *IEEE Journal on Selected Areas in Communications*, 17, 1369–1379, 1999.
3. M.S. Corson and A. Ephremides, A distributed routing algorithm for mobile wireless networks, *ACM/Baltzer Wireless Networks*, 1, 61–81, 1995

4. D.B. Johnson, Routing in Ad Hoc Networks of Mobile Hosts, *Proceedings of the IEEE Workshop on Mobile Computing Systems and Applications (WMCSA)*, IEEE Computer Society, Santa Cruz, CA, Dec. 1994, pp. 158–163.

5. C.E. Perkins, Ed., *Ad Hoc Networking*, Addison-Wesley, Reading, MA, 2000.

6. C.-K. Toh, *Ad Hoc Mobile Wireless Networks*, Prentice Hall PTR, Upper Saddle River, NJ, 2002.

7. R. Prakash, Unidirectional Links Prove Costly in Wireless Ad-Hoc Networks, *Proceedings of the 3rd International Workshop on Discrete Algorithms and Methods for Mobile Computing and Communications — Dial M '99*, Seattle, Aug. 20, 1998, pp. 15–22.

8. F. Stajno and R.J. Anderson, The Resurrecting Duckling: Security Issues for Ad Hoc Wireless Networks, *Proceedings of Security Protocols 17th International Workshop*, vol. 1796 of Lecture Notes in Computer Science, Cambridge, U.K., April 19–21, 1999, pp. 172–194.

9. P. Papadimitratos and Z.J. Haas, Secure Routing for Mobile Ad Hoc Networks, *SCS Communication Networks and Distributed Systems Modeling and Simulation Conference (CNDS 2002)*, San Antonio, TX, Jan. 27–31, 2002.

10. L. Zhou and Z.J. Haas, Securing Ad Hoc Networks, *IEEE Network Magazine*, Vol. 13, no. 6, Nov./Dec. 1999.

11. Z.J. Haas and M.R. Pearlman, The Zone Routing Protocol (ZRP) for Ad Hoc Networks, Internet Draft, draft-ietf-manet-zone-zrp-02.txt, June 1999.

12. R.V. Boppanam and S.P. Konduru, An Adaptive Distance Vector Routing Algorithm for Mobile Ad Hoc Networks, *Proceedings of IEEE INFOCOM 2001*, Anchorage, AK, April 22–26, Vol. 3, 2001, pp. 1753–1762.

13. C.-C. Chiang, H.-K. Wu, W. Liu, and M. Gerla, Routing in Clustered Multihop, Mobile Wireless Networks with Fading Channel, *IEEE Singapore International Conference on Networks* (SICON 97), Kent Ridge, Singapore, April 1997, pp. 197–211.

14. R. Sivakumar, P. Sinha, and V. Bharghavan, CEDAR: a core-extraction distributed ad hoc routing algorithm, *IEEE Journal on Selected Areas in Communications*, 17, 1454–1465, 1999.

15. S. Giordano, I. Stojmenovic, and L. Blazevic, http://www.site.uottawa.ca/~ivan/routing-survey.pdf.

16. C. Hedrick, The Routing Information Protocol, RFC 1058, June 1988.

17. J. Moy, OSPF version 2, RFC 1247, July 1991.

18. C.E. Perkins and P. Bhagwat, Highly dynamic destination-sequenced distance-vector routing (DSDV) for mobile computers, *Computer Communications Review*, Vol. 24, no. 4, Oct. 1994, pp. 234–244.

19. T. Clausen et al., Optimized Link State Routing Protocol, IETF MANET Working Group Internet Draft, draft-ietf-MANET-olsrr-06.txt, Sep. 2001.

20. A. Qayyum, L. Viennot, and A. Laouiti, Multipoint Relaying: An Efficient Technique for Flooding in Mobile Wireless Networks, *35th Annual Hawaii International Conference on System Sciences (HICSS '2001)*, Maui, 2001.

21. M. Gerla, X. Hong, and G. Pei, Fisheye State Routing Protocol (FSR) for Ad Hoc Networks, draft-ietf-MANET-fsr -02.txt, IETF MANET Working Group — Internet Draft, Dec. 2001.

22. T.-W. Chen and M. Gerla, Global State Routing: A New Routing Scheme for Ad-hoc Wireless Networks, *Proceedings of IEEE ICC*, Atlanta GA, June 8–11, 1998, pp. 171–175.

23. P.A. Humblet, Another adaptive shortest-path algorithm, *Trans. Commun.*, 39, 995–1003, 1991.

24. S. Murthy and J.J. Garcia-Luna-Aceves, An Efficient Routing Protocol for Wireless Networks, *ACM Mobile Networks and Applications Journal*, Oct. 1996.

25. C. Cheng, R. Reley, S.P.R. Kumar, and J.J. Garcia-Luna-Aceves, A loop-free extended Bellman–Ford routing protocol without bouncing effect, *ACM Computer Communications Review*, 19(4), 224–236, 1989.

26. D.B. Johnson and D.A. Maltz, Dynamic source routing in ad hoc wireless networks, in *Mobile Computing*, T. Imielinski and H. Korth, Eds., Kluwer Academic Publishers, Dordrecht, 1996, pp. 153–181.

27. D.B. Jhonson, D.A. Maltz, and Y.-C. Hu, The Dynamic Source Routing Protocol for Mobile Ad Hoc Networks, draft-ietf-manet-dsr-07.txt, IETF MANET working group, Feb. 2002.

28. D.-K. Kim, C.-K. Toh, and Y.-H. Choi, RODA: A New Dynamic Routing Protocol Using Dual Paths to Support Asymmetric Links in Mobile Ad Hoc Networks, *Proceedings of IEEE IC3N*, October 2000.

29. S.Y. Ni, Y.C. Tseng, Y.S Chen, and J.P. Sheu, The Broadcast Storm Problem in Mobile Ad Hoc Networks, *Proceedings of ACM/IEEE MobiCom 99*, Seattle, WA, Aug. 15–20. 1999, pp. 151–162.

30. Y.-C. Hu and D.B. Johnson, Caching Strategies in On-demand Routing Protocols for Wireless Ad Hoc Networks, *Proceedings of IEEE/ACM MobiCom 00*, Aug. 2000, Boston, MA, pp. 231–242.

31. M.K. Marina and S.R. Das, Performance of Route Cache Strategies in Dynamic Source Routing, *Proceedings of ICDCS-2001*, Scottsdale, AZ, Apr. 2001.

32. C.E. Perkins and E.M. Royer, Ad Hoc On Demand Distance Vector Routing, *Proceedings of the 2nd IEEE Workshop on Mobile Computing Systems and Applications*, New Orleans, Feb. 1999, pp. 90–100.

33. C.E. Perkins, E.M. Belding-Royer, and S.R. Das, Ad Hoc On-Demand Distance Vector (AODV) Routing, draft-ietf-manet-aodv-10.txt, IETF MANET working group, Jan. 2002.

34. V.D. Park and M.S. Corson, A Highly Adaptive Distributed Routing Algorithm for Mobile Wireless Networks, *Proceedings of IEEE INFOCOM '97*, Kobe, Japan, Apr. 1997.

35. C.K. Toh, Associativity based routing for ad hoc mobile networks, *Wireless Personal Communications*, Vol. 4, no. 2, Mar. 1997, pp. 1–36.

36. R. Dube, C. Rais, K. Wang, and S. Tripathi, Signal stability based adaptive routing (SSA) for ad hoc mobile network, *IEEE Personal Communications*, 4, 36–45, 1997.

37. M.R. Pearlman and Z.J. Haas, Determining the optimal configuration for the zone routing protocol, *IEEE Journal on Selected Areas in Communications*, 17, 1395–1414, 1999.

38. Z J. Haas, M.R. Pearlman, and P. Samar, The Interzone Routing Protocol (IERP) for Ad Hoc Networks, draft-ietf-manet-zone-ierp-01.txt, IETF MANET working group, Dec. 2001.

39. Z J. Haas, M.R. Pearlman, and P Samar, The Intrazone Routing Protocol (IARP) for Ad Hoc Network, draft-ietf-manet-zone-iarp-01.txt, IETF MANET working group, Dec. 2001.

40. N. Nikaein, H. Labiod, and C. Bonnet, Distributed Dynamic Routing Algorithm for Mobile Ad Hoc Networks, *Proceedings of ACM MobiHoc 2000*, 2000, Boston, MA, pp. 19–27.

41. Y.-B. Ko and N.H. Vaidya, Location-Aided Routing (LAR) in mobile ad hoc networks, *ACM/IEEE MobiCom 98*, Dallas, 1998.

42. Y-B. Ko and N.H. Vaidya, Location-Aided Routing (LAR) in mobile ad hoc networks, in *ACM/ Baltzer Wireless Networks (WINET) Journal*, Vol. 6–4, 2000.

43. S. Corson and J. Macker, Mobile Ad Hoc Networking (MANET): Routing Protocol Performance Issues and Evaluation Considerations, RFC2501, Request for Comments, Jan. 1999.

44. G. Holland and N.H. Vaidya, Analysis of TCP performance over mobile ad hoc networks, *Proceedings of ACM/IEEE MobiCom '99*, Seattle, Aug. 15–20, 1999, pp. 219–230.

45. M. Gerla, K. Tang, and R. Bagrodia, TCP Performance in Wireless Multi-hop Networks, in *Proceedings of IEEE Workshop on Mobile Computing Systems and Applications (WACSA)*, Feb. 25–26, 1999, pp. 41–50.

46. D. Kim, C.-K. Toh, and Y. Choi, TCP BuS: Improving TCP performance in wireless ad hoc networks, *Journal of Communication and Networks*, 3, 2001.

47. F. Wang and Y. Zhang, Improving TCP Performance over Mobile Ad-Hoc Networks with Out-of-Order Detection and Response, in *Proceedings of ACM MobiHoc 02*, Lausanne, Switzerland, June 2002.

48. http://www.monarch.cs.cmu.edu/cmu-ns.html.

49. http://may.cs.ucla.edu/projects/glomosim/.

50. J. Broch, D.A. Maltz, D.B. Johnson, Y.-C. Hu, and J. Jetcheva, A Performance Comparison of Multi-Hop Wireless Ad Hoc Network Routing Protocols, *Proceedings of the Fourth Annual ACM/IEEE International Conference on Mobile Computing and Networking*, Dallas, Oct. 1998.

51. S.R. Das, C.E. Perkins, and E.M. Royer, Performance Comparison of Two On-Demand Routing Protocols for Ad Hoc Networks, *IEEE INFOCOM*, vol. 1, Mar. 2000, pp. 3–12.

52. S.-J. Lee and M. Gerla, A Simulation Study of Table-Driven and On-Demand Routing Protocols for Mobile Ad Hoc Networks, *IEEE Network*, Vol. 3, no. 4, Jul./Aug. 1999, pp. 48–54.

8

Satellite
Communications

Matthew N.O. Sadiku
Prairie View A&M University

Abstract

Wireless communication is undergoing explosive growth, and satellite-based delivery is a major player. With the introduction of satellite personal communication services in the near future, an important step will be made toward the implementation of a global communication infrastructure.

Satellite communications were first deployed in the 1960s and have their roots in military applications. Since the 1965 launch of the Early Bird satellite (the first commercial communication satellite) by the U.S. National Aeronautics and Space Administration (NASA) proved the effectiveness of satellite communications, satellites have played an important role in both domestic and international communication networks. They have brought voice, video, and data communications to areas of the world that are not accessible with terrestrial lines. By extending communications to the remotest parts of the world, satellites have helped to allow virtually everyone to be part of the global economy.

Communication by satellite is not a replacement for the existing terrestrial systems but rather an extension of wireless systems. However, satellite communication has the following merits over terrestrial communication:

- *Coverage:* Satellites can cover a much larger geographical area than traditional ground-based systems can. Satellites have the unique ability to cover the globe.
- *High bandwidth:* A Ka-band (27–40 GHz) can deliver throughput of gigabits per second rate.
- *Low cost:* A satellite communications system is relatively inexpensive because there are no cable-laying costs and one satellite covers a large area.
- *Wireless communication:* Users can enjoy untethered mobile communication anywhere within the satellite coverage area.
- *Simple topology:* Satellite networks have simpler topology, which results in more manageable network performance.
- *Broadcast/multicast:* Satellites are naturally attractive for broadcast/multicast applications.

TABLE 8.1   Advantages and disadvantages of satellites

Advantages	Disadvantages
Wide-area coverage	Propagation delay
Easy access to remote sites	Dependency on a remote facility
Costs independent of distance	Less control over transmission
Low error rates	Attenuation due to atmospheric particles (e.g., rain) severe at high frequencies
Adaptable to changing network patterns	Continual time-of-use charges
No right-of-way necessary; earth stations located at premises	Reduced transmission during solar equinox

Source: D.J. Marihart, *IEEE Transactions on Power Delivery,* 16, 181–188, 2001.

- *Maintenance:* A typical satellite is designed to be unattended, requiring only minimal attention by customer personnel.
- *Immunity:* A satellite system will not suffer from disasters such as floods, fire, and earthquakes and will therefore be available as an emergency service should terrestrial services be knocked out.

Of course, satellite systems do have some disadvantages. These are weighed against their advantages in Table 8.1 [1]. Some of the services provided by satellites include fixed satellite service (FSS), mobile satellite service (MSS), broadcasting satellite service (BSS), navigational satellite service, and meteorological satellite service.

This chapter explores the integration of satellites with terrestrial networks to meet the demands of highly mobile communities. After looking at the fundamentals of satellite communication, we will discuss its various applications

8.1   Fundamentals

A satellite communications system may be viewed as consisting of two parts: the space and ground segments. The space segment consists of the satellites and all their on-board tracking and control systems. The ground segment comprises the earth terminals, their associated equipment, and the links to terrestrial networks [2].

8.1.1   Types of Satellites

There were only 150 satellites in orbit by September 1997. The number is expected to be roughly 1700 by the year 2002. With this increasing trend in the number of satellites, there is a need to categorize them according to the height of their orbits and their "footprint" or coverage on the earth's surface. They are classified as follows [3]:

Geostationary Earth Orbit (GEO) Satellites

These satellites are launched into a geostationary or geosynchronous orbit, 35,786 km above the equator. (Raising a satellite to such an altitude, however, required a rocket, so the achievement of a GEO satellite did not take place until 1963.) A satellite is said to be in geostationary orbit when the satellite is matched to the rotation of the earth at the equator. A GEO satellite can cover nearly one-third of the earth's surface, i.e., it takes three GEO satellites to provide global coverage. Due to their large coverage, GEO satellites are ideal for broadcasting and international communications. (GEO are sometimes referred to as high earth orbits [HEO]). Examples of GEO satellite constellations are Spaceway, designed by Boeing Satellite Systems, and Astrolink, by Lockheed Martin. Another example is Thuraya, designed by Boeing Satellite Systems to provide mobile satellite services to the Middle East and surrounding areas.

There are at least three major objections to GEO satellites [4]:

1. The propagation delay (or latency) between the instant a signal is transmitted and when it returns to earth is relatively long (about 240 milliseconds). This is caused by speed-of-light transmission delay and signal processing delay. This may not be a problem if the signal is going only one way. However, for signals such as data and voice, which go in both directions, the delay can cause problems. GEO satellites, therefore, are less attractive for voice communication.
2. Coverage is lacking at far northern and southern latitudes. This is unavoidable because a GEO satellite is below the horizon and may not provide coverage at latitudes as close to the equator as 45°. Unfortunately, many of the European capitals, including London, Paris, Berlin, Warsaw, and Moscow, are north of this latitude.
3. Both the mobile unit and the satellite of a GEO system require a high transmit power.

In spite of these objections, the majority of satellites in operation today are GEO satellites, but that may change in the near future.

Middle Earth Orbit (MEO) Satellites

These satellites orbit the earth at 5000 to 12,000 km. GEO satellites do not provide good coverage for places far north, and satellites in inclined elliptical orbits are an alternative. Although the lower orbit reduces propagation delay to only 60 to 140 milliseconds round trip, it takes 12 MEO satellites to cover most of the planet. MEO systems represent a compromise between Low Earth Orbit (LEO) and GEO systems, balancing the advantages and disadvantages of each. (MEO are sometimes referred to as intermediate circular orbit [ICO] satellites).

Low Earth Orbit (LEO) Satellites

These satellites circle the earth at 500 to 3000 km. For example, the Echo satellite circled the earth every 90 minutes. To provide global coverage may require as many as 200 LEO satellites. Latency in a LEO system is comparable with terrestrial fiber optics, usually less than 30 milliseconds round trip. LEO satellites are suitable for personal communication service (PCS). However, LEO systems have a shorter life space of 5–8 years (compared with 12–15 years for GEO systems) due to the increased amount of radiation in low earth orbit. LEO systems have been grouped as Little LEO and Big LEO. The Little LEOs have less capacity and are limited to nonvoice services such as data and message transmission. An example is OrbComm designed by Orbital Corporation, which consists of 36 satellites, each weighing 85 pounds. The Big LEOs have larger capacity and voice transmission capability. An example is Loral and Qualcomm's Globalstar, which will operate in the L-band frequencies and employ 48 satellites organized in eight planes of six satellites each.

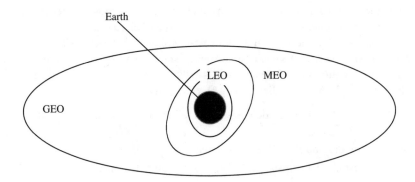

FIGURE 8.1   The three common types of satellites — GEO, MEO, and LEO.

The arrangement of the three basic types of satellites is shown in Fig. 8.1. Both MEO and LEO satellites are regarded as non-GEO satellites. As seen from the earth station, the GEO satellite never appears to move any significant distance. As seen from the satellite, the earth station never appears to move. The MEO and LEO satellites, on the other hand, as seen from the ground, are continuously moving. Likewise, the earth station, as seen from the satellites, is a moving target. GEO systems are well suited for the delivery of broadcast services, while LEO systems are very efficient for the delivery of interactive services. LEO satellites tend to be smaller, lighter, and cheaper than MEO, which in turn are likely to be less expensive than GEO. The evolution from GEO to MEO and LEO satellites has resulted in a variety of global satellite systems. The convenience of GEO was weighed against the practical difficulty involved with it and the inherent technical advantages of LEO, such as lower delay and higher angles of elevation. While it has been conceded that GEO is in many respects theoretically preferable, LEO or MEO systems would be preferred for many applications. Although a constellation (a group of satellites) is required instead of only one for hemispheric coverage, the loss of individual satellites would cause only gradual degradation of the system rather than a catastrophic failure.

Several attractive services can be offered by utilizing the small satellite technology. These include:

- PCS on a global scale
- Digital audio broadcasting (DAB)
- Environmental data collection and distribution
- Remote sensing/earth observation
- Several military applications

When planning a stand-alone satellite or constellation satellite system, the following geometric parameters should be considered:

- Number of satellites
- Number of orbital planes
- Inclination angle of each orbital plane
- Relative spacing between orbital planes
- Number of satellites in each orbital plane
- Relative spacing between orbital planes
- Relative phasing of satellites between adjacent planes
- Satellite altitude and elevation angle for ground users

8.1.2   Frequency Bands

Every nation of the world has the right to access the satellite orbit, and no nation has a permanent right or priority to use any particular orbit location. Without a means for nations to coordinate use of satellite frequency bands, the satellite services of one nation could interfere with those of another, thereby creating a chaotic situation in which neither country's signals could be received clearly.

To facilitate satellite communications and eliminate interference between different systems, international standards govern the use of satellite frequencies. The International Telecommunication Union (ITU) is responsible for allocating frequencies to satellite services. As far as frequency allocation is concerned, the entire world has been divided into three regions:

- *Region 1:* Europe, Africa, the Middle East, Mongolia, and the Asian region of the former Soviet Union
- *Region 2:* North and South America and Greenland
- *Region 3:* The remainder of Asia, Australia, and the southwest Pacific

TABLE 8.2   Satellite Frequency Allocations

Frequency Band	Range (GHz)
L	1–2
S	2–4
C	4–8
X	8–12
Ku	12–18
K	18–27
Ka	27–40

Since the spectrum is a limited resource, the ITU has reassigned the same parts of the spectrum to many nations and for many purposes throughout the world.

The frequency spectrum allocations for satellite services are given in Table 8.2. Notice that the assigned segment is the 1–40 GHz frequency range, which is the microwave portion of the spectrum. As microwaves, the signals between the satellites and the earth stations travel along line-of-sight paths and experience free-space loss that increases as the square of the distance.

Satellite services are classified into 17 categories [5]: fixed, intersatellite, mobile, land mobile, maritime mobile, aeronautical mobile, broadcasting, earth exploration, space research, meteorological, space operation, amateur, radiodetermination, radionavigation, maritime radionavigation, and standard frequency and time signal. The Ku band is presently used for broadcasting services and also for certain fixed satellite services. The C band is exclusively for fixed satellite services, and no broadcasting is allowed. The L band is employed by mobile satellite services and navigation systems.

A satellite band is divided into separation portions: one for earth-to-space links (the uplink) and one for space-to-earth links (the downlink). Separate frequencies are assigned for sending to the satellite (the uplink) and receiving from the satellite (the downlink). Table 8.3 provides the general frequency assignments for uplink and downlink satellite frequencies. We notice from the table that the uplink frequency bands are slightly higher than the corresponding downlink frequency bands. This is to take advantage of the fact that it is easier to generate RF power within a ground station than it is onboard a satellite. In order to direct the uplink transmission to a specific satellite, the uplink radio beams are highly focused. In the same way, the downlink transmission is focused on a particular footprint or area of coverage

All satellite systems are constrained to operate in designed frequency bands depending on the kind of earth station used and service provided. The satellite industry, particularly in the U.S., is subject to several regulatory requirements, domestically and internationally, depending upon which radio services and frequency bands are proposed to be used on the satellite. In the U.S., the Federal Communications Commission (FCC) is the independent regulatory agency that ensures that the limited orbital/spectrum resource allocated to space radiocommunications services is used efficiently. After receiving an application for a U.S. domestic satellite, FCC initiates the advance publication process for a U.S. satellite. This is to ensure the availability of an orbit position when the satellite is authorized. FCC does not guarantee international recognition and protection of satellite systems unless the authorized satellite operator complies with all coordination requirements and completes the necessary coordination of its satellites with all other administrations whose satellites are affected [6].

TABLE 8.3   Typical Uplink and Downlink Satellite Frequencies (GHz)

Uplink Frequencies	Downlink Frequencies
5.925–6.425	3.700–4.200
7.900–8.400	7.250–7.750
14.00–14.50	11.70–12.20
27.50–30.00	17.70–20.20

8.1.3 Multiple Access Technologies

Multiple access allows different users to utilize a satellite's resources of power and bandwidth without interfering with each other. Satellite communication systems use different types of multiple access technology including frequency division multiple access (FDMA), time division multiple access (TDMA), and code division multiple access (CDMA). The access technology can vary between the uplink and downlink channels.

The ability of multiple earth stations or users to access the same channel is known as frequency division multiple access (FDMA). In FDMA, each user signal is assigned a specific frequency channel. One disadvantage of FDMA is that once a frequency is assigned to a user, the frequency cannot be adjusted easily or rapidly to other users when it is idle. The potential for interference from adjacent channels is another major shortcoming.

In time division multiple access (TDMA), each user signal is allotted a time slot. A time slot is allocated for each periodic transmission from the sender to a receiver. The entire bandwidth (frequency) is available during the time slot. This access scheme provides priority to users with more traffic to transmit by assigning those users more time slots than it assigns to low-priority users. Satellite providers will extend the capability and will employ multiple frequencies (MF-TDMA). If there are N frequencies, each offering M Mb/sec of bandwidth, then the total available bandwidth during a time slot is $N \times M$ Mb/sec. Although FDMA techniques are more commonly employed in satellite communications systems, TDMA techniques are more complex and are increasingly becoming the de facto standard.

In code division multiple access (CDMA), users occupy the same bandwidth but use spread spectrum signals with orthogonal signaling codes. This technique increases the channel bandwidth of the signal and makes it less vulnerable to interference. CDMA operates in three modes: direct sequence (DS), frequency hopping (FH), and time hopping (TH).

8.1.4 Basic Satellite Components

Every satellite communication involves the transmission of information from a ground station to the satellite (the uplink), followed by a retransmission of the information from the satellite back to the earth (the downlink). Hence, the satellite must typically have a receiver antenna, a receiver, a transmitter antenna, a transmitter, some mechanism for connecting the uplink with the downlink, and a power source to run the electronic system. These components are illustrated in Fig. 8.2 and explained as follows:

- *Transmitters:* The amount of power that a satellite transmitter is required to send out depends on whether it is GEO or LEO satellite. The GEO satellite is about 100 times farther away than the LEO satellite. Thus, a GEO satellite would need 10,000 times as much power as a LEO satellite. Fortunately, other parameters can be adjusted to reduce this amount of power.

- *Antennas:* The antennas dominate the appearance of a communication satellite. The antenna design is one of the more difficult and challenging parts of a communication satellite project. The antenna geometry is constrained physically by the design and the satellite topology. A major difference between GEO and LEO satellites is their antennas. Since all the receivers are located in the coverage area, which is relatively small, a properly designed antenna can focus most of the transmitter power within that area. The easiest way to achieve this is simply to make the antenna larger. This is one of the ways the GEO satellite makes up for the apparently larger transmitter power it requires.

- *Power generation:* The satellite must generate all of its own power. The power is often generated by large solar cells, which convert sunlight into electricity. Since there is a limit to how large the solar panel can be, there is also a practical limit to the amount of power that can be generated. Satellites must also be prepared for periods of eclipse, when the earth is between the sun and the satellite. This necessitates having batteries on board that can supply power during eclipse periods and recharge later.

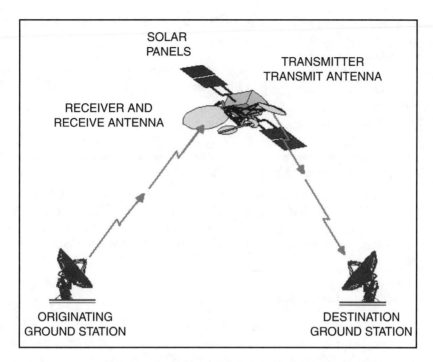

FIGURE 8.2 Basic components of a communication satellite link. (With permission of Regis Leonard, NASA Lewis Research Center.)

- *Transponders:* These are the devices each satellite must carry. They receive radio signals at one frequency, amplify them, and convert them to another frequency for transmission, as shown typically in Fig. 8.3. For example, a GEO satellite may have 24 transponders with each assigned a pair of frequencies (uplink and downlink frequencies).

8.1.5 Effects of Space

Space has two major effects on satellite communications:

1. The space environment, with radiation, rain, and space debris, is harsh on satellites. The satellite payload, which is responsible for the satellite communication functions, is expected to be simple and robust. Traditional satellites, especially GEOs, serve as bent pipes and act as repeaters between

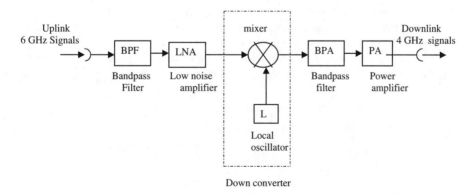

FIGURE 8.3 A simplified block diagram of a typical transponder.

communication points on the ground. There is no onboard processing (OBP). However, new satellites allow OBP, including decoding/recoding, demodulation/remodulation, transponder, beam switching, and routing [7].

2. The second effect is that of wave propagation. Light-of-sight and attenuation due to atmospheric particles (rain, ice, dust, snow, fog, etc.) are not significant at L-, S-, and C-bands. Above 10 GHz, the main propagation effects are [8,9]:

   • *Tropospheric propagation effects:* attenuation by rain and clouds, scintillation, and depolarization.

   • *Effects of the environment on mobile terminals:* shadowing, blockage, and multipath caused by objects in the surroundings of the terminal antenna.

The troposphere can produce significant signal degradation at the Ku-, Ka-, and V-bands, particularly at lower elevation angles. Most satellite systems are expected to operate at an elevation angle above roughly 20°. Rain constitutes the most fundamental obstacle encountered in the design of satellite communication systems at frequencies above 10 GHz. The resultant loss of signal power makes for unreliable transmission.

8.2   Orbital Characteristics

Since a satellite is a spacecraft that orbits the earth, an intuitive question to ask is, "What keeps objects in orbit?" The answer to the question is found in the orbital mechanical laws governing satellite motion. Satellite orbits are essentially elliptical and obey the same laws of Johannes Kepler that govern the motion of planets around the sun. Kepler's three laws are stated as follows [10]:

   • *First law:* The orbit of each planet follows an elliptical path in space with the sun serving as the focus.

   • *Second law:* The line linking a planet with the sum sweeps out equal areas in equal time.

   • *Third law:* The square of the period of a planet is proportional to the cube of its mean distance from the sun.

Besides these laws, Newton's law of gravitation states that any two bodies attract each other with a force proportional to the product of their masses and inversely proportional to the square of the distance between them, i.e.

$$\mathbf{F} = -\frac{GMm}{r^2}\mathbf{a}_r \tag{8.1}$$

where M is the one body (earth), m is the mass of other body (satellite), $\mathbf{F}$ is the force on m due to M, r is the distance between the two bodies, $\mathbf{a}_r = \mathbf{r}/r$ is a unit vector along the displacement vector $\mathbf{r}$, and $G = 6.672 \times 10^{-11}$ Nm/kg^2 is the universal gravitational constant. If M is the mass of the earth, the product $GM = \mu = 3.99 \times 10^{14}$ m^3/sec^2 is known as Kepler's constant.

Kepler's laws, in conjunction with Newton's laws, can be used to completely describe the motion of the planets around the sun or that of a satellite around the earth. Newton's second law can written as

$$\mathbf{F} = -m\frac{d^2r}{dt^2}\mathbf{a}_r \tag{8.2}$$

Equating this with the force between the earth and the satellite in Eq. (8.1) gives

$$\frac{d^2r}{dt^2}\mathbf{a}_r = -\frac{\mu}{r^2}\mathbf{a}_r \tag{8.3}$$

or

$$\ddot{r} + \frac{\mu}{r^3}\mathbf{r} = 0 \tag{8.4}$$

where is the vector acceleration. The solution to the vector second-order differential equation, Eq. (8.4), is not simple, but it can be shown that the resulting trajectory is in the form of an ellipse given by [10,11]

$$r = \frac{p}{1 + e\cos\theta} \tag{8.5}$$

where r is the distance between the geocenter and any point on the trajectory, p is a geometric constant, e ($0 \le e < 1$) is the eccentricity of the ellipse, and θ (known as the true anomaly) is the polar angle between r and the point on the ellipse nearest to the focus. These orbital parameters are illustrated in Fig. 8.4. The point on the orbit where the satellite is closest to the earth is known as the *perigee*, while the point where the satellite is farthest from the earth is known as the *apogee*. The fact that the orbit is an ellipse confirms Kepler's first law. If a and b are the semimajor and semiminor axes (see Fig. 8.4), then

$$b = a\sqrt{(1 - e^2)} \tag{8.6}$$

$$p = a(1 - e^2) \tag{8.7}$$

Thus, the distance between a satellite and the geocenter is given by

$$r = \frac{a(1 - e^2)}{1 + e\cos\theta} \tag{8.10}$$

Note that the orbit becomes a circular orbit when e = 0.

The apogee height and perigee height are often required. From the geometry of the ellipse, the magnitudes of the radius vectors at apogee and perigee can be obtained as:

$$r_a = a(1 + e) \tag{8.8}$$

$$r_p = a(1 - e) \tag{8.9}$$

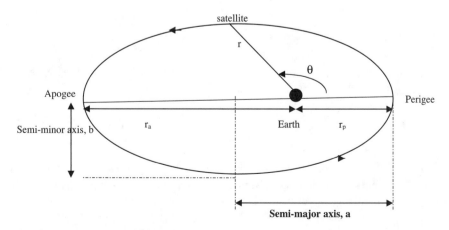

FIGURE 8.4 Orbital parameters.

To find the apogee and perigee heights, the radius of the earth must be subtracted from the radii lengths. The period T of a satellite is related to its semimajor axis a using Kepler's third law as

$$T = 2\pi \sqrt{\frac{a^3}{\mu}} \tag{8.10}$$

For a circular orbit to have a period equal to that of the earth's rotation (a sidereal day 23 hours, 56 minutes, 4.09 seconds), an altitude of 35,803 km is required. In this equatorial plane, the satellite is "geostationary."

The velocity of a satellite in an elliptic orbit is obtained as

$$v^2 = \mu \left(\frac{2}{r} - \frac{1}{a} \right) \tag{8.11}$$

For a synchronous orbit (T = 24 h), r = a = 42,230 km, and v = 3074 m/sec or 11,070 km/h.

A constellation is a group of satellites. The total number N of satellites in a constellation depends on the earth central angle γ and is given by [12]

$$N \approx \frac{4\sqrt{3}}{9} \left(\frac{\pi}{\gamma} \right)^2 \tag{8.12}$$

To determine the amount of power received on the ground due to satellite transmission, we consider the power density

$$\psi = \frac{P_t}{S} \tag{8.13}$$

where P_t is the power transmitted and S is the terrestrial area covered by the satellite. The value of P_t is a major requirement of the spacecraft. The coverage area is given by

$$S = 2\pi R^2 (1 - \cos \gamma) \tag{8.14}$$

where R = 6,378 km is the radius of the earth. S is usually divided into a cellular pattern of spot beams, thereby enabling frequency reuse. The effective area of the receiving antenna is a measure of the ability of the antenna to extract energy from the passing electromagnetic wave and is given by

$$A_e = G_r \frac{\lambda^2}{4\pi} \tag{8.15}$$

where G_r is the gain of the receiving antenna and λ is the wavelength. The power received is the product of the power density and the effective area. Thus,

$$P_r = \psi A_e = \frac{G_r \lambda^2}{4\pi S} P_t \tag{8.16}$$

This is known as Friis equation, relating the power received by one antenna to the power transmitted by the other. We first notice from this equation that for a given transmitted power P_t, the received power P_r is maximized by minimizing the coverage area S. Second, mobile terminals prefer having nondirectional antennas, thereby making their gain G_r fixed. Therefore, to maximize P_r encourages using as long a wavelength as possible, i.e., as low a frequency as practicable within regulatory and technical constraints.

The noise density N_0 is given by

$$N_0 = kT_0 \tag{8.17}$$

where $k = 1.38 \times 10^{-23}$ Wsec/K is Boltzmann's constant and T_0 is the equivalent system temperature, which is defined to include antenna noise and thermal noise generated at the receiver. Shannon's classical capacity theorem for the maximum error-free transmission rate in bits per second (b/sec) over a noisy power-limited and bandwidth-limited channel is

$$C = B \log_2 \left(1 + \frac{P_r}{N_0} \right) \tag{8.18}$$

where B is the bandwidth and C is the channel capacity.

8.3 Applications

Satellite communication services are uniquely suited for many applications involving wide area coverage. Satellites provide the key ingredient in the development of broadband communications and information processing infrastructure. Here, we consider five major applications of satellite communications: the use of very small aperture terminals (VSATs) for business applications; fixed satellite service (FSS), which interconnects fixed points; and mobile satellite (MSAT) service (MSS), which employs satellites to extend cellular networks to mobile vehicles, satellite radio, and satellite-based Internet.

8.3.1 VSAT Networks

A very small aperture terminal (VSAT) is a dish antenna that receives signals from a satellite. (The dish antenna has a diameter that is typically in the range of 1.2 to 2.8 m, but the trend is toward smaller dishes, not more than 1.5 m in diameter.) A VSAT may also be regarded as a complete earth station that can be installed on the user's premises and provide communication services in conjunction with a larger (typically 6–9 m) earth station acting as a network management center (NMC), as illustrated in Fig. 8.5.

VSAT networks arose in the mid-1980s as a result of electronic and software innovations that allowed all the required features to be contained in an affordable package about the size of a personal computer (PC). VSAT technology brings features and benefits of satellite communications down to an economical and usable form. VSAT networks have become mainstream networking solutions for long distance, low density voice and data communications because they are affordable for both small and large companies.

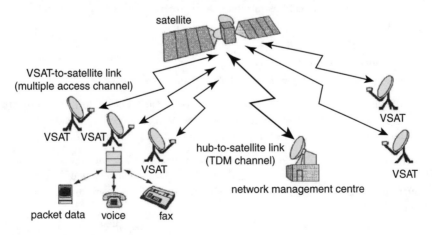

FIGURE 8.5 A typical VSAT network.

Other benefits and advantages of VSAT technology include lower operating costs, ease of installation and maintenance, ability to manage multiple protocols, and ability to bring locations where the cost of leased lines is very expensive into the communication loop.

Satellite links can support interactive data applications through two types of architectures [3,13]: mesh topology (also called point-to-point connectivity) and star topology (also known as point-to-multipoint connectivity). Single-hop communications between remote VSATs can be achieved by full-mesh connectivity. Although the mesh and star configurations have different technical requirements, it is possible to integrate the two if necessary.

The star network employs a hub station. The hub consists of an RF terminal, a set of baseband equipment, and network equipment. A VSAT network can provide transmission rates of up to 64 kb/sec. As is common with star networks, all communication must past through the hub. That is, all communication is between a remote node and the hub; no direct node-to-node information transfer is allowed in this topology. This type of network is highly coordinated and can be very efficient. The point-to-multipoint architecture is very common in modern satellite data networks and is responsible for the success of the current VSAT. Its simple mode of communication makes it useful for businesses; the hub station is located at the company headquarters while the remote VSATs are located at the branches. Most VSAT star networks employ the TDMA access method for the inroute (from VSAT to hub) and TDM outroute (from the hub to the VSAT).

A mesh network is more versatile than star network because it allows any-to-any communications. Also, the star network can provide transmission rates up to 64 kb/sec per remote terminal, whereas the mesh network can have its data rates increased to 2 Mb/sec or more. Mesh topology was used by the first satellite networks to be implemented. With time, there was a decline in the use of this topology but it remains an effective means of transferring information with least delay. Mesh topology applies to either temporary connections or dedicated links to connect two earth stations. All full-duplex point-to-point connectivities are possible and provided, as typical of a mesh configuration. If there are N nodes, the number of connections is equal to the permutation of $N(N - 1)/2$. Mesh networks are implemented at C and Ku bands. The transmission rate ranges from about 64 kb/sec to 2.048 Mb/sec (E1 speed). Users have implemented 45 Mb/sec.

Transponder resources are assigned to VSATs on the network using transponder access protocols. Three such protocols are frequency division multiple access (FDMA), time division multiple access (TDMA), and code division multiple access (CDMA). FDMA is the most popular access method because it allows the use of comparatively low-power VSAT terminals. TDMA is not efficient for low-density uplink traffic from the VSAT, which is mostly data transfer of bursty nature.

Although TV used to dominate domestic satellite communications, the transmission of data has grown tremendously with the advent of VSAT. VSAT technology has the following advantages [14]:

- Ability to extend communications to remote areas where provision of terrestrial facilities can be very expensive
- Ability to configure or reconfigure a network quickly
- Ability to manage multiple protocols
- Single-vendor support for equipment installation and maintenance

The major drawbacks of VSAT networks are the high costs, the tendency toward optimizing systems for large networks (say with 500 VSATs), and lack of direct VSAT-to-VSAT connectivity.

In spite of these drawbacks, VSAT technology has found applications in the following areas [2,15]:

- *PC integration with VSAT:* A PC serves as a direct user interface with the VSAT in applications where online information delivery is required. The most popular service of this kind is DirecPC, an offering of hardware, software, and satellite service delivery from Hughes Network Systems (HNS).
- *Integrating LANs with VSAT networks:* A VSAT star network can create an efficient WAN environment for interconnection of legacy LANs. The functionality provided is similar to that of LAN-

to-LAN interconnection using bridges, routers, and gateways. For example, LANAdvantage is a specific implementation of LAN connectivity with a VSAT star network by HNS.

- *Television broadcasting:* Much of the popularity of direct-broadcast satellite (DBS) can be attributed to the quality of TV communications provided by digital technology. For example, DirecTV provides personalized TV through programming capabilities. It was the first high-power DBS service to deliver up to 175 channels of digital-quality programming.
- *Others:* VSATs are mainly being employed as a replacement for terrestrial data networks using analog private lines in various industries such as teleconferencing, training, retailing, insurance, credit card checking, reservation systems, interactive inventory data sharing, automobile sales and distribution, banking, travel reservation, lodging, and finance. For example, Walmart employed VSATs to extend its reach to thousands of remote towns in rural America. Chrysler Corporation has provided every one of its 6000 U.S. dealers with a VSAT to be used for order entry. Space will not permit mentioning similar applications by General Electric, Holiday Inn, Toyota of America, etc.

Several types of VSAT networks are now in operation, both domestically and internationally. There were over 1000 VSATs in operation at the beginning of 1992. Today, there are over 100,000 two-way Ku band VSATs installed in the U.S. and over 300,000 worldwide. Almost all of these VSATs are designed primarily to provide data for private corporate networks, and almost all two-way data networks with more than 20 earth stations are based on some variation of an ALOHA protocol for access [16,17]. The price of a VSAT started at around $20,000 and dropped to around $6000 in 1996.

8.3.2 Fixed Satellite Service

Several commercial satellite applications are through earth stations at fixed locations on the ground. The international designation for such an arrangement is *fixed satellite service* (FSS). The FSS provides communication service between two or more fixed points on earth, as opposed to mobile satellite services (MSS) (to be discussed later), which provide communication for two moving terminals. Although ITU defined FSS as a space radiocommunication service covering all types of satellite transmissions between given fixed points, the borderline between FSS and Broadcasting Satellite Service (BSS) for satellite television is becoming more and more blurred [18]. FSS applies to systems that interconnect fixed points such as international telephone exchanges. It involves GEO satellites providing 24 hour per day service.

Table 8.4 shows the WARC (World Administrative Radio Conference) frequency allocations for FSS. The table only gives a general idea and is in no way comprehensive.

The FSS shares frequency bands with terrestrial networks in the 6/4 and 14/12 GHz bands. Thus, it is possible that a terrestrial network could affect a satellite on the uplink or that a terrestrial network may be affected by the downlink from a satellite.

As exemplified by Intelsat, FSS has been the most successful part of commercial satellite communications. Early applications were point-to-point telephony and major trunking uses. Current applications

TABLE 8.4   Frequency Allocations for FSS (Below ~30 GHz)

Downlinks (in GHz)	Uplinks (in GHz)
3.4–4.2 and 4.5–4.8	5.725–7.075
7.25–7.75	7.9–8.4
10.7–11.7	
11.7–12.2 (Region 2 only)	12.75 13.25 and 14.0–14.5
12.5–12.75 (Region 1 only)	
17.7–21.2	27.5–31.0

of the FSS can be classified according to frequency (from about 3 MHz to above 30 GHz), the lowest frequency being the high frequency (HF) band. They include:

- *High frequency service:* The (HF) bands have been crowded due to the fact that this is the only technology that could provide very long range coverage with a minimum investment in infrastructure. Since HF signals are reflected back to earth by the ionosphere, they can travel long distances. Because of the long-range capability using inexpensive equipment, HF is valuable for many long-range fixed applications. However, HF communication users must consider the constantly changing nature of the ionosphere, high levels of ambient noise, interference, and the need for relatively large antennas. The DOD and many other federal agencies use HF fixed service to support priority communications after natural disasters such as earthquakes and hurricanes or to maintain many HF links to overseas bases, embassies, and offices. Many private industries use HF links to communicate with their foreign offices.

- *Private fixed services:* These microwave services are licensed by the FCC. They include services operated by organizations mainly to carry signals for their own purposes. Major users include private companies, utilities, transportation providers, and local and state governments.

- *Auxiliary broadcasting (AUXBC) services:* These include applications that support the TV and AM/FM broadcasting industry. Electronic news gathering (ENG) uses transportable microwave links to provide live coverage of events. AM and FM stations may use such services for temporary live coverage of events as well as for studio-to-transmitter links.

- *Cable relay service (CARS):* This supports the TV industry. CARS and AUXBC use ENG in identical ways to provide temporary real time coverage of events outside the studio.

- *Federal government fixed services:* The U.S. federal government uses fixed services for many functions. Federal civilian agencies such as the Department of the Interior and Department of Agriculture use microwave networks to support mobile radiocommunication sites in federally controlled remote areas such as national parks and national forests. The Departments of Justice and Treasury maintain extensive urban radio nets to support national law enforcement and security. Numerous federal agencies make extensive use of fixed services to communicate with a wide range of sensors that keep track of airways, weather, stream flow, etc. Military operations and training make extensive use of fixed satellite microwave terminals to support range safety and security, relay data received from airborne platforms to central control sites, provide closed circuit TV for safety, provide radar tracking and air traffic control information, and support a wide range of logistics and administrative support activities. NASA's Advanced Communications Technology Satellite (ACTS) is a major development for FSS. Launched in September 1993, ACTS is designed to verify new FSS technologies.

Although the telecommunications industry as a whole is growing rapidly, the FSS industry is not. The market trend is toward the replacement of long-haul microwave systems with fiber. Fiber provides much greater capacity than microwaves.

8.3.3 Mobile Satellite Service

There is the need for global cellular service in all geographical regions of the world. The terrestrial cellular systems serve urban areas well; they are not economical for rural or remote areas where the population or tele-density is low. Mobile satellite (MSAT) systems can complement the existing terrestrial cellular network by extending communication coverage from urban to rural areas. Mobile satellite services (MSS) are not limited to land coverage but include marine and aeronautical services [5,19]. Thus, the coverage of mobile satellites is based on geographical and not on population coverage as in terrestrial cellular systems and could be global.

MSAT or satellite-based PCS/PCN is being developed in light of the terrestrial constraints. The low cost of installation makes satellite-based PCS simple and practical. The American Mobile Satellite Cor-

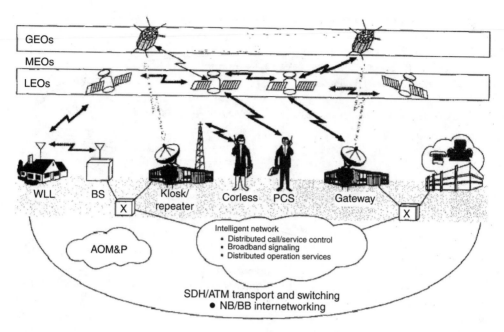

FIGURE 8.6 MSAT concept.

poration (AMSC) and Telesat Mobile of Canada are designing a geosynchronous MSAT to provide PCS to North America. The concept of MSAT is illustrated in Fig. 8.6.

Satellite communication among mobile earth stations is different from cellular communication. First, the cells move very rapidly over the earth, and the mobile units, for all practical purposes, appear stationary — a kind of inverted cellular telephone system. Second, due to different designs, use of a handheld is limited to the geographical coverage of a specific satellite constellation, and roaming of handheld equipment between different satellite systems will not be allowed. With personal communication systems (PCS), there will be a mix of broad types of cell sizes: picocell for low power indoor applications, microcell for lower power outdoor pedestrian applications; macrocell for high power vehicular applications; and supermacrocell with satellites, as shown in Fig. 8.7. For example, a microcell of a PCS has a radius of 1 to 300 meters.

There are two types of constellation design approaches to satellite-based PCS. One approach is provide coverage using three GEO satellites at approximately 36,000 km above the equator. The other approach involves using the LEO and MEO satellites at approximately 500 to 1500 km above the earth's surface. Thus, MSS are identified as either GEO or nongeostationary orbit (NGSO) satellites. The LEO and MEO satellites provide lower attenuation to the uplink and downlink signals in addition to lower signal delays because they operate at a lower altitude than the GEO satellites. Therefore, the NGSO satellites are emerging as major players in the world of wireless and personal communications [20].

The main purpose of MSAT or MSS is to provide data and/or voice services into a fixed or portable personal terminal, close to the size of today's terrestrial cellular phones, by means of interconnection via satellite. LEO and MEO satellites have been proposed as an efficient way to communicate with these handheld devices. The signals from the handheld devices are retransmitted via a satellite to a gateway (a fixed earth station), which routes the signals through the public switched telephone network (PSTN) to their final destination or to another handheld device.

Satellite systems designed for personal communications include the Iridium, Globalstar, and ICO systems [21–24]. All are global systems covering everywhere on earth. Each of these is characterized by two key elements: a constellation of non-geosynchronous satellites (LEO or MEO) arranged in multiple planes as shown in Fig. 8.8, and a handheld terminal (handset) for accessing PCS.

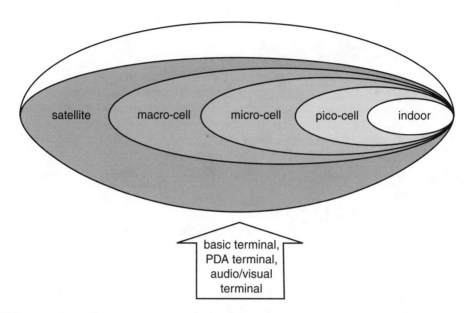

FIGURE 8.7   Various cell sizes.

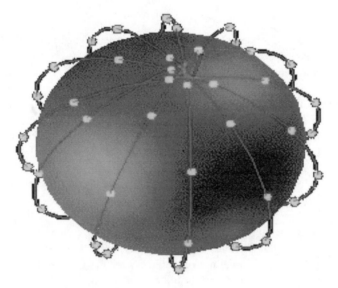

FIGURE 8.8   A typical satellite constellation.

Iridium

Mobile satellite service (MSS) had a turning point in 1992 when Motorola introduced the concept of a LEO satellite system capable of directly serving handheld terminals. Motorola based the design of its big LEO constellation, known as Iridium, on the classical work of Adams and Ryder [25]. Iridium (www.iridium.com), which began in 1990, is the first mobile satellite telephone network to offer voice and data services to and from handheld telephones anywhere in the world. It uses a network of intersatellite switches for global coverage and GSM-type technology to link mobile units to the satellite network. The Iridium system has three antenna arrays, each producing 16 cellular spot beams (or cells or rather megacells), and the beams are juxtaposed within the coverage area. The 48 spot beams travel very rapidly along the surface of the earth, and the mobile units will appear stationary.

Several modifications have been made to the original idea, including reducing the number of satellites from 77 to 66 by eliminating one orbital plane. (The name Iridium was based on the fact that Iridium is the element in the periodic table with an atom with 77 electrons.) Some of the key features of the current Iridium satellite constellation are [26–29]:

- Number of (LEO) satellites: 66 (each weighing 700 kg or 1500 lb)
- Number of orbital planes: 6 (separated by 31.6° around the equator)
- Number of active satellites per plane: 11 (uniformly spaced, with one spare satellite per plane at 130 km lower in the orbital plane)
- Altitude of orbits: 780 km (or 421.5 nmi)
- Inclination: 86.4°
- Period of revolution: 100 minutes
- Design life: 8 years

Satellites in planes 1, 3, and 5 cross the equator in synchronization, while satellites in planes 2, 4, and 6 also cross in synchronization but out of phase with those in planes 1, 3, and 5. Collision avoidance is built into the orbital planning such that the closest distance between two satellites is 223 km. Each satellite covers a circular area roughly the size of the U.S. with a diameter of about 4400 km. The coverage area is divided into 48 cells. The satellites can project 48 spot beams using the L-band frequency assignments. Each of the spot beams is roughly 372 nautical miles in diameter.

The Iridium network uses FDMA/TDMA to produce efficient use of the spectrum; it provides voice at 4.8 kb/sec and data at 2.4 kb/sec. With FDMA, the available spectrum is subdivided into smaller bands assigned to individual users. Iridium extends this multiple access scheme further by using TDMA within each FDMA sub-band. Each user is assigned two time slots (one for sending and the other for receiving) within a repetitive time frame. During each time slot, the digital data are burst between the mobile handset and the satellite. The total spectrum of 5.15 MHz is divided into 120 FDMA channels with each satellite having about 1100 channels. Within each FDMA channel, there are four TDMA slots in each direction (uplink and downlink). Each TDMA slot has length 8.29 msec in a 90 msec frame. The coded data burst rate with QPSK modulation and raised cosine filtering is 50 kb/sec. All this is designed to use less spectrum, keep channels closer without undue interference, and allow for acceptable levels of inter-modulation.

The Iridium handsets are built by Motorola and Kyocera, a Japanese manufacturer of cellular telephones. The handsets will permit both satellite access and terrestrial cellular roaming capability. Paging options are also available. The price of a typical handset is around $3000. The satellite service is about $3.00 per minute, which is about 25 percent more than the normal cellular roaming rate. The expected break-even market for Iridium is about 600,000 customers globally.

Outside the U.S., Iridium must obtain access rights in each country where service is provided. Altogether, Iridium is seeking access to some 200 nations through a negotiating process. In spite of some problems expected of a complex system, Iridium is already at work. Its 66 LEO satellites have been fully commercial as of November 1, 1998. But on August 13, 1999, Iridium filed for bankruptcy and was later bought by Iridium Satellite LCC. Vendors competing with Iridium include Aries, Ellipso, Globalstar, and ICO.

Globalstar

The second system is Globalstar (www.globalstar.com), which is a satellite-based cellular telephone system that allows users to talk from any place in the world. It will serve as an extension of terrestrial systems worldwide except for the polar regions. The constellation is capable of serving up to 30 million subscribers.

Globalstar is being developed by the limited partnership of Loral Aerospace Corporation and Qualcomm with ten strategic partners. A functional overview of Globalstar is presented in Fig. 8.9. The key elements are [30–32]:

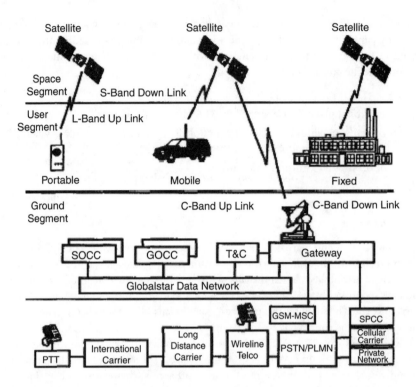

FIGURE 8.9   Globalstar system architecture.

- *Space segment:* It comprises a constellation of 48 active LEO satellites located at an altitude of 1414 km and equally divided in eight planes (six satellites per plane). The satellite orbits are circular and are inclined at 52° with respect to the equator. Each satellite illuminates the earth at 1.6 GHz L-band and 2.5 GHz S-band with 16 fixed beams with service links, assignable over 13 FDM channels.
- *User segment:* This includes mobile and fixed users.
- *Ground segment:* This consists of gateways (large ground station), ground operations control center (GOCC), satellite operations control center (SOCO), and Globalstar data network (GDN). The gateway enables communications to and from handheld user terminals (UTs), relayed via satellite, with Public Switched Telephone Network (PSTN). A gateway with a single radio channel transmits on a single frequency. The GOCC allocates capacity among gateways and collects operational control and billing information. The SOCC provides satellite maintenance and directs orbital maneuvers of the satellites.

The Globalstar satellites employ "bent pipe" transponders with the feeder link at C-band. Each satellite weighs about 704 pounds and has a capacity of 2800 full-duplex circuits. It covers the earth with only 16 spot beams. Access to and from the satellite is at L- and S-band, respectively, using CDMA in channels that are 1.25 MHz in bandwidth. The Qualcomm CDMA waveform employed in the user terminals and gateways spreads RF energy evenly over the allotted spectrum.

The Globalstar system offers the following services:

- One phone for both cellular and satellite calls
- Data and fax transmission
- Global roaming

- Short messaging service (SMS)
- Internet access

Since Globalstar plans to serve the military with commercial subscriptions, it employs signal encryption for protection from unauthorized calling parties. Unlike Iridium, which offers a global service, Globalstar's business plan calls for franchising its use to partners in different countries.

ICO

In Europe, Inmarsat in 1997 spun off a commercial organization known as Intermediate Circular Orbit (ICO) Global. The motivation was to use MEO rather than GEO as initially used by Inmarsat, thereby reducing the link propagation loss. The ICO system (originally called Inmarsat-P) was built by Hughes Space and Communications (now Boeing Satellite Systems). The ICO constellation is made up of [33–35]:

- Ten operational MEO satellites with five in each of the two inclined circular orbits at an altitude of 10,355 km
- One spare satellite in each plane, making 12 total launched
- A total of 163 spot beams on each satellite
- An integrated C- and S-band payload to be carried on each satellite
- Twelve Satellite Access Nodes (SANs) located globally

The inclination of the orbits is 45° — making this the lowest of the systems described. Although this reduces the coverage at high latitudes, it allows for the smallest number of satellites. The configuration is designed to provide coverage of the entire surface of the earth at all times and to maximize the path diversity of the system. ICO differs from Iridium and Globalstar in that it adopts a TDMA scheme for the service links. Each satellite is designed to support at least 4500 telephone channels using TDMA.

The ICO system (www.ico.com) is designed to provide the following services:

- Global paging
- Personal navigation
- Personal voice, data, and fax

The satellites will be linked on a terrestrial network known as the ICONET, which interconnects 12 ground stations, referred to as Satellite Access Nodes (SANs). Each SAN consists of earth stations with multiple antennas for communicating with satellites, associated switching equipment, and a database to support mobility management. The orbital pattern of ICO is designed for significant coverage overlap to ensure that usually two (but sometimes three or four) satellites will be in view of a user and a SAN at any time. Each satellite will cover roughly 25 percent of the earth's surface at any given time. The satellites will communicate with the terrestrial network through the ICONET, a high-bandwidth global Internet Protocol (IP) network. The ICONET will include a system for managing global user mobility based upon the existing digital cellular standard, GSM. The lifespan of ICO constellation is approximately 12 years. ICO is in the process of obtaining approvals from national and international regulatory authorities covering operation of the system and service provision. The three constellations are compared in Table 8.5.

Applications of MSS include:

- *International travelers:* Satellites will provide services to people on the move. There is virtually no limit to the number of services that can be provided to the traveler, whether it be by land, sea, or air. The only possible constraint is that of limited spectrum.
- *Global PCS:* MSS is designated to provide personal communication services (PCS) to those who need mobile communications in their own countries but who travel beyond the reach of terrestrial cellular systems.
- *Government agencies:* These include law enforcement, fire, public safety, and other services.

TABLE 8.5 Characteristics of Satellite PCS Systems

Parameter	Iridium	Globalstar	ICO
Company	Motorola	Loral/Qualcomm	ICO-Global
No. of satellites	66	48	10
No. of orbit planes	6	8	2
Altitude (km)	780	1414	10,355
Weight (lb)	1100	704	6050
Bandwidth (MHz)	5.15	11.35	30
Frequency up/down (GHz)	30/20	5.1/6.9	14/12
Spot beams/satellite	48	16	163
Carrier bit rate (k/sec)	50	2.4	36
Multiple access	TDMA/FDMA	CDMA/FDMA	TDMA/FDMA
Cost to build ($billion)	4.7	2.5	4.6
Service start date	1998	1999	2003

- *Broadband services:* Broadband satellite networks are the new generation of satellite networks in which Internet-based services will be provided to users regardless of their degree of geographical mobility. Still voice, e-commerce, and low bit rate applications are among the services that will be provided.

As more and more customers sign up for satellite mobile service (MSS), information will flow more freely, the world will grow smaller, and the global economy will be stimulated. The dawning age of global personal communication will bring the world community closer together as a single family.

8.3.4 Satellite Radio

Satellite radio is broadcasting from satellite. With satellite radio, one can drive from Washington, DC to Los Angeles without changing the radio station and without static interference. Satellite radio eliminates localization, which is the major weakness of conventional radio. Satellite radio will permanently change radio just as cable changed television. It is regarded as radio beyond AM, beyond FM, or radio to the power of X. It transforms radio from a local medium into a national one.

Figure 8.10 displays a typical architecture of satellite radio. Satellite radio is based on digital radio, which produces a better sound from radio than analog radio does. Digital radio systems are used extensively in communication networks. Digital radio offers CD quality sound, efficient use of the spectrum, more programming choices, new services, and robust reception even under the most challenging conditions.

Satellite radio is both a new product and a service. As a product, it is a new electronic device that receives the satellite signal. As a service, it will provide consumers with 100 national radio stations, most of which will be brand-new, comprising various music, news, sports, and comedy stations.

Satellite radio service is being provided by two companies: XM Satellite Radio (also known as XM Radio), based in Washington, and Sirius Satellite Radio, based in New York. The two companies obtained FCC licenses to operate a digital audio radio service (DARC) system coast-to-coast throughout the continental U.S.A. To avoid competition with terrestrial radio broadcasters, both satellite broadcasters will carry advertisements of nationally branded products.

XM Satellite Radio is made possible by two satellites, officially named "Rock and Roll," placed in geostationary orbit, one at 85 degrees West longitude and the other at 115 degrees West longitude. Rock and Roll are Boeing 702 satellites, built by industry leader Boeing Satellite Systems. The satellites will be positioned above the United States. The two powerful satellites will have a footprint that covers the entire country. They will transmit up to 100 channels of revolutionary radio programming nationwide. XM Satellite Radio service will be uplinked to the satellites and transmitted directly to auto, home, and portable radios. In September 2001, XM Satellite Radio started to broadcast. Subscribers will pay as little as $9.95 per month after they purchase an AM/FM/XM radio.

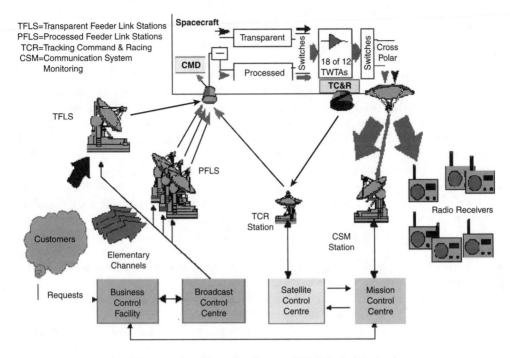

FIGURE 8.10 A typical architecture of satellite radio. (Source: EBU Technical Review.)

Sirius Satellite Radio, on the other hand, does not use GEO satellites. Rather, it is flying three satellites, which are equally spaced in an elliptical 47,000 × 24,500 km orbit that takes 24 hours to complete. This ensures that each satellite spends about 16 hours a day over the continental U.S., with at least one satellite over the country at any time. It also means that Sirius will be higher in the sky than XM, which is at the zenith only at the equator. This in turn implies that fewer ground stations will be needed for the Sirius system. Programs will be beamed to one of the three satellites, which in turn will transmit the signal to your radio receiver or to ground repeaters for listeners in urban areas where the satellite signal can be interrupted. Sirius charges $12.95 a month for their service. The systems of the two satellite radio companies are compared in Table 8.6

Of paramount importance to the two broadcasters-to-be are early agreements from major U.S. car manufacturers. General Motors, Acura, American Honda Motor, Saab, Suzuki, and Isuzu have agreed to install XM radio receivers in some of their car and truck models. Sirius has exclusive alliances to install AM/FM/XM radios in Ford, Chrysler, Mercedes, BMW, Jaguar, and Volvo cars.

TABLE 8.6 Comparison of the XM and Sirius Systems

	XM	Sirius
Constellation	2 satellites	3 satellites
Satellite type	Boeing 702	SS/Loral FS-3000
Terrestrial repeaters	1500 in 70 areas	105 in 46 areas
Satellite costs	$439 m	$120 m
Transmission rate	4 Mb/sec	4.4 Mb/sec
Uplink frequencies	7.05–7.075 GHz	7.06–7.0725 GHz
Downlink frequencies	2.3325–2.345 GHz	2.32–2.3325 GHz

Both companies need the support of radio receiver manufacturers. XM-ready radio receivers produced by Pioneer are now available for sale by some retailers such as Best Buy, Circuit City, Sears, Tweeter stores, and Radio Shack. XM Satellite Radio is now extending its market to airlines. The Sirius system is being developed by manufacturers such as Panasonic, Visteon, Kenwood, Clarion, and Jensen.

Besides XM Satellite Radio and Sirius Satellite Radio that operate in the U.S., WorldSpace is another radio satellite broadcasting company already broadcasting in Africa and Asia. With a constellation of three satellites (AfriStar to cover Africa and Middle East, AmeriStar to serve Latin America and the Caribbean, and AsiaStar to serve nearly all Asia), WorldSpace intends to touch all or parts of four continents, especially those areas of the world that most conventional radio stations cannot reach.

As a new technology, satellite radio is not without its own peculiar problems. First, people are not yet used to paying for radio programming. If the programming of the satellite broadcasters is not better than what people are getting free from regular, terrestrial radio, they will be reluctant to pay. So the real question is: How many people are going to subscribe? Before one can receive from satellite radio, one needs a car stereo costing between $200 and $400 and a monthly payment of $9.95. Some critics think this is asking for too much. Others think that most Americans are dissatisfied with conventional radio (having to endure 18 minutes of commercials each hour and radio signals that may begin to fade away 30 miles from the source), and if people are offered better service than regular radio, the satellite broadcasting companies can transform radio into a subscription service in much the same way cable has changed television. Also, satellite broadcasting requires near-omnidirectional receive antennas for cars, which in turn require a powerful signal from the satellite. In addition, some believe that the two companies will face a big hurdle in transforming radio from a local medium into a national one. The many-pie-in-the-sky companies are faced with great risks ahead of them.

Satellite radio will transform the radio industry, which has seen little technological change since the discovery of FM, some 40 years ago. Receiving digital-quality music from a radio satellite is a major technical milestone. It is as revolutionary to the entertainment industry as was the invention of radio itself. The future of radio-by-satellite is exciting but uncertain [36–38].

8.3.5 Satellite-Based Internet

The Internet is becoming an indispensable source of information for an ever-growing community of users. The thirst for Internet connectivity and high performance remains unquenched. This has led to several proposals for integrating satellite networks with terrestrial ISDN and the Internet [39–42].

Several factors are responsible for this great interest in IP-over-satellite connectivity:

1. Satellites cover areas where land lines do not exist or cannot be installed. Satellites can serve as an access link between locations separated by great distances.
2. Developments in satellite technology allow home users to receive data directly from a geostationary satellite channel at a rate 20 times faster than of an average telephone modem. For example, the majority of today's PC-based modems operate at a 28.8 kb/sec or less, while the DirectPC satellite system (developed by Hughes Network Systems in 1996) provides a direct 400 kb/sec link. With more power transponders utilizing wider frequencies, commercial satellite links can now deliver up to 155 Mb/sec.
3. The unique positioning of satellites between sender and receivers lends itself to new applications such as IP multicast, streaming data, and distributed Web caching.
4. Satellite connectivity can be rapidly deployed because trenches and cable installation are unnecessary. Moreover, satellite communication is highly efficient for delivering multimedia content to businesses and homes [43].

As an inherently broadcast system, a satellite is attractive to point-to-multipoint and multipoint-to-multipoint communications, especially in broadband multimedia applications. The asymmetrical nature of Web traffic suggests a good match to VSAT systems since the VSAT return link capacity would be much smaller than the forward link capacity.

A typical network architecture for a satellite-based Internet service provider (ISP) is shown in Fig. 8.11, which has been simplified to focus on the basic functionality. It includes its own satellite network

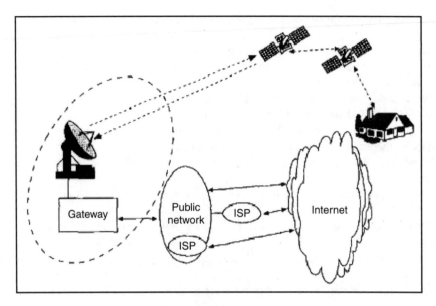

FIGURE 8.11 A typical configuration for satellite-based Internet.

and a network of ground gateway stations. The ground gateway stations interface with the public network through which access to the Internet is gained. The number of satellites may vary from dozens to hundreds, and they may be GEO, MEO, or LEO. Thus, the satellite-based Internet has several architectural options due to the diverse designs of satellite systems, orbit types, payload choice, and intersatellite link designs [44].

There is ongoing research into various aspects of implementation and performance of TCP/IP over satellite links. Related issues include the slow start algorithm, the ability to accommodate large bandwidth-delay products, congestion control, acknowledgment, and error recovery mechanisms [45].

More information about satellite communications systems can be obtained from [46–48].

References

[1] D.J. Marihart, Communications technology guidelines for EMS/SCADA systems, *IEEE Transactions on Power Delivery*, 16, 181–188, 2001.

[2] D. Calcutt and L. Tetley, *Satellite Communications: Principles and Applications*, Edward Arnold, London, 1994, pp. 3–15, 321–387.

[3] B.R. Elbert, *The Satellite Communication Applications Handbook*, Artech House, Norwood, MA, 1997, pp. 3–27, 257–320.

[4] W. Pritchard, Geostationary versus non-geostationary orbits, *Space Communications*, 11, 205–215, 1993.

[5] T.T. Ha, *Digital Satellite Communications*, McGraw-Hill, New York, 1990, pp. 1–30, 615–633.

[6] T.S. Tycz, Fixed satellite service frequency allocations and orbit assignment procedures for commercial satellite systems, *Proceedings of IEEE*, 78, 1283–1288, 1990.

[7] Y. Hu and V.O.K. Li, Satellite-Based Internet: A Tutorial, *IEEE Communications Magazine*, March 2001, pp. 154–162.

[8] Propagation special issue, *International Journal of Satellite Communications*, 19(3), May/June 2001.

[9] D.C. Hogg and T.S. Chu, The role of rain in satellite communication, *Proceedings of IEEE*, 63, 1308–1331, 1975.

[10] M. Richharia, *Satellite Communication Systems*, McGraw-Hill, New York, 1995, pp. 16–49.

[11] T. Pratt, *Satellite Communications*, John Wiley & Sons, New York, 1986, pp. 11–51.

[12] W.W. Wu et al., Mobile satellite communications, *Proceedings of IEEE*, 82, 1431–1448, 1994.

[13] B.R. Elbert, *Introduction to Satellite Communication*, Artech House, Norwood, MA, 1999, pp. 390–395.

[14] N.J. Muller, *Mobile Telecommunications Factbook*, McGraw-Hill, New York, 1998, pp. 336–363.

[15] M.H. Hadjitheodosiou et al., Next generation multiservice VSAT networks, *Electronics and Communication Engineering Journal*, June 1997.

[16] N. Abramson, Internet Access Using VSATs, *IEEE Communications Magazine*, July 2000, pp. 60–68.

[17] N. Abramson, VSAT data networks, *Proceedings of IEEE*, 78, 1267–1274, 1990.

[18] J.C. Raison, Television via satellite: convergence of the broadcasting-satellite and fixed-satellite service — the European experience, *Space Communications*, 9, 129–141, 1972.

[19] P. Wood, Mobile Satellite Services for Travelers, *IEEE Communications Magazine*, Nov. 1991, pp. 32–35.

[20] F. Abrishamkar, PCS Global Mobile Services, *IEEE Communications Magazine*, Sep. 1996, pp. 132–136.

[21] G. Comparetto and R. Ramirez, Trends in Mobile Satellite Technology, *Computer*, Feb. 1997, pp. 44–52.

[22] J.V. Evans, Satellite Systems for Personal Communications, *IEEE Antennas and Propagation Magazine*, June 1997, pp. 7–20.

[23] Anon., Satellite systems for personal communications, *Proceedings of the IEEE*, vol. 39, no. 3, June 1997, pp. 7–20.

[24] Anon., Satellite communications — a continuing revolution, *IEEE Aerospace and Electronic Systems Magazine*, Oct. 2000, pp. 95–107.

[25] W.S. Adams and I. Rider, Circular polar constellations providing continuous single or multiple coverage above a specified latitude, *The Journal of the Astronautical Sciences*, 35(2), Apr./June 1967.

[26] B. Pattan, *Satellite-Based Cellular Commuications*, McGraw-Hill, New York, 1998, pp. 45–88.

[27] P. Lemme et al., Iridium: Aeronautical Satellite Communications, *IEEE AES Systems Magazine*, Nov. 1999, pp. 11–16.

[28] Y.C. Hubbel, A Comparison of the Iridium and AMPS Systems, *IEEE Network*, Mar./Apr. 1997, pp. 52–59.

[29] R.J. Leopold and A. Miller, The Iridium Communications System, *IEE Potentials*, Apr. 1993, pp. 6–9.

[30] E. Hirshfield, The Globalstar system: breakthroughs in efficiency in microwave and signal processing technology, *Space Communications*, 14, 69–82, 1996.

[31] F.J. Dietrich et al., The Globalstar cellular satellite system, *IEEE Transactions on Antennas and Propagation*, 46, 935–942, 1998.

[32] R. Hendrickson, Globalstar for the military, *Proceedings of MILCOM*, 3, 808–813, 1998.

[33] P. Poskett, The ICO System for Personal Communications by Satellite, *Proceedings of the IEE Colloquim (Digest)*, part 1, 1998, pp. 211–216.

[34] L. Ghedia et al., Satellite PCN — the ICO system, *International Journal of Satellite Communications*, 17, 273–289, 1999.

[35] M. Werner, Analysis of system parameters for LEO/ICO-satellite communication networks, *IEEE Journal of Selected Areas in Communications*, 13, 371–381, 1995.

[36] M.N.O. Sadiku, XM Radio, *IEEE Potentials*, Feb./Mar. 2002.

[37] D. Wood, Digital Radio by Satellite, *EBU Technical Review*, Summer 1998, pp. 1–9.

[38] D.H. Layer, Digital Radio Takes to the Road, *IEEE Spectrum*, July 2001, pp. 40–46.

[39] T. Otsu et al., Satellite communication system integrated into terrestrial ISDN, *IEEE Transactions on Aerospace and Electronics Systems*, 36, 1047–1057, 2000.

[40] C. Metz, TCP over Satellite …The Final Frontier, *IEEE Internet Computer*, Jan./Feb. 1999, pp. 76–80.

[41] H.K. Choi, Interactive Web Service via Satellite to the Home, *IEEE Communications Magazine*, Mar. 2001, pp. 182–190.

[42] Y. Hu and V.O.K. Li, Satellite-based Internet: A Tutorial, *IEEE Communications Magazine*, Mar. 2001, pp. 154–162.

[43] C. Metz, IP-over-Satellite: Internet Connectivity Blasts Off, *IEEE Internet Computer,* July/Aug. 2000, pp. 84–89.

[44] P.W. Cooper and J.F. Bradley, A space-borne satellite-dedicated gateway to the Internet, *IEEE Communications Magazine,* Oct. 1999, pp. 122–126.

[45] J. Farserotu and R. Prasad, A Survey of Future Broadband Multimedia and Satellite Systems, Issues, and Trends, *IEEE Communications Magazine,* June 2000, pp. 128–133.

[46] Special issue, Satellite communications, *Proceedings of the IEEE,* 65(3), 1977.

[47] Special issue, Satellite communication networks, *Proceedings of the IEEE,* 72(11), 1984.

[48] Special issue, Global satellite communications technology and system, *Space Communications,* 16, 2000.

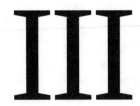

III

Wireless Communication Systems and Protocols

9

Wireless Communication Protocols

Pascal Lorenz
University of Haute Alsace

Abstract

In a few years, there will be more wireless communications through mobile telephones and wireless Internet accesses. Therefore, many different wireless communication protocols have been developed [1,3,12,16]. In this chapter, a distinction has been made between the cellular networks used by the telecommunication world, the satellite networks, and private wireless communications. The Mobile IP and Cellular IP protocols are introduced, which should be widely used in the future wireless communication architectures [2,10,15,22,25,27].

9.1   Cellular Networks

Cellular networks are composed of cells that cover an operator territory [6,19]. The cells are partially stackable to efficiently cover the territory. Today, three generations of cellular networks are defined:

- In the first generation, the radio interface was analogical.
- In the second generation, the radio interface became numerical. For the second generation (2G) of cellular networks:

- Europe has developed the GSM network.
- The United States has developed the IS95 (Interim Standard 95) and IS136 (Interim Standard 136) networks.
- Japan has developed the PDC (Pacific Digital Cellular) network.
- The third generation will integrate multimedia applications [7,8]. For third generation (3G) networks:
    - Europe will use the UMTS (Universal Mobile Telecommunication System) network.
    - The United States will use the CDMA2000 (Code Division Multiple Access 2000) and UWC136 (Universal Wireless Communications 136) networks.
    - Japan will use the W-CDMA (Wideband-Code Division Multiple Access) network.

These different protocols will be explained in the next sections.

The different functionalities used by a cellular network are:

- Mobility Management (MM) for the localization of the users
- Connection Management (CM) for the establishment of the connection
- Radio Resource (RR) management to offer the best channel to a communication

9.1.1 GSM (Global System for Mobile Communications)

9.1.1.1 Introduction

The GSM protocol was created in 1980 in France. It uses the 890–915 MHz radio band for the upload traffic and the 935–960 MHz radio band for the download traffic. GSM is a 2G system, based on FD-TDMA (Frequency Division–Time Division Multiple Access) radio access, which offers a 9.6 kb/sec rate. Today, more than 450 million subscribers in the world use the GSM system for their wireless cellular communications.

The problem is that GSM will not be able to satisfy news services such as data networks.

In 1990, the DSC (Digital Communication System) was developed to offer additional bands:

- DSC1800 uses the 1710–1785 MHz radio band for the upload traffic and the 1805–1880 MHz radio band for the download traffic.
- DSC1900 uses the 1850–1910 MHz radio band for the upload traffic and the 1930–1990 MHz radio band for the download traffic.

9.1.1.2 Architecture

The GSM equipment is based on the PLMN (Public Land Mobile Network), which is composed of the:

- BSS (Base Station Subsystem) to manage the radio interface
- NSS (Network SubSystem) to manage the interconnections with the fixed network
- OSS (Operation Support Subsystem) to supervise the PLMN

9.1.1.2.1 BSS (Base Station Subsystem)

The BSS is composed of the:

- MS (Mobile Station) used by the final users
- BTS (Base Transceiver Station) used by the MS for the connections to the cellular network. The BTS functionalities include:
    - Modulation
    - Demodulation
    - Error detection and correction
- BSC (Base Station Controller), which gathers several tens of BTS

The BSC manages the radio channels, the calls admission control, and the handovers/handoffs. It is the Abis interface that is used to connect the BSC with the BTS [21].

9.1.1.2.2 NSS (Network SubSystem)

The NSS is composed of the:

- MSC (Mobile service Switching Center), which is connected to the BSC through the A interface. The MSC manages the interconnections between the PLMN and the PSTN (Public Switched Telephone Network).
- VLR (Visitor Location Register), which is connected to the MSC through the B interface. The VLR is a local database avoiding a connection to the HLR.
- HLR (Home Location Register), which is connected to the MSC through the C interface and to the VLR through the D interface. The HLR stores all information concerning the characteristics of a subscriber to the cellular network.

The GSM architecture can be represented as shown in Fig. 9.1.

Inside the NSS, all the communications are managed by the MAP (Mobile Application Part) protocol, which is an upgraded version of the SS7 (Signaling System number 7) telecommunication protocol.

9.1.1.3 SMS (Short Message Service) and WAP (Wireless Application Protocol)

The SMS (Short Message Service) is a service enabling the sending of short messages of 160 octets length. Today, more than 15 billion SMS messages are exchanged every month in the world. In the near future, the MMS (Multimedia Messaging Service) will enable the sending of not only text but also movies and sounds.

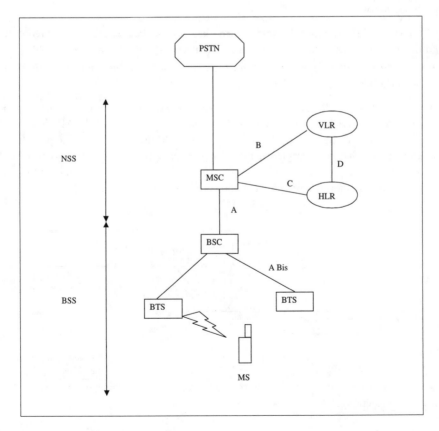

FIGURE 9.1 Architecture of a GSM network.

The WAP (Wireless Application Protocol) is a light browser for cellular phones in which the information is only textual. WAP is based not on HTML language but on the WML (Wireless Markup Language) language. A WAP gateway is used to translate the IP packets into binary format.

9.1.2 IS95 (Interim Standard 95) AND IS136 (Interim Standard 136)

IS95 (Interim Standard 95) is a 2G network based on CDMA (Code Division Multiple Access) radio access. IS95, also called CDMAOne, has been developed by Qualcomm in the 800 and 1900 MHz radio bands. It offers two different rates: 8 and 13 kb/sec.

The IS95 architecture is similar to the GSM, but without BSC equipment. The signalization in the core network is done not by the MAP protocol but by the IS41 protocol. IS95 will be used for the development of CDMA2000 networks.

IS136 (Interim Standard 136) is a 2G network, based on TDMA.

9.1.3 GPRS (General Packet Radio Service) and EDGE (Enhanced Data for GSM Evolution)

GRPS (General Packet Radio Service) and EDGE (Enhanced Data for GSM Evolution) are 2.5 generation networks enabling the offering of a 170 kb/sec data transfer rate. They will use the IP protocol for data transport [14]. For these networks, the billing for data communication is based on the data exchanged and not on the duration of the communication.

GPRS networks can work in parallel with GSM networks: GSM is used for voice communications and GPRS for data communications. Therefore, a mobile station can be connected at the same time to the GSM and GPRS networks.

GPRS is based on two routers:

- SGSN (Serving GPRS Support Node) for the packets transfer between the wireless radio subsystem and the fixed network. The SGSN routers are connected to the BSS and to the GGSN routers.
- GGSN (Gateway GPRS Support Node) for the management of the public data, for example, for the QoS negotiation.

The GTP (GPRS Tunneling Protocol) protocol is used for the encapsulation of data between the SGSN and GGSN, by using the TCP and UDP protocols. Between the SGSN and the MS, it is the SNDCP (SubNetwork Dependent Convergence Protocol) protocol that is used to manage the packets.

The HLR equipment has been improved in order to offer point-to-point and point-to-multipoint services.

The GPRS architecture can be represented as shown in Fig. 9.2.

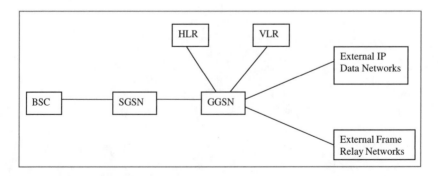

FIGURE 9.2   GPRS architecture.

The EDGE networks will offer a 348 kb/sec rate by introducing new coding schemes. EDGE phase 1 is also called E-GPRS (Enhanced-GPRS), which is often considered as a 3G network and is used by the UWC136 networks. In EDGE networks, the BSS is renamed E-RAN (EDGE Radio Access Network).

9.1.4   The Third Generation (3G) Systems

The third generation system IMT 2000 (International Mobile Telecommunication system) is based on the 2000 MHz radio band and is composed of:

- UTRA (Universal Mobile Telecommunication System Terrestrial Radio Access), proposed by the ETSI (European Telecommunication Standard Institute)
- CDMA2000 (Code Division Multiple Access 2000), proposed by TIA (Telecommunications Industry Association), which is an evolution of the IS95 standard
- UWC136 (Universal Wireless Communications 136), proposed by TIA (Telecommunications Industry Association), which is an evolution of the IS136 standard
- W-CDMA (Wideband–Code Division Multiple Access), proposed by ARIB (Association of Radio Industries and Business), which is an evolution of the PDC standard.

W-CDMA is based on the FDD (Frequency Division Duplex) mode. In the FDD mode, the uplink (reverse direction) and the downlink (forward direction) traffic use different frequencies. In the TDD (Time Division Duplex) mode, the uplink and the downlink traffic are separated in time.

9.1.4.1   UMTS (Universal Mobile Telecommunication System)

The 3GPP (Third Generation Partnership Project) is composed of national standardization committees (ETSI, ARIB, TTC, TTA, T1P1, CWTS). It works on the development of UMTS (Universal Mobile Telecommunication System) standards through five technical committees (Radio Access Network, Core Network, Service and System Aspects, Terminals, and GSM Enhanced Radio Access Network). In 2010, the UMTS Forum foresees 2 billion users for 3G systems.

The UTRA (UMTS Terrestrial Radio Access) standard proposes five different accesses to the radio resources:

- W-CDMA (Wideband CDMA), used by the FDD mode
- OFDMA (Orthogonal Frequency Division Multiplexing)
- TD-CDMA (Time Division-CDMA), used by the TDD mode
- W-TDMA (Wideband TDMA)
- ODMA (Opportunity Driven Multiple Access), based on the ad hoc networks

UMTS integrates the TD-CDMA and the W-CDMA systems.

The RNS (Radio Network Subsystem) (called BSS in GSM) is composed of the:

- UE (User Equipment) (called MS in GSM)
- Node B (called BTS in GSM)
- RNC (Radio Network Controller) (called BSC in GSM)

The RNC is the Iub interface (called Abis interface in GSM), which is used to connect the RNC to the NodeB.

The UMSC (UMTS MSC) is connected to the RNC through the Iu interface (called A interface in GSM).

The UMTS architecture can be represented as shown in Fig. 9.3.

UMTS will offer a 2 Mb/sec; 384 kb/sec in urban areas and 144 kb/sec in rural areas.

9.1.4.2   CDMA2000 (Code Division Multiple Access 2000)

CDMA2000 (Code Division Multiple Access 2000) is an evolution of the IS95 standard in which the packet mode is more efficient. The two protocols, CDMA2000 and IS95, use the same frequency band.

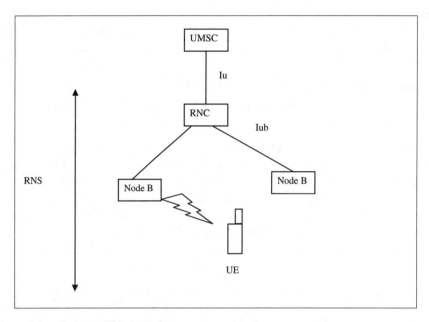

FIGURE 9.3   Architecture of an UMTS network.

Rhe PPP (Point-to-Point Protocol) protocol does the interconnection between the IP protocol and the CDMA2000 standard.

9.1.5   Conclusion

As is true today, in the near future many different systems will be used for cellular communications. The major spectrum allocations are summarized in Fig. 9.4.

Therefore, in the future, software radio solutions will be developed to enable dynamic reconfiguration (for all layers) and to offer a multifrequency and multimode system [20].

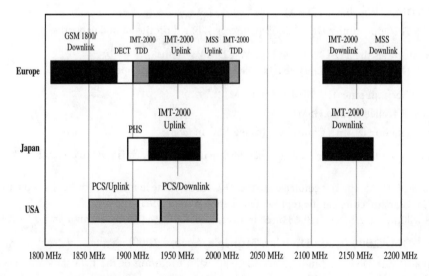

FIGURE 9.4   Spectrum allocation.

9.2 Satellite Communications

Wideband satellites will increasingly be used for the transmission of multimedia information [11,23]. There are three categories of satellites:

- LEOS (Low Earth Orbital Satellite) located at a distance of 1000 km from the earth. The round trip delay between the earth and the satellite is equal to 0.01 sec. It is necessary to have many satellites to cover all the world, and many handovers (soft-handover or hard-handover) should be performed to maintain a communication. The same frequency can be reused 20,000 times. For example, Globalstar is a 48-satellite constellation system at an altitude of 1410 km, which was designed to provide mobile telephony services between 69° of north and south latitude. The constellation of satellites can be represented as shown in Fig. 9.5.
- MEOS (Medium Earth Orbital Satellite) located at a distance of 13,000 km from the earth. The round trip delay between the earth and the satellite is equal to 0.1 sec. About ten satellites can cover all the earth. For example, ICO is a 12-satellite system at an altitude of 10,335 km, which is designed to provide mobile telephony services.
- GEOS (Geostationary Earth Orbital Satellite) located at a distance of 36,000 km from the earth. The round trip delay between the earth and the satellite is equal to 0.25 sec. Three satellites can cover all the earth, but some parts of the earth cannot be covered (over 75° of latitude), and it is difficult to reuse the frequencies. The power transmission of the terminals should be important.

Access to satellites can be performed through FAMA (Fixed Assignment Multiple Access), RA (Random Access), PR (Packet Reservation), or DAMA (Demand Assignment Multiple Access).

9.3 Private Networks

In the private networks, we can find WPAN (Wireless Personal Area Networks) and WLAN (Wireless Local Area Networks) networks [9]. The standardization of WPAN is performed mainly by the IEEE 802.15 group [17,18,24,28].

9.3.1 Bluetooth

The WPAN Bluetooth is inexpensive technology that uses the 2450 MHz frequency to interconnect equipment. The range of the transmission in the Bluetooth piconets is about 10 m maximum at a rate

FIGURE 9.5 Constellation of satellites.

of 1 Mb/sec. The communication can be synchronous (through the SCO Synchronous Connection Oriented link) or asynchronous (through the ACO Asynchronous Connection Oriented link). Bluetooth is based on the Master/Slave transmission mode.

To avoid spontaneous communications, for example, between two PDA, some authentication mechanisms have been introduced.

In the future, version 2.0 of Bluetooth will offer a rate of 10 Mb/sec.

9.3.2   Home RF (Home Radio Frequency)

The WPAN Home RF (Home Radio Frequency) networks are used for home automation [13]. The Home RF systems use the same frequency as IEEE 802.11 and Bluetooth.

Home RF can interconnect computers and DECT (Digital Enhanced Cordless Telecommunications) phones at a rate of 1.6 Mb/sec.

9.3.3   IEEE 802.11

The WLAN cellular IEEE 802.11 networks are based on the CSMA/CA (Carrier Sense Multiple Access/ Collision Avoidance) protocol. The IEEE 802.11 protocol offers two types of service:

- BSS (Basic Set Service). Several terminals establish wireless connections with the AP (Access Point)
- IBSS (Independent Basic Set Service). This is an ad hoc mode enabling the interconnection of the different terminals without any infrastructure and without AP.

Security can be handled by the WEP (Wired Equivalent Privacy) protocol.

The cellular IEEE 802.11b network uses the 2450 MHz frequencies and can cohabit with Bluetooth. The range of the transmission of a IEEE 802.11b network is about 150 meters in the office and 600 meters when there is no obstacle. The 802.11b network offers a rate of 11 Mb/sec and is beginning to be widely used in public locations, such as airports, stations, etc.

The IEEE 802.11a networks use the 5000 MHz frequency and offer a rate of 54 Mb/sec.

With the IEEE 802.11e networks, the security and the QoS of the IEEE 802.11a networks are improved.

9.3.4   HiperLAN

The WLAN HiperLAN networks have been specified by the ETSI standardization committees. There are four different types of HiperLAN networks:

- HiperLAN type 1. The transmission has a range of about 50 meters at a rate of 23 Mb/sec and uses the 5 GHz radio frequency. The access network technique is the same as the one used by the IEEE 802.11 networks.
- HiperLAN type 2. The transmission has a range of about 200 meters for wireless ATM networks at a rate of 23 Mb/sec and uses the 5 GHz radio frequency.
- HiperLAN type 3. The transmission has a range of about 5000 meters for wireless ATM networks at a rate of 20 Mb/sec and uses the 5 GHz radio frequency. This network can be used for WLL (Wireless Local Loop) communications.
- HiperLAN type 4. The transmission has a range of about 200 meters for wireless ATM connections at a rate of 155 Mb/sec and uses the 17 GHz radio frequency.

9.3.5   LMDS (Local Multipoint Distribution Service)

The LMDS (Local Multipoint Distribution Service) network uses radio frequencies higher than 25 GHz and offers a rate of 20 Mb/sec. The range of the transmission is about 10 km. Today, the available LMDS products are based on IP over ATM protocols. Each transmitter in a cell serves a relatively small area, about 5 km in diameter. This small cell size means that the LMDS network requires a large number of antennas.

The majority of system operators will use point-to-multipoint wireless access designs, although point-to-point systems and TV distribution systems can be provided within the LMDS system. It is expected that the LMDS services will be a combination of voice, video, and data. Due to its limited range of transmission, LMDS is not a good choice to provide wide area coverage of digital television service.

The DAVIC (Digital Audio-Visual Council) group will use LMDS networks for the transmission of video.

9.3.6 MMDS (Multi-Channel Multipoint Distribution Service)

The MMDS (Multi-Channel Multipoint Distribution Service) networks are used to offer a 10 km range for the unidirectional transmission of cable TV in rural areas, by using the 2.7 GHz radio frequency.

Repeater stations can be used to redirect MMDS signals to screened areas. The range of a transmitting antenna can reach 50 km depending on the broadcast power.

MMDS networks are characterized by the limited number of channels available in the low radio frequency bands. Only a 200 MHz spectrum is allocated for MMDS; this constraint reduces the effective number of channels in a single MMDS system. For TV signals with 6 MHz bandwidth, only 33 channels can be fit into the spectrum.

9.3.7 DVB (Digital Video Broadcasting) and DAB (Digital Audio Broadcasting)

DAB (Digital Audio Broadcasting) will replace the analogical FM radio, and DVB (Digital Video Broadcasting) can be used for high definition television. DVB is based on MPEG-2 transmissions.

There are three standards for DVB:

- DVB-S: DVB via satellite
- DVB-T: DVB via terrestrial hertzian, which uses the same frequencies as analogical television
- DVB-C: DVB via cable

The integration of video in mobile terminals can use the MPEG-4 standard, which introduces flow grading mechanisms.

The MPEG-7 standard will enable the integration of video with databases and movie diffusion in 3G terminals through the Web.

In the future, the MPEG-12 standard will improve the compression technologies of the MPEG-7 and MPEG-4 standards.

9.4 Mobile IP and Cellular IP

9.4.1 Mobile IP

The Mobile IP protocol will manage the mobility in Internet networks. It will offer to users connections to the Internet anywhere in the world without contacting local providers and without needing to send a new address to correspondents.

When a mobile node reaches a new domain, the Foreign Agent (FA) of the new domain offers a temporary address (called the Care-Of Address) to the mobile node. In the local domain of the mobile node, the Home Agent (HA) does the association between the IP address of the mobile node and the new Care-Of Address. Thus, a tunnel is created between the HA and the FA.

When the mobile node is moving to another domain, there is a modification of the Care-Of Address: this handoff is called smooth-handoff.

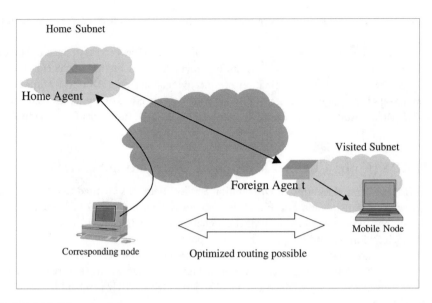

FIGURE 9.6   Mobile IP mechanisms.

In future IPv6 networks, the auto configuration process will enable the removal of the FA because the mobile node can itself find the Care-Of Address. Thus, in this case a tunnel is created between the HA and the Care-Of Address.

The Mobile IP mechanisms can be represented as shown in Fig. 9.6.

9.4.2   Cellular IP

The Cellular IP protocol is more adapted for mobility inside a domain. The different stations send an empty packet called paging-update to update the different paging caches.

MANET (Mobile Ad hoc NETwork) is a protocol developed by IETF. In MANET, there are two types of protocols: reactive protocols (using the flooding mechanism to find another station) and proactive protocols (using routing tables built by the exchanges of signals).

The Distance Vector routing is based on AODV (Ad hoc On demand Distance Vector) for the reactive routing technique and DSDV (Destination Sequence Distance Vector) for the proactive routing technique [4,5,26].

The Link State routing is based on OLSR (Optimized Link State Routing Protocol) for the reactive routing technique.

The Source Routing is based on DSR (Dynamic Source Routing) for the proactive routing technique.

9.5   Conclusion

The fourth generation of mobile networks will introduce QoS (Quality of Service) and the integration of different protocols in the same device. For computer and telephony communications, WPAN and WLAN will be used in buildings; outside, IMT2000 solutions will be used; and when there is no connectivity, satellites will be used.

The IP protocol will be used by all types of terminals and by all networks. The 4G terminals will be a mobile and a wireless terminal integrated by Mobile IP and Cellular IP protocols.

In the fifth generation, the satellites will be replaced by HAPS (High Altitude Stratospheric Platform) solutions to offer high-speed wireless communications. This system, located at 20 km from earth, should be based on automatic airships or airplanes (such as Awacs).

References

[1] Aljadhai, A. and Znati, T.F., Predictive mobility support for QoS provisioning in mobile wireless environments, *IEEE Journal on Selected Areas in Communications*, 19, 1915–1930, 2001.

[2] Becchetti, L. et al., Enhancing IP Service Provision over Heterogeneous Wireless Networks: A Path Toward 4G, *IEEE Communications Magazine*, Aug. 2001, pp. 74–81.

[3] Carmel, B. et al., PiNet: Wireless Connectivity for Organizational Information Access Using Lightweight Handheld Devices, *IEEE Personal Communications*, Aug. 2001, pp. 18–23.

[4] Chakrabarti, S. and Mishra, A., QoS Issues in Ad Hoc Wireless Networks, *IEEE Communications Magazine*, Feb. 2001, pp. 142–148.

[5] Chandran, K. et al., A Feedback-Based Scheme for Improving TCP Performance in Ad Hoc Wireless Networks, *IEEE Personal Communications*, Feb. 2001, pp. 34–39.

[6] Chi-Jui H. et al., Call admission control in the microcell/macrocell overlaying system, *IEEE Transactions on Vehicular Technology*, 50, 992–1003, 2001.

[7] Dixit, S. et al., Resource Management and Quality of Service in Third Generation Wireless Networks, *IEEE Communications Magazine*, Feb. 2001, pp. 125–133.

[8] Frodigh, M. et al., Future-Generation Wireless Networks, *IEEE Personal Communications*, Oct. 2001, pp. 10–17.

[9] Gutierrez, J.A. et al., IEEE 802.15.4: A Developing Standard for Low-Power Low-Cost Wireless Personal Area Networks, *IEEE Network*, Sep./Oct. 2001, pp. 12–19.

[10] Huston, G., TCP in a Wireless World, *IEEE Internet Computing*, Mar./Apr. 2001, pp. 82–84.

[11] Jamalipour, A. and Tung, T., The Role of Satellites in Global IT: Trends and Implications, *IEEE Personal Communications*, June 2001, pp. 5–11.

[12] Kleine-Ostmann, T. and Bell, A.E., A data fusion architecture for enhanced position estimation in wireless networks, *IEEE Communications Letters*, 5, 343–345, 2001.

[13] Lansford, J. and Bahl, P., The design and implementation of HomeRF: a radio frequency wireless networking standard for the connected home, *Proceedings of the IEEE*, 88, 1662–1676, 2000.

[14] Leung, K.K. et al., Link adaptation and power control for streaming services in EGPRS wireless networks, *IEEE Journal on Selected Areas in Communications*, 19, 2029–2039, 2001.

[15] Marczynski, J. and Tabak, D., A Wireless Interconnection Network for Parallel Processing, *Proceedings of the Euromicro Symposium on Digital Systems Design*, 2001, pp. 386–389.

[16] Misra, A. et al., Autoconfiguration, Registration, and Mobility Management for Pervasive Computing, *IEEE Personal Communications*, Aug. 2001, pp. 24–31.

[17] Ramjee, R. et al., IP-Based Access Network Infrastructure for Next-Generation Wireless Data, *IEEE Personal Communications*, Aug. 2000, pp. 34–41.

[18] Rong-Hou W. et al., Planning system for indoor wireless network, *IEEE Transactions on Consumer Electronics*, 47, 73–79, 2001.

[19] Sari, H., A multimode CDMA with reduced intercell interference for broadband wireless networks, *IEEE Journal on Selected Areas in Communications*, 19, 1316–1323, 2001.

[20] Varshney, U. and Jain, R., Issues in Emerging 4G Wireless Networks, *Computer*, June 2001, pp. 94–96.

[21] Wang, W. and Akyildiz, I.F., A new signaling protocol for intersystem roaming in next-generation wireless systems, *IEEE Journal on Selected Areas in Communications*, 19, 2040–2052, 2001.

[22] Webb, W., Broadband Fixed Wireless Access as a Key Component of the Future Integrated Communications Environment, *IEEE Communications Magazine*, Sep. 2001, pp. 115–121.

[23] Wood, L. et al., Effects on TCP of Routing Strategies in Satellite Constellations, *IEEE Communications Magazine*, Mar. 2001, pp. 172–181.

[24] Worrall, S.T. et al., Optimal Packetisation of MPEG-4 Using RTP over Mobile Networks, *Communications, IEE Proceedings*, Aug. 2001, pp. 197–201.

[25] Tseng, Y.-C. and Tan, C.-C., Termination detection protocols for mobile distributed systems, *IEEE Transactions on Parallel and Distributed Systems*, 12, 558–566, 2001.

[26] Tseng, Y.-C. et al., Location Awareness in Ad Hoc Wireless Mobile Networks, *Computer*, June 2001, pp. 46–52.

[27] Zhang, T. et al., IP-based Base Stations and Soft Handoff in All-IP Wireless Networks, *IEEE Personal Communications*, Oct. 2001, pp. 24–30.

[28] Zheng, H. and Boyce, J., An improved UDP protocol for video transmission over Internet-to-wireless networks, *IEEE Transactions on Multimedia*, 3, 356–365, 2001.

10

An Integrated Platform for Ad Hoc GSM Cellular Communications

George N. Aggélou[1]
Institute of Technology, Greece

Abstract

The latest developments and experimentation in mobile ad hoc networks (MANETs) show that MANETs will be an alternative candidate in many private and public multimedia networks. Current interest in MANET systems has grown considerably because they can rapidly and economically extend the boundaries of any terrestrial network; integrating MANET and GSM (Global System for Mobile Communication) offers a great number of benefits (e.g., increasing capacity, improving coverage) at the cost of increasing the complexity of the mobile terminal and its battery consumption. The objective of this article is to address new concepts in the GSM system, dealing with both standardized features and theoretically and technologically feasible improvements, which contribute to evolutionary changes in general. Dynamic evolution of GSM presents a platform for the Universal Mobile Telecommunication System (UMTS) introduction. Major trends in GSM development will be addressed in this chapter, in particular progress towards a generic platform to accommodate relaying capability in GSM cellular networks. A GSM simulation tool has been constructed for quantifying the integrated system characteristics.

10.1 Introduction

GSM (Global System for Mobile Communication) is undoubtedly the most successful second-generation digital mobile radio system. One of the key factors for this exceptional performance is the constant

---

[1] The work is supported by Lucent Technologies, UK, under grant 33/1/PRS/LUC22.

evolution of the GSM system and its derivatives DCS-1800 and PCS-1900. The continued introduction of GSM wireless systems represents significant capital investments by network operators around the globe. GSM networks are operational or planned in almost 60 countries in many parts of the world including Europe, the Middle East, the Far East, Africa, South America, and Australia. By the beginning of 1997, there were over 70.2 million subscribers. The mobile user population is expected to increase to 300 million by the end of 2001.

Given the worldwide trend for connectivity, the market for inexpensive mobile telephony, inspired by the catalytic presence of GSM, is expected to increase significantly during the next decade. This is also facilitated by the emergence of a new type of terminals ("smart phones") that combine cellular data capabilities with personal digital assistant devices and multimode operation over GSM.

For the past years, the emphasis has been on utilizing means to extend and enhance coverage; simply put, how to achieve the desired level of coverage by employing the minimum of costly infrastructure equipment (base stations, base site controllers, switching centers, etc.). At the same time, customers are becoming more discerning; they are beginning to demand better quality in terms of intelligibility, clarity, and absence of the artifacts traditionally associated with radio communications systems. This, allied to the stated aim of many operators to attract customers away from rational wire line telecomm service providers, has placed momentum behind moves to improve quality, too.

10.2 Towards Ad Hoc GSM Communications

Where the future UMTS or IMT-2000 shall offer a global mobility throughout the world, GSM offers similar mobility but has some capacity and coverage constraints when addressing the residential market. The latest developments and experimentation in the GSM cellular network have shown that an integrated satellite–GSM communication platform could provide global roaming, whereas a DECT–GSM integrated network platform could improve indoor coverage. However, even if any of the above or any other stand-alone or integrated system could achieve acceptable coverage at (ideally) all locations within a GSM cell, there are still places where any present communication platform would fail to provide successful communication. These places often referred in the literature to as *dead spots* [1]. Dead spots include subway train platforms, indoor environments, and basements. The research community faces formidable technical challenges in providing reliable wireless communication systems and networks that provide efficient communication performance in dead spot locations. This is because it is not economical (commercially at least) to install additional antennas and extensive antenna processing and directionality at each dead spot location.

Relaying is one of the enhancements that GSM may use to improve coverage and robustness against network or radio link failures, and to increase capacity by lowering transmission powers and associated intercell interference with negligible increase to the mobile station's complexity or cost.

To this extent, the standard GSM radio interface needs to be extended with sufficiently flexible capabilities to support relaying [9,10]. Conceptually, this implies that the radio access part of an ad hoc GSM (A-GSM) protocol would comprise two segments: the GSM radio access and the wireless multihop (or else, ad hoc) access part. The basic idea is similar to the Opportunity Driven Multiple Access (ODMA) TDMA/WB-CDMA proposal [1]. Figure 10.1 illustrates different scenarios of a GSM cell using relaying to improve indoor as well as outdoor coverage and to transparently extend communication at dead spot locations.

In this chapter, the networking requirements for the relaying component of an A-GSM cellular network are examined, and an architecture that best satisfies the requirements of an integrated solution with the present GSM cellular system is identified [9,10]. The aim is to bring GSM functionality closer to the user terminal, without violating its connection-oriented nature, as well as to design an A-GSM system that will make use of the existing GSM system entities with minimal changes.

In order to support the rapidly deployable, wireless, multimedia network requirements, several problems will need to be addressed:

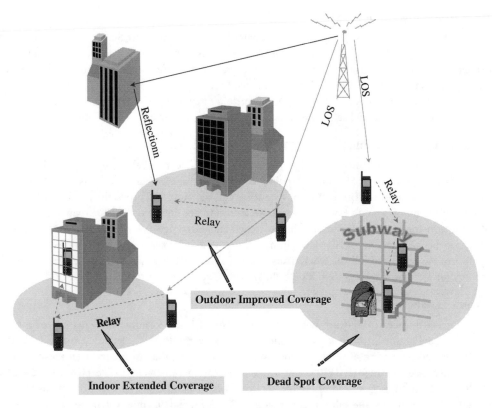

FIGURE 10.1 Scenario for GSM with A-GSM enhancement.

1. The protocols should provide the necessary functionality including reconfiguration and dynamic control over the topology to instantly set up an infrastructure when new nodes start up and to reconfigure the network when nodes move or fail.
2. The network must use available bandwidth efficiently and must control interference in order to support real_time traffic.
3. Protocols should adapt to the rapidly changing propagation conditions that characterize the mobile radio channel. The phenomenon of shadowing, which occurs when the communication path is obstructed by obstacles (e.g., buildings, trees, tracks, etc.), may cause the signal to be weak or even fall below a minimum threshold level, in which case the call may be dropped.

Since there is always a penalty to be paid due to the transmission overheads on the air interface, alternative solutions for the A-GSM protocol layering and the impact on higher layer protocols is further investigated.

10.3 Ad Hoc GSM (A-GSM) Cellular System

10.3.1 A-GSM Network Entities

Most of the future GSM enhancements will share common characteristics with the present GSM cellular architecture [8], supported protocol standards, user terminals, access scheme, and interconnection to terrestrial networks. Therefore, in a typical Ad Hoc GSM (A-GSM) system the following network entities are considered.

Base Station Subsystem (BSS)

The Base Station Subsystem is composed of two parts, the Base Transceiver Station (BTS) and the Base Station Controller (BSC). The Base Transceiver Station houses the radio transceivers that define a cell and handles the radio link protocols with the Mobile Station. In a large urban area, there will potentially be a large number of BTSs deployed; thus, the requirements for a BTS are ruggedness, reliability, portability, and minimum cost. The Base Station Controller manages the radio resources for one or more BTSs. It handles radio channel setup, frequency hopping, and handovers.

Mobile Services Switching Center

The central component of the Network Subsystem is the Mobile services Switching Center (MSC). MSCs link groups of neighboring BSSs through point_to_point landline or microwave-based E1 trunks. The MSC acts as the nerve center of the system. It controls call signaling and processing, and coordinates the handover of the mobile connection from one base station to another as the mobile roams around. Each MSC is in turn connected to the local public switched telephony network (PSTN, or ISDN) to provide connectivity between mobile and fixed telephony users, as well as the necessary global connectivity among the MSCs of the cellular mobile network. Thus, the global connectivity provided by the existing landline telephony infrastructure is used to link up cellular mobile subscribers throughout the world.

Network Databases

GSM defines a number of network databases that are used in performing the functions of mobility management and call control in a public land mobile network (PLMN). These elements include the location registers consisting of the home location register (HLR), the visiting location register (VLR), the equipment identity register (EIR), and the authentication center (AC). The HLR maintains and updates the mobile subscriber's location and service profile information. The VLR maintains the same information locally, where the subscriber is roaming. The VLR is defined as a stand-alone function but is usually viewed by vendors as part of the MSC. The EIR is used to list the subscribers' equipment identities, which are used for identification of unauthorized subscriber equipment, and hence denial of service by the network. The AC provides the keys and algorithm for maintaining the security of subscriber identities and for encrypting information passed over the air interface.

The Home Location Register (HLR) and Visitor Location Register (VLR), together with the MSC, provide the call routing and roaming capabilities of GSM. The HLR contains all the administrative information of each subscriber registered in the corresponding GSM network, along with the current location of the mobile. The location of the mobile is typically in the form of the signaling address of the VLR associated with the mobile station. The actual routing procedure will be described later. There is logically one HLR per GSM network, although it may be implemented as a distributed database.

Integrated Dual Mode Terminals

Several issues arise from the integration of GSM and ad hoc mobile networks. First, in such a system with dissimilar network and air interfaces, the issues of integration and/or interworking, as well as the level of integration, affect almost all the signaling transfer, radio resource, mobility, security, and communication management procedures. Special care has to be taken to design optimally a call routing or rerouting (handover) algorithm, as this algorithm could significantly affect the performance of the system and thereby the quality of offered services in the combined system. The capability and complexity of a dual mode handset will also impact design of the signaling load, delay, required level of modification, implementation complexity, and cost as well as operators' requirements.

As far as handover in an interworking system is concerned, a dual-mode handset that is capable of inter-GSM/MANET handover must function as two handsets with a direct interface (Fig. 10.2). While one A-GSM MT serves an active call either in the GSM or the MANET system, the other system is on the watch for information indicating the presence of an alternative system. This allows the handset to access the alternative system in case handover needs to be performed. Thus, a very attractive feature

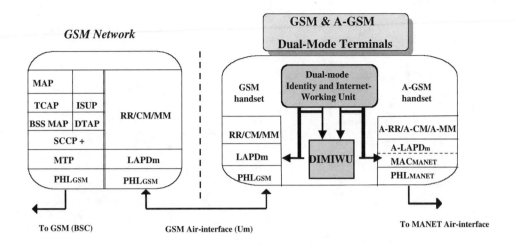

- **PHL:** Physical Layer
- **MTP:** Message Transfer Part
- **SCCP:** Signalling Connection Control Part
- **DTAP:** Direct Transfer Application Part
- **BSS:** Base Station Subsystem
- **MAP:** Mobile Application Part
- **LAPDm:** Link Access Protocol for the D
- **DIMIWU:** InterWorking Function

- **RR:** Radio Resource Management
- **CM:** Communication Management
- **MM:** Mobility Management
- **A-LAPD m:** GSM-MANET tailored Link Layer
- **MAC**MANET**:** MANET specific Medium Access Layer
- **A-CM:** A-GSM Communication Management
- **A-MM:** A-GSM Mobility Management channel
- **A-RR:** A-GSM Radio Resource Management

FIGURE 10.2　GSM-MANET dual mode terminal. (Sources: G. Aggélou, dissertation, University of Surrey, 2001 and G Aggélou and R Tafazolli, *IEEE Personal Communications,* Feb. 2001, pp. 6–13.)

of this type of terminal would be service mobility between modes by means of initiating a service in one mode, then handing over the service to a more suitable system in terms of system capability and application.

Some interworking functionality would be required in dual-mode handset to translate the information coming from two different systems and make some decision based on a predefined algorithm.

Internet Working Unit (DIMIWU)

This is a nonstandard, specially designed unit, responsible for providing access to a GSM/A-GSM network. It performs all the necessary user terminal protocol adaptations to the GSM/A-GSM protocol platform. DIMIWU also includes all physical layer functionalities such as channel coding, modulation, and demodulation and the radio frequency parts. All of the supported terminals share the same access scheme and protocol stacks.

10.3.2 Protocol Layering

The ad hoc segment of an A-GSM communication system has different characteristics from the GSM system, and this explains why modifications of the existing management modules are needed. The A-GSM protocol architecture used for the exchange of signaling messages pertaining to mobility, radio resource, and connection management functions comprises two broad segments: the GSM radio access and fixed network part, and the A-GSM radio access part. Practically, A-GSM layers should inherit the semantics and roles of their peer GSM layers.

A prototype protocol architecture of an A-GSM communication system is presented in Fig. 10.3.

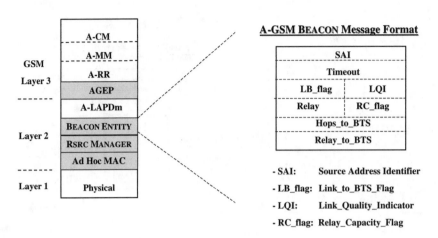

FIGURE 10.3   Proposed A-GSM protocol stack.

Link Layer Protocol

The data link layer over the radio link (connecting two MTs) is based on the GSM LAPDm protocol [6], labeled A-LAPDm, which is designed for operation within the constraints and functional requirements of an A-GSM protocol.

In particular, the Link Access Protocol for the D channel should be enhanced to support the following procedures:

Beaconing

Beaconing is a mechanism used to indicate mobile activity in an A-GSM network. During operation a MT *may* (see discussion in the *Resource Manager* subsection) offer connectivity by broadcasting local beacon messages as follows. At periodic intervals, nodes generate a broadcast message, called BEACON message. The message fields are illustrated in Fig. 10.3.

The LINK_TO_BTS flag is set to state ON if the sender of the BEACON has a direct link to BTS. If LINK_TO_BTS is set to OFF, the ID of the relay node through which the sender of the BEACON can reach the BTS (RELAY_TO_BTS) along with the number of hops (HOPS_TO_BTS), are provided. If, however, the RELAY field is set to negative one, this means that the sender has neither a direct link nor a multihop connection to BTS, and the message is silently discarded. The significance and use of the RELAY_CAPACITY flag are illustrated in the following subsection.

Resource Manager

As capacity is a scarce resource in wireless environments, the protocol should ensure that the relaying of calls does not degrade the performance of the relay nodes. To this avail, the resource manager entity is responsible for coordinating the allocation of resources of a relay node. An A-GSM–to–GSM connection routed through a mobile utilizes resources of this mobile. These resources include link bandwidth, buffer space, and processing time. Thus, a certain number of connection requests can be established in parallel via a relay mobile. When a MT receives a request to set up a connection, the Resource Manager will execute a function called the Connection Admission Control (CAC) to define whether or not it can accept the connection. If the connection is accepted, the resources are reserved.

Furthermore, the resource manager is responsible to inform the BEACON entity when no resources are available for relaying. The lack of resources is in turn indicated in the BEACON messages, using the RELAY_CAPACITY flag, so that nodes that receive the message and currently do not have a connection through the sender silently discard the message.

Assuming that MTs are capable of relaying multiple calls, a protocol parameter then is the relaying (or forwarding) capacity per node; that is, the number of calls that a MT can simultaneously relay.

An alternative for relaying for a mobile radio with no resources would be to declare itself as "busy" instead of broadcasting BEACONS with the RELAY_CAPACITY flag set to OFF. Following this approach, the mobile radio will refrain from sending BEACONS until sufficient resources for forwarding are regained. By doing so, the neighbor mobiles of this radio would not unnecessarily be interrupted from their schedule to receive and process these otherwise "useless" BEACONS. Therefore, neighbor mobiles would avoid extra processing overhead, as the BEACONS from this radio do not serve any purpose in the A-GSM system because the radio cannot be of any help for the salvage of calls in case relay is needed from one of its neighbors.

However, it is argued that a mobile should continuously transmit beacons even when the number of calls that it is currently relaying exceeds its forwarding capacity threshold. That is because its neighbor nodes should constantly assess their link to this node so that when resources are found and the mobile node is again able to relay calls, its neighbor nodes would have a good quality indicator of their link to this radio over a long period of time.

A-GSM Encapsulation Protocol (AGEP)

Protocol encapsulation is a simple and easy-to-implement technique for passing arbitrary information through network entities. In this scenario, the AGEP protocol platform is designed to transparently support different user terminal standards through a proprietary A-GSM–specific interface.

In addition, AGEP can be used to improve overall network performance by reducing the number of network control packet broadcasts through encapsulation and aggregation of multiple A-GSM–related control packets (e.g., routing protocol packets, acknowledgments, link status sensing packets, "network-level" address resolution, etc.) into larger AGEP messages. The encapsulation protocol could also provide an architecture for MANET router identification, interface identification, and addressing [12]. The AGEP will run at the network layer (Fig. 10.3) and will be an adjunct to whichever network protocol is using it.

Usage of the AGEP seems to be desirable because per-message, multiple access delay in contention-based schemes such as the IEEE 802.11 standard [3] is significant and thus favors the use of fewer, larger messages. It also may be useful in reservation-based, time-slotted access schemes where smaller packets must be aggregated into appropriately sized network layer packets for transmission in a given time slot. Another purpose of AGEP concerns the commonality of certain functionality in many network-level control algorithms. Many algorithms intended for use in a A-GSM will require common functionality such as link status sensing, security authentication with adjacent routers, one-hop neighbor broadcast (or multicast) reliability of control packets, etc. This common functionality can be extracted from these individual protocols and put into the AGEP protocol thus serving as a unified, generic protocol useful to all.

10.3.3 Call Rerouting Phases

During a call session, the system has to ensure the continuity of its provision when the mobile moves across different base stations as well as different network vicinities within the same cell. This feature is generally achieved by the provision of call rerouting. The rerouting (or handover) process involves three successive phases:

- Radio measurement
- Initiation and trigger
- Handover control

Measurements

A MT participating in an A-GSM protocol has to perform two types of radio link measurements: one for the link to BTS and one to each of its neighbor nodes. As specified in the GSM specification [7], the measurements associated with the BTS radio link are sent to BTS over the SACCH, which can be associated with a TCH or SDCCH. The latter facilitates the handover process while the MT is in the signaling state. A complete SACCH block of data is received by the BS every 480 msec, approximately. Each measurement is averaged over the SACCH block period. In addition to reports relevant to the downlink from the serving cell, the MT also reports the received level of the six strongest surrounding or neighbor cells, as well as their Base Station Identity Codes (BSIC), which can be decoded from the SCH on the BCCH of the neighboring cells.

However, measuring the signal strength of the surrounding MTs (through the beaconing process) raises one major issue: when can MTs perform this beaconing process? In GSM, MTs measure the characteristics of the neighbor cells during the interval between the transmission of an uplink burst and the reception of a downlink burst. The uplink direction is derived from the downlink one by a delay of three burst periods (BPs). However, this approach presents fundamental problems if it is to be applied for the A-GSM beaconing. These problems arise from the fact that the beaconing process requires the MTs be synchronized during the measurement period. On the one hand, different nodes have different transmission/reception patterns, thus making it difficult to achieve synchronization of nodes between the uplink and downlink bursts, whereas on the other hand the intervals during the transmission of an uplink burst and the reception of a downlink burst are of various lengths, depending on the dedicated channel type. These factors prevent this approach from being a solid solution to the problem of beaconing transmission timing.

Within each of these cycles, 24 slots are used for the TCH, one slot for the corresponding SACCH, and there is one slot where nothing is sent. These 26 small intervals between the transmission of an uplink burst and the reception of a downlink burst could then be used by the A-GSM MTs in order to perform signal strength measurements on their surrounding MTs through the transmission and reception of beacon messages.

A second solution could be to lower the TCH rate and use one of the 24 slots to perform the measurements on the surrounding MTs. Alternatively, the idle frame could be shared for both GSM and A-GSM measurements, although this scheme may have some implications on the strength measurements of the surrounding BTSs.

Handover Initiation

Based on signal measurement, a handover decision is made based on both absolute and relative signal strength measurements, and in particular, those measurements taken at the mobile rather than at the base stations (received signal strength at BSs may not be reliable for handover decisions, especially in systems employing power control). Additional parameters such as the bit error rate (BER) or the carrier-to-interference ratio (C/I) could be used as alarm condition indicators in order to increase the efficiency of the handover mechanism.

A-GSM handover (GSM–to–A-GSM and A-GSM–to–A-GSM) occurs when a high probability exists that the call will be lost or the quality of the ongoing connection will be seriously degraded if the current link to BTS (direct or multihop) is not changed. In GSM, this is interpreted as changing the serving cell, whereas in A-GSM, depending on the type of handover, this is interpreted as changing the serving relay (A-GSM–to–A-GSM case) or the serving base station (GSM–to–A-GSM case).

The A-GSM handover may be triggered because of:

1. Serving BTS failure
2. Change of base station, frequency band, or time slot due to signal quality degradation and interference, in order to provide a service with better quality of service, but no neighbor BTS is accessible with adequate signal quality
3. Change of base station, frequency band or time slot due to the mobility of the user in a multicell environment (For example, lack of continuous coverage could initiate the handover procedure, where the user with a call in progress crosses the border of coverage area, but similar to the first two cases, the neighbor BTSs are either not accessible or their communication link quality is below acceptable thresholds.)

Let us assume a mobile that is moving towards its serving BTS while it is also crossing a number of relay mobiles. As shown in Fig. 10.4, due to path loss, the received signal level in the A-GSM pico-cell

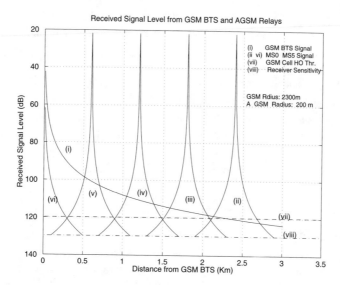

FIGURE 10.4 Received signal levels from GSM BTS and A-GSM relays.

shows about 95 dB variation, whereas within the same distance in the GSM micro-cell, the variation is less than 15 dB.

Many variations in the handover algorithm are possible; the choice of each method would depend on the characteristics of the system. As the author in [11] pointed out in a study on efficient handover techniques in micro-/pico-cell integrated systems, the signal level received in the dual-mode terminal is dependent on the location of the pico-cell base station relative to the position of the micro-cell base station. The study concludes that hysteresis level may not be suitable as a handover initiation criterion in an integrated micro-/pico-cell system.

In A-GSM, one could assume that the coverage of beacon transmissions from a single user forms a virtual pico-cell such that the sender of the beacon is the base station of the pico-cell whereas the radius of the pico-cell equals the transmitting range of the node. Thus the A-GSM terrain can be viewed as a platform that consists of a micro-cell and many virtual "mobile" pico-cells. Similar to the study in [11], it is also concluded here that a hysteresis-based handover is not appropriate in an A-GSM integrated system as prior knowledge of the crossover signal strength between two base stations is required to define the hysteresis margin. Hence, in a GSM–A-GSM system, a threshold-based handover could be applicable.

Thus, in current implementation a threshold-based handover algorithm is proposed and modeled as:

$$\left\{r_{GSM}(x) < T_{GSM}\right\} \text{ or } \left\{r_{GSM}(x) fails\right\} \text{ and } \left\{A-GSM \text{ relay exists}\right\} \text{ and}$$

$$\left\{r_{RELAY}(x) > T_{A-GSM}\right\} \text{ and } \left\{r_{GSM\_NEIGHBOUR}(x) < T_{GSM}\right\}$$

where r_{GSM} and r_{A-GSM} are the average signal levels received from the serving BTS and neighbor mobile relay, respectively. T_{GSM} and T_{A-GSM} are the GSM and A-GSM handover threshold levels, respectively. Therefore, handover is initiated when r_{GSM} falls below the threshold level, there exists a neighbor relay with its signal (r_{A-SM}) strongest among the mobile relay ranking list and higher than the A-GSM handover threshold level, and also, no GSM coverage is available to perform GSM-to-GSM cell handover (the last probability is assumed to be zero in the following discussions as the analysis is focused on a single GSM cell).

The proposed criteria for different A-GSM handover cases could be modeled as:

- GSM–to–A-GSM handover is performed if the following conditions are simultaneously fulfilled:

A. The averaged signal level of the serving BS falls below a threshold T_{GSM} (dBm).

B. (*Optional*) No averaged signal level of any other BS is greater than that of the serving BS by a hysteresis of h (dB).

C. No averaged signal level of any other BS is greater than the threshold T_{GSM} (dBm).

D. The averaged signal level of a neighbor MT is greater than a threshold $T_{A\text{-}GSM}$ (dBm).

- A-GSM–to–A-GSM handover is performed if the following conditions are simultaneously fulfilled:

A. No averaged signal level of any BS is greater than the threshold T_{GSM} (dBm).

B. The averaged signal level of the serving MT falls below the threshold $T_{A\text{-}GSM}$ (dBm).

C. The averaged signal level of a neighbor MT is greater than the threshold $T_{A\text{-}GSM}$ (dBm).

- A-GSM–to–GSM handover is performed if the average signal level of a BTS is greater than the threshold T_{GSM}(dBm).

The different handover decision algorithms can be seen in Fig. 10.5.

Handover Control

The basic type of handover protocol followed in the present GSM system is similar to the mobile-assisted handover (MAHO) [5], in which both network and MT make measurements on radio link parameters. The downlink measurements made by the MT are reported to the network periodically based on both uplink and downlink measurements. The network is then responsible for carrying out the handover procedure.

In an A-GSM system, however, a hybrid handover scheme should be applied. For the cases of MANET-to-GSM and GSM-to-GSM handover, the handover type is MAHO. For all other types of handovers, including GSM-to-MANET and MANET-to-MANET, the mobile-controlled handover (MCHO) is to be applied. In an MCHO scheme, the MT itself makes both the radio link measurement and the handover decision. In terms of handover process, the network will be under the instructions of the relay MTs. As the handover decision is made totally by the MT, the handover can be initiated very quickly. In addition, since the handover decision is made without assistance from the network, handover-related radio link signaling is low during a call. However, there are concerns about the performance aspect of this scheme as the handover decision may not be reliable if the radio link characteristics on multihop uplink and downlink are not correlated.

Furthermore, two cases need to be distinguished with regard to participation of network components in the handover, depending on whether the signaling sequences of a handover execution also involve an MSC. Since the Resource Reservation (RR) module of the network resides in the BSC, the BSS can

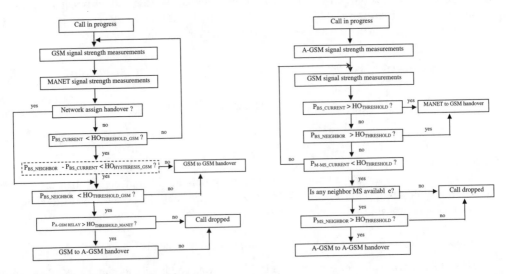

FIGURE 10.5 Handover design algorithms.

perform the handover without participation of the MSC. Such handovers occur between cells that are controlled by the same BSC and are called internal handovers. They can be performed independently by the BSS; the MSC is informed only about the successful execution of internal handovers. All other handovers require participation of at least one MSC, or their BSSMAP and MAP parts, respectively. These handovers are known as external handovers.

As the MAHO scheme is similar to the GSM handover protocol, the internal handover is analyzed in this chapter. The basic signaling structure for a GSM-to-MANET *internal* handover is presented in Fig. 10.6.

In GSM–to–A-GSM handover protocol, a MT requests a handover from its BSS by sending an encapsulated HANDOVER REQUIRED message through its relay MT and starts the timer T3124, as specified in the GSM specification[1] [2].

The A-GSM HANDOVER REQUIRED message (Fig. 10.6) contains the following elements:

- Message type
- Cause (handover type)
- Cell identifier list
- Details of the resource that is required
- MT relay(s)

The recipient MTs along the multihop path try first to see if direct communication with the BTS can be established. If so, the message is sent directly to the BTS. Otherwise, the HANDOVER REQUIRED message is forwarded from each relay MT along the multihop path to the BTS. The forwarding process continues until either a relay MT found to have a direct link to BTS or a MT has neither a relay nor a direct link to BTS. In the latter case, the packet is silently discarded, and the handover phase resumes (with failure. If failure occurs, the node that reports the handover failure either sends an error message to the initiator of the handover or does not assume any action. In the former case, the node that initiated the handover, upon reception of the error message, may initiate a new handover phase by sending a HANDOVER REQUIRED message through a different mobile, while in the latter case the initiator of the handover will eventually react to the failure upon timeout of the time T3124.

If, however, the relaying process has successfully forwarded the encapsulated message to the BSS, the message is forwarded to the BSC which decapsulates it to read the contents of the GSM-compatible HANDOVER REQUIRED message.

On receipt of this message, the BSC shall choose a suitable idle radio resource. If a radio resource is available, then this will be reflected back to the MS in an A-GSM HANDOVER COMMAND within its "Layer 3 Information Element," which is in fact the RR-Layer3 A-GSM HANDOVER COMMAND, and the timer T3103, as specified in GSM specification, is started; again this timer is modified to A-GSM protocol semantics.

Information about the appropriate new channels and a handover reference number chosen by the new BSS are contained in the A-GSM HANDOVER COMMAND.

If, however, handover cannot be carried out, the BSS informs the MS by sending an A-GSM HANDOVER REQUIRED REJECT (A-HANDOVER REQUIRED REJECT) message.

The A-GSM HANDOVER REQUIRED shall be repeated by the MS periodically until:

- An A-GSM HANDOVER COMMAND is received from the MS.
- An A-GSM Handover Required Reject is received.
- The transaction ends, e.g., call clearing.

---

[1] Note, however, that all the timers specified by the GSM standards must be modified accordingly to account for the additional delay imposed from the multihop relaying. This implies that A-GSM timers may differ from the respective GSM timers. Timer resolution might prove to be a problem though for some switching complex manufacturers hoping to reuse existing products.

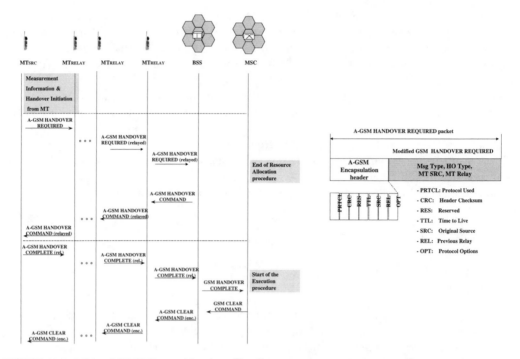

FIGURE 10.6   GSM-to-MANET internal handover signaling.

The sending of the A-GSM HANDOVER COMMAND by the BSS to the MS ends the handover Resource Allocation procedure. The Handover Execution procedure can now proceed.

Upon receipt of the A-GSM HANDOVER COMMAND message, the MT initiates the release of link-level connections, disconnects the physical channels, commands the switching to the assigned channels, and initiates the establishment of lower layer connections (this includes the activation of the channels, their connection, and the establishment of the data links).

After lower layer connections are successfully established, the MT returns a HANDOVER COMPLETE message. The sending of this message on the MT side and its receipt on the network side allow the resuming of the transmission of signaling layer messages other than those of RF management. When receiving this message, the network stops the A-GSM timer T3103 and releases the old channels. The BSS shall then take all necessary action to allow the MS to access the radio resource that the BSS has chosen. Since the new RR traffic connection is essentially an A-RR connection, the BSS shall then switch to the A-GSM mode.

Now, on the MT side, if timer T3124 times out or if a lower layer failure happens on the new channel before the HANDOVER COMPLETE message has been sent, the MT deactivates the old channels and disconnects the TCHs, if any. On the network side, if the timer T3103 elapses before the HANDOVER COMPLETE message is received on the new channels, the old channels are released and all contexts related to the connections with that MT are cleared. Finally, BSC informs the MSC at the completion of the process.

10.4 System Comparisons

To simulate an A-GSM system, a cellular GSM simulator was created [9]. The objective of this simulator was the investigation of the A-GSM network layer throughput performance with respect to the following factors:

- Number of dead spot locations
- Average dead spot size

• Mobile node population

Results from a single GSM cell are presented; that is, intercell handover as well as intercell interference were not considered in these experiments. The simulated cell has a radius $R = 1.5$ km and carrier frequency $f = 900$ MHz. The receiver sensitivity was set to -134 dBm and the transmitting power of base station and mobile station were 10.0 and -33.13 dBW, respectively. The propagation channel included inverse fourth law path loss and a lognormally distributed (correlated) shadowing component. The complete BCCH channel structure has been implemented for the measurements to BTS link (uplink). The beacon rate was set to 1 sec.

The system throughput is defined as the ratio of successfully delivered to BTS calls to the number of generated calls. The throughput results are illustrated in Fig. 10.7 for different scenarios.

A higher number of nodes results in a higher node diversity and average network connectivity. Therefore, the system improvements for higher node populations first plot in Fig. (10.7) are attributed to the fact that the A-GSM multihop connections are more responsive and robust to radio link failures such that a dynamic routing topology is reactively established if, during a call, an A-GSM handover is required. On average, an 8–17% improvement on system throughput is observed for different mobile and dead spot populations.

It is worth noting, however, that a higher number of users trades off user diversity with increased intraMDAL (MANET Distributed Access Layer) as well as inter-cell multiple access interference.

Furthermore, as expected, the average dead spot size also impacts the performance of the system as a higher number of users becomes "trapped"[2] in the dead spots. It is observed that significant improvements could be achieved by using the adaptive A-GSM routing protocol. As illustrated in the second plot of Fig. 10.7, a 10–12% improvement is achieved for different dead spot populations and different dead spot sizes, whereas similar improvements are reported in the third plot of the same figure.

Finally, it has to be mentioned that different values of the simulation parameters can be related to different GSM applications such as residential, public, and business applications. It has to be stressed then that the results are sensitive to assumptions on user mobility, cell size, and transmission powers, highlighting the need for careful investigation of these parameters in the specified application.

10.5 Conclusions

In this chapter, a generic platform for accommodating relaying in the GSM cellular network is described. Integrating two different network architectures involves many details and critical issues, and it is not an easy task to extract out all of the crucial concepts and design specifics. Instead, a network layer platform is presented and some of the functional requirements for enabling relaying of calls in GSM are highlighted. The A-GSM network protocol platform was proposed to improve the GSM system throughput performance. The benefits of an integrated ad hoc GSM protocol are attributed to the fact that dependency of the handover performance on the BTS availability is significantly reduced.

Finally, the complexity of a dual mode system will also impact the design and performance of the integrated system. The choice of each method and option is a tradeoff among the signaling load, the delay, the required level of modification, the implementation complexity and cost, and finally, the operators' and customer's requirements. For a complete description of this work and fundamental discussions on the subject, the reader is further referred to [9].

Acknowledgment

The author would like to thank Kanagasabapathy Narenthiran for his valuable comments on the design and implementation of the GSM radio propagation channel.

---

[2] Note that the average dead spot residence time does not depend on the average velocity of mobiles, as it does when a user crosses a GSM cell with different average velocities. The subway scenario is a representative example.

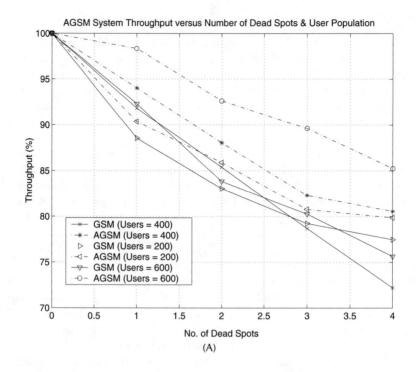

(A)

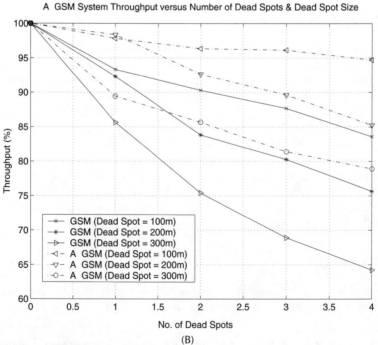

(B)

FIGURE 10.7 (A) (B)   A-GSM system throughput.

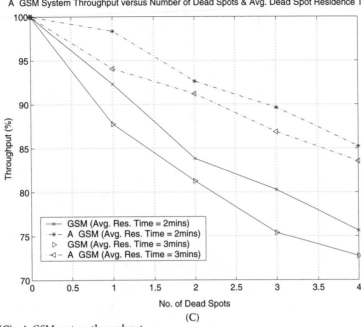

FIGURE 10.7 (C) A-GSM system throughput.

References

1. WB-TDMA/CDMA — System Description Performance Evaluation, Tdoc SMG 899/97.
2. J. Eberspacher and H. Vogel, *GSM Switching, Services and Protocols*, John Wiley & Sons, New York.
3. 3IEEE 802.11, Wireless LAN Medium Access Control (MAC) and Physical Layer (PHY) specification, Standard, IEEE, New York, November 1997.
4. C. Perkins, IP Encapsulation within IP, RFC2003, October 1996.
5. GSM-Rec. 03.09, Handover Procedures.
6. GSM-Rec. 04.06, MS-BSS Data Link Layer Specification.
7. GSM-Rec. 04.08, Mobile Radio Interface Layer 3 Specification.
8. GSM-Rec. 03.02, Network Architecture.
9. G. Aggélou, Dynamic IP Routing and Quality-of-Service Support in Mobile Multimedia Ad Hoc Networks, Ph.D. dissertation, University of Surrey, U.K., 2001.
10. G Aggélou and R. Tafazolli. On the Relaying Capability of Next Generation GSM Cellular Network, *IEEE Personal Communications*, Feb. 2001, pp. 6–13.
11. S. Niri, Advanced Mobility Management Techniques for Hierarchical Cell Structure, Ph.D. thesis, University of Surrey, U.K., May 1998.
12. S. Corson, S. Papademetriou, P. Papadopoulos, V. Park, and A. Qayyum, An Internet MANET Encapsulation Protocol (IMEP) Specification, Internet-draft, draft-ietf-manet-imep-spec01.txt, August 1998, work in progress.

11

IEEE 802.11 and Bluetooth: An Architectural Overview

Sal Yazbeck
Barry University

Abstract

IEEE 802.11 wireless local area networks (WLANs) and Bluetooth wireless technology have positioned themselves as technologies with varied industry applications and as having distinct technical features. IEEE 802.11 applications focus on high-speed WLAN scenarios, whereas Bluetooth focuses on wireless personal area network (PAN) and cable replacement applications. This chapter highlights some of these applications and presents an overview of the architectural differences that make up these technologies, including network configurations, and market trends.

11.1 Introduction

Over the last several years, the wireless world has seen significant developments with a new generation of radio frequency (RF) networking products. The goal of these RF products is to simplify and transform the way we communicate in business and conduct our everyday lives. Third generation (3G) wireless networks are being developed to enable personal high-speed interconnectivity with wide area networks (WANs) through the infrastructure of a wireless Carrier. IEEE 802.11a and b working groups are working to bring wireless high-speed connectivity of computers to corporate or university network infrastructures that interconnect with WAN services. On the personal area side, technologies such as Bluetooth are striving to serve as a cable replacement solution for interconnecting devices and to serve as an ad hoc network service where personal area networks (PANs) between devices can be spontaneously set up to serve a specific function and collapsed when no longer needed.

IEEE 802.11b has been making inroads into universities as a high-speed campus network, while it is also being tested in other markets such as in coffee shops and airports to provide wireless access to networked services. Wireless Carriers are also taking a closer look as to how wireless local area network (WLAN) developments can augment their aspirations for 3G network implementations. The interest here is twofold. One is that wireless LANs utilizing the IEEE 802.11 family of specifications provide a practical high-bandwidth solution to wireless networking. The other is that WLAN operability is license free utilizing the 2.4 GHz Industrial/Scientific/Medical (ISM) and the 5 GHz Unlicensed National Information Infrastructure (UNII) frequency bands.

The original IEEE 802.11 standard, in terms of its Standards roots, presented specifications for a Medium Access Control (MAC) layer and for a Physical (PHY) layer where the use of both Spread Spectrum (SS) and Diffuse Infrared (Ir) technologies is specified.[3] Whereas both techniques of Spread Spectrum technology, Direct Sequencing (DS) and Frequency Hopping (FH), were included in first generation IEEE 802.11 PHY specifications, later versions of the standard, IEEE 802.11a and b, specify compatibility with only Direct Sequencing Spread Spectrum (DSSS). The MAC layer specified in the original IEEE 802.11 standard remains the same for IEEE 802.11a and b. The IEEE 802.11 family of specifications supports a variety of networking functions including roaming, ad hoc communications, and infrastructure based networking.

Bluetooth wireless networking brings a new perspective to the personal dimension. Although it does not provide many of the functions that IEEE 802.11 specifications offer, it does provide service discovery capabilities that enable ad hoc networking and is perceived to provide a cable replacement solution for consumer devices. Bluetooth technology offers the freedom of wirelessly connecting a computer to a cellular phone, a personal digital assistant (PDA) to a computer, or a PDA to a cellular phone, among many other consumer electronics devices. When Bluetooth is used to interconnect a cellular phone to another computing device, the cellular phone can in effect operate as a personal gateway to send and receive information over a local or wide area network. It can also be used to synchronize personal information located on both devices such as schedules and contact lists. These scenarios can occur using Bluetooth wireless networking without the requirement for line-of-sight communications and, in the case of the cellular phone, it can be located within a closed briefcase and can easily interconnect with a PDA or laptop computer. Thus, Bluetooth brings a new level of interconnectivity to the personal dimension.

Bluetooth is also competing in a wireless space where the Infrared Data Association (IrDA) was projected to reign. In its earlier days, the IrDA was unregulated, and it struggled with issues of interoperability and vendor incompatible devices. Even though these issues were addressed to some degree, and Infrared (Ir) ports became available on many consumer electronics devices, IrDA never became an industry or cross-industry standard.[4] This technology has proved to be frustrating to users through its difficulty to configure, requirements of having an exact setup with appropriate system information, and need to aim products at close range at one another since misalignment causes connectivity problems.

Bluetooth aims to provide several key features to the PAN industry. These include offering an air interface that is universal, low cost, user friendly, and capable of replacing the variety of proprietary cables that consumers need to carry and use to connect their personal devices.[1]

Whereas the wireless industry has been struggling the past few years to define the respective roles that the IEEE 802.11 family of specifications and Bluetooth may play in the networking realm, specific niche areas have emerged for each, namely the WLAN market for 802.11a and b and the PAN market for Bluetooth wireless technology.

In this chapter, an overview of key distinguishing features between IEEE 802.11a/b (including the original standard) and Bluetooth specifications is presented. Several sections that highlight the history of each respective technology are also presented, followed by a discussion on key protocol stack reference models and implementation models. The chapter concludes with trends in these technologies and with a summary of the chapter.

11.2 The History of the Technologies

11.2.1 IEEE 802.11 Family of Specifications

The IEEE 802.11 wireless local area network (WLAN) standard is the first WLAN standard that has gained market acceptance. Originally, it was conceived in 1987 by the Institute of Electrical and Electronics Engineers (IEEE) as part of the IEEE 802.4 token bus standard with a given name of 802.4L. In 1990, the 802.4L group was renamed the IEEE 802.11 WLAN Project Committee, which created an independent 802 standard tasked with defining three Physical (PHY) Layer specifications and one common Medium Access Control (MAC) layer for the lower portion of the Data-Link layer for WLANs.[4] Figure 11.1 shows an illustrative comparison between the Open System Interconnect (OSI) model and the IEEE 802 reference model. Whereas the OSI model presents a detailed and structured communications structure, the IEEE 802 model addresses the lower layer portion, specifically, the Data Link Layer and the Physical Layer.

The purpose of the IEEE 802.11 standard was to foster industry product compatibility between WLAN product vendors which, consequently, led to the approval of the IEEE 802.11 standard on June 27, 1997.[3] Since then, two additional IEEE standards have been ratified to extend the data rate of WLANs by enhancing the PHY layer specifications. These current specifications are IEEE 802.11b, ratified in 1999, and IEEE 802.11a, also ratified in 1999. Both standards, IEEE 802.11a and b, share the same MAC specifications with the original IEEE 802.11 standard.[3] The differences are evident in newer PHY specifications, where 802.11a utilizes orthogonal frequency-division multiplexing (OFDM) in the 5 GHz UNII band to achieve a higher data rate of up to 54 Mbps, while 802.11b utilizes complementary code keying (CCK) in the 2.4 GHz ISM band to achieve data throughput of up to 11 Mbps (refer to Fig. 11.2).

11.2.2 Bluetooth Wireless Technology

In July 1999, version 1.0 was published by the Bluetooth Special Interest Group (SIG); the specification is currently at version 1.1. However, Bluetooth started five years earlier in 1994 when Ericsson Mobile

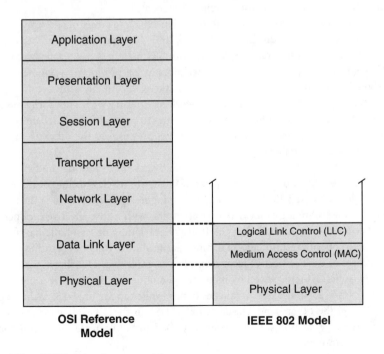

FIGURE 11.1 OSI and IEEE 802 reference models.

Industry Standards	Roaming Support	Supported PHY Technology	Data Rate (in Mbps)	ISM Band (in GHz)	UNII Band (in GHz)	Network Classification
IEEE 802.11	Yes	DSSS, FHSS, Diffuse Ir	1, 2	2.4 - 2.48	N/A	WLAN
IEEE 802.11a	Yes	OFDM	6, 9, 12, 18, 24, 36, 48, 54	N/A	5.15 - 5.25 5.25 - 5.35 5.72 - 5.87	WLAN
IEEE 802.11b	Yes	DSSS	1, 2, 5.5, 11	2.4 - 2.48	N/A	WLAN
Bluetooth	No	FHSS	1	2.4 – 2.48	N/A	WPAN

FIGURE 11.2 Industry standards and key features.

Communications began a study to find out how wireless technology can be used effectively as a cable replacement to link cellular phones with accessories. The study focused on radio links since radio is not directional and does not require line of sight. The choice of using radio had an obvious advantage over infrared technology, which was previously used for the same purpose. Of the many requirements for Bluetooth, support for voice and data were key since the purpose of this technology was to connect phones to headsets and accessories.[4]

Out of this study came the Bluetooth specification for Bluetooth wireless technology. The name Bluetooth comes from the Danish King Harald Blatand (Blatand is Danish for Bluetooth). King Bluetooth is credited with uniting the Scandinavian people during the tenth century. Likewise, the Bluetooth wireless technology aims to unite personal computing devices. The name was selected temporarily pending development of a formal name to apply to this technology. However, the selection of a new name never materialized, and thus the Bluetooth name became permanent.

The Bluetooth special interest group (SIG) was formed in 1998 by a group of core corporate promoters, specifically, Ericsson Mobile Communications AB, Intel, IBM, Toshiba, and Nokia. Today, the core group has expanded significantly, and the number of Bluetooth SIG members has reached into the thousands since only SIG members are entitled to use the technology for product development. However, the introduction of Bluetooth wireless technology to the marketplace initially sent a confusing signal as to how it would compete for a niche in the wireless networking industry where primarily the IEEE 802.11b standard had been introduced. This phenomenon continues to prompt heated debates in industry with regards to which standard, Bluetooth or IEEE 802.11b, will prevail over the other as an internationally viable wireless networking technology.

Accordingly, the wireless industry has taken a competitive stance regarding the two technologies — one stance in favor of 802.11b, and the other in favor of Bluetooth wireless technology. Both technologies operate within the global 2.4 GHz ISM band. In this regard, one argument is among those in the wireless industry stating that the technology that goes to market first with a cost effective product is inherently assured a market winner — a position that the IEEE 802.11 family of standards has already attained. Though, upon analysis of both technologies, a distinct clarification is evident regarding the technical and application differences of each respective technology. For example, Fig. 11.2 shows that the IEEE 802.11 family of products is increasingly viewed by its technical abilities for supporting a high-speed wireless network in the unlicensed 2.4 and 5 GHz frequency bands, while supporting DSSS technology. Bluetooth wireless technology, on the other hand, supports only FHSS in the 2.4 GHz band and is increasingly perceived as a short-range ad hoc solution primarily used for cable replacement on computers and consumer electronics products.

11.3 Network Configuration Models

11.3.1 Key IEEE 802.11 MAC-Layer Features

In this section, we will discuss key characteristics of the common MAC layer that is shared by the original IEEE 802.11 as well as the enhanced IEEE 802.11a and b specifications.

The common MAC published in the original IEEE 802.11 specification works with two basic network configurations, as represented by Fig. 11.3. These are:

1. *Infrastructure configuration:* Computers connect directly through Access Points (AP). As part of a wider network infrastructure, APs provide devices with access to a range of networks for an extended coverage area (Fig. 11.3a).
2. *Ad hoc configuration (also called peer-to-peer networking):* Computers interact with one another independent of any infrastructure support. Devices communicate directly with one another and thus provide a very limited coverage area (Fig. 11.3b).

In a typical application, laptops carry a PC card that connects the laptops with the wired network via the AP, which also contains a PC card. The PC card in the laptops and APs supports the MAC and PHY of the IEEE 802.11. The AP in this case acts as a bridge between the laptops and the wired network to convert the IEEE 802.11 protocol to the appropriate MAC and PHY layers operating on the wired network, which is usually 802.3 Ethernet LAN.[4]

The IEEE 802.11 MAC layer provides a basic access mechanism that supports several characteristics such as clear channel assessment, link setup, authentication, roaming, power management, and channel synchronization.[2] Roaming, a key feature of IEEE 802.11 WLANs, allows mobile users to move between Basic Service Areas (BSA). Essentially, a BSA is the coverage area of one access point. Figure 11.4 illustrates how two access points are interconnected with the wired backbone infrastructure while a mobile user seamlessly moves between two BSAs.

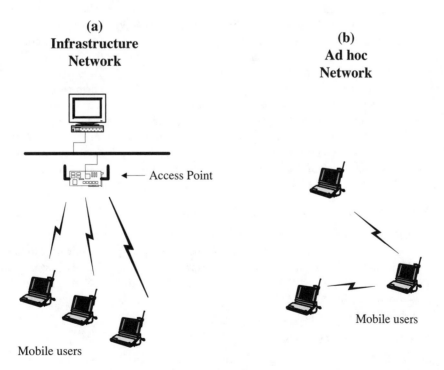

FIGURE 11.3 Topologies for IEEE 802.11 WLANs: (a) infrastructure network; (b) ad hoc network.

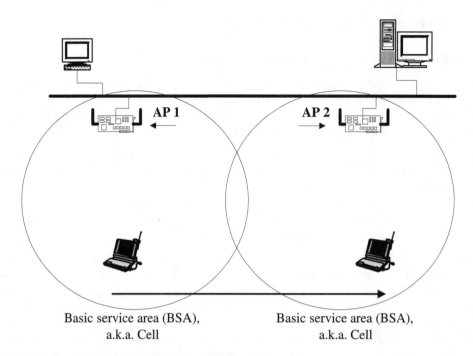

FIGURE 11.4   Roaming feature for IEEE 802.11 WLANs.

This roaming capability is a key feature of the IEEE 802.11 family of specifications that Bluetooth does not support.

The IEEE 802.11 MAC layer also specifies the use of the Carrier Sense Multiple Access (CSMA) protocol with Collision Avoidance (CA) to reduce packet collisions on the network. The CSMA used in wireless networks is similar to the CSMA scheme used in wired LANs. However, the collision detection (CD) technique for wired LANs cannot be used effectively for wireless LANs since nodes cannot detect over-the-air collisions when they occur. The absence of detection is caused by the strong signals present at the transmitters that also serve to drown out other communicating signals.[7] Accordingly, collision avoidance (CA) adds many features to the CSMA scheme on WLAN systems to help reduce the number of over-the-air collisions that can occur. A wireless node utilizes CSMA to listen to or sense another carrier (or the presence of another transmission) before attempting to transmit its own data packets. If another transmission is sensed, the terminal delays its own transmission until the medium becomes available. The CA scheme in CSMA/CA further provides a random back-off delay feature before a new transmission attempt is executed. This random delay helps avoid collisions from simultaneous multi-user transmissions, since other wireless nodes could also be waiting to send data over the network.[7]

11.3.2 Key IEEE 802.11 PHY-Layer Features

The original 802.11 standard specifies the use of three different PHY layers, any of which can utilize the same MAC layer. These PHY layers include two spread spectrum techniques, Frequency Hopping Spread Spectrum (FHSS) and Direct Sequence Spread Spectrum (DSSS). The third PHY layer is an optical technique utilizing Diffuse Infrared (DFIR). The rest of this section discusses issues that are important to the application of the IEEE 802.11 PHY layer. Specifically, spread spectrum technology, the ISM band, and the UNII band are presented.

Spread-Spectrum Technology

The original IEEE 802.11 standard and the revised IEEE 802.11b specification utilize spread spectrum technology to operate within the unlicensed Industrial/Scientific/Medical (ISM) bands located at 2.4 to 2.4835 gigahertz (GHz). IEEE 802.11a wireless LANs operate at the 5 GHz Unlicensed National

Information Infrastructure (UNII) bands, which became available for use in 1985. Both ISM and UNII frequency bands present an important market opportunity in terms of the IEEE 802.11 family of specifications since no licensing is required by the Federal Communications Commission (FCC) to implement WLAN networks utilizing their frequencies.

The spreading function in spread spectrum (SS) is achieved by using a nonspread signal (such as narrowband microwave) and combining it with one of the following two schemes: direct sequence (DS) or frequency hopping (FH).[6] Direct sequence transmission is supported by the original IEEE 802.11 family of specifications and utilizes the entire available bandwidth to transmit a signal over the air. This transmission uses a spreading function to multiply the bit stream with higher-frequency signals, thus achieving a spread signal. As the signal is received, the original data stream is recovered by correlating it with that same spreading function.

Frequency hopping (FH), on the other hand, operates by different technical means and is used only by the original IEEE 802.11 specification. In order to enable the narrowband signal to operate on a much wider bandwidth, the spreading effect is achieved by moving the transmitting and receiving mechanisms between frequencies (78 hopping channels in the 2.4 GHz band) in discrete time. Then, as data come in, the receiver takes on the task of recovering the original signal, which was initially sent by the transmitting end.

Diffused Infrared (IR)-Based WLANs

The IEEE 802.11 original specification calls also for an optical PHY layer utilizing Diffused Infrared (DFIR) as a carrier medium. Diffused IR is another method available to network a multi-user environment. The only way to use IR in radio-like form is to implement DFIR technology within a room environment.[7] This type of IR floods a room with IR radiation which then allows many users to connect to a wired network from anywhere within that room. The IEEE 802.11 standard is the industry standard that defines diffused IR LAN technology. This standard specifies a Basic Access Rate (BAR) data speed of 1 Mb/sec at 16 PPM (pulse per minute) modulation and an Enhanced Access Rate (EAR) of 2 Mb/sec at 4 PPM.[3]

As with all infrared products, good security measures are achieved with optical LANs since IR signals cannot penetrate through walls. This nonpenetration feature helps protect transmissions against external data detection and tampering.

11.3.3 Key Bluetooth Communication Features

Bluetooth is the first popular technology for short-range voice and data communication. Unlike the IEEE 802.11 WLAN standards, Bluetooth supports several unique features including a lower data rate and lower power consumption, and it has a wireless PAN (WPAN) designation. This section will present an overview of the protocol architecture of Bluetooth and its network topologies.

Bluetooth and IEEE 802.15 WPAN

In June 1997, a WPAN project effort was initiated by the IEEE as part of the IEEE 802.11 project group. Subsequently, the first WPAN specifications were published in January 1998. In 1998, Bluetooth responded to an IEEE invitation for participation in WPAN standardization development. In March 1999, the IEEE 802.15 project committee was approved to handle WPAN standardization. Bluetooth has since been selected as the base specification for IEEE 802.15 WPANs.[4]

Bluetooth Protocol Stack

A key feature of the Bluetooth specification is to foster device interconnectivity from a variety of vendor products. In this regard, Bluetooth does not only define a radio system, but also introduces a layered protocol stack that enables applications to discover other Bluetooth devices in an area, discover what services they present, and contain the capability to use those services.[2] In an effort to understand this communication process of Bluetooth, it is useful to draw a comparison between Bluetooth and the familiar OSI standards model for protocol stacks, even though the layers do not exactly match. Figure 11.5

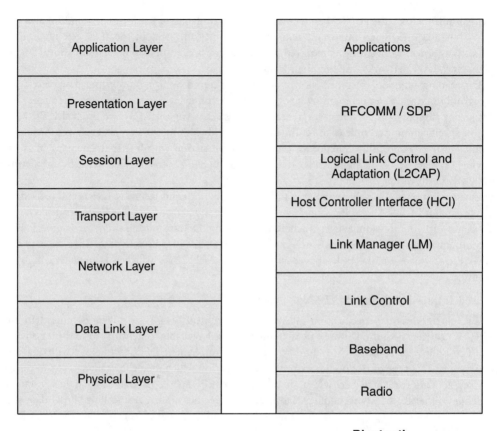

OSI Reference Model   **Bluetooth**

FIGURE 11.5 OSI and Bluetooth reference models.

shows a side-by-side illustration of the OSI model and Bluetooth with their respective communication hierarchies. The Bluetooth protocol stack represents eight layers, as compared to the OSI reference model, which represents seven layers. Layer one of the Bluetooth protocol stack is called the Radio Layer and is responsible for the electrical interface to the communications media, coding/decoding, and modulation/demodulation of data for transmission. Here, the license-free ISM band is at 2.4 GHz and is based on the FHSS scheme. The operating band is divided into 1 MHz–spaced channels with each signaling data at 1 Megasymbol per second (Ms/sec) thus attaining maximum channel bandwidth availability.[5] Utilizing the chosen modulation scheme of Gaussian Frequency Shift Keying (GFSK), data throughput of 1 Mb/sec is achieved.

The Baseband Layer and Link Controller overlap to cover the functionalities of the Data Link Layer. Together they control the physical links via the Radio Layer by assembling packets, controlling frequency hopping, and performing error checking and correction. Above these levels, a direct comparison between the two reference models becomes less clear. The OSI Network Layer is responsible for data transfer across the network. This process occurs independent of the network topology and types of media traversing it. This role compares with the Link Manager (LM) Layer of Bluetooth, which controls and configures links to other Bluetooth devices. The LM is responsible for connecting Slaves to a piconet and creating their active member addresses.[5] Additionally, LM serves to establish asynchronous connectionless (ACL) data and synchronous connection-oriented (SCO) voice links and is capable of putting connections into low-power mode. The Transport Layer is responsible for the reliability and multiplexing of data transfer across a network. Thus, it overlaps with the LM and the Host Controller Interface (HCI) Layers, where HCI handles communications between a separate host and a Bluetooth module.

The Session Layer is responsible for the management and data flow control services. Accordingly, the Logical Link Control and Adaptation (L2CAP) Layer multiplexes data from higher layers and converts between different packet sizes. The Presentation Layer provides a common representation for Presentation Layer data, which RFCOMM/Service Discovery Protocol (SDP) handles by emulating serial connections similar to RS232 serial ports, since Bluetooth involves mainly point-to-point links. The SDP portion of RFCOMM allows Bluetooth devices to discover what services other Bluetooth devices support. Finally, the Application Layer manages communications between host computers.[2]

Bluetooth Configuration

Bluetooth networks can operate in one of two network configurations: as a Master, or as a Slave. The Master is responsible for setting the frequency hopping sequence that the Slaves will tune into. Thus, Slaves synchronize to the Master in time and frequency by applying the master's hopping sequence.[2] Bluetooth specifies a 10 meter (about 30 feet) radio range and supports up to seven devices in a piconet. Within a piconet, point-to-point full-duplex communication is used between the Master and Slave. Frequency Hopping (FH) is used to combat interference presence and fading. Bluetooth can support three full-duplex voice channels concurrently in a piconet at a data rate of 721 kb/sec while transmitting at a power rate of 800 microamps.[5]

Every Bluetooth device contains a unique Bluetooth device address and a Bluetooth clock. The Baseband part of the Bluetooth protocol stack, layer two, contains an algorithm that can calculate a frequency hop sequence from a Bluetooth device address and clock of the Master. Slaves can use this to calculate the frequency hop sequence. In addition, since all Slaves use the Master's clock and address, they are all synchronized to the Master's frequency hop sequence.

Data traffic is also controlled by the Master. The Master permits Slaves to transmit by allocating slots for voice or data traffic utilizing Time Division Multiplexing (TDM). The Master further controls the total available bandwidth and how it will be divided among the Slaves. Multi-hop communications are achieved through the scatternet concept, where several Masters from different piconets must establish links with each other. In this context, the Master becomes the bottleneck.[5]

Bluetooth Topologies

Bluetooth wireless networks can be implemented in two network topologies: piconets and scatternets. A piconet is a collection of Slave devices operating together with one Master device. They can take the form of a point-to-point design where only one Slave and a Master exist in a network, or they can take on a point-to-multipoint design where one Master is connected to many Slaves in a network. Figure 11.6 shows a point-to-multipoint architecture where the Master becomes the head of the piconet and also serves as the central controller. The Bluetooth specification limits the number of slaves to seven within a piconet. Should other devices be present within a piconet, they should not be active or will be considered Parked.[5]

All devices in a piconet adhere to the same frequency hopping and timing provided by the Master, and a direct link is only made between the Master and Slave, not directly between Slaves. Thus, communication between Slaves must be routed through the Master.

The overlap of one piconet over another results in the formation of a scatternet. As Fig. 11.7 illustrates, when such an overlap occurs, a Master of one piconet has to serve as a Slave of the other piconet. No device can serve as a Master of two piconets. When a Slave from one piconet wishes to communicate with another Slave from another piconet, both Masters from each piconet must be involved in the relay of packets across the piconets.[2] As additional piconets overlap, it is possible for one Master to serve as a Slave of two piconets. In such a scenario, this Master/Slave acts as a network bridge and router across piconets. However, such a multi-hop scenario poses performance degradation issues due to the presence of time switching among piconets, as well as potential signal interference from adjacent piconets.

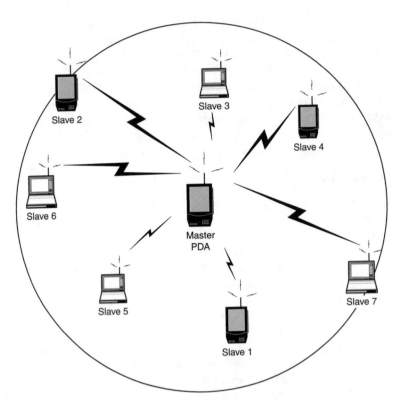

FIGURE 11.6   Bluetooth piconet architecture.

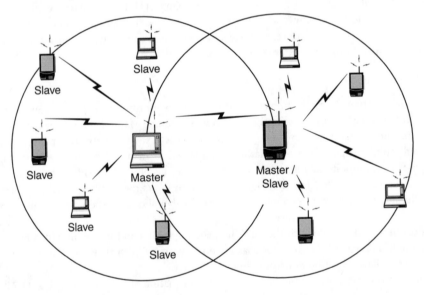

FIGURE 11.7   Bluetooth scatternet architecture.

11.4 Trends in Wireless Networking

As highlighted earlier, there has been much discussion regarding the existence of both IEEE 802.11b and Bluetooth. Both compete in the same frequency band and in the same market, although the IEEE 802.15

project group is handling this issue through coexistence developments. Consider the following scenario regarding coexistence. A traveler in an airport might use a Bluetooth enabled device such as a PDA or cellular phone via associated access points to check flight schedules and to book hotel reservations. Then, the traveler would use a WLAN (i.e., IEEE 802.11b) to surf the Web and to view video e-mails via other WLAN access points. The difference in power consumption requirements (i.e., laptop vs. smart phone or PDA) as well as the cost of the product used would contribute to such a pattern in network usage.

Additionally, when product costs eventually drop to a level that is attractive to consumers, Bluetooth might become an attractive technology for the small office network. In such a case, small clustered piconets could conceivably interconnect common office printing and disk sharing functions. Bluetooth access points can further provide interconnectivity with backbone networks or to the WAN.[2]

Given its specification guidelines, Bluetooth does not compete directly with WLANs. The Bluetooth SIG is already working on efforts that can lead to coexistence between other WLAN standards in the ISM band. For example, a user might want to insert an IEEE 802.11b PC card into a notebook that already contains a Bluetooth module. Considering this case, manufacturers developing Baseband and radio devices that can work with more than one ISM band communications protocol would create possibilities of bridging devices that could link Bluetooth to 2.4 GHz WANs.

Already, several developments are underway that can significantly enhance the communications experience. For example, the existing work that is evident by the IEEE 802.11 family of standards has delivered the IEEE 802.11a standard, which offers WLAN data throughput of up to 54 Mb/sec in the 5 GHz frequency bands. A perception exists that eventually all WLAN networking functions would migrate towards this 5 GHz ISM band. Bluetooth, on the other hand, is foreseen to provide specific cable replacement roles such as enabling a phone to work with hands-free devices in cars, for instance, where usage applications can be many and varied. This is because a car can communicate internally with the occupant as well as between its automotive parts. Additionally, a car can communicate with its external surrounding environment such as with a parking meter for payment. Similarly, it can also communicate with diagnostic machines in an auto repair shop. Accordingly, coexistence is seen as a key trend in Bluetooth and IEEE 802.11b communications.

11.5 Chapter Summary

The IEEE 802.11 family of specifications represents the original IEEE 802.11 WLAN standard as well as the newer standards IEEE 802.11a and b. All three versions operate in the global ISM band. The original standard permits data rates of 1 Mb/sec with an optional 2 Mb/sec rate in the 2.4 GHz band, IEEE 802.11a permits data rates of up to 54 Mb/sec in the 5 GHz band, and IEEE 802.11b has a data rate of up to 11 Mb/sec in the 2.4 GHz band. WLANs offer infrastructure-based and ad hoc networking among computing devices.

Bluetooth belongs to the WPAN market, operates in the ISM 2.4 GHz band, and has a data rate of 1 Mb/sec. It is a specification for short-range, low-cost, and small form factors that enables user-friendly connectivity between handheld devices. It was developed with the notion of replacing cable connectivity among such devices. Bluetooth offers ad hoc network communication among Bluetooth devices and supports both voice and data capabilities.

Both IEEE 802.11b and Bluetooth have been much debated as to which technology would eventually succeed the other. However, coexistence is projected for these technologies as each is seen to support a niche in the market — IEEE 802.11 supporting the WLAN segment, while Bluetooth supports the WPAN segment. Accordingly, Bluetooth wireless technology is currently the base specification for the IEEE 802.15 WPAN workgroup.

References

1. Bisdikian, C., An Overview of the Bluetooth Wireless Technology, *IEEE Communications Magazine*, Dec. 2001, 86.
2. Bray, J. and Sturman, C., *Bluetooth: Connect Without Cables*, 2nd ed., Prentice Hall, New York, 2002.
3. IEEE, Wireless LAN Medium Access Control (MAC) and Physical Layer (PHY) Specifications, IEEE 802.11, The Institute of Electrical and Electronics Engineers, New York, 1997.
4. Pahlavan, K. and Krishnamurthy, P., *Principles of Wireless Networks, A Unified Approach*, Prentice Hall, New York, 2002.
5. Toh, C.-K., *Ad Hoc Mobile Wireless Networks, Protocols and Systems*, Prentice Hall, New York, 2002.
6. Venkatraman, S., Anytime, anywhere — the flexible networking environment of wireless LANs, *Journal of Systems Management*, 45, 6, 1994.
7. Wickelgren, I.J., Local-area networks go wireless, *IEEE Spectrum*, 33, 34, 1996.

IV

Routing Techniques in Ad Hoc Wireless Networks, Part I

12

Position-Based Routing in Ad Hoc Wireless Networks

Jörg Widmer
University of Mannheim

Martin Mauve
University of Mannheim

Hannes Hartenstein
NEC Europe Ltd.

Holger Füßler
University of Mannheim

Abstract

In ad hoc networks, autonomous nodes collaborate to route information through the network. Commonly, nodes are end-systems and routers at the same time. In cases in which nodes have a notion of their geographic position, position-based routing protocols can be used; these protocols' properties of statelessness and fast adaptability to changes in the topology match very well with the characteristics of ad hoc networks.

Position-based routing does not require the maintenance of routes; instead, forwarding decisions are made locally, based only on the node's own position, the positions of its neighbors, and the position of the destination. The routing algorithms are complemented by location services through which a node can obtain the position of a packet's destination.

We will introduce the main components of position-based routing, discuss position-based routing algorithms as well as location services, and present an application scenario where position-based routing can be used for intervehicle communication.

12.1 Introduction

Position-based routing protocols use the geographic position of nodes to make routing decisions in ad hoc wireless networks. In order to use a position-based approach, the nodes forming the network must be able to determine their own positions. This can be done by means of GPS or some other type of positioning mechanism [7,16]. Such positioning mechanisms are beyond the scope of this chapter; a survey regarding this topic can be found in [12]. The routing decision at each node is based on the destination's position and the position of the forwarding node's neighbors. Typically, the packet is forwarded to a neighbor that is closer to the destination than the forwarding node itself, thus making progress toward the destination. In order to inform all neighbors in radio range about its own position, a node transmits beacons at regular intervals. The position of a packet's destination can be obtained from a so-called location service.

Position-based routing does not require the establishment or maintenance of routes. The nodes neither have to store routing tables nor transmit messages to keep routing tables up to date. This is an important advantage if the topology of a network changes fast, as is the case in many ad hoc wireless networks. Also, when the path from the source to the destination changes while a packet is en route, the packet does not have to be discarded as would often be the case with non–position-based routing mechanisms. As a further advantage, position-based routing supports the delivery of packets to all nodes in a given geographic region in a natural way. This type of service is called geocasting [22].

Generally, position-based routing can be separated into two parts: the location service and the actual routing of data packets. The location service maps the ID of a node to its geographical position. The location service is required by the sender of a data packet to find the location of the destination. The resulting location is usually included in the packet header. Intermediate nodes may or may not consult the location service again to obtain a more accurate position of the destination. During the routing process, a node needs to determine the neighbor to which the packet should be forwarded. This includes the handling of cases where no direct neighbor exists that is closer to the destination than the forwarding node itself.

In the remainder of this chapter, we will first investigate existing location services and their characteristics. Then, different routing strategies are discussed in detail. Finally, we give an example of how position-based routing can be used in vehicular networks to enable communication in a highly dynamic network.

12.2 Location Services

Position-based routing aims to forward packets in the geographic direction of the packet's destination. This requires that nodes can query the location of a packet's destination through a so-called location service. Since a main concept of ad hoc networking is the independence of a fixed infrastructure, a key design principle for such a location service is to utilize a distributed algorithm. The failure of single nodes of the network should not disrupt the service. Furthermore, scalability to large networks is a desirable feature.

12.2.1 Classification

Location services for mobile ad hoc networks can be classified according to the following criteria:

- Architecture: distributed vs. centralized
- Update strategy: reactive vs. proactive
- Structure: hierarchical vs. flat
- Type: some-for-some vs. some-for-all vs. all-for-some vs. all-for-all

As discussed above, centralized location services are not well suited for mobile ad hoc networks. Hence, we will concentrate our discussion on mechanisms with a distributed architecture.

The update strategy decides when the position information is updated. Reactive services only take action when they receive a location query and then try to find out the location of the node in question. Proactive services continuously disseminate up-to-date information about the locations of the nodes. Thus, proactive services usually have a shorter response time to queries, but the maintenance of the location information causes a higher communication overhead when queries are infrequent. It is possible to combine reactive and proactive elements to reach a compromise between response time and update complexity.

By introducing a hierarchy into the dissemination or storage of the location information, the scaling characteristics of a location service can be improved. A location service operating on a flat set of nodes is less complex but may be limited in the number of nodes it can support. Location services can further be classified according to how many nodes participate in providing the service. These can either be some specific nodes or all nodes of the network. Furthermore, each node that does participate may maintain the position information of some specific nodes or all nodes of the network. We abbreviate the four possible combinations as some-for-some, some-for-all, all-for-some, and all-for-all.

The location services differ in update and query complexity. In order to assess these criteria, we determine the number of one-hop transmissions required to perform updates and queries with respect to the number of nodes (n) in the network. For this purpose, we assume that the density of nodes remains constant when the number of nodes increases and that the average distance between two uniformly sampled participants increases proportional to the square root of the increase in nodes [13].

The more nodes participate in the location service and the higher the number of nodes that store location information for a particular node, the higher the update overhead. A large amount of information is invalidated when a node moves to a different location, and a large number of update messages may be necessary. The advantages of redundant information are a reduced overhead for location queries and a higher resilience against network failures. By limiting the number of nodes that store information, update overhead is reduced at the cost of a higher location query complexity. Table 12.1 classifies the location services discussed in detail in the next section according to the above criteria.

12.2.2 Reactive Location Services

A simple reactive location service is RLS, as described in [9]. RLS stores information about the location of a node only at the node itself. Querying the location of a node is equivalent to reaching that node with the query, and the node can then respond with its location. A node's position is queried by flooding the query packet. Instead of immediately flooding with the maximum hop distance (i.e., the diameter of the network), it is possible to gradually increment the flood radius until the corresponding node is reached. The characteristics of the algorithm are largely determined by the chosen method of incrementing the search radius (e.g., linear or exponential) and the time intervals between successive attempts. RLS does not require any position updates. The overhead of a single position query scales with $O(n)$.

The mechanism is fairly simple to implement and very robust against node failure or packet loss. Furthermore, the location service only consumes resources when data packets have to be sent. Since only the node itself maintains its location, storage requirements are minimal. Nevertheless, the overhead caused by flooding location requests makes such a mechanism unsuitable for scenarios where location queries are frequent or the network is large.

12.2.3 Proactive Location Services

12.2.3.1 Homezone

The Homezone [11] and Home Agent [23] location services introduce the concept of a virtual home region of a node. All nodes within the virtual home region of a certain node have to maintain up-to-date position information for that node. Through a well-known hash function, the identifier of a node is hashed to a position, and the virtual home region is formed by all the nodes within a certain radius of that position. The radius has to be chosen such that the virtual home region contains a sufficient number of nodes. To obtain the position of a node, the same hash function is applied to the node identifier, and a location query can then be sent to the resulting position of the home region. Any node within the home region can answer the query.

The schemes operate on a flat set of nodes without any hierarchy. The reduction in complexity comes at the expense of increased inflexibility and inefficiency. Nodes can be hashed to a far away home region leading to high response delays. Furthermore, if only one home region exists per node, it is possible that the home region of a node cannot be reached (e.g., because of network partitioning). If the home region

TABLE 12.1 Classification of the Discussed Protocols

Mechanism	Update Strategy	Structure	Type
RLS	Reactive	Flat	All-for-some
Homezone	Combined	Flat	All-for-some
DREAM	Proactive	Flat	All-for-all
GLS	Combined	Hierarchical	All-for-some
GRSS	Proactive	Hierarchical	All-for-all

Note: RLS, Reactive Location Service; DREAM, Distance Routing Effect Algorithm for Mobility; GLS, grid location service; GRSS, Geographical Region Summary Service

is sparsely populated, its size has to be increased. As a consequence, several tries with increasing radius around the center of the home region may be necessary for location update messages as well as queries. Since only the home region is queried and updated, the overall communication overhead of this scheme scales with $O(\sqrt{n})$.

12.2.3.2 Distance Routing Effect Algorithm for Mobility

For the location service used in the Distance Routing Effect Algorithm for Mobility (DREAM) approach [1], each node stores location information for each other node of the network. It can therefore be classified as an all-for-all approach.

A node broadcasts position update packets to update the position information maintained by the other nodes. A node can control the accuracy of its position information available to other nodes by (1) modifying the frequency with which it sends position updates and (2) indicating how far a position update packet is allowed to travel before being discarded. The temporal resolution of the updates is coupled with the mobility rate; the higher the speed of a node, the more frequent the updates it sends. Location updates with a high maximum hop count are sent less frequently than updates that only reach nearby nodes. Thus, a node provides accurate location information to its direct neighborhood and less accurate information (because of fewer updates) for nodes farther away. The reasoning for this update strategy is that "the greater the distance separating two nodes, the slower they appear to be moving with respect to each other" (termed the distance effect) [1]. The distance effect is a reasonable paradigm when intermediate hops are allowed to update the position information contained in the destination address of a packet. The closer the packet gets to its final destination, the more accurate the position information contained in the packet header.

Compared to periodically flooding the network with location information, DREAM achieves a substantial reduction in the communication overhead it produces. Nevertheless, nodes need to flood the network occasionally to provide faraway nodes with their location, and thus the update complexity is $O(n)$. Since a node that wishes to communicate with another node already knows approximately where the target node is located, there is no need to send location queries. However, the storage requirements for keeping a list with entries for each node at all the nodes are very high. This, together with the necessary flooding of the whole network, limits the scalability of DREAM to small networks.

12.2.3.3 Grid Location Service

The Grid Location Service (GLS) [20] divides the area covered by the ad hoc network into a hierarchy of squares. In this hierarchy, n-order squares contain exactly four $(n - 1)$–order squares, constructing a so-called quadtree. The lowest order squares typically have a size comparable to the radio range of a node whereas the highest order square covers the whole network. Each node maintains a table of all other nodes within the local first-order square. The table is constructed with the help of periodic position broadcasts, which are scoped to the area of the first-order square.

We demonstrate the mechanism using a simple example (Fig. 12.1). To determine where to store position information, GLS establishes a notion of near node IDs, defined to be the least ID greater than a node's own ID within a given n-order square. (ID numbers wrap around after the highest possible ID.) When node 10 in the example wants to distribute its position information, it sends position updates to

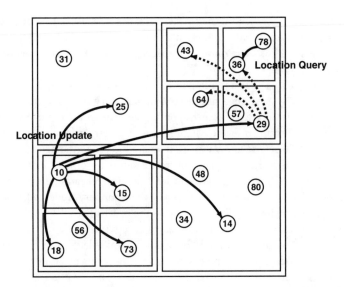

FIGURE 12.1 Grid location service.

the node with the nearest ID in each of the three surrounding first-order squares. Thus, the position information is available at nodes 15, 18, 73 and all nodes that are in the same first-order square as 10 itself. In the surrounding three second-order squares, again the nodes with the nearest ID are chosen to host the node's position; in the example these are nodes 14, 25, and 29. This process is repeated until the area of the ad hoc network is covered. The "density" of the position information decreases logarithmically with the distance to the node that updates its position.

Let us assume that node 78 wants to obtain the position of node 10. It therefore should locate a "nearby" node that knows about the position of node 10. An ideal candidate in the example would be node 29. It is therefore useful to have a look at the position servers for node 29. Its position is stored in the three surrounding first-order squares in nodes 36, 43, and 64. Note that each of these nodes and node 29 are automatically also the ones in their respective first-order squares with the nearest ID to 10. Thus, there exists a "trail" with descending node IDs to the correct position server from each of the squares of all orders. Position queries for a node can now be directed to the node with the nearest ID the querying node knows of. In our example this would be node 36. The node with the nearest ID does not necessarily know the sought-after node, but it will know a node with a nearer node ID (node 29, which already is the sought-after position server). The process continues until a node that has the position information available is found.

Note that a node does not need to know the IDs of its position servers, which makes a bootstrapping mechanism that discovers a node's position servers unnecessary. Position information is forwarded to nodes with nearer IDs in a process closely resembling position queries, only that information is written instead of read, ensuring that the position information reaches the correct node, where it is then stored.

Since GLS requires that all nodes store the information of some other nodes, it can be regarded as an all-for-some approach. The burden of maintaining position information is distributed evenly among all nodes. The hierarchical structure of GLS allows scaling to large networks, since update and query complexities both scale with $O(\sqrt{n})$.

12.2.3.4 GRSS

Another approach using a grid-like hierarchy of the network is proposed in [14]. In this location service, called GRSS (Geographical Region Summary Service), the network is logically partitioned into overlapping squares of different orders, as described in the previous section on GLS. This structure is a priori known such that each node knows in which square of each order it resides.

All nodes know their neighbors within the smallest square (order-0). A local routing protocol (using link-state routing) generates exact paths to each of them. The size of order-0 squares should be chosen small enough so that the local link-state routing protocols provide valid routes even when node mobility is high (e.g., two-hop radio range).

Using the knowledge of all node IDs residing in the same order-0 square, the nodes at the border of an order-0 square periodically generate a summary. This summary is a bit vector with one bit for each node ID in the network. A bit is set if the corresponding node is in the order-0 square, otherwise it is not set. The summary is transmitted to the adjacent order-0 square, where the information is flooded. Thus the nodes in an order-0 square know which nodes are located in the neighboring order-0 squares. The nodes at the border of the order-1 square generate a summary of all nodes in that order-1 square and transmit this information to the neighboring order-1 squares, repeating the process described above. A boundary node of order n (i.e., the shared order of the boundary node and its siblings) generates summaries for each of the squares it is located in up to the order of n.

At the end of this process, a union of all summaries covers the whole network except for the own order-0 square, for which complete routing information is available. Similar to DREAM, the closer a node is located to a given node, the more precise the knowledge of its position gets. For example, if two nodes are located in the same order-0 square, they know the exact position of each other. If they are located in different order-0 squares but in the same order-1 square they know in which order-0 square the other node resides and so on.

Besides the "exact summary" generation that uses one bit for each node in the network, GRSS also supports "imprecise summary" generation. In order to decrease the length of the bit vector, imprecise summary generation uses the technique of bloom filters [4]. To generate an imprecise summary, t hash functions (for $t \geq 1$), which generate t bit positions in the vector, are applied to the node's ID. If a node resides in the generating node's own order-0 square, all bits at the t positions are set in the vector. If one wants to know if a certain node (ID) is located in the corresponding square, the t hash functions with the node's ID as an input are applied to the bit vector. If all the corresponding bits are set, the node is located in that square with a very high probability. However, there is the possibility of a set of different nodes setting the bits belonging to the node ID ("false positive"). The consequence of a false positive is that there are two or more alternatives for the location of the node. Thus, one would have to either duplicate the packet, resulting in additional overhead, or just choose one alternative, which can result in a detour of the packet. Typically one would choose the latter option to remain scalable.

As all nodes have some information about the location of all other nodes, GRSS is an all-for-all approach. When used with precise summary generation, GRSS has similar characteristics to DREAM: position updates scale with O(n) and position lookups are free. The imprecise summary generation may be used to reduce the overhead of position updates at the cost of collisions.

12.3 Routing

In position-based routing, the forwarding decision of a node is primarily based on the position of a packet's destination and the position of the node's immediate one-hop neighbors. The position of the destination is contained in the header of the packet. If a node knows a more accurate position of the destination, it may choose to update the position in the packet before forwarding it. The position of the neighbors is typically learned through one-hop broadcasts. These beacons are sent periodically by all nodes and contain the position of the sending node.

We can distinguish three main routing strategies for position-based routing: greedy forwarding, directed flooding, and hierarchical routing. For the first two strategies, a node forwards a given packet to one (greedy forwarding) or more (directed flooding) one-hop neighbors that are located closer to the destination than the forwarding node itself. The third forwarding strategy is to form a hierarchy in order to scale to a large number of mobile nodes. Hierarchical mechanisms use different types of ad hoc routing protocols at different levels of the hierarchy (e.g., a non–position-based ad hoc routing protocol at one level and a position-based protocol at a different level).

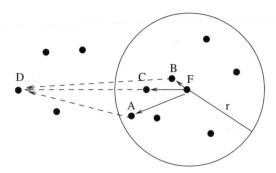

FIGURE 12.2 Greedy routing strategies.

In order to assess the individual routing strategies, we use the following metric: the communication complexity is the average number of one-hop transmissions that is required to send a packet from one node to another node under the assumption that the position of the destination is known. Ideally, the communication complexity should be equal to the number of hops on the shortest path between the source and the destination of a packet.

12.3.1 Greedy Routing

Using greedy routing, an intermediate node forwards the packet to a neighbor lying in the general direction of the packet's recipient. Ideally, this process is repeated until the recipient is reached.

Often, more than one neighbor is located closer to the destination than the forwarding node. There are different strategies to decide to which of those neighbors a given packet should be forwarded. These strategies are illustrated in Fig. 12.2, where F is the forwarding node and D is the destination of the packet. The circle with radius r indicates the maximum transmission range of node F. An intuitive strategy is for F to forward the packet to the node that makes the most progress towards D. In the example, this would be node A. This strategy is known as most forward within radius (MFR) [24]; it tries to minimize the number of hops a packet has to traverse in order to reach D.

When the signal strength of the transmission can be adapted to the distance between a sender and the receiving neighbor, then the transmission radius caused by MFR will be usually be fairly large. As a consequence, the transmission is likely to interfere with the transmissions of other nodes. The larger the radius, the higher the probability that such an interference occurs and that it prevents one or more receivers from decoding the transmission. This observation has led to the development of a different strategy called nearest with forward progress (NFP) [13]. For NFP, the packet is transmitted to the nearest neighbor of the sender that is closer to the destination. In Fig. 12.5, this would be node B. If all nodes use NFP, the probability of packet collisions is significantly reduced. Therefore, the average progress of the packet, calculated as the product of the packet's progress and the likelihood that it is not destroyed by a collision with other transmissions, is higher than with MFR.

Yet another strategy for forwarding packets is compass routing, which selects the neighbor closest to the straight line between sender and destination [19]. In the example, this would be node C. Compass routing tries to minimize the spatial distance that a packet travels.

Unfortunately, greedy routing may fail to find a path between sender and destination, even though one does exist. An example of this problem is depicted in Fig. 12.3. In this figure the sender of the packet is node S, and node D is the destination. Using greedy forwarding for two hops, the packet arrives at an intermediate node F. The half-circle around D has the radius of the distance between F and D, and the circle around F shows the transmission range of F. Note that a valid path from S over F to D exists. However, F is closer to the destination D than any of the nodes in its transmission range. Greedy routing has reached a local optimum from which it cannot recover.

For early approaches to greedy routing, it has been suggested that in these situations the packet should be forwarded to the node with the least backward (negative) progress [24]. However, this raises the

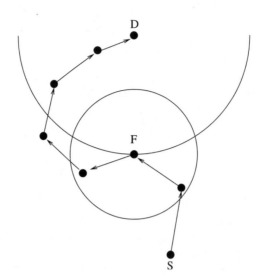

FIGURE 12.3 Greedy routing failure.

problem of routing loops, which do not occur when packets are forwarded only towards the destination with positive progress. Other researchers proposed to discard packets that have reached a local maximum [13]. This is justifiable since it can be expected that the problem is only short lived due to the dynamic nature of ad hoc wireless networks. Packet drops may later be repaired by a retransmission at the transport layer or above.

In the following we describe two greedy routing approaches that have recently been proposed. Both use the same recovery strategy for local optima.

12.3.1.1 face-2 and Greedy Perimeter Stateless Routing

Both face-2 [5] and Greedy Perimeter Stateless Routing (GPSR) [17] use greedy routing as described above. For the selection of the next hop MFR is used, assuming that the strength of the radio signal cannot be adapted to the distance between sender and receiver. The algorithm used to recover from local optima is performed on a per packet basis and does not require the nodes to store and maintain any additional information. The key idea of this approach is to perform a planar graph traversal upon reaching a local optimum. A packet continues to be forwarded in greedy mode when it reaches a node closer to the destination than the node where the packet entered the recovery mode.

A set of nodes in an ad hoc network can be considered a graph in which the nodes are vertices and an edge exists between two vertices if they are close enough to communicate directly with each other. Planar graphs are graphs with no intersecting edges. The graph formed by an ad hoc network is generally not planar, but well-known algorithms to construct planar subgraphs from nonplanar graphs exist [25]. These algorithms remove edges from the nonplanar graph so that it becomes planar without being partitioned. They can be applied locally by each node without any global information.

The traversal of the planar graph is then based on the left hand rule. Figure 12.4 illustrates the application of the left hand rule to the scenario shown in Fig. 12.3. First, an imaginary line is drawn from the node where the local optimum was reached (F) to the destination (D) of the packet. The packet is then transmitted over the first edge counterclockwise from that line. The next hop checks whether it is closer to the destination than the node where the local optimum was encountered. If this is the case, greedy routing resumes. Otherwise, the packet is forwarded over the next edge counterclockwise from the edge where the packet arrived. This is repeated until one of the following occurs:

1. The destination is reached.
2. Greedy mode is resumed.
3. An edge is to be traversed in the same direction for the second time.

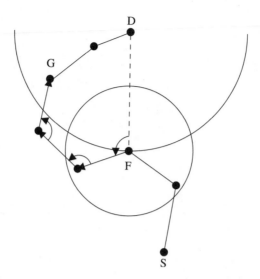

FIGURE 12.4 Left hand rule.

In the last case, the destination is unreachable and the packet should be discarded. In the example, the left hand rule will be used until node G is reached, where the second condition holds true and the packet is therefore forwarded further in greedy mode.

GPS and face-2 use a somewhat more sophisticated recovery strategy than plain left-hand rule. This recovery strategy is based on the traversal of faces around the area without connectivity.

Greedy routing has a low communication complexity of $\sqrt{n}$, since a single copy of the packet is forwarded from the source to the destination. The recovery mode of face-2 and GPSR is used only to recover from local optima. The resulting overhead does not dominate the communication complexity as long as the network is relatively dense. Greedy routing can be used together with any of the location services described in the previous section. It can make use of more accurate position information as it approaches the destination, but this is not required.

It is frequently stated in the relevant literature that greedy routing is inherently loop free and does not cause packet duplication. However, this is only true in a static network. As described in [18], in a dynamic network routing loops may occur because of outdated position information about a node's neighbors. Consider the case where A and B are neighbors and D is the destination. At a certain point in time, A is closer to D than B. At this time A and B transmit a beacon. Then A moves away from D such that now B is closer to D. At this point A receives a packet headed for D. Since A knows its current position and the position of B at the time the beacons were sent, it concludes that B is closer to D than A and forwards the packet to B. Due to the outdated position information B has about A, B thinks that A is closer to D and forwards the packet to A, causing a routing loop. While greedy routing protocols can prevent infinitely looping packets by including a maximum hop count in the packet header, even temporary loops are undesirable. The resulting overhead may cause congestion in an entire area of the ad hoc network by consuming most of the available network capacity. This remains an open problem for greedy routing based on inaccurate information.

Packet duplication can occur when a packet is forwarded from a node to a neighbor close to the maximum transmission range. A transmission is typically acknowledged in wireless networks (e.g., refer to IEEE 802.11). When the sending node does not receive an acknowledgment, it resends the packet. If this remains unsuccessful, this is considered a link failure and the node may select a new neighbor to forward the packet to. Duplication of packets occurs when the initial data transmission is successful but the acknowledgment is lost and the retransmissions remain without answer, too. The sender assumes that the packet was lost and selects a new neighbor. However, the original recipient of the packet forwards it as well thus causing a duplication of the packet.

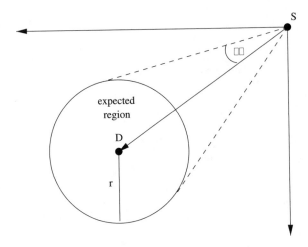

FIGURE 12.5 Directed flooding in DREAM.

12.3.2 Directed Flooding

In directed flooding, packet duplication does not occur by accident but is part of the standard forwarding algorithm. A node will forward a packet to all neighbors that are located in the direction of the destination. Directed flooding is very robust at the cost of heavy network load.

12.3.2.1 DREAM

In DREAM, the direction toward the destination is determined by means of a so-called expected region. As depicted in Fig. 12.5, the expected region is a circle around the position of D as it is known to a forwarding node N. Since this position information may be outdated, the radius r of the expected region is set to $(t_1 - t_0)v_{max}$, where t_1 is the current time, t_0 is the timestamp of the position information that N has about D, and v_{max} is the maximum speed with which a node may travel in the ad hoc network.

Given the expected region, the "direction towards D" for the example given in Fig. 12.5 is defined by the line between N and D and the angle φ. The neighboring hops repeat this procedure using their information on D's position. If a node does not have a one-hop neighbor in that direction, a recovery strategy is necessary. This procedure is not part of the DREAM specification.

Even though directed flooding limits the flooding to the direction toward the destination, the communication complexity has the same order as pure flooding, $O(n)$; it is just better by a constant factor. Directed flooding therefore does not scale to large networks with a high volume of data transmissions. On the other hand, it is highly robust against the failure of individual nodes and position inaccuracy, and it is very simple to implement. This qualifies it for applications that require a high reliability and fast message delivery for very infrequent data transmissions. DREAM works best in combination with an all-for-all location service that provides more accurate information close to the destination. This reduces the size of the expected region and thus the area in which the packet is flooded.

12.3.3 Hierarchical Routing

The introduction of a hierarchy is a well-known method in traditional networks to reduce the complexity each node has to handle. It allows networks to scale to a very large number of nodes. Currently there exist two approaches that introduce a two-layer hierarchy to routing in ad hoc wireless networks: Terminodes Routing and Grid Routing. Both approaches combine the use of a non–position-based approach on one level of the hierarchy with a position-based approach at the other level.

12.3.3.1 Terminodes Routing

In Terminodes routing [2], packets are routed according to a proactive distance vector scheme if the destination is close (in terms of hops) to the sending node. Over long distances, position-based routing is used. Once a long-distance packet reaches the area close to the recipient, it is forwarded by a local routing protocol from there on. The authors of [3] show by means of simulation that the introduction of a hierarchy can significantly improve the ratio of successfully delivered packets and the routing overhead compared to reactive ad hoc routing algorithms.

In order to prevent greedy forwarding for long distance routing from reaching a local optimum, the sender includes a list of positions in the packet header. These are called geodesic anchors. On its way to the destination, a packet must traverse the areas in which the geodesic anchors are located. The packet forwarding between the areas is done on a purely greedy basis. This approach can be thought of as position-based source routing. It requires that the sender know about appropriate positions leading to the destination. In Terminodes routing, the sender requests this information from nodes it is already in contact with (e.g., the nodes that are reachable using the local routing protocol). Once a sender has the information, it needs to check at regular intervals whether the path of positions is still valid or can be improved. Therefore, Terminodes long-distance routing contains elements of reactive ad hoc routing approaches.

Terminodes routing for long distances is based on greedy forwarding. However, due to the usage of a non–position-based approach at the local level, Terminodes routing can tolerate a higher degree of position inaccuracy regarding the destination.

12.3.3.2 Grid Routing

Another method for position-based ad hoc routing containing hierarchical elements is based on so-called location proxies [8] and is part of the Grid [27] project. As in Terminodes routing, a proactive distance vector routing protocol is used at the local level, while position-based routing is employed for long-distance packet forwarding. For Grid routing, the hierarchy is not only introduced to improve scalability, but also to allow nodes that do not know their own position to participate in the ad hoc network. If at least one position-aware node exists in each area where the proactive distance vector protocol is used, the position-aware node may be used as a proxy: position-unaware nodes use the position of a position-aware node as their own position. Packets addressed to a position-unaware node therefore arrive at a position-aware proxy and are then forwarded according to the information of the proactive distance vector protocol.

As a repair mechanism for greedy long-distance routing, Intermediate Node Forwarding (INF) is proposed [8]. If a forwarding node has no neighbor with forward progress, it discards the packet and sends a notification to the sender of the packet. The sender of the packet then chooses a single intermediate position randomly for a circle around the midpoint of the line between the sender and the receiver. As in Terminodes source routing, packets have to traverse that intermediate position. If the packet is discarded again, the radius of the circle is increased and another random position is chosen. This is repeated until the packets are delivered to the destination or until a predefined value has been reached and the sender assumes that the destination is unreachable.

The main characteristics of Grid routing are similar to those of Terminodes routing. However, as an additional benefit in Grid routing, it is acceptable that some nodes are not able to determine their own geographic position.

12.4 Application Scenario

Communication between vehicles is one prime area where wireless ad hoc networks are likely to be deployed in the near future. Emergency warnings, the distribution of traffic information, and cooperative driving are just three examples of the vast number of applications that may become feasible with intervehicle communication. In order to push development in this area, vehicle manufacturers and their suppliers are actively supporting research on integrating ad hoc wireless networks into vehicles [21,26].

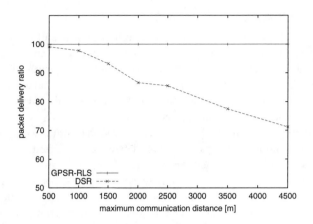

FIGURE 12.6 Percentage of successfully delivered packets for GPSR/RLS and DSR.

As a routing strategy, position-based routing seems to be very well suited for vehicular ad hoc networks. The reasons for this are twofold. First, the network topology of a vehicular network changes frequently, which is problematic for non–position-based approaches. Secondly, it is realistic to assume that in the near future all vehicles will be equipped with a positioning device such as a GPS receiver or even with sophisticated navigation systems; thus each node will know its own position at virtually no additional cost.

We performed a simulation study with the network simulator ns-2 comparing a position-based routing approach with a reactive non–position-based scheme for a highway traffic scenario. Specifically, we compared greedy position-based forwarding as implemented in GPSR with RLS (see Section 12.2.1) to Dynamic Source Routing (DSR) [15], an on-demand ad hoc routing protocol that shows very good performance in standard mobility scenarios based on the random waypoint mobility model [6]. The vehicular movement patterns were generated using a well-validated microscopic traffic simulation tool of the DaimlerChrysler AG. The generated highway scenario is 10 in km length with about 300 vehicles and has two lanes per direction. The traffic density is six vehicles per kilometer and lane, which matches weekday traffic on German highways. The transmission range was set to 250 meters. In our study we measured the packet delivery ratio for GPSR/RLS and DSR by conducting simulation runs with varying maximum distance between communication partners. Communication partners were randomly selected within the fixed maximum communication distance range and exchange packets for 5 seconds with eight packets per second.

The results, as given in Fig. 12.6, show that the position-based approach performs very well for the investigated distances, while DSR's performance decreases with increasing distances between communication partners. For the tested scenario, the protocol overhead of GPSR/RLS did not exceed the overhead produced by DSR. The detailed simulation study is given in [9]. The study shows the benefit of the "forwarding on-the-fly" behavior of a position-based routing approach compared to route setup and maintenance problems of a non–position-based method in the context of highly dynamic ad hoc networks, such as a vehicular network.

12.5 Conclusions

Position-based routing can be separated into two distinct tasks:

1. Discovering the position of the destination
2. The actual forwarding of packets, based on this information

Location services can be proactive or reactive, depending on whether the position of nodes is updated within the service at all times or only when it is needed. Furthermore, they can be classified as to how many nodes of the ad hoc network participate in providing the service and how many locations each of these nodes manages. RLS is the only example of a reactive all-for-some approach, where the querying node floods the network with a location request while the destination itself replies to this request. DREAM and GRSS are active all-for-all services, where information about the location of all nodes is flooded in the network. The GLS approach works by hashing the ID of a node on the IDs of so-called location servers. These location servers are updated by the destination node with regard to its own position and are queried by the source nodes that want to contact the destination node. It is thus a proactive all-for-some approach. Finally, the Homezone algorithm requires that the ID of a node be hashed on a position. All nodes close to this position are informed about the position of the node and provide this information to sources that want to contact it. As with GLS, this is proactive and belongs to the all-for-some category.

The actual routing of packets based on position information was separated into three distinct areas. Greedy routing works by forwarding packets in the direction of the destination. If a local maximum is encountered, a repair strategy such as face-2 or GPSR's perimeter routing can be used to avoid dropping the packet. In directed flooding, as used by DREAM, the packets are broadcasted in the general direction of the destination. On the way, the position information in the packets may be updated if a node has more current information about the destination's position. In the Terminodes and Grid projects, routing is done hierarchically by means of a position-independent protocol at the local level and a greedy variant at the long-distance level.

References

[1] S. Basagni, I. Chlamatac, V. Syrotiuk, and B. Woodward, A Distance Routing Effect Algorithm for Mobility (DREAM), *Proceedings of the 4th Annual ACM/IEEE International Conference on Mobile Computing and Networking (MobiCom) '98*, Dallas, TX, 1998, pp. 76–84.

[2] L. Blazevic, L. Buttyan, S. Capkun, S. Giordano, J. Hubaux, and J. Le Boudec, Self-Organization in Mobile Ad Hoc Networks: The Approach of Terminodes, *IEEE Communications Magazine*, June 2001.

[3] L. Blazevic, S. Giordano, and J.Y. Le Boudec, Self-Organizing Wide-Area Routing, *Proceedings of SCI 2000/ISAS 2000*, 2000.

[4] B. Bloom, Space/time trade-offs in hash coding with allowable errors, *Communications of the ACM*, 13, 422–426, 1970.

[5] P. Bose, P. Morin, I. Stojmenovic, and J. Urrutia, Routing with Guaranteed Delivery in Ad Hoc Wireless Networks, *Proceedings of 3rd ACM International Workshop on Discrete Algorithms and Methods for Mobile Computing and Communications (DIAL M99)*, 1999, pp. 48–55.

[6] J. Broch, D. Maltz, D. Johnson, Y.-C. Hu, and J. Jetcheva, A Performance Comparison of Multi-Hop Wireless Ad Hoc Network Routing Protocols, *Proceedings of the 4th ACM/IEEE International Conference on Mobile Computing and Networking (MobiCom) '98*, Dallas, TX, 1998, pp. 85–97.

[7] S. Capkun, M. Hamdi, and J. Hubaux, GPS-free Positioning in Mobile Ad-Hoc Networks, *Proceedings of the Hawaii International Conference on System Sciences*, Maui, 2001.

[8] D. De Couto and R. Morris, Location Proxies and Intermediate Node Forwarding for Practical Geographic Forwarding, Technical Report MIT-LCS-TR-824, MIT Laboratory for Computer Science, June 2001.

[9] H. Füßler, M. Mauve, M. Käsemann, D. Vollmer, and H. Hartenstein, A Comparison of Routing Strategies for Vehicular Networks, Technical Report TR-03-2002, Department of Computer Science and Mathematics, University of Mannheim, Germany, 2002.

[10] B. Ghosh, Random distances within a rectangle and between two rectangles, *Bull. Calcutta Math. Society*, 43, 17–24, 1950.

[11] S. Giordano and M. Hamdi, Mobility Management: The Virtual Home Region, Technical Report SSC/1999/037, October 1999.

[12] J. Hightower and G. Borriello, Location Systems for Ubiquitous Computing, *IEEE Computer*, Aug. 2001, pages 57–66.

[13] T. Hou and V. Li, Transmission range control in multihop packet radio networks, *IEEE Transactions on Communications*, 34, 38–44, 1986.

[14] P. Hsiao, Geographic Region Summary Service for Geographical Routing. *ACM Mobile Computing and Communication Review*, Oct. 2001, pages 25–39.

[15] D. Johnson and D. Maltz, *Mobile Computing*, Kluwer Academic Publishers, Dordrecht, 1996, pp. 153–181.

[16] E. Kaplan, *Understanding GPS*, Artech House, Norwood, MA, 1996.

[17] B. Karp and H. Kung, Greedy Perimeter Stateless Routing for Wireless Networks. *Proceedings of the 6th Annual ACM/IEEE International Conference on Mobile Computing and Networking (MobiCom 2000)*, 2000, Boston, MA, pp. 243–254.

[18] M. Käsemann, H. Hartenstein, H. Füßler, and M. Mauve, Analysis of a Location Service for Position-Based Routing in Mobile Ad Hoc Networks, Accepted at WMAN 2002, Ulm, 2002.

[19] E. Kranakis, H. Singh, and J. Urrutia, Compass Routing on Geometric Networks, *Proceedings of 11th Canadian Conference on Computational Geometry*, Aug. 1999.

[20] J. Li, J. Jannotti, D. De Couto, D. Karger, and R. Morris, A Scalable Location Service for Geographic Ad Hoc Routing, *Proceedings of the 6th Annual ACM/IEEE International Conference on Mobile Computing and Networking (MobiCom 2000)*, Boston, MA, 2000, pp. 120–130.

[21] R. Morris, J. Jannotti, F. Kaashoek, J. Li, and D. De Couto, CarNet: A Scalable Ad Hoc Wireless Network System. *Proceedings of the 9th ACM SIGOPS European Workshop: Beyond the PC: New Challenges for the Operating System*, Sep. 2000.

[22] J. Navas and T. Imielinski, Geographic Addressing and Routing, *Proceedings 3rd ACM/IEEE International Conference on Mobile Computing and Networking (MobiCom '97)*, Budapest, Sep. 1997.

[23] I. Stojmenovic, Home Agent Based Location Update and Destination Search Schemes in Ad Hoc Wireless Networks, Technical Report TR-99–10, Computer Science, SITE, University of Ottawa, Canada, Sep. 1999.

[24] H. Takagi and L. Kleinrock, Optimal Transmission Ranges for Randomly Distributed Packet Radio Terminals, *IEEE Transactions on Communications*, 32, 246–257, 1984.

[25] G. Toussaint, The relative neighborhood graph of a finite planar set, *Pattern Recognition*, 12, 261–268, 1980.

[26] The Fleetnet Homepage, http://www.fleetnet.de.

[27] The Grid Project Homepage, http://www.pdos.lcs.mit.edu/grid/.

13

Structured Proactive and Reactive Routing for Wireless Mobile Ad Hoc Networks

Ahmed M. Safwat
Queen's University, Canada

Hossam S. Hassanein
Queen's University, Canada

Hussein T. Mouftah
Queen's University, Canada

Abstract

On-demand routing schemes suffer from the initial route setup latency, introduced by their discovery phase. This degrades the performance of interactive and/or multimedia applications. Table-driven routing, on the other hand, responds more quickly to call/route requests, but it wastes a large portion of the scarce wireless bandwidth on routing table broadcasts. Structured routing can be utilized to reduce the route setup time and increase throughput. In our Virtual Base Stations (VBS) architecture, a mobile node is elected from a set of nominees to act as a mobile base station within its zone. Based on our VBS routing protocol, a mobile station that wishes to send a packet to another mobile node in the network sends the packet to its VBS, which forwards it to the proper next hop. In this chapter, we also propose a novel routing scheme based on the VBS architecture, namely, the Virtual Base Stations Proactive-Reactive (VBS-PR) routing protocol. VBS-PR promises to achieve considerable gains in terms of increasing the capacity of the wireless ad hoc network, reducing the call setup times and end-to-end delays and increasing the routing efficiency. VBS-PR can operate as a purely on-demand scheme and utilize the wireless mobile infrastructure to increase efficiency.

13.1 Introduction

In ad hoc networks, the nonexistence of a centralized authority complicates the routing and medium access problems. Third generation wireless networks use a variation of the time division multiple access (TDMA) and code division multiple access (CDMA) protocols [4]. Protocols such as TDMA/TDD [5],

MASCARA [6], DS/CDMA [7], and others [4] all have slot reservation capabilities for handling multi-media ATM constant bit rate (CBR) and variable bit rate (VBR) traffic. However, such protocols cannot be extended to ad hoc networks since they assume a centralized access technique that is dependent on the base stations. For ad hoc networks, the IEEE 802.11 CSMA/CA distributed access protocol [8] is assumed. CSMA/CA is a random access protocol and cannot support bandwidth reservation or provide QoS routing.

Ensuring QoS communications in a wireless mobile ad hoc network is, apparently, highly dependent upon routing and medium access control. Obviously, conforming to QoS measures, such as delay bounds, depends on the chosen route from the source mobile station to the destination mobile station. In addition, complying with QoS guarantees imposes the use of a MAC method that guarantees the successful transmission of packets under high mobility and/or heavy load circumstances and hence meets the QoS constraints dictated by the communications application. This can never be made feasible without a central, rather than the typically distributed, administration of the ad hoc network. Hence, an efficient centralized mechanism must be devised for wireless mobile ad hoc networks and deployed by a central body. As a consequence, a central regulatory authority is needed to carry out the significant functions of routing and MAC in wireless mobile ad hoc networks. Intuitively, this places more emphasis on the indispensability of infrastructure-based communications and structured routing for ad hoc networks.

In view of the foregoing, developing an infrastructure for infrastructure-less wireless mobile ad hoc networks is of utmost importance. Such an infrastructure reduces the problem of wireless mobile ad hoc communications from a multi-hop problem to a single-hop problem, as in conventional cellular communications. The previous apprehension originates from the intuitive realization of the problems of medium access and routing in wireless mobile ad hoc networks and their effects on QoS-guaranteed communications. Infrastructure-based communications enable the use of simple variations of the widely used cellular protocols in ad hoc networks. Such a dynamic infrastructure will form the basis for developing MAC protocols and routing algorithms, which can utilize it to conform to different QoS requirements. Infrastructure-based wireless mobile ad hoc communications will help circumvent routing and medium access control issues. Therefore, in this chapter, a wireless mobile infrastructure is developed for ad hoc networks. We propose an infrastructure-creation protocol for wireless mobile ad hoc networks, namely, the *Virtual Base Stations (VBS)* protocol [2]. Nevertheless, the developed infrastructure is, essentially, a mobile infrastructure. The wireless mobile infrastructure acts as the executive regulatory authority that carries out mobility tracking, and hence routing, in wireless mobile ad hoc networks. The proposed infrastructure formation scheme demonstrates quick response to topological changes in the ad hoc network. Additionally, the protocol is scalable to networks with large populations of mobile stations. It outperforms current infrastructure-creation schemes in stability and load balancing among the mobile stations forming the infrastructure. We also propose a novel routing scheme that is both reactive and proactive. In *VBS Proactive-Reactive (VBS-PR) routing*, only VBSs and border mobile terminals (BMTs) participate in the routing process.

13.2 Previous Work

Several hierarchal architectures for ad hoc networks have been proposed [9–11]. A mobile infrastructure is developed in [9] to replace the wired infrastructure and base stations in conventional cellular networks. Moreover, an adaptive hierarchical routing protocol is devised. The routing protocol utilizes the mobile infrastructure in routing packets from a source mobile terminal (MT) to a destination MT. The algorithm divides the wireless mobile ad hoc network into a set of clusters, each of which contains a number of MTs that are at most two hops away. However, clusters are only a logical arrangement of the nodes of the mobile network. Thus, unlike clusterheads, cluster centers may not coordinate transmissions. Nodes that relay packets between clusters are called repeaters. Repeaters play a major role in performing the routing function. Since route maintenance was not studied by means of simulation, it is not possible to evaluate its efficiency.

In [10], the Random Backbone (RB) algorithm requires that the clusterheads form a dominanting and independent set in the network graph. RB is used to derive a solution for the basic problem of connecting the clusterheads into a virtual backbone (VB) using fewer connections as compared to the deterministic case. The scheme provides a probabilistic connectivity guarantee. RB constructs a virtual backbone that is connected with a probability that approaches 1 for large and dense networks. It is apparent that for this type of approach to address the communications problem in wireless mobile ad hoc networks, it would add great overhead to the MTs in the network, since mobile stations might have to switch roles from time to time. Every time this happens, clusterheads will have to be determined from the beginning and the virtual backbone must be rebuilt from scratch. Also, route construction and maintenance were not addressed and, as a result, the effect of the proposed architecture on packet delivery was not studied.

A heuristic, called max-min, is proposed in [11] and forms d-hop clusters. For the sake of determining the set of clusterheads, each run of max-min uses two-dimensional rounds of flooding. It is when the mobile stations with the largest node IDs are spaced d hops apart, that max-min's solution is optimal; a condition which cannot be guaranteed in a dynamic environment such as that of an ad hoc network. In addition, since max-min forms d-hop clusters, it does not suit the dynamic nature of ad hoc networks, where topology changes due to node mobility are unpredictable. Therefore, determining (or predicting) the time as to when the heuristic should be best run is significant.

13.3 The Virtual Base Stations (VBS) Architecture

In our scheme, some of the MTs, based on an agreed-upon policy, become in charge of all the MTs in their neighborhood or a subset of them. This can be achieved by electing one to be a *Virtual Base Station (VBS)*. If a VBS moves or stops acknowledging its presence via its so-called *hello* packets for a period of time, a new one is elected. A number of issues have to be solved. The first issue is the way in which the VBSs are to be chosen. Electing a single VBS from a set of nominees should be done in an efficient way. Another issue to be addressed is the handing of responsibilities of a VBS over from one VBS to another. Every MT (*zone_MT/VBS*) has a *sequence number* that reflects the changes that occur to that MT, and a *my_VBS* variable, which is used to store the VBS in charge of that MT. If an MT has a VBS, its my_VBS variable will be set to the ID number of that VBS, else if the MT is itself a VBS, then the my_VBS variable will be set to 0, otherwise it will be set to −1. Hello messages sent by VBSs contain their current knowledge of the ad hoc network, i.e., the whole ad hoc network. A VBS accumulates information about all other VBSs and their lists of MTs and sends this information in its periodic hello messages. On the other hand, unlike VBSs, zone_MTs accumulate information about the network from their neighbors between hello messages, and their network information is cleared afterwards.

After being chosen as a VBS, a node stores information about all other nodes in the network. An MT is chosen by one or more MTs to act as their VBS based on an agreed-upon rule, viz., the MT with the smallest ID number. A noteworthy remark is that a node with a smaller ID number than another may be thought of as one that is more capable (in terms of processing speed, battery capacity, or any other criterion) than the latter. MTs announce their ID numbers with their periodic hello messages. An MT sends a *merge-request* message to another MT if the latter has a smaller ID number. The receiver of the merge-request responds with an *accept-merge* message, increments its sequence number by 1 to indicate that some change took place, and sets its my_VBS variable to 0. When the MT receives the accept-merge, it increments its sequence number by 1 and sets its my_VBS variable to the ID number of its VBS. If an MT hears from another MT whose ID number is smaller than that of its VBS, it sends a merge-request message to the former. When it receives an accept-merge message, it increments its sequence number by 1 and updates its my_VBS field. The MT then sends a *dis-join* message to its previous VBS (the dis-join message is only sent if an accept-merge was received, otherwise it would not be sent). When the old VBS receives the dis-join, it removes the sender from its list of MTs, which it is in charge of, and it increments its sequence number by one.

Unlike other single-hop infrastructure creation protocols, such as the least cluster change (LCC) protocol proposed in [12], VBS puts more emphasis on a node becoming a VBS rather than being supervised by a VBS. Hence, if a node receives a merge request, it responds by sending an accept-merge message, even if it was being supervised by a VBS. Moreover, it is noteworthy here that this degrades neither intracluster nor intercluster communications by any means. This is because if the node that became a VBS was originally acting as a gateway for its previous VBS, it will still be a gateway, besides being a VBS. This becomes of great significance if the criterion upon which VBSs were elected was one that relies on the assets possessed by the MTs of the ad hoc network. Processing speed, main and secondary storage, MAC contention experienced in the neighborhood of the MT, and more can be among such assets. Consequently, if an MT chooses not to become a clusterhead, as in [12], only because it is under the supervision of another MT, even though it possesses the required assets to become one, the node requesting to merge might experience demoted communications to other nodes in the ad hoc network because it does not have the proper resources, nor is it able to be associated with a VBS.

13.3.1 VBS Illustrated

This section explains the operation of the VBS infrastructure-creation protocol by means of illustrations. Each of the following examples (see Figs. 13.1 through 13.3) represents a scenario and the set of corresponding actions performed by the mobile stations running the VBS protocol.

In Fig. 13.1A, all the MTs are within radio range of one another, except 1 and 4. Due to the asynchronous transmission of the hello messages, MT 2 may broadcast its hello message before MT 1 does. Therefore, MTs 3 and 4 receive MT 2's hello message first. They send merge-request messages to MT 2 [a] and it sends accept-merge messages back to each one of them [b]. The scenario in Fig. 13.1B starts with MT 1 sending its hello message. MT 3, realizing that it heard from an MT whose ID number is smaller than that of its VBS, sends a merge-request message to MT 1 [a]. MT 1 sends back a merge-accept message [b]. After receiving the merge-accept messages, 3 sends a dis-join message to 2 [c], which

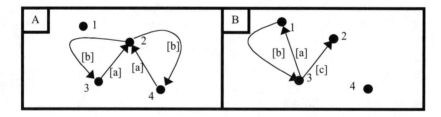

FIGURE 13.1 Finding the VBSs.

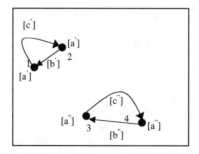

FIGURE 13.2 Motion of a VBS out of the transmission range of its zone MTs.

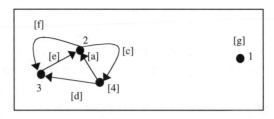

FIGURE 13.3 Terminal mobility and choosing the VBS.

removes 3 from its list of supervised MTs. MT 2 is still the VBS of 4, which did not send a dis-join message to MT 2 since 4 is not within the wireless transmission range of MT 1. The my_VBS variable of 3 becomes 1, and the my_VBS variable of 4 remains 2. The my_VBS variable of 1 and 2 is equal to 0, reflecting that they are both VBSs.

In Fig. 13.2, 1 and 2 move such that the other MTs no longer hear their transmissions. MTs 3 and 4 will each try to find another VBS, as they can no longer hear from their previous VBSs (i.e., their VBSs' timers expire [a"]). MT 3 will not send any merge-request messages, as it has the smallest ID number in its neighborhood. MT 4 sends a merge-request message to 3 [b"] that will, in response, send a merge-accept message [c"], indicating that 3 is willing to become the VBS of 4. In quite the same manner, 1 will become 2's VBS. Notice, however, that the zone MT timers in case of 1 and 2 expire [a'], and both try to find a VBS. MT 1 acts as the VBS of itself until 2 sends a merge-request message to it [b']. MT 1 becomes 2's VBS as well [c'].

As shown in Fig. 13.3, when 2 moves such that its hello transmissions are heard by 3 and 4, 3 will not be able to send a merge-request message to 2 because it is currently acting as a VBS for 4. MT 4 sends a merge-request message to 2 [a]. MT 2, being under the supervision of 1 (see Fig. 13.2), sends a dis-join message to 1 (not shown) [b]. The dis-join message is not received by 1, which is out of the transmission range of 2. Once the merge-request messages are received by 2, it will send a merge-accept message to 4 [c]. Having received the merge-accept messages from 2, 4 will send a dis-join message to 3 [d] that will, as a result, find out that it is no longer the VBS for any MT. Its my_VBS variable will be assigned the value of −1 to reflect this situation. MT 3 sends a merge-request message to 2 [e] and receives a merge-accept message [f]. The my_VBS variable of 2's zone MTs will be set to two. MT 1's timer for 2 expires [g], and it will set its my_VBS variable to −1 indicating that it is a VBS of itself.

13.3.2 Some VBS Properties

This section contains a description of the main properties of the VBS scheme. The validity of each of these properties is also proved.

Lemma 1: If an MT, X, has a VBS then the VBS is unique, else X is itself a VBS (of itself or other MTs).

Proof:

To prove the first part of Lemma 1, let Z be an MT that has two VBSs, X and Y. Without loss of generality, the following scenario is assumed. Z merges with X and increments its sequence number by one to become z_x. Let Y have a smaller ID than X. Z merges with Y and increments its sequence number by one to become z_y.

Now, there are two cases. X and Y can hear each other's hello messages (case I), or X and Y cannot hear each other's hello messages (case II).

Case I:

X has Z on its list with sequence number z_x, and Y has Z on its list with sequence number z_y. Z, after merging with X, having noticed that Y has a smaller ID number, sends a merge-request message to Y, which, in return, sends a merge-accept message to Z. Afterwards, Z sends a dis-join message to X.

It is, therefore, impossible for Z to have more than one VBS at the same time. But X and Y might both have Z as one of their supervised MTs. However, when Y receives X's hello, Y takes no action and it keeps Z on its list of supervised MTs.

On the other hand, when X receives Y's hello, X removes Z from its list of supervised MTs and adds it to its copy of Y's list of supervised MTs. This is because $z_y > z_x$. Therefore, Z has one and only one VBS.

Case II:

In this case, X and Y cannot hear each other's hello messages, but they can hear Z's hello messages. Similar to case I above, after receiving Z's merge request, Y will remove Z from its copy of X's list of supervised MTs, which it received earlier from Z, and will instead add Z to its own list of supervised MTs. Since $z_y > z_x$, when X receives Z's first hello message after merging with Y, it removes Z from its list of supervised MTs. Therefore, Z has one and only one VBS.

The second part of Lemma 1 above is equivalent to: *If the MT is not a VBS, then it must be under the supervision of a VBS or the VBS of its own.* If the MT is not a VBS, then one of the following holds:

(A) X is the MT with the greatest ID number in its neighborhood, and in this case, it will send a merge-request to the MT whose ID number is the smallest among its neighbors.

(B) There are MTs with smaller ID numbers than X's, and X can hear at least one of these MTs, e.g., Y with $y_{ID} < x_{ID}$. In this case, X will send a merge-request message to Y and will receive an accept-merge message from Y, and X will be under the supervision of Y, and the my_VBS variable of X will be equal to y_{ID}.

(C) X cannot hear any of these MTs. The my_VBS variable of X will be set to −1 indicating that it is a VBS of its own.

From (A), (B), and (C), it follows that X is either under the supervision of a VBS or is a VBS of its own.

Let the *Global Convergence Time (GCT)* be the time needed for every VBS to know its list of supervised MTs and the other VBSs' lists of supervised MTs. Likewise, let a *Border Mobile Terminal (BMT)* be an MT that can hear the transmissions of more than one VBS (i.e., it lies in the transmission range of more than one VBS).

Lemma 2: One hello period ≤ GCT ≤ (number of VBSs + number of BMTs) hello periods.

Proof:

Lower bound for GCT:

There is only one VBS in the best case. Therefore, in order to come up with the lower bound, we assume that the MT that will be chosen as the VBS is the first node to send its hello message. Hence, the lower bound for every VBS (and there is only one in the best case) to know its list of supervised MTs and the other VBSs' (there are not any in the best case) lists of supervised MTs = 1 hello period.

Upper Bound for GCT:

To find the upper bound, we must determine the worst-case scenario for information update. This occurs when the time, Δt, between the periodic hello of any announcer and the periodic hello of the corresponding receiver(s) is zero (meaning that they send their hellos at the same time, i.e., if an MT sends its hello containing its knowledge of the network, the receiver[s] will

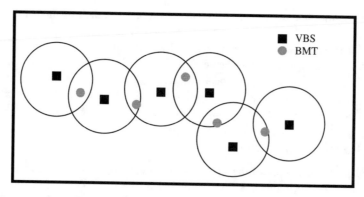

FIGURE 13.4 A linear array of zones.

have sent their hello messages [to their MAC layer] by the time they receive the hello message of the announcer, which implies that they will have to wait for another hello period to send their updated list). In addition, no two VBSs can hear each other's transmissions directly, but rather through BMTs. Every BMT connects exactly two VBSs (i.e., its own VBS and another VBS within its wireless transmission range, see Fig. 13.4), or its own VBS and a BMT of another VBS. This is equivalent to a *linear array of zones.*

From Lemma 1, two hello periods are needed for the BMTs to know their VBSs' lists. Moreover, for each BMT, one hello period (i.e., in the worst case) is needed to transmit its knowledge of the network to the other VBS/BMT that can hear it. Therefore, the upper bound = 2 + (number of VBSs – 2) + (number of BMTs) = (number of VBSs + number of BMTs) hello periods. The upper bound ensures that the VBS at one side of the linear array will obtain the list of the VBS at the other side of the linear array.

Hence, one hello period < total time for every VBS to know its list of supervised MTs and the other VBSs' lists of supervised MTs ≤ (number of VBSs + number of BMTs) hello periods. That is, one hello period ≤ GCT ≤ (number of VBSs + number of BMTs) hello periods.

13.4 VBS Routing

The VBS routing protocol [1,3] utilizes the VBS infrastructure-creation protocol described in Section 13.3 to route packets from a source node to a destination node. The VBS protocol is used in successfully developing a fast and reliable routing scheme that provides a solution to the intricate problem of routing in wireless mobile ad hoc networks.

All the MTs run the VBS protocol, and hence, some are elected to act as VBSs. The wireless mobile ad hoc network relies on the wireless mobile infrastructure created by VBS to provide communications between the different MTs. Each and every VBS is in charge of a set of MTs, including itself. In VBS routing, only certain nodes, i.e., the nodes that form the dynamic infrastructure, are eligible to acquire the knowledge of full network topology. Route requests are not flooded to the rest of the ad hoc network due to the existence of the wireless mobile infrastructure. As explained earlier, a BMT is an MT that lies within the wireless transmission range of one or more nodes that are being supervised by different VBSs, or are VBSs themselves, other than its own. When an MT wishes to send a packet, it sends it to its VBS, which forwards the packet to the VBS in charge of the destination or to the correct BMT. The sent packet contains the address of the destination. When the VBS receives the message, it looks up the destination address in its table. If the destination is found, the VBS of the source MT will forward the packet to the VBS in charge of the destination. This is done by consulting the BMT field of that VBS. The message is then forwarded to the MT (which might also be a VBS), the ID number of which is stored in the BMT field. The BMT, after receiving the message, forwards it to its own VBS (if it was not itself a VBS). This process is repeated until the message reaches the destination. Thus, path maintenance is an on-the-fly, built-in process.

Apparently, MTs are responsible neither for the burden of discovering new routes nor for maintaining existing ones. Hence, this novel routing scheme eliminates the initial search latency, introduced by the discovery phase in on-demand routing protocols, which degrades the performance of interactive and/or multimedia applications, due to the lack of a wireless mobile infrastructure that can efficiently handle routing. In addition, unlike table-driven protocols, MTs do not store the whole network topology.

13.4.1 VBS Routing Illustrated

This section explains the operation of the VBS routing protocol by means of illustrations. In Figs. 13.5 to 13.9, inclusive, a subscript represents the ID number of the VBS in charge of the MT. A VBS that wishes to send a packet to one of its zone MTs sends the packet directly to that MT. The example in Fig. 13.5 illustrates a similar scenario, where MT 1 is the VBS for MTs 2, 3, and 4. MT 1 wishes to send a packet to 2. It finds out from its *VBS table* that 2 is currently one of its zone MTs. Therefore, it sends the message to 2 directly (as indicated by the arrow from 1 to 2).

Likewise, a zone MT communicates directly with its VBS whenever it wants to send a packet to that VBS. Figure 13.6 shows four MTs. MT 2 wishes to send a message to MT 1. Since 2 is a zone MT, 2 figures out that it has to send the message through its VBS, which is 1 (as indicated by the arrow from 2 to 1). After receiving the message from its zone MT, 1 finds out that the message is destined for itself. If the message is destined for an MT that is under the supervision of another VBS, the VBS of the source sends the message

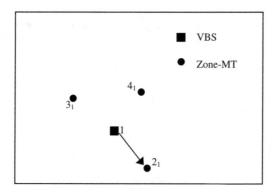

FIGURE 13.5 Sending a message from a VBS to one of its zone MTs.

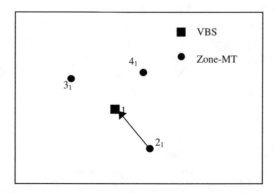

FIGURE 13.6 Sending a message from a zone MT to its VBS.

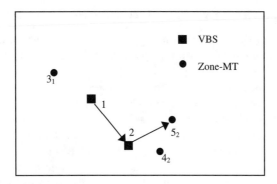

FIGURE 13.7   Sending a message from a VBS to a zone MT of another VBS.

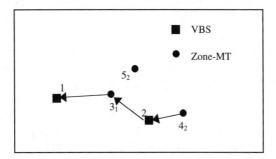

FIGURE 13.8   Sending a message from a zone MT to a VBS other than its own VBS.

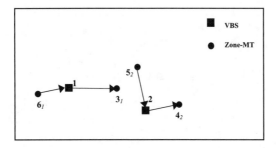

FIGURE 13.9   Sending a message from a zone MT to another zone MT of the same VBS.

to the VBS of the destination. Figure 13.7 demonstrates such a scenario. MT 1 decides to send a message to 5. Being a VBS, it consults its VBS table. It finds out that 5 is a zone-MT of 2. As a result, it forwards the message to 2 (as indicated by the arrow from 1 to 2). Consequently, 2 consults its VBS table to find out that 5 is one of its zone MTs and, therefore, sends the message directly to it (as indicated by the arrow from 2 to 5).

A zone MT that wishes to send a message to any other MT in the wireless ad hoc network sends that message to its VBS first. In Fig. 13.8, MT 4 decides to send a message to 1. Since 4 is a zone MT, 4 forwards the message to its VBS, 2 (as indicated by the arrow from 4 to 2). MT 2, after learning about 1 from 3, forwards the message to 3 (as indicated by the arrow from 2 to 3). MT 3 forwards the message directly to 1 as it is 3's VBS (as indicated by the arrow from 3 to 1). Finally, 1 receives the message.

In an effort to put the least amount of routing load on MTs other than those acting as VBS's, a zone MT wishing to send any message forwards it directly to its VBS even if the message was destined for a zone MT which is under the supervision of the same VBS. MT 6 of Fig. 13.9 sends a packet to its VBS, 1. Consequently, 1 forwards the packet directly to 3. A similar scenario takes place in the case when the source is 5 and the destination is 4.

13.4.2 VBS Proactive-Reactive (VBS-PR) Routing

We have proposed [13] a novel approach to routing in wireless ad hoc networks. Our approach utilizes the VBS wireless backbone in efficiently routing packets to their destinations. The new scheme is called *VBS Proactive-Reactive (VBS-PR)* routing [13]. As in other on-demand routing schemes, a source node broadcasts a RREQ packet whenever it wishes to communicate with another wireless node. Any of the VBSs or BMTs that are neighbors to the source node may reply with a RREP packet. Only VBSs and BMTs may rebroadcast the received RREQ. This ensures that MAC-level contention will be less likely to occur due to the RREQ and RREP storms that result in conventional on-demand schemes. If a wireless station does not overhear the transmission of the RREQ by its VBS or receive a RREP within a specific timeout period, it reelects a new one or becomes a VBS itself and rebroadcasts the RREQ with the *broadcastByAll* flag set to true. In Fig. 13.10, only VBSs and BMTs take part in the routing process. The source, S, is only required to broadcast the RREQ packet. All the nodes in Fig. 13.10 take part in the routing process except the destination, D, because its VBS will send a RREP packet as soon as it receives the RREQ packet. Nevertheless, in Fig. 13.11, the destination responds with a RREP packet since it receives the RREQ packet fromB.

Our algorithm will find a path if and only if at least one exists. It can be shown that our scheme performs better than any on-demand routing scheme given that the same routing criterion had been adopted by both protocols [13]. Since only VBSs and BMTs are required to broadcast the RREQs, VBS-PR may fail only if the source or an intermediate node that receives the RREQ is neither a neighbor of a VBS nor a BMT. One such example is node I in Fig. 13.12. However, this can never take place in a wireless ad hoc network and it can be shown that node 1 will either become a BMT or D's VBS, depending on its ID number. VBS-PR nodes are allowed to exchange routing tables only under light to modest network conditions.

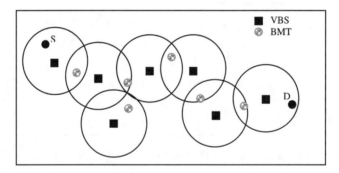

FIGURE 13.10 The RREP is sent by the VBS of the destination.

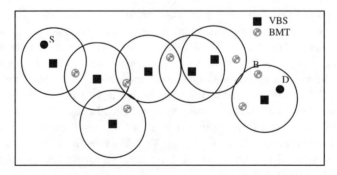

FIGURE 13.11 The RREP is sent by the destination itself.

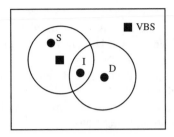

FIGURE 13.12 VBS-PR failure.

13.5 Conclusions

The Virtual Base Stations (VBS) architecture mimics and maintains the operation of the conventional fixed infrastructure in cellular networks. In the VBS protocol, a mobile node is elected from a set of nominees to act as a mobile base station within its zone. We have utilized the VBS infrastructure-creation protocol in successfully developing a fast and reliable routing scheme, i.e., VBS-based routing, which provides a solution to the intricate problem of routing in wireless mobile ad hoc networks. Moreover, it provides some form of mobility management for users of the ad hoc network by tracking mobile users within the ad hoc network and associating them with VBSs, so as to be able to route packets destined for any user, anywhere, anytime.

We also presented a novel routing scheme based on the VBS architecture, namely, the Virtual Base Stations Proactive-Reactive (VBS-PR) routing protocol. VBS-PR promises to achieve considerable gains in terms of increasing the capacity of the wireless ad hoc network, reducing the call setup times and end-to-end delays, and increasing the routing accuracy. VBS-PR can operate as a purely on-demand routing scheme with the added advantage of the wireless mobile infrastructure.

We have provided a thorough explanation of the operation of our proposed VBS-based routing protocols. We have also provided illustrative comprehensive examples on VBS and VBS-PR routing.

References

1. A. Safwat and H. Hassanein, Infrastructure-based routing in wireless mobile ad hoc networks, *Journal of Computer Communications*, 25(3), 210–224, 2002.
2. H. Hassanein and A. Safwat, Virtual base stations for wireless mobile ad hoc communications: an infrastructure for the infractructure-less, *International Journal of Communication Systems*, 14, 763–782, 2002.
3. A. Safwat and H. Hassanein, Structured Routing in Wireless Mobile Ad Hoc Networks, *Proceedings of the 6th IEEE Symposium on Computers and Communications (ISCC) 2001*, July 2001, pp. 332–337.
4. I.F. Akyildiz, J. McNair, L. Carrasco, R. Puigjaner, and Y. Yesha, Medium Access Control Protocols for Multimedia Traffic In Wireless Networks, *IEEE Network*, July/Aug. 1999, pp. 39–47.
5. D. Raychaudhri et al., WATMnet: a prototype wireless ATM system for multimedia personal communication, *IEEE JSAC*, 15, 83–95, 1997.
6. N. Passas, S. Paskalis, D. Vali, and L. Merakos, Quality of Service–Oriented Medium Access Control for Wireless ATM Networks, *IEEE Communications Magazine*, Nov. 1997, pp. 42–50.
7. R. Pichna and Q. Wang, A medium access control protocol for cellular packet CDMA carrying multirate traffic, *IEEE JSAC*, 14, 1728–1736, 1996.
8. IEEE LAN/MAN standards committee, Wireless LAN Medium Access Protocol (MAC) and Physical Layer (PHY) Specifications, IEEE Standard 802.11, 1999.
9. C.R. Lin and M. Gerla, A Distributed Architecture for Multimedia in Dynamic Wireless Networks, *Proceedings of IEEE GLOBECOM*, New York, 1995, pp. 1468–1472.

10. I. Chalmtac and A. Farago, A new approach to the design and analysis of peer-to-peer mobile networks, *ACM-Baltzer J. Wireless Networks,* 149–156, 1999.
11. A.D. Amis, R. Prakash, T.H.P. Vuong, and D.T. Huynh, Max-min d-cluster Formation in Wireless Ad Hoc Networks, *19th IEEE INFOCOM.*
12. C.-C. Chiang, H.-K. Wu, W. Liu, and M. Gerla, Routing in clustered multihop, mobile wireless networks with fading channel, *Sicon,* 1997.
13. A. Safwat, H. Hassanein, and H. Mouftah, Virtual Base Stations Proactive-Reactive (VBS-PR) Routing, Submitted for publication, March 2002.

14

Hybrid Routing: The Pursuit of an Adaptable and Scalable Routing Framework for Ad Hoc Networks*

Prince Samar
Cornell University

Marc R. Pearlman
Cornell University

Zygmunt J. Haas
Cornell University

Abstract

Advances in ad hoc network research have opened the door to an assortment of promising military and commercial applications for ad hoc networks. However, because each application has unique characteristics (such as traffic behavior, device capabilities, mobility patterns, operating environments, etc.), routing in such a versatile environment is a challenging task, and numerous protocols have been developed to address it. While many protocols excel for certain types of ad hoc networks, it is clear that a single basic protocol cannot perform well over the entire space of ad hoc networks. To conform to any arbitrary ad hoc network, the basic protocols designed for the edges of the ad hoc network design space need to be integrated into a tunable framework.

* This work has been supported in part by the DoD Multidisciplinary University Research Initiative (MURI) program administered by the Office of Naval Research under grant number N00014-00-1-0564 and by National Science Foundation grant numbers ANI-9980521 and ANI-0081357.

The Zone Routing framework demonstrates how multi-scoping can provide the basis for a hybrid routing protocol framework. Zone Routing proactively maintains routing information for a local neighborhood called the routing zone (local scope), while reactively acquiring routes to destinations beyond the routing zone. In this paper, we review the Zone Routing concept and propose Zone Routing with independently sized routing zones capability. With this capability, each of the nodes in the network can adaptively configure its own optimal zone radius in a distributed fashion. We show that the performance of Zone Routing is significantly improved by the ability to provide fine-tuned adaptation to local and temporal variations in network characteristics.

14.1 Ad Hoc Networks Overview

An ad hoc network is a self-organizing wireless network made up of mobile nodes and requiring no fixed infrastructure. The limitations on power consumption imposed by portable wireless radios result in a node transmission range that is typically small relative to the span of the network. To provide communication throughout the entire network, nodes are designed to serve as routers. The result is a distributed multi-hop network with a time-varying topology.

Because ad hoc networks do not rely on existing infrastructure and are self-organizing, they can be rapidly deployed to provide robust communication in a variety of hostile environments. This makes ad hoc networks very appropriate for providing tactical communication for the military, law enforcement, and emergency response efforts. Ad hoc networks can also play a role in civilian forums, such as the electronic classroom, convention centers, and construction sites. With such a broad scope of applications, it is not difficult to envision ad hoc networks operating over a wide range of coverage areas, node densities, and node velocities.

This potentially wide range of ad hoc network operating configurations poses a challenge for developing efficient routing protocols. On one hand, the effectiveness of a routing protocol increases as network topology information becomes more detailed and up to date. On the other hand, in an ad hoc network, the topology may change quite often, requiring large and frequent exchanges of control information among the network nodes. This is in contradiction with the fact that all updates in the wireless communication environment travel over the air and are, thus, costly in resources.

14.2 Brief Survey of Basic Ad Hoc Routing Protocols

Existing routing protocols can be classified either as *proactive* or as *reactive*. Proactive protocols attempt to continuously evaluate the routes within the network, so that when a packet needs to be forwarded, the route is already known and can be immediately used. Early applications of proactive routing schemes for ad hoc networks were Distance Vector protocols based on the Distributed Bellman–Ford (DBF) algorithm [2]. Modifications to the basic DBF algorithm (i.e., [4,7,27]) were proposed to address the inherent problems of convergence and excessive traffic (both can be quite severe problems in ad hoc networks, where bandwidth is scarce and topologies are often very dynamic). The convergence problem has also been addressed by the application of Link State protocols such as TBRPF [1], STAR [8], ALP [9], and GSR [3] to the ad hoc environment. In general, Link State protocols converge faster than Distance Vector protocols do, but may lead to a lot of control traffic. Motivation to both, improve protocol convergence and to reduce control traffic, has led to the development of proactive path finding algorithms, which combine the features of the Distance Vector and Link State approaches. Realizations of the path finding algorithms, such as the Wireless Routing Protocol (WRP) [19,20], are able to eliminate the "counting-to-infinity" problem and to reduce the occurrence of temporary loops, often with less control traffic than traditional Distance Vector schemes.

In contrast, reactive protocols[1] invoke a route determination procedure on an on-demand basis. The reactive route discovery is usually based on a query–reply exchange, where the route query uses some

---

[1] *Reactive protocols* are also referred to as *on-demand protocols.*

flooding-based process to reach the desired destination. In the case of the Temporally Ordered Routing Algorithm (TORA) [21], the route replies are also flooded, in a controlled manner, distributing routing information in the form of directed acyclic graphs (DAGs) rooted at the destination. In contrast, the Dynamic Source Routing (DSR) [16] and Ad hoc On Demand Distance Vector (AODV) [28] protocols unicast the route reply back to the querying source along a path constructed during the route query phase. In the case of DSR, the routing information is accumulated in the query packet and the complete sequence of nodes is returned to the source (to be used for source routing of the actual user data). AODV, on the other hand, distributes the discovered route in the form of next-hop information stored at each node in the route. The on-demand discovery of routes can result in much less traffic than the standard, proactive Distance Vector or Link State schemes, especially when innovative route maintenance schemes are employed. However, the reliance on flooding of the reactive schemes may still lead to considerable control traffic in the highly versatile ad hoc networking environment. Moreover, due to the large increase in control traffic at the times of route discovery, the delay of the route discovery process in reactive protocols can be significant.

14.3 Multi-Scope Routing

All else being equal, the value of information decreases with respect to the distance from the information source. For example, a node cannot compute routes without knowing any of its neighbors. On the other hand, a node may make near-optimal forwarding decisions even in spite of outdated or missing *distant* state information.

This relationship between information value and distance is extremely valuable for routing protocol design. Simply distributing the same information, at the same rate, to all nodes in the network does not provide the most "bang for the buck." There is more value in providing nearby nodes with more frequent updates and/or detailed information, at the expense of keeping more distant nodes less informed.

To some extent, most basic protocols exhibit some degree of multi-scope behavior. Many proactive routing protocols monitor the status of neighbor connectivity through neighbor broadcast HELLO beacons, which occur at a faster rate than the global link state (or distance vector) advertisements. In many reactive routing protocols, route discovery is based on querying on a global scale, whereas subsequent route repair utilizes local querying, constrained by a time-to-live (TTL) packet hop counter.

The high quality local route information provided by multi-scope routing can be used to provide an assortment of new and enhanced services. By identifying overlaps in local connectivity, broadcast messages can be distributed to all nodes more efficiently (e.g., OLSR's multipoint relay [6]). Moreover, local exchange of route information can be further exploited to provide a *bordercast* query distribution service, in which only a subset of the network's nodes needs to be queried. Such a service can be applied to global route discovery, name-address translation, and general database lookups. In the case of global reactive protocols, once a route has been discovered, changes in local connectivity can be quickly identified, allowing for either proactive route repair or proactive route shortening [23]. Local multi-hop feedback of link layer acknowledgments can be used to discover and reliably use unidirectional links [24]. Intelligent node participation/sleep-mode algorithms can use local route information to determine if a node's absence would compromise the network connectivity.

Perhaps the most familiar examples of multi-scope routing are the various flavors of hierarchical routing [5,15,26]. In basic clustered routing, nodes are aggregated into subnets. Each node knows the topology of its subnet through a local proactive protocol. On a global scale, each subnet is represented by a clusterhead, which knows the connectivity to other subnets' clusterheads, but not the details of the other subnets' topologies. Nodes are located in the hierarchy through relative addressing that associates nodes with clusterheads. This two-level example can be easily extended by grouping clusterheads into intermediate level subnets, thus creating a deeper hierarchy.

A variation on clustered routing is *landmark routing* (e.g., LANMAR [10]). As in the previous example, nodes are organized into local subnets and assume hierarchical addresses. However, the clusterheads are replaced with globally visible landmarks. A global distance vector routing protocol is used to provide all nodes with routes to the landmarks. The role of the landmark is to identify the general location of the associated subnet. Data packets are forwarded toward the landmark until they reach the subnet, at which point the subnet nodes can forward the packet directly to the destination.

Another hierarchical routing approach is based on the concept of a *core*. In these schemes, local topology is proactively monitored for the purpose of selecting a set of core nodes, such that every node has at least one core node neighbor. The purpose of the core nodes is to determine routes on behalf of the nodes that they cover. This is generally accomplished through global route discovery that is carried through the core. Although the route discovery occurs in the core, the core nodes apply knowledge of their local topology to construct routes that do not necessarily pass through the core. In addition to this basic operation, the CEDAR protocol [29] introduces an interesting local scoping behavior by advertising higher quality links (e.g., high capacity links) over greater distances. In the Dynamic Virtual Backbone scheme [17], which employs a similar concept of the core, the route queries are restricted within the virtual backbone.

Specialized node roles and regional node addressing help hierarchical routing protocols to scale with network size, especially when there is a structure in the underlying network connectivity (for example, group mobility) that can be exploited [18]. However, as network behavior becomes less coordinated, the overhead of the hierarchy maintenance (e.g., clusterhead election, node readdressing) becomes a limiting factor for scaling. In addition, hierarchical routing may introduce uneven resource utilization, traffic hotspots, and in some cases, suboptimal routing.

It is also possible to provide multi-scope routing without the limitations and overhead of hierarchy management. For example, in FSR [25], a node's link state is distributed over various distances (scopes), with longer distance updates occurring at lower frequencies. This provides each node with a fresh view of the surrounding topology but a more dated view of farther network regions. The less accurate distant views effectively serve as landmarks, getting data packets forwarded in the right general direction. As the data packets approach the destination, the path to the destination becomes more accurate and the forwarding more refined.

Additional benefits of multi-scope routing can be realized when larger scope protocols are able to exploit the information provided by a smaller scope. Two protocols that exhibit this kind of scope integration are OLSR and Zone Routing. In OLSR [6], an extended neighbor discovery provides each node with the topology of its surrounding two hops. This local information is used to provide an efficient global link state broadcast, based on *multipoint relay*. Multipoint relay identifies a "minimal" subset of neighbors needed to relay a message, such that all nodes two hops away will receive the message. In the case of Zone Routing, a proactive routing protocol is used to provide each node with a view of its surrounding "routing zone" topology. This local information enables an efficient query distribution service (bordercasting), which is used by a global reactive route discovery protocol. (Zone Routing is described in more detail later in this chapter). For OLSR and Zone Routing, global protocol efficiency increases with the size of the local "zone." The cost of local vs. global scope can be traded off, and ultimately optimized, through the adjustment of a single parameter — the zone radius.

14.4 Protocol Hybridization

The diverse applications of ad hoc networks pose a challenge for a single protocol that operates efficiently across a wide range of operational conditions and network configurations. Each of the purely proactive or purely reactive protocols described above performs well in a limited region of this range. For example, reactive routing protocols are well suited for networks where the "call to mobility" ratio is relatively low. Proactive routing protocols, on the other hand, are well suited for networks where this ratio is relatively high. Figure 14.1 shows the ad hoc network design space with node mobility and call rate as the two dimensions and the approximate regions where each of these two kinds of protocols performs well. The performance of either class of protocols degrades when the protocols are applied to regions of ad hoc network space between the two extremes.

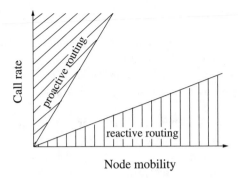

FIGURE 14.1 The ad hoc network design space and the applicability of proactive vs. reactive routing.

Given multiple protocols, each suited for a different region of the ad hoc network design space, it makes sense to capitalize on each protocol's strengths by combining them into a single framework (i.e., hybridization). In the most basic hybrid framework, one of the protocols would be selected based on its suitability for the specific network's characteristics. Although not an elegant solution, such a framework would perform as well as the best suited protocol for any scenario and outperform either protocol over the entire ad hoc network design space. However, by not using both protocols together, this approach fails to capitalize on the potential synergy that would make the framework perform as well or better than either protocol *for any given scenario.*

A more promising approach for protocol hybridization is to have the base protocols operate simultaneously, but with different "scopes." For the case of a two-protocol framework, protocol A would operate locally, while the operation of protocol B would be global. The key to this framework is that the local information acquired by protocol A is used by protocol B to operate in a more efficient manner. This framework can be tuned to network behavior simply by adjusting the size of the protocol A's scope. In one extreme configuration, the scope of protocol A is reduced to nothing, leaving protocol B to run by itself. As the scope of protocol A is increased, the information provided to protocol B increases as well, thereby decreasing protocol B's overhead. At the other extreme, protocol A is made global, eliminating the load of protocol B altogether. So, at either extreme, the framework defaults to the operation of an individual protocol. In the wide range of intermediate configurations, the framework performs better than either protocol on its own.

14.5 Framework Tuning

The motivation behind hybrid routing and multi-scope routing is to provide a framework that can be configured to match the network's operational condition and configuration. Therefore, an integral component of the framework is a tuning mechanism. In particular, the three basic ingredients for tuning are a means for measuring relevant network characteristics, a mapping of these measurements to a framework configuration, and a scheme to update the configurations of affected nodes.

The most basic approach to tuning is to determine the network characteristics and proper configuration offline, prior to the network deployment. Typically, the configuration would be determined through network simulation and subsequent parameter optimizations. The nodes are loaded with the proper configuration and then activated. When it is not possible to preconfigure all nodes individually, a small number of nodes may be configured, and this configuration can be shared with other nodes as part of an automatic configuration procedure.

The main advantages of preconfiguration are that it requires limited network intelligence and low real-time processing overhead and it ensures stable and consistent configuration. However, for many applications, preconfiguration is not an option. Preconfiguration requires a central configuration authority, which may not exist for distributed applications. In addition, the network characteristics may not be known *a priori*, or may vary over time, preventing the offline analysis and reducing the effectiveness of the static configuration.

Ad hoc networks naturally lend themselves to dynamic reconfiguration. Through the course of normal operation, nodes directly measure (or infer) local network characteristics. Each node may use its own local measurements for independent self-configuration. Alternatively, the measurements could be relayed to a central configuration node or shared with surrounding nodes for a distributed configuration approach.

At first glance, centralized dynamic reconfiguration may appear to prevent inconsistent configuration, as is the case for centralized static configuration. However, the multi-hop nature of ad hoc networks makes it impossible to reliably perform tightly synchronized configuration updates for all nodes. This means that, for some period of time, the network could be in an inconsistent state. As this also affects distributed and independent reconfiguration, it is necessary that a dynamically tunable routing framework be able to deal with, and potentially exploit, nonuniform node configurations. The way in which a routing framework supports nonuniform configuration depends on its particular design. We will later see how nonuniform configuration is supported in the Zone Routing framework.

With support for nonuniform configuration, reconfiguration decisions and the associated measurement/control traffic can be kept local (or eliminated altogether in the case of independent reconfiguration), making the tuning mechanism scalable. Furthermore, the framework can be fine tuned to adapt to regional changes or even to nodal behavior, rather than broadly tracking average network behavior. This can lead to significant performance improvements, especially in the case of networks where node behavior has regional dependencies.

14.6 The Zone Routing Framework

Protocol hybridization, multi-scope operation, and dynamic reconfiguration, the key features of adaptable and scalable routing, form the basis of the Zone Routing framework. At a local level, a proactive routing protocol provides a detailed and fresh view of each node's surrounding local topology (*the routing zone*). The knowledge of local topology is used to support services such as proactive route maintenance, unidirectional link discovery, and guided message distribution. One particular message distribution service, called *bordercasting*, directs queries throughout the network across overlapping routing zones. Bordercasting is used in place of traditional broadcasting to improve the efficiency of a global reactive routing protocol.

The benefits provided by routing zones, compared with the overhead of proactively tracking routing zone topology, determine the optimal framework configuration. As network conditions change, the framework can be dynamically reconfigured through adjustment of each node's routing zone.

In the following sections, we describe the routing and dynamic configuration components of the Zone Routing framework in more detail. (See also Fig. 14.2.)

14.6.1 Local Proactive (Intrazone) Routing

In Zone Routing, the Intrazone Routing Protocol (IARP) proactively maintains routes to destinations within a local neighborhood, which we refer to as a routing zone. More precisely, a node's routing zone is defined as a collection of nodes whose minimum distance in hops from the node in question is no greater than a parameter referred to as the *zone radius*. Note that each node maintains its own routing zone. An important consequence, as we shall see, is that the routing zones of neighboring nodes overlap.

Figure 14.3 illustrates the routing zone concept with a routing zone of radius two hops. This particular routing zone belongs to node S, which we refer to as the central node of the routing zone. Nodes A through K are members of S's routing zone. Node L, however, is three hops away from S and is, therefore, outside of S's routing zone. An important subset of the routing zone nodes is the collection of nodes whose minimum distance to the central node is exactly equal to the zone radius. These nodes are aptly

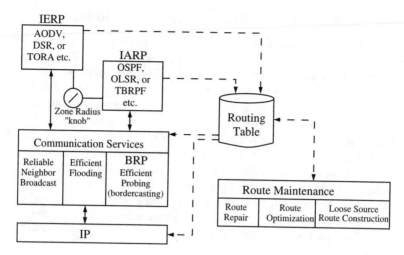

FIGURE 14.2 Architecture of the Zone Routing framework.

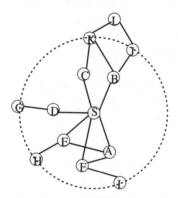

FIGURE 14.3 A Routing Zone of radius two hops. (Reprinted with permission from [22] © 1999 IEEE.)

named *peripheral nodes*. In our example, nodes G–K are peripheral nodes of node S. We typically illustrate a routing zone as a circle centered around the central node. However, one should keep in mind that the zone is a description not of physical distance, but rather of nodal connectivity (hops).

The construction of a routing zone requires a node to know first who its neighbors are. A neighbor is defined as a node with whom direct (point-to-point) communication can be established and is, thus, one hop away. Identification of a node's neighbors may be provided directly by media access control (MAC) protocols, as in the case of polling-based protocols. In other cases, neighbor discovery may be implemented through a separate Neighbor Discovery Protocol (NDP). Such a protocol typically operates through the periodic broadcasting of "hello" beacons. The reception (or quality of reception) of a "hello" beacon can be used to indicate the status of a connection to the beaconing neighbor.

Neighbor discovery information is used as a basis for the IARP. IARP can be derived from globally proactive link state routing protocols that provide a complete view of network connectivity (for example, the Shortest Path First [OSPF]). The base protocol needs to be modified to ensure that the scope of the route updates is restricted to the radius of the node's routing zone [13]. In this paper, IARP is based on a simple, timer-based, link state protocol. To track the topology of R-hop routing zones, each node periodically broadcasts its link state for a depth of R hops (controlled by a time-to-live [TTL] field in the update message).

14.6.2 Bordercast-Based Global Reactive (Interzone) Routing

Route discovery in the Zone Routing framework is distinguished from standard broadcast-based route discovery through a message distribution service known as *bordercasting* [14]. Rather than blindly broadcasting a route query from a neighbor to a neighbor, bordercasting allows the query to be directed outward, toward regions of the network (specifically, toward peripheral nodes) that have not yet been "covered" by the query. (A covered node is one that belongs to the routing zone of a node that has received a route query). The query control mechanisms reduce route query traffic by directing query messages outward from the query source and away from covered routing zones, as illustrated in Fig. 14.4.

A node can determine local query coverage by noting the addresses of neighboring nodes that have forwarded the query. In the case of multiple channel networks, a node can only detect query packets that have been directly forwarded to it. For single channel networks, a node may be able to detect any query packet forwarded within the node's radio range (i.e., through eavesdropping in promiscuous reception mode). When a node identifies a query forwarding neighbor, all known members of that neighbor's routing zone (i.e., those members that belong to both the node's and neighbor's routing zones) are marked as covered.

When a node is called upon to relay a bordercast message, it again uses its routing zone topology to construct a bordercast tree, which is rooted at itself and spans its uncovered peripheral nodes. The message is then forwarded to those neighbors in the bordercast tree. By virtue of the fact that this node has forwarded the query, all of its routing zone members are marked as covered. Therefore, a bordercasting node will not forward a query more than once.

Query detection can be enhanced by introducing a random delay prior to construction of the bordercast tree. During this time, the waiting node benefits from the opportunity to detect the added query coverage from other bordercasting neighbors. This, in turn, promotes a more thorough pruning of the bordercast tree. Increasing the average delay can significantly improve performance, up to a point. Once the bordercast delays are sufficiently spread out, further increases in delay have only a negligible impact on query efficiency.

The use of random delays does not necessarily result in extra route discovery delay. Many route discovery protocols use random pretransmission jitter to dilute the "instantaneous" channel load of neighboring query retransmissions. This forwarding jitter may be scheduled any time between query packet reception and query packet retransmission, including just prior to bordercast tree construction.

Given the implementation of an underlying bordercast service, the operation of Zone Routing's global reactive Interzone Routing Protocol (IERP) is quite similar to standard route discovery protocols. An IERP route discovery is initiated when no route is locally available to the destination of an outgoing data packet. The source generates a route query packet, which is uniquely identified by a combination of the source node's address and request number. The query is then relayed to a subset of neighbors as

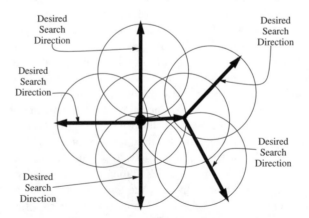

FIGURE 14.4 Guiding the search in desirable directions. (Reprinted with permission from [22] © 1999 IEEE.)

determined by the bordercast algorithm. Upon receipt of a route query packet, a node records its ID in the route query packet. The sequence of recorded node IDs specifies an accumulated route from the source to the current node. If a valid route for the destination is not known (i.e., the destination is not in the node's routing zone and an active route does not appear in the node's route cache), then the node bordercasts the query. This process continues until the query reaches a node that has a valid route to the destination or until the query reaches the destination itself. In that case, a route reply is sent back to the source, along the path specified by reversing the accumulated route. The operation of the IERP is sufficiently general so that many existing reactive protocols can be used as an IERP with minimal modification [12].

14.7 Motivation for Independent Zones

Intuition, confirmed by simulation results of Zone Routing, suggests that high mobility and/or low call rates favor a smaller zone radius. And vice versa, low mobility and/or high call rates favor a larger zone radius. Now consider a network where different parts of the network have different mobility and call rate patterns. Due to these differences, it may turn out that the different sections may have different optimal zone radii. This motivates the development of the Zone Routing framework with independent zones capability, such that different nodes are possibly assigned different zone radii. It is quite likely that such a framework would perform better than the single zone size case, as it can be fine tuned to the local conditions of the network. Furthermore, if the network characteristics change over time, such a framework can easily and quickly adapt to the changing conditions of the network.

According to the description of Zone Routing in the previous section, every node participating in network routing should have the same value of the zone radius. This means that before the network becomes operational, all the nodes in the network should come to a consensus on the optimal value of the zone radius by some extraneous means. Also, any node joining the network later or undergoing rebooting should be able to infer the correct value of the zone radius with which the rest of the network is operating.

Many applications of ad hoc networks require that the network be formed and be operational quickly and that nodes be free to join and leave the system at their own will, without the need for any external configuration. In such networks, the constraint of having a uniform zone radius for all the nodes may not be desirable. Having independent zones capability within the Zone Routing framework would allow nodes to dynamically, distributedly, and automatically configure their optimal zone radii, making the framework truly flexible.[2]

All these points motivate the development of Zone Routing with the capability to have independent routing zones. Such a framework would help concoct a hybrid mix of proactive and reactive components, which is just right for the specific characteristics and operational conditions of the network. The proportion of proactive and reactive components in this hybrid mix can easily be changed over time and location by changing a single parameter — the zone radius of each node. We expect that such a framework would not only reduce the routing overhead, but would be responsive to the needs of the network traffic as well.

14.8 IZR Introduction

In the Independent Zone Routing (IZR) framework, different nodes may have different sized "routing zones." What does it mean for the nodes to have independent routing zones and how does such a routing protocol operate? Before exploring these questions, we begin by re-defining some terms in the IZR context, as follows:

---

[2] We will see later how a node distributedly determines the optimal value of its zone radius based on its measurements of the routing control traffic.

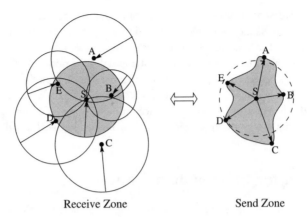

Receive Zone   Send Zone

FIGURE 14.5 The receive zone is regular in shape, but the send zone may not be.

- *Routing zone or receive zone:* The *neighborhood* around each node about which a node proactively maintains routing information is called its routing zone. A node maintains this information by receiving proactive updates from these nodes in the neighborhood; hence, this zone is also called its receive zone. This neighborhood consists of the set of all nodes, whose minimum distance, in hops, from the node is not more than the *zone radius, R.*
- *Send zone:* All the nodes that require proactive updates from the node in question, in order to maintain their intrazone routing information, belong to the node's send zone. A node is expected to broadcast proactive updates to the members of its send zone.
- *Peripheral nodes:* The farthest members of a node's routing zone, whose minimum distance from the node is *R* hops, are called its peripheral nodes.

We have seen that in the case of equally sized routing zones, a node broadcasts proactive routing information to all the members of its zone and also receives the same information from each one of them. Thus, the send zone of a node is the same as its receive zone, when all nodes have equal zone radius. However, when the nodes in the network are allowed to have independent routing zones, this may not be the case.

In IZR, the routing zone or the receive zone is also *regular* in shape — that is, it can be represented by a circle of radius proportional to the zone radius of the node. All nodes with a lesser number of hops from the node lie inside this circle, and the peripheral nodes lie on the circle. In contrast, in IZR, the send zones may not have such a regular shape. The members of the send zone of a particular node S consist of all nodes of which S is a routing zone member (so that they expect to receive a routing update from S). With possibly different routing zone sizes for different nodes, S may be a member of routing zones of nodes that may possibly form an irregularly shaped area. The send zone may not even be a connected (contiguous) area.

Figure 14.5 shows the routing or receive zones of nodes S, A, B, C, D, and E, which are regular in shape. As S is a routing zone member of A, B, C, D, and E, they belong to its send zone, which is irregular in shape.

14.9 IZR Details

The basic operation of IZR is similar to Zone Routing as discussed above. If a source node has a packet to send to a destination node that is not a member of its routing zone, it bordercasts a route query packet. However, due to the presence of unequal routing zones in the network, a somewhat different bordercasting scheme is used. As unequal routing zones imply that the send zone of a node may be irregular in shape, the Intrazone Routing Protocol (IARP) has to be modified in order to distribute the proactive updates in such a send zone. Below, we discuss the IARP and Bordercast Resolution Protocol (BRP) for the Independent Zone Routing framework.

14.9.1 Intrazone Routing Protocol (IARP)

Each node maintains proactive routing information about the members of its routing or receive zones. For this to happen, each node needs to broadcast its proactive updates to the members of its sendzone. As the send zone may be irregular in terms of the distance in hops to the "boundary" nodes, a node first needs to infer the size and the shape of its send zone.

Consider a scenario where each node broadcasts "zone building packets" to all the members of its routing zone. As the routing zones are regular in shape, this can easily be done by setting the time-to-live (TTL) field of the update packet equal to the zone radius R. The value in the TTL field is decremented by one each time the packet travels one hop. If the TTL value reaches zero, the packet is dropped, else it is rebroadcasted. In Fig. 14.5, nodes A, B, C, D, E, and S broadcast their zone building packets to the members of their routing zones (marked by a circle around each of them). Thus, each node will receive a zone building packet from all those nodes to whose routing zones it belongs. In particular, S would receive A, B, C, D, and E's zone building packets, as it is a member of each of their routing zones. Note that all these nodes, whose zone building packets are received by a node, will belong to that node's send zone — A, B, C, D, and E belong to S's send zone, as in Fig. 14.5.

Based on the above discussion, in the general case of independent zone radius, a node can find out the size and extent of its send zone. The following scheme is used by the nodes to distribute the proactive updates in their send zones. Along with the zone radius field and the dynamic time-to-live TTL field, an update packet also has a field that contains the initial value of TTL at the source, the TTL_0 field. The source node sets the value of the TTL and TTL_0 fields equal to the distance in hops to the farthest member of its send zone.

Initializing the TTL field to the above value makes the updates reachable to all the members of a node's send zone. For example, in Fig. 14.6, B sets the TTL (and TTL_0) field equal to the distance in hops to one of the farthest members of its send zone (node C, E, or G). However, this may lead to some extra overhead, due to the updates being broadcasted in the area marked by the horizontal lines in the figure (which lies outside B's send zone). In order to reduce this overhead, each of the peripheral nodes of B maintains information about members of B's send zone that lie further away from B than itself. That is, A maintains information about C, D about E, and F about G. B's other peripheral nodes, H, J, and K, do not have any such nodes. Hence, when A, D, and F receive B's proactive updates, they send them toward C, E, and G, respectively. When H, J, and K receive B's updates, they do not forward the update packets, thus reducing the extra overhead.

This information to reduce the overhead is maintained as follows. A node, A, maintains a list of all nodes for whom it serves as a peripheral node. For each node B in this list, A maintains another list called the *expecting_nodes_list*, which consists of all nodes C whose "zone building packets" are received by A such that the value of TTL in the packet is not less than the zone radius of B. (This implies that B lies in C's routing zone, or equivalently, C lies in B's send zone.)

Now a peripheral node A of a node B does not forward a proactive update packet originated at B if A has no nodes in the *expecting_nodes_list* for node B. This reduces unnecessary traffic going beyond the peripheral nodes, if there are no nodes in that region that have B in their routing zone. Note that all these conditions can be checked by using the TTL, TTL_0, and the zone radius values available in the update packets or the zone building packets.

Using the above scheme, each node in the network broadcasts proactive update packets by initializing the values of TTL and TTL_0 as above. Propagation of unnecessary update packets may be terminated by the peripheral nodes, if no nodes beyond them are expecting these packets, as found by examining the maintained lists.

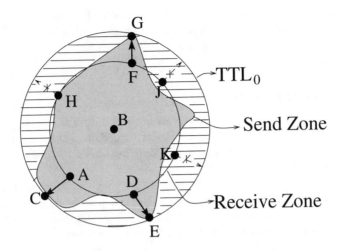

FIGURE 14.6   The irregular send zone of node B is indicated by the shaded area.

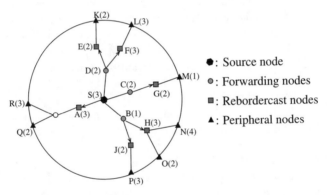

FIGURE 14.7   The bordercast tree of the source node S. Numbers in parentheses next to the node labels indicate the zone radii of the nodes.

The "zone building packets'"may be combined with the proactive update packets to reduce the overhead. The broadcasting of the proactive updates by IARP can be based on one of the strategies proposed in [30] for more efficient performance.

14.9.2 Bordercast Resolution Protocol (BRP)

With independently sized routing zones in the network, it is possible that some of the nodes in the bordercast tree of the source node have a routing zone that is small, so that it lies completely within the source node's routing zone. Such nodes may not be able to reconstruct the source node's bordercast tree and may not be able to correctly judge whom to forward the packets to. In order to deal with situations like these, a different bordercasting mechanism is used.

Figure 14.7 shows the bordercast tree of the source node S, which has a zone radius of three. Nodes A, B, C, and D are the bordercast tree neighbors of S. The zone radius of A is three and its zone extends beyond S's routing zone. However, as the zone radii of B, C, and D are small (one, two, and two, respectively), their routing zones lie completely inside S's routing zone. So, S examines the nodes that are two hops from it in the bordercast tree downstream from B, C, and D. It finds that the routing zones of E, F, G, H, and J include newer regions outside its own routing zone, and so they get selected. S then sends the route query packet to A, E, F, G, H, and J (via forwarding through one-hop neighbors, if needed), which could, in turn, query unexplored regions of the network.

Formally, we define the following two kinds of nodes that a bordercasting node identifies after constructing the bordercast tree to its *uncovered* peripheral nodes:

- *Rebordercast Node:* The node closest to the source node on the bordercast path from the source node to a peripheral node, such that its routing zone extends beyond the source node's routing zone, is called a rebordercast node of the source node corresponding to that peripheral node. For example, in Fig. 14.7, *H* is a rebordercast node corresponding to *O* and *N*, *J* is a rebordercast node corresponding to *P*, etc.

- *Forwarding Node:* Nodes lying on the bordercast-path between the source node and a rebordercast node belong to the set of forwarding nodes corresponding to that rebordercast node. For example, *B* is a forwarding node corresponding to *J* and *H*, while *A* does not have any forwarding node. Note that it is possible that this set may be empty.[3]

The following bordercasting mechanism is used by the nodes in order to guide a route query "outwards," towards unexplored regions of the network:

1. Source node *S* constructs the bordercast tree to *uncovered* peripheral nodes.
2. *S* chooses rebordercast nodes corresponding to each of its *uncovered* peripheral nodes.
3. *S* then sends the query packet to each of these rebordercast nodes via the forwarding nodes, if any.
4. When the rebordercast nodes receive the query packet, they become the bordercast nodes and go back to step 1.

For query control, a node marks certain members of its zone as *covered* and tries to steer the query away from such nodes. The following rules are used by a node for identifying such covered regions of its routing zone:

- A forwarding node marks all the members of its zone as covered.
- A rebordercast node marks:
  - the nodes lying in the intersection of its zone with the zone of the bordercasting node as covered, if the bordercasting node is a member of its zone[4]
  - the nodes lying in the intersection of its zone with the zone of the last forwarding node as covered, if the bordercasting node does not lie in its zone[5]

The above mechanism ensures that the query always gets bordercasted by nodes whose routing zones cover newer, unexplored regions.

14.9.3 Zone Radius Determination Algorithm

In order to determine the optimal zone radius of each node in the network independently, the algorithm for zone radius determination should depend only on the local measurements made at the node. A hybrid of *Min Searching* and *Adaptive Traffic Estimation* schemes, described in [22], is used to determine the optimal zone radius of each node dynamically and independently.

The Min Searching scheme involves iteratively searching for the minima of the routing control traffic curve by incrementally increasing or decreasing the routing zone radius of a node by one hop. During each estimation interval, the amount of routing traffic is measured. If the amount of routing traffic in the current estimation interval is less than that in the previous interval, the zone radius is further

---

[3] The routing zone of a forwarding node does not extend beyond that of the source node, otherwise it itself would be a rebordercast node.

[4] As the bordercasting node is a member of the rebordercasting node's routing zone, the rebordercasting node knows the bordercasting node's position relative to the other members of its routing zone and thus can infer the intersection of their routing zones.

[5] The last forwarding node will be a member of the rebordercasting node's routing zone and, thus, it has the required information to mark the intersection of their routing zones as covered.

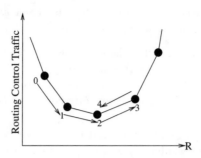

FIGURE 14.8 Min Searching to determine the minima of the routing control traffic curve. The algorithm starts at t = 0 and converges to the optimal value at t = 4. (Reprinted with permission from [22] © 1999 IEEE.)

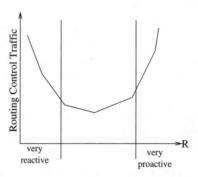

FIGURE 14.9 The location of optimal zone radius is in the region where the routing control raffic is neighter very reactive nor very proactive (Reprinted with permission from [22] © 1999 IEEE.)

incremented/decremented in the same direction. Else, the direction of zone radius change is reversed. The process continues until a minimum is detected, as shown in Fig. 14.8. The Min Searching scheme converges to the local minima, provided that the network characteristics do not change substantially during the search and that the estimation interval is long enough to provide a good measurement of the routing control traffic.

The optimal zone radius (Fig. 14.9) of a node lies in a region where the total routing control traffic is neither predominantly reactive nor predominantly proactive. The Adaptive Traffic Estimation scheme exploits the fact that close to the minimum the amount of these two traffic types is relatively equal. The scheme involves iteratively changing the routing zone radius based on the relative proportions of reactive and proactive components in the total routing overhead generated in each estimation interval. Let $\Gamma(R)$ be the ratio of reactive (IERP) traffic to proactive (IARP) traffic at zone radius R during a certain estimation interval. Adjustments to the zone radius are made by comparing this ratio with a predetermined threshold, Γ_{thres}. If $\Gamma(R) > \Gamma_{thres}$, increase the zone radius; if $\Gamma(R) < \Gamma_{thres}$, decrease the zone radius. However, possibly changing the zone radius after each estimation interval could lead to too frequent adaptation of the zone radius and to possible network instability. Hence, a triggering mechanism is introduced by a hysteresis term, so that if $\Gamma(R) > \Gamma_{thres} \cdot H$, increase the zone radius; if $\Gamma(R) < \Gamma_{thres}/H$, decrease the zone radius. In this scheme, the decision to change the zone radius is based on the measurements made in the current estimation interval only. This is desirable as the scheme can better track the changes in the network, and the dependence on the need for correlation between successive intervals is reduced.

A hybrid zone radius determination scheme is used in IZR, where control can switch between the Min Searching and the Adaptive Traffic Estimation schemes. Initially the Min Searching scheme is under control. The Min Searching scheme provides useful statistics, which could be used to improve the estimates of Γ_{thres} and H for the Adaptive Traffic Estimation scheme [22]. Once the minimum of the control traffic curve is reached, the control switches to the Adaptive Traffic Estimation scheme. Also, whenever the zone radius R reaches one, the Min Searching scheme assumes control again [22]. This is because the Adaptive Traffic Estimation scheme could be misled by the highly reactive traffic at a zone radius of one ($\Gamma(1) = \infty$).

14.10 Performance Evaluation

The OPNET™ simulation environment was used to simulate the Independent Zone Routing framework. Link-state based IARP, described in [13], was used as the proactive component, and a source-route based IERP, described in [12], was used as the reactive component.

The network consists of 50 nodes, spread randomly in an area of 1000×1000 *meter*^{2}. The transmission radius of the nodes is set at 225 meters. A node moves at a constant speed v and is assigned an initial direction θ, which is uniformly distributed between 0 and 2π. When a node reaches an edge of a simulation region, it is reflected back into the network. A node's session with a randomly chosen destination consists of sending an average of 25 packets. The interarrival times between sessions are exponentially distributed. As different simulation runs were performed for different zone radius settings, the network behavior was made to remain exactly the same; i.e., the nodes moved in exactly the same path and started sessions with exactly the same nodes at the very same instants.

Figure 14.10a shows the amount of routing control traffic generated during a simulation duration of 180 seconds. The scenario consists of 50 nodes, half of which (Set I) move at the constant speed (v) of 14 m/sec and have the mean session interarrival delay (MSID) of 1 sec. The other half (Set II) move at the speed of 1 m/sec and have the mean session interarrival delay of 0.1 sec. From the plot, it can be seen that Independent Zone Routing with dynamic radius configuration leads to about a 50% reduction in routing control traffic, as compared to regular Zone Routing. The plot also shows the amount of routing control traffic for IZR with fixed but different zone radius (R) assignments for the two sets of nodes. The reduction in control traffic for this case reinforces our belief that different zone radii may be preferable for nodes with different characteristics.

The curves in Fig. 14.10b show the amount of routing control traffic for a scenario consisting of 50 nodes. Half of these nodes (Set I) move at the speed of 7 m/sec and have the mean session interarrival delay of 1 sec. The rest (Set II) move at the speed of 1 m/sec and have the mean session interarrival delay of 0.1 sec. The plot shows the reduction in routing control traffic as compared to Zone Routing, when IZR with dynamic zone radius configuration or IZR with fixed but different zone radii for the two sets of nodes are used.

In both Figs. 14.10a and b, the values for IZR with fixed but different zone radii are the representative combinations of the zone radii for the two sets of nodes. The routing control traffic generated for IZR with dynamic zone radius configuration is between the different values for IZR with fixed but different zone radii combinations. The reason for this behavior is that the dynamic zone radius configuration algorithm tries to estimate the state of the network based on the traffic measurements in an estimation interval. Making the estimation interval small makes the traffic estimates during that interval inaccurate, whereas making it large makes the algorithm slow in adaptively tracking the changing conditions of the network. Thus choosing a value of the estimation interval implies a tradeoff between the accuracy of the zone radius determination scheme and the speed with which it adapts to the changing network conditions.

Figure 14.11 shows how the routing control traffic generated in the network varies as time progresses. The points on the curves represent the amount of control traffic generated over a window of the last 25 seconds. Initially, the amount of traffic generated is high, as the network is operating at suboptimal routing zone radii values. Soon, the zone radius determination algorithm is able to find optimal values of the zone radii for the nodes, and the routing overhead decreases to below the level of regular Zone Routing. The routing control traffic values for the case of IZR with dynamic zone radius configuration in Figs. 14.10a and b do not include this initial overhead of the scheme during its stabilization. Γ_{thres} has

(a)

(b)

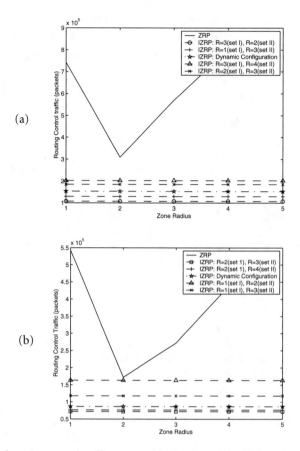

FIGURE 14.10 Total routing control traffic generated in the network consisting of 50 nodes divided into two sets. (a) Set I: $v = 14$ m/sec, MSID = 1 sec; Set II: $v = 1$ m/sec, MSID = 0.1 sec. (b) Set I: $v = 7$ m/sec, MSID = 1 sec; Set II: $v = 1$ m/sec, MSID = 0.1 sec.

been set equal to one (corresponding to equal proactive and reactive components) for these curves. The curves for different hysteresis (H) values show that the dependence on H is not very strong.

Thus, we have demonstrated that IZR enhances the Zone Routing framework by enabling each node in the network to independently and adaptively configure its optimal zone radius. Furthermore, it can lead to reduction in routing control traffic as well, as observed from the simulation results. IZR enables setting the zone radius of each of the nodes to its optimal value over time and space. This can improve the efficiency and increase the scalability of the routing protocol.

14.11 Conclusions

Hybridization, multi-scope operation, and local tuning form the basis for scalable, adaptable routing, as demonstrated by the Zone Routing framework. Zone Routing provides a flexible solution to the challenge of discovering and maintaining routes in a wide variety of different and differing ad hoc networking environments. The proportions of proactive and reactive components can be changed depending on the network conditions. With independent zones capability, Zone Routing can be fine tuned to the local conditions of the network. Each of the nodes in the network can dynamically, distributedly, and automatically configure its zone radius to the temporally and locally optimal value. This configuration is done at each node by analyzing just the local route control traffic, making the tuning mechanism itself

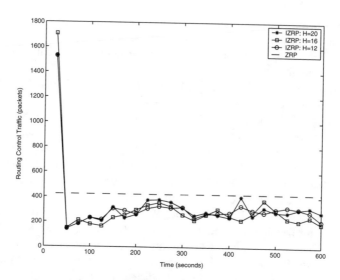

FIGURE 14.11 The variation in routing control traffic for IZR with dynamic zone radius configuration. The points correspond to routing overhead generated over a window of the last 25 seconds.

scalable. All these factors lead to significant performance improvements and increase the scalability and robustness of the routing protocol.

Possible future directions consist of extending the hybrid routing framework into additional dimensions, for example balancing the tradeoffs between bandwidth efficiency and local processing/storage requirements. Hybrid frameworks for other layer 3+ global operations, such as security and database management, can also be developed.

References

1. Bellur, B. and Ogier, R.G., A Reliable, Efficient Topology Broadcast Protocol for Dynamic Networks, *IEEE INFOCOM,* Mar. 1999.
2. Bertsekas, D. and Gallager, R., *Data Networks,* 2nd ed., Prentice Hall, Inc., Englewood Cliffs, NJ, 1992.
3. Chen, T.-W. and Gerla, M., Global State Routing: A New Routing Scheme for Ad-hoc Wireless Networks, *IEEE ICC '98,* Atlanta, GA, June 1998.
4. Cheng, C., Reley, R., Kumar, S.P.R., and Garcia-Luna-Aceves, J.J., A loop-free extended Bellman–Ford routing protocol without bouncing effect, *ACM Computer Communications Review,* 19, 224–236, 1989.
5. Chiang, C.-C., Routing in Clustered Multihop, Mobile Wireless Networks with Fading Channel, *IEEE SICON '97,* Apr. 1997.
6. Clausen, T., Jacquet, P., Laouiti, A., Minet, P., Muhlethaler, P., Qayyum, A., and Viennot, L., Optimized Link State Routing Protocol (OLSR), IETF MANET, Internet Draft, Oct. 2001.
7. Garcia-Luna-Aceves, J.J., Loop-free routing using diffusing computations, *IEEE/ACM Transactions on Networking,* 1, 130–141. 1993.
8. Garcia-Luna-Aceves, J.J. and Spohn, M., Efficient Routing in Packet-Radio Networks Using Link-State Information, *IEEE WCNC 99,* Aug. 1999.
9. Garcia-Luna-Aceves, J.J. and Spohn, M., Scalable Link-State Internet Routing, *IEEE International Conference on Network Protocols (ICNP 98),* Austin, TX, Oct. 14–16, 1998.
10. Gerla, M., Hong, X., and Pei, G., Landmark Routing for Large Ad Hoc Wireless Networks, *IEEE GLOBECOM 2000,* San Francisco, CA, Nov. 2000.
11. Haas, Z.J. and Pearlman, M.R., The performance of query control schemes for the zone routing protocol, *ACM/IEEE Transactions on Networking,* 9, 427–438, 2001.

12. Haas, Z.J., Pearlman, M.R., and Samar, P., Interzone Routing Protocol (IERP), IETF MANET, Internet Draft, June 2001.
13. Haas, Z.J., Pearlman, M.R., and Samar, P., Intrazone Routing Protocol (IARP), IETF MANET, Internet Draft, June 2001.
14. Haas, Z.J., Pearlman, M.R., and Samar, P., Bordercast Resolution Protocol (BRP), IETF MANET, Internet Draft, June 2001.
15. Iwata, A., Chiang, C.-C., Pei, G., Gerla, M., and Chen, T.-W., Scalable routing strategies for ad hoc wireless networks, *IEEE JSAC*, special issue on ad hoc networks, 17(8), August 1999.
16. Johnson, D.B. and Maltz, D.A., Dynamic source routing in ad hoc wireless networking, in *Mobile Computing*, Imielinski, T. and Korth, H., Eds., Kluwer Academic Publishing, Dordrecht, 1996.
17. Liang, B. and Haas, Z.J., Hybrid Routing in Ad Hoc Networks with a Dynamic Virtual Backbone, to appear in *Transactions on Wireless Communications*.
18. McDonald, A.B. and Znati, T., Predicting Node Proximity in Ad-Hoc Networks: A Least Overhead Adaptive Model for Electing Stable Routes, *MobiHoc 2000*, Boston, Aug. 4th, 2000.
19. Murthy, S. and Garcia-Luna-Aceves, J.J., A Routing Protocol for Packet Radio Networks, *Proc. of ACM Mobile Computing and Networking Conference, MOBICOM '95*, Nov. 14–15, 1995.
20. Murthy, S. and Garcia-Luna-Aceves, J.J., An efficient routing protocol for wireless networks, *MONET*, 1, 183–197, 1996.
21. Park, V.D. and Corson, M.S., A Highly Adaptive Distributed Routing Algorithm for Mobile Wireless Networks, *IEEE INFOCOM '97*, Kobe, Japan, 1997.
22. Pearlman, M.R. and Haas, Z.J., Determining the Optimal Configuration for the Zone Routing Protocol, *IEEE JSAC*, special issue on ad hoc networks, 17(8), August 1999.
23. Pearlman, M.R., Haas, Z.J., and Mir, S.I., Using Routing Zones to Support Route Maintenance in Ad Hoc Networks, *IEEE WCNC 2000*, Chicago, Sep. 2000.
24. Pearlman, M.R., Haas, Z.J., and B.P. Manvell, Discovering and Reliably Communicating over Unidirectional Links in Ad Hoc Networks, *IEEE WCNC 2000*, Chicago, Sep. 2000.
25. Pei, G., Gerla, M., and Chen, T.-W., Fisheye State Routing: A Routing Scheme for Ad Hoc Wireless Networks, *ICC 2000*, New Orleans, June. 2000.
26. Pei, G., Gerla, M., Hong, X., and Chiang, C.-C., A Wireless Hierarchical Routing Protocol with Group Mobility, *IEEE WCNC '99*, New Orleans, Sep. 1999.
27. Perkins, C.E. and Bhagwat, P., Highly Dynamic Destination-Sequenced Distance-Vector Routing (DSDV) for Mobile Computers, *ACM SIGCOMM*, 24, 234–244, 1994.
28. Perkins, C.E. and Royer, E.M., Ad Hoc On-Demand Distance Vector Routing, *IEEE WMCSA'99*, New Orleans, Feb. 1999.
29. Sivakumar, P., Sinha, R., and Bharghavan, V., CEDAR: A Core-Extraction Distributed Routing Algorithm, *IEEE JSAC*, special issue on ad-hoc networks, 17(8), Aug. 1999.
30. Samar, P. and Haas, Z.J., Strategies for Broadcasting Updates by Proactive Routing Protocols in Mobile Ad Hoc Networks, *IEEE MILCOM 2002*, Anaheim, CA, Oct. 2002.

V

Routing Techniques in Ad Hoc Wireless Networks, Part II

15

Adaptive Routing in Ad Hoc Networks

Yantai Shu
Tianjin University

Oliver Yang
University of Ottawa

Lei Wang
Tianjin University

Abstract

The dynamics of an ad hoc network are a challenge to protocol design because mobility inevitably leads to unstable routing, and consequently flows encounter fluctuations in resource availability on various paths during the lifetime of a session. This has become serious, especially for those protocols based on single-path reservation, as frequent reservation and restoration of reservation-based flows increase the instability of connections. Advances in wireless research are focusing more and more on the adaptation capability of routing protocols due to the interrelationship among various performance measures such as those related to topological changes (link breakages, node mobility, etc.) and quality of service (QoS) parameters (load, delay, etc). After giving a more detailed discussion of the existing work in adaptive routing, we summarize our work on Multipath Source Routing (MSR) in order to introduce our latest work on QoS-MSR.

Multipath Source Routing (MSR) is an extension of DSR (Dynamic Source Routing) that incorporates the multipath mechanism into DSR. MSR is an adaptive routing for ad hoc networks. It considers the two fundamental issues in its design. MSR may adapt to topology changes by retaining the route discovery and route maintenance mechanism of DSR. In addition, MSR employs a probing-based load-balancing mechanism. Simulation results show that MSR can improve the packet delivery ratio and the throughput of TCP and UDP, and it reduces the end-to-end delay and the average queue size while adding little overhead. As a result, MSR decreases network congestion and increases the path fault tolerance quite well.

As a multipath QoS routing protocol, QoS-MSR can collect QoS information through a route discovery mechanism and establish a QoS route with reserved bandwidth by using Multipath Bandwidth Splitting Reservation (MBSR). MBSR allows a bandwidth request to be split into several smaller bandwidth requests among multiple paths. According to our simulation results, QoS-MSR with MBSR decreases network congestion and improves the packet delivery ratio and end-to-end delay of all connections. In addition, the reserved packet ratio indicates that MBSR can improve QoS of reservation-based flows and can be made adaptive to ad hoc networks with high mobility.

15.1 Introduction

An ad hoc mobile network is a collection of mobile nodes that are dynamically and arbitrarily located in such a manner that the interconnections between nodes are capable of changing on a continual basis. In order to facilitate communication within the network, a routing protocol is used to discover routes between nodes. The primary goal of the routing protocol is to obtain correct and efficient route establishment between a pair of source and destination nodes so that messages may be delivered in a timely manner. The general performance measures are the delay and throughput of information. Due to the limited bandwidth associated with most wireless channels, it is obvious that route construction should be done with a minimum of overhead and bandwidth consumption. And due to the nature of node distribution, another performance measure is path reliability, which distinguishes ad hoc networks from other types of networks. Much work has appeared in these areas, but advances in wireless research are focusing more and more on the adaptation capability of routing protocols due to the interrelationship among these performance measures. In general, adaptive routing protocols can be classified by how they address the two fundamental issues: topology changes and load changes.

There are many papers addressing the topology change issues. Some, such as Dynamic Source Routing (DSR)[1], Multipath Source Routing (MSR) [2], Temporally Ordered Routing Algorithm (TORA) [3], and Source-tree On-demand Adaptive Routing (SOAR) [4], deal with topologies. Various factors can affect topologies, including link breakages and node mobility, such as the Associativity-Based Routing (ABR) [5]. Other researchers seek methods to combat the topology change issues such as the adaptive clustering method [6–8] to achieve more efficient routing management under topological changes, and adaptive multicasting methods [9–11].

Other papers focus more on traffic load issues with the idea of combating congestion. As traditionally performed, these approaches fall into two categories:

1. Prevention methods such as System and Traffic dependent Adaptive Routing Algorithm (STARA) [12] and the Adaptive Distance Vector (ADV) [13]
2. The avoidance method [14,15]

Some of these protocols have considered the two fundamental but related issues in their design. For example, STARA has considered load changes due to topology changes, and ADV has also considered mobility issues in addition to the load issue. As a matter of fact, recent advancement in the adaptive approach has focused more on the QoS issues by combining these two related issues. Typical works are end-to-end delay based, such as Distributed Dynamic Routing (DDR) [16], or bandwidth-based, such as the Core-Extraction Distributed Routing Algorithm (CEDAR) [17] and Multipath Source QoS Routing (QoS-MSR) [18].

In the following, Section 15.2 gives a more detailed discussion of the existing work in adaptive routing. Section 15.3 summarizes our work on MSR to introduce our latest work on QoS-MSR in Section 15.4. Section 15.5 provides some concluding remarks.

15.2 A Survey Of Adaptive Routing Protocols

This section surveys and classifies the existing protocols. As mentioned earlier, some of them are inter-related.

15.2.1 Topology-Related Protocols

In addition to dealing with topological changes, the protocols here consider related issues of node mobility and methods to combat these issues.

15.2.1.1 Adaptation to Topological Changes

The Dynamic Source Routing (DSR) protocol proposed in [1] adapts quickly to routing changes when host movement is frequent, and it requires little or no overhead during periods in which hosts move less frequently. When a host wants to send packets, the routing cache is first consulted to find the route to the destination. If there is no entry for that particular destination, a route discovery process is initiated. When a node detects that the packets are not received by the next hop, it first consults its route cache to find another route to the destination. If there is no other route available, the node sends a route error message to the source node, and a new route discovery process is launched by the source. More details will be provided in Section 15.3.

The Multipath Source Routing (MSR) protocol proposed in [2,19] is a multipath routing extension of DSR. A scheme to distribute load among multiple paths based on measurement of the round-trip time of every path is proposed. Simulation results show that the new approach improves the packet delivery ratio and the throughput of TCP and UDP and reduces the end-to-end delay and the average queue size, while adding little overhead. As a result, MSR decreases network congestion and increases the path fault tolerance quite well. More details will be provided in Section 15.3.

TORA [3] is a distributed protocol with source-initiated routing based on the link reversal algorithm. The scope of TORA is to limit the amount of control information, especially in topological changes, to isolate the propagation of control messages to a small set of nodes near the event. Route creation is launched by a host by broadcasting a query with the node ID of the destination. When the query packet reaches a node that has a "height value" for the destination, an update is sent as a response, with the "height" of the node attached in the packet. The offset metric of the height is incremented in the receiving node and sent to the neighbors. In this way, a directed acyclic graph is constructed from the source to the destination. When there are changes in the topology, a new reference level is generated and forwarded to the network. When routes are no longer valid, they are erased using clear messages.

The Source-tree On-demand Adaptive Routing (SOAR) protocol [4] is an on-demand link-state protocol based on partial link-state information, in which a wireless router communicates to its neighbors the link states of only those links in its source tree (i.e., those links that belong to the paths a router chooses to advertise link-state information for reaching destinations with which it has active flows). A little bit different from ABR, SOAR does not require periodic link-state advertisements when there are no link connectivity changes in the network. Minimal source trees can be updated incrementally or automatically, and updates to source trees are validated using sequence numbers. Thus, SOAR requires fewer control packets and has a better performance.

15.2.1.2 Adaptation to Mobility

Associativity-Based Routing (ABR) [5] is a distributed routing protocol that can adapt to the mobility of neighboring nodes. This protocol employs a new routing scheme where a route is selected based on nodes having associativity states that imply periods of stability. The key idea behind the ABR protocol is the ability to take into account the stability of the links when making a routing decision. The metric that is used for links is the degree of association stability, and is determined by beacons that are generated periodically. Each node makes a record of the received beacons, and when the node or one of its neighbors moves, it deletes irrelevant entries. By comparing the number of beacons heard per link, a node can draw conclusions about the mobility of the neighbor: if the metric is high, the node is considered stable; if the metric is low, the node is considered to be highly mobile. In this manner, the routes selected are likely to be long-lived and hence there is no need to restart frequently, resulting in higher attainable throughput.

15.2.1.3 Adaptive Clustering

In [6], an adaptive approach for hierarchical routing management is proposed. At first, the network infrastructure is constructed by several communication groups, which are called routing groups. A routing group communicates with other routing groups via the boundary mobile hosts as forwarding nodes. In a routing group, the mobile hosts are divided by means of dominating values into two groups — one positive cluster and several nonpositive clusters. The nodes in the positive cluster maintain the topology information of the routing group. Under such a construction environment, intra-group routing performs unicasting and gets multiple paths, while inter-group routing performs on the group level by propagating route requests to the boundary clusters, which are called bridge clusters. This routing scheme massively reduces message complexity, and as far as the dynamic topology characteristics of ad hoc networks are concerned, this approach also provides a more efficient infrastructure update.

The (a, t)-cluster framework in [7] defines a strategy for adaptively organizing ad hoc networks into clusters in which the probability of path availability between nodes is bounded over time. The purpose of this dynamic arrangement is to support an adaptive hybrid approach to routing that is efficient under all conditions yet can achieve more optimal routing when mobility patterns favor it.

In [8], a distributed algorithm is presented that partitions the nodes of a fully mobile network (ad hoc network) into clusters, thus giving the network a hierarchical organization. The algorithm is proven to be adaptive to changes in the network topology due to nodes' mobility and their addition or removal. A new weight-based mechanism, not available in previous solutions, is introduced for efficient cluster formation/maintenance; it allows the cluster organization to be configured for specific applications and is adaptive to changes in the network status. Simulation results demonstrate that up to an 85% reduction of the communication overhead associated with cluster maintenance with respect to clustering algorithms previously proposed.

15.2.1.4 Adaptive Multicasting

The Adaptive Demand-Driven Multicast Routing protocol (ADMR) [9] is a multicast routing method that can adapt its behavior based on application sending pattern, allowing efficient detection of link breaks and expiration of routing states that are no longer needed. In addition, ADMR can detect when mobility in the network is too high to allow timely multicast state setup, without requiring GPS (or other positioning information) or any additional control traffic. When such high mobility is detected, an ADMR source can switch to flooding for some period of time, after which it may attempt to operate efficiently with multicast again in case the mobility in the network has decreased.

The Core-Assisted Mesh Protocol (CAMP) [10] is another multicast routing that can function in highly dynamic topologies; it is based on building and maintaining a multicast mesh. The mesh is practically a subgraph of the multihop topology that provides at least one path connecting any pair of source and receiver within the multicast group. The use of a mesh for the distribution of data within the group implies that the simplicity of multicasting based on minimum spanning trees is sacrificed.

The multicast protocol in [11] is tree-based and relies on the idea of assigning dynamically increasing identity numbers; these numbers represent the "logical height" of the node and thus allow the rapid formation of the shared tree. The protocol targets fast adaptation to topological changes, while the tree maintenance related control overhead is limited in general to the region of the link breakage.

15.2.2 Protocols that Are Adaptive to Traffic Load

15.2.2.1 Traffic Intensity Prevention

STARA [12] presents a highly adaptive routing algorithm that uses a more appropriate distance measure given by the expected delay along a path instead of the number of hops, which is the measure used in most of the existing proposals. This metric allows the algorithm to adapt to changes not only in the topology of the network, but also in the traffic intensity.

Adaptive Distance Vector (ADV) [13] is a distance vector routing algorithm that exhibits some on-demand characteristics by varying the frequency and the size of the routing updates in response to the network load and mobility conditions. It has been shown via simulations, that ADV outperforms AODV and DSR by having significantly higher peak throughputs and lower packet delays in high-mobility cases. Furthermore, ADV uses fewer routing and control overhead packets than AODV and DSR do, especially at moderate to high loads.

15.2.2.2 Traffic Congestion Avoidance

In [14], a new distributed routing algorithm is presented to perform dynamic load balancing for wireless access networks. The algorithm constructs a load-balanced backbone tree, which simplifies routing and avoids per-destination state routing and per-flow state for reservation. We evaluate the performance adaptation to mobility, degree of load-balance, bandwidth blocking rate, and convergence speed. We find that the algorithm achieves better network utilization by lowering bandwidth-blocking rates than other methods do.

In [15], a mechanism for adaptive computation of multiple paths is used to transmit a large volume of data packets from a source to a destination. Two aspects are considered: The first is to perform preemptive route rediscoveries before the occurrence of route errors while transmitting a large volume of data. This helps to find out dynamically a series of multiple paths in temporal domain to complete the data transfer. The second aspect is to select multiple paths in spatial domain for data transfer at any instant of time and to distribute the data packets in sequential blocks over those paths in order to reduce congestion and end-to-end delay.

15.2.3 QoS-Based Adaptive Routing

15.2.3.1 End-to-End Delay

Distributed dynamic routing (DDR) [16] is a simple loop-free bandwidth-efficient distributed routing algorithm that uses classical concepts such as zone and forest. It tries to achieve several goals at the same time. First, it provides a different mechanism to drastically reduce routing complexity and improve delay performance. Second, it does not even require physical location information. Finally, zone naming is performed dynamically, and broadcasting is reduced noticeably.

15.2.3.2 Bandwidth

CEDAR [17] is a distributed algorithm that can identify a group of nodes called the core of the network, which can help in providing routes to applications with minimum bandwidth requirements. It has the distinction of being one of the first few algorithms proposed to provide QoS routing for MANETs. CEDAR is claimed to be robust and adaptive, using only local state for route computation at each core node. Two important methods are used to find the shortest–widest path (the width being the bandwidth of the link) between two core nodes that represent the source and the destination: the maintenance of an approximate dominating set called the Core to reduce the complexity of routing, and an algorithm to perform QoS (bandwidth) routing by local propagation of control messages.

The multipath QoS routing (QoS-MSR) protocol proposed in [18] can collect QoS information through the route discovery mechanism of MSR [2] and establish a QoS route with reserved bandwidth. In order to reserve bandwidth efficiently, we present a novel bandwidth reservation approach called the multipath bandwidth splitting reservation (MBSR), under which the overall bandwidth request is split into several smaller bandwidth requests among multiple paths. Through extensive simulation, our results show that the QoS-MSR routing protocol with the MBSR algorithm can improve the call admission ratio of QoS traffic, the packet delivery ratio, and the end-to-end delay of both best-effort traffic and QoS traffic. Therefore, QoS-MSR with MBSR is an efficient mechanism that supports QoS for ad hoc networks.

15.3 Case Study I: DSR vs. MSR

DSR is a fundamental but important piece of work that leads to much extension. In this section, we summarize DSR and introduce MSR, an extension of DSR. We compare the performance of MSR with DSR using simulation.

15.3.1 The DSR (Dynamic Source Routing) Protocol

DSR [1,20] uses source routing instead of hop-by-hop packet routing. Each data packet carries the complete path from source to destination as a sequence of IP addresses. The main benefit of source routing is that intermediate nodes need not keep route information because the path is explicitly specified in the data packet. DSR is on-demand based; that is, it does not require any kind of periodic message to be sent. The source routing mechanism, coupled with the on-demand nature of this protocol, eliminates the needs for periodic route advertisement and neighbor detection packets that are characteristic of other protocols.

The DSR protocol consists of two mechanisms: Route Discovery and Route Maintenance. Route discovery is initiated by a source whenever the source has a data packet to send but does not have any routing information to the destination. To establish a route, the source floods the network with request messages carrying a unique request ID. When a request message reaches the destination or a node that has route information to the destination, the node sends a route reply message containing path information back to the source. In order to reduce overhead generated, the "route cache" at each node records routes that a node has learned and overheard during this route discovery phase.

Route Maintenance is the mechanism by which a sender S of a packet detects network topology changes that render useless its route to the destination D (e.g., when two nodes listed in the route have moved out of range of each other). When Route Maintenance indicates a source route is broken, S is notified with a ROUTE ERROR packet. The sender S can then attempt to use any other route to D already in its cache or can invoke Route Discovery again to find a new route.

15.3.2 MSR (Multiple Source Routing) Protocol

By using source routing, MSR [2,19] can improve performance by giving applications the freedom to use multiple paths within the same path service. However, maintaining alternative paths requires more routing table space and computational overhead. Fortunately, some characteristics of DSR can suppress these disadvantages. First, source routing is so flexible that messages can be forwarded on arbitrary paths, which makes it very easy to dispatch messages to multiple paths without any demanding path calculation at the intermediate hops. Second, the on-demand nature of DSR helps to reduce the routing storage and routing computation greatly. In the following, we shall provide the details on the implementation of MSR.

15.3.2.1 Path Finding

Each route discovered is stored in the route cache with a unique route index. So it is easy to pick multiple paths from the cache. In multipath routing, path independence is an important property because a more independent path set can offer more aggregate physical resources between a node pair. (When those resources are not shared, it is less likely that the performance of one path would affect the performance of others). To achieve high path independence, disjoint paths are preferred in MSR.[1] There is no looping problem in MSR, as the route information is contained inside the packet itself; routing loops, either short- or long-lived, cannot be formed as they can be immediately detected and eliminated.

15.3.2.2 Packet Forwarding and Load Balancing

Since MSR uses source routing, intermediate nodes need not do anything except to forward the packet as indicated by the route in its header, thus adding no more processing complexity than that in DSR. All the work for path calculation is done in the source hosts. In MSR, source nodes are responsible for load balancing. In our protocol, we implement a "mul_dest table," which contains the multiple path information to the specific destination. This can be illustrated by the following data structure:

```
struct mul_dest
{
    int index ;
    ID Dest;
    float Delay;
    float Weight;          ...
}
```

Here, "Dest" is the destination of a route. "Index" is the current index of the route in DSR's route cache that has a destination to "Dest." "Delay" is the current estimate of the round-trip time. "Weight" is a per-destination based load distribution weight among all the routes that have the same destination. "Weight" is the number of packets to be sent consecutively on the same route every time. The choice of "Weight" is an interesting and challenging task, and we can make the following observation.

Within an ad hoc network, which is always an autonomous system acting as a stub network, there is less heterogeneity in some sense when compared to WAN and MAN. For instance, in WAN or MAN, the maximal bandwidths that every node can obtain vary little; so do the round-trip delays. Therefore, we assume the bandwidth–delay product is a constant. Thus, the available bandwidth is inversely proportional to the RTT (Round Trip Time), so the traffic can be distributed among multiple paths proportional to the available bandwidth. The principle is inherently simple but reasonable in wireless networks. In wireline networks, due to the very different bandwidths, delay cannot be a definite indicator of the available bandwidth.

From our above observation, we propose to choose the weight W_i^j (i is the index of the route to j) according to a heuristic equation, Eq. (15.1):

$$W_i^j = \mathrm{Min}\left(\left\lceil \frac{d_{max}^j}{d_i^j} \right\rceil, U\right) \times R \qquad (15.1)$$

where d_{max}^j is the maximum delay of all the routes to the same destination, d_i^j is the delay of route with index i, and U is a bound to ensure that W_i^j should not to be too large. The factor R controls the switching frequency between routes. The larger the value of R, the less frequently the switching would happen and the less processing overload would result from searching and positioning an entry in the mul_dest table. When choosing R, the interface priority queue (IFQ)[2] buffer's size should also be taken into consideration. Unlike the work done in [23–25], note that we have done extensive experiments beyond the R = 1 work in [19], and we found R = 3 to be better in reducing the out-of-order deliveries

---

[1] In this chapter, we address the multipath routing problem in the context of single channel. For the disjoint path problem in a multiple channel environment, refer to [21].

[2] The network stack for a mobile node consists of a link layer (LL), an ARP module connected to the LL, an interface priority queue (IFQ), a MAC layer (MAC), and a network interface (netIF), all connected to the channel [26].

in TCP. So in our experiment, R is set to 4 for an IFQ size of 50. When distributing the load, the weighted-round-robin scheduling strategy is used.

To aid load balancing and to decouple the interlayer dependence of delay measurement, a network layer probing mechanism is employed. Probing is also an enhancement to the DSR route maintenance mechanism. Normally, in DSR, a link breakage can be noticed only when a Route Error message is returned. However, in the wireless mobile environment, there is a nontrivial chance that the Route Error message cannot reach the original sender successfully. Although, "as a last resort, a bit in the packet header could be included to allow a host transmitting a packet to request an explicit acknowledgment from the next-hop receiver"[1], we found that probing one path constantly only to test its validity is not cost effective. Therefore, the function of probing in our MSR is twofold: to obtain the path delay status and to test the validity of active paths.

15.3.2.3 Toward Gratuitous Mode

In DSR, when a data packet is received as the result of operating in promiscuous receive mode, the node checks whether the Routing Header packet contains its address in the unprocessed portion of the source route. If so, the node knows that packet could bypass the unprocessed hops preceding it in the source route. The node then sends what is called a gratuitous Route Reply message to the packet's source, giving it the shorter route without these hops [1]. Since in MSR, there are always routes that are not the shortest ones, the GRAT (GRATuitous) packets increase greatly, which takes too much IFQ and ARP buffer space. Thus, we have to turn off the gratuitous options in our simulations.

15.3.3 Performance Evaluation

We evaluated the performance of our algorithm using ns-2 simulation [26]. CMU has extended ns-2 with some wireless supports, including new elements at the physical, link, and routing layers of the simulation environment [27]. Using these elements, it is possible to construct detailed and accurate simulations of ad hoc networks. For scenario creation, two kinds of scenario files are used to specify the wireless environment. The first is a movement pattern file that describes the movement that all nodes should undergo during the simulation. The second is a communication pattern file that describes the packet workload that is offered to the network layer during the simulation. To obtain the performance of MSR at different moving speeds, we use two simulation sets with speeds of 1 and 20 m/sec, respectively. Our simulations model a network of 50 mobile hosts placed randomly within a 1500 m × 300 m area, both with zero pause time. To evaluate the performance of MSR, we experimented with different application traffic, including CBR and file transfer protocol (FTP). CBR uses UDP as its transport protocol, and FTP uses TCP. The channel is assumed error free except for the presence of collision. For other simulation detail, please refer to [28].

We chose the following metrics for our evaluation:

- Queue size: The size of the IFQ (Interface Priority Queue) object at a node
- Packet delivery ratio: The ratio between the number of packets originated by the "application layer" CBR sources and the number of packets received by the CBR sink at the final destination
- Data throughput: The total number of packets received during a measurement interval divided by the measurement interval
- End-to-end delay
- Packet drop probability

For TCP, another issue is the out-of-order problem described in [23]. To present the packet dynamics clearly, the ack time-sequence plot is given.

15.3.3.1 Performance of UDP Traffic

We look at CBR traffic implemented with UDP agents at a maximum moving speed of 20 m/sec. A scenario with 20 CBR connections is created. Since UDP has no feedback control mechanism, all the

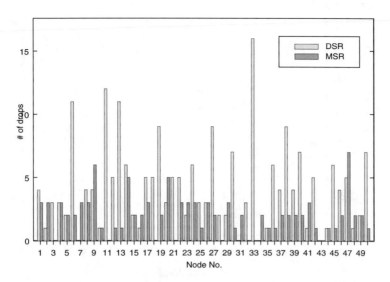

FIGURE 15.1 CBR packets dropped at each node.

TABLE 15.1 CBR Packet Drops Summary

CBR Drops Summary	DSR	MSR
Total packets	217	114
Total bytes	119,976	63,096
No route	129	90
TTL expired	0	0
RTR queue full	0	0
Timeout	0	0
Routing loop	0	0
IFQ full	30	9
ARP full	53	14
MAC callback	0	0
SIM end	5	1

CBR traffic generated is constant no matter how the network runs. We shall use it as a reference for comparing with other routing protocols. Figure 15.1 shows that fewer CBR packets are dropped in MSR than in DSR. Table 15.1 shows the drop summary in detail; the main reasons for dropping are "No route" and "IFQ full." Figure 15.2 provides the end-to-end packet delivery ratio of every connection, and the comparison shows that MSR is better than DSR.

From Fig. 15.3, we can see that MSR achieves higher throughput than DSR on almost every connection, just as we expected. This can be attributed to the fact that the multipath routing effectively utilizes currently unallocated network resources. Figure 15.4 shows the end-to-end delay of every connection. Figure 15.5 presents the average queue size for all 50 hosts. From Fig. 15.5, we can see that, in MSR, the packets that should have been queued in the IFQ have been redistributed to other nodes that have light load, through which the traffic is balanced. Balancing the route load in MSR shortens the delay, as the chance of congestion is reduced.

Table 15.2 shows the routing overheads in DSR and MSR, respectively. We can see that the number of routing messages in MSR is only a little more than that of DSR. However, the packet drops probability is lower than that of DSR. The main drop reasons are still "No route" and "IFQ full."

15.3.3.2 Performance of TCP Traffic

For TCP traffic, we take a scenario with 30 FTP connections, with the network rather heavily loaded but with the same maximum moving speed of 20 m/sec. Since TCP has an AIMD (Additive Increase

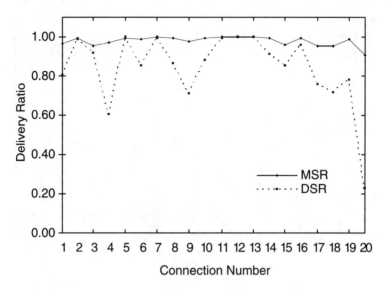

FIGURE 15.2   Packet delivery ratio of every connection.

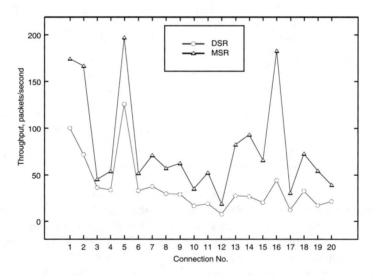

FIGURE 15.3   End-to-end throughput.

Multiplicative Decrease) feedback control mechanism, the statistics at every node have less meaning than those of UDP. Therefore, we focus on the end-to-end packet dynamics instead. Figures 15.6 and 15.7 show that multipath routing can also be used to reduce the end-to-end network latency and message drop probability or to increase the likelihood of message delivery for TCP connections.

From Figs. 15.8 and 15.9, we can see there are not many out-of-order deliveries in MSR. On the contrary, the end-to-end throughput of TCP in MSR has increased substantially due to the smooth increase of sequence number. Figure 15.9 also implies that MSR recovers more quickly than DSR does when the connection meets severe packet drops (e.g., at time 90 sec). It illustrates that our load-balancing method achieves a good switching granularity.

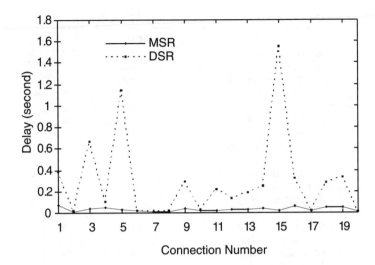

FIGURE 15.4 End-to-end delay.

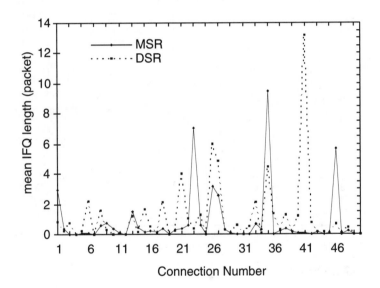

FIGURE 15.5 IFQ queue size at each node.

15.3.4 Discussion

In our initial experiments, we found that the major statistics of Routing Packets of MSR are comparable to those of DSR, except that the GRAT packets count in MSR is too large compared to DSR. Thus, we turn off the gratuitous options, and the results become better. The simulation results show that the main reasons for packet drops in DSR are "No route" and "IFQ full," while these two factors improve substantially in MSR.

We have also examined the scenario under a maximum speed of 1 m/sec; the throughput and end-to-end delay of MSR are also better than those of DSR. We found there is no significant difference in packet drops between DSR and MSR. Therefore, we can conclude that one of the main gains we get from MSR is attributed to fewer "No route" drops. In other words, multipath routing compensates for route failures efficiently in high-speed movement. This is consistent with the results in Tables 15.1 and 15.2.

TABLE 15.2 Routing Level Statistics for CBR (in Packets)

Type	DSR	MSR
DSR TOTALS transmitted	2200	2313
DSR TOTALS received	21,786	25,970
DSR TOTALS forwards	2740	3392
DSR TOTALS drops	56	58
IFQ len above 25	100	73
DSR REQUEST transmitted	199	189
DSR REQUEST received	18,255	19,389
DSR REQUEST forwards	1145	1232
DSR REQUEST drops	0	1
DSR REPLY transmitted	72	89
DSR REPLY received	202	318
DSR REPLY forwards	137	233
DSR REPLY drops	5	4
DSR ERROR transmitted	596	676
DSR ERROR received	1085	1261
DSR ERROR forwards	492	593
DSR ERROR drops	6	2

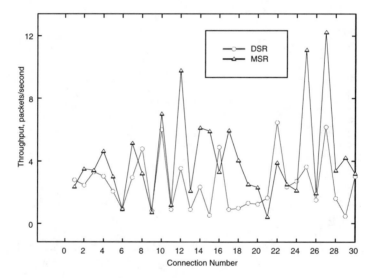

FIGURE 15.6 End-to-end throughput of each connection.

When evaluating a network routing protocol, control load should also be considered. There is no more control load in MSR than in DSR, except for probing packets transmitted in networks. Since we use SRping (which is unicast), rather than flooding, to test the validity of paths currently used, and the probing interval we choose is very conservative, little overload is added.

15.4 The QoS-MSR Protocol

With the development of ad hoc networks, there are ever-increasing demands on the transmission of various types of services. In addition, in certain real-time applications (for instance, the transmission of information such as voice, video, and images), not only correct routing information but also guaranteed QoS are required. Although DSR performs very well in storing a route by using route cache [20,28–30], the cache mechanism lacks the capability to store additional QoS information (e.g., available bandwidth and delay) that may change at any moment, even if the path may remain stable during a long time. A natural idea is to increase the frequency of route discovery. However, if flooding is used as in DSR,

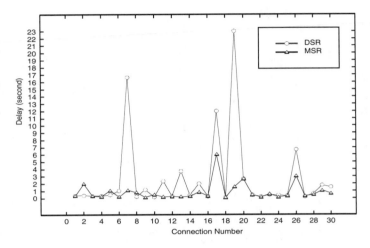

FIGURE 15.7 End-to-end delay of each connection: (a) DSR, (b) MSR.

increasing the frequency of route discovery excessively may result in unacceptable control overhead. Therefore, a different mechanism is necessary to maintain QoS information.

As discussed above, MSR is a multipath extension of DSR. Based on the measurement of the round-trip time of every path, MSR can adaptively distribute load among multiple paths. Therefore, MSR can decrease network congestion and increase the path fault tolerance quite well. So MSR is excellent for best-effort traffic in ad hoc networks, and it is quite natural to extend it to support QoS. In order to implement multipath QoS routing efficiently, we have extended MSR to support the QoS route discovery mechanism. Recall that MSR [2] uses the route discovery mechanism of DSR to discover and return multiple paths, each of which is stored in the route cache with a unique route index. So it is easy for us to pick multiple paths from the cache. In the route discovery mechanism in DSR, some fields are inserted into a route request packet. When the route request packet passes one node, the node's information such as bandwidth and delay will be recorded in the fields designated for them in order for a link's QoS information to be retrieved easily later. In order to ensure that the route information is timely and accurate, we use periodic probing approach to measure the path. Therefore, we can improve the route discovery mechanism in DSR so that QoS information of every link/node may be obtained during the route discovery phase. Then the maintenance mechanism for QoS status can be introduced. In the following, we shall discuss the implementation in detail.

15.4.1 The Implementation of QoS Route Discovery

In order to append bandwidth information to the route request packet, each route request packet must contain a bandwidth record. When any node processes a route request packet, in addition to appending route information to the route record, it also appends its available bandwidth to the bandwidth record in the route request packet. Thus a source node can obtain not only a path but also the bandwidth information of every node at the same time.

Since the route cache of MSR does not store any QoS information, in order to keep the QoS information of different paths, the bandwidth field is added into the route cache in QoS-MSR. As there is no fixed link connection in a wireless network, the bandwidth discussed here is node oriented. Every route record in the route cache contains the available bandwidth of each node in a path. So the available (bottleneck) bandwidth of a path can be obtained from the minimum of bandwidths among all the nodes in the path.

In QoS-MSR, during the process of route discovery, an intermediate node may not be able to provide enough resources to satisfy the QoS request. One way to deal with this problem may be to terminate this route discovery, but the information obtained in the route discovery process would be wasted. In order

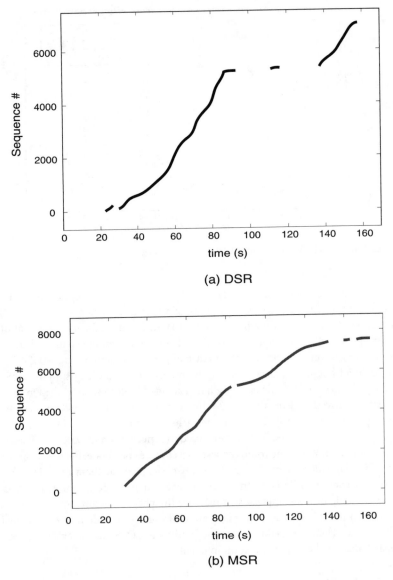

FIGURE 15.8 TCP segments time–sequence plot of a heavy connection, node 49 to node 50.

to avoid this disadvantage, we would like to allow the route discovery to continue even when an intermediate node cannot provide enough resources to satisfy the QoS request. Eventually, the information on this path will go back to the source node and be recorded in the route cache. Although the current QoS request cannot use this path, some other QoS requests later may be able to use this path directly, thus taking full advantage of all the information accumulated in a route discovery.

15.4.2 QoS Route Maintenance

In this protocol, we shall use a probing mechanism to maintain QoS information. There are two types of probing used: on demand and periodic. When there is no required route information in the route cache, a flooding-based on-demand route discovery will be initiated. If there is already a path in the route cache (but its QoS information is outdated), a unicast probing will be initiated to measure QoS information of this path. This method will increase the network overhead a little bit because it is based on unicast.

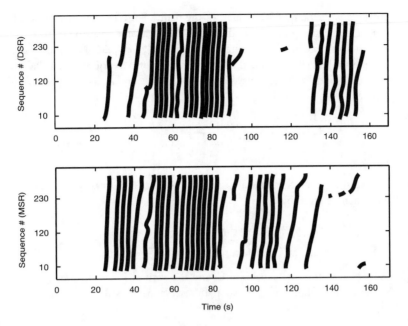

FIGURE 15.9 Time–sequence plot with sequence # mod 300.

In addition to active probing, passive piggybacking is used in QoS-MSR to obtain QoS information for every node. When a source node or an intermediate node forwards a data packet, the node will record in the data packet some QoS information, such as bandwidth and length of the interface queue. When any other nodes receive the data packet, they may record into its route cache this piggybacked QoS information.

To guarantee the continuity of one path, QoS-MSR adds an additional multipath route table. As the route discovery mechanism only appends a new route to the route cache, a synchronization mechanism between the multipath route table and route cache must be introduced. Besides, the route status of the path in use must be measured in as timely a manner as possible. In practice, before a node initiates probing for a path or when the node receives a probing packet, a timestamp is added to the path, and a time-out timer is initiated at the same time. If the timestamp is not refreshed during the time-out period, it shows that the probing packet may be dropped and the path may be broken; the path should be deleted from the route cache and the multipath route table of the source node.

15.4.3 Multiple Bandwidth Splitting Reservation (MBSR)

Since the available bandwidth of a network can be limited, it would take a considerable amount of time to smooth out bursty traffic. During this time, if an application requests a large bandwidth, its probability of success will be very low. This becomes more apparent in ad hoc networks because of the very constrained bandwidth. Therefore, we propose the MBSR (Multipath Bandwidth Splitting Reservation) algorithm. In this algorithm, the overall bandwidth request is split into several smaller bandwidth requests on multiple paths. Since the amount of requested bandwidth on each path is reduced, the probability of a successful reservation on each path will be increased. As a result, the success probability of the overall reservation will be increased.

Mathematically, for each request of bandwidth B, the algorithm finds a group of paths P = {P1, P2, P3, ..., Pn} from all the paths between the source and destination so that

$$bandwidth(P) = \sum_i bandwidth(P_i) \geq B$$

where $bandwidth(P_i)$ is the bandwidth request of the path i. For a request of bandwidth R between a source node S and a destination node D, let the number of available paths be n (n > 1), the capability of each path be C, the available bandwidth of each path be B_{avl} ($B_{avl} \leq C$), and the success probability of a bandwidth request R be P(R). Assume that P(R) has a uniform distribution on [0,C]. Then splitting R into n independent bandwidth requests R_i (i = 1, 2, ..., n) among n paths would give a success probability $P_m(R)$ below.

$$P_m(R) = \prod_{i=1}^{n} P_{s'}(R_i) \tag{15.2}$$

where $P_{s'}$ is the successful reservation probability of each path. The success probability of a single-path reservation P_s may then be given by

$$P_s = \frac{C-R}{C} = 1 - \frac{R}{C} \tag{15.3}$$

Assume the n paths have the same capability C, and the request R is split averagely on n paths, i.e., $R_i = R/n$. The success probability of every subpath reservation in MBSR is

$$P_{s'} = 1 - \frac{R}{nC} \tag{15.4}$$

Using Eqs. (15.2) and (15.4), we obtain

$$P_m = \left(P_{s'}\right)^n = \left(1 - \frac{R}{nC}\right)^n \tag{15.5}$$

Let $\eta = R/C$, then $0 \leq \eta \leq 1$, and we obtain

$$P_m = \left(1 - \frac{\eta}{n}\right)^n \text{ and } P_s = 1 - \eta$$

Now, let

$$\gamma = \left(1 - \frac{\eta}{n}\right)^n - (1 - \eta)$$

i.e., $\gamma = P_m - P_s$; this is the difference between the success probability of MBSR and the success probability of single-path reservation. Then we can show that

$$\frac{d\gamma}{d\eta} = 1 - \left(1 - \frac{\eta}{n}\right)^{n-1} > 0$$

Apparently, when $0 \leq \eta \leq 1$, γ is an increasing function. When $\eta = 0$, then $\gamma = 0$, and when $\eta > 0$, then $\gamma > 0$. Also R > 0 leads to $\eta > 0$, which gives $P_m > P_s$. So one can conclude that the probability of a successful bandwidth request is higher than that of single-path reservation.

Figure 15.10 shows the relation between the success probability of MBSR and the success probability of single-path reservation. The abscissa denotes η, and the ordinate denotes γ. Besides the known fact that the success probability of MBSR is always higher than that of single-path reservation, it is interesting to see that the higher η is, the more obvious the advantage of MBSR is; and the larger n is, the higher

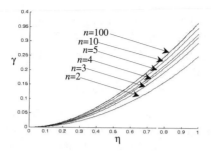

FIGURE 15.10 The gain of the conservative success probability of MBSR compared to single-path reservation.

the gain. Most gain improvement occurs when the number of multiple paths is increased to two or three. After that, the gain improvement diminishes. For example, when n > 100, the gain improvement is negligible. From this observation, it can be seen that the number of multiple paths need not be too large, and it is appropriate to choose just two or three.

To look further, let the arrival rate of MBSR be λ', and let the sum of connection reservation time and connection continuance time have an exponential distribution with a mean of $1/\mu'$. Let $\rho' = \lambda'/\mu'$, and the success probability of reservation be P'_{suc}. Although the requests for bandwidth are split in n paths, their aggregate arrival rate λ' should be the same as the arrival rate λ in [22], i.e., $\lambda' = \lambda$. However, since MSR can adaptively distribute load among multiple paths, the request of bandwidth on each path is decreased; so the mean sum of connection reservation time and connection continuance time is decreased, i.e., $\mu' > \mu$, $\rho' < \rho$. Therefore, according to the equation

$$P_{suc} = \frac{n}{n+\rho}$$

in [22], we have $P'_{suc} > P_{suc}$, i.e., the success probability of MBSR is higher than the success probability of multipath reservation in [22]. Hence, the success probability of MBSR is higher than that of single-path reservation.

15.4.4 Performance Evaluation

We are interested in comparing QoS-MSR using MBSR with DSR using single-path reservation.[3] Through comparison, we shall show that QoS-MSR using MBSR can improve QoS of reservation-based flows and can be made adaptive to ad hoc networks with high mobility.

We used ns-2 [26] to conduct the simulation in an environment similar to that for the comparison of MSR and DSR in Section 15.3 except that the node speed was fixed at 20 m/sec. We ran our simulations using five different pause times (0, 30, 60, 120, and 300 seconds) and 10 movement patterns for each value of pause time. Each simulation ran for 300 seconds of simulated time. (A pause time of 0 seconds corresponds to continuous motion, and a pause time of 300 seconds [the length of the simulation] corresponds to no motion.)

Also in the communication pattern file, we chose our traffic sources to be constant bit rate (CBR) sources this time. We experimented with a sending rate of 4 packets per second, a packet size of 512 bytes, and a network with 20 CBR sources (including six QoS sources). The simulation of QoS traffic was implemented by UDP flow. We used different flow ports to distinguish between QoS flows (Ports #1 to 4) and best-effort flows (using Port #0).

The following performance metrics were used for the comparison:

---

[3] The QoS-MSR and DSR mentioned in the discussion and figures hereinafter denote QoS-MSR with MBSR and DSR with single-path reservation, respectively.

- *Reserved packet ratio:* This is the ratio of the number of packets that received QoS service to the number of total packets in QoS traffic. When used as a metric to QoS admission, the higher the ratio is, the more QoS service is accepted.
- *Packet delivery ratio:* This is the ratio of the number of packets received by the destination to the number of packets sent by the sources.
- *End-to-end delay:* This is an important performance parameter for QoS traffic.

Figure 15.11 shows the reserved packet ratios of QoS-MSR and DSR, respectively. One can see that both ratios are decreasing with increasing mobility (i.e., as pause time decreases from 300 to 0 sec), but the reserved packet ratio of QoS-MSR is higher than that of DSR and is more stable at different pause times (mobility). Under high mobility conditions (i.e., pause time is 0 or 30 sec), QoS-MSR provides more performance improvement over DSR (at least 7 percent improvement in the reserved packet ratio). Under low to moderate mobility conditions (i.e., pause time is longer than 60 sec), the benefit of QoS-MSR over DSR is less because DSR can provide higher reserved packet ratio under lower mobility conditions.

In summary, the reserved packet ratio of QoS-MSR is higher than that of DSR. The main reason is that MBSR has increased the probability of successful reservation for every request, so that the ratio of reserved packets in every subflow also increases. Moreover, MBSR can keep the path for a longer lifetime, thus decreasing the frequency of path change.

The packet delivery ratio and end-to-end delay are shown in Figs. 15.12 and 15.13, respectively. According to these figures, QoS-MSR is more stable than DSR in different mobility patterns. Similar to the reserved packet ratio, under high mobility conditions, QoS-MSR provides obvious performance improvement over DSR: the packet delivery ratio is increased and the end-to-end delay is decreased. However, under low to moderate mobility conditions, the paths are relatively stable, so both DSR and QoS-MSR perform very well.

We have also obtained results for different numbers of sources (10, 30, and 40 sources). Our experiments show that the benefits of QoS-MSR over DSR are not significant for 10 sources or for more than 30 sources. With 10 sources, using a single path performs quite well already because the load is very light; therefore the gain of using multipath is minimal. With more than 30 sources, the load is so heavy that there are not even enough vacant resources for multipath. In view of the above, one can see why under a moderate load condition (20 sources), QoS-MSR with MBSR performs better than DSR with single-path reservation.

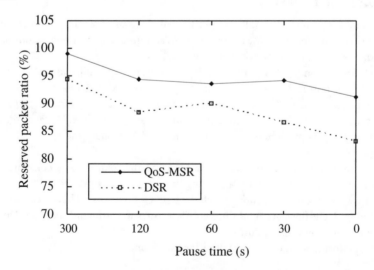

FIGURE 15.11 Comparison between QoS-MSR with MBSR and DSR with single-path reservation of the fraction of packets successfully reserved (reserved packets ratio) as a function of pause time. Pause time 0 represents constant mobility.

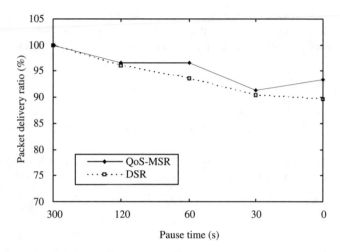

FIGURE 15.12 Comparison between QoS-MSR with MBSR and DSR with single-path reservation of the packet delivery ratio as a function of pause time. Pause time 0 represents constant mobility.

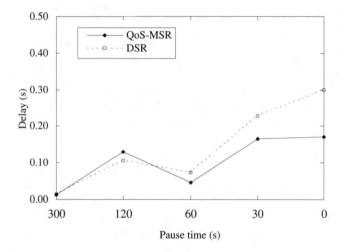

FIGURE 15.13 Comparison between QoS-MSR with MBSR and DSR with single-path reservation of the end-to-end delay as a function of pause time. Pause time 0 represents constant mobility.

15.5 Concluding Remarks

The dynamics of an ad hoc network are a challenge to protocol design because mobility inevitably leads to unstable routing, and consequently flows encounter fluctuations in resource availability on various paths during the lifetime of a session. This has become serious, especially for those protocols based on single-path reservation, as frequent reservation and restoration of reservation-based flows increase the instability of connections. Therefore, adaptive protocols are important for ad hoc networks facing increasing traffic and service demands.

MSR meets this challenge by incorporating the multipath mechanism into DSR and employing a probing based load-balancing mechanism. Consequently, the throughput, end-to-end delay, and drop probability have been improved significantly. Furthermore, by using Multipath Bandwidth Splitting Reservation (MBSR), one can support a QoS guarantee in ad hoc networks. According to the simulation results, MBSR decreases network congestion and improves the packet delivery ratio and end-to-end delay of all connections. In addition, the reserved packet ratio indicates that MBSR can improve QoS of reservation-based flows and can be made adaptive to ad hoc networks with high mobility.

Acknowledgment

This research was supported in part by the National Natural Science Foundation of China (NSFC) under grant No. 90104015, the Engineering Research Council of Canada (NSERC) under grant No. OGP0042878, and Cisco Systems.

The following people have contributed to the discussion and reading of the manuscripts and their help is acknowledged here. From Tianjin University, Lianfang Zhang, Yongjie Fan, and Guanghong Wang, and from the University of Ottawa, Song Guo.

References

[1] Johnson, D. and Maltz, D.A., The Dynamic Source Routing Protocol for Mobile Ad Hoc Networks, Internet Engineering Task Force Internet-Draft, http://www.ietf.org/internet-drafts/draft-ietf-manet-dsr-03.txt, Internet standards organization protocol specification contribution, 1999.

[2] Wang, L., Shu, Y.T., Yang, O.W.W., Dong, M., and Zhang, L.F., Adaptive Multipath Source Routing in Wireless Ad Hoc Networks, *Proceedings of the IEEE International Conference on Communications*, Helsinki, 2001, p. 867.

[3] Park, V.D. and Corson, M.S., A Highly Adaptive Distributed Routing Algorithm for Mobile Wireless Networks, *Proceedings of IEEE INFOCOM*, Kobe, Japan, Apr. 1997, pp. 1405–1413.

[4] Roy, S. and Garcia-Luna-Aceves, J.J., Using Minimal Source Trees for On-Demand Routing in Ad Hoc Networks, *Proceedings of IEEE INFOCOM*, Anchorage, 2001, p. 1172.

[5] Toh, C.K., A Novel Distributed Routing Protocol to Support Ad Hoc Mobile Computing, *Proceedings of 15th IEEE Annual International Phoenix Conference on Computers and Communications*, Phoenix, 1996, p. 480.

[6] Chang, Y.L. and Hsu, C.C., Routing in wireless/mobile ad hoc networks via dynamic group construction, *ACM/Kluwer Mobile Networks and Applications*, 5, 27, 2000.

[7] McDonald, A.B. and Znati, T.F., A mobility-based framework for adaptive clustering in wireless ad hoc networks, *IEEE Journal on Selected Areas in Communications*, 17, 1466, 1999.

[8] Basagni, S., Distributed and Mobility-Adaptive Clustering for Multimedia Support in Multi-Hop Wireless Networks, *Proceedings of the IEEE Vehicular Technology Conference, (VTC-Fall)*, Amsterdam, September 19–22, 1999, vol. 2, pp. 889–893.

[9] Jetcheva, J.G. and Johnson, D.B., The Adaptive Demand-Driven Multicast Routing Protocol for Mobile Ad Hoc Networks (ADMR), Internet Engineering Task Force Internet-Draft, http://www.ietf.org/internet-drafts/draft-jetcheva-manet-admr-00.txt, Internet standards organization protocol specification contribution, 2001.

[10] Garcia-Luna-Aceves, J.J. and Madruga, E.L., The core-assisted mesh protocol, *IEEE Journal on Selected Areas in Communications*, 17, 1380, 1999.

[11] Wu, C.W. and Tay, Y.C., AMRIS: A Multicast Protocol for Ad Hoc Wireless Networks, *Proceedings of IEEE MILCOM*, Atlantic City, NJ, 1999, pp. 25–29.

[12] Gupta, P. and Kumar, P.R., A System and Traffic Dependent Adaptive Routing Algorithm for Ad Hoc Networks, *Proceedings of the 36th IEEE Conference on Decision and Control*, San Diego, CA, 1997, pp. 2375–2380.

[13] Boppana, R.V. and Konduru, S.P., An Adaptive Distance Vector Routing Algorithm for Mobile Ad Hoc Networks, Proceedings of the IEEE INFOCOM, Anchorage, 2001, pp. 1753–1762.

[14] Hsiao, P.H., Hwang, A., Kung, H.T., and Vlah, D., Load-Balancing Routing for Wireless Access Networks, *Proceedings of the IEEE INFOCOM*, Anchorage, 2001, pp. 986–965.

[15] Das, S.K., Mukherjee, A., et al., Improving Quality-of-Service in Ad Hoc Wireless Networks with Adaptive Multipath Routing, *Proceedings of the IEEE Globecom,* San Francisco, CA, 2000, pp. 261–265.

[16] Nikaein, N., Labiod, H., and Bonnet, C., DDR-Distributed Dynamic Routing Algorithm for Mobile Ad Hoc Networks, *Proceedings of the IEEE MobiHoc,* Boston, August 2000, pp. 19–27.

[17] Sivakumar, R., Sinha, P., and Bharghavan, V., CEDAR: a core-extraction distributed ad hoc routing algorithm, *IEEE Journal on Selected Areas in Communications,* 17, 1454, 1999.

[18] Wang, G.H., Shu, Y.T., Wang, L., Yang, O.W.W., Fan, Y.J., and Zhang, L.F., Multipath QoS Routing and Bandwidth Reservation in Ad Hoc Networks, Computer Science Department Report, Tianjin University, China, Jan. 2002.

[19] Wang, L., Zhang, L.F., Shu, Y.T., Dong, M., and Yang, O.W.W., Multipath Source Routing in Wireless Ad Hoc Networks, *Proceedings of the IEEE CCECE,* Halifax, 2000, p. 479.

[20] Johnson, D.B. and Maltz, D.A., Dynamic source routing in ad hoc wireless networks, in *Mobile Computing,* Imielinski, T. and Korth, H., Eds., Kluwer Academic Publishers, Dordrecht, 1996, chap. 5.

[21] Pearlman, M., Haas, Z., Sholander, P., and Tabrizi, S.S., On the Impact of Alternate Path Routing for Load Balancing in Mobile Ad Hoc Networks, *Proceedings of MobiHoc 2000,* Boston, MA, Aug. 11, 2000, pp. 3–10.

[22] Cidon, I., Rom, R., and Shavitt, Y., Analysis of multipath routing, *IEEE Trans. on Networking,* 7, 885, 1999.

[23] Gogate, N. and Panwar, S.S., Supporting Applications in a Mobile Multihop Radio Environment Using Route Diversity I. Non–Real Time Data, *Proceedings of the IEEE ICC,* Atlanta, June 1998, pp. 802–880.

[24] Nasipuri, A., Castaneda, R., and Das, S.R., Performance of multipath routing for on-demand protocols in mobile ad hoc networks, *ACM/Baltzer Mobile Networks and Applications (MONET) Journal,* 6, 339, 2001.

[25] Nasipuri, A. and Das, S.R., On-Demand Multipath Routing for Mobile Ad Hoc Networks, *Proceedings of IEEE ICCCN '99,* Boston, MA, Oct. 1999.

[26] ns simulation software that can be downloaded from http://mash.cs.berkeley.edu/ns.

[27] Extensions to the ns-2 simulation software that can be downloaded from http://www.monarch.cs.cmu.edu/cmu-ns.html

[28] Broch, J., Maltz, D., Johnson, D., Hu, Y.C., and Jetcheva, J., A Performance Comparison of Multi-Hop Wireless Ad Hoc Network Routing Protocols, *Proc. IEEE/ACM MobiCom,* Dallas, TX, 1998, pp. 85–97.

[29] Maltz, D.A., Broch, J., Jetcheva, J., and Johnson, D.B., The effects of on-demand behavior in routing protocols for multihop wireless ad hoc networks, *IEEE Journal on Selected Areas in Communications,* 17, 1439, 1999.

[30] Perkins, C.E., Royer, E.M., Das, S.R., and Marina, M.K., Performance comparison of two on-demand routing protocols for ad hoc networks, *IEEE Personal Communications,* 8, 16, 2001.

16

Position-Based Ad Hoc Routes in Ad Hoc Networks

Silvia Giordano
LCA-IC-EPFL

Ivan Stojmenovic
University of Ottawa

Abstract

Highly dynamic topology is a fundamental characteristic of mobile ad hoc networks. For this reason, maintaining a consistent state for routing purposes can be a very difficult task. The introduction of node position information can significantly simplify the routing task. Additionally, small, inexpensive low power Global Positioning System (GPS) receivers and techniques for finding relative coordinates based on signal strengths have recently become available. A number of position-based routing methods for ad hoc networks have been developed. As an inherent part of the position-based approach to routing, different schemes for location updates have been presented. This chapter surveys routing and location updates methods, with the main focus on schemes that are loop free and localized and follow a single-path strategy, all of which are desirable characteristics of scalable routing protocols.

16.1 Introduction

16.1.1 Ad Hoc Networks and the Routing Task

Ad hoc networks consist of wireless hosts that communicate with each other in the absence of a fixed infrastructure. They have received significant attention in recent years. Networks of this type can be used in disaster relief and conference and battlefield environments. Other contexts include rooftop networks — static networks with nodes placed on top of buildings — to be used when wired networks fail. Sensor networks are a class of wireless ad hoc networks. Wireless networks of sensors are likely to be widely deployed in the near future because they greatly extend our ability to monitor and control the physical environment from remote locations and improve the accuracy of information obtained via collaboration among sensor nodes and online information processing at those nodes. Networking these sensors (empowering them with the ability to coordinate among themselves on a larger sensing task) will revolutionize information gathering and processing in many situations.

In an ad hoc network, a message sent by a node reaches all neighboring nodes that are located at distances up to the transmission radius. Because the transmission radius is limited, the routes between nodes are normally created through several hops in such multi-hop wireless networks. In the widely accepted *unit* graph model, two nodes A and B in the network are neighbors if the distance between them is at most R, where R is the transmission radius, which is equal for all nodes in the network. Variations of this model include unit graphs with obstacles (or subgraphs of unit graphs) and minpower graphs where each node has its own transmission radius and links are allowed only when bidirectional communication is possible. Little credible research has been published on any model other than the unit graph model (one important exception is in [5]). However, in power and cost savings and congestion aware methods, nodes may adjust their transmission power to merely reach the intended receiver.

In this chapter, we consider the global routing task, in which a message is to be sent from a source node to a destination node in a given wireless network. The task of finding and maintaining routes in sensor and ad hoc networks is nontrivial since host mobility and changes in node activity cause frequent unpredictable topological changes. The destination node is known and addressed by means of its location. Packet forwarding is performed by a scheme based on this information that is generally classified as a *position-based scheme.*

The traditional routing approach based on IP addressing has one critical aspect: the addressing. In fact, the addressing approach used in wired networks, as well as its adaptation for mobile IP [23], would drastically increase the control overhead. Therefore, a new addressing approach for such networks is required. Moreover, given that in the foreseen topology of the addressing approach, interaction among different routing protocols can easily happen, a common addressing approach is necessary. This issue is still a matter of ongoing research. The IETF document describing Internet mobile ad hoc networks states that the development of an approach that will permit routing through a multi-technology fabric, permit multiple hosts per router, and ensure long-term interoperability through adherence to the IP addressing architecture is underway. Supporting these features appears only to require identifying host and router interfaces with IP addresses, identifying a router with a separate router ID, and permitting routers to have multiple wired and wireless interfaces [8]. Geographical location of nodes, i.e., node coordinates in two- or three-dimensional space, has been suggested for simplifying the addressing issue in combination with the Internet addressing scheme, among other purposes. The existing position-based routing protocols propose to use location information for reducing the propagation of control messages, for reducing the intermediate system functions, or for making packet forwarding decisions. Geographical routing allows nodes in the network to be nearly stateless; the information that nodes in the network have to maintain is about their one-hop neighbours.

The use of the nodes' position for routing poses evident problems in terms of reliability. The accuracy of the destination's position is an important problem to consider. In some cases the destination is a fixed node (such as a monitoring center known to all nodes or the geographic area that is monitored), and some networks are static. The problem of designing location update schemes to provide accurate destination information and enable efficient routing in mobile ad hoc networks appears to be more difficult than routing itself and will be discussed in Sections 16.4 and 16.5.

16.1.2 Advantages of Using Position in Routing Decisions: Localized Ad Hoc Routes for Scalability

The distance between neighboring nodes can be estimated on the basis of incoming signal strengths. Relative coordinates of neighboring nodes can be obtained by exchanging such information between neighbors. Alternatively, the location of nodes may be available directly by communicating with a satellite, using GPS, if nodes are equipped with a small low-power GPS receiver. The position-based approach in routing becomes practical due to the rapidly developing software and hardware solutions for determining absolute or relative positions of nodes in ad hoc networks [7,16].

The routing algorithms should perform well for wireless networks with an arbitrary number of nodes. Sensor and rooftop networks, for instance, have hundreds or thousands of nodes. While other characteristics of each algorithm are easily detected, *scalability* is sometimes judgmental and/or dependent on the performance evaluation outcome. A scalable solution is one that performs well in a large network. It has been experimentally confirmed [17,21] that routing protocols that do not use geographic location in their routing decisions, such as *AODV, DSDV* or *DSR* (their recent surveys are given in [25,34]) are not scalable. For instance, [21] describes *GLS* (a scalable location service), similar to the doubling circle method independently proposed by Amouris, Papavassiliou and Li [1]. Experiments using the *ns* simulator for up to 600 mobile nodes show that the storage and bandwidth requirements of *GLS* grow slowly with the size of the network. Furthermore, *GLS* tolerates node failures well: query performance degrades gracefully as nodes fail and restart and is relatively insensitive to node speeds [21]. Simple geographic forwarding [13] combined with *GLS* compares favorably with *DSR*, delivering more packets and consuming fewer network resources. Similar conclusions were made in [13], where the depth-first search based *GRA* scheme was compared with *DSDV* protocol. Therefore, only position-based approaches provide satisfactory performance for large networks. We shall now elaborate on other properties and reasons for difference in scalability.

Localized algorithms are distributed in nature and resemble greedy algorithms, where simple local behavior achieves a desired global objective. In a localized routing algorithm, each node makes a decision to which neighbor to forward the message based solely on the location of itself, its neighboring nodes, and the destination. In the shortest (weighted) path based nonlocalized algorithms, each node maintains accurate topology of the whole network. In addition, since nodes change between active and sleep periods, the activity status for each node is also required. Although routing table (typical nonposition) based solutions merely keep the best neighbor information on a route toward the destination, the communication overhead for maintenance of routing tables due to node mobility and topology changes is quadratic in network size (each change in edge or node status may trigger routing table modifications in a large portion of the network). On the other hand, position-based localized algorithms avoid that overhead by requiring only accurate neighborhood information and a rough idea of the position of the destination. For example, edge and node changes in one part of the network have no immediate impact on almost any route, since all routes are ad hoc routes. That is, ad hoc networks require ad hoc routes to provide scalable solutions. Clearly, only localized algorithms provide such solutions, especially for networks with critical power resources at nodes (e.g., sensor networks).

16.1.3 Path Strategies, Metrics, Memorization, Guaranteed Delivery, Location Updates, and Robustness

Desirable qualitative properties of routing protocols include: distributed operation, loop freedom (to avoid a worst case scenario of a small fraction of packets spinning around in the network), demand-based operation, and "sleep" period operation (when some nodes become temporarily inactive).

The shortest path route is an example of a *single path* strategy, where one copy of the message is in the network at any time. Arguably, the ideal localized algorithm should follow a single path. On the other extreme are *flooding* based approaches, where messages are flooded through the whole network area or a portion of the area. The "compromise" is *multipath* strategy, that is, a route composed of a few single recognizable paths. Since power and bandwidth are two main limitations in wireless networks, single path strategies are preferred.

The metrics that are used in simulations normally reflect the goal of the designed algorithm, and are naturally decisive in the route selection. Most routing schemes use *hop count* as the metric, where hop count is the number of transmissions on a route from a source to a destination. However, if nodes can adjust their transmission power (knowing the location of their neighbors), then the constant per hop metric can be replaced by a *power* metric that depends on distance between nodes. The goal is to minimize the energy required per each routing task. Some nodes participate in routing packets for many source–destination pairs, and the increased energy consumption may result in their failure. Thus, a pure power consumption metric may be misguided in the long term, and longer paths that pass through nodes that have plenty of energy may be a better solution. The *cost* metric (a rapidly increasing function of decreasing remaining energy at a node) is used with the goal of maximizing the number of routing tasks that a network can perform. Current congestion and other metrics can also be used.

Solutions that require nodes to memorize routes or past traffic are sensitive to node queue size, changes in node activity, and node mobility while routing is ongoing (e.g., monitoring environment). It is better to avoid memorizing past traffic at any node, if possible. However, the need to memorize past traffic is not necessarily a demand for significant new resources in the network for several reasons. First, a great deal of memory space is available on tiny chips. Next, the memorization of past traffic is needed for short periods of time, while an ongoing routing task is in progress, and therefore after a timeout outdated traffic can be safely removed from memory. Finally, the creation of a Quality-of-Service (QoS) path, that is, a path with bandwidth, delay, and connection time requirements, requires that the path be memorized in order to optimize the traffic flow and satisfy QoS criteria.

The delivery rate is the ratio of numbers of messages received by the destination and sent by senders. The primary goal of every routing scheme is to deliver the message, and the best assurance one can offer is to design a routing scheme that will *guarantee delivery*. Wireless networks normally use a single frequency communication model where a message intended for a neighbor is heard by all other neighbors within the transmission radius of the sender. Collisions normally occur in medium access schemes, such as IEEE 802.11. The guaranteed delivery property for routing assumes that the medium access layer is always able to transmit a message between any two neighboring nodes, possibly with retransmissions.

Robust strategies handle the position deviation due to the dynamicity of the network. Another aspect of robust algorithms is their ability to deliver a message when the communication model deviates from the unit graph, due to obstacles or noise. Robust variants of algorithms described here are given in [2,5] (due to space limitations, they will not be discussed here).

In this chapter, we start by presenting (in Section 16.2) routing schemes for *greedy mode*, when the node currently holding the message may advance it toward the destination. The "advance" may be defined in different ways (e.g., distance to destination) or may not be measured at all, leading sometimes to non–loop-free schemes. The basic distance, progress, direction, power, cost, power–cost, congestion, and fading channel methods belong to this group. Greedy mode routing was shown to nearly guarantee delivery for dense graphs but to fail frequently for sparse graphs. Several of these schemes were further developed for integrating a *recovery mode*. The routing process is converted from the greedy mode to recovery mode at a node where greedy mode fails to advance the message toward the destination (refered to as the concave node in the sequel). All the schemes allow the return from recovery mode to greedy mode and aim at guaranteed delivery.

In Section 16.3, we cover schemes that address the routing task by requesting the help of other nodes in the network. They differ in the way the nodes are selected and for the collaborative tasks performed.

Section 16.4 covers schemes that act in proactive way; the data message is directly sent toward the destination location without previous exploration. In these schemes, the main information used for the routing task is the neighborhood, and that this has to be maintained consistent. Therefore, part of Section 16.4 is dedicated to methods for neighborhood updates.

In Section 16.5 we cover the reactive schemes; the destination location is searched before starting the forwarding. These schemes are in general a combination of a simple routing strategy and a location updates scheme. We present both schemes that explicitly address the routing issue and schemes that focus more on the location updates aspects.

16.2 Routing Toward a Destination at a Given Position

16.2.1 Greedy Routing Schemes

In a localized routing scheme, node *S*, currently holding the message, is aware only of the position of its neighbors within the transmission radius and destination *D* (indicated by black circles in Fig. 16.1). Takagi and Kleinrock [33] proposed the first position-based routing scheme, based on the notion of progress. Given a transmitting node *S*, the *progress* of a node *A* is defined as the projection onto the line connecting *S* and *D*. In the *MFR (Most Forward within Radius)* scheme [33], the packet is forwarded to a neighbor with maximal progress, such as node *M* in Fig. 16.1. Nelson and Kleinrock also discussed the *random* progress method (choosing at random one of the nodes with progress and adjusting the transmission radius to reach up to that node), arguing that there is a tradeoff between the progress and transmission success. Hou and Li discussed the *NFP (Nearest Forward Progress)* method (selecting node *N* in Fig. 16.1).

Finn [13] proposed the greedy routing scheme based on geographic distance. *S* selects neighboring node *G* (see Fig. 16.1) that is closest to the destination among its neighbors. Only neighbors closer to the destination than *S* are considered. Otherwise there is a lack of advance, and the method fails. A variant of this method is called the *GEDIR (GEographic DIstance Routing)* scheme [27]. In this variant, applied to other schemes as well, all neighbors are considered, and the message is dropped if the best choice for a current node is to return the message to the node the message came from (stoppage criterion indicating lack of advance). The *Nearest Closer (NC)* method was proposed in [26] (node *N* in Fig. 16.1).

In the *compass routing* method (referred also as the *DIR* method) proposed by Kranakis, Singh, and Urrutia [18], the message *m* is forwarded to the neighbor *A* (see Fig. 16.1), such that the direction *SA* is closest to the direction *SD* (that is, the angle ∠*ASD* is minimized).

The *MFR* and greedy/*GEDIR* methods, in most cases, provide the same path to the destination, and are loop free [27]. The hop count for the *DIR* method is somewhat higher than for the *greedy* scheme, while the success rate is similar. All methods have high delivery rates for dense graphs, and low delivery rates for sparse graphs (about half the messages at average degrees below 4 are not delivered) [27]. When successful, hop counts of greedy and *MFR* methods nearly match the performance of the shortest path algorithm. The *DIR* method, and any other method that includes forwarding the message to a neighbor with closest direction, such as *DREAM* [4] and *LAR* [19], are not loop free [27].

Hop count was traditionally used to measure the energy requirement of a routing task, thus using a constant metric per hop. However, if nodes can adjust their transmission power, then the constant metric can be replaced by a power metric $u(d) = d^\alpha + c$ (for some constants α and c) that depends on distance

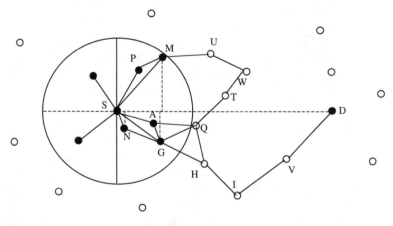

FIGURE 16.1 *S* selects *M* in *MFR* path *SMUW* that fails to deliver, *G* in greedy path *SGHIVD*, *A* in direction based path *SAQTW*, *P* in power path *SPMUW*, *N* in NFP/NC path *SNGQHIVD*.

d between nodes. The value of *c*, which includes energy lost due to startup, collisions, retransmissions, and acknowledgments, is relatively significant, and protocols using any kind of periodic hello messages are extremely energy inefficient [14].

A localized *power aware routing algorithm* is described in [26]. It is based on a formula and an intuition. For two given nodes, [26] described a formula for the optimal power needed for transmission between the two nodes, assuming that additional nodes can be placed at will, the desirable position of which is described. Of course, such nodes are not available in a given ad hoc network, but nevertheless the result is used to attempt to find the most promising forwarding neighbor. It should be as close to the destination as possible, but also as close as possible to the optimal position of a forwarding node for the optimal power transmission. The current node *S* will forward the packet to a neighbor *B* for which the sum of power for transmitting from *S* to *B* and ideal (optimal) power from *B* to *D* is minimized. The algorithm proceeds until the destination is reached, or no closer node to the destination exists.

Singh, Woo, and Raghavendra [31] proposed a cost metric as a function proportional to the inverse of the remaining battery power. Power and cost are combined into a single metric in order to choose power efficient paths among cost optimal ones. Localized cost and power–cost efficient algorithms were proposed in [26], and their performance, when successful, was surprisingly competitive with shortest weighted path algorithms.

Other metrics for choosing best forwarding neighbor in localized routing schemes were considered recently. Yeh [36] proposed several variable-radius routing protocols for achieving higher throughout, smaller latency at a given traffic load, and/or lower power consumption in ad hoc networks. Larsson [20] described a forwarding method for routing in multi-hop networks that takes into account Raleigh fading and nonfading channels. Candidate nodes, addressed in the data packet header, which successfully receive the data packet return acknowledgments in consecutive order, i.e., as their addresses are listed in the packet header. The first neighbor to respond is the forwarding neighbor.

16.2.2 Guaranteed Delivery with Memorization

Greedy schemes encounter difficulties when packets reach concave nodes (normally defined as nodes that have no closer neighbor to destination than themselves). In this condition, some schemes switch from greedy to recovery mode. A simple *greedy/flooding* method is proposed in [27]; concave nodes flood their neighbors and then reject further copies of the same message. Each neighbor then continues with greedy routing, except that nodes that announced their concavity are ignored in forwarding decisions. For each message seen by a node, a list of concave neighbors to be avoided is memorized. If a node is left without a "willing" neighbor, it does not forward the packet further. The method was subsequently improved so that one neighbor in each connected component of the neighborhood subgraph receives a forwarding message from the concave node.

The *Geographic Routing Algorithm (GRA)* by Jain, Puri, and Sengupta [17] requires nodes to partially store routes toward certain destinations (those for which they are concave) in routing tables. *GRA* applies the greedy strategy in forwarding messages. However, concave nodes start the route discovery protocol if the information in routing tables in outdated. The route discovery finds a path from *S* to *D* and updates the routing tables toward *D* at any node on the path with this information. After the route discovery protocol is successfully completed, the stuck packet can be routed from *S* to *D*. The authors propose two route discovery strategies: *breadth first search* (which is equivalent to flooding) and *depth first search (DFS)*. *DFS* yields a single acyclic path from *S* to *D*. Each node puts its name and address on the route discovery packet *p*. Then it forwards *p* to a neighbor that has not seen *p* before. This neighbor is one of all the neighbors that minimize $d(S,y) + d(y,D)$, where $d(x,y)$ is the distance between nodes *x* and *y*. If a node has no possibilities to forward the packet, it removes its name and address from the packet and returns the packet to the node from which it originally received it. Route discovery packets are kept for some time. If a node receives the same packet twice, it refuses it. *DFS* can alternatively be used to deliver the packet without route discovery and routing tables, as proposed in [30] (with applications for the construction of quality of service paths).

16.2.3 Stateless Routing with Guaranteed Delivery

Stateless routing schemes are localized schemes where nodes do not need to memorize past traffic. All decisions are based on the location of neighboring nodes, location of the destination, position of the neighboring node that forwarded the message in the previous step, and the information that "arrives" with the message. Bose, Morin, Stojmenovic, and Urrutia described the *face* and *GFG (Greedy-Face-Greedy)* routing schemes [6], with subsequent improvements, such as the use of two-hop neighborhood information and dominating set concept [11]. Most importantly, Barriere, Fraigniaud, Narajanan, and Opatrny [5] made GFG robust against interferences by addressing instability in the transmission ranges of the host.

In order to ensure message delivery, the *face* algorithm [6] constructs a planar and connected so-called Gabriel subgraph (*GG*) of the unit graph and then applies routing along the faces of the subgraph that intersect the imaginary line between the source and the destination.

GG is a spanning subgraph of the original network. It is defined as follows: given any two adjacent nodes *U* and *V* in the network, the edge *UV* belongs to *GG* if and only if no other node *W* of the network is located in the disk with *UV* as its diameter. This test is fully localized and requires no additional information other than the position of all neighboring nodes.

16.3   Assisted Routing Algorithms

Terminode routing [3] is a combination of two routing protocols: Terminode Local Routing *(TLR)* and Terminode Remote Routing *(TRR)*. TLR is a mechanism that allows for destinations to be reached in the vicinity of a terminode and does not use location information for making packet forwarding decisions. It uses local routing tables that every terminode proactively maintains for its close terminodes. In contrast, *TRR* is used to send data to remote destinations and uses geographic information; it is the key element for achieving scalability and reduced dependence on intermediate systems.

TRR default method is Geodesic Packet Forwarding *(GPF)*. GPF is basically a greedy method [13] and does not perform well if the source and the destination are not well connected along the shortest geodesic path. If the source estimates that *GPF* cannot successfully reach the destination, it uses *anchored paths*. In contrast with traditional routing algorithms, an anchored path does not consist of a list of nodes to be visited for reaching the destination. An anchored path is a list of fixed geographic points, called anchors. In traditional paths made of lists of nodes, if nodes move far from where they were at the time when the path was computed, the path cannot be used to reach the destination. Given that geographic points do not move, the advantage of anchored paths is that an anchored path is always "valid."

In order to forward packets along an anchored path, *TRR* uses the method called Anchored Geodesic Packet Forwarding *(AGPF)*. AGPF is a loose source routing method designed to be robust for mobile networks. A source terminode adds to the packet a route vector made of a list of anchors, which is used as loose source routing information. Between anchors, geodesic packet forwarding is employed. When a relaying terminode receives a packet with a route vector, it checks whether it is close to the first anchor in the list. If so, it removes the first anchor and sends the packet towards the next anchor or the final destination using geodesic packet forwarding. If the anchors are correctly set, then the packet arrives at the destination with a high probability. Simulation results show that the introduction of the anchored paths is beneficial for the packet delivery rate [3].

The *GRID* assisted geographic forwarding is composed of the location proxy and intermediate node forwarding techniques [10]. The assisted geographic forwarding is an extension of the greedy routing scheme proposed by Finn [13], in conjunction with a location service such as *GLS* [21]. The intermediate node forwarding technique is a limited-radius variant of the *DSDV* protocol, propagating route entries for a fixed number of hops, which is used for increasing the number of neighbors each node can use for making geographic forwarding decisions. When forwarding a packet, a node first checks its local *DSDV* neighbor table for a route to the destination. If this exists, the packet is routed to the next hop indicated. Otherwise the next hop is searched still in the list of neighbors of the *DSDV* table using the geographic forwarding rules.

16.4 Proactive Routing Algorithms and Neighborhood Updates

The proactive approach is similar to the connectionless approach of traditional datagram networks, which is based on a constant update of the routing information. To maintain consistent and up-to-date routes between each source–destination pair, the propagation of a large amount of routing information, regardless if this is needed or not, is required. As a consequence, with proactive protocols, a route between any source–destination pair is always available, but such protocols could not perform properly when the mobility rate in the network is high or when the number of nodes in the network is large. In fact, the control overhead, in terms of both traffic and power consumption, is a serious limitation in mobile ad hoc networks, where bandwidth and power are scarce resources [25]. The proactive approaches are more similar in design to traditional IP routing protocols; thus, they are more likely to retain the behavior features of present routing protocols used in practice. Existing transport protocols and applications are more likely to operate as designed over proactive routing approaches than over on-demand routing approaches [23].

Specifically, in proactive position-based approaches the data message is directly sent toward the destination location, without previous exploration. This implies that the main information used for the routing task is the neighborhood, and that this has to be maintained consistent.

16.4.1 Location Updates Between Neighboring Nodes

One of the most important ingredients in all location update schemes is the update between neighboring nodes. Several methods were proposed in the literature to establish when a node decides to send a message to all its neighbors announcing its new location. In principle, location updates are more effective when done as reaction to a topological change. As a basic ("bonus") update, nodes may update their location information with each exchange of routing messages between them.

16.4.2 Request Zone Routing

A distance routing effect algorithm for mobility (*DREAM*) is described in [4]. The source or any intermediate node *A* calculates the direction of destination *D* and, based on the mobility information about *D,* chooses an angular range. The message *m* is forwarded to all neighbors whose direction belongs to the selected range. The range is determined by the tangents from *A* to the circle centered at *D* and with radius equal to a maximal possible movement of *D* since the last location update. Reference [4] proposes recovery procedures, based on partial or full flooding, to start flooding if the given algorithm fails to find the route within a timeout interval. In *DREAM,* the moving nodes send location update messages, which are limited to the two-hops neighborhood if the node remains local or are flooded if the node moves further.

16.4.3 Doubling Circles Routing

Amouris, Papavassiliou, and Li [1] presented a position-based multi-zone routing protocol for wide area mobile ad hoc networks. Their algorithm is based on position updates within circles of increasing radii. Each node updates its location to all nodes located within circles of radii $P, 2P, 4P, 8P,\dots$ (each subsequent circle has twice larger radius than previous one). Whenever a given node *A* moves outside one of these circles of radius 2^tP for some *t,* node *A* broadcasts its location update to all nodes located inside the circle centered at the current node position, and with radius $2^{t+1}P.$ The routing toward the destination then follows these circles of last updates. Source nodes send messages toward the last reported position of the destination (using the *DIR* method), which since the last report has moved within the circle of some radius. As the routing message moves closer to the destination, the information about the position of the destination becomes more precise, and nodes are able to send messages toward the center of circles with twice smaller radius than previously, until the node is eventually reached. This method is very interesting and certainly competitive. We observe that the radii of larger circles may encompass almost all nodes of the network, and that the routing paths discovered by the algorithm do not have near optimal hop counts (which may be important in quality of service applications). However, if the path quality is important, one can consider this algorithm only as the destination search step in the three-phase routing algorithm described above. A similar algorithm, using squares instead of circles and additional sophisticated techniques, is proposed in [21].

The location update techniques discussed so far include occasional flooding of location information to all or a large portion of nodes in the network. In the next two sections, methods that never use such flooding are discussed.

16.5 Reactive Routing Algorithms and Location Updates

A reactive protocol creates and maintains routes between a source–destination pair only when necessary, in general when requested by the source (on-demand approach). Location updates are still performed, both locally and beyond, but to a lesser degree than in the proactive approach, therefore with smaller control overhead. However, similar to connection-oriented communications, a route is not initially available and this generates a latency period due to the route discovery procedure. The on-demand design is based on:

1. The observation that in a dynamic topology routes expire frequently
2. The assumption that not all the routes are used at the same time

Therefore, the overhead expended to establish and/or maintain a route between a given source–destination pair will be wasted if the source does not require the route prior to its invalidation due to topological changes. Note that this assumption may not hold true in all architectures, but it may be suitable for many wireless networks. The validity of this design decision is dependent, in part, on the traffic distribution and the topology dynamics in the network. More specifically, in the case of the position-based approach, a scheme is classified as reactive if the destination location is searched before starting the forwarding. Some schemes do that with an exchange of short messages, others assume some location update schemes. If a message is reasonably "short," it can be broadcasted (that is, flooded), using an optimal broadcasting scheme (nonblind broadcasting schemes are discussed in section 16.5.2). If a message is relatively "long," then destination search (or route discovery) can be initiated, which is a task of broadcasting a short search message. The destination then reports back to the source by routing a short message containing its position. The source then is able to route the full message toward the accurate position of the destination. We present several schemes that explicitly address the destination search and the location update issues.

16.5.1 Request Zone Routing

Several request zone routing schemes, as opposed to *DREAM* [4], are reactive. The main difference (as for example in *LAR scheme 1* [19]) includes sending route requests before the message itself. Note that a route request may be considered as a routing of short messages, as already discussed above. Additionally, the authors propose *LAR scheme 2*. In this scheme, the source or each intermediate node A will forward the message to all nodes that are closer to the destination than A is.

The definition of the request zone [4,19] was modified in [29] to provide a uniform framework with the corresponding notions in *GEDIR* and *MFR* methods. Reference [29] discusses the *V-GEDIR*, *CH-MFR*, and *R-DIR* methods, in which m is forwarded to exactly those neighbors that may be best choices for a possible position of the destination (using the appropriate criterion). The request zone in the *R-DIR* method [29] may include one or two neighbors that are outside of angular range because they can have the closest direction for the tangents to the circle. In the *V-GEDIR* method, these neighbors are determined by intersecting the Voronoi diagram of neighbors with the circle (or rectangle) of possible positions of the destination, while the portion of the convex hull of neighboring nodes is analogously used in the *CH-MFR* method.

16.5.2 Intelligent Flooding for Destination Search

Intelligent and scalable flooding solutions are based on the concept of dominating sets. A set of nodes in a network is dominating if all the nodes in the system are either in the set or neighbors of nodes in the set. If only nodes in a connected dominating set retransmit, all nodes will still receive the message in a collision-free environment. Clusterheads and gateway nodes in a cluster structure define such a set, and were the first "intelligent" flooding solution proposed in the literature. However, the node mobility either worsens the quality of the structure dramatically or otherwise causes a chain reaction. (Local changes in the structure could trigger global updates.)

Localized connected dominating set concepts, proposed recently, avoid such chain reactions and have similar or better rebroadcast savings. Their maintenance does not require any communication overhead in addition to maintaining positions of neighboring nodes or information about two-hop neighbors. One such concept is based on coverage of all two-hop neighbors by a minimal size set of one-hop neighbors [9,22,24]. The other [32] (to be elaborated below) is based on creating a fixed dominating set, where nodes that do not have two unconnected neighbors and nodes that are "covered" by one or two connected neighbors (each neighbor of a covered node is a neighbor of one of nodes that cover it) are eliminated. Neighbor elimination was also applied (solely or in conjunction with other concepts) [31], where nodes that do not have any neighbor not already receiving one of the arriving message copies give up retransmitting.

It is desirable, in the context of broadcasting, to create a dominating set with minimal possible size. Wu and Li [35] proposed a simple and efficient distributed algorithm for calculating a connected dominating set in ad hoc wireless networks. They introduced the concept of an *intermediate* node. A node A is an *intermediate* node if there exist two neighbors B and C of A that are not direct neighbors themselves. For example, nodes C and K in Fig. 16.2 are not intermediate nodes, while other nodes are. The concept is simple, but not many nodes are eliminated from the dominating set.

Consider two intermediate neighboring nodes v and u. If every neighbor of v is also a neighbor of u, and $id(v) < id(u)$, then node v is not an *inter-gateway* node [35]. We may also say that node v is "covered" by node u. Observe that retransmission by v, in this case, is covered by retransmission of u, since any node that might receive a message from v will receive it instead from u. Reference [31] proposed to replace node *id*s with a record *key = (degree, x, y)*, where *degree* is the number of neighbors of a node (and is the primary key in the comparison), and x and y are its two coordinates (that serve as secondary and ternary keys). This significantly reduces the size of the dominating set. Using such keys, consider the example in Fig. 16.2. Note that node J is forced by node K, for which it is the only neighbor, to be in the dominating set for all possible definitions of dominating sets that do not include node K. Nodes A and B are covered by node D, node H is covered by node F, and node L is covered by G. The remaining six nodes are inter-gateway nodes and are squared in Fig. 16.2.

Assume that u, v, and w are three inter-gateway nodes that are mutual neighbors. If each neighbor of v is a neighbor of u or w, where u and w are two connected neighbors of v, and v has lowest *id* among the three, then v can be eliminated from the list of gateway nodes [35]. Reference [31] again proposed to use above defined key instead of *id*. The reason for elimination of v is that any node that can benefit from retransmission by v will receive the same message instead from either u or w. All inter-gateway nodes in Fig. 16.2 remain gateway nodes. Although all neighbors of node I are neighbors of either F or G, it does not have the lowest *id* (in this example, x coordinate serves as *id*). If *id* is changed appropriately, node I may become covered. This suggests that further improvements to the gateway definition might be possible, but the enhancement may require informing neighbors about dominating set status. In the current definition, nodes may decide their own dominating set status without any message exchange but cannot decide the same for their neighbors. Experiments in [31] indicate that the percentage of gateway nodes decreases from 60 to 45% when the average graph degree increases from 4 to 10.

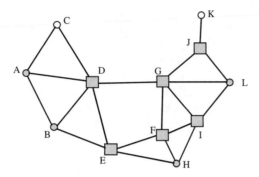

FIGURE 16.2 Nodes *C* and *K* are not intermediate; nodes A and H are not inter-gateway nodes.

16.5.3 Quorum Based Strategy

Quorum based approaches for information dissemination are based on replicating information at multiple nodes acting as repositories. The choice of repositories and the query search must be mutually compatible. Such a query and update strategy has been previously employed for location management in cellular networks. For example, 25 servers can be organized into five rows and five columns. Each column serves as a quorum. Thus each node (i,j) (located in the i-th row and j-th column) replicates its data to all servers (i',j) in its column. To extract the information from server (i,j), server $(i'j')$ may inquire within its i'-th row, and the server (i',j) will provide requested information. In [29], nodes in an ad hoc network do not stay in the same "column," and the distributed information may easily disperse due to node movement. Moreover, it is not clear what the "column" is, and how all the nodes in a column, once defined, will receive the latest updates. Nevertheless, this idea is worth pursuing.

The main location update method is to forward the new location information (and node's identifier) within a "column"' in the network, in the following way. Each node uses a counter to count the number of previously made changes in edge existence (the number of created or broken edges). When the counter reaches a fixed threshold value e, location information is forwarded along the "column," and e is reset to 0. The "column" may have arbitrary "thickness," but we shall assume, for clarity, thickness one here, which means that the created column is a single path in the north–south direction, including neighbors of nodes in that path. Moving node *A* initiates two routing messages, in the directions north and south, while other nodes follow only one of these directions. Each follows a variation of the *MFR* algorithm [33], with the destination always to the north (south, respectively) of the current node, as follows. Current node *B* transmits update information to all its neighbors, and indicates, in the same message, which of them is its northernmost (southernmost, respectively) neighbor. That node will, in turn, do the same, until a node is reached which does not have such a neighbor.

A frequent problem with the scheme is that the northernmost node as determined by the northward update may be only locally northernmost. A "horizontal" destination search can miss such a node, which can remain "below" it. To overcome this problem, each locally northernmost node may switch to *FACE* mode [6] until another node, more northern, is found on a face. It then converts back to the regular upward move. This switch can be repeated a few times. The *FACE* algorithm can be improved by applying a shortcut scheme [11]. The final result will be that all nodes at the outer face of the network will receive location updates. This method guarantees that the "horizontal" destination search and the "vertical" location update will intersect at one of the nodes on the outer face. The drawback is that nodes on the outer boundary will have more traffic demands.

The search for the destination is performed similarly, using the horizontal east–west direction instead. The search message includes the time of the last available information, and other nodes are requested to provide more up-to-date information, if they have it. Nodes that are not on the horizontal path, but that receive the search message and have more up-to-date information about the destination, will respond in the process with their information. Thus, the data link layer protocol should be efficient here to avoid collisions. When easternmost and westernmost nodes are reached (or the outer boundary is traversed, if the *FACE* algorithm is also incorporated), the search strategy changes. The message search is then oriented toward the destination, using the latest available information for each search message. Three searches are initiated. The first one originates at sender node *S*, using the best locally available information. This search does not need to wait for the results of searches in the east and west directions. The easternmost and westernmost nodes in a given "row" may initiate two other searches. Each of three search tasks follows a path toward the destination, using the greedy/*GEDIR* strategy [13,27] or *GFG* algorithm [6]. That is, at each step, the neighbor closest to the destination is selected to forward the message. Nodes that hear messages between two neighbors may react if they have better information about the destination's location.

16.5.4 Home Agent Based Strategy

A strategy where the location is searched with the support of other specific nodes is the Virtual Home Region (*VHR*) solution [15]. A virtual home region is a set of nodes located close to each other. The *VHR* is in charge of keeping some information generated by a node. This node is called the owner node (because it owns the information stored in the *VHR*). The *VHR* is defined by a point in the space; its center *C*, and a radius *R*. The nodes that belong to a *VHR* are those located in the disk *(C,R)*. Each node defines one *VHR*. When the owner node *A* moves to a new location, it stores its new location in its *VHR*, using a simple *SNMP*-like protocol. If a node *B* is willing to send packets to *A*, it sends a query to *A*'s *VHR* and retrieves its location. The *VHR* associated with *A* is known to *B* because it is known to all nodes. The advantage of this approach is that it has a reasonable amount of communications to find a new position, in addition to the fact that no external devices or servers are necessary. However, this solution has its main drawbacks in the communication costs when the network is large. Costs are proportional to the distance between a node and its *VHR* as well as to the distance between a node and the *VHR* of the selected destination. Additionally, estimating parameters such as the radius of the *VHR* or the timer for retransmission in case of messages lost, which are fundamental for the robustness of the approach, can be a very complex and long task.

Another strategy that uses a similar home agent based scheme is independently proposed in [28].

16.6 Conclusion

This chapter did not include a discussion about relevant issues such as physical requirements, experimental design, location updates, QoS, congestion, scheduling node activity, topology construction, broadcasting, and network capacity.

The successful design of localized single-path loop-free algorithms with guaranteed delivery is an encouraging start for future research. The search for localized routing methods that have excellent delivery rates, short hop counts, small flooding ratios, and power efficiency is far from over. In QoS applications, memorization does not appear to require additional resources and is therefore acceptable. However, the research on QoS position-based routing is scarce.

Further research is needed to identify the best position-based routing protocols for various network contexts. These contexts include nodes positioned in three-dimensional space and obstacles, nodes with unequal transmission powers, or networks with unidirectional links. Finally, the mobility caused loop needs to be further investigated and solutions to be found and incorporated into position-based routing schemes.

References

[1] K.N. Amouris, S. Papavassiliou, and M. Li, A Position Based Multi-Zone Routing Protocol for Wide Area Mobile Ad-Hoc Networks, *Proc. 49th IEEE Vehicular Technology Conference*, Amsterdam, 1999, pp. 1365–1369.

[2] L. Blazevic, L. Buttyan, S. Capkun, S. Giordano, J.-P. Hubaux, and J-Y. Le Boudec, Self-organization in Mobile Ad Hoc Networks: The Approach of Terminodes, *IEEE Communication Magazine,* June 2001, pp. 166–175.

[3] L. Blazevic, S. Giordano, and J.Y. Le Boudec, Self organized terminode routing, *Journal of Cluster Computing,* 5(2), 2002.

[4] S. Basagni, I. Chlamtac, V.R. Syrotiuk, and B.A. Woodward, A Distance Routing Effect Algorithm for Mobility (DREAM), *Proc. MobiCom,* Dallas, TX, 1998, pp. 76–84.

[5] L. Barriere, P. Fraigniaud, L. Narajanan, and J. Opatrny, Robust Position Based Routing in Wireless Ad Hoc Networks with Unstable Transmission Ranges, *Proc. of 5th ACM Int. Workshop on Discrete Algorithms and Methods for Mobile Computing and Communications DIAL M01,* 2001, pp. 19–27.

[6] P. Bose, P. Morin, I. Stojmenovic, and J. Urrutia, Routing with guaranteed delivery in ad hoc wireless networks, *3rd Int. Workshop on Discrete Algorithms and Methods for Mobile Computing and Communications,* Seattle, WA, Aug. 20, 1999, pp. 48–55; *ACM/Kluwer Wireless Networks,* 7, 609–616, 2001.

[7] S. Capkun, M. Hamdi, and J.P. Hubaux, GPS-free Positioning in Mobile Ad-Hoc Networks, *Proc. Hawaii Int. Conf. on System Sciences,* Maui, Jan. 2001.

[8] S.Corson and J. Macker, Mobile Ad Hoc Networking (MANET), *IETF RFC 2501,* January 1999.

[9] G. Calinescu, I. Mandoiu, P.J. Wan, and A. Zelikovsky, Selecting Forwarding Neighbors in Wireless Ad Hoc Networks, *Proc. DIAL M,* 2001.

[10] D.S.J. De Couto and R. Morris, Location Proxies and Intermediate Node Forwarding for Practical Geographic Forwarding, MIT Lab. for Computer Science Technical Report MIT-LCS-TR-824, June 2001.

[11] S. Datta, I. Stojmenovic, and J. Wu, Internal nodes and shortcut based routing with guaranteed delivery in wireless networks, *Cluster Computing,* Apr. 2002, pp. 169–178, to appear.

[12] M. Ettus, System Capacity, Latency, and Power Consumption in Multihop-Routed SS-CDMA Wireless Networks, *Proc. IEEE Radio and Wireless Conf.,* Colorado Springs, CO, Aug. 1998.

[13] G.G. Finn, Routing and Addressing Problems in Large Metropolitan-scale Internetworks, ISI Research Report ISU/RR-87–180, Mar. 1987.

[14] L.M. Feeney, An energy-consumption model for performance analysis of routing protocols for mobile ad hoc networks, *Mobile Networks and Applications,* 6, 239–249, 2001.

[15] S. Giordano and M. Hamdi, Mobility Management: the Virtual Home Region, EPFL, Mar. 3, 2000, TR DSC99–037.

[16] J. Hightower and G. Borriello, Location Systems for Ubiquitous Computing, *IEEE Computer,* Aug. 2001, pp. 57–66.

[17] R. Jain, A. Puri, and R. Sengupta, Geographical Routing Using Partial Information for Wireless Ad Hoc Networks, *IEEE Personal Communication,* Feb. 2001, pp. 48–57.

[18] E. Kranakis, H. Singh, and J. Urrutia, Compass Routing on Geometric Networks, *Proc. 11th Canadian Conference on Computational Geometry,* Vancouver, Aug. 1999.

[19] Y.B. Ko and N.H. Vaidya, Location-aided routing (LAR) in mobile ad hoc networks, *MobiCom,* 1998, 66–75; *Wireless Networks,* 6, 307–321, 2000.

[20] P. Larsson, Selection Diversity Forwarding in a Multihop Packet Radio Network with Fading Channel and Capture, *Proc. ACM MobiHoc,* Long Beach, CA, 2001, pp. 279–282.

[21] J. Li, J. Jannotti, D.S.J. De Couto, D.R. Karger, and R. Morris, A Scalable Location Service for Geographic Ad Hoc Routing, *Proc. ACM MobiCom,* Boston, MA, 2000, pp. 120–130.

[22] H. Lim and C. Kim, Flooding in Wireless Ad Hoc Networks, *Proc. ACM MSWiM Workshop at MOBICOM,* Aug. 2000; *Computer Communication J.,* 24, 353–363, 2001.

[23] J.P. Macker, V.D. Park, and M.S. Corson, Mobile and Wireless Internet Services: Putting the Pieces Together, *IEEE Communication Magazine,* June 2001.

[24] A. Qayyum, L. Viennot, and A. Laouiti, Multipoint Relaying: An Efficient Technique for Flooding in Mobile Wireless Networks, *Proc. IEEE Hawaii Int. Conf. System Sciences,* Maui, Jan. 2002.

[25] E. Royer and C.-K. Toh, A Review of Current Routing Protocols for Mobile Ad-Hoc Networks, *IEEE Personal Communications*, Apr. 1999.

[26] I. Stojmenovic and X. Lin, Power-aware localized routing in wireless networks, *IEEE Transactions on Parallel and Distributed Systems*, 12, 1122–1133, 2001.

[27] I. Stojmenovic and X. Lin, Loop-free hybrid single-path/flooding routing algorithms with guaranteed delivery for wireless networks, *IEEE Transactions on Parallel and Distributed Systems*, 12, 1023–1032, 2001.

[28] I. Stojmenovic, Home Agent Based Location Update and Destination Search Schemes in Ad Hoc Wireless Networks, Computer Science, SITE, University of Ottawa, TR-99–10, Sep. 1999.

[29] I. Stojmenovic, Location updates for efficient routing in ad hoc networks, in *Handbook of Wireless Networks and Mobile Computing*, Wiley, New York, 2002, pp. 451–471, www.site.uottawa.ca/~ivan.

[30] I. Stojmenovic, M. Russell, and B. Vukojevic, Depth First Search and Location Based Localized Routing and QoS Routing in Wireless Networks, *IEEE Int. Conf. On Parallel Processing*, Toronto, Aug. 21–24, , pp. 173–180.

[31] S. Singh, M. Woo, and C.S. Raghavendra, Power-aware Routing in Mobile Ad Hoc Networks, *Proc. MobiCom*, Dallas, TX, 1998, pp. 181–190.

[32] I. Stojmenovic, M. Seddigh, and J. Zunic, Dominating sets and neighbor elimination based broadcasting algorithms in wireless networks, *IEEE Transactions on Parallel and Distributed Systems*, 13, 14–25, 2002.

[33] H. Takagi and L. Kleinrock, Optimal transmission ranges for randomly distributed packet radio terminals, *IEEE Transactions on Communications*, 32, 246–257, 1984.

[34] Y.C. Tseng, W.H. Liao, and S.L. Wu, Mobile ad hoc networks and routing protocols, in *Handbook of Wireless Networks and Mobile Computing*, I. Stojmenovic, Ed., John Wiley & Sons, New York, 2002, pp. 371–392.

[35] J. Wu and H. Li, On Calculating Connected Dominating Set for Efficient Routing in Ad Hoc Wireless Networks, *Proc. DIAL M*, Seattle, WA, Aug. 1999, pp. 7–14.

[36] C. Yeh, Variable-Radius Routing Protocols for High Throughput, Low Power, and Small Latency in Ad Hoc Wireless Networks, *IEEE Int. Conf. Wireless LANs and Home Networks*, Dec. 2001.

17

Route Discovery Optimization Techniques in Ad Hoc Networks

Boon-Chong Seet
Nanyang Technological University

Bu-Sung Lee
Nanyang Technological University

Chiew-Tong Lau
Nanyang Technological University

Abstract

Efficient use of network bandwidth is an important concern in wireless networks. The prevailing use of flooding for on-demand route discovery is bandwidth expensive and can lead to serious degradation in network performance particularly when there are large numbers of concurrent peer-to-peer communications. This chapter discusses techniques for optimizing bandwidth efficiency of route discovery and also explores some ideas for possible future research.

17.1 Introduction

Existing on-demand routing protocols perform route discovery by flooding the network with a query message requesting a route to the destination. Flooding is used because of its simplicity, and because it has greater success in finding not only a route but also the best route between the source and destination available at that time of route discovery. However, as flooding involves querying all reachable network nodes, frequent flooding can rapidly deplete the energy reserves at each node, in addition to consuming significant portions of the available network bandwidth. Further, as the number of communicating nodes increases, more congestion, contention, and collisions can be expected. This chapter focuses on optimizations of route discovery. More specifically, we discuss techniques for limiting the extent and effects of query flooding. Optimizations of other aspects of on-demand routing such as route maintenance are beyond the scope of this chapter and are thus omitted to allow a more focused discussion of the issues relating directly to this topic.

The rest of this chapter is structured as follows. Section 17.2 presents an overview of the concepts of on-demand routing and two such representative protocols, the Ad-Hoc On-Demand Distance Vector (AODV) [1] and Dynamic Source Routing (DSR) [2] protocols. Section 17.3 describes and assesses the proposed optimizations for flooding-based route discovery. Section 17.4 provides a brief discussion of the implications and explores some ideas for possible future research. Finally, concluding remarks are made in Section 17.5.

17.2 Revisiting On-Demand Routing

On-demand routing creates routes only when needed for communication. The route is maintained until the communication ends or when the destination becomes unreachable. When a route to some previously unknown destination is needed, the source floods the network with a query packet to discover a route to that destination, which upon receiving sends a reply back to the source. But since multiple copies of the same query may arrive at the destination via alternate paths, the destination *may* send more than one reply to the source, each giving a different route. The source may then select one it deems optimal (using its own definition of optimal). Once established, the source uses the route to send data packets to the destination. Should the route break at any time during the communication, i.e., due to node movements, the node that notices the break informs the source, which in turn invokes route discovery to set up another route to the destination. When the communication ends, the source does not attempt to further maintain the connectivity to the destination. Hence, network resources are not expended to maintain unneeded routes.

AODV and DSR are two representatives of such on-demand routing protocols submitted to the Internet Engineering Task Force (IETF) Mobile Ad-hoc Networks (MANET) Working Group [3] for consideration as proposed standards. Both protocols perform route discovery in a similar way as described above. The main difference is that AODV uses forwarding tables at each node while DSR uses source routing. Hence, the size of query packets in DSR can be larger than in AODV since the routing path information is appended to the header of the query packet as it traverses along the potential route. In AODV, this information is individually stored in each node between the source and destination. However, source routing has its own advantage in that the source can learn more routing information, including routes to all intermediate nodes, since it possesses the complete knowledge of the route leading to the destination. Such information can be saved for future use by the source, obviating the need for some route discovery processes. Another aspect of difference is that AODV does not attempt to use any unidirectional links encountered during the route discovery. Links suspected to be unidirectional are only blacklisted and ignored. In DSR, provisions are made to permit the use of such links. This can be critical particularly when a unidirectional link is the only link available to reach a destination, and thus failure to use it may result in repeated unsuccessful route discoveries. Other significant differences between the two protocols are the optimizations to route discovery, which will be the subject of discussion in the next section.

17.3 Optimizations for Flooding-Based Route Discovery

Excessive bandwidth overhead produced by query flooding during route discovery has prompted protocol designers to seek new ways to reduce the frequency and spread of route discovery. In the following sections, we will discuss these optimization techniques, some of which are already part of the AODV and DSR specifications. We classify optimizations common in both AODV and DSR as basic techniques. We also discuss optimizations specific to each of these protocols. Other optimizations proposed by individual researchers that can be applied to these protocols are also discussed.

17.3.1 Basic Techniques

17.3.1.1 Suppressing Duplicate Route Requests

One common assumption in many existing ad hoc routing protocols is the use of omni-directional antennas. Due to the broadcast nature of transmissions using these antennas, i.e., a node's transmission may be heard by several nearby nodes, it is possible that a node may receive multiple copies of the same route request (RREQ) message rebroadcast by its neighbors. For reasons of bandwidth efficiency, processing and forwarding of duplicate RREQs are normally not desirable.[1] Intermediate nodes process and forward only the first copy of the RREQ and discard other copies of the same RREQ that arrive later.[2] To ascertain whether a RREQ received has been previously processed, nodes keep track of each RREQ's source IP address and RREQ ID, which is a sequence number generated by the source to uniquely identify a particular RREQ.

17.3.1.2 Intermediate Nodes Replying to Route Requests

During the process of route discovery, any intermediate node receiving the RREQ and having a route to the destination may send a route reply (RREP) message to the source, instead of continuing to forward the RREQ to the destination. This, in effect, "quenches" (or stops the spreading of) the query flooding at the intermediate node. To a large extent, this may reduce the route discovery overhead and latency. However, it does not guarantee that the route obtained will be the "best" route (i.e., fewest hops, most long-lived, least congested, etc.), or that the path between the intermediate node and the destination will still be intact. The source node may actually be prevented from discovering a better route even if it exists. Besides, when the freshness of routes is more important, or when some up-to-date, end-to-end information must be collected to assist in proper route selection, i.e., end-to-end bandwidth availability and individual node's energy reserve, this option often has to be disabled.

17.3.2 AODV-Specific Techniques

17.3.2.1 Expanding Ring Search

In the previous section, we discussed two techniques to suppress and quench RREQs *after* a network-wide flooding has been triggered. We now discuss the expanding ring search technique currently employed by AODV to *delay* the triggering of such flooding. Concisely put, expanding ring search is a technique that searches in increasingly larger neighborhoods, i.e., by sending out successive RREQs, each with a larger "time-to-live" (TTL) that limits how far (how many hops) a RREQ can traverse from the source. Below, a generalized description of the technique is given without dwelling on the implementation details in AODV.

In an expanding ring search, several attempts at route discovery may be conducted. The source initially conducts a route discovery that searches only the region within some limited hops from itself. When a route to the destination cannot be found after some timeout, the source attempts another route discovery, but with a greater search scope than the preceding attempt by increasing the TTL of the RREQ. This process continues until the TTL reaches a maximum threshold, after which the RREQ is flooded throughout the network. This technique is effective only if the destination can be found within the initial attempts. If not, the route discovery overhead can be even higher than with flooding on the first attempt, in addition to increasing route discovery latency due to multiple timeouts.

17.3.2.2 Local Repair

AODV includes an additional provision to allow an intermediate forwarding node (instead of the source node) to initiate a route rediscovery (or route recovery) to the destination upon detection of a link failure. This process is typically known as "local repair," in which the intermediate node upstream of the link failure sends out a RREQ, the TTL of which is set to the remaining hop-distance to the destination, added with an increment value. Data packets are buffered at this node during the route rediscovery and sent as soon as the route is repaired. If after some timeout, no route to the destination could be found, the buffered data packets would be dropped, and a Route Error (RERR) message would be sent to the source, which then attempts a source-initiated route discovery.

---

[1] Unless there is a specific purpose, such as to facilitate the discovery of disjoint routes for multipath routing [6].

[2] DSR additionally specifies that the RREQ should be discarded if the receiving node's address is already listed in the route record.

Local repair is useful, in particular in large networks where routes could be long and thus more susceptible to link failures. By performing route rediscovery from the "point of failure," rather than from the source, the amount of control traffic generated could be reduced, especially if the link failure occurs close to the destination. One downside of this technique is that after local repair, the route may no longer be the shortest route available to the destination and thus may degrade performance in terms of packet transfer time and delivery rate. Further, if the initial local repair by the intermediate node fails, the source has to perform a new route discovery, which would further increase the delay and overhead.

17.3.3 DSR-Specific Techniques

17.3.3.1 Caching Overheard Routing Information

One distinct feature of DSR is the caching of overheard routing information available from the headers of passing-by packets. Nodes with network interface operating in promiscuous mode are able to eavesdrop on packets that may not be addressed to them. Routing information learned from these packets can be stored for future use by the nodes themselves. Hence, nodes may perform less route discovery since the desired routes may be already available in their caches. However, the promiscuous learning of routes may incur high processing overhead, as well as requiring large memory storage.

17.3.3.2 Preventing Route Reply Storm

Another feature of DSR is the deliberate introduction of delay to an intermediate node that is sending a RREP message. Due to the nature of broadcast transmission, many nodes around the broadcasting node may receive the RREQ and send RREPs simultaneously. This may result in what is dubbed a RREP "storm," in which a great number of nodes attempt to send RREPs from their caches simultaneously, causing local congestion and excessive packet collisions. Having some nodes delay sending their RREPs may mitigate this problem. The delay time is specified to be: $d = H \times (h - 1 + r)$, where h is the number of hops of the returned route, r is a random number between 0 and 1, and H is a small constant delay to be introduced per hop. Notice that shorter the route, the earlier the node giving this route would reply.

17.3.3.3 Nonpropagating Route Requests

Earlier, we discussed the expanding ring search technique in AODV. There also exists in DSR a variant of such an approach, but it limits the number of route discoveries to two attempts. In the initial attempt, the source sends a "nonpropagating" RREQ with a hop limit of 1 (i.e., TTL = 1) to look for either the destination or some node with a route to the destination within its immediate neighborhood. If a route cannot be found, the source node sends a "propagating" RREQ with no hop limit, which essentially floods the network. It should be relatively straightforward to extend this dual-phase search to an expanding ring search by allowing the hop limit to increase in incremental steps. Perhaps more challenging is finding good techniques to estimate a minimum hop limit for RREQ to acquire a route on the first attempt.[3]

17.3.3.4 Rate Limiting of Repeated Route Discovery

To avoid the possible generation of an overwhelming amount of control traffic due to repeated unsuccessful route discoveries to some (temporarily) unreachable destinations because of network partitioning, DSR uses an exponential back-off scheme to limit the rate at which the source node may repeat route discovery to the same destination.

17.3.4 Other Techniques

17.3.4.1 Solving the Broadcast Storm Problem

Recall that in Section 17.3.1, we discussed the suppression of duplicate RREQs, in which nodes avoid rebroadcasting copies of the same RREQ they have already seen. However, even if nodes only rebroadcast

---

[3] We noted that a similar issue has been investigated in [7].

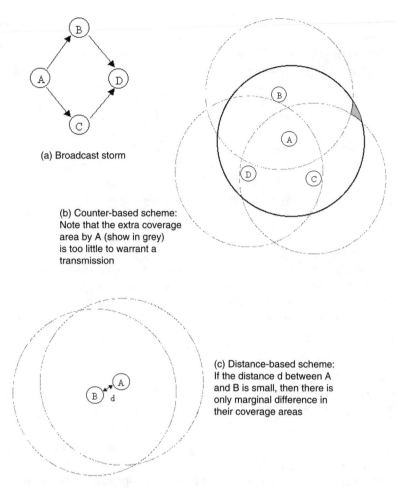

(a) Broadcast storm

(b) Counter-based scheme:
Note that the extra coverage
area by A (show in grey)
is too little to warrant a
transmission

(c) Distance-based scheme:
If the distance d between A
and B is small, then there is
only marginal difference in
their coverage areas

FIGURE 17.1 (a) Broadcast storm; (b) Counter-based scheme: Note that the extra coverage area by A (shown in gray) is too little to warrant a transmission; (c) Distance-based scheme: If the distance d between A and B is small, then there is only marginal difference in their coverage areas.

RREQs they receive for the first time, some of the rebroadcasts may be redundant. As shown in Fig. 17.1a, suppose Node A broadcasts a new RREQ to Nodes B and C, which in turn rebroadcast to Node D. Hence, Node D receives two copies of the same RREQ, one of which is *redundant*. Moreover, if Nodes B and C are close to each other and both transmit at the same time, channel *contention* could occur. Further, RTS/CTS exchange is not used in broadcast transmission. If the underlying MAC does not provide collision detection capability (i.e., CSMA/CA cannot listen while sending), packet *collisions* could be damaging. The resulting redundancy, contention and collisions constitute what is called the "broadcast storm" problem [8]. Several schemes are proposed by the authors to alleviate this problem:

1. Probabilistic
2. Counter-based
3. Distance-based
4. Location-based
5. Cluster-based

Below is a concise description of these schemes.

In the *probabilistic* scheme, each node rebroadcasts the message it received for the first time with some fixed probability *p*.

In the *counter-based* scheme, each node rebroadcasts the message only if the same message has not
been heard for more than C times, before it itself can transmit. The assumption made by the node
is that if the message has been rebroadcast several times by its neighbors, then the extra coverage
contribution from its own rebroadcast is probably too low to be worth transmitting. This is
illustrated in Fig. 17.1b.

In the *distance-based* scheme, each node rebroadcasts the message only if the physical distance between
itself and the node from which it received the message is not less than d (Fig. 17.1c). The node
uses the signal strength of the received message to estimate this distance.

In the *location-based* scheme, the message is rebroadcast only if the extra area expected to be covered
from this broadcast is greater than A. The node uses the location information from GPS to
determine the area of this extra coverage.

As for the *cluster-based* scheme, only cluster-heads and gateway nodes are able to rebroadcast the
message. The nodes may use any of the other schemes to determine whether or not to rebroadcast
the message.

The schemes proposed in the above are effective, in particular for densely populated networks, in
which nodes are communicating in close proximity of each other. One problem, however, is that all the
threshold values are *fixed*, which may result in some messages not being broadcast to the destination
under certain conditions, i.e., when the network is sparse. Thus, some improvements have been proposed
to adapt the threshold values to changing node density [13].

17.3.4.2   Query Localization

Query localization [9] is a technique that exploits the knowledge of some previously known route to
restrict the query flooding to a specific region of the network. The basic premise behind such technique
is that the topology has not changed drastically soon after a link failure and thus many of the nodes on
the previous route may be used to reconstruct a new route to the destination. Two schemes for query
containment are proposed:

The first scheme assumes that the new route cannot be very different from an older route, with at
most k nodes different (path locality).

The second scheme assumes that the destination is within k hops away from any nodes on the older
route (node locality).

In both schemes, every query packet carries a counter that is initialized to zero and then incremented
each time the query encounters a node that was not on the previous route to the destination. When the
query does encounter a node on the previous route, only the second scheme resets the counter to zero.
Once the counter exceeds the threshold value k, the query is dropped.

This technique should be useful for the source to initiate route rediscovery soon after a link failure.
For such cases, the initial value of k is given some small value, i.e., two hops. If a route to the destination
cannot be found with this value, then k is increased, and the process repeats until the maximum threshold
for k is reached. For a new route discovery in which no previous route to the destination is known, the
initial value of k is set to the network diameter, thus flooding the query over the entire network. As with
local repair, the query localization shares the possibilities of reconstructing a longer path, as well as
increasing the route discovery latency when the initial route rediscovery fails.

17.3.4.3   Location-Aided Routing

Location-Aided Routing (LAR) [10] is a technique that proposes using location information obtained
from GPS to confine the route search to a region where the destination is likely to be found. Two variants
of this protocol are proposed. Figure 17.2 illustrates the concepts of LAR1. By knowing the physical
location L and average speed v of the destination at time $t0$, the source defines at time $t1$ a circular region
of radius $v\ (t1 - t0)$ called "expected zone." This is the region in which the destination may be found. In
addition, the source defines the smallest rectangle that includes the expected zone and itself as the "request

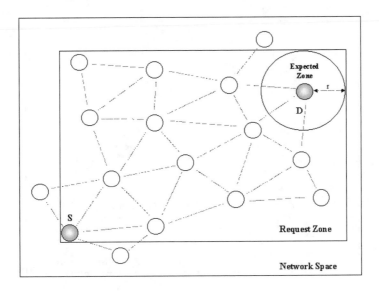

FIGURE 17.2 LAR's request and expected zone concepts.

zone," in which only nodes that reside in this zone can forward the RREQ. The source attaches this information on request zone to the RREQ.

In LAR2, the source uses location information to compute the distance between the destination and itself and then attaches this distance value to the RREQ. When a node receives the RREQ, it computes its own distance to the destination and then forwards the RREQ only if it is closer to the destination than the node from which the RREQ is received. Hence, the RREQ will only get progressively closer to the destination after each relay. In both schemes, nodes may attach their location information onto any packets they are sending (i.e., RREQ) in order to allow other nodes to learn about their location. Moreover, if a RREP is not received after some timeout period, the source initiates a new route discovery using flooding.

One potential weakness of the protocol is the dependence on GPS for obtaining one's location, since direct line-of-sight access to GPS satellites may not always be possible due to blockage by objects such as buildings and foliage. Further, prior knowledge of the destination's location may not always be available at the source. For the former, the problem may be remedied by using some non-GPS techniques as proposed in [4]. For the latter, the protocol may require more mobile nodes to communicate their locations more frequently, or alternatively enlist the aid of a distributed location service [5] if necessary. Savings from the reduced flooding of RREQs when LAR is performed may far outweigh the costs of retrieving location information.

17.3.4.4 Unicast Query Mechanism

We now discuss another location-based optimization, which is the unicast query mechanism [11,12]. This is a mechanism that can be used to improve the overhead performance of LAR. Consider the case when the source and target are not in proximity: a significant portion of the network may be flooded with RREQs, i.e., due to a larger request zone. The unicast query mechanism can help to mitigate this problem by allowing the source to use location information to select an existing route for *unicasting* its RREQ to a node in the neighborhood of the target. This node, which is known as the "target neighbor," in turn broadcasts the RREQ to the nearby target, i.e., one or two hops away, as shown in Fig. 17.3.[4] Hence, the RREQ is broadcast near the target and not at the source as in LAR, which helps reduce the

---

[4] The request and expected zones are not used by this mechanism but are shown to highlight its potential to improve LAR.

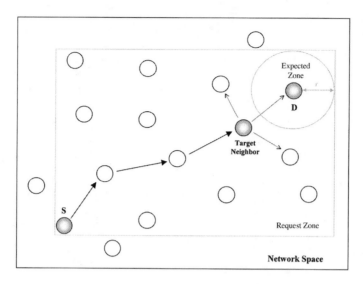

FIGURE 17.3   The unicast query mechanism.

number of RREQ and RREP messages generated. Besides, if any intermediate nodes along the query path are allowed to respond to the RREQ when they have a route to the target, then a broadcast of the RREQ may not be required, and the overhead can be further reduced.

As with expanding ring search, one potential drawback of this mechanism is the increase in route discovery latency when the initial attempt to discover a route using this mechanism fails, i.e., when the unicast query path has been invalidated as a result of node movements. Extra latency thus can be incurred when the source retries route discovery, either using a different unicast query path (if available), or by broadcast.

Another potential source of latency is introduced when the intermediate nodes are allowed to respond to RREQs. Though network-wide broadcast is expensive, it enables the source to discover multiple routes to the destination. But more importantly, it allows many nodes, i.e., intermediate nodes that forward RREPs to the source, to discover a route to the destination, as well as to other nodes along the route (i.e., by virtue of source routing). This greatly increases a node's ability to respond to others' RREQs. Conversely, this ability diminishes when the searching space of many route discoveries is constrained, thus increasing the RREQ's traversal time.[5]

Further, since the RREQs are not broadcast end-to-end, i.e., from source to destination, the routes constructed by the mechanism may not always be the shortest. But this shortcoming can be remedied with route maintenance features such as "automatic route shortening" from DSR, which makes possible the self-optimization of path length over time.

17.4 Thoughts and Suggestions for Future Research

One of the key objectives of most optimization techniques is to minimize the amount of control traffic generated in a route discovery. But in the process, they often impact other aspects of performance in ways that are not always desired. Expanding ring search, for example, compromises packet latency for bandwidth efficiency, and so do other techniques that confine the search space for routes or limit the query to only a subset of nodes. Early quenching of route requests by intermediate nodes may result in fewer control packets and a shorter query time. However, the routes obtained can be obsolete or non-optimal, which results in both increased packet loss and latency. Inherently, there exists a tradeoff between

---

[5] We expect this phenomenon to occur as well (though to different extents) in LAR and other similar optimizations.

overhead of route discovery and other performance areas. This is not unexpected, but a question arises as to whether such tradeoffs in performance can be averted. For example, if the message to be sent is urgent, or the current network utilization is low, then flooding may be used to discover a better route in a shorter time. If not, better resource conserving techniques such as the unicast query mechanism may be employed for route discovery. Conceivably, some adaptive methods of optimizing route discovery would be useful to allow a flexible tradeoff between efficiency and performance. This may be interesting for future investigation.

In the previous section, we also discussed techniques that utilize geographic location information for directional route discovery. Underlying these techniques is the notion that a route to the destination can be found by searching in the general direction of the destination. Terrain features such as buildings, hills, and foliage are currently not considered in these techniques. The presence of such objects can obstruct or substantially weaken the transmission of radio signals, making communication across them difficult if not impossible even though the communicating nodes may be close physically. Lack of terrain awareness may render the use of these techniques less effective in real-life scenarios.

As an example, we examine a case where route discovery using LAR may be problematic when obstructions are not considered. In Fig. 17.4, we represent the obstructions by rectangular objects in gray and assume them to be impenetrable by radio waves. Suppose that Node S initiates a route discovery to Node D by broadcasting a query message. Node S floods this query to its request zone only to find that Node D is unreachable because no queries rebroadcast by other nodes in the request zone have reached Node D due to obstructions. In fact, a route does exist through Node K. However, this route is not discovered since Node K is lying beyond Node S's request zone. If the obstructions are known *a priori*, then Node S may (for example) increase the search space around edges of the obstruction at Node D, so that Node K can be encompassed within Node S's request zone. There are, of course, other solutions possible. But in general, knowing the terrain over which communication is to take place is expected to yield greater success in route discovery.

It is also possible that some obstructions are semipenetrable (i.e., forested areas), where radio signals are weakened but not completely obstructed. If a direct route is desired over one that makes a detour around the obstruction, then nodes may instead increase their transmit power to get the queries across, i.e., Node C may increase its transmit power to send to Node D. However, increasing transmit power would cause greater interference with surrounding nodes. Hence, the use of directional antennas can be envisaged.

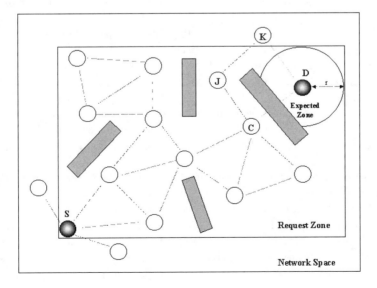

FIGURE 17.4 Route discovery with LAR in the presence of obstructions.

Surface terrain information can be obtained from computerized terrain databases such as geographic information systems (GIS). Intuitively, nodes may also learn about their physical environment by examining other nodes' geographic location and their logical connectivity, i.e., though Node C and Node D are close in physical space, Node C requires two hops to reach Node D via Nodes J and K. If the type of radio interface used is the same between the nodes (having equal transmit power and bi-connected links), then one may infer that an obstruction exists in the region bounded by Nodes C, J, K, and D. This is a simplistic inference that has not considered other important factors such as node distribution and density, and clearly much work remains to be done. By itself, this may be another interesting topic that is worth exploring.

17.5 Summary and Concluding Remarks

In this chapter, we discussed several techniques to limit the extent and effects of query flooding during a route discovery. However, as we noted earlier, there is always a tension between the conflicting goals of efficiency and performance. Hence, the gain in bandwidth efficiency is not without its costs. We also explored some ideas for future research, including adaptive methods of optimizing route discovery and LAR with terrain awareness.

References

1. Perkins, C.E., Royer, E.M., and Das, S.R., Ad Hoc On-Demand Distance Vector (AODV) Routing, Internet-Draft, draft-ietf-manet-aodv-10.txt, Jan. 2002, work in progress.
2. Johnson, D.B., Maltz, D.A., Hu, Y.-C., and Jetcheva, J. G., The Dynamic Source Routing Protocol for Mobile Ad Hoc Networks (DSR), Internet-Draft, draft-ietf-manet-dsr-06-txt, Nov. 2001, work in progress.
3. Internet Engineering Task Force (IETF) Mobile Ad Hoc Networking (MANET) Working Group, http://www.ietf.org/html.charters/manet-charter.html.
4. Capkun, S., Hamdi, M., and Hubaux, J.-P., GPS-free Positioning in Mobile Ad-Hoc Networks, *Proc. 34th Hawaii Int. Conf. System Sciences*, Maui, 2001, p. 3481.
5. Li, J., Jannotti, J., De Couto, D.S.J., Karger, D.R., and Morris, R., A Scalable Location Service for Geographic Ad Hoc Routing, *Proc. 6th Int. Conf. Mobile Computing and Networking*, Boston, MA, 2000, p. 120.
6. Lee, S.-J. and Gerla, M., Split Multipath Routing with Maximally Disjoint Paths in Ad Hoc Networks, *Proc. IEEE Int. Conf. Communications*, Helsinki, 2001, p. 3201.
7. Sucec, J. and Marsic, I., An Application of Parameter Estimation of Route Discovery by On-Demand Routing Protocols, *Proc. 21st Int. Conf. Distributed Computing Systems*, Phoenix, 2001, p. 207.
8. Ni, S.-Y., Tseng, Y.-C., Chen, Y.-S., and Sheu, J.-P., The broadcast storm problem in mobile ad hoc networks, *Proc. 5th ACM/IEEE Int. Conf. Mobile Computing and Networking*, Seattle, 1999, p. 151.
9. Castaneda, R. and Das, S.R., Query localization techniques for on-demand routing protocols in ad hoc networks, *Proc. 5th ACM/IEEE Int. Conf. Mobile Computing and Networking*, Seattle, 1999, p. 186.
10. Ko, Y. and Vaidya, N., Location-aided routing (LAR) in mobile ad hoc networks, *Proc. 4th ACM/IEEE Mobile Computing and Networking*, Dallas, 1998, p. 66.
11. Seet, B.-C., Lee, B.-S., and Lau, C.-T., Route discovery optimisation for dynamic source routing in mobile ad hoc networks, *IEE Electronic Lett.*, 36, 1963, 2000.
12. Seet, B.-C., Lee, B.-S., and Lau, C.-T., Study of a unicast query mechanism for dynamic source routing in mobile ad hoc networks, *Lecture Notes on Computer Science*, 2094, Springer-Verlag, Berlin, 2001, p. 168.

13. Tseng, Y.-C., Ni, S.-Y., and Shih, E.-Y., Adaptive Approaches to Relieving Broadcast Storms in a Wireless Multihop Mobile Ad Hoc Network, *Proc. 21st Int. Conf. Distributed Computing Systems,* Phoenix, 2001, p. 481.

VI

Applications of Ad Hoc Wireless Networks

18

Location-Aware Routing and Applications of Mobile Ad Hoc Networks

Yu-Chee Tseng
National Chiao-Tung University

Chih-Sun Hsu
National Central University

18.1 Introduction

Wireless communications have made great progress recently. Computing technologies have also advanced quickly as we see a variety of portable, small, light devices appearing on the market. These together have made computing and communication anytime, anywhere possible. One of the promising wireless network architectures that can realize communication anytime, anywhere is the *mobile ad hoc network (MANET)*. A MANET consists of a set of mobile hosts without the support of base stations. It is attractive since it can be quickly deployed and operated by batteries only.

We have observed that wireless networks typically operate in a three-dimensional real space because wireless communications must rely on signals traveling in the space. On the contrary, in traditional wireline networks, cables may interconnect hosts into (ideally) any kind of topology. Thus, we may say that wireline networks are not limited to humans' three-dimensional world. This interesting observation has led to many researchers working on *location-aware* MANETs. By location awareness, we mean that a host is capable of knowing its current physical location in the three-dimensional world. In traditional networks, hosts only have logical names (such as IP addresses) and do not know exactly what their current physical locations are.

GPS (Global Positioning System) is the most widely used tool to calculate a device's physical location. GPS is a worldwide, satellite-based radio navigation system. The GPS system consists of 24 satellites in six orbital planes. The satellites transmit navigation messages periodically. Each navigation message contains the satellite's orbit element, clock, and status. After receiving the navigation messages, a GPS receiver can determine its position and roaming velocity. To determine the receiver's longitude and latitude, we need at least three satellites. If we also want to determine the altitude, another satellite is needed. More satellites can increase the positioning accuracy. The positioning accuracy of GPS ranges in about a few tens of meters. GPS receivers can be used almost anywhere near the surface of the Earth.

By connecting to a GPS receiver, a mobile host will be able to know its current physical location. This can greatly help the performance of a MANET, and it is for this reason that many researchers have proposed to adopt GPS in MANETs. For example, mobile hosts in a MANET can avoid using naïve flooding to find routes; neighbors' or destinations' locations may be used as a guideline to find routing paths efficiently. Several works have addressed location-aware routing protocols for MANETs [Jain et al., 2001; Karp and Kung, 2000; Ko and Vaidya, 1998; Lin and Stojmenovic, 1999; Mauve and Widmer, 2001; Stojmenovic and Lin, 2001]. Proposals that partition the physical area into nonoverlapping zones to facilitate routing have also been proposed [Joa-Ng and Lu, 1999; Liao et al., 2001]. One interesting feature of such zone-based protocols is that a host can easily decide which zone it belongs to, and only one representative host needs to be active to collect routing-related information. The route search cost can be reduced significantly too since nonrepresentative hosts will not flood the route request packets.

The applications of location information are not limited to routing protocols. Navigation systems, which already incorporate GPS, can further combine MANET for group communications. Geocast, the goal of which is to deliver a message to a target area, is another potential service [Ko and Vaidya, 1999; Liao et al., 2000]. A computer-assisted tour guide system may take advantage of location information as well as the wireless communication capability of ad hoc networks.

The rest of the chapter is organized as follows. Section 18.2 discusses several location-assisted routing protocols. Section 18.3 reviews two zone-based routing protocols. Section 18.4 presents some location-aware applications. Conclusions are presented in Section 18.5.

18.2 Location-Assisted Routing Protocols

In this section, we review some routing protocols for MANETs that take advantage of location information of the hosts [Jain et al., 2001; Karp and Kung, 2000; Ko and Vaidya, 1998; Lin and Stojmenovic, 1999; Mauve and Widmer, 2001; Stojmenovic and Lin, 2001]. Different levels of knowledge are assumed to be known in advance. Generally, these works assume that a source host knows the destination's location or all its one-hop neighbors' locations. Some assume that each mobile host knows the locations of all its two-hop neighbors [Stojmenovic and Lin, 2001]. The location-aided routing (LAR) protocol also exploits roaming speeds of destination hosts [Ko and Vaidya, 1998]. Most of the routing protocols mentioned here do not need to go through the route discovery procedure before sending packets. Mobile hosts can forward packets directly to next hops according to local location information. Greedy approaches are widely adopted by using distance [Jain et al., 2001; Karp and Kung, 2000; Lin and Stojmenovic, 1999] or direction [Lin and Stojmenovic, 1999] as the metric to pick the next host to forward packets. However, greedy solutions may fall into the dilemma of running into a local maximum host (such as a dead end). When trying to avoid local maximum hosts, loops may occur. Solutions are proposed in [Stojmenovic and Lin, 2001].

18.2.1 LAR (Location-Aided Routing)

The *location-aided routing (LAR)* protocol [Ko and Vaidya, 1998] assumes that the source host (denoted as S) knows the recent location and roaming speed of the destination host (denoted as D). Suppose that S obtains D's location, denoted as (Xd, Yd), and speed, denoted as v, at time t_0 and that the current time is t_1. We can define the *expected zone* in which host D may be located at time t_1 (refer to the circle in Fig. 18.1). The radius of the expected zone is $R = v(t_1 - t_0)$.

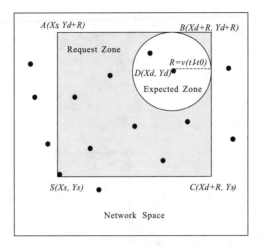

FIGURE 18.1 Request and expected zones in the LAR protocol.

From the expected zone, we can define the *request zone* to be the shaded rectangle as shown in Fig. 18.1 (surrounded by corners *S, A, B,* and *C*). The LAR protocol basically uses restricted flooding to discover routes. That is, only hosts in the request zone will help forward route-searching packets. Thus, the searching cost can be decreased. When *S* initiates the route-searching packet, it should include the coordinates of the request zone in the packet. A receiving host simply needs to compare its own location to the request zone to decide whether or not to rebroadcast the route-searching packet.

After *D* receives the route-searching packet, it sends a route reply packet to *S*. When *S* receives the reply, the route is established. If the route cannot be discovered in a suitable timeout period, *S* can initiate a new route discovery with an expanded request zone. The expanded request zone should be larger than the previous request zone. In the extreme case, it can be set as the entire network. Since the expanded request zone is larger, the probability of discovering a route is increased with a gradually increasing cost.

18.2.2 GPSR (Greedy Perimeter Stateless Routing)

The *greedy perimeter stateless routing (GPSR)* protocol [Karp and Kung, 2000] assumes that each mobile host knows all its neighbors' locations (with direct links). The location of the destination host is also assumed to be known in advance. Different from the LAR protocol, the GPSR protocol does not need to discover a route prior to sending a packet. A host can forward a received packet directly based on local information. Two forwarding methods are used in GPSR: *greedy forwarding* and *perimeter forwarding.*

Figure 18.2 shows an example of greedy forwarding. When host *S* needs to send a packet to host *D*, it picks from its neighbors one host that is closest to the destination host and then forwards the packet to it. In this example, host *A* is the closest one. After receiving the packet, host *A* follows the same greedy forwarding procedure to find the next hop. This is repeatedly used until host *D* or a local maximum host is reached.

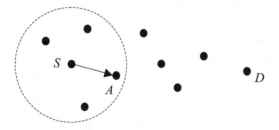

FIGURE 18.2 An example of greedy forwarding in the GPSR protocol.

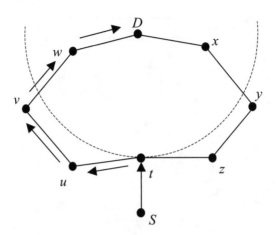

FIGURE 18.3 An example of perimeter forwarding in the GPSR protocol.

A *local maximum host* is one that finds no other hosts that are closer to D than itself. In the example in Fig.18.3, host t is a local maximum because all its neighbors are farther from D than itself. Therefore, the greedy forwarding method will not work here. When this happens, the perimeter forwarding method is used to forward the packet. The perimeter forwarding method works as follows. The local maximum host first "planarizes" the graph representing the network topology. A graph is said to be *planar* if no two edges cross. The graph may be transformed into a *relative neighborhood graph (RNG)* or a *Gabriel graph (GG)*. Both RNG and GG are planar graphs. After the graph is planarized, the local maximum host can forward the packet according to a *right-hand* rule to guide the packet along the perimeter of a plane counterclockwise. For example, in Fig. 18.3 at t, we can forward the packet along the perimeter of the plane *Dxyztuvw* counterclockwise. As the packet is forwarded to host w, we know that we are closer to D (as opposed to the location of host t). Then the greedy forwarding method can be applied again, and the packet will reach destination D. Overall, these two methods are used interchangeably until the destination is reached. The GPSR is a stateless routing protocol since it does not need to maintain any routing table.

18.2.3 GRA (Geographical Routing Algorithm)

The *geographical routing algorithm (GRA)* [Jain et al., 2001] is also derived based on location information. To send or forward a packet, a host first checks route entries in its routing table. If there is one, the packet is forwarded according to the entry. Otherwise, a greedy approach is taken, which will try to send the packet to the host closest to the destination. If the packet runs into a local maximum host, GRA will initiate a route discovery procedure to discover a route from the host to the destination. This is done by flooding. After the route reply comes back, the route entry will be stored in the host's routing table to reduce possible flooding in the future.

18.2.4 GEDIR (Geographic Distance Routing)

The *geographic distance routing (GEDIR)* protocol [Lin and Stojmenovic, 1999] also assumes that each host has the locations of its direct neighbors. Similar to GPSR, the GEDIR protocol also directly forwards packets to next hops without establishing routes in advance. There are two packet-forwarding policies: *distance approach* and *direction approach*. In the distance approach, the packet is forwarded to the neighbor whose distance is nearest to the destination. However, in the direction approach, the packet is forwarded to the neighbor whose direction is closest to the destination's direction. The latter can be formulated by the angle formed by the vector from the current host to the destination and to the next hop.

The distance approach may lead a packet to a local maximum host, while the direction approach may lead a packet into an endless loop. To resolve these problems, several variations are proposed, such as the f-GEDIR ("f" stands for flooding) and *c*-GEDIR (i.e., concurrently sending from the source to *c* hosts). These mechanisms are used to help the packet leave the local maximum host or the loop.

To further improve the performance of GEDIR, [Stojmenovic and Lin, 2001] recommends that hosts collect the locations of their two-hop neighbors. A host, on requiring to send/forward a packet, first picks a host (say *A*) from its two-hop neighbors whose distance (or direction) is nearest (or closest) to the destination. If host *A* is a one-hop neighbor, the packet is directly forwarded to *A*. Otherwise, the packet is forwarded to the host that is *A*'s one-hop neighbor. The protocol is called *2-hop GEDIR*. This protocol can also be combined with flooding to discover a route. Both GEDIR and 2-hop GEDIR have been proven to be loop free.

18.3 Zone-Based Routing Protocols

Below, we discuss two protocols that are derived based on partitioning the network space into nonoverlapping zones. In the zone-based routing protocol [Joa-Ng and Lu, 1999], the network space is partitioned into squares. Each host can decide which zone it belongs to according to its current location. The two-level hierarchy can decrease the route discovery cost. To send a packet, a host only needs to know the destination's zone ID and host ID. This result is a mobility-tolerant protocol proper for networks with changing topologies. Another protocol called GRID is proposed in [Liao et al., 2001]. The protocol enjoys a fully location-aware routing capability since it utilizes location information in route discovery, packet relay, and route maintenance. These two routing protocols are discussed in more detail below.

18.3.1 Zone-Based Routing Protocol

In the *zone-based routing protocol* [Joa-Ng and Lu, 1999], the geographic area of the MANET is divided into squares in advance. Since the partitioning plan is known in advance and each mobile host knows it own location, a host can easily compute its current zone. An example is shown in Fig. 18.4.

This protocol is a table-driven one. Therefore, each mobile host needs to spread its link state throughout the network from time to time. However, to save bandwidth, two types of link-state packets are sent in the two-level hierarchy: *intra-zone* and *inter-zone*. When there is any link state change inside a zone, the change will be propagated through a link-state protocol. However, the propagation is limited only within the zone itself. For example, in Fig. 18.4a, if link (*A*,*B*) is broken, only hosts in zone 2 need to be informed.

A *gateway* is a host that is connected to host(s) in other zone(s). The existence of gateways defines the connectivity between two zones. For example, the inter-zone connectivity in Fig. 18.4a is reflected in Fig. 18.4b. Only when there is any change of connectivity between two zones will the information be broadcasted throughout the whole network from zone to zone by gateways. Therefore, a local change of link states will not cause a global flooding unless it changes the inter-zone connectivity. For example, in Fig. 18.4a, if link (*A*,*C*) becomes broken, the inter-zone connectivity is unchanged. But when both links (*A*,*C*) and (*B*,*C*) are broken, the inter-zone connectivity will change, and the information needs to be propagated to other zones.

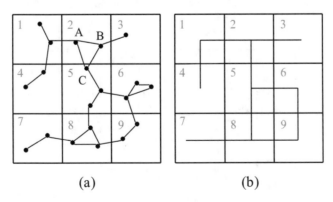

(a)   (b)

FIGURE 18.4 An example of the zone-based protocol: (a) intra-zone and (b) inter-zone connectivity.

By exchanging link states, each host maintains an inter-zone routing table and an intra-zone table. To discover a route, the source host first searches its intra-zone routing table. If an entry exists, data packets will be routed locally. Otherwise, a location request packet will be broadcast to all other zones through gateways to search the zone where the destination currently resides. Once the zone is discovered, data packets can be sent first through inter-zone routing. After reaching the destination's zone, data packets can be sent through intra-zone routing.

18.3.2 GRID

Observe that most routing protocols need to resolve three problems: route discovery, packet relay, and route maintenance. The GRID protocol [Liao et al., 2001] is claimed to be a *fully location-aware* one since it exploits location information in dealing with these three issues. The geographic area is partitioned into squares called *grids*. In each nonempty grid, one mobile host is elected as the *leader* of the grid. Routing is then performed in a grid-by-grid manner. Only the grid leaders have the responsibility to relay data packets. A routing example in GRID is shown in Fig. 18.5.

Location information is utilized in GRID in this way:

- *Route Discovery:* The concept of the request zone, similar to that in LAR [Ko and Vaidya, 1998], is used to confine the route-searching area. In addition, only grid leaders are responsible for forwarding route-searching packets. Note that nonleaders' route-searching packets are likely to be redundant since hosts in the same grid are close to each other (and so are their neighbors). Therefore, GRID can significantly save route-searching packets.

- *Packet Relay:* In GRID, a route is not denoted by host ID. Instead, it is denoted by a sequence of grid ID's. Each entry in a routing table records the next grid leading to a particular destination. For example, in Fig. 18.5, host *B* will record grid (3,1), instead of the MAC address of host *C*, as the next hop leading to host *D*. This provides an interesting "handoff" capability in the sense that if *C* roams away, the next leader (if any) in the same grid can take over and serve as the relay host without breaking the original route. Thus, GRID has been shown to be more resilient to host mobility.

- *Route Maintenance:* In GRID, routes are maintained by reelecting a new leader if the previous leader moves away. For example, in Fig. 18.5, when host *A* roams away, another host in grid (1,1) will be elected as the new leader to take over host *A*'s job of relaying packets. Therefore, the route is still alive. On the contrary, in most other protocols, such as DSR, AODV, LAR, and ZRP, once any intermediate host in a route roams away, the route is considered broken. Further, even if the source *S* roams into another grid, the route may still remain alive. For example, in Fig. 18.5, after *S* moves from grid (0,0) to grid (0,1), the route is still alive.

In each grid, hosts have to run a leader election protocol to maintain its leader. When a leader roams off its original grid, a "handoff" procedure needs to be executed to pass its routing table to the newly

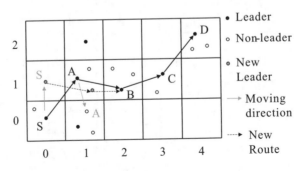

FIGURE 18.5 A routing example in the GRID protocol.

elected leader. In most other routing protocols, such a handover procedure is not possible. Thus, routes in GRID can survive for a longer lifetime. Therefore, GRID is less vulnerable than most other routing protocols to host mobility. In addition, the amount of control traffic is quite insensitive to host density. These merits make GRID quite scalable.

18.3.3 Comparison

The routing strategies and required information of the aforementioned routing protocols are compared and summarized in Table 18.1. Table 18.2 compares how location information is utilized in different stages of routing

18.4 Location-Aware Applications of MANET

Location information, when integrated into MANETs, may provide many potential services:

- *Navigation:* When location information and wireless communication capability are integrated into navigation systems, users will be able to talk to each other in an ad hoc manner. Quick wireless communication links can be established whenever needed. A user will be able to find out who is at what location. Location-dependent emergency rescue and law enforcement services would be possible.

- *Geocast:* The goal of geocast is to send messages to all hosts in a specific area. When urgent events (such as fires, traffic accidents, or natural disasters) occur in a specific area or we want to advertise some information to people in certain areas, geocasting can be a convenient way to achieve this goal.

TABLE 18.1 Comparison of Routing Protocols

Scheme	Routing Strategy	Required Information
LAR [Ko and Vaidya, 1998]	Discover route by flooding request packets in request zone	Destination's location and roaming speed
GPSR [Karp and Kung, 2000]	Greedy forwarding (distance-based) and perimeter forwarding	Destination's location and all neighbors' locations
GRA [Jain et al., 2001]	Greedy forwarding (distance-based) and flooding	Destination's location and some neighbors' locations
GEDIR [Lin and Stojmenovic, 1999]	Greedy forwarding (distance- or direction-based) and flooding	Destination's location and all neighbors' locations
Zone-Based [Joa-Ng and Lu, 1999]	Intra-zone: table-driven Inter-zone: zone-by-zone, table-driven	Intra-zone and inter-zone routing tables
GRID [Liao et al., 2001]	Intra-grid: direct transmission Inter-grid: grid-by-grid, on-demand	Destination's grid *ID*

TABLE 18.2 Comparison of Routing Protocols on How Location Information Is Used

Scheme	Route Discovery	Packet Relay	Route Maintenance
DSR, AODV	No	No	No
LAR	Yes	No	No
GPSR	Stateless	Yes	Stateless
GRA	Yes (on-demand)	Yes	Yes
GEDIR	Connectionless	Yes	Connectionless
Zone-Based	Yes (table-driven)	No	Yes (table-driven)
GRID	Yes (on-demand)	Yes	Yes

- *Tour guide:* Tour guide systems can provide location-dependent information to tourists (such as map, traffic, and site information). The effort needed to search for tourism information can be significantly reduced with the help of positioning.

Below we discuss three location-aware applications: Geocast, location-assisted broadcast, and location-assisted tour guide.

18.4.1 Geocast

Geocast is a location-based multicast. The goal of geocast is to send messages to all mobile hosts within a specified geographical region. Different from traditional multicast, the destination address is not a multicast IP, but instead a geographic area/coordinate. The first geocast work is in [Ko and Vaidya, 1999], where two different approaches are proposed. The first approach is similar to LAR [Ko and Vaidya, 1998]. The goal is to forward geocast packets to a region called the *geocast region*. It confines the propagation of geocast packets within a certain region called the *forwarding zone*, which is the smallest rectangle that includes the location of the sender and the geocast region. For example, as Fig. 18.6 shows, the rectangle *SABC* is the forwarding zone and the rectangle *OPQB* is the geocast region. After sender *S* broadcasts a geocast packet, host *I* will forward the packet because it is within the forwarding zone. However, host *J* will not relay the packet. In the second approach, a host decides whether it will forward the geocast packet or not according to its distance to the "center" of the geocast region. A host *X*, on receiving a geocast packet from *Y*, will forward the packet only if *X* is closer than *Y* to the center.

The GeoGRID [Liao et al., 2000] protocol is also for geocast. It is modified from the GRID protocol. As mentioned earlier, the GRID protocol divides the network area into several nonoverlapping squares called grids. Geocasting messages are sent in a grid-by-grid manner through grid leaders. However, in GeoGRID, no spanning tree or routing path needs to be established before geocasting. Instead, a connectionless mode is adopted. Two approaches are suggested to propagate geocast packets. The first approach is *flooding-based*. Every grid leader in the forwarding zone will forward the geocast packets. The second approach is *ticket-based*. Only the grid leader that holds a ticket will forward the geocast packet. The purpose of issuing tickets is to avoid blind flooding. The source needs to decide how many tickets will be issued. On their way to the geocast region, tickets may be split to different grids. The number of tickets issued may affect the arrival rate of geocast packets. The GeoGRID protocol can reduce network traffic and achieve a high data arrival rate.

Most of the previous works assume that geocasting protocols are operated in an obstacle-free area. In practice, obstacles blocking the way might be inevitable. To overcome obstacles, an *obstacle-free single-destination geocasting protocol (OFSGP)* is proposed in [Chang et al., undated]. Interestingly, the protocol also extends the definition of geocasting such that there may be more than one target geocast region.

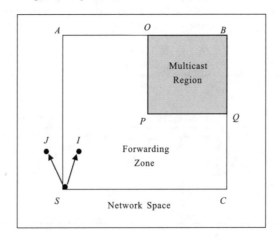

FIGURE 18.6 Geocast.

18.4.2 Location Services

The above reviewed location-aware routing protocols all rely on certain knowledge of the destination host's location. The definition of *location service* is to provide one or more hosts that constantly collect all hosts' locations. Other hosts can contact these hosts and request location-related information. This is similar to the location-tracking problem in personal communication systems. However, the difficulty is that the location service providers may also be members of the MANET and may roam around. In systems such as GSM, the home location registers (HLRs) cannot be mobile hosts. As a result, this problem is more challenging, and it is preferred that multiple location service providers exist to tolerate mobility. Here, we review several such solutions.

The *DREAM (distance routing effect algorithm for mobility)* framework [Basagni et al., 1998] proposes that each host maintain a position database that stores location information of each other host. An entry in the database may include a host's ID, location, and the time when this entry was created. Each host regularly floods packets to update its database. Two concepts called *temporal resolution* and *spatial resolution* are proposed to control the accuracy of location information. Since each host knows the other hosts' locations, this can be classified as an *all-to-all* approach [Mauve and Widmer, 2001].

The *quorum-based scheme* [Haas and Liang, 1999] intends to provide location service using the concept of quorum that is widely used in distributed database design. A number of hosts are designated as the location service providers. These hosts are partitioned into a number of quorum sets $Q_1, Q_2,..., Q_k$. The design of quorums should guarantee that for each $1 \leq i, j \leq k$,

$$Q_i \cap Q_j \neq \phi$$

When a host changes its location, it can pick any nearest quorum Q_i to update its location (based on any optimization criteria). When a host needs any other host's current location, it can query any nearest quorum Q_j. Since the intersection of Q_i and Q_j must be nonempty, the most up-to-date location information can be obtained.

Another distributed way to store location information is the *Homezone* mechanism [Giordano and Hamdi, 1999]. A host X can choose a position as its homezone by computing a globally known hashing function. Any host with a distance of R to the homezone point has responsibility to store X's current location. Host X, when changing location, should update its location (say, by geocasting) with all hosts in its homezone. In this way, the location database is distributed among all hosts, and the communication bottleneck problem can be relieved. When a host needs other hosts' locations, it simply queries their homezones by computing the hashing function. This solution is classified as an *all-for-some* approach [Mauve and Widmer, 2001].

18.4.3 Location-Assisted Broadcasting in MANET

Broadcasting is a common operation in any kind of network, including MANETs. It is shown in [Ni et al., 1999] that broadcasting by naïve flooding will cause many redundant rebroadcasts, collisions, and contentions. This phenomenon is called the *broadcast storm problem*. A location-based scheme is proposed in [Ni et al., 1999] to alleviate the broadcast storm problem. It is suggested that a host can decide whether or not it should rebroadcast a received broadcast message depending on its own and neighbors' locations. Before rebroadcasting the received packet, a host (say X) may have heard other neighbors rebroadcasting the same packet already. Host X can measure the size of the area that has not been covered by its neighbors' rebroadcasts. If the uncovered area is greater than a certain threshold (say 30%), host X will rebroadcast the packet. Otherwise, its rebroadcast will be prohibited to save bandwidth.

For example, in Fig. 18.7a, if hosts A, B, and C have already rebroadcast the broadcasting packet and this has been heard by host X, then X will prohibit its retransmission. However, in Fig. 18.7b, if only host A has rebroadcast the packet, then X's retransmission can still cover a substantial amount of area (the shaded area), and it is beneficial for X to rebroadcast. Based on this mechanism, the location-based broadcast is shown to be able to save considerable traffic while keeping the packet reachability ratio relatively high, thus improving the efficiency [Ni et al., 1999].

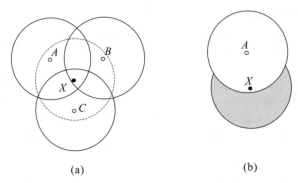

(a)   (b)

FIGURE 18.7 Examples of location-based broadcast: (a) All of host X's transmission area is covered by hosts A, B, and C. (b) Only part of host X's transmission area is covered by its neighbor A.

18.4.4 Location-Assisted Tour Guide

To demonstrate the application of location-aware ad hoc wireless networks, we have implemented a campus tour guide system at the National Chiao-Tung University. The system is targeted at serving tourists visiting the campus. The hardware consists of a Compaq iPAQ PDA, which is connected to a GPS and a wireless LAN card. The system is shown in Fig. 18.8.

The system is developed on WinCE operating system. The software consists of the following components:

- *Map component:* This consists of a map associated with some geographic information. On the background is the campus map. A user can point to any building, and text information associated with the building will be shown. The user's current location will also be shown on the map. Figure 18.9 shows the map component.
- *Positioning component:* The part communicates to the GPS through RS-232 to get the current location of the device. The information is fed to the map component so that a roaming user can see his/her moving path on the map.
- *Ad hoc networking:* A group of tourists who use the tour guide system is regarded as a mobile ad hoc network. The communication is through wireless LAN cards. IEEE 802.11b wireless LAN cards are used in the implementation. We have developed a short message system on which tourists can talk to each other by sending short messages. This component is shown in Fig. 18.10.

FIGURE 18.8 The campus tour guide system.

FIGURE 18.9 The map component.

FIGURE 18.10 The short message system.

- *Location database:* Each PDA will update its current location with other PDAs. As a result, each PDA is able to know all other PDAs' locations. Here we adopt a simple all-to-all approach to support the location service. These locations can be shown on the map component. A tourist is thus able to find where the other tourists are and identify a particular tourist's current location.

Our goal is to demonstrate how location information and wireless communication can facilitate highly mobile people, such as tourists. This tool can provide up-to-date location-dependent information to a tourist. Without knowing his/her current location, a tourist may find a self-guided trip to be a very boring one.

One difficulty that we experienced in the implementation is the lack of UDP function in the version of the development platform we used. As a result, TCP has to be used to simulate UDP. Using UDP is important since in a mobile environment, a device has to discover approaching and leaving neighbors from time to time. Thus, connectionless types of communication, such as UDP, are essential for the efficiency of such systems.

18.5 Conclusions

We have explained some location-aware routing protocols and applications of mobile ad hoc networks in this chapter. With the assistance of location information, routing protocols can be improved significantly. We have discussed several ways to apply location information in different stages of routing, including route discovery, packet relay, and route maintenance. Several location-aware routing protocols, including LAR, GPSR, GRA, GEDIR, zone-based routing, and GRID, have been discussed. Comparisons of these protocols are given. We have also presented several location-dependent applications and services, including geocast, location service, and location-based tour guide. We expect that more and more location-aware applications will appear in the future to facilitate human life.

Acknowledgments

Yu-Chee Tseng's research was funded by the Lee and MTI Center for Networking Research at NCTU, the Ministry of Education (contract number 89-H-FA07-1-4), and the National Science Council, Taiwan (contract number NSC90-2213-E009-154).

References

S. Basagni et al., A Distance Routing Effect Algorithm for Mobility (Dream), *Proc. 4th Annual ACM/IEEE Int'l Conf. Mobile Computing and Networking (MobiCom '98)*, Dallas, TX, 1998, pp. 76–84.

C.-Y. Chang, C.-T. Chang, and S.-C. Tu, Obstacle-Free Geocasting Protocols for Single/Multi-Destination Short Message Services in Ad Hoc Networks, ACM Wireless Networks.

S. Giordano and M. Hamdi, Mobility Management: The Virtual Home Region, Tech. Report, Oct. 1999.

Z.J. Haas and B. Liang, Ad hoc mobility management with uniform quorum systems, *IEEE/ACM Trans. on Networking*, 7, 228–240, 1999.

R. Jain, A. Puri, and R. Sengupta, Geographical Routing Using Partial Information for Wireless Ad Hoc Networks, *Personal Communications*, Feb. 2001, 48–57.

M. Joa-Ng and I.-T. Lu, A peer-to-peer zone-based two-level link state routing for mobile ad hoc networks, *IEEE Journal on Selected Areas in Communications*, 17, 1415–1425, 1999.

B. Karp and H.T. Kung, GPSR: Greedy Perimeter Stateless Routing for Wireless Networks, *MobiCom*, Boston, MA, 2000, pp. 243–254.

Y.-B. Ko and N.H. Vaidya, Location-Aided Routing (LAR) in Mobile Ad Hoc Networks, *MobiCom*, Dallas, TX, 1998, pp. 67–75.

Y.-B. Ko and N.H. Vaidya, Geocasting in Mobile Ad Hoc Networks: Location-Based Multicast Algorithms, *IEEE Workshop on Mobile Computing Systems and Applications*, New Orleans, LA, Feb. 1999.

W.-H. Liao, Y.-C. Tseng, K.-L. Lo, and J.-P. Sheu, GeoGRID: a geocasting protocol for mobile ad hoc networks based on GRID, *Journal of Internet Technology*, 1, 23–32, 2000.

W.-H. Liao, Y.-C. Tseng, and J.-P. Sheu, GRID: a fully location-aware routing protocol for mobile ad hoc networks, *Telecommunication Systems*, 18, 61–84, 2001.

X. Lin and I. Stojmenovic, GEDIR: Loop-Free Location Based Routing in Wireless Networks, *Proc. IASTED International Conference on Parallel and Distributed Computing and Systems*, 1999, pp. 1025–1028.

M. Mauve and J. Widmer, A Survey on Position-Based Routing in Mobile Ad Hoc Networks, *IEEE Networks*, Nov./Dec. 2001, pp. 30–39.

S.Y. Ni, Y.C. Tseng, Y.S. Chen, and J. P. Sheu, The Broadcast Storm Problem in a Mobile Ad Hoc Network, *Proceedings of the Fifth Annual ACM/IEEE International Conference on Mobile Computing and Networking*, Seattle, WA, Aug. 1999, pp. 151–162.

I. Stojmenovic and X. Lin, Loop-free hybrid single-path/flooding routing algorithms with guaranteed delivery for wireless network, *IEEE Transactions on Parallel and Distributed Systems*, 12, 2001.

Y.-C. Tseng, S.-L. Wu, W.-H. Liao, and C.-M. Chao, Location Awareness in Ad Hoc Wireless Mobile Networks, *IEEE Computer*, 34, 46–52, 2001.

19

Mobility over Transport Control Protocol/ Internet Protocol (TCP/IP)

José Ferreira de Rezende
Federal University of Rio de Janeiro

Michele Mara de Araújo Espíndula Lima
State University of Paraná West

Nelson Luis Saldanha da Fonseca
State University of Campinas

Abstract

Advances in wireless communications have enabled access to the Internet for mobile users, but the Transport Control Protocol/Internet Protocol (TCP/IP) protocol stack of the Internet, designed to be independent of the link layer technology, involves mechanisms that lead to poor performance when used for mobile networks. This chapter identifies some of the important issues of mobility over TCP/IP and outlines the major proposals for overcoming them.

19.1 Introduction

The availability of wireless communication devices with increased processing capabilities allows the connectivity of mobile users to the global Internet. This will certainly change the way we communicate, but in mobile computing and communication many challenges are yet to be met. The major hindrances are related to mobility management and the poor performance of legacy protocols over wireless networks; these problems restrict large-scale deployment of such technologies.

The TCP/IP architecture employed by the Internet was designed for fixed node networks. Various changes have been proposed to overcome the difficulties of using it for mobile networks. In this chapter, various proposals for dealing with mobility in the Internet Protocol (IP) supported by underlying layers are presented. These new architectures aim at guaranteeing transparency without increasing signaling load and performance degradation due to handoffs of user access point. These solutions involve scalability issues raised by the huge number of mobile nodes expected in cellular networks.

The Transport Control Protocol (TCP) was developed to operate regardless of the link layer technologies of a network. It assumes that packet losses are due only to congestion, which is true for wired networks but not for wireless networks, where losses tend to be due to bit errors. The proposals suggested to overcome these difficulties differ largely in their awareness of the existence of wireless links. These proposals will be presented here.

19.2 Architectures for IP Mobility

The IP protocol allows the interconnection of heterogeneous networks and is used to form the so-called Internet, in which a vast set of client/server and peer-to-peer computing applications exist, thus enabling a global system of interpersonal communication, e-commerce, and source information. IP, however, was designed for traffic exchange between fixed nodes. The development of wireless devices with built-in high-speed packet radios makes wireless access to the Internet possible but requires modifications of the IP protocol to be feasible. To meet these demands, the Internet Engineering Task Force (IETF) proposed the Mobile IP [1], aimed at providing continuous access to the Internet without disruption in computing activities due to changes in the point of attachment of the user. This "mobile computing" is different from the "portable computing" in use today.

The main features of mobile computing involve application transparency and seamless handoffs [2]. The former implies the continuity of the use of present-day technology in a future mobile environment, since mobile users should not have to buy new versions of current applications for mobile environments. Seamless handoff, on the other hand, implies that no disruption in services should occur during handoff, thus leading to the enhancement of mobility transparency. Moreover, high-speed wireless access requires smaller cells, due to the increase in handoff frequency.

The preferred solution for mobility involves network layers as these provide more benefits. In recent years, various proposals for enhancing Mobile IP have been submitted to IETF working groups. In this section, these proposals for mobile architectures are presented and discussed. Most of them were designed for micromobility scenarios where scalability and security issues are important. Moreover, most were designed for the third generation cellular networks in which service disruption during handoff is a major concern.

19.2.1 Mobile IP (MIP)

The original Mobile IP (MIP) architecture was specified in the Request for Comments — RFC 3220 [1]. This architecture treats mobility at the network layer, thus avoiding changes in existing nonmobile hosts and applications. In MIP, mobility is thus transparent to layers higher than that of the network layer.

The mobility support provided by the IP layer can be achieved by modifying the path followed by datagrams addressed to the mobile node so that they will arrive at the actual point of attachment of the user. This routing process is based on the IP address destination incorporated in the IP datagram itself. IP addresses are hierarchical, with topological numbers assigned to nodes according to their location, so that nodes located in a given IP subnetwork will have the same higher-order bits (*network number*). Normal IP routing tables store next-hop information for each network as a whole rather than for an end-system, thus leading to a substantial gain in routing scalability. In order to preserve this feature, each time a mobile node is attached to a new point, it receives a new network number and, hence, a new IP address. On the other hand, it is highly desirable that mobile users keep their IP addresses. Such conflicting requirements are addressed by MIP.

In MIP, mobile nodes have two addresses: the *home address* and the *care-of address*. The *home address* is a static address that the mobile user maintains regardless of location; it is the IP subnetwork to which the mobile user (node) originally belongs. The *home address* identifies the mobile node so that it continually receives data in its *home network*. The *care-of address* is an address temporarily assigned to a node to reflect a new point of attachment. This address guarantees the correct delivery of packets to a mobile node when it is located in a *foreign network*, outside its original network. The key idea of Mobile IP is thus to allow a node to change its point of attachment while retaining the same IP address [2].

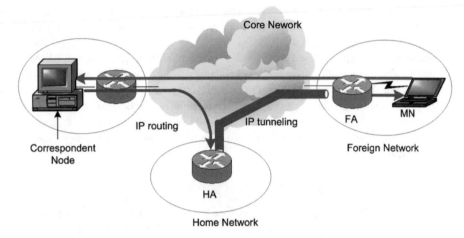

FIGURE 19.1 Mobile IP architecture.

In MIP architecture (Fig. 19.1), two routers or mobility agents, the *home agent* (HA) and the *foreign agent* (FA), cooperate to deliver packets to the mobile node. These agents perform three related functions: agent discovery, registration, and tunneling [2]. In agent discovery, performed when the mobile node moves to a *foreign network* or returns to its *home network*, the mobile node receives the *care-of address* in periodic advertisement messages that are broadcast by the FA as ICMP (Internet Control Message Protocol) messages into the network where it is attached. Information about special features, such as encapsulation and header compression algorithms available, is also provided by the agent in these messages. The mobile node registers with the HA in a way that prevents malicious users from making false registrations to redirect the traffic destined to reach it.

Whenever a packet arrives at the *home address* of the mobile node after a successful registration, this HA sends the packet via a tunnel to the corresponding *care-of address,* which does not necessarily need to be collocated with the mobile node. If the mobile node is not collocated with its *care-of address,* the FA receives the packet, deencapsulates it, and forwards it to the mobile node. Otherwise the mobile node receives the encapsulated packet. The routing of packets sent by the mobile node is performed normally, even if the receiving node is located in a *foreign network*.

The process just described can be optimized if the sender is an MIP node. The mobile node sends the sender of a packet a *binding update,* which contains its *care-of address.* Further packets are sent directly to this address instead of to the *home address.* This functionality, called route optimization, however, raises security issues related to the authenticity of updates, and the HA must be responsible for providing authenticated binding updates to the corresponding nodes.

One problem that commonly arises in mobile environments is a change in point of attachment. The period of time between the moment a mobile node detaches from its old FA and that in which its HA is informed about the new FA can be quite long; consequently, many packets directed to the old FA may be lost. To limit such losses, the transition process of *handoff* must be as smooth as possible. One way of effecting this is by route optimization, which solves the problem by allowing cooperating FAs to exchange and maintain bindings for their former mobile visitors.

Whenever a mobile node is located in a foreign network, MIP faces several problems generally related to a lack of changes in the routing infrastructure, thus leading to the need for indirect routing, or triangulation, through the HA. This triangulation may cause a significant increase in end-to-end transmission delay, however, especially when the mobile node receives data originating from a foreign network in which the mobile agent is currently located. Both triangulation and IP tunneling complicate the integration with IP QoS (Quality of Service) models such as Integrated Services. Moreover, MIP relies on frequent reports to the HA that can create an unreasonably high signaling load if the mobile node moves frequently. Furthermore, unless all routers have HA and FA functionality, the implementation of MIP is somewhat restricted. Nevertheless, there is a trend to provide macromobility for mobile IP in future wireless networks.

Several solutions have been proposed to provide IP micromobility to tackle the problems caused by frequent handoffs. The main goal of these proposals is to provide fast, seamless local handoffs for mobile nodes, thus leading to shorter delays, decreases in packet losses, and lower signaling loads.

Another relevant issue in mobility management is the tracking of the location of mobile nodes, which is important in some applications. As the number of mobile users grows, this task can consume power and increase signaling. The maintenance of information when a node is idle (_paging_) is an important functionality; however, MIP does not support it.

19.2.2 Hierarchical Mobile IP (HMIP)

Hierarchical Mobile IP (HMIP) [3] architecture, proposed by the research centers of Ericsson and Nokia, uses a hierarchy of FAs to handle Mobile IP registration locally. In MIP, a mobile node registers with its HA each time it changes its _care-of address_. In HMIP, in contrast, registrations are local to the regional domain, thus reducing signaling load and the delay incurred in registration when the foreign network is located a long way from the home network [3].

Registration messages establish tunnels between neighboring FAs along the path from the mobile host to a gateway FA (GFA). Therefore, when a node sends packets to a mobile node, these packets traverse as many tunnels as there are intermediate nodes between the GFA and FA located at the access point where the mobile node is connected. In this model, the _care-of address_ does not change when the mobile node attachment changes from one FA to another, as long as both are under the same GFA. In fact, the GFA address is registered at the HA as the _care-of address_ for the mobile node. Only when changing GFAs must a mobile node perform a home registration. The proposed regional registration protocols support one level of FA hierarchy below the GFA and may be utilized to support several levels of hierarchy, as presented in [3].

An extension to HMIP has been proposed to support regional paging [4]. An idle node can save power since it does not need to perform subsequent registrations when it moves through the IP subnetworks of an area. When a packet addressed to an idle mobile node is received by a FA in areas with such paging, the FA pages the mobile node to reestablish a path to the current point of attachment.

19.2.3 HAWAII

The Handoff-Aware Wireless Access Internet Infrastructure (HAWAII) (Fig. 19.2) was proposed to improve the quality of service and reduce the inefficiency of MIP [5]. In HAWAII, user mobility is restricted to a given administrative domain. Mobile nodes implement MIP as before, whereas host-based forwarding entries are installed on selected routers, thus creating routes to the mobile nodes in the domain. These nodes support intra-domain mobility while using traditional MIP for inter-domain mobility. The nodes thus maintain their _home address_ without any triangulation or IP tunneling.

HAWAII segregates the network into a hierarchy of domains, as in the autonomous system hierarchy of the Internet. Each domain owns a root gateway called the _domain root router_, which takes on the role of the HAs. Each node has an IP address and a home domain. As long as a node moves within its home domain, the mobile node retains its IP address, and packets destined to the mobile node are routed to the home domain root router by using the IP subnet address of the domain. The packets received are then forwarded to the node by using special dynamically established paths. The establishment of such paths is triggered by the mobile node via the usual MIP registration messages whenever it moves between two access points (APs). In this way, APs behave as different FAs. Within a home domain, these messages create direct routing entries at the intermediate nodes they cross.

When a node moves to a foreign domain, the usual MIP procedures are used. The foreign domain root router is now the FA and is responsible for assigning a care-of address and forwarding packets

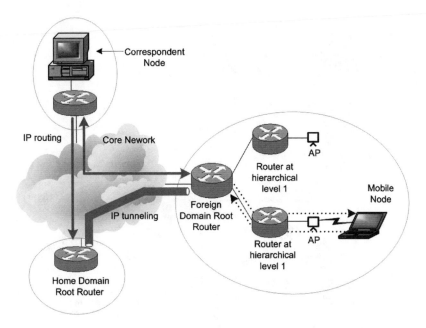

FIGURE 19.2   HAWAII architecture.

to or from the mobile node. As long as the mobile node is moving within this foreign domain, it retains its care-of address unchanged. The connectivity is maintained using dynamically established paths defined by a protocol operating in the access network. The path state is maintained in the routers as a soft state, which increases the robustness of the protocol under conditions of failure of the router or of a link. When the mobile host moves from one base station to another, four path setup schemes are available for avoiding data disruption during handoff. As in MIP, HAWAII nodes are IP routers. HAWAII supports seamless mobility, passive connectivity, and paging.

19.2.4 Cellular IP (CIP)

Cellular IP (CIP) architecture (Fig. 19.3) was proposed by Columbia University and the Ericsson Research Center [6]. It employs two different handoff techniques, as well as supporting paging. In contrast to HAWAII and HMIP, however, Cellular IP uses regular data packets transmitted by the mobile nodes to maintain reverse path routes. CIP nodes snoop packets originated at mobile nodes in order to maintain a distributed hop-by-hop location database used to route packets to the mobile nodes. The handoff schemes are hard handoff and semisoft handoff. The former is a simple operation that provides a tradeoff between packet losses and minimal signaling. The latter, however, makes use of layer-2 information or access point signal strength to predict handoff, thus permitting earlier triggering of layer-3 procedures, and consequently minimizing packet losses.

Each CIP domain is composed of a number of CIP nodes structured in a tree topology, with a MIP gateway as the root node. These nodes can route IP packets inside the CIP network, as well as communicate with mobile nodes through the wireless interface.

The CIP nodes maintain both routing and paging caches. The latter are those records that are created by mobile nodes that do not send or receive many packets, and are maintained by paging-update packets sent to the nearest access point each time the mobile node moves. Routing caches, on the other hand, are used to locate roaming mobile nodes. These caches are updated by the information in IP packets being transmitted by that node. At these CIP nodes, a chain of temporary cached records is created to provide information on the downlink path for packets destined to that node. Whenever a packet destined

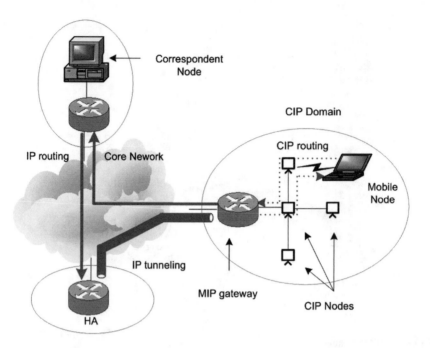

FIGURE 19.3   Cellular IP architecture.

to a mobile node arrives at a CIP node, it is sent to all interfaces mapped in the routing cache, but these mappings must be refreshed periodically by the node. Otherwise, they expire and are deleted.

When a mobile node approaches a new access point in a CIP domain, it redirects all outgoing packets from the old access point to the new access point, thus updating the routing caches all the way up to the gateway. All packets destined to that mobile node will then be forwarded to both access points for a time interval equal to the routing cache timeout. Once the old path expires, packets destined to the mobile node are forwarded only to the new path. Therefore, even if the node has no packets to send during handover, it must generate route-update messages in order to allow correct updating of the routing caches. Normal MIP procedures are used for macromobility among CIP domains.

It should be noted that in CIP all packets generated within a CIP domain must be routed through the gateway, even if the destination is adjacent to the source.

19.2.5 TIMIP

Terminal Independent for Mobile IP (TIMIP) architecture uses a completely different approach to support IP mobility [7]. It avoids the need for special IP layer signaling between the mobile node and the AP by adding layer-2 handoff mechanisms at the AP to the IP layer. In this way, mobile nodes are not required to use a mobility-aware protocol stack, and nodes with legacy IP protocol stacks can take advantage of mobility. As with HAWAII and CIP, TIMIP relies on MIP to support macromobility.

19.3 TCP and Mobility

TCP, the most popular Internet transport layer protocol, was developed to offer reliable services in networks where congestion is the main cause of packet loss. It provides mechanisms that enable connections to share a link fairly by dynamically estimating the available bandwidth and avoiding collapse due to congestion.

TCP was designed to operate regardless of the link layer technology employed and does not take into consideration any specific link technology characteristics. TCP performance, however, can degrade severely in the presence of wireless links and mobile nodes.

In order to avoid TCP-caused bottlenecks, certain modifications have been proposed to improve TCP's performance for mobile networks. Prior to the presentation of these specific solutions for TCP performance problems, however, the congestion control mechanism of TCP and its impact on performance are discussed.

19.3.1 TCP Reno Congestion Control Mechanism

The congestion control mechanism of TCP Reno, the most popular TCP implementation, is composed of four algorithms: Slow Start, Congestion Avoidance, Fast Retransmit, and Fast Recovery [8]. The former two algorithms are used by TCP senders to control the amount of data injected into the network, while the latter two are used to recover from packet losses quickly without the need for Retransmission TimeOuts (RTOs).

At the beginning of a transmission, a TCP sender uses the Slow Start algorithm to probe the capacity available, so that congestion can be avoided by not sending large bursts of data that the network cannot absorb. While executing Slow Start, a TCP sender increments its congestion window ($cwnd$) for each acknowledgment (ACK) received, doubling the number of packets that can be sent in a window. When the congestion window exceeds a certain threshold value ($ssthresh$), the execution of Slow Start ends, and the execution of the Congestion Avoidance algorithm begins and continues its execution until congestion is observed by the detection of a packet loss, signaled by the occurrence of a timeout or the receipt of duplicate ACKs (DupACKs). During the execution of this algorithm, the congestion window $cwnd$ has a linear growth, instead of the exponential growth during Slow Start phase.

The receiver will signal the receipt of an out-of order packet by immediately sending a duplicate ACK. The sender interprets the receipt of three of these DupACKs as an indication that a packet was lost and uses the Fast Retransmit algorithm to reduce the congestion window by half and to retransmit the lost packet immediately, without waiting for the retransmission timer to expire. After the retransmission of such a packet, the Fast Recovery algorithm is activated to control the transmission of new packets until a nonduplicate ACK is received. It thus interprets the receipt of duplicate ACKs not only as an indication of the loss of a packet but also as the receipt of some other packets. Therefore, the sender will continue sending data to the receiver, but it will do so at a slower transmission rate. If the packet loss was detected by the occurrence of a timeout, the congestion window is reduced to its initial value, the lost packet is retransmitted, and the Slow Start algorithm is started to increase the congestion window value.

Since the losses in wireless networks are mostly due to transmission errors and handoff, the reduction of the TCP transmission rate upon detection of a packet loss leads to unnecessary degradation of the performance of these networks.

Four main problems result from the use of TCP in wireless networks. The first of these is the waste of available capacity during the Slow Start phase. The problem is especially acute in long-delay networks, where transmissions are short in comparison to the delay * bandwidth product, which defines the amount of data that a protocol can send while acknowledgments are still pending.

The second problem involves the high rate of errors, typical of the radio and infrared wave transmission used in these networks. The high bit error rate (BER) values cause packets to be corrupted, resulting in their discard. TCP, however, interprets such losses as congestion, consequently reducing the throughput values [9].

The third problem results from the disconnections that are so common in mobile networks, whether due to signal blocking by physical obstacles or due to the movement of the mobile node. Whenever a mobile node moves, the handoff leads to a brief disconnection, resulting in lost packets and ACKs, thus forcing the TCP sender to decrease its congestion [9].

The final problem is a consequence of the third, since frequent disconnections lead to serial timeouts, which result from numerous consecutive retransmissions of the same packet. Since TCP uses an exponential back-off mechanism to retransmit packets, serial timeouts can lead to the end of a connection or inactivity, even when the mobile host is reconnected [9].

19.3.2 TCP Extensions for Mobile Networks

Enhancements of TCP for mobile networks fall into three categories: those incorporated into TCP itself, those that can be applied to link-layer protocols, and those involving application-layer protocols [10]. In this section, only the proposals that fall into the first category are presented, and they vary basically in terms of the awareness of mobility [11].

19.3.2.1 Mobile Unaware TCP

In this approach, the TCP at fixed hosts remains unmodified. The basic idea is that local problems must be solved locally, whereas TCP must be generic and independent of link technologies [11].

Connection Segmentation

Connection Segmentation avoids modifications to the TCP at the fixed host by treating the TCP connection between the mobile host and the fixed host as two concatenated connections, one between the mobile host (MH) and an intermediate node (base station), and another between the base station and the fixed host (FH).

The connection between the base station and the fixed host is a standard TCP connection, whereas that between the base station and the mobile host is a modified version of TCP that is aware of the presence of a wireless link. Figure 19.4 illustrates a segmented connection between a mobile host and a fixed host.

Two proposals for segmented connections will be presented here.

Indirect TCP — Indirect TCP (I-TCP) is based on the Indirect Protocol Model of Badrinath and Bakre [12]. The existence of two separate connections makes it possible to distinguish the functionality of flow and congestion control on wireless links from that on fixed links, thus allowing the mobile support router (MSR) to manage the communication overhead for a MH without interfering in the communication with the FH.

Figure 19.5 presents the operation of I-TCP. When a MH wishes to communicate with a FH, a request is sent to the current MSR, which opens a standard TCP connection with the FH on behalf of the MH. The communication to the MSR, however, uses a modified version of TCP, specially tuned for wireless networks, which is aware of mobile nodes. If the MH moves, the state of the two sockets associated with the connection at MSR-1 is handed over to the new MSR. Since neither of the connection endpoints changes after a move, it is not necessary to reestablish the connection at the new MSR, thus ensuring transparency to the fixed host [12].

Since in I-TCP no need to change the fixed network exists, the protocol is left unchanged, and all current optimizations to TCP still work. Moreover, the mobile host communication overhead is managed by the MSR, so the common transmission errors of the wireless link are not propagated, and the MSR can use more aggressive strategies to recover from errors.

The disadvantage of this approach, however, is that the end-to-end semantic of TCP is lost. This means that the receipt of an ACK by the FH does not indicate that the MH has actually received a packet. Moreover, latency may increase due to the buffering of data at the MSR and the forwarding of data to a new MSR during a handoff. Furthermore, in a mobile environment subject to frequent disconnections, I-TCP only localizes the problem but does not solve it [9].

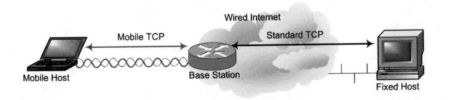

FIGURE 19.4 Segmented TCP connection.

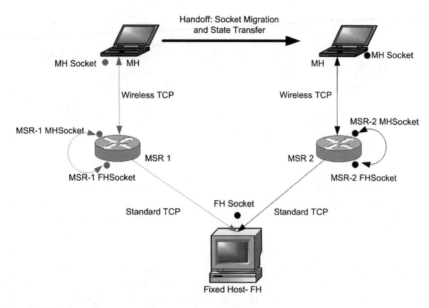

FIGURE 19.5 I-TCP connection setup and connection transfer.

M-TCP — M-TCP proposes a TCP extension for modern cellular networks, which can support multimedia services over high bandwidth channels. This extension was designed to work well in the presence of frequent disconnections and over low bit-rate wireless links subject to dynamic changing of bandwidths.

This architecture can be viewed as a three-level hierarchy as in Fig. 19.6. In the lowest level are the mobile hosts (MH), which communicate with the mobile support station (MSS) nodes in each cell. Each

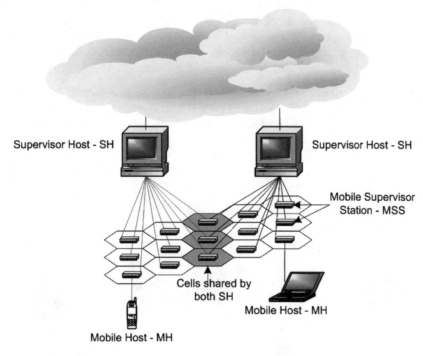

FIGURE 19.6 M-TCP architecture.

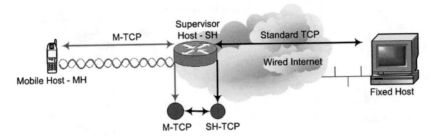

FIGURE 19.7 Setting up a TCP connection.

group of MSS nodes is controlled by a supervisor host (SH), which is connected to the wired network and handles most of the routing and other protocol details. If a MH moves from one cell to another, there is no need for the MSSs involved to perform state transfer, as long as they are controlled by the same SF [9].

The TCP at the sender is unchanged, although that at the SH does use a modified version of the TCP called SH-TCP. The communication between the MH and the SH uses M-TCP. Figure 19.7 shows how the TCP connection is split.

When the TCP client at the SH receives data from the TCP sender, this is passed to the M-TCP client for delivery of the data to the MH, but no ACK is sent to the sender until an ACK from the wireless host is received. If the size of the advertised windows at SH-TCP is greater than the number of ACKs received from the wireless host, then the SH-TCP acknowledges the same number of bytes less one and delays the sending of ACK for the last byte. If, after this, the MH disconnects, the M-TCP client assumes that the MH has been temporarily disconnected and sends an indication of this to the SH-TCP, which then advertises a window size of zero, signalling that the sender should move to a persistent state, in which all timers are frozen, i.e., TCP does not back off its timers.

When a MH reestablishes the connection, it notifies the SH, which in turn notifies the M-TCP client, which passes the information on to the SH-TCP. This host now sends a duplicate ACK to the sender, which will reopen the receiving window. This approach provides a solution to the problem of frequent disconnections by maintaining the end-to-end semantic of TCP and not using a buffer for data forwarding. Nonetheless, it propagates wireless losses into the fixed network, and the SH complexity is high [11].

TCP Snooping

TCP Snooping is a TCP extension that requires modification of the base station (BS) as well as of TCP software of the mobile host, although it preserves the end-to-end semantics of TCP. The modification at the BS involves the inclusion of a snoop module, which serves as a cache for packets at the base station, which were sent by the fixed host (FH). The snoop module keeps track of all the acknowledgments sent from the MH and uses this information to perform local retransmission back to it. For this, it uses a link level retransmission mechanism to overcome the problems caused by high bit error rates [13]. Figure 19.8 illustrates the behavior of a snoop module.

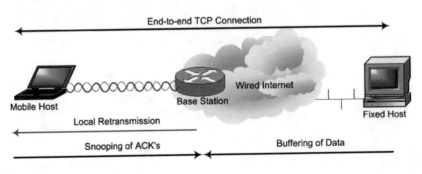

FIGURE 19.8 Snoop module behavior.

When a FH sends a packet to the MH, the snoop module in the BS adds a copy to its cache and forwards the packet to the MH. The snoop module also keeps track of all the acknowledgments sent by the MH. When the BS receives a duplicate ACK (DupACK) with sequence number P from the MH, it retransmits the packet containing this sequence number, provided a copy of that packet is cached [13], thus avoiding the trigger of the Fast Retransmit algorithm by the sender.

When a packet sent by the MH to the FH is lost, however, there is no way for the mobile host to know whether this packet was lost in the wireless link or elsewhere in the network. To eliminate this uncertainty, the BS keeps track of all packets that are lost and generates negative acknowledgments (NACKs) to the MH, which uses these NACKs to retransmit packets that were lost [13]. The support for a TCP selective acknowledgment scheme at the fixed host is required to inform the BS which packets were lost, so that communication from the mobile host to the fixed host can be enhanced. One such scheme is presented in Section 19.4.3.

The BS can also identify groups of BSs to which the MH can move and broadcasts packets sent to that MH by all of these BSs. Thus, when the MH moves to a new BS, the overhead of state transfer during a handoff can be minimized. Movement to a BS outside of this group, however, will require some time to build up its cache [11].

In a slow wireless link, the round trip time may be long, which can cause the TCP sender to enter into timeout while a retransmission is still on the wireless link. Moreover, if the MH host is subject to frequent disconnections, the snoop module may not be able to prevent the TCP sender from executing Fast Retransmit. In such environments, Snoop-TCP does not perform well.

A final problem is that Snoop-TCP will be useless if encryption is used because it needs to be able to verify the packet header.

Delayed Duplicate ACKs

The Delayed Duplicate ACKs (Delayed DupACKs) scheme is an attempt to approximate the behavior of Snoop-TCP without the need to make changes in the TCP. The base station, however, must implement a mechanism to perform link level retransmissions, although it does not need to look at the TCP header [14].

Delayed DupACKs tries to reduce the interference between TCP retransmissions and link level retransmissions by implementing a link level retransmission scheme for packets that are lost on the wireless link. The difference between Delayed DupACKs and Snoop-TCP is that the former uses link level acknowledgments to trigger link level retransmissions instead of DupACKs, thus reducing the interference between TCP retransmissions and link level retransmissions by delaying the third and subsequent DupACKs for a certain period of time instead of discarding these acknowledgments [14].

Whenever a receiver receives packets out of order, it must send DupACKs to signal this situation to the TCP sender. But when this transmission is delayed, there is time for the next-in-order packet to be received so that the delayed DupACKs are not sent. After this delay time, the DupACKs are released normally. Since Delayed DupACKs simulates Snoop-TCP behavior, it presents the same disadvantages.

19.3.2.2 Mobile Aware TCP

In this approach, the TCP sender recognizes the existence of both the wireless link and the mobile host; thus, it is able to deal with the problems characteristic of mobile networks, therefore improving its performance.

This section presents and discusses some of these mobile aware TCP proposals. First, three proposals that try to distinguish losses due to congestion from those due to bit errors are discussed. Then, the fast retransmit proposal is presented, and finally, the transmission and timeout freezing proposal is introduced.

Distinguishing Congestion Losses from Those Due to Bit Error

If the TCP recognizes the existence of a wireless link, it is able to infer the cause of a packet loss and to trigger the congestion control mechanism only when a loss is actually due to congestion. This will improve the performance of TCP. Three proposals designed to discriminate congestion losses from those due to bit errors are presented. The first identifies the cause of packet loss using end-to-end bandwidth estimation. The second uses a heuristic based on packet inter-arrival times at the receiver, with this information used by the receiver to inform the sender of the cause of packet loss. The third uses two connections to distinguish the cause of packet loss, one with the mobile host, and the other with the base station. The latter, the control connection, estimates congestion on the wired network.

Identifying Loss Type by Using End-to-End Bandwidth Estimation — TCP Westwood (TCPW) is an extension used in the sender side TCP to improve the performance of the TCP Reno in a wireless network by using end-to-end bandwidth estimation to identify the cause of packet losses [15]. The main idea is to take advantage of the information available in acknowledgment packets to estimate end-to-end bandwidth. This estimation is the result of the average of the measurements of the rates of arrival of ACKs at the sender. TCPW uses this bandwidth estimate (BWE) to set the congestion window (*cwnd*) and the slow start threshold (*ssthresh*) after a packet loss. When a loss occurs, TCPW adjusts the congestion window by considering the available network capacity. In other words, TCPW selects values of *cwnd* and *ssthresh* consistent with the effective bandwidth used by the connection when a loss was experienced, rather than blindly halving the congestion window as in TCP Reno. This modified mechanism for loss recovery is called "Faster Recovery."

TCPW maintains the end-to-end semantics of TCP and requires no modifications at the TCP receiver. Moreover, it uses only the feedback that comes from the network to estimate the bandwidth. It neither needs to inspect nor intercept TCP packets at any intermediate node, nor does it need to probe the network.

Since TCPW does not address handoffs or black periods, however, its performance degrades as the loss bit error rate (BER) increases as in TCP New Reno (a TCP Reno enhancement that solves the problem of recovering from lost packets when multiple losses occur in the same window of data presented by TCP Reno).

Identifying Loss Types by Using Receiver Inter-arrival Times — In TCP-aware, it is assumed that the TCP sender is incorporated in a wired network, with only the last hop involving a mobile host. Moreover, this wireless link is the bottleneck for the connection, and the sender performs bulk data transfer.

The central idea of the TCP-aware is explicit in the fact that the TCP receiver knows exactly which packets were lost. This mechanism uses the inter-arrival time between packets at the receiver to distinguish between losses due to congestion and those due to link errors. The scheme assumes that a certain number of packets has been lost due to bit errors if the time between the arrival of an out-of-order packet and the time of arrival of the last in order packet is greater than the value obtained by multiplying the minimum inter-arrival time observed in the connection by the number of packets with bit errors plus one and if it is also less than the minimum inter-arrival time multiplied by the number of packets lost plus two [16].

This heuristic is more effective if the bandwidth available on the wireless link is much narrower than that available on the wired network and if the overall rate of loss is small [16].

Identifying Loss Type by Making Two Connections — The assumption underlying this proposal is that the source host is connected to a base station by a wired network, while the mobile host is only a single hop away from the base station. Moreover, the mobile host receives all its packets from the base station. Therefore, any packets lost between the source host and the base station can be assumed to be lost due to congestion, whereas those lost in the wireless link will be due to link errors [17].

The key idea to distinguish between losses is to create two connections when the sender wants to communicate with the mobile host. One connection is with the base station and the other is with the mobile host. The former is called the control connection and is used to estimate congestion on the wired network.

If the mobile host receives all its packets from the base station, then packets sent to both connections must be routed through the same route. The source host thus sends packets to the control connection at regular intervals and compares the percentage of packets that are acknowledged by both connections within the timeout. If there is a significant difference, it can be concluded that the losses are due to wireless link errors, rather than to congestion, and the source host will continue to increment its congestion window at the same rate, rather than triggering its congestion control algorithm. If the difference is small, however, losses can be assumed to be due to congestion, and congestion control mechanisms are triggered [17].

This proposal, however, does not address the problems caused by handoffs. Moreover, if a mobile host is disconnected for a long time or if disconnections are frequent, this scheme does not improve performance significantly.

Fast Retransmit

During handoff, the packets sent to the mobile host are lost. In [18], it was proposed that the wireless host resume communication immediately after the handoff and send three duplicate ACKs to force the TCP sender to execute the Fast Retransmit algorithm. This small modification improves the performance of TCP, since timeout for lost packets is avoided.

This solution combines well with the effects of short disconnections, although if disconnections are long or frequent, this scheme will not improve performance significantly.

Transmission and Timeout Freezing

One of the major problems of TCP over mobile networks is the frequent disconnections during which no data are exchanged. Fortunately, this situation can be detected by most of the link layers of wireless systems [10].

It was proposed that the link layer inform the TCP sender whenever disconnection occurs, so that the TCP sender can freeze the connection, stopping the time counter, retaining packets, and leaving the congestion window unmodified, since the sender will not be making an incorrect assumption about the existence of congestion. As soon as the service resumes, the link layer informs the TCP sender, which then restarts the transmission [19].

19.3.3 Generic TCP Extensions

Some TCP extensions were designed to improve the performance of TCP regardless of the type of network being used. Such extensions, however, require changes at both sender and receiver TCP.

19.3.3.1 Larger Initial Window

The situation created when the Slow Start algorithm probes the available bandwidth at a low rate, especially in long-delay networks, can be overcome by increasing the initial value of the congestion window (*cwnd*). A TCP extension presented in [20] allows the initial size of *cwnd* to be increased from one segment to the value given by the following equation: *min* [*4*MSS, max (2*MSS, 4380 bytes)*], where MSS is the maximum segment size.

Using this value allows more packets to be sent during the first RTT (Round Trip Time), thus triggering more ACKs, which allow the congestion window to grow more rapidly. Moreover, when sending more than one segment at the beginning, there is no need to wait for the delayed ACK timer to expire. This increase in *cwnd* can save up to three RTTs and delay the ACK timeout in comparison with an initial value of *cwnd* of one segment.

19.3.3.2 Transaction TCP

The connection establishment/close phases of TCP require at least three messages (three-way handshake), which can be an overhead if the application has a small quantity of data to be transferred. RFC 1644 [21] defines a transaction TCP for this type of application that avoids these messages by combining connection establishment and connection-close packets with data packets.

19.3.3.3 Selective Acknowledgment TCP

A TCP connection that uses the TCP Reno Fast Recovery Algorithm may experience poor performance when multiple packet losses occur in a single window of data, since it can only recover from these losses by using the timeout mechanism. One attempt to overcome this limitation involves the use of the Selective ACKnowledgment (SACK) mechanism, which was proposed in RFC 2018 [22]. In this mechanism, the TCP receiver can inform the TCP sender which packets were received correctly, so that the TCP sender can implement a selective retransmission policy, with only the missing segments being retransmitted.

When SACK is adopted, the sender can generally recover from losses in a single RTT, avoiding the costly Slow Start algorithm as well as unnecessary retransmissions.

19.3.3.4 Explicit Congestion Notification

Active queue management (AQM) is a mechanism designed to detect congestion before overflow occurs by providing an indication of congestion to the end nodes. One of the ways in which congestion is indicated is to drop packets. Another way is for routers to set the Congestion Experienced (CE) bit in the IP header of packets from ECN-capable senders [23]. This extra information in the IP header can help the sender distinguish the cause of packet losses, thus enabling the use of the correct error recovery strategy.

19.4 Conclusions

The independence of the TCP/IP protocol stack from the link layer technology has allowed excessive spread of the use of this stack and has led to what is now known as the Internet. It is not surprising that the generality of this spread does not perfectly match the characteristics of specific link layer technologies. Especially in mobile networks, performance degradation due to changes in access point and the interpretation of packet losses as a signal of congestion have motivated the development of proposals to improve the system. In the present chapter, a general overview of these proposals has been provided. Despite the engineering beauty of these proposals, however, only those that truly provide seamless access to the Internet in a cost-effective way will survive.

References

[1] C. Perkins, *IP Mobility Support for IPv4*, RFC 3220, Jan. 2002.

[2] C. Perkins, Mobile IP, *IEEE Communications Magazine*, May 1997.

[3] E. Gustafsson, A. Jonsson, and C.E. Perkins, Mobile IPv4 Regional Registration, Internet draft, draft-ietf-mobileip-reg-tunnel-06.txt, work in progress, Mar. 2002.

[4] H. Haverinen and J. Malinen, Mobile IP Regional Paging, Internet draft, draft-haverinen-mobileip-reg-paging-00.txt, work in progress, June 2000.

[5] R. Ramjee, T. La Porta, S. Thuel, and K. Varadhan, HAWAII: A Domain-Based Approach for Supporting Mobility in Wide-Area Wireless Networks, *Proc. of Seventh International Conference on Network Protocols*, Toronto, 1999, pp. 283 –292.

[6] A.T. Campbell, J. Gomez, S. Kim, A.G. Valko, C-Y. Wan, and Z. Turanyi, Design, Implementation, and Evaluation of Cellular IP, *IEEE Personal Communications*, Aug. 2000.

[7] A. Grilo et al., Terminal Independent Mobility for IP (TIMIP), *IEEE Communications Magazine*, Dec. 2001.

[8] M. Allman, V. Paxson, and W. Stevens, TCP Congestion Control, RFC 2581, Apr. 1999.

[9] K. Brown and S. Singh, M-TCP: TCP for mobile cellular networks, *ACM Computer Communications Review*, 27(5), 1997.

[10] J.D. Solomon. *Mobile IP: The Internet Unplugged*, Prentice Hall, New York, 1998.

[11] N. Deshpande, TCP Extensions for Wireless Networks, http://www.cis.ohio-state.edu/~jain/cis788–99/tcp_wireless/index.html.

[12] B.R. Badrinath and A. Bakre, I-TCP: Indirect TCP for Mobile Hosts, *Proc. IEEE 15th International Conference on Distributing Computing Systems (ICDCS)*, May 1995.

[13] H. Balakrishna, S. Srinivasan, and R.H. Katz, Improving Reliable Transport and Handoff Performance in Cellular Wireless Networks, *ACM Wireless Networks*, Dec. 1995.

[14] N. Vaidya and M. Mehta, Delayed Duplicate Acknowledgments: A TCP-Unaware Approach to Improve Performance of TCP over Wireless, Texas A&M University, Technical Report 99–003, Feb. 1999.

[15] C. Casetti, M. Gerla, S. Mascolo, M.Y. Sanadidi, and R. Wang, TCP-Westwood: Bandwidth Estimation for Enhanced Transport over Wireless Links, *Proceedings of ACM MobiCom 2001*, Rome, July 16–21, 2001, pp. 287–297.

[16] S. Biaz and N. Vaidya, Discriminating Congestion Losses from Wireless Losses using Inter-Arrival Times at the Receiver, *IEEE Symposium ASSET '99*, Richardson, TX, Mar. 1998.

[17] S. Banerjee and J. Goteti, Extending TCP for Wireless Networks, Department of Computer Science, University of Maryland, College Park, May 1997.

[18] R. Caceres and L. Iftode, Improving the Performance of Reliable Transport Protocols in Mobile Computing Environments, *IEEE Transactions on Selected Areas in Communication*, June 1995.

[19] B. Bakshi, P. Krishna, N. Vaidya, and D. Pradhan, Improving Performance of TCP over Wireless Networks, *17th International Conference on Distributed Computing Systems*, May 1997.

[20] M. Allman, S. Floyd, and C. Partridge, Increasing TCP's Initial Window, RFC 2414, Sep. 1998.

[21] R. Braden, T-TCP — TCP Extensions for Transaction: Functional Specification, RFC 1644, July 1994.

[22] M. Mathis, J. Mahdavi, S. Floyd, and A. Romanow, TCP Selective Acknowledgment Options, RFC 2018, Oct. 1996.

[23] K.K. Ramakrishnan, S. Floyd, and D. Black, The Addition of Explicit Congestion Notification (ECN) to IP, RFC 3168, Sep. 2001.

20

An Intelligent On-Demand Multicast Routing Protocol in Ad Hoc Networks

Kuochen Wang
National Chiao Tung University

Chaou-Tang Chang
National Chiao Tung University

Abstract

In this paper, we present an *intelligent on-demand multicast routing protocol* (IOD-MRP), which is suited for rapidly changing network environments, such as ad hoc networks. This protocol simplifies and enhances the existing *core-assisted mesh protocol* (CAMP). The main difference between our protocol and CAMP is that we have eliminated the cores from CAMP. We remove the cores from CAMP and apply an on-demand receiver-initiated procedure to dynamically build routes and maintain multicast group membership. In addition, an intelligent mobility management procedure is used to maintain and optimize the multicast mesh. This procedure monitors multicast traffic and learns about link states of the mesh. As a result, the effect of mobility can be reduced, and the mesh structure can be optimized. With this procedure, control messages due to flooding can be reduced as well. Flooding is an important issue in ad hoc networks, and our protocol is able to reduce flooding messages. We compare the proposed IOD-MRP scheme with CAMP. Simulation results show that IOD-MRP reduces number of control messages by 3 to 13% compared to CAMP and still has slightly lower packet delay and packet loss rate than CAMP.

20.1 Introduction

We consider the problem of providing multicast services in a large Mobile Ad hoc NETwork (MANET) environment. There are several challenges to providing multicast services in MANET, including:

- Challenges in wireless networks
- Challenges in ad hoc networks
- Challenges of multicast in ad hoc networks

20.1.1 Challenges in Wireless Networks

Due to the global sharing of wireless medium, wireless networks have high interference, high error rates, and limited bandwidth. For mobile hosts, battery life is also an important issue. Therefore, power control, signaling traffic, and error detection and correction in wireless networks are more important than those in wireline networks. Due to mobility, the address and physical position of a mobile host are not associated anymore, hence traditional addressing and routing protocols must be modified to fit wireless networks.

20.1.2 Challenges in Ad Hoc Networks

An ad hoc network can be viewed as a set of mobile hosts that can communicate with each other without the help of wired infrastructure. In an ad hoc network, mobile hosts must act as routers and react to the network topology, which changes dynamically and frequently. Consequently, we need a simple, band-width-saving, and robust multicast protocol for ad hoc networks.

20.1.3 Challenges of Multicast in Ad Hoc Networks

The main challenge of multicast in ad hoc networks comes from a very high rate of topology changes in the networks. Consequently, traditional multicast protocols that employ periodic broadcast may cause the networks to be full of control messages. In the worst case, the update rate of the multicast delivery structure may not catch up with the rate of topology changes. As a result, some members of a multicast group may be lost, and the network will be filled with packets with invalid destination addresses.

The shared-based multicast tree is a well-established concept used in several multicast protocols for wireline networks, e.g., core-based trees (CBT) [1–3], PIM (protocol independent multicast) sparse mode [4], etc. However, the established tree structure may not be able to react well to the high mobility of ad hoc networks. Recently, a new multicast delivery structure, mesh [5], was employed to provide a more robust delivery structure than a shared-based tree. In this paper, we propose a multicast protocol based on the mesh structure to conquer the challenges of multicast in ad hoc networks. The protocol has the following features:

- Mesh type multicast delivery structure
- On-demand construction of multicast mesh
- No core
- Localized effect of changing multicast delivery structure
- Selection of a multicast mesh node based on node's mobility and link states
- Bandwidth saving by reducing number of control messages

In Section 20.2, we introduce the basic concepts of unicast and multicast in ad hoc networks. After examining some existing multicasting protocols in Section 20.3, we propose our design approach in Section 20.4. Simulation results and evaluation are described in Section 20.5. Finally, we present concluding remarks and discuss future work.

20.2 Background

20.2.1 Fundamentals of Ad Hoc Networks

Due to the rapid development of wireless communication technology, wireless networks have become more and more popular in the computer and communications industry. There are two types of mobile wireless networks [6]: infrastructured mobile networks and infrastructureless mobile networks. An infrastructured mobile network has a fixed and wired backbone network. A mobile host communicates with the network through the nearest base station (or access points) within its communication scope. To allow communication to continue seamlessly throughout the network, *handoff* occurs when a mobile host travels out of the range of one base station into the range of another base station; traffic for the mobile host will be redirected from the old base station to the new one. Typical backbones of infrastructured networks include public land mobile networks (PLMNs) and wireless local area networks (WLANs).

Infrastructureless mobile networks are commonly known as ad hoc networks. No fixed and wired backbone exists in an infrastructureless mobile network. To communicate with each other, some mobile hosts must act as routers that discover and maintain routes to other mobile hosts in the network. Example applications of ad hoc networks are emergent search-and-rescue operations, meetings or conventions in which people wish to quickly share information, data acquisition operations in inhospitable terrain [7], and mobile commerce applications [8].

20.2.2 Overview of Ad Hoc Unicast Routing Protocols

IOD-MRP utilizes an on-demand receiver-initiated procedure to dynamically build a multicast mesh. This on-demand concept is based on the demand-driven concept of ad hoc unicast routing protocols [7,9,10]. In ad hoc networks, unicast routing protocols may generally be categorized into two types. We introduce these two types of protocols to explain why we use an on-demand procedure to build a multicast mesh. They are categorized as follows:

- Table-driven: CGSR [11], DSDV [12], WRP [13]
- Source-initiated (demand-driven): AODV [14], DSR [15], TORA [9]

20.2.2.1 Table-Driven Routing Protocols

Table-driven unicast routing protocols attempt to maintain consistent, up-to-date routing information from each node to every other node in the network [7]. In order to maintain a consistent view of network topology, each mobile host propagates update messages throughout the network when the network topology changes. As a result, this kind of protocol requires each node to maintain a large table to store routing information for the entire network. In addition, the update messages occupy much network bandwidth.

20.2.2.2 Source-Initiated On-Demand Routing Protocols

To overcome the drawbacks of table-driven routing protocols, source-initiated on-demand routing protocols were proposed. The routing information is established only when needed, and hence there is no need for mobile hosts to maintain routing information of the entire network. A mobile host will initiate a route discovery procedure when it requires a route to a destination. After the route has been established, a route maintenance procedure is used to keep a connection between source and destination until either the destination becomes inaccessible from the source or the route is no longer desired.

20.2.3 Fundamentals of Multicast

Different from point-to-point communications (i.e., unicast), multicast refers to both point-to-multipoint and multipoint-to-multipoint communications. Multicast is also known as *selective broadcast*, which can simultaneously send data packets to a subset of specific end hosts. Communications among multiple hosts are now common, especially in conferencing applications, which involve sharing text, images, and even real-time data among multiple hosts.

A multicast group is defined as a set of senders and receivers who are involved in the same multicast communication session. The route established by a multicast protocol for transmitting multicast traffic is called a multicast delivery structure (e.g., tree, mesh). Most studies of multicast focus on how to build and maintain a multicast delivery structure because the delivery structure greatly influences the performance of a multicast protocol. Multicasting in wireless networks such as ad hoc networks is more complex than in traditional wireline networks because of the frequently changing network topology. Therefore, traditional multicast for wireline networks must be modified to suit the mobility, dynamically changing topology, and radio channel characteristics of wireless networks.

20.3 Existing Methods

20.3.1 Traditional Multicast Protocols

In wireline networks, there are several ways to support multicast. Figure 20.1a shows an example of a popular technique that maintains per-source trees. For every multicast sender, there is a multicast tree connecting every multicast receiver. A sender multicasts data packets to all destinations along the source-based tree using *reverse path forwarding*. This technique is implemented in the distance vector multicast routing protocol (DVMRP) [16] and the protocol independent multicast (PIM) dense mode [4]. Source-

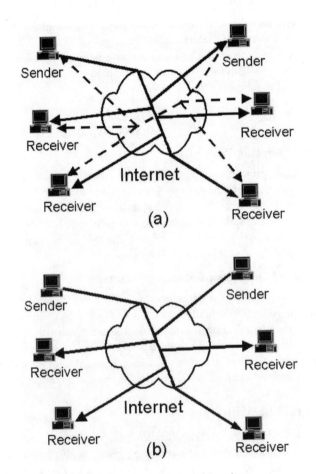

FIGURE 20.1   Multicasting using (a) source-based tree (b) shared-based tree.

based tree multicast has many attractive properties. The shortest path information to establish the source-based tree can easily be obtained from most common unicast routing protocols that use distance vector or link state. Furthermore, the multicast traffic is distributed evenly in the network. It does not rely on a central control point, and the algorithms to initialize and maintain the source-based tree are very simple.

However, the number of source-based trees depends on the number of senders in a multicast group. For example, in Fig. 20.1a, the multicast group has two senders and hence two source-based trees. Therefore, the source-based tree is not suited for multipoint-to-multipoint multicast, which has multiple senders. To overcome this problem, shared-based tree (Fig. 20.1b) multicast strategies were proposed [17,18]. Instead of many source-based trees per multicast group, a single tree rooted at a core is maintained. In the shared-based tree, the sender unicasts a data packet to the core, and the core then multicasts the data packet on the shared-based tree to the intended receiver. Examples of the shared-based tree approach are core-based tree (CBT) [1–3] and PIM sparse mode [4]. The shared-based tree approach also has some drawbacks. First, the paths are often nonoptimal, and packet transmission delay may increase. Second, traffic is concentrated in the backbone tree, and delivery of all multicast data packets will stop in the event of core failure.

20.3.2 Core-Assisted Mesh Protocol (CAMP)

In mobile networks, traditional multicast protocols may have a problem. For example, a receiver may move faster than the routing tables can track it. By the time packets reach their specified destination, the receiver may have already moved to another place. This means that the receiver may never receive some packets intended for it. A simple way to overcome this problem is to increase the routing update rate with mobility. However, this will increase control message overhead and will not scale well with network size. This is a very serious problem for the shared-base tree because high mobility of cores will cause cores to be unavailable frequently.

To offer a robust multicast protocol that fits ad hoc networks, a new multicast protocol named the core-assisted mesh protocol (CAMP) [5] was proposed. The main difference between CAMP and traditional multicast protocols is that it builds and maintains a *multicast mesh* instead of a tree structure. A multicast mesh is a subset of the network topology that provides at least one path from each source to each receiver in the multicast group [5]. A multicast mesh has high connectivity because a member router of the multicast mesh has multiple paths to any other routers in the same mesh. Due to high connectivity of mesh, topology changes will not interrupt the flow of multicast data.

CAMP ensures that the shortest paths from receivers to sender (called reverse shortest path [19]) are part of a group's mesh [5]. To join a multicast group, a mobile host first determines the address of the group. The mobile host then sends a join request towards a core if none of its neighbors are members of the group. If cores are not reachable from a router that wishes to join the multicast group, the router broadcasts its join request using an *expanded ring search* (ERS) that eventually reaches some group members. This is the worst case joining procedure and the network will flood with this join request message. Eventually, one or multiple responses are sent back to the joining router, and it will choose any of these responses to use as a reverse path to the mesh.

Examples of other nontree-type multicast protocols are FGMP [20] and ODMRP [21–23]. FGMP and ODMRP are built from a variation of meshes. The difference between these two protocols involves who starts flooding control messages. In FGMP, the receivers start flooding, while in ODMRP, the senders do.

20.4 Design Approach

20.4.1 The Aim of the Proposed Protocol

To propose a multicast protocol suited for ad hoc networks, we must give careful consideration to the challenges described in Section 20.1. We found that most challenges and problems come from the rapidly changing network topology. To overcome this, our main goal is to propose a multicast protocol suited to the rapidly changing network topology. Our multicast protocol has the following features:

- Mesh type multicast delivery structure
- On-demand construction of multicast mesh
- No core
- Localized effect of changing multicast delivery structure
- Selection of a multicast mesh node based on the node's mobility and link states
- Bandwidth saving by reducing number of control messages

20.4.1.1 On-Demand Construction of Multicast Mesh

In ad hoc networks, table-driven routing algorithms may cause high control message overhead. Hence, our multicast protocol is based on an on-demand unicast routing protocol, and the on-demand concept is used to avoid control message storm.

20.4.4.2 Localized Effect of Changing Multicast Delivery Structure

When a multicast delivery structure must change, localizing the effects of changing mesh structures would make the whole multicast mesh more stable and efficient.

20.4.4.3 Selection of a Multicast Mesh Node Based on the Node's Mobility and Link States

To obtain a stable multicast delivery structure, we select nodes with low mobility and stable links as multicast intermediate nodes.

20.4.2 Overview of the Proposed Approach

The proposed approach is an on-demand receiver-initiated mesh type multicast routing protocol. The basic idea of our approach is to simplify the core-assisted mesh protocol (CAMP) by eliminating the role of cores and to provide an intelligent mobility management procedure to handle the multicast mesh. The role of cores in CAMP is to reduce flooding messages in the joining phase by directly unicasting JOIN-REQUEST messages to the cores. However, the unicast routing information to the cores is not always correct (or available) because of stale routing information. Hence, this will result in a JOIN-REQUEST failure for a node, and the node has to join the mesh by flooding. In our approach, we overcome this problem by directly applying on-demand receiver-initiated flooding to join the mesh and use the intelligent procedure to learn and monitor multicast memberships in order to efficiently maintain the multicast mesh and to reduce flooding traffic on the whole network. By using the on-demand receiver-initiated technology, we guarantee that a node can join the multicast mesh through the shortest path (hence, in the shortest time) and that the node can also have multiple paths to the mesh. By using the intelligent procedure, we guarantee that there is at least one stable and optimal path between each multicast sender and receiver. Therefore, we can avoid using unstable links and reduce the frequency of flooding. In sum, our protocol is more robust and can save more bandwidth in ad hoc networks with high mobility nodes.

20.4.3 Types of Membership and Node Classification

There are three types of membership:

- *Sender:* sends out multicast data packets
- *Receiver:* receives multicast data packets
- *Sender and receiver:* sends and receives multicast data packets

There are four types of nodes [24]:

- *Interested node (I-node):* A node that is interested in a specific multicast session and wishes to join. It does not matter if it is interested in joining as a sender, receiver, or both [24].
- *Uninterested node (U-node):* A node that is not interested in a specific multicast session but is forced to be part of the session because it is required to be a relay/intermediate node on the multicast delivery path [24].
- *Leaf node:* A node at the edge of the multicast mesh. A leaf node may be a sender or a receiver or both [24].
- *Intermediate node:* A node in the internal branches of the multicast mesh. It may be either an I-node or U-node [24].

20.4.4 Initializing Multicast Mesh

There is no special control message when opening a new multicast session. The procedure of initializing a multicast session is similar to the procedure of joining a multicast session. The only difference between joining and opening is that a node *s* must first geerate a new multicast-id when opening a new multicast session. Figure 20.2 shows the flowchart of the multicast session initialization and join procedure. When a new multicast session is opened, no information about this session is present in the multicast routing table (MRT). Hence, node *s*, which wishes to open a new multicast session, will enter the Broadcast-JOIN-REQ mode. After that, every node in the network will have the routing information of the new multicast session in the MRT. When an I-node receives the Broadcast-JOIN-REQ, it will send a Unicast-JOIN-REQ to the sender of the Broadcast-JOIN-REQ and try to join this multicast session. Eventually, the multicast session will be built. This new multicast session information will exist in the network for a while. In Broadcast-JOIN-REQ mode, the join request will time out if there is no other node joining this multicast session, and the protocol will inform node *s* about this. Node *s* can keep advertising this new multicast session message or just wait for other nodes to join this session.

20.4.5 Joining Multicast Session

We use the on-demand receiver-initiated procedure to allow a node to join the multicast mesh as soon as possible and to reduce control message overhead. There are two modes of joining: Unicast-JOIN-REQ mode and Broadcast-JOIN-REQ mode. If an I-node already has the routing information of the multicast session in its MRT, it will use the Unicast-JOIN-REQ mode to join this multicast session. Otherwise, it will use the Broadcast-JOIN-REQ mode instead.

If a node *x* wants to join a multicast group, it will become an I-node and enter the join procedure, as shown in Fig. 20.2. At first, node *x* will check its MRT to see if there is any routing information about this multicast group. In the following situations, node *x* will have this information:

- Node *x* belongs to the mesh structure of this multicast group.
- Node *x* has at least one neighbor that belongs to the mesh structure of this multicast group.
- Node *x* has received the Broadcast-JOIN-REQ from the sender of this multicast group.

In the above three situations, there is enough unicast routing information from node *x* to the multicast mesh. Hence, node *x* will enter the Unicast-JOIN-REQ mode and try to join the multicast group by unicasting join request messages to every sender in its MRT. Note that the join request message will be sent only once if two or more entries in the MRT have the same relay.

As shown in Fig. 20.3, node *x* will wait for an acknowledgment (ACK), and any acknowledgment from the same multicast group (with the same multicast session *id*) will be accepted. If there is at least one positive ACK received, node *x* successfully joins this multicast mesh. Node *x* will set the active flag of this multicast session in its MRT as active. Otherwise, node *x* will enter the Broadcast-JOIN-REQ mode. When a node receives a Unicast-JOIN-REQ message from node *x*, it will check whether it is the destination of the message and whether it is a mesh member (the entry in MRT is set to active) first. If so, it will reply a positive ACK back to node *x*. Otherwise, it will try to deliver this message to the destination or send a negative ACK back to node *x*. Figure 20.4 shows the detailed flowchart.

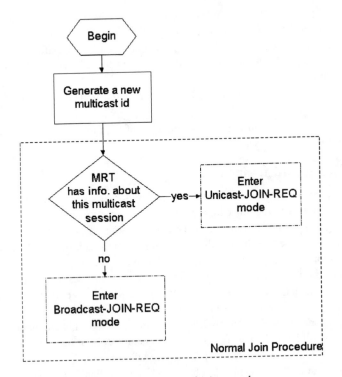

FIGURE 20.2   Flowchart of multicast session initialization and join procedure.

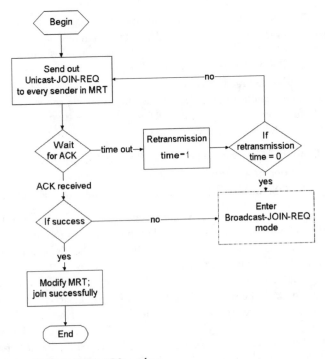

FIGURE 20.3   Flowchart of Unicast-JOIN-REQ mode.

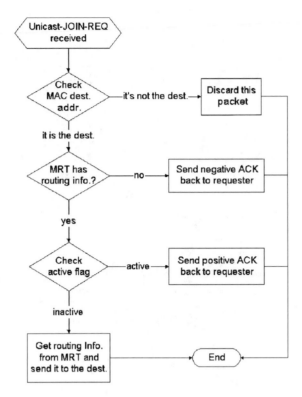

FIGURE 20.4 Node reaction when receiving a Unicast-JOIN-REQ message.

Node x will enter the Broadcast-JOIN-REQ mode if it does not have the routing information about this multicast group. The Broadcast-JOIN-REQ mode invokes an expanded ring search (ERS) that eventually reaches at least one group member in the mesh. When a node that does not have the routing information of the multicast group receives a nonduplicate Broadcast-JOIN-REQ message, it will record the message in its MRT and then rebroadcast the message out until the message reaches a node that has the routing information of the multicast group or until time-to-live (TTL) expires. Note that if a node has the routing information of the multicast group but the corresponding active flag in MRT is inactive, this node will set itself as an intermediate node and invoke the Unicast-JOIN-REQ mode to join the multicast group. This procedure can prevent broadcast messages from flooding all over the network. Figure 20.5 shows the detailed flowchart.

20.4.6. Maintaining Multicast Mesh

We propose an intelligent procedure to maintain the multicast mesh after the joining phase. This procedure can monitor multicast traffic and learn the link states of the mesh from it. This intelligent procedure has three main functions:

- Handling the problems of mobility
- Optimizing the multicast mesh
- Reducing control message overhead

Our protocol is different from other multicast protocols in the following aspect. When a U-node receives a multicast packet, the U-node will learn about this multicast session from the header of the multicast packet and record the header on the MRT (the active flag will be set OFF to indicate that it is not a multicast member). After that, the multicast packet is dropped (as in other protocols). In this way,

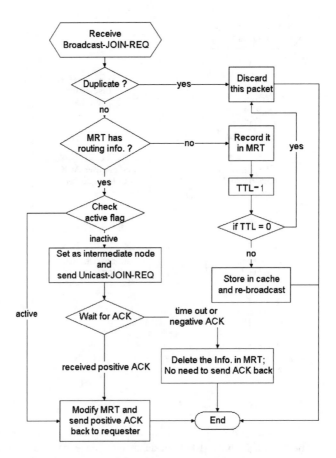

FIGURE 20.5 Node reaction when receiving a Broadcast-JOIN-REQ message.

node joining is more efficient because an I-node can unicast a Unicast-JOIN-REQ message without the help of the unicast protocol.

After successfully joining a multicast mesh, the intelligent procedure will monitor the multicast traffic and decide the best path for each sender. It finds out the best path for each sender by the following steps:

- If most of the multicast data packets from sender *S* come from a specific path, it means that this path is the shortest path between the receiver and sender *S*.
- Furthermore, if this path also has a low packet loss rate, it means that this path has good link states.

When a receiver finds out such a path, it will unicast a SET-ANCHOR to sender *S* through this path. When a node receives this control message, it will set the anchor flag in the anchor table (AT) ON, and this will force the node to stay in the mesh. Hence, this path is more stable for routing the traffic from *S* to the receiver. If a receiver finds another path better than the current path, it will send a RESET-ANCHOR control message to the sender through the old path and send SET-ANCHOR through the better path; the anchor path will be changed to the better path. Our intelligent procedure guarantees that one anchor path exists between each pair of sender and receiver.

In the worst case, when it is concluded that there is no good path to sender *S*, our intelligent procedure will trigger the unicast routing protocol to construct a new unicast path between *S* and the receiver. Then the receiver will unicast a Direct-JOIN-REQ to *S* through this path to construct a new multicast delivery path. When a node receives the Direct-JOIN-REQ packet, it will check its unicast routing table and forward the packet to the next relay. The node will also modify its MRT and directly set the anchor flag ON. If the unicast protocol reports that there is no path to *S*, it means that network partition occurs.

20.4.7 Forwarding Policy

After finishing the group establishment and mesh route construction process, a source can multicast packets to receivers via the mesh. The basic idea of the packet forwarding scheme is to try to forward the first arrival multicast data packet from the sources along the paths in the mesh. The header of the multicast packet consists of:

- The address of the intended multicast group
- The address of the sending router
- A sequence number to ensure that it is not a duplicate packet
- A time to live (TTL) used to limit the time that each packet is allowed to remain in the network

A node receiving a multicast packet without errors from a neighbor node will accept the packet only if:

- The node is a member of the multicast group specified in the packet.
- The packet's sequence number is not in the packet-forwarding cache.
- The packet's TTL did not expire.

When a router accepts a packet, it adds a sequence number and the identifier of the source to its packet forwarding cache. This step prevents the same packet from being accepted more than once by a node, provided that the entries in the cache persist longer than the time it takes for packets to revisit a node [24]. A node forwards an accepted multicast packet only if the node has more than one child node. Whether a node forwards a packet or not, the node updates its MRT with the fact that the sending router belongs to the multicast group addressed by the packet. This can allow the router to join the group through simple announcement if it is required in the future.

20.4.8 Terminating a Multicast Session

Nodes are of course free to leave a multicast session. Any sender in the multicast session can send a SESSION-LEAVE message when it wants to leave this multicast session. This SESSION-LEAVE message will be forwarded to every node on the mesh. All nodes that receive the SESSION-LEAVE message purge any entry/state associated with the sender. Other multicast members who miss the SESSION-LEAVE message will eventually notice the expiration of any entry/state associated with the sender. If all the senders in the multicast session have left, the SENDER TABLE will be empty, and the protocol ends the multicast session.

20.5 Evaluation and Discussion

20.5.1 Simulation Setup

We use Borland C++ builder on Microsoft Windows 98 to simulate our protocol (IOD-MRP) and CAMP. The simulation model is based on the ad hoc network model in *ns-2* [25], which is summarized as follows. Each mobile host is an independent entity that is responsible for computing its own position and velocity. Each mobile host can have one or more network interfaces, each of which is attached to a channel. Channels are the conduits that carry packets between mobile hosts [25]. When a mobile host transmits a packet onto a channel, the channel distributes a copy of the packet to all the other network interfaces on the channel. These interfaces then use a radio propagation model [25] to determine if they are actually able to receive the packet. There are five parameters for simulation:

- NUM_OF_NODE: Total number of mobile hosts in the network
- MAX_MNODE: Maximum number of multicast members
- MAX_SPEED: Maximum speed that a mobile host can choose
- SIM_TIME: Total simulation time
- RADIO_RANGE: Transmission scope of mobile hosts

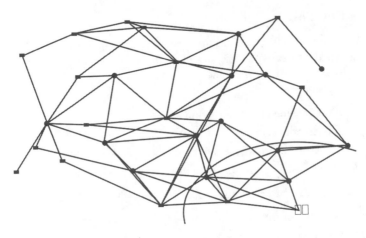

FIGURE 20.6 The initial topology of the ad hoc network for simulation.

20.5.2 Experimental Results

IOD-MRP simplifies CAMP, a representative multicast routing protocol based on the mesh. Therefore, for evaluation, we implemented CAMP with random core selection. The objective of the experiments is to prove that IOD-MRP based on the on-demand procedure and intelligent mobility management can save more bandwidth than CAMP. We compare the performance of IOD-MRP vs. CAMP in terms of average packet delay, packet loss rate, number of control messages, and scalability. The packet loss rate at a receiver is simply the percentage of packets sent by the traffic source that were not seen by the specific receiver.

We ran a number of experiments to compare IOD-MRP with CAMP. Figure 20.6 shows the initial topology of the ad hoc network used in the simulation. The network has 30 mobile hosts, and they all are members of the same multicast session. All 30 mobile hosts can send and receive multicast packets. The links shown in the diagram represent the constructed mesh. The circle shows the radio range of the node with a cross mark. The simulation is based on a single broadcast channel, so the transmission of a node is received by all the nodes within the radio range. Experiments were run for 300 seconds, and the same conditions were applied to the simulation runs for both protocols.

Figure 20.7 shows that IOD-MRP has a smaller number of control messages than CAMP. Both protocols use increasing numbers of control messages as average moving speed increases. IOD-MRP reduces more control messages than CAMP does, especially when the average moving speed is above 72 km/h. This is due to our Unicast-JOIN-REQ mode and intelligent mobility management. The Unicast-JOIN-REQ mode can reduce the number of flooding (control) messages because the node that received a Broadcast-JOIN-REQ control message has the routing information of the multicast session. Our intelligent mobility management procedure can select the most stable route. Therefore, IOD-MRP has a lower packet loss rate and packet delay, as shown in Fig. 20.8 and Fig. 20.9, respectively, than CAMP. Finally, we evaluate the scalability of IOD-MRP and CAMP. The network has 30 mobile hosts, and the number of multicast members grows from five to 30. As shown in Fig. 20.10, the numbers of control messages needed for IOD-MRP and CAMP increase linearly as the number of multicast members increases. From Fig. 20.10, we can say that IOD-MRP and CAMP have the same scalability. We conclude that IOD-MRP has slightly better performance in terms of packet delay and packet loss rate than CAMP. In addition, IOD-MRP reduces the number of control messages by 3 to 13% compared to CAMP.

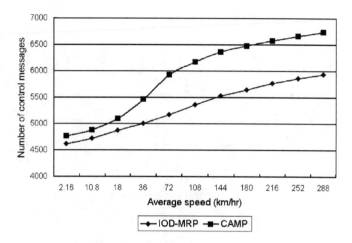

FIGURE 20.7 The effect of mobility on number of control messages.

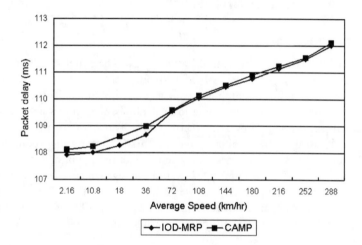

FIGURE 20.8 The effect of mobility on packet delay

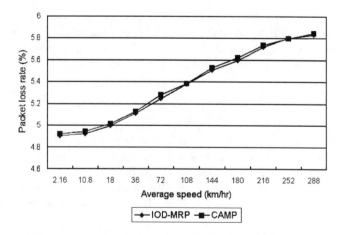

FIGURE 20.9 The effect of mobility on packet loss rate.

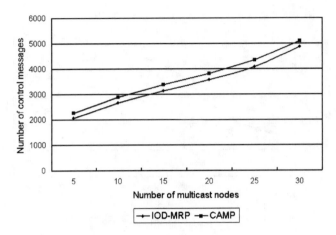

FIGURE 20.10 The effect of number of multicast nodes on number of control messages.

20.6 Conclusions and Future Work

20.6.1 Concluding Remarks

We have presented an intelligent on-demand multicast routing protocol (IOD-MRP) in ad hoc networks. IOD-MRP simplifies and enhances the core-assisted mesh protocol (CAMP) by totally removing the role of cores, and it has an intelligent mobility management procedure to handle the multicast mesh. We apply an on-demand receiver-initiated procedure to dynamically build routes and maintain multicast group membership instead of using cores. The intelligent mobility management procedure can maintain and optimize the multicast mesh by monitoring multicast traffic and learning about link states of the mesh. As a result, control messages due to flooding can be reduced significantly. Experimental results show that IOD-MRP can reduce the number of control messages more effectively than CAMP can, especially when the average moving speed is above 72 km/h. Our IOD-MRP reduces the number of control messages by 3–13% compared to CAMP and still has slightly better performance in terms of packet delay and packet loss rate than CAMP. In sum, our IOD-MRP is suited for ad hoc networks with high-mobility nodes.

20.6.2 Future Work

The intelligent mobility management procedure in IOD-MRP can take some quality of service (QoS) parameters into account such that real time traffic can be handled. In addition, the on-demand route discovery procedure can be integrated with source-initiated on-demand unicast routing protocols, such as AODV [14]. With the help of the on-demand unicast protocol, our multicast protocol can have better performance.

References

[1] A. Ballardie, Core Based Trees (CBT version 2) Multicast Routing, *RFC-2189*, Sep. 1997.
[2] A. Ballardie, Core Based Trees (CBT) Multicast Routing Architecture, *RFC-2201*, Sep. 1997.
[3] A. Ballardie, P. Francis, and J. Crowcroft, Core Based Tree (CBT): An Architecture for Scalable Inter-Domain Multicast Routing, Proc. ACM SIGCOMM '93, Oct. 1993, pp. 85–95.
[4] Protocol Independent Multicast (pim), http://www.letf.org/html.charters/pim-charter.html, Sept. 2002.
[5] J.J. Garcia-Luna-Aceves, A Multicast Routing Protocol for Ad-Hoc Networks (CAMP), *Proc. IEEE INFOCOM '99*, vol. 2, 1999, pp. 784–792.

[6] C.-K. Toh, *Wireless ATM and Ad-Hoc Networks*, Kluwer Academic Publishers, Dordrecht, 1997.

[7] E.M. Royer, A Review of Current Routing Protocols for Ad Hoc Mobile Wireless Networks, *IEEE Personal Communications*, Apr. 1999, pp. 46–55.

[8] U. Varshney, Multicast Support in Mobile Commerce Applications, *IEEE Computer*, Feb. 2002, pp. 115–117.

[9] V.D. Park and M.S. Corson, A Highly Adaptive Distributed Routing Algorithm for Mobile Wireless Networks, *Proc. IEEE INFOCOM '97*, Kobe, Japan, Apr. 1997, pp. 1040–1048.

[10] E. Bommaiah, M. Liu, A. McAuley, and R. Talpade, AMRoute: Ad Hoc Multicast Routing Protocol, Internet Draft, http://www.ietf.org/internet-drafts/draft-talpade-manet-amroute-00.txt, Aug. 1998.

[11] C.C. Chiang, Routing in Clustered Multihop, Mobile Wireless Networks with Fading Channel, *Proc. IEEE SICON '97*, Apr. 1997, pp. 197–211.

[12] C.E. Perkings and P. Bhagwat, Highly Dynamic Destination-Sequenced Distance-Vector Routing (DSDV) for Mobile Computers, *Proc. ACM SIGCOMM '94*, London, Oct. 1994, pp. 234–244.

[13] S. Murthy and J.J. Garcia-Luna-Aceves, An Efficient Routing Protocol for Wireless Networks, *ACM Mobile Networks and Application Journal*, Oct. 1996, pp. 183–197.

[14] C.E. Perkins and E.M. Royer, Ad Hoc On-Demand Distance Vector Routing, *Proc. 2nd IEEE Workshop Mobile Computing Systems and Applications*, Feb. 1999, pp. 52–61.

[15] D.B. Johnson and D.A. Maltz, The Dynamic Source Routing in Ad Hoc Wireless Networks, in *Mobile Computing*, T. Imielinski and H. Korth, Eds., Kluwer, Dordrecht, 1996, pp. 153–181.

[16] S. Deering, C. Partridge, and D. Waitzman, Distance Vector Multicast Routing Protocol, *RFC-1075*, Nov. 1988.

[17] L. Ji and M.S. Corson, A Lightweight Adaptive Multicast Algorithm, *Proc. IEEE GLOBECOM '98*, vol. 2, 1998, pp. 1036–1042.

[18] C.C. Chiang, M. Gerla, and L. Zhang, Adaptive Shared Tree Multicast in Mobile Wireless Networks, *Proc. IEEE GLOBECOM '98*, vol. 3, 1998, pp. 1036–1042.

[19] Y.K. Dalal and R.M. Metcalfe, Reverse path forwarding of broadcast packets, *Communications of the ACM*, 21, 1040–1048, 1978.

[20] C. Chiang and M. Gerla, On-Demand Multicast in Mobile Wireless Networks, *Proc. IEEE ICNP 98*, Austin, TX, Oct. 1998, pp. 262–270.

[21] M. Gerla et al., On-Demand Multicast Routing Protocol, Internet Draft, http://www.ietf.org/internet-drafts/draft-ietf-manet-odmrp-00.txt, Nov. 1998.

[22] S.H. Bae et al., The design, implementation, and performance evaluation of the on-demand multicast routing protocol in multihop wireless networks (ODMRP), *IEEE Network*, 14, pp. 70–77, 2000.

[23] S.H. Bae, S.-J. Lee, and M. Gerla, Multicast Protocol Implementation and Validation in an Ad Hoc Network Testbed, *Proc. IEEE ICC*, 2001, pp. 3196–3200.

[24] C.-K. Toh, C.W. Wu, and Y.C. Tay, Ad Hoc Multicast Routing Protocol Utilizing Increasing Id-numbers (AMRIS) Functional Specification, Internet Draft, http://www.ietf.org/internet-drafts/draft-ietf-manet-amris-spec-00.txt, Nov. 1998.

[25] The CMU Monarch Project's Wireless and Mobility Extensions to *ns*, http://www.monarch.cs.cmu.edu/, 1999.

21

GPS-Based Reliable Routing Algorithms for Ad Hoc Networks

Young-Joo Suh
Pohang University of Science and Technology

Won-Ik Kim
Electronics & Telecommunications Research Institute

Dong-Hee Kwon
Pohang University of Science and Technology

Abstract

The routing protocols designed for wired networks can hardly be used for mobile ad hoc networks due to unpredictable topology changes, and thus several routing protocols for mobile ad hoc networks have been proposed. The goal of this chapter is to select the most reliable route that is impervious to failures due to topological changes caused by host mobility. To select a reliable route, we introduce the concept of stable zone and caution zone and then apply it to the route discovery procedure of the existing ad hoc routing protocols. The concept of the stable zone and caution zone, which are located in a mobile node's transmission range, is based on a mobile node's location and mobility information received by the Global Positioning System (GPS). We evaluated the proposed algorithms by simulation in various conditions, and we obtained an improved performance in route maintenance time, the number of route disconnections, and packet delivery ratio.

21.1 Introduction

Mobile multi-hop wireless networks, called ad hoc networks, are networks with no fixed infrastructure, such as underground cabling or base stations, where mobile nodes are connected dynamically in an arbitrary manner. Thus, nodes in such networks function as routers, which discover and maintain routes to other nodes. A central challenge in the design of ad hoc networks is the development of dynamic routing protocols that can efficiently find routes between the source and destination. The routing protocols must be able to keep up with the high degree of node mobility, which often changes the network topology drastically and unpredictably [6,15].

Routing protocols in conventional wired networks generally use either distance vector or link state routing protocols, both of which require periodic routing advertisements to be broadcast by each router [7]. However, such protocols do not perform well in dynamically changing ad hoc network environments. The limitations of mobile networks, such as limited bandwidth, constrained power, and host mobility, make designing ad hoc routing protocols particularly challenging. To overcome these limitations, several source-initiated on-demand routing protocols, including Dynamic Source Routing (DSR) [8] and Ad-hoc On-demand Distance Vector (AODV) [9,10], have been proposed. These protocols create routes only when the source node has data to transmit. When a node requires a route to a destination, it initiates a route discovery procedure. This procedure is completed once a route has been found or all possible route permutations have been examined. Once a route has been established, it is maintained by a route maintenance procedure until either the destination becomes inaccessible or the route is no longer desired.

During the route discovery procedure, some of the existing source-initiated routing protocols such as DSR and AODV attempt to choose a route having the minimum number of hops among available routes. However, the route having the minimum number of hops is not always the best routing path. Although the route having the minimum number of hops may be faster than other routes in packet delivery, it may be highly probable that the spatial distance between any two intermediate nodes in the route may be larger than those in other routes. The larger distance between neighboring nodes may give rise to shorter link maintenance time, which in turn shortens the route maintenance time [13,16]. If there are frequent route failures due to host mobility, it will require additional time to reconfigure the route from the source to destination, which results in increased amounts of control packet flooding. Therefore, it may not be said that a route with the smallest hop count is necessarily optimal. The goal of our work is to select the most reliable route that is impervious to link failures by topological changes by host mobility, where a route discovery is performed with the location and mobility information received by Global Positioning System (GPS). To accomplish this goal, we propose new route discovery algorithms referred to as *Reliable Route Selection (RRS)*.

In this chapter, we assume that each node is aware of its current location through the use of GPS receivers with which each node is equipped. GPS has been successfully employed for determining a mobile node's position and speed. It is expected that the proliferation of GPS-based positioning technology will proceed at a fast pace, and the accuracy of this technology will be dramatically enhanced [12,14].

The remainder of this chapter is organized as follows. Section 21.2 briefly describes related works — DSR and AODV routing protocols. Section 21.3 provides some background on GPS. Section 21.4 describes the proposed GPS-based reliable route selection algorithms. The performance of the proposed algorithms is evaluated and compared in Section 21.5. We summarize our results in Section 21.6.

21.2 Ad Hoc Routing Protocols

21.2.1 Dynamic Source Routing (DSR) Protocol

The DSR protocol [8,18] uses source routing, where a packet carries in its header the complete list of nodes through which the packet must pass [6,8]. DSR is composed of two mechanisms: *route discovery* and *route maintenance*. Route discovery and route maintenance operate on demand.

Route discovery is the mechanism by which a node S wishing to send a packet to a destination D obtains a source route to D. When node S does not know a route to node D, S initiates a route discovery by transmitting a Route Request (RREQ) message as a single local broadcast packet, which is received by all nodes currently within transmission range of S. Each RREQ message contains a unique request ID and a record listing the addresses of each intermediate node through which the RREQ has been forwarded. When a node receives a RREQ message, if it is the target of the route discovery, it replies with a Route Reply (RREP) message to node S. Otherwise, if the node receiving the RREQ message recently saw another RREQ from the same initiator having the same request ID, or if it finds that its own address is already listed in the route record in the RREQ message, it discards the RREQ message. If not, the node appends its own address to the route record in the RREQ message and forwards it. On receiving the RREP message that includes the route from node S to node D, node S caches the route included in the RREP message. When node S sends a data packet to node D, the entire route is included in the packet header (hence source routing). Intermediate nodes use the source route included in the packet to determine to whom the packet should be forwarded.

Route maintenance is the mechanism by which the sender node S detects whether the network topology has been changed such that it can no longer use its route to D because a link along the route no longer works. Each forwarding node is responsible for confirming the receipt of each packet by the next hop node by a link-layer acknowledgment. If a packet is retransmitted the maximum number of times and no receipt confirmation is received, the node returns a Route Error message to original sender S, identifying the link over which the packet could not be forwarded. Node S then removes the broken link from its cache. If node S has another route to D in its route cache, S can send packets using the new route immediately. Otherwise, S may perform a new route discovery.

In the DSR protocol, a node forwarding or overhearing any packet may add the routing information to its own route cache, which can speed up a route discovery and reduce the propagation of route requests. But stale caches can adversely affect the performance of DSR. With passage of time and host mobility, cached routes may become invalid. A sender host may try several stale routes before finding a good route [11].

21.2.2 Ad Hoc On-Demand Distance Vector (AODV) Routing Protocol

The AODV routing protocol [9,10,18] supports multi-hop routing among mobile nodes for establishing and maintaining an ad hoc network. Unlike DSR, which uses source routing, AODV uses hop-by-hop routing. AODV is based on the Destination Sequence Distance Vector (DSDV) routing protocol [7]. The main difference is that AODV is reactive, while DSDV is proactive. AODV requests a route only when it is needed, but it does not require mobile nodes to maintain routes to the destination that are not actively used. AODV retains the desirable feature of DSR that routes are maintained only between nodes that need to communicate. When node S wants to send a packet to node D, it checks its route table to determine whether it has a route to D. If S has a route to D, S forwards the packet to the next-hop node toward D. If S does not have a route to D, S initiates route discovery. Source node S floods a Route Request (RREQ) message, which contains source address, destination address, sequence number, and broadcast ID. After sending the RREQ message, S sets a timer to wait for a reply. When a node receives a RREQ message, it checks whether it has already seen the RREQ message by noting the source address/ broadcast ID pair. If so, it discards the message. Otherwise, it sets up a reverse path pointing towards the source.

A node can send a Route Reply (RREP) message provided that it has an unexpired entry for the destination in its route table, and it knows a more recent path than the one previously known to sender S. To determine whether the path known to an intermediate node is more recent, a destination sequence number is used. A new RREQ message by a node is assigned a higher destination sequence number. If the node cannot send a RREP message, it increments the RREQ's hop count and then broadcasts the RREQ message to its neighbors.

When the intended destination receives a RREQ message, it replies by sending a RREP message. When an intermediate node receives the RREP message, it sets up a forward path entry to the destination in its route table. The RREP travels along the reverse path set up when the RREQ message was forwarded. Source node S can begin data packet transmissions as soon as the first RREP is received and later update its routing info if it discovers a better route.

Figure 21.1 shows an example of a route discovery by the AODV routing protocol. As shown in Fig. 21.1a, a RREQ message from node S is flooded through the network until it reaches node D. On its way through the network, the RREQ initiates the creation of temporary route table entries for the reverse path at all the nodes it passes, as shown in Fig. 21.1b. Next, a RREP message from node D is transmitted back to node S along the temporary reverse path, as shown in Fig. 21.1c. When the RREP message is routed back along the reverse path, all nodes on this route set up a forward path by pointing to the node that transmitted the RREP message.

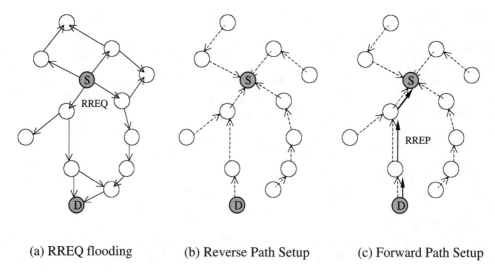

(a) RREQ flooding   (b) Reverse Path Setup   (c) Forward Path Setup

FIGURE 21.1   Route discovery procedure of AODV.

A route selected by the route discovery procedure is maintained as follows. If a route is broken due to the movement of the source, the source reinitiates route discovery to find a new route to the destination. If a route is broken due to the movement of either the destination or an intermediate node on the route, the node upstream of the break transmits a Route Error (RERR) message to the source node. When the source node receives the RERR message, it can reinitiate a route discovery if a route is still needed.

Neighborhood information is obtained from hello messages periodically transmitted by neighboring nodes. Each time a node receives a hello message from a given neighbor, it updates the info associated with the neighbor in its route table. Hello messages can be used to maintain the local connectivity of a node. When a mobile node uses its shared link layer protocol such as IEEE 802.11, instead of using hello messages, the node may listen to the retransmission of data packets to ensure that the next-hop node is still within its transmission range [9,10].

21.3 Global Positioning System

The Global Positioning System (GPS) is a worldwide radio navigation system formed from a constellation of satellites and their ground stations [12]. The system provides accurate, continuous, worldwide, three-dimensional position and velocity information to GPS receivers. The satellite constellation consists of 24 satellites arranged in six orbital planes with four satellites per plane. The ground stations are spread worldwide and monitor the status of the satellites. The satellites are used as reference points to calculate positions accurate to a matter of meters. GPS can provide service to an unlimited number of users since the user receivers operate passively, much like television sets.

GPS utilizes the concept of one-way time of arrival (TOA) ranging. Satellite transmissions are referenced to highly accurate atomic frequency standards onboard the satellites, which are synchronized with an internal GPS system time base. The satellites broadcast ranging codes and navigation data on two frequencies using a technique called code division multiple access (CDMA). That is, there are only two frequencies in use by the system, called L1 (1575.42 MHz) and L2 (1227.6 MHz). Each satellite transmits on these frequencies, but with different ranging codes from those employed by other satellites. These codes are also called "pseudo random number," and they have low cross-correlations with respect to one another. The navigation data provide the means for the receiver to determine the location of the satellite at the time of signal transmission, whereas the ranging code enables the user's receiver to determine the propagation delay of the signal from the satellite to a GPS receiver. With that information, a GPS receiver can determine the satellite-to-user range. This technique requires that the user receiver also contain a clock. Utilizing this technique to measure the receiver's three-dimensional location requires that TOA ranging measurements be made to four satellites. If the receiver clock is synchronized with the satellite clocks, only three measurements would be required. The results of TOA ranging measurements are applied to the "triangulation method," and then the location information with some error can be calculated. The errors are due to the fact that the signal is deteriorated by ionospheric and tropospheric effects, noisy channel, and clock inaccuracy during its travel from the satellite to the receiver.

Basically, GPS provides two services: the Standard Positioning Service (SPS) and the Precise Positioning Service (PPS). The SPS is designated for the public and provides a predictable accuracy of at least 100 m (2 drms, 95%) in the horizontal plane and 156 m (95%) in the vertical plane. The distance root mean square (drms) is a common measure used in navigation. The value of 2 drms is the radius of a circle that contains at least 95% of all possible fixes that can be obtained with a system at any one place. The PPS is used for military purposes and provides predictable accuracy of at least 22 m (2 drms, 95%) in the horizontal plane, 27.7 m (95%) in the vertical plane.

Since SPS and PPS do not provide exact location information, the Differential GPS (DGPS) system has been introduced to provide more accurate location information. In the basic form of DGPS, a reference station with a precisely known location is used. The reference station also performs GPS signal calculation. By comparing the result of location information obtained by GPS signal with the preknown location information, the reference station can produce error correction information. The error correction information is broadcast by the reference station and used for error correction of DGPS receivers, which can hear signals both from the satellites and the reference station. Some DGPSs can provide exact location information with no more than 1 m error.

21.4 Reliable Route Selection Algorithms

In AODV, the source node transmits its data through the route determined by the first RREQ that arrived at the destination. The selected route generally has the lowest hop count, which means that any two intermediate nodes on the route may be remotely located from each other, and thus the route maintenance time may not last long, causing a link failure in an ad hoc network environment where mobile nodes frequently move. If a link in the selected route is broken, a route reconstruction procedure is initiated by on-demand routing protocols. During the time a route reconstruction procedure is being performed, the source cannot transmit its data, and control messages generated to reconstruct the route may degrade the network performance. Therefore, selecting a reliable route during route discovery is important.

To achieve reliability of the selected route, we propose new route selection algorithms using the concept of a *stable zone* and a *caution zone* based on a mobile node's position, speed, and direction information obtained from GPS. The proposed algorithm is first applied to the AODV routing protocol and then applied to other ad hoc routing protocols.

21.4.1 Stable Zone and Caution Zone

For the proposed algorithms, we introduce the concept of virtual zone, where the transmission range of a node can be divided into two zones: *stable zone* and *caution zone*. Stable zone is the area in which a mobile node can maintain a relatively stable link with its neighbor node since they are located close to each other. Caution zone is the area in which a mobile node can maintain an unstable link with its neighbor nodes since they are located relatively far from each other. These zones are used for deciding whether or not the link state between any two nodes is reliable. The stable zone and the caution zone change dynamically depending on a mobile node's speed and direction. As mentioned previously, we know the position, speed, and direction of mobile nodes using the GPS information (i.e., latitude, longitude, and altitude). For simplicity, we assume that all mobile nodes have the same altitude value and transmission range. Figure 21.2 shows the two zones. In the figure, the radius of the transmission range is R, the stable zone is a smaller inner circle with a radius of r in the transmission range, and the caution zone is the transmission range excluding the stable zone. The inner circle that indicates the stable

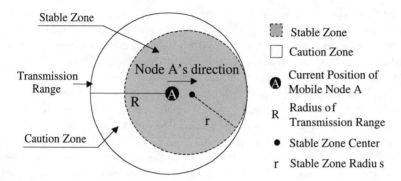

FIGURE 21.2   Stable zone and caution zone.

zone is inscribed in an outer circle that indicates the transmission range. The stable zone radius (r) is determined from the speed of a mobile node. If a mobile node's speed is zero, r will be the same as R, and it becomes smaller when a mobile node starts to move. If the mobile node increases its speed, then the value of r becomes smaller in proportion to the speed of the mobile node, and vice versa. In the proposed algorithm, adequate selection of stable zone radius (r) for a node's speed is very important. This will be discussed in detail later.

In addition to a mobile node's speed, its moving direction is also important. Even when a mobile node is located in the border range of a neighbor node, if the two nodes progress to the face-to-face direction, the link between the two nodes will be stable. In Fig. 21.2, since node A moves to the right, the location of the stable zone center is to the right of node A's current location and on the line of the mobile node's movement direction. Thus, the moving direction of a node will be the direction pointing from the node's current location to the stable zone center.

21.4.2 AODV-RRS Protocol Description

Now we describe the proposed protocol called AODV-RRS, which is obtained by applying the Reliable Route Selection (RRS) algorithm using the concept of stable zone and caution zone to the AODV routing protocol. In AODV-RRS, the following additional GPS information is included in each RREQ control packet:

- mobile_node_position (x,y)
- stable_zone_center (x',y')
- stable_zone_radius (r)

The *mobile_node_position (x,y)*, *stable_zone_center (x',y')*, and *stable_zone_radius (r)* indicate the current position of a mobile node, the center of the stable zone, and the radius of the stable zone, respectively. The GPS information is used to decide whether or not a link between two nodes is stable.

The route discovery mechanism of AODV-RRS is very similar to that of AODV, but AODV-RRS requires the following additional steps:

1. The source node or an intermediate node (e.g., node S) floods a RREQ message, which includes its own GPS information, to all nodes within its transmission range.
2. If a node (e.g., node M) receives the RREQ message, it calculates whether or not it is in the stable zone of node S, using its own GPS information and the GPS information included in the received RREQ message (i.e., GPS information of node S).
   - If node M is located in the stable zone of the node that transmitted the RREQ message and it is not the final destination, node M inserts its own GPS information in the RREQ message and then floods it.
   - Otherwise, node M drops the RREQ message.

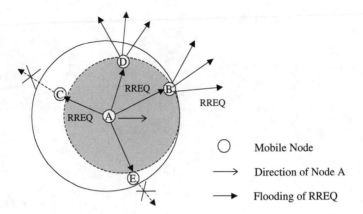

FIGURE 21.3 Route discovery using AODV-RRS.

Figure 21.3 shows an example of route discovery by AODV-RRS, where node A is the source node or an intermediate node entitled to forward a RREQ message, and the shaded area indicates the stable zone of node A. Four nodes — B, C, D, and E — are located in node A's transmission range, with nodes B and D located in the stable zone and nodes C and E located in the caution zone. When node A transmits a RREQ message, B, C, D, and E receive the message. Then each node calculates whether or not it is in the stable zone of node A. Since nodes B and D are in the stable zone, they insert their own GPS information into the RREQ message and then flood the message. Although nodes C and E receive the RREQ message, they drop the message since they are located in the caution zone of node A.

Figures 21.4a and 21.4b show how a RREQ message is flooded by AODV and AODV-RRS, respectively. When a route discovery is performed by AODV, an intermediate node floods the RREQ to other nodes as soon as it receives the message, except when a duplicated RREQ is received or when the node is the destination. If two neighboring nodes on a selected route are located near the border of each other's transmission range (that is, caution zone) as shown in Fig. 21.4a, then a small movement of a node may cause the node to be out of the other node's transmission range and thus cause a route failure. Frequent route failures cause overheads such as time delay and flood of control packets to reconstruct a new route. Thus, it is not desirable for a mobile node to set up a link with a node located near the border of each other's transmission range. With the proposed algorithm, on the other hand, when a mobile node receives a RREQ message from a neighbor node located in its caution zone, the node ignores the message and thus does not flood the RREQ. This provides a reliable route that is not easily broken, even though the two nodes move in opposite directions as shown in Fig. 21.4b. Thus, the proposed algorithm reduces the control overhead required to reconstruct a new route.

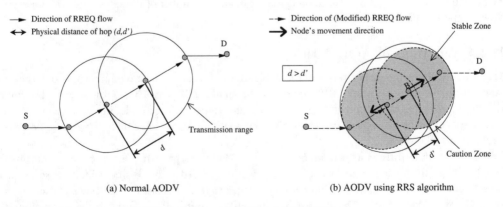

FIGURE 21.4 Comparison of RREQ flow between AODV and AODV-RRS.

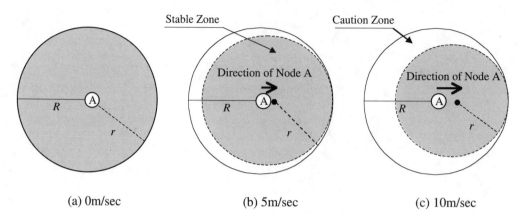

FIGURE 21.5 Variation of stable zone radius and center according to the mobile node's speed.

21.4.3 Effect of Stable Zone Radius

In the proposed algorithm, determining the optimum value of the stable zone radius variation rate due to node mobility is very important since it affects several performance parameters such as the number of hops between the source and destination, the time delay until a route is constructed, and route reliability. Figures 21.5a, b, and c show the stable zone size (or stable zone radius) when the speed of node A is 0, 5, or 10 m/sec, respectively. When node A does not move, the stable zone extends to the whole transmission range as shown in Fig. 21.5a. But the stable zone shrinks with the increase of node A's speed as shown in Figs. 21.5b and c.

The relation between the speed of a mobile node (S), transmission range radius (R), and stable zone radius (r) can be expressed as $r = R - \beta S$, where β is the parameter determining the variation rate of the stable zone radius due to host mobility. If we use a very high value of β, then the stable zone size shrinks drastically with a small increase of node speed, which can construct a more reliable route that cannot be easily broken. However, the cost of reliability is the increased number of hops between the source and destination. As shown in Fig. 21.4, the average length of a link in a selected route by AODV-RRS is shorter than that by AODV (that is, $d' \leq d$), which increases the number of hops in a route between the source and destination. Another problem is that if there are a small number of nodes in a network, a very small stable zone size due to a large value of β may cause no route construction. However, AODV-RRS has the important feature that the maintenance time of the selected route lasts considerably longer (and the route is thus more reliable) than the route selected by AODV. Thus, a tradeoff between reliable route construction and the optimal number of hops in each source–destination pair is needed, and our goal is finding the optimum value of β without causing much increase in the number of hops. We studied it for several β values, and it is discussed in Section 21.5.

21.4.4 Variations of AODV-RRS

As mentioned above, the variation rate of the stable zone radius due to host mobility (β) is an important parameter. In the proposed AODV-RRS algorithm, the value of β is fixed. Thus, a route may not be found by using a large value of β, although there are less reliable routes through nodes in caution zones. A less-reliable route may be better than no route in several applications. To solve this problem, we propose a variation of AODV-RRS.

In the algorithm, there is a "*β value field*" in the RREQ message. When a source node transmits a RREQ message, the predetermined value of β (β_max) is recorded at the β value field. If the source fails to find a route to the destination within a predetermined time, it decreases the β value and then starts a route discovery process again. This process repeats until the source finds a route to the destination. If a route is found with $\beta = 0$, then the route is the same route found by AODV.

AODV-RRS may require more time delay until the source node constructs a route to the destination. However, this variation of AODV-RRS can find the best route ever possible (although it may not actually be stable) to the destination. In the worst case, the route found by the algorithm is the same route that will be found by AODV. Thus, this scheme using carefully selected values of β_max and the number of levels between β_max and zero may show better performance than AODV and AODV-RRS.

When there are a small number of nodes in a network, another variation of AODV-RRS may be a good solution. When a source node initiates a route discovery process, the route is searched by AODV-RRS. If a route cannot be found within a predetermined time or predetermined number of hops, the algorithm is switched to AODV, and then another route discovery process starts.

21.4.5 Extensions to Other Ad Hoc Routing Protocols

The proposed RRS algorithm can be applied to other ad hoc routing protocols. Here, we give examples of such extensions of the RRS algorithm to existing protocols.

Destination-Sequence Distance Vector (DSDV) [7] is one of the well-known proactive ad hoc routing protocols. The RRS algorithm can easily be applied to DSDV. When a node updates its routing table from the routing control packets from its neighbor node, it can determine whether or not to update the routing table based on the RRS algorithm. If the neighbor node that sent the routing control packet is in the caution zone, the packet can be silently discarded without any update to its routing table. This makes sense because if a link between any of two neighboring nodes is likely to break easily, then it may be desirable to think that those nodes are not within each other's transmission range. Thus, RRS for DSDV can reduce the number of route update control packets while providing a reliable route to a destination.

Dynamic Source Routing (DSR) is one of the well-known reactive ad hoc routing protocols. Since its basic route discovery mechanism is very similar to that of AODV, the proposed RRS algorithm can easily be applied to the route discovery procedure of DSR except for some optimizations such as the promiscuous mode. The promiscuous mode is an on-the-fly packet overhearing technique used for efficient route cache update for a node. But we can apply the RRS algorithm to the technique, i.e., modify the overhearing rule such that a node decides according to the RRS algorithm whether or not it learns route information or updates its route cache from the route information contained in the overheard packet.

21.5 Performance Evaluation

We studied the proposed AODV-RRS protocol by simulation in various situations and compared it to AODV. For the simulation study, we used the Network Simulator (*ns*) [1] and a mobility extension of ns (i.e., *ns-2*) [2].

We assumed that initially 75 mobile nodes are distributed randomly in a flat area of 2250 m × 450 m. Each node in a location moves to a randomly selected location (we call it the target location) with a predetermined speed. Once a node reaches the target location, another random target location is selected. We ran our simulations with movement patterns generated by five different speeds (2.5, 5, 7.5, 10, and 12.5 m/sec). The reason why we limited the node's speed to 2.5 ~ 12.5 is that an ad hoc network is not applicable to the extremely high or low speed environment. Each simulation was executed for 300 seconds. Among 75 nodes, 30 nodes were randomly selected as source nodes, and they generated continuous bit rate (CBR) traffic. The packet size was 64 bytes, the packet generation rate was 4 packets/sec, and the bandwidth of each link was 2 Mb/sec. The radio transmission range (*R*) of each node was 250 meters. To study the effect of β value, we used three values of β: $\beta = 2$, $\beta = 3$, and $\beta = 4$. Table 21.1 summarizes the parameters used in our simulation study.

The mobility pattern of a mobile node followed a randomly selected scenario file. Multiple runs with different seed numbers were conducted for each scenario, and output data were averaged over those runs. These simulations of random scenarios are very similar to the approaches in [3–5]. For fair comparison, identical mobility and traffic scenarios were used for AODV and AODV-RRS. Both protocols detect a link breakage using a feedback from the MAC layer, and no additional network layer mechanism such as hello messages is used.

TABLE 21.1 Simulation Environments

Parameters	Value
Transmission range	250 m
Network size	2250 m × 450 m
Simulation time	300 sec
Number of mobile nodes	75
Number of traffic sources	50
Bandwidth	2 Mb/sec
Packet transmission rate	4 packets/sec
Traffic type	Constant bit rate
β	2, 3, and 4

Figure 21.6 shows the average route maintenance time per source node as a function of node mobility (speed). The average route maintenance time is an average period of time from the time when a route is established to the time when the route is broken. As shown in the figure, AODV-RRS shows longer route maintenance time than AODV, which indicates that the route established by AODV-RRS is more reliable than that established by AODV. Note that AODV-RRS shows longer route maintenance time when β values are large. This is due to the fact that the stable zone size becomes smaller with a large value of β, and thus a route lasts a longer period of time before it breaks. The average route maintenance time gets smaller as the speed is increased. For all classes of speed, AODV-RRS outperforms AODV in route maintenance time.

Figure 21.7 shows the average route recovery latency per source node as a function of speed. The average route recovery latency is an average period of time from the time when a route is broken to the time when a route is reconstructed, which includes the route discovery latency. The average route recovery latency of AODV-RRS is longer than that of AODV. This is because the number of hops in a route constructed by AODV-RRS is generally larger than that in a route constructed by AODV, and the probability that no route is found by AODV-RRS is higher than that of AODV, especially when the number of nodes in a network is very small. The average route recovery latency of AODV-RRS increases as the β value increases, since a high value of β generally causes an increased number of hops in a route.

From Figs. 21.6 and 21.7, we can see that there are tradeoffs between the average route maintenance time and the average route recovery latency. However, the differences between the average route maintenance time of AODV-RRS and that of AODV are on the order of seconds, while the differences in the average route recovery latencies of the two algorithms are on the order of milliseconds.

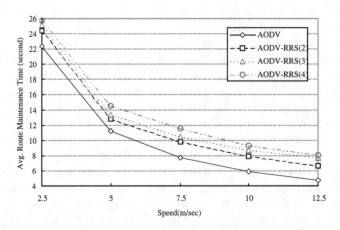

FIGURE 21.6 Average route maintenance time per source node.

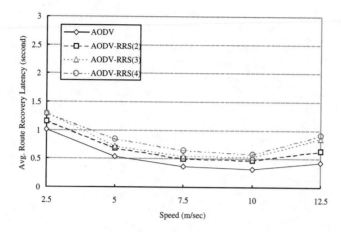

FIGURE 21.7   Average route recovery latency per source node.

Figure 21.8 shows the number of route disconnections per source node during the whole simulation time. As illustrated in the figure, AODV-RRS shows a smaller number of route disconnections than AODV. This is due to the fact that AODV-RRS constructs very stable routes, and thus the routes are not easily broken when nodes move to other locations. Although the number of route disconnections increases with the increase of node speed, the difference in the number of disconnections of the protocols becomes larger when node speed becomes higher because AODV-RRS constructs routes through nodes located very close to each other even when nodes move fast, and thus the routes last a longer period of time, while routes constructed by AODV are fragile even with a small movement of nodes. According to an analysis of the effects of mobility on TCP performance in mobile ad hoc networks [17], mobile nodes' movements result in frequent route disconnections, which in turn cause significant TCP throughput degradation. Although we did not conduct a TCP performance study, we expect that AODV-RRS also improves the TCP performance since it reduces the number of route disconnections.

From Figs. 21.6–21.8, we can expect that the characteristics of AODV-RRS with a reasonably chosen β value are especially good for multimedia applications that require constant delay jitters without frequent disconnections. While it is generally true that the delay bound for packet delivery is critical to multimedia applications, in ad hoc networks, how long a route remains connected will also be an important factor for determining service quality.

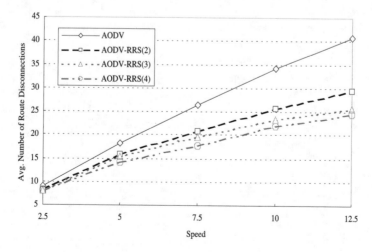

FIGURE 21.8   Number of route disconnections per source node.

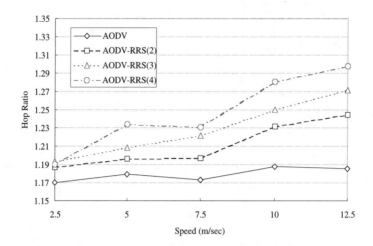

FIGURE 21.9 Normalized number of actual hops of a route.

Figure 21.9 shows the normalized number of the actual hops of a route. It is an average value of the ratio of the actual number of hops to the optimal number of hops for successfully arrived packets at destinations. As shown in the figure, as the β value increases, the number of hops of AODV-RRS also increases. This is due to the fact that AODV-RRS tries to maintain a more reliable route at the sacrifice of an increased number of hops. But the difference between AODV-RRS and AODV is not so significant, at most about 0.13 hops even when a mobile node's speed is 12.5 m/sec and $\beta = 4$.

Figure 21.10 shows the average packet delay as a function of speed. The average packet delay is the average period of time from when the sender transmits a packet to when the packet arrives at the destination. As shown in the figure, AODV and AODV-RRS show very comparable packet delays. This can be explained as follows. If there are no link failures, a route constructed by AODV will have a shorter packet delay than a route constructed by AODV-RRS, since the route constructed by AODV has the smaller number of hops. However, the increased number of link failures in a route constructed by AODV lengthens the packet delay. Thus, frequent link reconstructions by AODV lengthen the average packet delay.

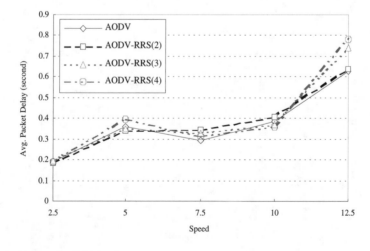

FIGURE 21.10 Average packet delay.

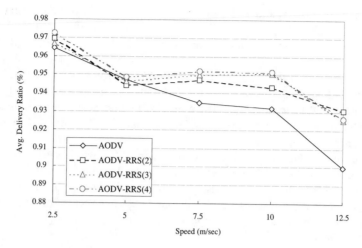

FIGURE 21.11 Average packet delivery ratio.

Figure 21.11 shows the average packet delivery ratio as a function of speed. This ratio is the number of packets that arrived at the destination successfully divided by the number of packets transmitted by the sender. The packet delivery ratios of the protocols generally become smaller as the node speed increases. When a mobile node's speed is low (e.g., 2.5 or 5 m/sec), the average delivery ratios of the protocols are comparable. But when the speed of a mobile node becomes higher, the average delivery ratio by AODV-RRS is better than that by AODV since the number of link disconnections increases significantly with host mobility.

21.6 Conclusions

In ad hoc networks, network topology changes drastically and unpredictably due to host mobility. Existing protocols for ad hoc networks have the problem of a fragile route. Consequently, a selected route comes to have a short route maintenance time, which causes the overhead of reestablishing a new route. To solve the problem, we propose new route selection algorithms to establish a reliable route that does not break easily due to the frequent topological changes caused by mobile nodes' mobility. We have introduced the concept of stable zone and caution zone. These zones change dynamically depending on a mobile node's speed and direction. Our simulation study shows that a route selected by AODV-RRS lasts a considerably longer period of time than that selected by AODV, and the number of route disconnections by AODV-RRS is smaller than that by AODV.

References

[1] K. Fall and K. Varadhan, ns notes and documentation, The VINT project, UC Berkeley, LBL, USC/ ISI, and Xerox PARC, May 1998.
[2] The CMU Monarch Project, The CMU Wireless and Mobility Extensions to ns, URL: http:// www.monarch. cs.cmu.edu/.
[3] S.-J. Lee, M. Gerla, and C.-K. Toh, A Simulation Study of Table-driven and On-Demand Routing Protocols for Mobile Ad Hoc Networks, *IEEE Network*, 1999, pp. 48–54.
[4] S.R. Das, C.E. Perkins, and E.M. Royer, Performance Comparison of Two On-Demand Routing Protocols for Ad-hoc Networks, *Proceedings of the Nineteenth Annual Joint Conference of the IEEE Computer and Communications Societies (INFOCOM 2000)*, Tel Aviv, Mar. 2000, pp. 3–12.

[5] P. Johansson, T. Larsson, and N. Hedman, Scenario-Based Performance Analysis of Routing Protocols for Mobile Ad-hoc Networks, *Proceedings of the Fifth Annual ACM/IEEE International Conference on Mobile Computing and Networking (MobiCom '99)*, Seattle, WA, Aug. 1999, pp. 195–206.

[6] E.M. Royer and C.-K. Toh, A Review of Current Routing Protocols for Ad-Hoc Mobile Wireless Networks, *IEEE Personal Communications*, Apr. 1999, pp. 46–55.

[7] C.E. Perkins and P. Bhagwat, Highly Dynamic Destination-Sequenced Distance-Vector Routing (DSDV) for Mobile Computers, *ACM Comput. and Commun. Rev. (ACM SIGCOMM '94)*, Oct. 1994, pp. 234–244.

[8] D. Johnson and D. Maltz, Dynamic source routing in ad-hoc wireless networks, in T. Imielinski and H. Korth, Eds., *Mobile Computing*, Kluwer Academic Publishers, Dordrecht, 1996, Chap. 5.

[9] C.E. Perkins, Ad Hoc On Demand Distance Vector (AODV) Routing, Internet Draft, Nov. 1997.

[10] C.E. Perkins and E.M. Royer, Ad-hoc On-demand Distance Vector Routing. *Proceedings of the 2nd IEEE Workshop on Mobile Computing Systems and Applications (WMCSA '99)*, Feb. 1999, pp. 90–100.

[11] S. Basagni, I. Chlamtac, and V.R. Syrotiuk, Dynamic Source Routing for Ad-Hoc Networking Using the Global Positioning System, *Proceedings of the IEEE Wireless Communications and Networking Conference 1999 (WCNC '99)*, Sep. 1999, pp. 301–305.

[12] E.D. Kaplan, *Understanding the GPS: Principles and Applications*, Artech House, Norwood, MA, Feb. 1996.

[13] S. Basagni et al., Route Selection in Mobile Multimedia Ad-hoc Networks, *Proceedings of the IEEE International Workshop on Mobile Multimedia Communications (MoMuC '99)*, Nov. 1999, pp. 97–103.

[14] Y.-B. Ko and N.H. Vaidya, Location-Aided Routing (LAR) in Mobile Ad Hoc Networks, *Proceedings of ACM/IEEE International Conference Mobile Computing and Networking Conference (MobiCom '98)*, Dallas, TX, Oct. 1998, pp. 66–75.

[15] C.-K. Toh, *Wireless ATM and Ad-hoc Networks: Protocols and Architectures*, Kluwer Academic Publishers, Dordrecht, 1997.

[16] W.-I. Kim, D.-H. Kwon, and Y.-J. Suh, A Reliable Route Selection Algorithm Using Global Positioning Systems in Mobile Ad-hoc Networks, *Proceedings of the IEEE International Conference on Communications (ICC '2001)*, June 2001.

[17] G. Holland and N.H. Vaidya, Analysis of TCP Performance over Mobile Ad Hoc Networks, *Proceedings of the Fifth Annual ACM/IEEE International Conference on Mobile Computing and Networking (MobiCom '99)*, Seattle, WA, Aug. 1999, pp. 219–230.

[18] C.E. Perkins, Ed., *Ad Hoc Networking*, Addison-Wesley, Reading, MA, 2001.

VII

Power Management in Ad Hoc Wireless Networks

22

Power-Aware Wireless Mobile Ad Hoc Networks

Ahmed M. Safwat
Queen's University, Canada

Hossam S. Hassanein
Queen's University, Canada

Hussein T. Mouftah
Queen's University, Canada

Abstract

Power-aware protocol design is important due to the limited battery capacity of the mobile devices making up the ad hoc wireless network. The performance of the medium access control (MAC) scheme not only has a significant effect on the performance of the routing method employed, but also on the energy consumption of the wireless network interface card (NIC). Thus, a thorough energy-based comparative and performance study is essential to any bandwidth-based study. We explore the shortcomings of the MAC schemes proposed for ad hoc wireless networks in the context of power awareness herein. In addition, we investigate the potential energy consumption pitfalls of non–power-based and power-based routing schemes. Moreover, we introduce a thorough energy-based performance study of power-aware routing protocols for wireless mobile ad hoc networks. Our energy consumption model is based on an implementation of the IEEE 802.11 physical layer convergence protocol (PLCP) and MAC sublayers. We also present the statistical performance metrics measured by our simulations.

22.1 Introduction

Most of the research performed in the field of wireless ad hoc networks has focused on the problems of routing and medium access control (MAC). Nevertheless, the limited battery capacity of the mobile devices making up the ad hoc network draws our attention to the importance of power awareness in wireless ad hoc network design. Hence, ad hoc routing and MAC protocols ought to be energy conservative. A thorough energy-based comparative and performance study is essential to any bandwidth-based study. The simulation studies carried out for ad hoc networks fall short of examining essential power-based performance metrics, such as average node and network lifetime, energy-based protocol fairness, average dissipated energy per protocol, and standard deviation of the energy dissipated by each individual node.

In wireless ad hoc networks, the nonexistence of a centralized authority complicates the problem of medium access control. The centralized medium access regulation procedures, undertaken by base stations in cellular networks, have to be enforced in a distributed, and hence collaborative, fashion by mobile stations in the ad hoc network. Mobile stations may contend simultaneously for medium access. Consequently, transmissions of packets from distinct mobile terminals are more prone to overlap, resulting in packet collisions and energy losses. Likewise, the performance of the MAC scheme has a great effect on the performance of the routing method employed and on the energy consumption of the wireless network interface card (NIC).

In a multi-hop ad hoc network, or in a wireless LAN operating in independent basic service set (IBSS) mode, wireless stations must always be in standby mode to be able to receive incoming traffic from their neighbors. Due to the nonexistence of a fixed infrastructure, the wireless stations must always be awake. On the contrary, in a wide area or local area cellular environment, wireless nodes may be scheduled to sleep. Base stations and wireless access points, being central controllers, will be in charge of buffering all incoming packets to sleeping nodes. Thus, in a wireless ad hoc network, wireless stations may not sleep. All of the wireless nodes will consume power unnecessarily due to overhearing the transmissions of their neighboring nodes. Although this obviously wastes an extensive amount of the total consumed energy throughout the lifetime of the wireless station, on-demand routing protocols require that nodes remain powered on at all times so as to participate in on-the-fly route setup using route request broadcasts and route reply packets. Besides, table-driven protocols also require constant operation in the active state in order to exchange periodic updates and participate in packet routing. Utilizing a set of clusterheads, or *virtual base stations*, so as to reduce the idle time power consumption by having a fraction of the nonbackbone nodes sleep at a time has not been studied yet. There is an obvious tradeoff between energy conservation and routing accuracy. As in other cellular systems, the basic service set (BSS) and the extended service set (ESS) modes of operation in wireless LANs enable the mobile terminals to reduce their network interface–related energy consumption. Due to the large channel acquisition overhead, small packets have disproportionately high energy costs. However, in wireless ad hoc networks, other means need to be exercised so as to obtain power-efficient operation. Idle power consumption reflects the cost of listening to the wireless channel.

This chapter is organized as follows. Section 22.2 surveys the proposed MAC schemes for ad hoc networks and discusses their shortcomings in the context of power awareness. Section 22.3 presents a thorough comparative study of non–power-based and power-based routing schemes for ad hoc networks. In Section 22.4, two proposed application-level energy conservation techniques are described, and their potential for power conservation is examined. Finally, Section 22.5 presents the chapter summary and conclusions. Recommendations for power-efficient protocol design in ad hoc networks are also discussed.

22.2 Medium Access and Energy Conservation

The wireless stations in an ad hoc network must cooperatively resolve the problem of simultaneous medium access. Likewise, wireless ad hoc stations must resolve the problem of hidden terminals. A neighbor of the destination that is out of the wireless range of the source may interfere with the transmissions of the source. In Fig. 22.1, H is the hidden terminal; either H or S may transmit to D at a time, otherwise a collision will take place. It is noteworthy that neither the source nor the interferer will be aware of the collision. It is only the lack of a positive acknowlegment (ACK) from the destination that triggers the source to retransmit. Accordingly, the source will be aware of the occurrence of a collision or data corruption. On the other hand, a node that is in range of the sender but not the receiver is called an exposed terminal. Nevertheless, an exposed terminal can transmit at the same time as the sender, without causing a collision to occur. In spite of that, using the traditional carrier sense multiple access (CSMA) scheme, the exposed terminal will defer from accessing the channel. As a result, the capacity of the wireless ad hoc network will be reduced. Hence, energy is wasted groundlessly.

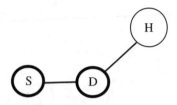

FIGURE 22.1 The hidden terminal problem in wireless networks.

In [1], a scheme that solves the hidden terminal problem using a busy tone is proposed. The protocol, named busy tone multiple access (BTMA), was developed for cellular networks. A busy tone is sent only while the base station is receiving. Thus, transmitters are prevented from accessing the channel. All nodes, including hidden terminals, within the cell site receive the busy tone and back off. Wireless nodes are allowed to transmit only in the absence of busy tones. Once a node detects a busy tone on the secondary channel, it cannot use the data channel even if no signal was detected on it. This implies that a node is allowed to use the data channel even while another node's transmissions are being heard. Apparently, the BTMA protocol cannot be used in a multi-hop wireless ad hoc network since there are no base stations. In addition, BTMA uses two widely separated channels instead of one and increases the hardware costs and complexity of the NIC.

In contrast with BTMA, dual busy tone multiple access (DBTMA) [2] uses two busy tones instead of one. In particular, the hidden terminal problem is resolved using the receive busy tone signal, whereas the exposed terminal problem is overcome by using the transmit busy tone signal. DBTMA wastes battery capacity by forcing the wireless node to continuously sense the medium for the transmit busy tone and the receive busy tone signals. A more energy-efficient MAC protocol would consider turning off the transceiver during standby time to save power.

In multiple access collision avoidance (MACA) [3], a node wishing to transmit a data packet to a neighbor first sends a request-to-send (RTS) frame to the neighbor. All nodes that receive the RTS are not allowed to transmit. Upon reception of the RTS, the neighbor that the RTS was sent to replies with a clear-to-send (CTS) frame. Also, any node that hears the CTS transmission is prevented from using the channel. Hence, the RTS-CTS message exchange clearly alleviates the hidden terminal problem present in wireless networks. Due to this scheme, data frames are, at least in theory, delivered collision-free. As a result, collisions can only affect control packets. In this case, the IEEE exponential backoff is used to resolve MAC contentions for control packets. In practice, however, collisions may still affect data frames. Nodes that do not properly receive a CTS frame are eligible to use the medium and their transmissions might overlap with those of the source. Hence, this MAC scheme will only decrease the probability of data collisions and the data remains vulnerable to corruption. From the standpoint of energy consumption, corrupted CTS frames will result in idle time energy losses for those neighbors that successfully receive the CTS frame. The neighborhood-wide energy loss is proportional to the number of neighbors of the sending station. On the downside also, MACA does not use link-layer positive or even negative ACKs, but rather end-to-end ACKs. It is worthy of notice, however, that the transmission of an RTS frame considerably reduces the energy costs of data collisions. RTS frames result in a favorable reduction in the consumed energy in case of a collision, as compared with the consumed energy in addition to the larger delay due to collision time, if it were the actual data frame for which the RTS is being sent. These delay and energy savings will be achievable in most cases. The use of the RTS frame is nullified whenever the size of data is comparable to that of the RTS frame. A threshold is used to specify the size of the data frames for which an RTS frame ought to be sent.

MACAW [4], on the other hand, uses link layer ACKs to increase data throughput. The IEEE 802.11 MAC and PHY standard [5–7], as in MACAW, utilizes link layer positive ACKs for all unicast traffic. Manifestly, the ACK frame allows the conveyance of fast ACKs, and fast recovery in the event of its absence. ACK frames may only be used with the unicast traffic. ACK frames sent in response to a broadcast message will have a large collision probability and will waste unnecessary energy and network bandwidth.

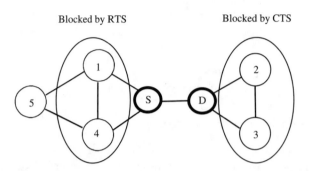

FIGURE 22.2 MACAW suffers from the exposed terminal problem.

In a wireless ad hoc network, almost all the proposed routing protocols rely to a large extent on broadcasts. Periodic updates in table-driven schemes, and route discovery, route setup, and route maintenance messages in on-demand protocols, are all examples of network-level broadcasts. An extensive amount of battery capacity would be wasted on ACK transmissions and ACK collisions throughout the wireless network. Apparently, because of the use of explicity ACKs, the exposed terminal problem was reintroduced, and thus the energy supply of the wireless nodes is gradually depleted by idle time power consumption. If node 1 or node 4 in Fig. 22.2 wants to send a data packet to node 5, it will needlessly wait until the end of the transmission. This contributes to increasing the total idle time energy.

In MACA by invitation (MACA-BI) [8], data must be foreseen beforehand by the receivers. Hence, MACA-BI may only be used in periodic and not unpredictable traffic. In the presence of bursty traffic, more efficient protocols have to be used. This is true since it will be impossible for the receivers to predict the instances at which the transmitters will send their data frames. Incorrect predictions will cause all the neighboring nodes to waste an amount of energy that is proportional to the size of the ready-to-receive (RTR) frame, which is sent by the receiver in place of the CTS frame. The sender then responds with the actual data, and the use of the RTS frame is rendered obsolete. In a wireless ad hoc environment, in which nodes are allowed to move freely at all times, anticipation of a source's transmissions would be extremely difficult due to the unpredictable patterns of contention for medium access by nodes neighboring to the source. Therefore, it would still not help the receiver to know the transmission schedule of the sender. Complex application-level and contention prediction schemes would be required. The authors claim that the RTR-DATA dialogue achieves improved performance over MACAW. As previously mentioned, the efficiency of this scheme relies on the ability to predict when the sources have data to send.

In all the above MAC schemes, nodes will consume most of their battery capacity while in their idle states, that is, while doing nothing. No special MAC-based energy conservation measures were adopted by any of them.

In [9], measurements show standby:receive:transmit ratios of 1:1.05:1.4. Ratios of 1:2:2.5 [10] and 1:1.2:1.7 [11] are reported elsewhere. Consider, for example, a single-hop ad hoc network (such as the one shown in Fig. 22.3) consisting of k nodes and a MAC scheme that uses an RTS-CTS-DATA-ACK message exchange sequence. Any transmission by one node will be heard by all the $k - 1$ other nodes. Consequently, a single successful packet transmission and reception would consume $E\_RTS_{Tx} + (k - 1)$ $E\_RTS_{Rx} + E\_CTS_{Tx} + (k - 1)E\_RTS_{Rx} + E\_P-2-P\_Data_{Tx} + (k - 1)E\_P-2-P\_Data_{Rx} + E\_ACK_{Tx} + (k - 1)$ E_ACK_{Rx}. This is in addition to the energy dissipated by all k nodes during the defer/back-off period.

Unlike other approaches that employ information from above the MAC layer to control radio power, the power-aware multi-access protocol with signaling (PAMAS) [11] is an energy conservative MAC protocol proposed for wireless ad hoc networks. In addition to the primary data channel, PAMAS uses a secondary channel, called the signaling channel, to carry control traffic. The neighbours of the traffic pair are prohibited from using the channel during an ongoing transmission via an RTS-CTS dialogue. A node sets its wireless NIC to sleep mode in case it overhears a neighbor's transmission. It is not clear if the energy-conserving behavior of PAMAS would negatively affect the end-to-end delay or not. In this

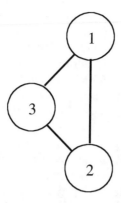

FIGURE 22.3 Wireless NIC energy consumption in a single-hop scenario.

scheme, a current receiver can force a neighbor to defer using the medium by sending a busy tone over the signaling channel. This can be used to force the sender of an RTS to defer. All nodes that successfully receive the busy tone are prevented from using the medium for the period of the data transmission over the primary channel. Even if the RTS is detected by the receiver, its CTS transmission will be corrupted by that of the busy tone since they will both be sent simultaneously over the control channel. Since most of the consumed energy in a wireless ad hoc network is attributed to the idle time, PAMAS will obtain high energy-conservation gains only in networks with high traffic loads. In networks with low to modest traffic, the energy-based performance of PAMAS will be close to that of other CSMA protocols. Therefore, nodes running PAMAS will still consume most of their battery capacity while in their idle states. Based on the aforementioned, PAMAS takes no measures so as to reduce idle time energy consumption unless there is an ongoing transmission.

22.3 Routing and Energy Efficiency in Wireless Ad Hoc Networks

Routing protocols in wireless mobile ad hoc networks can be classified as table-driven and on-demand. In table-driven routing protocols, all the mobile stations are required to have complete knowledge of the network through their periodic and incremental/triggered updates. Unnecessary periodic updates may have a negative effect on energy conservation. Table-driven routing is also known as proactive routing.

Examples of table-driven routing protocols include global state routing (GSR) [12] and destination-sequenced distance vector (DSDV) [13]. Nodes running GSR exchange link states packets (LSPs) with their neighbors. In DSDV, sequence numbers are used to provide a means to enable mobile nodes to distinguish old routes, as a result of mobility, from new ones. Thus, the formation of loops is avoided. On-demand routing requires that routes be built between nodes only as desired by source nodes. Hence, the terms on-demand and reactive can be used interchangeably. On-demand routing has two major components: route discovery and route maintenance. In route discovery, a source uses flooding to acquire a route to its destination. This degrades system-wide energy conservation. The transit nodes, upon receiving a query, learn the path to the source and enter the route in their forwarding tables. The destination node responds using the path traversed by the query. Route maintenance is responsible for reacting to topological changes in the network, and its implementation differs from one algorithm to the other.

On-demand protocols include ad hoc on demand distance vector routing (AODV) [14] and dynamic source routing (DSR) [15]. In on-demand protocols, route discovery and maintenance may become inefficient under heavy network load since intermediate nodes will have a higher probability of moving due to the delay in packet transmissions attributed to MAC contention. Hence, routes will also have a higher probability of breaking as a result of mobility. This wastes battery power, and thus the lifetime of the wireless nodes decreases. Moreover, flooding of route request and route reply packets in on-demand routing protocols may result in considerable energy drains under a realistic energy consumption model that takes idle time and promiscuous mode power into account. Every station that hears the route request broadcasts will consume an amount of energy proportional to the size of the broadcast packet. In addition, stations that hear a corrupted version of a broadcast packet will still consume some amount of energy. Besides, the simulation studies carried out for table-driven and on-demand routing protocols fall short of providing necessary power-based performance metrics, such as average node and network lifetime, average dissipated energy per protocol, standard deviation of the energy dissipated by each individual node, etc.

In cluster-based routing protocols, the nonclusterhead wireless nodes rely on the elected clusterheads to route the data packets to their destinations. Our virtual base stations (VBS)-based routing architecture in [16] is one example. The wireless mobile ad hoc network relies on the wireless mobile infrastructure created by VBS to provide communications between any pair of wireless stations. In VBS routing, only the nodes that form the dynamic infrastructure participate in the process of routing. Hence, those nodes will be more vulnerable than others to unfairly losing their remaining battery capacity. In [17], we propose a novel infrastructure for wireless mobile ad hoc networks. In our proposed architecture, namely, the Power-Aware Virtual Base Stations (PA-VBS) architecture, a mobile node is elected from a set of nominees to act as a base station within its zone based on its residual battery capacity. A wireless station is chosen by one or more stations to act as their PA-VBS based on a couple of energy thresholds. Unlike other clustering schemes, PA-VBS allows the mobile nodes to use their valuable battery power fairly; it does not drain all the battery capacity of the clusterheads since it introduces the concept of service denial, and, hence, it attains fair clustering. Moreover, it balances the load among the wireless nodes regardless of the carried routing load. Consequently, PA-VBS can be utilized as a basis for routing in ad hoc networks. Moreover, PA-VBS can form the cornerstone for the wise distribution of the network load among all the viable paths between a source and destination pair. It is shown that the proposed scheme attains load balancing and fair clustering. PA-VBS is also shown, in our simulation experiments, to react positively, from an energy-aware perspective, to different routing loads. PA-VBS may be utilized for routing and medium access regulation. Any routing protocol may nonetheless be used with it.

A number of routing proposals for ad hoc networks took energy conservation into consideration so as to extend the lifetime of the wireless nodes by wisely using their battery capacity. In [18], minimum total power routing (MTPR) is proposed. If the total transmission power for route R is P_R, then the route can be obtained from $P_{MTPR} = \min_{R \in S} P_R$, where S is the set containing all possible routes. On the downside, this routing approach will in most cases tend to select routes with more hops than others. This is realizable due to the fact that transmission power is inversely proportional to distance [18]. Thus, more energy may be wasted network-wide since a larger number of nodes is now involved in routing, as all nodes that are neighbors to the intermediate nodes will also be affected, unless they were in sleep mode.

Minimum battery cost routing (MBCR) [19] utilizes the sume of the inverse of the battery capacity for all intermediate nodes as the metric upon which the route is picked. However, since it is the summation that must be minimal, some hosts may be overused because a route containing nodes with little remaining battery capacity may still be selected. Min-max battery cost routing (MMBCR) [19] treats nodes more fairly from the standpoint of their remaining battery capacity. Smaller remaining battery capacity nodes are avoided and ones with larger battery capacity are favored when choosing a route. The route is found using the equation

$$P_{MMBCR} = \min_{R \in S} [\max_{n \in R} (1/\text{Battery\_Capacity}_n)]$$

However more overall energy will be consumed throughout the network because minimum total transmission power routes are no longer favored.

In [20], MTPR is used when all the nodes forming a path (note that one path is sufficient) have remaining battery capacity that is above a so-called battery protection threshold, and MMBCR is used if no such path exists. The combined protocol is called conditional max-min battery capacity routing (CMMBCR). The performance evaluation study carried out in [20] does not use a realistic power

TABLE 22.1 Power Consumption Figures for a 2-Mbps Wireless NIC

Mode	Reported Measurement
Transmit	1.33 W
Receive	967 mW
Idle	843 mW

TABLE 22.2 Interframe Spaces for DSSS Used in Simulations

Interframe Space	Value Used in Simulations (μsec)
Short interframe space (SIFS)	10
Random backoff slot	20
DCF Interframe space (DIFS)	50

consumption model. Instead, it assumes that the power drained from the batteries of the wireless nodes is only proportional to the amount of traffic transmitted and the amount of traffic relayed by the node. Promiscuous mode operation and the time spent by a node in idle mode drastically affect its drained power. Also, the power consumed by a frame transmission is not equivalent to the power spent receiving a frame. The conducted performance study also lacks detailed packet-level simulations. Also, the IEEE 802.11 MAC scheme was not considered. Consequently, the influence of the MAC layer on energy dissipation was not examined. One routing scheme may be vulnerable to MAC-level contention more than another resulting in larger defer durations and, possibly, more collisions. All of this adversely affects the performance of the routing protocol in a way that is unpredictable. A thorough packet-level performance evaluation study would help make the design of a power-based protocol more pellucid. Thus, the reported results may change significantly under a realistic IEEE 802.11 wireless ad hoc network setting.

In [21], we present a thorough energy-based performance study of power-aware routing protocols for wireless mobile ad hoc networks. Our energy consumption model is based on an implementation of the IEEE 802.11 physical layer convergence protocol (PLCP) and medium access control (MAC) sublayers. To our knowledge, this is the first such detailed performance study. Our energy model adopts the results reported in [22] and is shown in Table 22.1 for a 2 Mb/sec wireless NIC. However, unlike the linear model in [22], our study takes into account contention and transmission failures. This is critical to the proper quantification of energy loss in an ad hoc network

In our simulations, we provide a simple implementation for the PLCP sublayer. This is useful for our energy-based study so as to quantify most of the sources of energy loss since PLCP preambles and headers are transmitted at lower rates than those used for data frames. The PLCP preamble and a PLCP header are always attached to the start of each packet for synchronization purposes. Also, the length of the data packet and the rate are both obtained from the PLCP header by the receiver. To enable a lower-rate network interface card to communicate with a higher-rate one, the PLCP preambles and headers are transmitted at 1 Mb/sec. Taking the 1 Mb/sec PLCP rate into acount in the performance experiments is of utmost importance because usualy the lower the rate the more energy consumed. Various IEEE 802.11 parameters were considered in the study. Table 22.2 lists the IEEE 802.11 delay parameters for direct sequence spread spectrum (DSSS) used in our study.

The simulators measured the following noteworthy statistical performance metrics:

1. *Average Dissipated Energy* — This is measured in W.sec (joules). The smaller this value, the more energy-efficient the routing protocol is, and vice versa.
2. *The Standard Deviation of the Dissipated Energy* — This is also in joules. The computed value reflects whether the routing scheme overused any number of nodes. This is a very important performance measure since it is a measure of protocol fairness. Therefore, the sought value for this metric is one that is as close as possible to zero. The smaller this value, the fairer the routing criterion is.

3. *Energy Consumption Shares for Control Frame Transmissions and Receptions* — This shows the percentage of the total energy consumed throughout the simulation by all stations for RTS transmissions, RTS receptions, CTS transmissions, CTS receptions, ACK frame transmissions, and ACK frame receptions separately. These experiments show which frame type dominates energy consumption at the control frame level.

4. *Energy Consumption Shares for Defer and Data Transmissions and Receptions* — This shows the percentage of the total energy consumed throughout the simulation by all stations for defer events (including SIFS, random backoff, and DIFS), point-to-point data transmissions, point-to-point data receptions, broadcast data transmissions, and broadcast data receptions separately.

22.4 Application-Level Energy Conservation

Energy-conserving algorithms that operate above on-demand ad hoc routing protocols, such as DSR and AODV introduce a tradeoff btween routing fidelity and extended network lifetimes. Such algorithms promise minor modifications to the underlying routing protocols. The major contribution of such algorithms lies in turning off the wireless transceiver while being idle.

Two application-level energy conservation algorithms designed to run on top of existing on-demand schemes were proposed in [23]. Mobile nodes running the basic energy-conserving algorithm (BECA) power-off as long as possible. BECA does not take node density into consideration. A node is interrupted from its sleeping state only if it has data to send. The on-demand routing protocol must retry route requests a number of times in order to increase th eprobability of getting at least one route request through. This is the main disadvantage of this algorithm as it obviously becomes a source of energy loss and MAC contention. Evidently, route requests may collide, and none may ever successfully get through. Consider, for example, if the neighbor had been sleeping, and the route requests sent while it was listening were corrupted by another neighbor's transmissions.

The second scheme is the adaptive fidelity energy conserving algorithm (AFECA). In contrast with BECA, AFECA utilizes node density to switch off radios for longer time durations. Therefore, the number of neighbors must be estimated by each node. This scheme does not measure neighborhood size accurately. As a result, the neighborhood size may be underestimated by a node. Underestimation of neighborhood size causes a node to sleep for shorter periods and hence consume power needlessly. On the other hand, its overestimation will increase the node's sleep time and will in many cases translate to routing latency or even packet losses.

22.5 Summary and Conclusions

In a wireless multi-hop ad hoc network, the lack of a fixed infrastructure dictates that the wireless stations must always be awake. On the contrary, in a wide area or local area cellular environment, wireless stations may be scheduled to sleep. Base stations and wireless access points, being central controllers, will be in charge of buffering all incoming packets to sleeping nodes. Thus, in a wireless ad hoc network, wireless stations may not sleep. Consequently, all of the wireless nodes iwll consume power unnecessarily due to overhearing the transmissions of their neighboring nodes.

The hidden terminal problem in multi-hop wireless ad hoc networks is alleviatd using the RTS-CTS dialogue, which only represents a parial solution. As a result, savings in energy and delay are achieved, but collisions may still take place. Nevertheless, the use of the FTS frame is nullified whenever the size of data is comparable to that of the FTS frame. Although an exposed terminal may transmit at the same time as the sender, without causing a collision, the use of link-layer ACKs in the current IEEE 802.11 standards inhibits exposed terminals as well from using the medium.

Unnecessary periodic updates in table-driven routing protocols may have a negative effect on energy conservation. Nevertheless, the flooding of route request and route reply packets in on-demand routing protocols also results in considerable energy drains under a realistic energy-consumption model that takes idle time and promiscuous mode power into account. Every station that hears the route request broadcasts will consume an amount of energy proportional to the size of the broadcast packet. In addition, stations that hear a corrupted version of a broadcast packet will still consume some amount of energy.

Our MAC-based investigations [21] reveal that battery capacity alone may not be used to satisfy our requirements for a power-efficient routing scheme. One routing scheme may be vulnerable to MAC-level contention more than another resulting in larger defer durations and, possibly, more collisions. All of this adversely affects the performance of the routing protocol in a way that is unpredictable.

Utilizing the energy-efficient wireless infrastructure formed by PA-VBS [17] so as to reduce the standby power consumption by having a fraction of the non-backbone nodes turn off their wireless transceivers, promises considerable energy conservation gains. Unlike other infrastructure formation schemes, PA-VBS allows the wirelesss mobile nodes to use their valuable battery power fairly; it does not drain all the battery capacity of the clusterheads since it introduces the concept of service denial, and, hence, it attains fair clustering. Moreover, it balances the load among the wireless nodes regardless of the carried routing load. Consequently, PA-VBS can be utilized as a basis for medium access regulation and routing in ad hoc networks.

References

[1] F. Tobagi and L. Kleinrock, Packet Switching in Radio Channels: Part II - The Hidden Terminal Problem in CSMA and Busy-Tone Solution, *IEEE Transactions on Communications COM-23*, Dec. 1975, pp. 1417–1433.

[2] Z. Haas and J. Deng, Dual Busy Tone Multiple Access (DBTMA): A New Medium Access Control for Packet Radio Networks, *IEEE International Conference on Universal Personal Communications*, Oct. 1998.

[3] P. Karn, MACA — A New Channel Access Method for Packet Radio, *Proceedings of the 9th ARRL/CRRL Amateur Radio Computer Networking Conference*, Sep. 1992.

[4] V. Bharghavan, A. Demers, S. Shenker, and L. Zhang, MACAW: A Media Access Protocol for Wireless LAN's, *ACM SIGCOMM*, 1994.

[5] ETSI TC-RES, Radio Equipment and Systems (RES); HIgh PErformance Radio Local Area Network (HIPERLAN); Functional Specification, ETSI, 06921 Sophia Antipolis Cedex — France, July 1995, draft prETS 300 652.

[6] European Telecommunications Standards Institute, ETSI HIPERLAN/1 standard, Oct. 2000. Available at http://www.etsi.org/technicalactiv/hiperlan1.htm.

[7] IEEE LAN/MAN standards committee, Wireless LAN Medium Access Protocol (MAC) and Physical Layer (PHY) Specifications, IEEE Standard 802.11, 1999.

[8] F. Talucci, M. Gerla, and L. Fratta, MACABI (MACA By Invitation): A Receiver Oriented Access Protocol for Wireless Multiple Networks, in PIMRC '97, Helsinki, Sep. 1–4, 1997.

[9] Stemm, M. and Katz, R.H., Measuring and reducing energy consumption of network interfaces in handheld devices, *IEICE Transactions on Fundamentals of Electronics, Communications, and Computer Science*, 80, 1125–1131, Aug. 1997.

[10] O. Kasten, Energy Consumption, Swiss Federal Institute of Technology. http://www.inf.ethz.ch/~kasten/research/bathtub/energy_consumption.html, 2001.

[11] S. Singh and C.S. Raghavendra, Power-Efficient MAC Protocol for Multihop Radio Networks, *Proc. of IEEE PIRMC '98 conf.*, Sep. 1998, vol. 1, pp. 153–157.

[12] T.-W. Chen and M. Gerla, Global State Routing: A New Routing Scheme for Ad Hoc Wireless Networks, *Proceedings of IEEE ICC '98*.

[13] C. Perkins and P. Bhagwat, Highly Dynamic Destination-Sequenced Distance-Vector Routing (DSDV) for Mobile Computers, *Computer Communications Review*, Oct. 1994, pp. 234–244.

[14] C. Perkins and E. Royer, Ad Hoc On Demand Distance Vector Routing, *Proceedings of the 2nd IEEE Workshop on Mobile Computing Systems and Applications*, New Orleans, LA, Feb 1999, pp. 90–100.

[15] D. Johnson and D. Maltz, Dynamic source routing in ad hoc wireless networks, *Mobile Computing*, Kluwer Academic Publishers, Dordrecht, 1996.

[16] A. Safwat and H. Hassanein, Infrastructure-based routing in wireless mobile ad hoc networks, *Journal of Computer Communications*, 25, 210–224, 2002.

[17] A. Safwat, H. Hassanein, and H. Mouftah, Power-Aware Fair Infrastructure Formation for Wireless Mobile Ad hoc Communications, *Proceedings of IEEE Globecom 2001*, Vol. 5, pp. 2832–2836.

[18] K. Scott and N. Bambos, Routing and Channel Assignment for Low Power Transmission in PCS, *Proc. ICUPC '96*, Oct. 1996, vol. 2, pp. 498–502.

[19] S. Singh, M. Woo, and C. Raghavendra, Power-Aware Routing in Mobile Ad Hoc Networks, *Proc. MobiCom '98*, Dallas, TX, Oct. 1998.

[20] C.K. Toh, Maximum Battery Life Routing to Support Ubiquitous Mobile Computing in Wireless Ad Hoc Networks, *IEEE Communications Magazine*, June 2001.

[21] A. Safwat, H. Hassanein, and H. Mouftah, Energy-Aware Routing in Wireless Mobile Ad Hoc Networks, *Proceedings of the International Conference on Wireless Networks 2002 (ICWN '02)*, to appear.

[22] L. Feeney and M. Nilsson, Investigating the Energy Consumption of a Wireless Network Interface in an Ad Hoc Networking Environment, *Proceedings of IEEE Infocom*, Anchorage, AK, 2001.

[23] Y. Xu, J. Heidemann, and D. Estrin, Adaptive energy-conserving routing for multihop ad hoc networks, Technical Report TR-2000–527, USC/Information Sciences Institute, Oct. 2000. Available at ftp://ftp.isi.e.,du/isi~pubs/tr-527.pdf

23

Energy Efficient Multicast in Ad Hoc Networks

Hee Yong Youn
Sungkyunkwan University

Chansu Yu
Cleveland State University

Ben Lee
Oregon State University

Sangman Moh
ETRI

Abstract

In mobile ad hoc networks (MANETs), energy efficiency is as important as general performance measures such as delay or packet delivery ratio since it directly affects the network lifetime. In this chapter, we introduce two different approaches for energy efficient multicast protocols developed for MANETs. The first group of energy efficient multicast protocols is based on the assumption that the transmission power is controllable. Under this assumption, the problem of finding a tree with the least consumed power becomes a conventional optimization problem on a graph where the weighted link cost corresponds to the transmission power required for transmitting a packet between the two nodes of the link. The second approach focuses on maximizing sleep mode operation supported by the lower level protocol. A mobile node in tree-based protocols can safely put itself into low power sleep mode for conserving energy if it is not a designated receiver under the employed broadcast-based mesh protocol. It is shown that mesh-based protocols are more robust to mobility, but tree-based protocols may be preferable when energy is a primary concern.

23.1 Introduction

Wireless connectivity with mobility support has become an important issue in the modern computing infrastructure. Especially, *mobile ad hoc networks* (MANETs) [9,11] attract a lot of attention with the advent of inexpensive wireless LAN solutions such as *IEEE 802.11* [12], *HIPERLAN* [31], and *Bluetooth* [4] technology. Since they do not need communication infrastructure in their basic forms and utilize the unlicensed *ISM (Industrial, Scientific, and Medical)* band, they are highly likely to be rapidly adopted. Applications of MANETs encompass various areas including home-area wireless networking, on-the-fly conferencing, disaster recovery, wireless sensor networks [20], and *GSM (Global System for Mobile Telecommunications)* service extension covering dead spots [1]. For an extensive description of MANET, refer to [19].

This article investigates energy efficient multicast for MANETs. Multicasting has been extensively studied for MANETs because it is fundamental to many ad hoc network applications requiring close collaboration of the member nodes. A multicast packet is delivered to multiple receivers along a network structure such as a *tree* or *mesh*, which is constructed once a multicast group is formed. However, the network structure is fragile due to node mobility and, thus, some members may not be able to receive the multicast packet. In order to improve the *packet delivery ratio*, multicast protocols for MANETs usually employ control packets to refresh the network structure periodically. It has been shown that *mesh-based protocols* are more robust to mobility than *tree-based protocols* [15], due to many redundant paths between mobile nodes in the mesh. However, multicast mesh may perform worse in terms of energy efficiency because it uses costly broadcast-style communication involving more forwarding nodes than multicast trees. Another important aspect of energy efficiency is balanced energy consumption among all participating mobile nodes. In order to maximize the lifetime of a MANET, care has to be taken not to unfairly burden any particular node with many packet-relaying operations. Node mobility also needs to be considered along with energy balancing.

The rest of the chapter is organized as follows. Multicasting for MANETs is discussed in Section 23.2. Section 23.3 investigates energy efficient multicast protocols proposed for MANETs and analyzes energy efficiency assuming a static ad hoc network. Finally, concluding remarks are in Section 23.4.

23.2 Multicast Protocols for MANETs

This section briefly overviews the research efforts for multicast protocols targeting MANETs. They can be largely categorized into two types, *tree-based multicast* and *mesh-based multicast*, based on the multicast structure. Tree-based multicast is generally used in wired and infrastructured mobile networks (i.e., mobile networks with base stations), as well as in MANETs. Depending on the number of trees per multicast group, tree-based multicast can be further classified as *per-source tree multicast* and *shared tree multicast*.

A new approach unique to MANETs is the *mesh-based multicast*. A mesh is different from a tree since each node in a mesh can have multiple parents. Using a single mesh structure spanning all multicast group members, multiple paths exist and other paths are immediately available when the primary path is broken. This avoids frequent network reconfigurations, resulting in the minimization of disruption of ongoing multicast sessions and reduction of the overhead in implementing the protocol. However, care must be taken to avoid forwarding loops when multicast data are forwarded in a multicast mesh.

23.2.1 Tree-Based Multicast

As mentioned earlier, there are two versions of tree-based multicast in a MANET: *per-source tree* and *shared tree multicast*. Per-source based tree is established and maintained for each multicast source node of a multicast group. The advantage is that each multicast packet is forwarded along the most efficient path from the source node to each and every multicast group member. However, this method incurs a lot of control overhead and cannot quickly adapt to the movements of the nodes in a MANET.

On the other hand, shared tree multicast is a more scalable approach than the per-source tree approach. Instead of building multiple trees for each multicast group, a single shared tree is used for all multicast source nodes. Multicast packets are distributed along this shared tree to all members of the multicast group. To establish a shared tree, a special node is designated as a *core node*, which is responsible for creating and maintaining the shared tree. Hence, a *core selection algorithm* is needed. The established shared tree can be either *unidirectional* or *bidirectional*. In a unidirectional shared tree, multicast packets must be unicast to the core node, which is the root of the tree. From the core node, the multicast packets are distributed along the shared tree until they reach all the multicast group members. However, in a

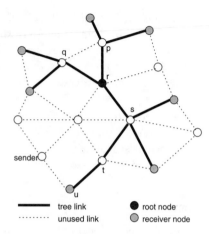

—————— tree link	● root node
············· unused link	○ receiver node

FIGURE 23.1 An example of tree-based multicast.

bidirectional shared tree, multicast packets can enter the shared tree at any point, and they are distributed along all the branches of the shared tree. The shared tree approach has lower control overhead, but the path is not necessarily optimal, i.e., the path from a multicast source to a receiver is not necessarily the shortest. Furthermore, in a dynamic network, throughput can be deteriorated dramatically unless the core node and shared tree quickly adapt to the node mobility.

Figure 23.1 shows an example of a shared unidirectional tree multicast. The tree consists of a root node *(r)*, four intermediate forwarding nodes *(p, q, s,* and *t)*, seven receiver nodes of a multicast group (gray-colored nodes), and eleven tree links. In the shared tree scheme, receiver nodes periodically send *join requests* to the root node, and the root updates the multicast tree using the path information included in the join request messages [3]. Joining a multicast group causes reports (i.e., join messages) to be periodically sent [15], while leaving a multicast group does not lead to any explicit action. The period must be carefully chosen to balance the overhead associated with tree update and the delay caused by the tree not having timely updates when the nodes move [26]. Various tree-based multicast protocols have been proposed, and here some representative ones are briefly reviewed:

- *Adhoc Multicast Routing Protocol (AMRoute)* [2] creates a bidirectional shared tree per multicast group. The tree contains only the group members, and multicast tunnels *(virtual links)* are assumed to exist between each pair of group members based on an underlying routing protocol. Therefore, the tree need not be reconstructed even though the network topology changes as long as routes between the group members exist.

- *Ad-hoc On-Demand Distance Vector Multicast Protocol (AODV)* [21] is another bidirectional shared tree multicast protocol. Here, if the sender does not belong to the multicast group, it first finds the nearest group member and lets it become a root for delivering the multicast packets.

- *Ad hoc Multicast Routing protocol utilizing Increasing-idS (AMRIS)* [32] is a shared tree multicast approach. Each node has *multicast session member id (msm-id)*. The msm-id provides each node with an indication of its "logical height" in the multicast delivery tree such that it increases as it radiates from the root of the delivery tree.

- *Lightweight Adaptive Multicast (LAM)* [10] builds a group-shared multicast routing tree centered at a preselected node called a *CORE*. LAM runs on top of *TORA (Temporally Ordered Routing)* protocol [18]; each node has information on its neighbors and the correct order of transmission path. Each member prepares a *JOIN* message containing the group id and the target *CORE* id, picks a neighbor with the lowest height as the receiver of the *JOIN* message, and sends the message. Since the *JOIN* message is supposed to travel along only a "downwards" path in the TORA DAG (directed acyclic graph) with respect to the target *CORE*, if a *JOIN* message is received over an upstream link, the tree is considered invalid and a valid one is constructed rooted at the *CORE*.

- In *Associativity-Based Multicasting Routing Protocol (ABAM)* [25], a multicast sender builds a per-source multicast tree with *i* messages sent by member receivers who received an *MBQ (multicast broadcast query)* message from the sender. The multicast sender decides a stable multicast tree based primarily on association stability, which refers to spatial, temporal, connection, and power stability of a node with its neighbors, and it generates an *MC-SETUP* message to establish a multicast tree.
- Multicast Routing Protocol based on Zone Routing (MZR) [5] is another per-source tree approach, in which a multicast delivery tree is created using a concept called the zone routing mechanism. A proactive protocol runs inside each zone, maintaining an up-to-date zone routing table at each node. A reactive multicast tree is created for inter-zone routing.

23.2.2 Mesh-Based Multicast

Tree-based protocols may not perform well in the presence of highly mobile nodes because the multicast tree structure is fragile and needs to be readjusted frequently as the connectivity changes. Mesh-based multicast protocols have been proposed to address the problem by constructing a mesh structure with redundant links between mobile nodes. Figure 23.2 shows an example of mesh-based multicast for the MANET of Fig. 23.1. Note that it includes three redundant links (marked in the figure) in addition to eleven tree links. As a result, even though the tree link from *s'* to *v'* is broken, node *v'* receives a multicast packet through the redundant link from *t'* to *v'*. Mesh-based protocols are more robust to mobility and thus allow better *packet delivery ratio*. We now present several mesh-based multicast protocols:

- *Multicast Core-Extraction Distributed Ad hoc Routing (MCEDAR)* [24] is an extension to the *CEDAR routing protocol* [23], and it provides the robustness of mesh-based routing protocols while approximating the efficiency of tree-based protocols. As CEDAR extracts core nodes, MCEDAR extracts a subgraph (called an *mgraph*) for each multicast group consisting only of core nodes as the routing infrastructure used for data forwarding.
- *Clustered Group Multicast (CGM)* [16] employs *advertising agents* to reduce traffic, which act as both a server and client for advertising join requests on behalf of their local clients. Multicast backbone is also used to reduce the control overhead. By implementing CGM over the multicast infrastructure, the clusterhead works as an advertising agent if one or more subscribers are within its cluster, and the inter-cluster routing approach lets the number of nodes in the backbone be smaller.

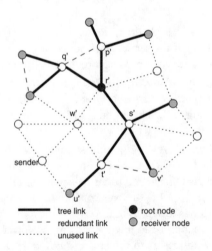

FIGURE 23.2 An example of mesh-based multicast.

- *Core-Assisted Mesh Protocol (CAMP)* [6] adopts the same basic architecture used in IP multicast. A node wishing to join a multicast mesh first consults a routing table to determine whether it has neighbors that are already members of the mesh. If so, the node announces its membership via a *CAMP UPDATE*. Otherwise, the node either propagates a *JOIN REQUEST* towards one of the multicast group "cores" or attempts to reach a member router by applying ring search of broadcast requests.

- *On-Demand Multicast Routing Protocol (ODMRP)* [13] employs on-demand routing techniques to avoid channel overhead and improve scalability. It uses the concept of *forwarding group*, a mesh of nodes responsible for forwarding multicast data on shortest paths between any pair of members. During the control message exchange between senders and group receivers (*JOIN REQUEST* and *JOIN TABLE*), a node realizes that it is part of the forwarding group when it is on the path from a receiver to the source.

- *Neighbor Supporting Multicast Protocol (NSMP)* [13] utilizes node locality to reduce the overhead of route failure recovery and mesh maintenance. A new source initially sends a *FLOOD REQ (FR)* packet containing an upstream node field. When an intermediate node receives it, it caches its upstream node and updates the field with its own address before forwarding it. When a receiver receives the *FR* packet, it sends an *REP* packet. The upstream node receives the *REP* packet and adds an entry for the group to its routing table, and the *REP* packet is forwarded eventually to the source node.

23.3 Energy Efficient Multicast Protocols

Two approaches have been proposed for energy efficient multicast in MANETs. The first is based on the assumption that the transmission power is controllable. Under this assumption, the problem of finding a tree with the least consumed power becomes a conventional optimization problem on a graph where the weighted link cost corresponds to the transmission power required for transmitting a packet between two nodes.

The second approach for energy efficiency comes from the difference of tree-based multicast from mesh-based multicast. One general idea of the power-saving mechanism is to put a mobile node in sleep (low power) mode while it is not sending or receiving packets. Since every mobile node in the mesh must not sleep and must be ready to receive packets during the entire multicast session, it would consume more energy. Even though data transmission through a wireless medium is broadcast in nature, this does not necessarily mean that all neighbor nodes have to receive the broadcast packets. Unicast transmission along the multicast tree is quite different from the intentional broadcast within the multicast mesh in that only the designated receiver needs to receive the transmitted data. A mobile node in tree-based protocols can safely put itself into a low power energy conserving sleep mode if it is not a designated receiver.

As mentioned in the introduction to this chapter, another important aspect of energy efficiency is balanced energy consumption among all participating mobile nodes. For example, consider a multicast tree shared by a number of multicast senders. In the shared tree, the root node of the tree consumes more battery energy and stops working earlier than other nodes. This affects the network connectivity and may lead to partitioning of the MANET and reduced network lifetime. A per-source tree-based multicast protocol alleviates this problem by using a separate tree per sender at the cost of increased tree management overhead [28,30]. Node mobility also needs to be considered along with energy balancing.

The two approaches are discussed in Sections 23.3.1 and 23.3.2, respectively. Section 23.3.3 quantitatively evaluates the multicast protocols in terms of energy efficiency.

23.3.1 Energy Efficiency via Adaptive Transmission Power Control

Network performance in a MANET greatly depends on the connectivity among nodes and the resulting topology. To create a desired topology for multicast, some multicast protocols adjust the nodes' transmission power assuming that it is controllable.

23.3.1.1 Broadcast Incremental Power (BIP) and Multicast Incremental Power (MIP) [28,29]

The object of BIP is the determination of the minimum-cost (in this case, minimum-power) tree, rooted at the source node, which reaches all other nodes in the network. The total power associated with the tree is simply the sum of the powers of all transmitting nodes. Initially, the tree consists of the source node. BIP begins by determining the node that the source node can reach with minimum power consumption, i.e., the source's nearest neighbor. BIP then determines which new node can be added to the tree at minimum additional cost (power). That is, BIP finds a new node that can be reached with minimum incremental power consumption from the current tree node. This procedure is repeated until there is no new (unconnected) node left. BIP is similar to *Prim's algorithm* in forming the *MST (minimum spanning tree)*, in the sense that new nodes are added to the tree one at a time on the basis of minimum cost until all nodes are included in the tree. Unlike Prim's algorithm, however, BIP does not necessarily provide minimum-cost trees for wireless networks.

To obtain the multicast tree, the broadcast tree is pruned by eliminating all transmissions that are not needed to reach the members of the multicast group. That is, the nodes with no downstream destinations will not transmit, and some nodes will be able to reduce their transmission power (i.e., if their distant downstream neighbors have been pruned from the tree). MIP is basically source-initiated tree-based multicasting of session (connection-oriented) traffic in ad hoc wireless networks. In both BIP and MIP, for simplifying trade-offs and evaluation of total power consumption, only the transmission energy is addressed, and it is assumed that the nodes do not move and that a large amount of bandwidth is available. Advantages over traditional network architectures come from the fact that the performance can be improved by jointly considering physical layer issues and network layer issues (i.e., by incorporating the vertical integration of protocol layer functions). That is, the networking schemes should reflect the node-based operation of wireless communications, rather than link-based operations originally developed for wired networks. The quantitative analysis of BIP in terms of approximation ratios can be found in [26].

23.3.1.2 Single-Phase Clustering (SPC) and Multi-Phase Clustering (MPC) [22]

The two distributed, time-limited energy conserving clustering algorithms for multicast, SPC and MPC, minimize the transmission power in two-tiered mobile ad hoc networks. In SPC, each master node pages the slave nodes at the same maximum power, and each slave node acknowledges the corresponding master node having the highest power level. The highest power at a slave node means that the paging master node is nearest to it; hence transmission power could be saved when the slave node selects the master node that provides the highest receive power. When slave nodes send acknowledgments to each master node, the master nodes set the transmission power level to support all acknowledged slave nodes.

MPC consists of the *dropping-rate-down phase* and the *power-saving phase*. In the dropping-rate-down phase, master nodes search the slave nodes that could receive the multicast packets from only one master node. The corresponding master nodes set the transmission power level to support those slave nodes, and then the searched slave nodes belong to the corresponding master node. In the subsequent power-saving phase, each master node pages the information about current power level. Paged slave nodes must have two or more candidate master nodes; hence each slave node selects one master node based on the difference of the current power (P_0) and the power to support the master node (P_n). When the master node is selected, the slave node acknowledges the master node with P_n, and each master node resets the transmission power level with the maximum value between the acknowledged P_n values.

The schemes are motivated by the fact that the most hierarchical networks such as *Bluetooth scatternet* are two-tier networks. The amount of energy consumption in two-tier mobile ad hoc networks could be varied with cluster configuration (e.g., the master node selection). However, an optimal cluster configuration cannot be obtained within a limited time for running a heuristic multicast algorithm. It is assumed that a slave node is connected to only one master node, and direct connection between the master node and a slave node is prohibited. MPC is desirable when energy conservation is more important than computation speed. Otherwise, SPC is preferable.

23.3.2 Energy Savings by Avoiding Broadcast-Based Multicast

As described in the introduction to this chapter, recent wireless LAN standards usually adopt sleep mode operation in order to reduce power consumption, i.e., a communication subsystem goes into energy conserving sleep mode if it has no data to send or receive. If a node sends a packet in unicast mode specifying a receiving node, other nodes except the receiver can continue to sleep. However, when a node sends a packet in broadcast mode, all neighbor nodes have to wake up and receive the packet even though they may eventually discard it. Since mesh-based multicast protocols depend on broadcast-style communication, they are not suitable in an energy-constraint environment. Based on this observation, the following multicast protocol employs a multicast tree but tries to improve the packet delivery ratio to the level achieved by mesh-based protocols.

23.3.2.1 Two-Tree Multicast (TTM) [17]

This protocol tries to reduce the total energy consumption while alleviating the energy balance problem without deteriorating the general performance. Since TTM is based on multicast trees, it inherits all the advantages of tree-based multicast protocols in terms of total energy consumption. TTM adopts shared-tree multicast rather than per-source tree multicast in order to avoid the tree construction overhead. It consumes less energy than mesh-based protocols by employing multi-destined unicast-based trees. As for the energy balance problem found in conventional single shared tree-based multicast (STM), TTM uses two trees called *primary* and *alternative* tree. When the primary tree becomes unusable or overloaded, the alternative tree takes the responsibility of the primary tree, and a new alternative tree is immediately constructed. By doing this, TTM maintains only two trees at a particular time instance, but, in fact, it uses many trees per multicast group as time advances. This is in contrast with a multicast mesh, which can be regarded as a superposition of a number of trees at a time instance.

TTM is similar to the relocation scheme [7], where the root node is periodically replaced with the one nearest the center location to achieve the shortest average hop distance from the root to all receiver nodes. In TTM, a group member with the largest remaining battery energy is selected to replace the root node, and the corresponding alternative tree is constructed and maintained to replace the primary tree. The selection of an alternative root is made in advance to provide a better quality of communication service. Using the example of Fig. 23.1, Fig. 23.3 shows the two trees constructed for a multicast group of eight members (one sender and seven receiver nodes). The primary tree consists of a primary root (r_p), four forwarding nodes $(p, q, s,$ and $t)$, and seven receiver nodes, while the alternative tree consists of an alternative root (r_a), four forwarding nodes $(p, r_p, s,$ and $t)$, and seven receiver nodes.

The TTM protocol performs as follows: Two trees are periodically reconstructed (e.g., every 3 seconds [15]) by periodic join messages (with the information on remaining battery energy) sent by all receiver

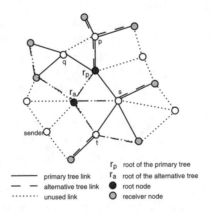

—— primary tree link	r_p root of the primary tree
— — alternative tree link	r_a root of the alternative tree
·········· unused link	● root node
	○ receiver node

FIGURE 23.3 An example of two trees in TTM.

nodes to r_p and r_a. The two root nodes independently construct multicast trees based on the forwarding paths that the join messages traverse. When a sender node intends to send a multicast message, it forwards the multicast message to r_p to be broadcast by the root node as in most shared multicast tree protocols [7,15].

23.3.3 Tree-Based vs. Mesh-Based Multicast Protocols

This subsection compares tree-based multicast protocols with mesh-based protocols as briefly introduced in Section 23.3.2, followed by quantitative evaluation in terms of energy efficiency. For the example of Fig. 23.1, receiver node u receives packets through nodes r, s, t, and u. It requires three transmissions and three receives. Now, consider the last transmission from node t to u. Even though it can be received by all neighbor nodes staying within node t's radio transmission range, those nodes except node u would not receive the multicast packet but stay in sleep mode because the packets are not addressed to them.

On the other hand, a multicast packet is broadcast within a multicast mesh as shown in Fig. 23.2. From node r' to u', it involves four transmissions and seventeen receives incurring much larger energy consumption than the tree-based multicast. For example, the transmission from node t' is received not only by node u' but also by nodes s', v', and w'. The neighbor nodes receive the data packet because the mesh-based protocol relies on broadcast-style communication for improved packet delivery ratio. The redundant link from node t' to v' may be useful when the path from node s' to v' is broken. Node w receives the multicast packet from node t because the packet is broadcast. However, the transmission from node t' to w' is of no use at all because node w' is neither a member nor an intermediate node (forwarding group) of the multicast group. Thus, it discards the packet but wastes energy to receive the packet (referred to as discarded links). Note here that node t' also sends the packet back to node s' since the packet is broadcast. Node s' will ignore the packet but waste additional energy for receiving it.

Based on the discussion above, we compare tree-based and mesh-based protocols with an analytic energy model. To simplify our analysis, static ad hoc networks are assumed.

23.3.3.1 Energy Model (First-Order Radio Model)

Let the total energy consumption per unit multicast message be denoted as E, which includes the transmission energy as well as the energy to receive the packet. We consider only data packets to analyze the total energy consumption for simplicity. According to the first-order radio model [8],

$$E = E_{TX} + E_{RX} = N_{TX} \times e_{TX} + N_{RX} \times e_{RX}$$

where N_{TX} and N_{RX} are the number of transmissions and receives, respectively, and e_{TX} and e_{RX} are the energy consumed to transmit and receive a unit multicast message via a wireless link, respectively. If e_{TX} and e_{RX} are assumed to be the same and denoted by e, the total energy consumption is simply $E = (N_{TX} + N_{RX})e$.

Thus, it is straightforward to show that in a multicast tree, N_{TX} is the number of tree nodes except the leaf receiver nodes (i.e., root and intermediate nodes) and N_{RX} is the number of tree links. In a multicast mesh, N_{TX} is the number of tree nodes (i.e., root, intermediate, and receiver nodes) for the multicast group and N_{RX} can be obtained by *(the number of tree links + the number of redundant links) × 2 + the number of discarded links*. Along a tree or a redundant link, two receives occur as exemplified in Fig. 23.2 (i.e., node t' receives a multicast packet from node s' and then, node s' receives the packet from node t' along the same tree link).

Example Network Model (Static Ad Hoc Network)

Consider a static ad hoc network consisting of k^2 nodes placed in a $k \times k$ grid. Figure 23.4 shows examples of tree-based multicast on an 8×8 grid network with node connectivity of 4 and 8. Figure 23.5 shows examples of mesh-based multicast on an 8×8 grid network. For upper bound analysis, we focus on complete multicast, where all the nodes in a network are member nodes as in Figs. 23.4a, 23.4d, 23.5a, and 23.5d. Figures 23.4b, 23.4e, 23.5b, and 23.5e show the worst cases where the total energy consumption is about the same as with complete multicast but with a smaller number of member nodes, i.e., member

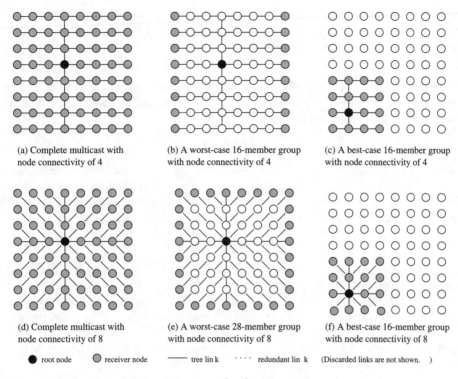

FIGURE 23.4 Examples of tree-based multicast on an 8×8 grid network.

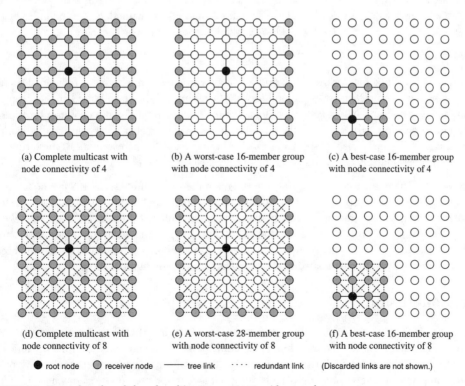

FIGURE 23.5 Examples of mesh-based multicast on an 8×8 grid network.

nodes reside at the edges of the network. Figures 23.4c, 23.4f, 23.5c, and 23.5f show the best cases where a multicast tree or mesh consists of only member nodes and, thus, the total energy consumption is the least with the given number of member nodes.

Quantitative Analysis

The following two theorems [17] formally analyze the upper and lower bounds of total energy consumption in a static ad hoc network as stated above. Theorem 1 analyzes the tree-based multicast, while Theorem 2 analyzes the mesh-based multicast.

Theorem 1

For a static ad hoc network of $k \times k$ grid topology with node connectivity of f, the total energy consumed to transfer a multicast message in a tree-based multicast method, E_{tree}, is bounded by $(2n - O[n^{1/2}])e \leq E_{tree} \leq (2k^2 - O[k])e$, where n is the number of member nodes and e is the energy consumed to transmit or receive a multicast message via a link.

Proof

Given a static ad hoc network of $k \times k$ grid topology with node connectivity of f, the total energy consumption of a tree-based multicast method for complete multicast can be regarded as the upper bound. In a complete multicast, $N_{TX} = k^2 - O(k)$, where $O(k)$ is mainly due to the boundary nodes having a smaller node connectivity than f, and $N_{RX} = k^2 - 1$ since, given a tree with n nodes, the number of edges is $n - 1$. Hence, $E_{tree} \leq (N_{TX} + N_{RX})e \leq (2k^2 - O[k])e$. In the best case, where a multicast tree consists of only member nodes, $N_{TX} = n - O(n^{1/2})$ and $N_{RX} = n - 1$. Hence, $(2n - O[n^{1/2}])e \leq E_{tree}$. **Q.E.D.**

Theorem 2

For a static ad hoc network of $k \times k$ grid topology with node connectivity of f, the total energy consumed to transfer a multicast message in a mesh-based multicast method, E_{mesh}, is bounded by $([f + 1]n - O[n^{1/2}])e \leq E_{mesh} \leq ([f + 1]k^2 - O[k])e$. n and e are defined as in Theorem 1.

Proof

Given a static ad hoc network of $k \times k$ grid topology with node connectivity of f, the total energy consumption in a complete mesh-based multicast can be regarded as the upper bound. In the complete multicast, $N_{TX} = k^2$ and $N_{RX} = fk^2 - O(k)$ since the mesh-based multicast protocol uses broadcast-style communication ($O[k]$ is due to the boundary nodes having smaller node connectivity than f). Hence, $E_{mesh} \leq (N_{TX} + N_{RX})e \leq ([f + 1] k^2 - O[k])e$. In the best case, where a multicast mesh consists of only member nodes, $N_{TX} = n$ and $N_{RX} = fn - O(n^{1/2})$. Hence, $([f + 1]n - O[n^{1/2}])e \leq E_{mesh}$. **Q.E.D.**

According to Theorems 1 and 2, $E_{mesh}/E_{tree} \approx (f + 1)/2$ in the worst and best cases. Since node connectivity, f, is usually much larger than two to avoid MANET partitioning, mesh-based multicast protocols consume around $(f + 1)/2$ times more energy than tree-based multicast protocols. The analysis above is based on the assumption that all nodes are located in a grid style. Even when the nodes are located in an arbitrary manner, if the variance of node connectivity is small enough, the analysis is still valid because the node connectivity is directly related to the tree structure and the number of transmissions.

23.4 Conclusion

We have discussed energy efficient multicast protocols proposed for MANETs. There are two different approaches. The first is based on the assumption that the transmission power is controllable. Under this assumption, the problem of finding a tree with the least consumed power becomes a conventional optimization problem. The second approach is to exploit low power sleep mode as much as possible by avoiding costly broadcast operations. Unicast transmission along the multicast tree is more energy efficient than the broadcast-style communication used in the multicast mesh. A mobile node in tree-based protocols can safely put itself into energy conserving sleep mode if it is not a designated receiver. Quantitative analysis is also presented to show that mesh-based protocols consume around $(f + 1)/2$ times more energy than tree-based protocols, where f is the node connectivity $(f >> 2)$.

The need for energy efficiency is due to the constraints imposed by battery capacity and heat dissipation. Battery and heat removal technologies have been improved at a slower pace than the rate of computation power increase and size of wireless device decrease. The key to energy efficiency in future wireless terminals will be at the higher levels of network protocols: low-energy protocols, energy-cognizant user interfaces, context dependent, and predictive shutdown management. The networked operation of a wireless device opens up additional techniques for increasing energy efficiency. Techniques for dynamically offloading computation from local terminals to remote, energy-rich nodes are also interesting. Other techniques include making various network protocols, such as links, MAC routing, and transport protocols, energy aware so that they continually strive to provide the most energy efficient transport of application data while meeting the desired QoS.

References

[1] Aggelou, G. and Tafazolli, R., On the Relaying Capability of Next-Generation GSM Cellular Networks, *IEEE Personal Communications*, Feb. 2001, pp. 40–47.

[2] Bommaiah, E., Liu, M., McAuley, A., and Talpade, R., AMRoute: Ad-hoc Multicast Routing Protocol, Internet-Draft, draft-talpade-manet-amroute-00.txt, Aug. 1998.

[3] Chiang, C., Gerla, M., and Zhang, L., Adaptive Shared Tree Multicast in Mobile Wireless Networks, *IEEE Global Telecomm. Conference (GlobeCom 1998)*, Nov. 1998, vol. 3, pp. 1817–1822.

[4] Complete Bluetooth Tutorial, http://infotooth.tripod.com/tutorial/complete.htm, 2000.

[5] Devarapalli, V. and Sidhu, D., MZR: A Multicast Protocol for Mobile Ad Hoc Networks, *IEEE International Conference on Communications*, Helsinki, Finland, 2001, vol. 3, pp. 886–891.

[6] Garcia-Luna-Aceves, J. and Madruga, E., The core assisted mesh protocol, *IEEE Journal on Selected Areas in Communications*, 17, 1380–1394, Aug. 1999.

[7] Gerla, M., Chiang, C., and Zhang, L., Tree multicast strategies in mobile, multihop wireless networks, *Baltzer/ACM Journal of Mobile Networks and Applications (MONET)*, 3, 193–207, 1999.

[8] Heinzelman, W., Chandrakasan, A., and Balakrishnan, H., Energy-Efficient Communication Protocols for Wireless Microsensor Networks, *Hawaii Int'l Conf. on System Sciences*, Maui, Jan. 2000, pp. 3005–3014.

[9] Internet Engineering Task Force (IETF) Mobile Ad Hoc Networks (MANET) Working Group Charter, http://www.ietf.org/html.charters/manet-charter.html, 2000.

[10] Ji, L. and Corson, M., A Lightweight Adaptive Multicast Algorithm, *IEEE Global Telecomm. Conference (GlobeCom 1998)*, 1998, vol. 2, pp. 1036–1042.

[11] Jubin, J. and Tornow, J., The DARPA packet radio network protocols, *Proceedings of the IEEE*, 75, 21–32, 1987.

[12] Kamerman, A. and Monteban, L., WaveLAN-II: A High-Performance Wireless LAN for the Unlicensed Band, *Bell Labs Technical Journal*, Summer 1997, pp. 118–133.

[13] Lee, S., Gerla, M., and Chiang, C., On-Demand Multicast Routing Protocol, *IEEE Wireless Communications and Networking Conference (WCNC '99)*, 1999, pp. 1298–1302.

[14] Lee, S. and Kim, C., Neighbor Supporting Ad Hoc Multicast Routing Protocol, *Workshop on Mobile Ad Hoc Networking and Computing (MobiHoc 2000)*, Boston, MA, Aug. 2000, pp. 37–44.

[15] Lee, S., Su, W., Hsu, J., Gerla, M., and Bagrodia, R., A Performance Comparison Study of Ad Hoc Wireless Multicast Protocols, *IEEE Infocom 2000*, Tel Aviv, Mar. 2000, vol. 2, pp. 565–574.

[16] Lin, C. and Chao, S., A Multicast Routing Protocol for Multihop Wireless Netorks, *IEEE Global Telecomm. Conference (GlobeCom 1999)*, 1999, pp. 235–239.

[17] Moh, S., Yu, C., Lee, B., and Youn, H., Energy Efficient Two-Tree Multicast for Mobile Ad Hoc Networks, Technical Report, Cleveland State University, Cleveland, OH, 2002.

[18] Park, V. and Corson, M., A Performance Comparison of the Temporally-Ordered Routing Algorithm and Ideal Link-State Routing, *IEEE Symposium on Computer and Communications*, July 1998.

[19] Perkins, C., Ed., *Ad Hoc Networking*, Addison-Wesley, Reading, MA, 2001.

[20] Pottie, G. and Kaiser, W., Wireless Integrated Network Sensors, *Communications of the ACM*, May 2000, pp. 51–58.

[21] Royer, E. and Perkins, C., Multicast Operation of the Ad-hoc On-Demand Distance Vector Routing Protocol, *MobiCom '99*, Seattle, WA, Aug. 1999, pp. 207–218.

[22] Ryu, J., Song, S., and Cho, D., A Power-Saving Multicast Scheme in Two-Tier Hierarchical Mobile Ad-Hoc Networks, *IEEE Vehicular Technology Conference (VTC 2000)*, Sep. 2000, vol. 4, pp. 1974–1978.

[23] Sinha, P., Sivakumar, R., and Bharghavan, V., Core Extraction Distributed Ad Hoc Routing (CEDAR) Specification, Internet Draft draft-ietf-manet-cedar-spec-00.txt, Sep. 1998.

[24] Sinha, P., Sivakumar, R., and Bharghavan, V., MCEDAR: Multicast Core Extraction Distributed Ad-hoc Routing, *IEEE Wireless Communications and Networking Conference (WCNC '99)*, 1999.

[25] Toh, C., Guichal, G., and Bunchua, S., ABAM: On-demand Associativity-based Multicast Routing for Ad Hoc Mobile Networks, *IEEE Vehicular Technology Conference (VTC Fall 2000)*, 2000, vol. 3, pp. 987–993.

[26] Varshney, U. and Chatterjee, S., Architectural Issues to IP Multicasting over Wireless and Mobile Networks, *IEEE Wireless Communications and Networking Conference (WCNC '99)*, Sep. 1999, vol. 1, pp. 41–45.

[27] Wan, P., Calinescu, G., Li, X., and Frieder, O., Minimum-Energy Broadcast Routing in Static Ad Hoc Wireless Networks, *IEEE Infocom 2001*, Apr. 2001, vol. 2, pp. 1162–1171.

[28] Wieselthier, J., Nguyen, G., and Ephremides, A., Algorithms for Energy-Efficient Multicasting in Ad Hoc Wireless Networks, *Military Communication Conference (MILCOM 1999)*, Atlantic City, NJ, Nov. 1999, vol. 2, pp. 1414–1418.

[29] Wieselthier, J., Nguyen, G., and Ephremides, A., Energy Efficiency in Energy-Limited Wireless Networks for Session-Based Multicasting, *IEEE Vehicular Technology Conference (VTC 2001)*, May 2001, vol. 4, pp. 2838–2842.

[30] Wieselthier, J., Nguyen, G., and Ephremides, A., On the Construction of Energy-Efficient Broadcast and Multicast Trees in Wireless Networks, *IEEE Infocom 2000*, Tel Aviv, Mar. 2000, vol. 2, pp. 585–594.

[31] Woesner, H., Ebert, J., Schlager, M., and Wolisz, A., Power-Saving Mechanisms in Emerging Standards for Wireless LANs: The MAC Level Perspective, *IEEE Personal Communications*, June 1998, pp. 40–48.

[32] Wu, C., Tay, Y., and Toh, C., Ad Hoc Multicast Routing Protocol Utilizing Increasing id-numberS (AMRIS) Functional Specification, Internet-Draft, draft-ietf-manet-amris-spec-00.txt, Nov. 1998.

[33] Xylomenos, G. and Polyzos, G., IP Multicast for Mobile Hosts, *IEEE Communications*, Jan. 1997, pp. 54–58.

24

Energy-Conserving Grid Routing Protocol in Mobile Ad Hoc Networks

Jang-Ping Sheu
National Central University

Cheng-Ta Hu
National Central University

Chih-Min Chao
National Central University

Abstract

The lifetime of a *mobile ad hoc network (MANET)* depends on the durability of the battery resources of the mobile hosts. Earlier research has proposed several routing protocols specifically for MANET, but most studies have not focused on the limitations of battery resources. The failure of battery resources severely impacts a communications system during natural disasters and in crucial communications environments, such as a battlefield. This study proposes a new energy-aware routing protocol that can increase the durability of the energy resource and, therefore, the lifetime of the mobile hosts and the MANET. The proposed protocol can conserve energy by shortening the idle period of the mobile hosts without increasing the probability of packet loss or reducing routing fidelity. Simulation results indicate that this new energy-conserving protocol can extend the lifetime of a MANET.

24.1 Introduction

A *mobile ad-hoc network (MANET)* is formed by a cluster of mobile hosts without any predesigned infrastructure of base stations. A host in a MANET can roam and communicate with other hosts at will. Two mobile hosts may communicate with each other either directly (if they are close enough) or indirectly, through intermediate mobile hosts that relay their packets because of transmission power limitations. A main advantage of a MANET is that it can be rapidly deployed since no base station or fixed network infrastructure is required. MANETs can be applied where predeployment of network infrastructure is difficult or impossible (for example, in fleets on the oceans, armies on the march, natural disasters, battlefields, festival grounds, and historic sites).

Many routing protocols have been proposed for MANETs [1–3,10,11,12,19]. Most of them concentrate on issues such as the packet delivery ratio, routing overhead, or shortest path between source and destination. In fact, energy constraints represent an equally important issue in MANET operations. Each mobile host that operates in a MANET has a limited lifetime due to its limited battery energy. Failure of one mobile host may disturb the whole MANET. Thus, battery energy should be considered to be a scarce resource, and an effective energy-conserving technique must be found to extend the lifetime of a mobile host and, hence, the whole MANET.

Several studies have addressed energy constraints. In [17,18], a minimum-power tree is established from source to a destination to support broadcast/multicast services. Rodoplu and Meng [8] proposed a distributed power-efficient transmission protocol to reduce the power consumption of a mobile host and thus increase the lifetime of the whole network. In [9,20], the topology of the whole network is controlled by adjusting the transmission power of mobile hosts. The goal is to maintain a connected network using minimum power. Wu, Tseng, and Sheu [6] proposed an energy-efficient MAC protocol to increase channel utilization and reduce both power consumption and co-channel interference. The issue of power controlled on the MAC layer is also addressed in [14]. Another protocol that addresses the conservation of energy consumption in the transmission mode can be found in [16].

Besides reducing transmission power, the energy of a mobile host can be conserved by occasionally turning off its transceiver [4,5]. As is well known, much power is consumed during transmission and reception by a mobile host. However, if the transceiver is powered on, then the power consumption is not reduced much even through the mobile host is idle. Table 24.1 shows the power consumption of the Lucent IEEE 802.11 WaveLan card in different modes [13].

A mobile host still consumes much energy even when idle. Turning off the transceiver and entering sleep mode whenever a mobile host is neither transmitting nor receiving is a better way to conserve energy. The problem with turning off the transceiver is that the mobile hosts may fail to receive broadcast or unicast packets. Two questions should be addressed in this energy conservation problem:

1. When should the transceiver be turned off?
2. How can packet loss be avoided when the destination host is in sleep mode?

A longer sleep is preferred to conserve energy. That is, the transceiver should be turned off as soon as possible when idle. However, long sleeping increases the probability of losing packets. This work considers these two issues together and seeks to maximize energy conservation without increasing the probability of packet loss.

The proposed protocol, *Energy-Conserving GRID (ECGRID)*, exploits the concept of a routing protocol called *GRID* [3] while considering the energy constraints. In GRID, each mobile host has a positioning device such as a Global Positioning System (GPS) receiver to determine its current position. The geographic area of the entire MANET is partitioned into two-dimensional logical *grids*. Routing is performed in a grid-by-grid manner. One mobile host will be elected as the *gateway* for each grid. This gateway is responsible for:

1. Forwarding route discovery requests to neighboring grids
2. Propagating data packets to neighboring grids
3. Maintaining routes for each entry and exit of a host in the grid

TABLE 24.1   Power Consumption (mA) of Lucent IEEE 802.11 WaveLan Card in Different Modes

Transmit Mode	Receive Mode	Idle Mode	Sleep Mode
284	190	156	10

Note: 11 Mb/sec, power = 4.74 V

No nongateway hosts are responsible for these jobs unless they are sources/destinations of the packets. For maintaining the quality of routes, we also suggest that the gateway host of a grid should be the one nearest to the physical center of the grid.

In ECGRID, grid partitioning and grid-by-grid routing are the same as in the GRID routing protocol. The main difference between these two protocols is that ECGRID considers the energy of mobile hosts but GRID does not. Moreover, ECGRID seeks to maximize the lifetime of the network. For each grid, one mobile host will be elected as the gateway and others can go into sleep mode. The gateway host is responsible for forwarding routing information and propagating data packets as in GRID. Sleeping nongateway hosts will return to active mode by the signaling of the gateway whenever data have been sent to them. (Accordingly, the transceiver is not periodically restarted to check whether data are to be received.)

The goal of this work is similar to that of Span [5] and GAF [4]. In Span, each mobile host switches between the coordinator and noncoordinator, according to a *coordinator eligibility rule*. Span coordinators stay awake continuously to perform packet routing. Span noncoordinators stay in sleep mode and wake up periodically to check whether any packets have been sent to them. In GAF, the geographic area is partitioned into grids as in GRID, and hosts in the same grid are defined as routing equivalent hosts. In a grid, one mobile host is active and others can sleep. A host will set the sleeping duration before it goes to sleep. After this period of sleeping, the host will wake up to check its activity.

ECGRID, Span, and GAF are compared as follows. ECGRID is superior since hosts need not periodically wake up from the energy-saving state, unlike in the other two schemes. In ECGRID, hosts can be awakened by a signal from the gateway host whenever packets must be sent to them. This signaling ensures that the probability of packet loss will not increase because of the power saving operations of ECGRID. In Span, an ATIM (Ad Hoc Traffic Indication Map) mechanism is proposed to solve this problem. GAF includes no way to ensure that a destination host is active when packets are sent to it. Thus, the packets cannot be delivered to the sleeping destination. In a location-aware scheme, such as ECGRID or GAF, more energy can be saved when host density is higher because only one host (gateway) in a grid is active. As the grid contains more hosts, each host can take turns to act as the gateway. Thus the saved power is proportional to host density. On the contrary, Span (not location aware) does not benefit from increasing host density [5].

The rest of this chapter is organized as follows. Section 24.2 presents our system environment. Section 24.3 presents our ECGRID protocol. Section 24.4 presents simulation results. Section 24.5 draws conclusions.

24.2 System Environment

As Fig. 24.1 illustrates, the geographic area of the MANET is partitioned into two-dimensional logical grids. This is exactly the same partition method as described in [3]. Each grid is a square area of size $d \times d$. Grids are numbered (x,y) following the conventional (x,y) coordinate system. Each host still has a unique ID (such as IP address or MAC address). Each mobile host is made location aware by being equipped with a positioning device, such as a GPS receiver, from which it can read its current location. A predefined mapping should exist from any physical location to its grid coordinate.

As mentioned above, each grid is a square of $d \times d$. Let r be the transmission distance of a radio signal. In this study, the value of d is chosen as

$$\frac{\sqrt{2}r}{3}$$

As shown in Fig. 24.2, this value setting means that a gateway located at the center of a grid can communicate with any gateway in its eight neighboring grids.

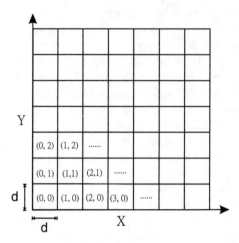

FIGURE 24.1 Logical grids to partition a physical area.

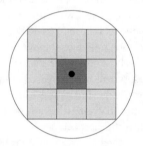

FIGURE 24.2 The relationship between d (the side length of grids) and r (the radio transmission

distance).$(d = \dfrac{\sqrt{2}r}{3})$

The mobile hosts are assumed to have the same power supply. That is, all the mobile hosts have the same maximum energy. Each mobile host periodically calculates its *ratio of battery remaining capacity* (R_{brc}). R_{brc} is defined as

$$R_{brc} = \frac{Battery \;\; remaining \;\; capacity}{Battery \;\; full \;\; capacity}$$

Battery remaining capacity represents the remaining energy of a mobile host while *Battery full capacity* means a host's maximum energy. Three energy levels are defined for each mobile host, as follows:

R_{brc} : *upper level* if $R_{brc} \geqq 0.6$; *boundary level* if $0.2 \leqq R_{brc} < 0.6$, and *lower level* if $R_{brc} < 0.2$. This classification of remaining energy is used in the *gateway election rule,* which will be presented in the next section.

Each host in a MANET is either in active mode (state S_0) or in sleep mode (state S_1). A mobile host in active mode can transmit and receive packets. Only one host (the gateway) is active in each grid. Other nongateway hosts can turn their transceivers off and enter sleep mode if they do not have packets to transmit or receive. These nongateway hosts need not periodically wake up to check whether any pending packet is at the gateway. Instead, whenever the gateway receives a packet that is sent to a host in sleep mode, the gateway will actively wake up the host. This method is made feasible by equipping every host with a device called a Remotely Activated Switch (RAS) [7], as depicted in Fig. 24.3. As its name reveals, an RAS module can be used remotely to activate a mobile host. This RAS is fulfilled by *radio frequency*

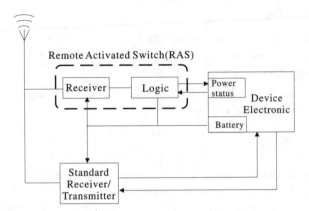

FIGURE 24.3 The architecture of a mobile host.

tags technology. When the RAS receives a correct *paging sequence* (a set of paging signals), it can turn on the transceiver and bring the host into active mode. Thus, if a unique paging sequence is assigned to each host, any host can be awakened by sending its paging sequence. Note that the power consumption of RAS is much lower than the transmitting/receiving power consumption and can thus be ignored in energy calculations.

24.3 Energy-Conserving Grid Routing Protocol

A routing protocol called *Energy-Conserving GRID (ECGRID)* is proposed. Every mobile host in the network must run ECGRID. Each host uses its unique ID as the paging sequence. One mobile host in each grid will be elected as the gateway and remains in active mode. Other nongateway hosts can sleep to conserve battery energy. The gateway host must maintain a *host table* that stores the host ID and status (transmit/sleep mode) of all the hosts in the same grid. Whenever the gateway receives a packet sent to a sleeping host, the gateway can wake up the mobile host by sending the paging sequence associated with that host.

A gateway may initiate a new gateway election process when it is leaving the grid or seeking to maintain the load balance. To elect a new gateway, all hosts in the same grid must be in active mode. Each grid has a unique *broadcast sequence*, which is defined as the *coordinate* of the grid. All hosts must move into active mode when they hear the broadcast sequence. That is, a gateway can wake up all the hosts in the same grid by sending the broadcast sequence. And then a new gateway can be elected according to the gateway election rules. This new gateway will inherit the routing table from the original gateway and remain continuously active.

The gateway is responsible for forwarding routing information and transmitting data packets. It plays the most important role in our protocol and consumes much energy. All hosts should take their turn as the gateway to prolong the lifetime of the network. The election of the gateway should take the remaining battery capacity and position into consideration. A gateway with remaining energy and nearer the center of a grid is preferred. These two principles prevent another gateway election from being triggered soon. The gateway election rules are as follows.

1. A host with higher level of remaining battery capacity has higher priority.
2. Given several hosts with the same highest energy level, the one that is closest to the center of the grid will be elected as the gateway. This rule allows a host that will stay in the grid for longer to be the gateway.
3. If no gateway can be elected according to steps 1 and 2, then the host with smallest ID (IP address or MAC address) will be elected as the gateway.

24.3.1 Gateway Election

This subsection describes the gateway election algorithm. This algorithm is executed distributively whenever a new gateway is needed, such as when the network is first initialized or when the gateway host runs out of energy or moves out of the grid. The algorithm is as follows:

1. Each host in active mode will periodically broadcast its *HELLO* message. The *HELLO* message contains the following five fields.
   A. *id*: host ID
   B. *grid*: grid coordinate
   C. *gflag*: gateway flag (set to one when the host is the gateway)
   D. *level*: remaining battery capacity level (upper, boundary, or lower)
   E. *dist*: distance to geographic center of the grid
2. After a *HELLO period*, which is predefined as the period for hosts to exchange their *HELLO* messages, all hosts are supposed to receive *HELLO* messages from neighboring hosts in the same grid. Then, each host will apply the gateway election rules to decide whether it becomes the gateway.
3. The host will declare itself as the gateway by sending a *HELLO* message with the *gflag* set. The gateway host is responsible for maintaining the host table, which is constructed from the *id* field of the *HELLO* messages.
4. All other nongateway hosts receiving the *HELLO* message from the gateway will move into sleep mode if they have no packets to transmit.

Figure 24.4 displays the state transition diagram of ECGRID. The gateway host must stay in active mode. The gateway can move into sleep mode only when it releases its gateway responsibility, as shown in Fig. 24.4a. In Fig. 24.4b, a nongateway host may move into sleep mode if it is not transmitting/receiving packets. A nongateway host in sleep mode must return to active mode if it receives its paging sequence or broadcast sequence.

24.3.2 Gateway Maintenance

In ECGRID, the gateway hosts must remain continuously active and are responsible for routing information and forwarding data packets. The correct operation of the gateway is critical in the protocol. In the following, two aspects of the selection of the gateway are discussed — the *mobility of mobile hosts* and the *load balance of the mobile host's battery energy*.

Before entering sleep mode, each mobile host will set a timer to wake up. This timer is set to the estimated dwell duration over which the host is expected to remain in its current grid. The estimation depends on the location and velocity of the host. These two parameters can be easily obtained since each host is equipped with a GPS device. When the timer expires, the host will wake up to see whether it is leaving the current grid. If it is leaving, the nongateway host will send a unicast message to the gateway host to update the routing and host tables. The host must remain active until it finds another gateway. If it is not leaving, it will recalculate the dwell duration, set the timer, and then enter sleep mode again.

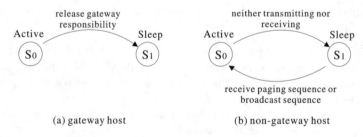

(a) gateway host   (b) non-gateway host

FIGURE 24.4   State transition diagram of ECGRID.

Two mobility situations should be addressed.

1. Hosts move into a new grid:

   Hosts will broadcast a *HELLO* message when they move into a new grid. The gateway host in each grid will also periodically broadcast its *HELLO* message. After the *HELLO* message from the gateway is received, the new incoming host will decide whether it should replace the gateway by applying the gateway election rule. In this situation, only a host with a battery level that is higher than that of the original gateway can replace the original gateway. This rule prevents frequent replacement of gateways: such replacement is an overhead of our protocol. If a gateway must be replaced, then the new gateway will declare itself by sending a *HELLO* message with the *gflag* set. The original gateway, receiving this *HELLO* message, will transmit the routing and host tables to the new gateway. If no gateway is replaced, then the new incoming hosts will enter sleep mode to conserve battery energy. If a new host does not receive any *HELLO* message during a *HELLO* period, the new host is in an empty grid and will declare itself as the gateway.

2. Hosts move out of a grid:

   The following considers the case of either a host or a gateway leaving one grid and entering another. A gateway must transfer its routing table to a new gateway before it leaves a grid. The gateway thus first sends a broadcast sequence, labeled (1) in Fig. 24.5, to wake up all the hosts in the same grid. The nongateway hosts in other grids will also receive this broadcast sequence, but they will simply ignore this signal because it is not the broadcast sequence associated with those hosts. Only sleeping hosts in the same grid will wake up, as shown in Fig. 24.5. After waiting for time, τ, the gateway will declare its departure by broadcasting a *RETIRE(grid, rtab)* message, labeled (2) in Fig. 24.5, where *grid* represents the grid coordinate of the gateway, and *rtab* represents the routing table. After receiving this *RETIRE* message, all nongateway hosts will store the routing table and apply the gateway election algorithm to elect a new gateway. Then, the new gateway transmits a *HELLO* message with the *gflag* set to inform all the hosts about its new status. If a nongateway host leaves a grid, it must notify the gateway about its departure by sending a unicast message to the gateway. The gateway will update its routing and host tables after receiving this unicast message.

In addition to host mobility, gateway replacement also takes place for load balance purposes. The gateway always remains in active mode, consuming the most battery energy. Without load balancing schemes, a host that acts as a gateway will rapidly run out of energy. All hosts in the same grid should share the job of the gateway to extend the lifetime of a mobile host and the network. The load balance scheme presented here releases a gateway from its gateway job when its battery level changes, i.e., from upper to boundary or from boundary to lower. The process by which a gateway quits its duty is the same process that is necessary when a gateway leaves a grid. A gateway will also quit its responsibility when it is going to exhaust its energy resource. That is, if a host with a lower battery level is elected as the gateway, it will act as the gateway until its battery is empty. Of course, the gateway will issue a broadcast sequence and a *RETIRE* message before its battery runs out.

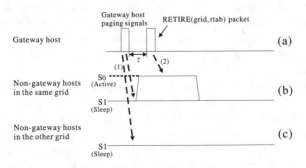

FIGURE 24.5   The gateway host wakes up all the hosts in the same grid.

24.3.3 Route Discovery and Data Delivery

In ECGRID, the routing table is established in a grid-by-grid manner, instead of in a host-by-host manner. Therefore, only the gateway is needed to maintain the routing table. Our ECGRID is an extension GRID (which is modified from AODV protocol [10]) by considering energy conservation. The gateway is the only host in a grid that is responsible for the routing discovery procedure. Nongateway hosts can enter sleep mode if they are neither transmitting nor receiving packets. Packets sent to a sleeping host are buffered at the gateway while the destination host is sleeping. Then, the gateway is responsible for waking up the destination host and forwarding these buffered data packets to it. If a sleeping host, S, must send data, it will wake up and send an acquire message $ACQ(gid)$ to identify the gateway. The gateway of S will respond with a *HELLO* message after receiving this *ACQ* message. Thus, host S can send data through the gateway to the destination. This handshaking is required since the gateway may be changed when a nongateway host is sleeping.

When a source host, S, needs a route to a destination host, D, it will broadcast a route request, $RREQ(S,s\text{-}seq,D,d\text{-}seq,id,range)$ packet to request a route to D. The pair (S,id) can be used to detect duplicate *RREQ* packets from the same source, S, avoiding endless flooding of the same request. The source sequence number, s_seq, represents the freshness of a reverse route from the destination to the source, and the destination sequence number, d_seq, indicates the freshness of the route from the source to the destination. The freshness information is used to determine whether a route is acceptable. The parameter *range* confines the area of search to where only the gateways within the area will participate in the route searching procedure from S to D. The searching area limits the broadcast packets and thus alleviates the *broadcast storm problem* [15]. Different searching areas involve different route search costs. Several ways of confining the searching area have been presented in [3]. Routes may fail to exist in the searching area. In such a situation, another round of route searching should be initialized to search all areas for a route. Notably, a global search for a route is also needed when the source does not have location information concerning the destination.

When a gateway receives an *RREQ* packet, the gateway will first check whether it is within the area defined by *range*. The gateway simply ignores this packet if the received packet is not within its *range*. Otherwise, the gateway checks the destination, D, from its routing table. If the destination, D, is in its routing table, then the gateway will rebroadcast the packet and set up a reverse pointer to the grid coordinate of the previous sending gateway. When D (or its gateway, if D is not a gateway) receives this *RREQ*, it will unicast a reply packet *RREP* $(S,D,d\text{-}seq)$ back to S through the reverse path. Notably, the reverse path is established when the *RREQ* packet is broadcast. A gateway that receives the *RREP* will add an entry to its routing table to specify that a route to D is available through the grid coordinate from which it received the *RREP* packet. A route from S to D is properly established when this *RREP* reaches S.

When destination D is a nongateway host, the gateway of D must wake D before forwarding data packets to it. This procedure is accomplished by transmitting the paging sequence and waiting for D's acknowledgment. On receiving the paging signals from the RAS, D will first compare the signals with its own paging sequence. If both signals match, D will change to the active state and send back an acknowledgment. After receiving the acknowledgment from D, the gateway can send all the data packets to D. Figure 24.6 depicts the process of data transmission from the gateway to a sleeping host. The order of transmission is labeled (1), (2), and (3). First the paging signal is sent to wake up the receiver. The gateway starts to transmit data after receiving the *ACK* from the receiver. Notably, although all hosts, other than the destination, will receive paging signals, no action is taken if the signals do not match these hosts' paging sequence.

Figure 24.7 shows an example to demonstrate how our protocol works. Suppose that the searching area is the smallest rectangle that can cover the grids of source S and destination D. Initially, all hosts are active and each of them will broadcast a *HELLO* message. After a *HELLO* period, hosts S, A, B, C, D, E, F, and I will be selected as the gateway of grids (1,1), (1,2), (2,2), (2,1), (5,3), (3,2), (4,2), and (0,2), respectively, by applying the gateway election protocol. After the gateway hosts are elected, nongateway hosts J, K, L, H, G, and M can enter sleep mode to conserve energy. Suppose that host S wants to

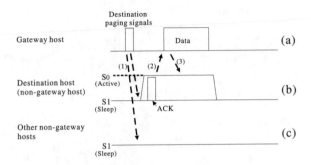

FIGURE 24.6   Data transmission from the gateway to a nongateway host.

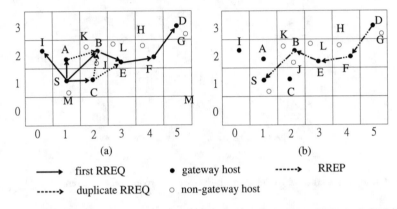

FIGURE 24.7   An example of routing discovery: (a) propagation of RREQ; (b) propagation of RREP packets.

communicate with destination host D; it will send an *RREQ* packet which specifies that the searching area is the rectangle bounded by grids (1,1), (1,3), (5,1), and (5,3). When host B receives this *RREQ* package for the first time, it will rebroadcast this packet, since host B is within the searching area. The reverse path that points to the grid (1,1), in which S belongs, is recorded in host B. Similar actions (rebroadcasting *RREQ* and recording a reverse path) will be performed at hosts E and F. The solid arrows in Fig. 24.7a show how the *RREQ* packets move. The packet finally reaches the destination, D, and a reverse path from D to S is also established. When destination D receives the *RREQ*, it will unicast an *RREP* packet as a reply to S through the reverse path that was established by the *RREQ* packets. Figure 24.7b shows the progress of the *RREP*.

In the above example, destination D is a gateway. Data transmission through *S-B-E-F-D* can properly take place. Assuming that the destination is a nongateway host, G, the route discovery process does not change and the path remains *S-B-E-F-D*. However, the data transmission path is *S-B-E-F-D-G*. The gateway, D, is responsible for waking G up, and buffer data packets are sent to G before G is ready to receive.

24.3.4 Route Maintenance

This subsection considers the maintenance of a route when the source or destination leaves its original grid. The purpose of route maintenance is to keep the route available. Notably, the source or destination host becomes a nongateway host when it moves into a new grid. Suppose that the source host roams from grid g_1 to another grid, and that grid g_2 is the next grid along the route to the destination, as shown in Fig. 24.8. Hosts A, B, C, and E are the gateways of grids (1,2), (2,2), (2,1), and (3,2), respectively. The following discusses four aspects of the route maintenance protocol according to the roaming direction of the source host:

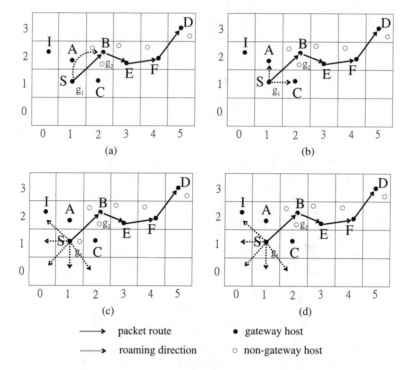

FIGURE 24.8   Route maintenance when the source host (gateway) roams off its current grid.

1. The source host moves into grid g_2, as shown in Fig. 24.8a. The route can still function correctly. When S roams into grid g_2, it decides whether it will replace host B as the gateway. If replacement occurs, then S inherits the routing table from B. Thus, the route through B is maintained. If S does not replace B as the gateway, then S's data will be transmitted with the help of its gateway, host B. The route can certainly work properly.

2. The source host S moves into a grid, g, that neighbors grid g_2, as shown in Fig. 24.8b. In this case, if S becomes the gateway of g, then no change is needed. Otherwise, S becomes a nongateway host, and all data packets are forwarded by the gateway of the new grid g, A in grid $(1,2)$ or C in grid $(2,1)$. In this case, a new *RREQ* is sent from A or C to B.

3. The source host S moves into a grid, g, which does not neighbor grid g_2, and a gateway exists in grid g_1, as shown in Fig. 24.8c. If S becomes the gateway of the new grid, g, then the source host changes its route entry for destination D to grid g_1. Then all data packets are forwarded to grid g_1. If S is a nongateway host in the new grid, then data packets of S are forwarded by the gateway of the new grid, g. In such a case, a new *RREQ* is sent to the gateway of grid g_1. In either case, the route is one hop longer than the original route.

4. The source host S moves into a grid that does not neighbor grid g_2, and a gateway does not exist in grid g_1, as depicted in Fig. 24.8d. Therefore, no host is available to forward the source host's packets, and thus the route will be considered broken. The source host S initiates a new route discovery procedure to request a route to destination host D.

The rules for the movement of the destination host are similar to those for the movement of the source host.

24.4 Simulation

The proposed ECGRID protocol is evaluated using an *ns-2* [21] simulator (CMU wireless and mobile extensions [22]). The simulation runs on a region of 1000 by 1000 square meters, a bandwidth of 2

Mb/sec and a transmission range of 250 meters. Hosts move according to the random way-point model, in which the hosts randomly choose a speed and move to a randomly chosen position. Then the hosts wait at the position for the pause time before they start to move to the next randomly chosen location and speed. Two kinds of movement speeds are selected — one uniformly distributed between 0 and 1 m/sec and one distributed between 0 and 10 m/sec. The energy consumption model described in [5], which uses measurements taken by the Cabletron Roamabout 802.11 DS High Rate network interface card that operates at 2 Mb/sec, is employed. The power consumption in transmit mode, receive mode, idle mode, and sleep mode is 1400, 1000, 830, and 130 mW, respectively. The cost of receiving paging signals is ignored. The grid size is set to 100; therefore, a gateway at the center of a grid can communicate with any gateway of its eight neighboring grids. Since each host is equipped with a GPS, we need to take into account the GPS power consumption. The GPS energy costs for GRID, ECGRID, and GAF are all 0.033 W [4].

Each source host sends a CBR (constant bit rate) flow with one or ten 512-byte packets per second. As mentioned in Section 24.1, a host running GAF protocol cannot be a destination host. Thus, our simulation requires two host models:

- *Model 1:* Ten infinite energy hosts act as sources or destinations. These hosts do not run GAF protocol nor do they forward traffic. Another 100 hosts, each with an initial energy of 500 joules, and which run GAF protocol, are used to evaluate energy consumption. This model applies only to GAF protocol because a source or destination host must always be active in GAF protocol.
- *Model 2:* Hosts with an initial energy of 500 joules are used to evaluate energy consumption. All hosts run ECGRID or GRID protocol. The source and destination hosts are randomly chosen. The number of hosts is 50, 100, 150, or 200. This model is used by ECGRID and GRID.

Three observations can be made:

1. *Effect of network lifetime:* In this experiment, the network lifetime of different protocols is observed. The network traffic load is 10 pkt/sec with constant mobility (pause time is 0 second). In Fig. 24.9a, the hosts' roaming speed is set to 1 m/sec. The network that runs GRID, which is not energy-conserving, is down when the simulation time = 590 seconds. Both ECGRID and GAF prolong the network lifetime. Since each active host in ECGRID must periodically send a *HELLO* message to maintain the host table, GAF is more energy conserving than ECGRID. For example, at speed = 1 m/sec and simulation time = 800 seconds, 85 and 81% of hosts are alive for GAF and ECGRID, respectively. However, the *HELLO* message exchanged in ECGRID guarantees successful data transmission at the expense of increased power consumption. Though the fraction of live hosts in GAF is higher than that of ECGRID, GAF assumes that the destination hosts are in active mode. Figure 24.9b presents the same simulation with a roaming speed of 10 m/sec. The results are similar to those in Fig. 24.9a.

2. *Effect of packet delivery:* Next, the *packet delivery rate* and *average packet delivery latency* are observed at different pause times. The packet delivery rate is defined as the number of data packets actually received by the destination, divided by the number of packets issued by the corresponding source host. The average packet delivery latency is defined as the average time elapsed between packet transmission and reception. The packet delivery qualities (packet delivery rate and latency) are compared for these three protocols with simulation time = 590 seconds, since the network hosts that run GRID exhaust all their energy at simulation time = 590 seconds. The results shown in Figs. 24.10 and 24.11 are based on a network traffic load of 10 pkt/sec at roaming speeds of 1 and 10 m/sec. Figure 24.10 reveals that the packet delivery rate exceeds 99% for all three protocols at a speed of 1 or 10 m/sec. Figure 24.10 shows that all three protocols have a similar average packet delivery latency, between 7.1 and 10.7 msec at a speed of 1 m/sec and between 8.5 and 12.5 msec at a speed of 10 m/sec. This experiment shows that our power-conserving protocol ECGRID can achieve its target without reducing the quality of delivered packets.

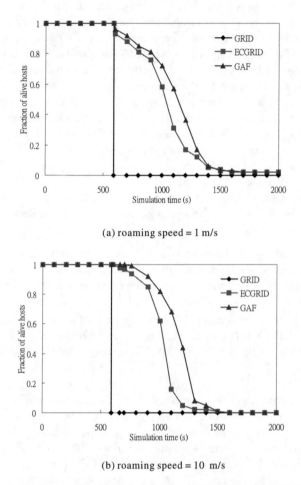

(a) roaming speed = 1 m/s

(b) roaming speed = 10 m/s

FIGURE 24.9 Fraction of alive hosts vs. the simulation time for GRID, GAF, and ECGRID: (a) roaming speed = 1 m/sec; (b) roaming speed = 10 m/sec. The number of hosts is 100, and the network traffic load is 10 pkt/sec with constant mobility (pause time 0).

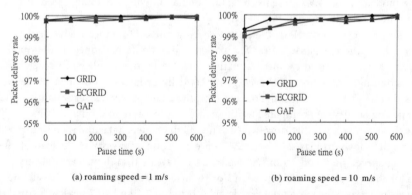

(a) roaming speed = 1 m/s   (b) roaming speed = 10 m/s

FIGURE 24.10 The packet delivery rate of GRID, GAF, and ECGRID for various pause times: (a) roaming speed = 1 m/sec; (b) roaming speed = 10 m/sec. The number of hosts is 100, and the network traffic load is 10 pkt/sec.

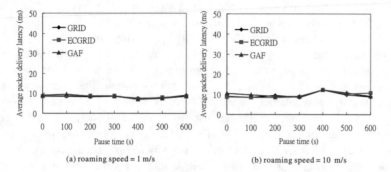

FIGURE 24.11 The packet delivery latency of GRID, GAF, and ECGRID for various pause times: (a) roaming speed = 1 m/sec; (b) roaming speed = 10 m/sec. The number of hosts is 100, and the network traffic load is 10 pkt/sec.

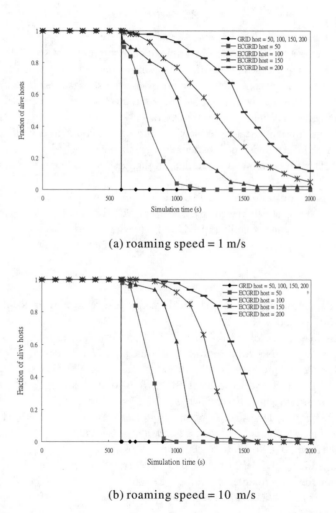

FIGURE 24.12 The fraction of alive hosts affected by host density for GRID and ECGRID: (a) roaming speed = 1 m/sec; (b) roaming speed = 10 m/sec. The host number is varied from 50 to 100, 150, and 200. The network traffic load is 10 pkt/sec with constant mobility (pause time = 0).

3. *Effect of host density:* In this experiment, the host density is varied to elucidate the relation between network lifetime and host density. The host number is set to 50, 100, 150, and 200. The network traffic load is 10 pkt/sec with constant mobility (pause time is 0 second). Figures 24.12a and b show the results at roaming speeds of 1 and 10 m/sec, respectively. The network lifetime in GRID is observed to be the same for various host densities because the network does not conserve energy. The network lifetime of our protocol increases with the host density. Comparing Figs. 24.12a and b shows that a higher roaming speed corresponds to better load balance between hosts. For example, with host number = 200, hosts start to run out of energy at simulation time = 660 seconds for a roaming speed of 1 m/sec (Fig. 24.12a) and at simulation time = 910 seconds for a roaming speed of 10 m/sec (Fig. 24.12b). The network lifetime is longer at a lower roaming speed because a higher roaming speed of each host leads to frequent gateway selection. The gateway selection will be restarted if a gateway host leaves its original grid, or a unicast message will be sent if a nongateway host leaves its original grid. Frequent gateway selection consumes much energy.

24.5 Conclusion

The issue of energy conservation is critical in a limited energy resource MANET. This study proposes a novel energy-aware routing protocol, ECGRID, for mobile ad hoc networks. ECGRID extends the GRID protocol to account for energy constraints. One host is elected as a gateway in each grid, according to the gateway election rule. A gateway is responsible for route discovery and packet delivery. Energy is conserved by turning the nongateway hosts' transceivers off when the hosts are idle. A gateway host can awaken sleeping hosts through RF tag technology. Accordingly, sleeping hosts need not wake up periodically. A load balance of the mobile host's battery energy scheme is applied to prolong the lifetime of all mobile hosts. Also, ECGRID eliminates the limitation that destination hosts must always be active (as is assumed for earlier protocols).

Simulation results demonstrate that ECGRID can not only prolong the lifetime of the entire network but also maintain good packet delivery quality. The proposed protocol outperforms in terms of prolonging the network lifetime at lower roaming speeds and achieving superior load balance at higher roaming speeds. Additionally, the lifetime is extended in proportion to the host density in the whole network.

References

[1] J. Broch, D.B. Johnson, and D.A. Maltz, The Dynamic Source Routing Protocol for Mobile Ad Hoc Networks (Internet draft), Feb. 2002.

[2] J. Broch, D.A. Maltz, D.B. Johnson, Y.-C. Hu, and J. Jetcheva, A Performance Comparison of Multi-Hop Wireless Ad Hoc Network Routing Protocols, *Proc. of 4th Annual ACM/IEEE International Conference on Mobile Computing and Networking (MobiCom '98)*, Dallas, TX, 1998, pp. 85–97.

[3] W.-H. Liao, Y.-C. Tseng, and J.-P. Sheu, GRID: a fully location-aware routing protocol for mobile ad hoc networks, *Telecommunication Systems*, 18, 37–60, 2001.

[4] Y. Xu, J. Heidemann, and D. Estrin, Geography-informed Energy Conservation for Ad Hoc Routing, *Proceedings of the 7th Annual ACM/IEEE International Conference on Mobile Computing and Networking (MobiCom '01)*, Rome, Italy, 2001, pp. 70–84.

[5] B. Chen, K. Jamieson, H. Balakrishnan, and R. Morris, Span: An Energy-Efficient Coordination Algorithm for Topology Maintenance in Ad Hoc Wireless Networks, *Proceedings of the 7th Annual ACM/IEEE International Conference on Mobile Computing and Networking (MobiCom '01)*, Rome, Italy, 2001, pp. 85–96.

[6] S.-L. Wu, Y.-C Tseng, and J.-P. Sheu, Intelligent medium access for mobile ad hoc networks with busy tones and power control, *IEEE Journal on Selected Areas in Communications*, 18, 1647–1657, 2000.

[7] C.F. Chiasserini and R.R. Rao, Combining Paging with Dynamic Power Management, *Proceedings of IEEE INFOCOM*, Anchorage, AK, vol. 2, 2001, pp. 996–1004.

[8] V. Rodoplu and T.H. Meng, Minimum Energy Mobile Wireless Networks, *Proceedings of the 1998 IEEE International Conference on Communications (ICC '98)*, vol. 3, June 1998, pp. 1633–1639.

[9] R. Ramanathan and R. Rosales-Hain, Topology Control of Multihop Wireless Networks Using Transmit Power Adjustment, *Proceedings of the IEEE INFOCOM*, Tel Aviv, vol. 2, 2000, pp. 404–413.

[10] C. Perkins and E.M. Royer, Ad Hoc on Demand Distance Vector (AODV) Routing (Internet draft), Jan. 2002.

[11] C.E. Perkins and P. Bhagwat, Highly Dynamic Destination-Sequenced Distance-Vector (DSDV) Routing for Mobile Computers, *ACM SIGCOMM Symposium on Communications, Architectures, and Protocols*, Sep. 1994, pp. 234–244.

[12] M. Royer and C.-K. Toh, A Review of Current Routing Protocols for Ad Hoc Mobile Wireless Networks, *IEEE Personal Communications*, Apr. 1999, pp. 46–55.

[13] L.M. Feeney and M. Nillsson, Investigating the Energy Consumption of a Wireless Network Interface in an Ad Hoc Networking Environment, *Proceedings of IEEE INFOCOM*, Anchorage, AK, vol. 3, 2001, pp. 1548–1557.

[14] J.P. Monks, V. Bharghavan, and W.-M.W. Hwu, A Power Controlled Multiple Access Protocol for Wireless Packet Networks, *Proceedings of IEEE INFOCOM*, Anchorage, AK, vol. 1, 2001, pp. 219–228.

[15] S.-Y. Ni, Y.-C. Tseng, Y.-S. Chen, and J.-P. Sheu, The Broadcast Storm Problem in a Mobile Ad Hoc Network, *Proceedings of the 5th ACM/IEEE International Conference on Mobile Computing and Networking (MobiCom '99)*, Seattle, WA, Aug. 1999, pp. 151–162.

[16] Q. Li, J. Aslam, and Daniela Rus, Online Power-Aware Routing in Wireless Ad-Hoc Networks, *Proceedings of the 7th Annual ACM/IEEE International Conference on Mobile Computing and Networking (MobiCom '01)*, Rome, Italy, 2001, pp. 97–107.

[17] J.E. Wieselthier, G.D. Nguyen, and A. Ephremides, On the Construction of Energy-efficient Broadcast and Multicast Trees in Wireless Networks, *Proceedings of IEEE INFOCOM*, Tel Aviv, vol. 2, 2000, pp. 585–594.

[18] P.-J. Wan, G. Galinescu, X.-Y. Li, and O. Frieder, Minimum-energy Broadcast Routing in Static Ad Hoc Wireless Networks, *Proceedings of IEEE INFOCOM*, Anchorage, AK, vol. 2, 2001, pp. 1162–1171.

[19] V. Park and S. Corson, Temporally-Ordered Routing Algorithm (TORA), Version 1 Functional Specification (Internet draft), July 2001.

[20] R. Wattenhofer, L. Li, P. Bahl, and Y.-M. Wang, Distributed Topology Control for Power Efficient Operation in Multihop Wireless Ad Hoc Networks, *Proceedings of IEEE INFOCOM*, Anchorage, AK, vol. 3, 2001, pp. 1388–1397.

[21] The Network Simulator — ns-2. http://www.isi.e.,du/nsnam/ns/.

[22] The CMU Monarch Project, http://www.monarch.cs.cmu.edu/.

25

Routing Algorithms for Balanced Energy Consumption in Ad Hoc Networks

Hee Yong Youn
Sungkyunkwan University

Chansu Yu
Cleveland State University

Ben Lee
Oregon State University

Abstract

In a mobile ad hoc network (MANET), a node communicates directly with the nodes within wireless range and indirectly with other nodes using a dynamically computed, multi-hop route via the other nodes of the MANET. In order to facilitate communication within the network, a routing protocol is used to discover routes between nodes. The primary goal of such an ad hoc network routing protocol is correct and efficient route establishment between a pair of nodes so that messages may be delivered in a timely manner. Although establishing efficient routes is an important goal, a more challenging goal is to provide energy efficient routing protocols, since a critical limiting factor for a mobile node is its operation time, restricted by battery capacity. However, the wireless link-only routing path in a MANET makes energy savings difficult to achieve. The corresponding reduction of nodes' lifetime directly affects the network lifetime since mobile nodes themselves collectively form a network infrastructure for routing in a MANET. This article surveys the energy aware routing mechanisms proposed for MANETs.

0-8493-1332-5/03/$0.00+$1.50
© 2003 by CRC Press LLC

25.1 Introduction

Recently, wireless technology has been one of the hottest topics in computing and communications. Since the late 1970s, consumer wireless applications such as mobile phones have begun to take off, and presently people are beginning to activate third-generation (3G) networks for commercial purposes. Wireless networking technology offering high data rates for mobile users will flourish, which will enable the handling of multimedia Web content, videoconferencing, e-commerce, etc. *Routing* is one of the key issues for supporting these demanding applications in a rather unstable and resource limited wireless networking environment.

There are two ways to implement mobile wireless networks — *infrastructured network* and *infrastructureless (ad hoc) network*. With an infrastructured network, mobile nodes communicate only with the base stations providing internode routing and fixed network connectivity. With the infrastructureless mobile network, each node communicates with other nodes directly or indirectly through intermediate nodes. Thus, all nodes are virtually routers participating in some protocol required for deciding and maintaining the routes.

A large number of routing protocols have been developed for *mobile ad hoc networks* (MANETs) [14], which are characterized by unpredictable network topology changes, high degree of mobility, energy-constrained mobile nodes, bandwidth-constrained intermittent connection, and memory-constrained. The routing problem has been well researched in infrastructured wireless networks, where the goals are efficient detection and adaptation to the network topology, scalability, and convergence. Even though these are equally valid for MANETs, the solutions are more difficult to find since MANETs are inherently more dynamic. In particular, energy efficiency may be the most important design criterion for mobile networks since a critical limiting factor for a mobile node is its operation time, restricted by battery capacity. In infrastructured wireless networks, where a wireless link is limited to one hop between an energy-rich base station and a mobile node, the goal of energy conservation can be largely achieved by relocating power intensive network operations to the base station.

However, the wireless link-only routing path in a MANET makes energy savings difficult to achieve. The corresponding reduction of nodes' lifetime directly affects the network lifetime since mobile nodes themselves collectively form a network infrastructure for routing in a MANET. To address this problem, many research efforts have been devoted to developing energy aware network protocols such as power saving *MAC (medium access control)* layer protocols, energy efficient routing algorithms, and power sensitive network architectures. Based on the aforementioned discussion, this chapter focuses on the energy-aware routing mechanisms proposed for MANETs.

The remainder of the chapter is organized as follows. Section 25.2 presents a general discussion on ad hoc routing protocols. Although the protocols discussed in this section do not consider energy consumption as a metric for routing, they provide the basis for energy-aware routing in MANETs. Section 25.3 surveys the routing protocols specifically designed for balanced energy consumption in MANETs. Finally, Section 25.4 provides a conclusion and a discussion on power issues.

25.2 Routing Protocols for Ad Hoc Networks

The routing protocols proposed for MANETs are generally categorized as *table-driven, source-initiated on-demand driven,* and *hybrid* based on the timing when the routes are updated. With the table-driven routing protocols, each node attempts to maintain consistent, up-to-date routing information to every other node in the network. With source-initiated on-demand routing, route discovery and maintenance are performed only when a source node desires them. The hybrid approach combines the two approaches to minimize the overhead incurred during route discovery and maintenance. In this section, the protocols belonging to each of the three aforementioned categories are discussed.

25.2.1 Table-Driven Routing Protocols

In table-driven routing protocols, each node maintains an up-to-date routing table by responding to changes in network topology and propagating the updates. Thus, it is *proactive* in the sense that when a packet needs to be forwarded, the route is already known and can be immediately used. As is the case for wired networks, each node in a MANET maintains a routing table containing a list of all the destinations, next hop, and the number of hops to each destination. The routing table is constructed using either link-state or distance vector algorithms. There are a number of protocols [5,6,7,12,19,22,23] that belong to this category, which are different in the number of tables manipulated for routing and the methods used for exchanging and maintaining routing tables.

Among the table-driven protocols, *Destination-Sequenced Distance Vector* (DSDV) [23], *Wireless Routing Protocol* (WRP) [19], and *Global State Routing* (GSR) [5] use destination sequence numbers to keep routes loop free and up to date. These sequence numbers are assigned by the destination node and allow the mobile nodes to distinguish invalid routes from new ones. GSR is similar to the DSDV scheme but uses the link state instead of the distance vector. Each node maintains a link-state table based on the information exchanged periodically with the neighbors. The update is selected based on the timestamp of the sequence numbers. In WRP, each node maintains a distance table, a routing table, a link-cost table, and a *Message Retransmission List* (MRL) table. MRL keeps a record of which updates in an update message need to be retransmitted and which neighbors should acknowledge the retransmission [19]. An update message is sent only between neighboring nodes and contains a list of updates (the destination, the distance to the destination, and the predecessor of the destination), as well as a list of responses indicating which mobile nodes should acknowledge (ACK) the update.

In contrast to DSDV and GSR, *Cluster Gateway Switching Routing* (CGSR) [6], *Hierarchical State Routing* (HSR) [7], and *Zone-based Hierarchy Link State* (ZHLS) [12] protocols use hierarchical routing schemes. The CGSR protocol extends DSDV by grouping nodes into clusters. Thus, each cluster is represented by a *clusterhead*, and two clusters can communicate via a *gateway* node that is within the communication range of the two clusters. Each node also maintains a cluster member table where the clusterheads' destinations are stored. Therefore, the cluster member table is used to perform intercluster routing, while the routing table is used to perform intracluster routing. The HSR protocol extends CGSR by forming a hierarchy of clusterheads. This is done by having nodes within a cluster broadcast their link information to each other. The clusterhead summarizes its cluster's information and sends it to neighboring clusterheads via gateway as done in CGSR. The hierarchy reduces the overhead associated with the link-state algorithm and the number of entries in the routing table.

In ZHLS, the network is divided into nonoverlapping zones without any *zone-head*. ZHLS defines two levels of topologies — node level and zone level. If any two nodes are within the communication range, a physical link exists. A virtual link exists between two zones if at least one node of a zone is physically connected to some nodes of the other zone. The node (zone) level topology provides the information on how the nodes (zones) are connected together by the physical (virtual) links. Thus, given the zone and node ID of a destination, the packet is routed based on the zone ID until it reaches the correct zone. Then, within that zone, it is routed based on node ID.

Fisheye State Routing (FSR) protocol [22] is another hierarchical routing scheme where information exchange is more frequent with closer nodes than with faraway nodes. FSR is an improvement over GSR in which the bandwidth overhead due to update messages is minimized. The FSR protocol scales well to large networks since the overhead is controlled.

25.2.2 Source-Initiated On-Demand Driven Protocols

These are reactive protocols where routes are created only when desired by the source node. The two basic procedures of source-initiated on-demand driven protocols are the *route discovery* process and the *route maintenance* process. The route discovery process involves sending *route-request* packets to neighbor nodes, which then forward the request to their neighbors, and so on. Once the route-request reaches the destination or the intermediate node with a "fresh enough" route, the destination/intermediate node responds by unicasting a *route-reply* packet back to the neighbor from which it first received the route-request. Once the route is established, it is maintained by some form of route maintenance process until either the destination becomes inaccessible along any path from the source or the route is no longer desired. In contrast to table-driven routing protocols, not all up-to-date routes are maintained at every node. This subsection discusses several source-initiated on-demand routing protocols [1, 8, 11, 13, 20, 24, 28].

The *Dynamic Source Routing* (DSR) protocol [13] is a typical example of the on-demand protocols, where each data packet carries in its header the complete ordered list of nodes the packet passes through. This is done by having each node maintain a *route cache* that learns and caches routes to destinations. Some on-demand routing protocols are extensions of table-driven protocols. For example, the *Ad-Hoc On-Demand Vector* (AODV) protocol [24] is an improvement on the DSDV protocol, where the number of required broadcasts is minimized by creating routes on an on-demand basis. Each node maintains its own sequence number, as well as a broadcast ID for the route-request. The broadcast ID is incremented for every route-request the node initiates, and together with the node's IP address it uniquely identifies a route-request. The *Cluster Based Routing Protocol* (CBRP) [11] is an extension of CGSR where nodes are divided into clusters. When a source has data to send, it floods route request packets only to the neighboring clusterheads. Upon receiving the request, a clusterhead checks to see if the destination is in its cluster. If so, the request is sent directly to the destination; otherwise, the request is sent to all its adjacent clusterheads.

Temporally Ordered Routing (TORA) [20] is a highly adaptive protocol that provides multiple routes for any desired source–destination pair and localizes the control messages to a very small set of nodes near the location of a topological change. To accomplish this, nodes maintain routing information on adjacent (one-hop) nodes and use a "height" metric to establish a *directed acyclic graph* (DAG) rooted at the destination. When the DAG route is broken during node mobility, route maintenance is necessary to reestablish a DAG rooted at the same destination. This is achieved using a *link reversal algorithm* at the site of the link failure to reestablish the path. The algorithm tries to localize the effect and gives many alternate paths to the destination. Thus, the algorithms not only save bandwidth in updates, but also provide alternate paths in case of path failures.

In contrast to aforementioned protocols that only use the shortest path as the routing metric, the *Associativity Based Routing* (ABR) [28] protocol uses the connection stability metric, called *associativity*, among mobile nodes to select the best route. In other words, a high degree of associativity may indicate a low state of node mobility, while a low degree may indicate a high state of node mobility. Associativity among nodes is determined by first having all nodes generate periodic beacons, and then the associativity tick of the receiving node with respect to the beaconing node is incremented. Thus, when packets arrive at the destination, the best route is selected by examining the associativity ticks along each of the paths. Associativity ticks are reset when the neighbors of the node or the node itself move out of proximity.

Similarly, the *Signal Stability Routing* (SSR) protocol [8] selects routes based on signal strength. SSR selects routes based on the signal strength between nodes and on a node's location stability, and it is divided into two cooperative protocols: the *Dynamic Routing Protocol* (DRP) and the *Static Routing Protocol* (SRP). DRP is responsible for maintaining the *Signal Stability Table* (SST) and the *Routing Table* (RT). SST records the signal strength of neighboring nodes as strong or weak using periodic beacons from each neighboring node. DRP passes a received packet to the SRP, which then forwards it using the RT. If there is no known route in RT, a route search is initiated by sending route-requests over only strong channels. The destination chooses the first arriving route-request packet to send back because it is most probable that the packet arrived over the shortest and/or least congested path. If no route-reply message is received by the source within a specific timeout period, the source node indicates that weak channels are acceptable, as these may be the only links over which the packet can be propagated.

The *Relative Distance Micro-Discovery Routing* (RDMAR) [1] protocol improves the ABR protocol by limiting the flooding of route-request packets to a certain radius. The estimate of the radius is based on the number of radio hops between two nodes. This protocol does not employ beaconing or a route cache.

25.2.3 Hybrid Routing Protocols

The hybrid approach combines the table-driven and source-initiated on-demand driven approaches such that the overhead incurred in route discovery and maintenance is minimized while the efficiency is maximized. Several protocols belonging to this approach are presented in this subsection [2,10,16,17,26].

The *Zone Routing Protocol* (ZRP) [10] partitions the network implicitly into zones, where a zone of a node includes all nearby nodes within the zone radius defined in hops. It applies proactive strategy inside the zone and reactive strategy outside the local zone. Each node may potentially be located in many zones. ZRP consists of two subprotocols. The proactive *intrazone routing protocol* (IARP) is an adapted distance-vector algorithm. When a source has no IARP route to a destination, it invokes a reactive *interzone routing protocol* (IERP), which is very similar to DSR.

The *Core Extraction Distributed Ad Hoc Routing* (CEDAR) protocol [26] is a hierarchical protocol that attempts to model the IP routing structure, with emphasis on QoS support, by identifying a subset of nodes called *core* nodes. Each node must be adjacent to at least one core node and picks one node as the leader or dominator. The core is determined by periodic exchange of messages between each node and its neighbors. Each core node maintains a path to the nearby nodes by issuing a limited broadcast. The core is dynamically extracted by approximating a minimum dominating set using local computation and local state, and it performs route computation on behalf of the nodes that belong to it. The bandwidth availability information is then propagated in the core subgraph. Each core node knows local links and nodes that are stable or having high bandwidth. When a source wants to send a packet to the destination, it informs its core. The core node then finds the path to the core node of the destination using some DSR-like probing. Finally, core nodes form a path using locally available link-state information.

The *Location-Aided Routing* (LAR) protocol [16] assumes that the sender has advance knowledge of the location and velocity of the destination node using the GPS. Based on the location and velocity of the destination node, the expected zone can be defined. Thus, LAR limits the search for a new route to a small zone resulting in fewer route discovery messages. The request zone is the smallest rectangle that encompasses the expected zone. The sender explicitly specifies the request zone in its route-request message to limit the boundary on the propagation of the route-request messages.

The *Distance Routing Effect Algorithm for Mobility* (DREAM) protocol [2] uses the fact that the greater the distance separating two nodes, the slower they appear to be moving with respect to each other. Accordingly, the location information in routing tables can be updated as a function of the distance separating the nodes without compromising the routing accuracy. DREAM sends the location updates by the moving nodes autonomously, based only on the node's mobility rate. This is because routing information on the slowly moving nodes needs to be updated less frequently than that for those with high mobility. This is done by sending messages in the "record direction" of the destination node, guaranteeing delivery by following the direction with a given probability.

The *Grid Location Service* (GLS) protocol [17] is a decentralized routing protocol. Each mobile node periodically updates a small set of other nodes (its *location servers*) with its current location. A node sends its position updates to its location servers without knowing their actual identities, assisted by a predefined ordering of node identifiers and a predefined hierarchy. Queries for a mobile node's location also use the predefined ordering and spatial hierarchy to find a location server for that node. For example, when node *A* wants to find the location of node *B*, it sends a request to the least node greater than or equal to node *B* for which it has location information. That node forwards the query in the same way, and so on. Eventually, the query will reach a location server of node *B*, which will then forward the query to node *B*. Since the query contains node *A*'s location, it can respond directly using geographic forwarding. Routing updates are carried out using either flooding based algorithm or link reversal algorithm.

25.3 Routing Protocols for Balanced Energy Consumption

This section surveys energy efficient routing protocols developed for MANETs. It is noted that direct comparison of these protocols is extremely difficult because these approaches have different goals with different assumptions and implementation levels. Nevertheless, there are three major issues involved in energy aware routing protocols. First, the goal is to find the path that either *minimizes* the absolute power consumed or *balances* the energy consumption of all mobile nodes. Balanced energy consumption does not necessarily lead to minimized energy consumption, but it keeps a certain node from being overloaded and thus ensures longer network lifetime. Since energy balance can be achieved indirectly by distributing network traffic, one such routing protocol is also discussed in this section. Second, energy awareness has been either implemented at purely routing layer or routing layer with help from other layers such as MAC or application layer. For example, information from the MAC layer is beneficial because it usually supports power saving features that the routing protocol can exploit to provide better energy efficiency. Third, some routing protocols assume that the transmission power is controllable and nodes' location information are available (e.g., via GPS). Under these assumptions, the problem of finding a path with the least consumed power becomes a conventional optimization problem on a graph where the weighted link cost corresponds to the transmission power required for transmitting a packet between the two nodes of the link.

25.3.1 PAR (Power Aware Routing) Protocol

The PAR protocol [25] is not a new routing protocol but suggests the use of different metrics when determining a routing path. The following energy-related metrics have been suggested instead of the shortest routing path between a source and a destination:

- Minimizing energy consumed/packet
- Maximizing time to network partition
- Minimizing variance in node power levels
- Minimizing cost/packet
- Minimizing maximum node cost

The first metric is useful for minimizing the overall energy consumption for delivering a packet. To this end, however, it is possible that some particular nodes are unfairly burdened to support many packet-relaying functions. These hot spot nodes may consume more battery energy and stop running earlier than other nodes do, resulting in link disconnection and network partitioning. A better routing path is the one where packets get routed through energy-rich intermediate nodes in spite of additional delay or hop count.

Maximizing the second metric, time to network partition, is considered an ultimate goal of a MANET because it directly addresses the network lifetime. However, since it is difficult to estimate the future network behavior, the next three metrics can be used to attempt to indirectly achieve the goal. For example, the third approach, minimizing variance in node power levels, is a direct approach to maintain the energy balance with information on all nodes' power levels. In the fourth and fifth approaches, each path is annotated with path cost measured by the accumulated battery life of all intermediate nodes and the minimal residual battery life among the intermediate nodes, respectively. The path with the maximum path cost is selected.

25.3.2 APR (Alternate Path Routing) Protocol

The APR protocol [21] indirectly balances energy consumption by distributing network traffic among a set of diverse paths for the same source–destination pair, called an *alternate route set*. APR's performance greatly depends on the quality of the alternate route set, which can be measured by *route coupling*, i.e., how many nodes and links two routes have in common. Since the movement of a common node breaks the two routes altogether, a good alternate route set consists of decoupled routes. A decoupled alternate route set can be constructed as shown in Fig. 25.1. When node S searches for a routing path to D, it may obtain three alternate routes: S→A→B→C→D, S→A→E→C→D, and S→E→B→D. Since they share some intermediate node(s), the alternate route set is not good enough. Each routing path is decomposed into constituent links, and additional alternate routes can be constructed with improved diversity and reduced length: S→A→B→D and S→E→C→D.

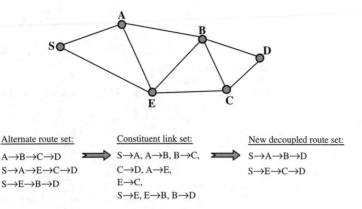

FIGURE 25.1   Construction of alternate route set in the APR protocol.

With proactive routing protocols (see Section 25.2.1), each node is provided with a complete and up-to-date view of the network connectivity and thus, it is capable of identifying the best alternate routes that exist in the network. However, in the presence of significant node mobility, tracking all the changes in network connectivity can be prohibitively expensive. With reactive routing protocols (see Section 25.2.2), the alternate route set is constructed during the route discovery process since a route query may produce multiple responses containing paths to the sought-after destination. Later, during the *reply phase*, the cached path information is used to redirect replies along more diverse paths back to the source.

25.3.3 LEAR (Localized Energy Aware Routing) Protocol

Unlike APR, the LEAR protocol [29] directly controls the energy consumption. In particular, it achieves balanced energy consumption among all participating mobile nodes. The LEAR protocol is based on DSR, where the route discovery requires flooding of route-request messages. When a routing path is searched, each mobile node relies on local information on *remaining battery level* to decide whether or not to participate in the selection process of a routing path. An energy-hungry node can conserve its battery power by not forwarding data packets on behalf of others. The decision-making process in LEAR is distributed to all relevant nodes, and the destination node does not need to wait or block itself in order to find the most energy-efficient path.

Upon receiving a route-request message, each mobile node has the choice to determine whether or not to accept and forward the route-request message depending on its *remaining battery power* (E_r). When it is higher than a *threshold value* (Th_r), the route-request message is forwarded; otherwise, the message is dropped. The destination will receive a route-request message only when all intermediate nodes along the route have good battery levels. Thus, the first arriving message is considered to follow an energy-efficient as well as a reasonably short path.

If any of the intermediate nodes along every possible path drop the route-request message, the source will not receive a single reply message even though one exists. To prevent this, the source will resend the same route-request message, but this time with an increased sequence number. When an intermediate node receives the same request message again with a larger sequence number, it adjusts (lowers) its Th_r to allow forwarding to continue. Table 25.1 describes the LEAR algorithm. In order to reduce the repeated request messages and to utilize the route cache, four routing-related control messages are introduced: DROP_ROUTE_REQ, ROUTE_CACHE, DROP_ROUTE_CACHE, and CANCEL_ROUTE_CACHE.

25.3.4 FAR (Flow Augmentation Routing) Protocol

The FAR protocol [3] maximizes network lifetime by balancing the traffic among the nodes in proportion to their energy reserves. The traffic balance, in turn, can be achieved by selecting the optimal transmission

TABLE 25.1 The LEAR Algorithm

Node	Steps
Source node	Broadcast a route-request; wait for the first arriving route-reply; select the source route contained in the message; ignore all later replies
Intermediate node	Upon receipt of a route-request message: If the message is not the first trial and $E_r < Th_r$, adjust (lower) Th_r by d; If it has the route to the destination in its cache, If $E_r > Th_r$, forward (unicast) ROUTE_CACHE & ignore all later requests; Else, forward DROP_ROUTE_CACHE & ignore all later requests; Else, If $E_r > Th_r$, forward (broadcast) route-request & ignore all later requests; Else, forward (broadcast) DROP_ROUTE_CACHE & ignore all later requests Upon receipt of a ROUTE_CACHE, If the message is not the first trial and $E_r < Th_r$, adjust (lower) Th_r by d; If $E_r > Th_r$, forward (unicast) ROUTE_CACHE & ignore all later requests; Else, forward (unicast) DROP_ROUTE_CACHE & ignore all later requests; and send backward (unicast) CANCEL_ROUTE_CACHE
Destination node	Upon receipt of the first arriving route-request or ROUTE_CACHE, send a route-reply to the source with the source route contained in the message

power levels and the optimal route. Given a static network topology, the selection problem turns out to be a conventional maximum flow optimization problem on a graph, where the transmission energy between two neighboring nodes corresponds to the link cost between them. Since there are multiple source–destination pairs with different data generation rates at each source, the solution can be obtained step by step with incremental data generation or data traffic. More specifically, FAR first solves the optimization problem with initial data traffic. It expends energy of the corresponding intermediate nodes. Then it augments data traffic at each source and solves the same problem again with the reduced energy reserves. The final and overall routing decision is obtained by repeatedly solving the optimization problem until any node runs out of its initial energy reserves.

The cost function of the optimization problem is the sum of link cost c_{ij} along the path, where c_{ij} is expressed as $e_{ij}^{x1} R_i^{-x2} E_i^{x3}$, e_{ij} is the energy cost for unit flow transmission over the link, and E_i and R_i are the *initial* and *residual energy* at the transmitting node i, respectively. Depending on the parameters x_1, x_2, and x_3, the corresponding routing algorithm $FA(x_1, x_2, x_3)$ achieves different goals. In $FA(0,0,0)$, the shortest cost path is the minimum hop path and, in $FA(1,0,0)$, it is the *minimum transmitted energy (MTE) path*. $FA(1,50,50)$ in the form of $FA(1,x,x)$ balances energy consumption and significantly improves the system lifetime over the conventional MTE routing algorithm. Table 25.2 summarizes those routing algorithms

25.3.5 OMM (Online Max-Min Routing) Protocol

The data transmission sequence (or data generation rate) is not usually known in advance. Without requiring that information, the OMM protocol [18] makes a routing decision that optimizes two different metrics: *minimizing power consumption* and *maximizing the minimal residual power* in the nodes of the network. Given the power level information of all nodes and the power cost between two neighboring nodes, this algorithm first finds the path that minimizes the power consumption (P_{min}) by using the

TABLE 25.2 FAR Routing Algorithms

Routing Algorithm	Optimization Objective
$FA(0, 0, 0)$	Minimum hop path
$FA(1, 0, 0)$	Minimum transmitted energy path
$FA(\cdot, x, x)$	Minimum normalized residual energy used
$FA(\cdot, \cdot, 0)$	Minimum absolute residual energy used

Dijkstra algorithm. Among the power efficient paths with some tolerance (less than zP_{min}, where $z \geq 1$), it selects the best path that optimizes the second metric by iterative application of the Dijkstra algorithm with edge removals.

The parameter z measures the tradeoff between the *max-min path* and the *minimum power path*. When $z = 1$, the algorithm optimizes only the first metric and thus provides the minimal power consumed path. When $z = \infty$, it optimizes only the second metric and thus provides the max-min path. Thus, proper selection of the parameter z is important in determining the overall performance. A perturbation method is used to compute z adaptively. First, the algorithm randomly chooses an initial value of z and estimates the lifetime of the most overloaded node. Then, z is increased by a small constant, and the lifetime is estimated again. The two estimates are compared, and the parameter z is increased or decreased accordingly. Since the two successive estimates are calculated during two different time periods, the whole process is based on the assumption that the message distributions are similar as time elapses. The algorithm steps are as follows:

1. Find the path with the least power consumption, P_{min}, using the Dijkstra algorithm.
2. Find the path with the least power consumption in the graph. If the power consumption is greater than $z \cdot P_{min}$ or no path is found, then the previous shortest path is the solution, stop.
3. Find the minimal residual power fraction on that path, and let it be u_{min}.
4. Find all the edges that have a residual power fraction smaller than u_{min} and remove them from graph.
5. Go to step (2).

OMM requires information about the power levels of all mobile nodes. In large networks, this requirement is not trivial. To improve the scalability, a *zone-based hierarchical routing mechanism* is used, where the area is divided into a small number of *zones*. A routing path usually consists of a global path from zone to zone and a local path (just a few hops) within the zone. With the extended OMM protocol, a node estimates the power level of each zone, computes a path across zones, and computes the best path within each zone.

25.3.6 PLR (Power-Aware Localized Routing) Protocol

MANET routing algorithms based on global information, such as data generation rate or power level information of other nodes, may not be practical because each node is provided with only the local information. The PLR protocol [27] is a localized, fully distributed energy aware routing algorithm. Assuming that the location information of its neighbors and the destination are available through GPS, each node selects one of its neighbors through which the overall transmission power to the destination is minimized.

Since the transmission power needed for direct communication between two nodes has super-linear dependency on distance, it is usually energy efficient to transmit packets via intermediate nodes. For example, direct transmission from node A to node D in Fig. 25.2 may consume more energy than indirect transmission via N_i provided that $|AD|$ is larger than $(c/(a(1 - 2^{1-\alpha})))^{1/\alpha}$, where the transmission and reception power between two nodes separated by a distance d is $u(d) = ad^\alpha + c$. It is also shown that the power consumption is minimized, which is denoted as $v(d)$, when $(n - 1)$ equally spaced intermediate nodes relay transmissions along the two end nodes, where $n = d[a(\alpha - 1)/c]^{1/\alpha}$ and $v(d) = dc[a(\alpha - 1)/c]^{1/\alpha} + da[a(\alpha - 1)/c]^{(1 - \alpha)/\alpha}$.

Therefore, the selection of an intermediate node among its neighbors requires evaluation of $u(d) + v(d)$. In other words, a node (A), whether it is a source or an intermediate node, selects one of its neighbors $(N_1, N_2, N_3,...)$ as the next intermediate node (N_i) to the destination node (D), which minimizes $u(|AN_i|) + v(|N_iD|)$. Note that A to N_i is a direct transmission, while N_i to D is an indirect transmission with some intermediate nodes between N_i and D. If the goal is to maximize the network lifetime, we only need to generalize the cost function by including the remaining lifetime of node N_i or all of N_i's neighbors.

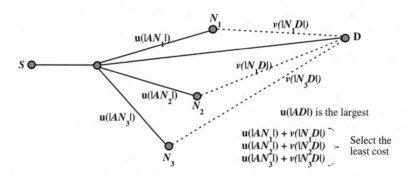

FIGURE 25.2   Transmission from node *A* to node *D*.

25.3.7 SPAN Protocol

Unlike other aforementioned routing protocols, the SPAN protocol [4] operates between the routing layer and the MAC layer. This is because SPAN tries to exploit the MAC layer's power-saving features in its routing decision. The basic idea of the MAC layer's power-saving mechanism is to power down (*sleep*) the radio device when it has no data to transmit or receive. This allows substantial energy savings since sleep operation consumes less power. For example, Lucent's WaveLAN-II, based on the IEEE 802.11 wireless LAN standard, consumes 250 and 300 mA when receiving and transmitting, respectively, while it consumes only 9 mA when it is in sleep mode [15].

In order to coordinate the sleep period operation in IEEE 802.11, one mobile node is selected as the *master*. The master node must be awake all the time and periodically sends a beacon packet to its slave nodes followed by a *TIM* (*Traffic Indication Map*) that indicates the desired receivers. Each slave wakes up at the beacon times and checks whether it is addressed or not. If the node is not addressed it sleeps again; otherwise, it stays awake to receive data. Figure 25.3 shows a simple power state diagram of the IEEE 802.11 standard.

The SPAN protocol makes the information on master nodes available to the network layer and lets them constitute a routing backbone to route most of the traffic in the MANET. All slave nodes need not wake up to forward traffic on behalf of other nodes; they conserve energy by sleeping most of time. On the other hand, master nodes must be awake all the time for routing. However, this does not expend any extra energy because they need to be up anyway for the MAC layer's sleep period coordination. To prevent overloading the masters and to ensure fairness, each master periodically checks whether it should withdraw as a master and give other neighbors a chance to become a master.

Selecting and replacing masters must be done in a distributed way. In SPAN, each node periodically determines whether it should become a master or not based on the following *master eligibility rule: If two of its neighbors cannot reach each other either directly or via one or two masters, it should become a master.* In Fig. 25.4, nodes *B* and *D* become masters. Node *H* would be eligible if either *B* or *D* does not

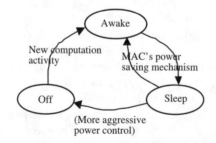

FIGURE 25.3   Power saving mechanism in IEEE 802.11 wireless LAN standard.

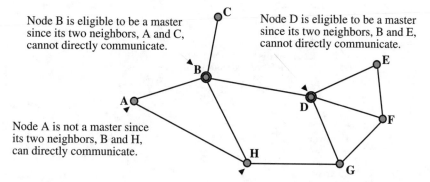

Node B is eligible to be a master since its two neighbors, A and C, cannot directly communicate.

Node D is eligible to be a master since its two neighbors, B and E, cannot directly communicate.

Node A is not a master since its two neighbors, B and H, can directly communicate.

Node H is not a master since all of its neighbors can communicate directly or via two masters. (If either B or D is not elected as a master yet, node H is eligible to be a master because of two neighbors A and G.)

FIGURE 25.4 Master eligibility rule in SPAN.

elect itself as a master when node *H* checks its eligibility (thus, the master selection process is not deterministic). This rule does not yield the minimum number of master nodes, but it provides robust connectivity with substantial energy savings.

25.3.8 GAF (Geographic Adaptive Fidelity) Protocol

Similar to SPAN, this protocol [30] identifies many redundant nodes with respect to routing and turns them off without sacrificing the routing fidelity. Each node uses location information based on GPS to associate itself with a *virtual grid*, where all nodes (except master nodes) in a particular grid square are redundant with respect to forwarding packets. Thus, these nodes switch between off and listening with the guarantee that one master node in each grid stays awake to route packets. For example, in Fig. 25.5, nodes 2, 3, and 4 in virtual grid B are equivalent, so one of them forwards packets between nodes 1 and 5 while the other two can sleep to conserve energy. The relationship between the grid size *r* and the radio range *R* can be easily deduced as $r^2 + (2r)^2 \leq R^2$ or $r \leq R\sqrt{5}$, since nodes 2 and 5 should be able to communicate directly.

In GAF, nodes are in one of three states as shown in Fig. 25.6: *sleeping, discovering,* and *active.* Initially, a node is in the *discovery* state and exchanges discovery messages including grid IDs to find other nodes within the same grid. A node becomes *active* if it does not hear any other discovery message for T_d. If more than one node is in the discovery state, the one with the longest expected lifetime becomes active. The active node remains active to handle routing for a predefined time duration, T_a. After T_a, the node changes state to discovery to give a chance to other nodes within the same grid to become active. In

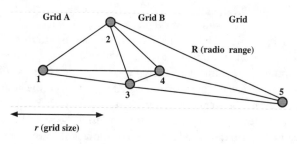

FIGURE 25.5 Virtual grid structure in the GAF protocol.

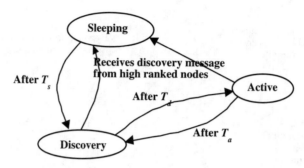

FIGURE 25.6 State transition in the GAF protocol.

scenarios with high mobility, *sleeping* nodes should wake up earlier to take over the role of an active node, where the sleeping time T_s is calculated based on the estimated time staying in the grid.

25.3.9 PEN (Prototype Embedded Network) Protocol

The PEN protocol [9] is designed for embedded networks where the rate of interaction is fairly low. It is thus more suited for control applications rather than data applications. Low power consumption is a key design criterion, which renders existing de facto protocols replaced by low power ad hoc protocol stack from the physical layer to the transport layer. As in SPAN and GAF, this protocol exploits the low duty cycle of communication activities and powers down the radio device when it is idle. Like SPAN, the PEN system has an additional layer between the MAC and the routing layer, called the *rendezvous layer*, which is responsible for scheduling and forecasting times of inactivity.

However, unlike SPAN, nodes interact asynchronously without master nodes and thus, costly master selection and cluster formation procedures can be avoided at the cost of extended delay. This asynchronous protocol is based on a *server beaconing* mechanism where each node periodically wakes up, broadcasts its routing capability as a server, and listens to the replies before powering down again. Any node wishing to send would wake up and listens to the beacons from such nodes. Route discovery and route maintenance procedures are similar to those in AODV (see Section 25.2.2), i.e., on-demand route search and routing table exchange between neighbor nodes. Due to its asynchronous operation, the PEN protocol minimizes the amount of active time and thus saves substantial energy.

25.4 Conclusion

A MANET consists of autonomous, self-organizing, and self-operating nodes. It is characterized by links with less bandwidth, nodes with energy constraints, and nodes with less memory and processing power, and it is more prone to security threats than fixed networks. However, it has many advantages and different application areas from fixed networks or infrastructured mobile networks. The field of ad hoc mobile networks is rapidly growing and changing, and while there are still many challenges that need to be met, it is likely that such networks will see widespread use within the next few years.

Routing is one of the main problems in MANETs. Numerous solutions to routing have been proposed, but energy efficient routing decision is more important than simple shortest path routing. In this chapter, we have provided descriptions of a number of energy aware routing schemes proposed for MANETs. While it is not clear that any particular algorithm or class of algorithms is the best for all scenarios, each protocol has definite advantages/disadvantages and is well suited for certain situations. Moreover, direct comparison of the energy efficient routing protocols is not possible because they are based on different assumptions such as location information availability and transmission power control. Instead, they must be carefully combined for extending the MANET lifetime.

References

[1] Aggelou, G. and Tafazolli, R., RDMAR: A Bandwidth-Efficient Routing Protocol for Mobile Ad Hoc Networks, *ACM Intl. Workshop on Wireless Mobile Multimedia (WoWMoM)*, Aug. 1999, pp. 26–33.

[2] Basagni, S., Chlamtac, I., Syrotiuk, V., and Woodward, B., A Distance Routing Effect Algorithm for Mobility (DREAM), *Intl. Conf. on Mobile Computing and Networking (MobiCom '98)*, Dallas, TX, 1998, pp. 76–84.

[3] Chang, J.-H. and Tassiulas, L., Energy Conserving Routing in Wireless Ad-hoc Networks, *Conf. on Computer Communications (IEEE Infocom 2000)*, Tel Aviv, 2000, pp. 22–31.

[4] Chen, B., Jamieson, K., Morris, R., and Balakrishnan, H., Span: An Energy-Efficient Coordination Algorithm for Topology Maintenance in Ad Hoc Wireless Networks, *Intl. Conf. on Mobile Computing and Networking (MobiCom '2001)*, Rome, Italy, Jul. 2001.

[5] Chen, T. and Gerla, M., Global State Routing: A New Routing Scheme for Ad Hoc Wireless Networks, *IEEE Int'l Conf. on Communications (ICC)*, 1998.

[6] Chiang, C.-C., Wu, H.K., Liu, W., and Gerla, M., Routing in Clustered Multihop, Mobile Wireless Networks with Fading Channel, *IEEE SICON '97*, Apr. 1997, pp. 197–211.

[7] Corson, M. and Ephremides, A., A distributed routing algorithm for mobile wireless networks, *ACM/Baltzer Wireless Networks Journal*, 1, 61–81, 1995.

[8] Dube, R., Rais, C.D., Wang, K.-Y., and Tripathi, S.K., Signal Stability Based Adaptive Routing (SSA) for Ad-Hoc Networks, *IEEE Personal Communications*, Feb. 1997, pp. 36–45.

[9] Girling, G., Wa, J., Osborn, P., and Stefanova, R., The Design and Implementation of a Low Power Ad Hoc Protocol Stack, *IEEE Wireless Communications and Networking Conference*, Sep. 2000.

[10] Haas, Z. and Pearlman, M., Performance of Query Control Schemes for the Zone Routing Protocol, *ACM SIGCOMM*, Aug. 1998.

[11] Jiang, M., Li, J., and Tay, Y.C., Cluster Based Routing Protocol, IETF Draft, Aug. 1999, http://www.ietf.org/internet-drafts/draft-ietf-manet-cbrp-spec-01.txt.

[12] Joa-Ng, M., and Lu, I., A peer-to-peer zone-based two-level link state routing for mobile ad hoc networks, *IEEE Journal on Selected Areas in Communications*, 1415–1425, 1999.

[13] Johnson, D. and Maltz, D., Dynamic source routing in ad hoc wireless networks, in *Mobile Computing*, T. Imielinski and H. Korth, Eds., Kluwer Academic Publishers, Dordrecht, 1996, pp. 153–181.

[14] Jubin, J. and Tornow, J., The DARPA packet radio network protocols, *Proceedings of the IEEE*, 75, 21–32, 1987.

[15] Kamerman A., and Monteban, L., WaveLAN-II: A High-Performance Wireless LAN for the Unlicensed Band, *Bell Labs Technical Journal*, Summer 1997, pp. 118–133.

[16] Ko, Y. and Vaidya, N., Location-aided routing (LAR) in mobile ad hoc networks, *Intl. Conf. on Mobile Computing and Networking (MobiCom '98)*, Dallas, TX, 1998, pp. 66–75.

[17] Li, J., Jannotti, J., De Couto, D., Karger, D., and Morris, R., A Scalable Location Service for Geographic Ad Hoc Routing, *Intl. Conf. on Mobile Computing and Networking (MobiCom '2000)*, Boston, MA, Aug. 2000.

[18] Li, Q., Aslam, J. and Rus, D., Online Power-aware Routing in Wireless Ad-hoc Networks, *Int'l Conf. on Mobile Computing and Networking (MobiCom '2001)*, Jul. 2001.

[19] Murthy, S. and Garcia-Luna-Aceves, J., An Efficient Routing Protocol for Wireless Networks, *ACM Mobile Networks and Applications Journal*, Oct. 1996, pp. 183–197.

[20] Park, V. and Corson, M., A Highly Adaptive Distributed Routing Algorithm for Mobile Wireless Networks, *The Conf. on Computer Communications (IEEE Infocom)*, Kobe, Japan, Apr. 1997.

[21] Pearlman, M., Hass, Z., Sholander, P., and Tabrizi, S., On the Impact of Alternate Path Routing for Load Balancing in Mobile Ad Hoc Networks, *The First Annual Workshop on Mobile Ad Hoc Networking & Computing (MobiHoc 2000)*, Boston, MA, Aug. 2000.

[22] Pei, G., Gerla, M., and Chen, T.-W., Fisheye State Routing: A Routing Scheme for Ad Hoc Wireless Networks, *IEEE Int'l Conf. on Communications (ICC)*, Jun. 2000, pp. 70–74.

[23] Perkins, C. and Bhagwat, P., Highly Dynamic Destination-Sequenced Distance-Vector Routing (DSDV) for Mobile Computers, *Computer Communications Review*, Oct. 1994, pp. 234–244.

[24] Perkins, C., and Royer, E., Ad-hoc On-Demand Distance Vector Routing, *2nd IEEE Workshop on Mobile Computing Systems and Applications*, Feb. 1999.

[25] Singh, S., Woo, M., and Raghavendra, C., Power-Aware Routing in Mobile Ad Hoc Networks, *Int'l Conf. on Mobile Computing and Networking (MobiCom '98)*, Dallas, TX, Oct. 1998.

[26] Sinha, P., Sivakumar, R., and Bhargavan, V., CEMAR: a Core Extraction Distributed Ad Hoc Routing Algorithm, *The Conf. on Computer Communications (IEEE Infocom)*, Mar. 1999.

[27] Stojmenovic, I. and Lin, X., Power-aware localized routing in wireless networks, *IEEE Transactions on Parallel and Distributed Systems*, 12, 1122–1133, 2001.

[28] Toh, C.-K., A Novel Distributed Routing Protocol To Support Ad-Hoc Mobile Computing, *IEEE Fifteenth Annual Int'l Phonetic Conf. on Computers and Communication*, Mar. 1996, pp. 480–486.

[29] Woo, K., Yu, C., Youn, H.Y., and Lee, B., Non-Blocking, Localized Routing Algorithm for Balanced Energy Consumption in Mobile Ad Hoc Networks, *Int'l Symp. on Modeling, Analysis and Simulation of Computer and Telecommunication Systems (MASCOTS 2001)*, Aug. 2001, pp. 117–124.

[30] Xu, Y., Heidemann, J., and Estrin, D., Geography-informed Energy Conservation for Ad Hoc Routing, *Int'l Conf. on Mobile Computing and Networking (MobiCom'2001)*, Rome, Italy, Jul. 2001.

VIII

Connection and Traffic Management in Ad Hoc Wireless Networks

26

Resource Discovery in Mobile Ad Hoc Networks

Jiang Chuan Liu
Hong Kong University of Science and Technology

Kazem Sohraby
Lucent Technologies

Qian Zhang
Microsoft Research

Bo Li
Hong Kong University of Science and Technology

Wenwu Zhu
Microsoft Research

Abstract

With the rising popularity of network-based applications and the potential use of mobile ad hoc networks in civilian life, an efficient resource discovery service is needed in such networks for quickly locating resource providers. In addition, to improve user experience, Quality of Service (QoS) aware-ness is also crucial. In this paper, we identify the challenges when basic resource discovery techniques for the Internet are used in mobile ad hoc networks. We then propose a framework that provides a unified solution to the discovery of resources and QoS-aware selection of resource providers. The key entities of this framework are a set of self-organized discovery agents. These agents manage the directory information of resources using hash indexing. They also dynamically partition the network into domains and collect intra- and inter-domain QoS information to select appropriate providers. Simulation results show that our framework improves the QoS delivered to clients, while the cost and response time are kept at low levels.

26.1 Introduction

An ad hoc network is generally formed by a set of wireless mobile nodes (hosts). Communication between two network nodes that are not in direct radio range of one another takes place in a multi-hop fashion, with other nodes acting as routers. Ad hoc networks can be used in military and rescue operations, as well as in meetings where people want to share information quickly.

Recently, the rising popularity of network-based applications among end users and the potential use of ad hoc networks in civilian life have led to research interests in resource sharing in large-scale ad hoc networks [25]. With the rapid increase of available resources and accessing requests, a crucial requirement here is that it should be possible to locate a resource without excessive overhead and long latency. In addition, providing desirable Quality-of-Service (QoS) is an important design objective. Specifically, when there are multiple/replicated providers for the same resource, the best one should be selected according to some QoS metrics to improve user experience. That is, an efficient and QoS-aware resource discovery system is needed.

Most previous work on resource discovery has focused on fixed-infrastructure networks, specifically, the Internet [2,4,6]. However, ad hoc networks have several distinct features that make resource discovery more challenging. The most important feature is that the topology of an ad hoc network changes with time. As a result, the design of the routing protocols for ad hoc networks is quite different from that for the Internet. For example, it has been shown that, in this case, reactive (on-demand) routing protocols are usually more efficient and scalable than traditional proactive (table-driven) protocols [12]. In addition, to be robust in the face of topology changes and node failures, applications for an ad hoc network generally prefer distributed and dynamic control mechanisms to centralized and static mechanisms, though the latter have proven to be efficient for many Internet applications or services, such as the Domain Name System (DNS) [20].

Furthermore, in previous resource discovery systems, the QoS to be delivered to a client is seldom considered. Some systems propose to use client-based probing techniques after discovery [8,18]. However, probing measures the QoS in a very short period. In our simulation, we find that it is not very effective in mobile ad hoc networks because of the mobility and wireless channel variations.

Some discovery standards have been proposed for ad hoc networks, such as the Service Discovery Protocol for Bluetooth [23]. However, they are limited to very small-scale networks and do not consider QoS. In this chapter, we propose a novel framework concerning resource discovery and provider selection in mobile ad hoc networks with cooperative nodes. This framework is targeted for large-scale networks. It provides a unified solution to the problems of the discovery of resources and the QoS-aware selection of resource providers. Furthermore, it has relatively low discovery latency and cost (in terms of the number of packets for each resource discovery query).

The key entities of our framework are a set of self-organized *Discovery Agents* (DAs), which efficiently integrate three functionalities that are specially designed for mobile nodes:

1. Directory information organization and query
2. Dynamic domain formation
3. Intra- and inter-domain QoS information monitoring

The effectiveness of our framework is demonstrated through simulation. The results show that it produces significant performance gain over the case where QoS is not considered. It also outperforms the case where QoS is considered but is estimated by probing. At the same time, it has relatively low cost and response time.

The rest of the chapter is organized as follows. Section 26.2 provides a brief review of existing resource discovery techniques and identifies their limitations when applied to mobile ad hoc networks. Section 26.3 proposes our QoS-aware resource discovery framework. Section 26.4 evaluates the performance of our framework. Finally, Section 26.5 concludes the paper and discusses some future directions.

26.2 Existing Work and Our Design Rationale

In this section, we review existing resource discovery and provider selection techniques for the Internet and identify their potential advantages and limitations when they are used for ad hoc networks. Most of these techniques can be classified into the following three approaches:

1. Query flooding and path probing [14]

   Query flooding is the most straightforward approach for resource discovery. In this approach, a discovery query is sent to all nodes using broadcast. Each node can determine how it will process the query and respond accordingly. The advantage of this approach is flexibility in query processing. However, the broadcast range and frequency need to be carefully controlled because broadcasting to the whole network consumes bandwidth and computation power; both are scarce in an ad hoc network.

   Path probing is also the basic way of measuring the path QoS between a resource provider and a client. *Ping* probes have been widely used in the Internet environment [29,30] to measure response time. Bandwidth can be measured by the packet-pair technique [31]. Nevertheless, as we said before, probing may not be effective in highly dynamic ad hoc networks as it measures path QoS for only a short time. In addition, with on-demand routing protocols, probing may initiate the route discovery process, incurring high cost.

2. Centralized directory service [6,7,14]

   In a centralized directory-based system, directory information of resources, such as meta data and addresses of resource providers, is registered at directory servers. To search the directory information of a requested resource, a client contacts its corresponding directory server. For the Internet, this approach has shown to be very efficient for resource discovery [2]. In fact, the Internet usually uses multiple directory servers that are hierarchically organized to improve query responsiveness and scalability [4,6,7]. QoS awareness can be easily incorporated into this hierarchy by statically partitioning the network into domains [7].

   Centralized server based techniques are also used in provider selection. One example is the use of the Domain Name System (DNS)-based server selection [27], which exploits the transparent nature of name resolution to redirect clients to an appropriate server. It relies on clients and their local name server being in close proximity, since redirection is based on the name server originating the request rather than the client. Another example is the IDMaps project [21], which aims at providing a distance map of the Internet from which relative distances between hosts on the Internet can be gauged, and the closest provider can be located based on the map. The architecture of IDMaps consists of a network of instrumentation boxes, called *Tracers*, distributed across the Internet. Tracers measure distances among themselves and between themselves and regions of the Internet to build the distance map. In [28], the issues of placing a given number of tracers in different topologies are addressed, and several heuristics are proposed to improve measurement accuracy in hierarchical topologies with partitioned domains.

   However, in mobile ad hoc networks, since there is no fixed topology, maintaining a hierarchal structure of directory or measurement servers is not an easy task. Moreover, statically configured domains do not reflect the dynamic relations of mobile nodes.

3. Decentralized hash indexing [15,16,26]

   Decentralized hash indexing has been proposed for resource discovery in peer-to-peer networks. In such a system, there is no special/centralized directory server. Instead, every node provides some directory service. A resource is given a unique key, and a hash function is used to build a deterministic mapping between the key and the node that stores the directory information of that resource. The network and peers are designed in such a way that, given a key, the corresponding resource can be located very quickly despite the network's size. However, in this approach, each network node could be involved in some queries. In an ad hoc network using on demand routing protocols, if a node has not communicated to other nodes for a certain time, a route discovery process is needed to find a route towards this node, which may incur high cost [10,11]. Furthermore, this approach does not address QoS issues.

Through analyzing the advantages and limitations of the existing approaches, we have arrived at the following design principles for QoS-aware resource discovery in mobile ad hoc networks. First, directory information should be distributed to only a small set of fault-tolerant directory agents. Most messages for discovery are exchanged among these agents to reduce the overhead of broadcast and route discovery. Only low-frequency or controlled broadcast is used to distribute some quasi-static or local information, such as the addresses or locations of the agents. Second, hash indexing can be applied to these agents to reduce query latency. Finally, QoS information should be monitored continuously using a distributed mechanism. These principles have led to our novel framework, described next.

26.3 A Novel Framework for QoS-Aware Resource Discovery

26.3.1 Framework Overview

Our framework is built on the application layer to provide generic and efficient tools for QoS-aware resource discovery.

In our framework, we assume that all nodes are cooperative and can communicate with each other via some single-hop or multi-hop path. Each node can take one or more of the following three roles[1]:

A **Client** that initiates a query for resource discovery and uses resources. There are two basic discovery modes:
   A *Browsing mode* where a client is looking for all resource providers that have the requested resource
   An *Accessing mode* where a client is looking for a resource provider that could provide the best quality of service
A **Resource Provider** *(RP)* that provides resources for clients. A RP is also responsible for registering the directory information of its resources and advertising its QoS information to discovery agents.
A **Discovery Agent** *(DA)* that performs many of the important operations in our framework, as follows:
   First, DAs collectively maintain directory information of the resources using hash indexing. This provides fault tolerance and fast query response.
   Second, DAs dynamically partition the whole network into *dynamic domains.* Each DA maintains a separate domain and acts as the *home DA* of that domain. It monitors the QoS information of the RPs in its domain and responds to discovery queries from clients in the domain.
   Third, all registration and query messages are exchanged between DAs. These frequently exchanged messages are also used to continuously estimate peer path QoS, such as the delay between two DA nodes.

26.3.2 DA Generation and Dynamic Domain Formation

Initially, there are no DA nodes in the network. They are generated through a bootstrapping process as follows. First, one node is elected as the *initial DA* using a procedure similar to the cluster head selection in the lowest-ID algorithm [22] for ad hoc networks. That is, all eligible nodes broadcast to the whole network about their existence to take part in the election, and the one with the smallest address will win the election. Suppose there are M DAs to be generated (the choice of M will be studied in the next section). The initial DA will then randomly select other $M - 1$ nodes to form the DA set and assign each of them a unique index in the set of $\{2,\ldots,M\}$. Specifically, the initial DA has index 1.

After the DAs are generated, their addresses are periodically broadcasted to the whole network at a low frequency. In addition, each non-DA node tries to find the nearest DA as its home DA and join that DA's domain.

Note that both DAs and other nodes move over time. Hence, the membership in a domain changes over time, and a dynamic domain formation process is periodically performed for a DA to update its domain members. Here, a nonnegative and additive metric is used to measure distance, which can be

---

[1] A node can have one or more functions. Specifically, a node that has Discovery Agent (DA) functions is called a DA node, or DA for short, and other nodes are called non-DA nodes.

the number of hops or delay in practice. Based on the properties of shortest paths with this type of metrics [13], we propose a simple distributed algorithm to form dynamic domains, as follows. A DA periodically broadcasts a formation announcement to its neighboring nodes, which includes the DA's index, expiration time of the announcement, and a distance field. The distance field records the distance between the DA and the node that receives the announcement. Upon receiving an announcement, a non-DA node first checks the value of the distance field; if it is greater than the distance to its current home DA, the node stops forwarding the announcement. Otherwise, it will set that DA as its home DA, and forward the announcement to all its neighboring nodes. To prove the correctness of this algorithm, we must show that:

1. Any non-DA node should be in a DA's domain.
2. The DA is the nearest DA to that non-DA node.

The first property can be proven as follows. Suppose a non-DA node is not covered by any DA's domain, then its neighboring nodes should not be covered by any DA's domain, either. Thus, by using induction on these nodes, we can conclude that either there is no DA in the network or there is a set of nodes that cannot communicate with the remaining part of the network. Both contradict our basic assumptions. The second property can be proved by the criterion for DA selection in the algorithm.

26.3.3 Directory Information Organization and Fault Recovery

In our framework, each resource has an attribute known to all intended clients in the network. To register a resource, its provider first issues a registration request to its home DA. The request includes the provider's address, attribute, expiration time, and other directory information. Assume the attribute of the resource is α, a hashing function $H(.)$ is used to produce an index $\beta = H(\alpha)$ in the set of $\{1,2,\ldots, M\}$. The home DA will then distribute the registration request to DA_β, $DA_{\beta + 1}, \ldots, DA_{\beta + K - 1}$, and the directory information of the resource will be registered to these DAs.

This organizational scheme has several advantages:

1. The replicated providers of the same resource always register to the same DAs; hence, we can obtain the full list of their directory information from only one DA.
2. The directory information of a resource can be quickly located by using hash indexing. Note that different resources may have the same attribute, and their directory information will thus be stored in the same DAs. Hence, our framework does not preclude the use of fuzzy search, such as wildcard-based search, in a DA.
3. This scheme provides fault tolerance if K is greater than 1.

Suppose the nodes are homogeneous with failure probability p; the number of replications, K, should be set to $|\log_{1-p} A|$ where A is the availability requirement for the directory information. When a DA is found failed by another DA in the discovery query process (this will be discussed in detail in Section 26.3.5), the latter will broadcast a DA selection message to the network. Non-DA nodes that are willing to take the place of the failed DA will respond to this message, and the one with the minimal last-known distance to the failed DA will be selected. Assume the index of the failed DA is i, the directory information can then be recovered from a subset of $DA_{i - K + 1}, DA_{i - K + 2}, \ldots, DA_{i + K - 1}$.

26.3.4 QoS Information Collection and Prediction

A DA is also responsible for QoS information collection and prediction. Note that the requirements of QoS are highly application specific. Hence, our framework provides generic QoS information to different applications to achieve a flexible solution. Specifically, the first type of QoS information is application-level QoS, including the CPU usage and available memory of a RP. A RP periodically provides this information to its home DA. The second type is path QoS between two nodes. In this chapter, we consider the path delay (packet latency) between two nodes, which is one of the most useful path QoS metrics for many applications [21]. However, other path metrics, such as bandwidth, can also be incorporated into the system. We assume the clocks of all DAs are synchronized by some global time service, such as the Universal Time Coordinate (UTC) service provided by the Global Position System (GPS) [24], and a message exchanging between DAs carries a timestamp. Thus, DAs can predict their peer path delay by an Autoregressive Moving Average (ARMA) predictor [17], which uses the packet latency calculated from those frequently exchanged messages.

For non-DA nodes, we do not directly measure their path QoS by exchanging probing messages between RP-Client pairs. This is because, first, probing may trigger high-cost route discovery operations if two nodes seldom communicate with each other, and second, the time for using a resource is usually much longer than the time for probing, and a short time probing may give a different estimate compared to the statistical behavior of a path. Hence, instead of using probing, we use an approximation method. We assume that the nodes in a dynamic domain are QoS-similar and use the home DA as a representative. The path QoS of two non-DA nodes is approximated by the path QoS of their home DAs. When there are enough DAs that move independently, the error of this approximation is expected to be small, as shown in Section 26.4.

26.3.5 Discovery Query

In our framework, resource discovery is done in two phases. The first is directory query for searching the resource directory information in the DA set. Figure 26.1 shows an example. Starting from the client's home DA (denoted as *hDA*) to which a query is submitted (Step 1 in Fig. 26.1), if no cached record matches the query, *hDA* will calculate the hashing index of the resource, β, to decide the qualified DA set, $DA_\beta, DA_{\beta+1}, ..., DA_{\beta+K-1}$. The query is then forwarded to the DA that is in the qualified DA set and is nearest to *hDA* (Step 2). If this DA fails, *hDA* will try to forward the query to the next nearest DA in the qualified set, until the query is successful. In the browsing mode, a full list of RP candidates (the providers that have the requested resource) is returned to *hDA* (Step 3) and then to the client that initiates the query with browsing mode (Step 4).

The second phase is QoS query, which is for accessing mode only. It needs to compare the QoS provided by all RP candidates and select the best one. Towards this end, *hDA* should query all DAs that are home DAs for the candidates. In the current version, we use a parallel search strategy (see Fig. 26.2). The DA

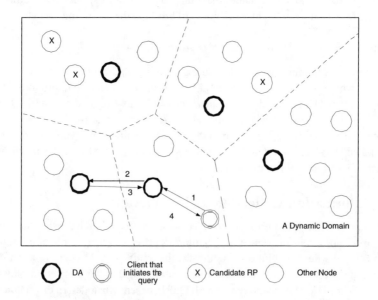

FIGURE 26.1 Operations for resource discovery, browsing mode.

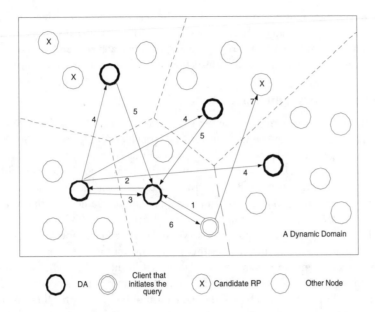

FIGURE 26.2 Operations for QoS-aware resource discovery, accessing mode.

that has the directory information sends a QoS query to all the DAs using multicast, or multiple-unicast if the underlying routing protocol does not support multicast (Step 4). The query includes the index of *hDA*, the list of the RP candidates, and the type of QoS of interest. If there are one or more candidates in a DA's dynamic domain, the DA will respond to *hDA* by providing the addresses of the candidates and the corresponding QoS information (Step 5). The QoS of all RP candidates is then compared by *hDA* according to the requirement of the client, and the result is returned to the client (Step 6). Finally, the directory information of the best candidate is returned to the client (Step 6), which accesses the resource using appropriate protocols (Step 7).

26.4 Performance Evaluation

In this section, we present our initial simulation results. Our main objective for the study is to investigate whether our framework enhances QoS awareness in resource discovery. Another objective is to see whether our framework exhibits satisfactory performance in terms of query responsiveness and overhead.

26.4.1 Simulation Environment

We simulate our framework using the LBNL network simulator *ns*-2 [19]. Our simulated network consists of 150 mobile nodes, whose initial positions are chosen from a uniform distribution over an area of 1000 m by 1000 m. Random waypoint [11] is used as the mobility model. The nodes' moving speeds are uniformly chosen from 10 to 72 km/h. The IEEE 802.11 protocol is used as the MAC layer protocol. Each wireless channel has a 2 Mb/sec bandwidth and a circular radio range with 250 m radius. For routing, we use the Ad hoc On-demand Distance Vector (AODV) protocol [10].

We assume that there are 100 different resources in the simulated environment. The popularity of each resource, measured in number of requests per minute, is randomly distributed between 1 and 5 requests per minute. For each discovery query, the client that initiates the query is randomly selected in the network. Each resource is served as a Constant Bit Rate (CBR) streaming application of 28 kb/sec and lasting 30 seconds. In the experiments, the path QoS of interest is the average packet latency. We chose this metric because it is relatively easy to estimate and luckily the most generally useful [21].

For the sake of comparison, we also simulated a traditional framework in which locating a resource and selecting a provider are considered as two separate issues. The resource discovery method is centralized directory–server based. When there are replicated providers, a client sends 15 consecutive packets to each provider to estimate path delay, and selects the one with the minimum average delay.

26.4.2 Query Latency and Cost

In the first set of simulation experiments, we investigate the query latency and cost of our framework and the traditional one. We vary the number of replicated providers, N_p, for each resource, from 1 to 10. We find that the average query latency is nearly independent of N_p in both frameworks, and also independent of M, the number of DAs in our framework. This is because QoS queries to different DAs or probes to different providers are sent simultaneously. Table 26.1 lists the query latency in different phases, including directory query and QoS query. It can be seen that, in our framework, the average latency for directory query is slightly higher than that in the traditional framework. This is because in the traditional framework, a client needs to contact a directory server only, while in our framework, a client needs to contact not only the home DA but also the DA that stores directory information. However, the latency of the QoS query in our framework is much lower than that in the traditional framework. As a result, the total latency in the traditional framework is about 2.2 times ours. This is because QoS query in our framework involves only querying all DAs, and the latency is thus bounded by the time-out factor between the home DA and the farthest DA. On the other hand, probing involves not only the transmission of a packet from a client to a provider, but also several cycles of this process.

Figure 26.3 shows the average cost in terms of the number of messages (packets) transmitted in the network for each discovery. We observe that, in our framework, the cost is also nearly independent of the number of replicated providers because the QoS queries are always sent to all the DAs regardless of the number of providers. On the contrary, in the traditional framework, since the client needs to probe every resource provider, the cost increases nearly linearly with the increasing number of replicated providers. When there are more than four replicated providers, the traditional framework incurs much higher cost than ours does. Note that in the QoS query phase of our framework, it is possible to use a heuristic algorithm to avoid querying DAs whose domain does not cover any qualified provider. We will study this approach in our future work.

26.4.3 Quality of Service

In this set of experiments, we studied the QoS awareness of the two frameworks. The metric of interest is our framework's performance gains over the traditional framework and the QoS-unaware case (the resource provider is randomly selected), where the gains are calculated by normalizing the reduced packet latencies.

Figure 26.4 shows the results with $N_p = 5$. The number of DAs varies from 10 to 20. We can see that performance gain increases with the increase of the number of DAs. This is because the expected area of a dynamic domain decreases when there are more DAs, and hence, in this case, it is more accurate to approximate the path QoS between two nodes using their home DAs as representatives. Specifically, when enough DAs are deployed, the gain of our framework is up to 45% compared to the QoS-unaware case, and 15% to that of the traditional framework. Moreover, the gain tends to saturate when there are more

TABLE 26.1 Comparison of Discovery Latencies

Framework	Latency (msec)		
	Directory Query	QoS Query	Overall
Traditional	287	1375	1662
Proposed	395	354	749

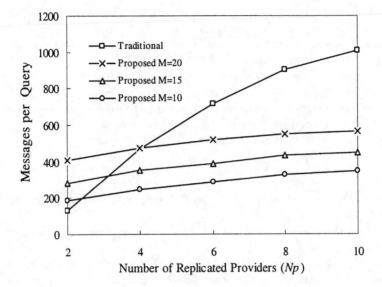

FIGURE 26.3 Comparison of discovery costs.

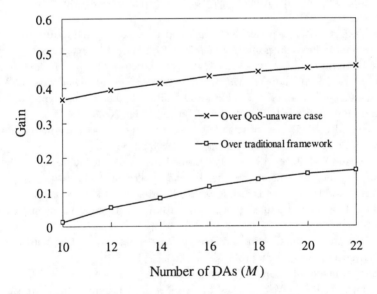

FIGURE 26.4 The performance gains of the proposed system over the traditional system and the QoS-unaware case.

than 20 DAs in this 150-node network. Hence, from Figs. 26.3 and 26.4, the choice of about 15 DAs provides satisfactory performance in terms of both discovery cost and accuracy. This number is much smaller than the total number of nodes.

26.5 Conclusions and Future Work

In this chapter, we first identified the limitations when basic resource discovery techniques are used in mobile ad hoc networks. We then proposed a novel framework that is specially designed for QoS-aware resource discovery in ad hoc networks. Our framework jointly considers the problems of resource discovery and QoS-based provider selection. The key entities in our framework are a set of self-organizing discovery agents. The agents adopt an efficient hash indexing method to store directory information. They also monitor QoS information continuously and predict path QoS on behalf of other nodes to reduce overall cost and improve accuracy.

Preliminary simulation results showed that our framework enhances QoS awareness compared to a traditional framework that uses centralized directory service and client-based probing. In addition, our framework incurs lower query latency and cost. We will conduct more experiments in our future work to investigate the behavior of this framework, such as potential oscillation among providers [18], scalability in large-scale networks, and performance with other QoS metrics.

References

[1] R. Iannella, Internet Resource Discovery Issues, http://archive.dstc.edu.au/RDU/reports/ QuestNet95.html.

[2] Decentralized Resource Discovery in Large Peer-based Networks, http://cubicmetercrystal.com/ alpine/discovery.html.

[3] Z. Fu and N. Venkatasubramanian, Combined Path and Server Selection in Dynamic Multimedia Environments, *Proceedings of ACM Multimedia 99*, Oct. 1999.

[4] R. van Renesse, Scalable and Secure Resource Location, *Proceedings of IEEE Hawaii International Conference on System Sciences*, Maui, Jan. 2000.

[5] M.S. Corson, Issues in Supporting Quality of Service in Mobile Ad Hoc Networks, *Proceedings of IWQOS 97*, May 1997.

[6] S. Czerwinski, B. Zhao, T. Hodes, A. Joseph, and R. Katz, An Architecture for a Secure Service Discovery Service, *Proceedings of ACM MobiCom 99*, Seattle, WA, Sep. 1999.

[7] D. Xu, K. Nahrstedt, and D. Wichadakul, QoS-Aware Discovery of Wide-Area Distributed Services, *Proceedings of IEEE International Symposium on Cluster Computing and the Grid*, May 2001.

[8] S.G. Dykes, C.L. Jeffery, and K.A. Robbins, An Empirical Evaluation of Client-Side Server Selection Algorithms, *Proceedings of IEEE INFOCOM*, Tel Aviv, Mar. 2000.

[9] E.M. Royer and C.-K. Toh, A Review of Current Routing Protocols for Ad Hoc Mobile Wireless Networks, *IEEE Personal Communications*, Apr. 1999, pp. 46–55.

[10] C.E. Perkins and E.M. Royer, Ad Hoc On-Demand Distance Vector Routing, *Proceedings of the 2nd IEEE Workshop on Mobile Computing Systems and Applications*, Feb. 1999.

[11] D.B. Johnson and D.A. Maltz, Dynamic Source Routing in Ad Hoc Wireless Networks, in *Mobile Computing*, T. Imielinski and H. Korth, Eds., Kluwer Publishing Company, Dordrecht, 1996, Chapter 5.

[12] S.-J. Lee, Routing and Multicasting Strategies in Wireless Mobile Ad Hoc Networks, Ph.D. Dissertation, University of California, Los Angeles, 2000.

[13] R.J. Wilson, *Introduction to Graph Theory*, 3rd ed., Longman, New York, 1987.

[14] E. Guttman, C. Perkins, J. Veizades, and M. Day, Service Location Protocol, Version 2, *RFC 2608*, June 1999.

[15] I. Stoica, R. Morris, D. Karger, F. Kaashoek, and H. Balakrishnan, Chord: A Scalable Peer-to-Peer Lookup Service for Internet Applications, *Proceedings of ACM SIGCOMM 01*, Sep. 2001.

[16] S. Ratnasamy, P. Francis, M. Handley, R. Karp, and S. Shenker, A Scalable Content-Addressable Network, *Proceedings of ACM SIGCOMM 01*, Sep. 2001.

[17] R. Wolski, Forecasting Network Performance to Support Dynamic Scheduling, *Proceedings of the 6th IEEE International Symposium on High Performance Distributed Computing*, 1997.

[18] Z. Fei, S. Bhattacharjee, E.W. Zegura, and M.H. Ammar, A Novel Server Selection Technique for Improving the Response Time of a Replicated Service, *Proceedings of IEEE INFOCOM 98*, San Francisco, CA, Apr. 1998.

[19] *The LBNL Network Simulator, ns-2*, available at http://www.isi.e.,du/nsnam/ns/.

[20] P. Mockapetris, Domain Names — Concepts and Facilities, *RFC 1034*, Nov. 1987.

[21] P. Francis, S. Jamin, C. Jin, D. Raz, Y. Shavitt, and L. Zhang, IDMaps: A global Internet host distance estimation service, to appear in *IEEE/ACM Transactions on Networking*, Oct. 2001.

[22] A. Ephremides, J. E. Wieselthier, and D.J. Baker, A design concept for reliable mobile radio networks with frequency hopping signaling, *Proceedings of the IEEE*, 75, 56–73, 1987.

[23] The Official Bluetooth Website: http://www.bluetooth.com.

[24] Notes on NAVSTAR GPS Operations, Available at website: http://tycho.usno.navy.mil/gpsinfo.html.

[25] K. Konishi, Report on Large Scale Networking Workshop, Mar. 2001, http://www.jp.apan.net/meetings/0103-NGI-workshop.

[26] R. Jain, Y.B. Lin, and S. Mohan, A caching strategy to reduce network impacts of PCS, *IEEE Journal on Selected Areas in Communications*, 12, 1434–1444, 1994.

[27] A. Shaikh, R. Tewari, and M. Agrawal, On the Effectiveness of DNS-based Server Selection, *Proceedings of IEEE INFOCOM*, Anchorage, AK, Apr. 2001.

[28] S. Jamin, C. Jin, Y. Jin, D. Raz, Y. Shavitt, and L. Zhang, On the Placement of Internet Instrumentation, *Proceedings of IEEE INFOCOM*, Tel Aviv, Mar. 2000.

[29] R. Carter and M. Crovella, Server Selection using Dynamic Path Characterization in Wide-Area Networks, *Proceedings of IEEE INFOCOM 97*, Kobe, Japan, Apr. 1997.

[30] M. Crovella and R. Carter, Dynamic Server Selection in the Internet, *Proceedings of the Third Workshop on the Architecture and Implementation of High Performance Communication Subsystems*, Aug. 1995.

[31] S. Keshav, A Control-Theoretic Approach to Flow Control, *Proceedings of ACM SIGCOMM*, Sep. 1991.

27

An Integrated Platform for Quality-of-Service Support in Mobile Multimedia Clustered Ad Hoc Networks

George N. Aggélou[1]
Institute of Technology, Greece

Abstract

One feature that distinguishes the ad hoc wireless network from traditional wired networks and PCS (Personal Communication Networks) is that all hosts in a mobile ad hoc network (MANET) are allowed to move freely without the need for static access points. This distinct feature, however, presents a great challenge to the design of the routing scheme and the support of multimedia services, since the link quality and the network topology may be fast changing as hosts roam around. In order to support the rapidly deployable, wireless, multimedia network requirements, Quality-of-Service (QoS) routing is the key to multimedia support. The goal is to find satisfactory paths that support the end-to-end QoS requirements of multimedia flows. This article presents fundamental issues in QoS routing in wireless networks; a QoS routing framework tailored for operation within a wireless ad hoc network is further proposed, and its operation is analyzed. The proposed QoS-based routing framework aims at providing a differentiated service treatment to multimedia

[1]The work was supported by Lucent Technologies, United Kingdom, under grant 33/1/PRS/LUC22.

traffic flows at the link level using novel techniques for channel assignment and end-to-end path QoS maximization. This is achieved by giving the routing mechanism access to QoS information, thus coupling the coarse grain (routing) and fine grain (congestion control) resource allocations. The system performance is examined through simulation experiments under various QoS traffic flows and mobility environments. The protocol's behavior and the changes introduced by variations on some of the mechanisms that make up the protocol are further investigated, examining which mechanisms have the greatest impact and exploiting the tradeoffs that exist between them.

27.1 Introduction to Multimedia Mobile Ad Hoc Communications

Mobile communications computing has enjoyed an explosive growth that began in the late 1990s and is expected to continue well into the new millennium. As more and more users are becoming mobile [23], mobile communications have to evolve at a faster rate than ever before to meet the increasing demands being placed on them. Observing the growing demands of roaming users, the mobile computing research community predicts that the next generation wireless networks will be burdened with bandwidth intensive traffic generated by personal multimedia applications such as video-on-demand, news-on-demand, web browsing, and traveler information systems. Multimedia services (voice, video, and text) support over wireless/mobile networks is the hottest telecommunications buzz word today. The current trend of connectivity anywhere, anytime, anyhow brings a new paradigm of accessing these services via wireless connectivity. However, the available bandwidth for supporting these applications is rather limited, and proper management of the bandwidth is necessary to accommodate the envisaged high-bandwidth applications and, thus, provide Quality of Service[1] (QoS) guarantees between the end systems.

Much work has been done in the past in cellular-like (single-hop) wireless access networks on efficient wireless bandwidth allocation. Specifically, most of the earlier research focused on the problem of optimizing frequency reuse. Some examples of today's existing public data networks are the Cellular Digital Packet Data [36], General Packet Radio Service [19], and High Speed Circuit Switched Data [37]. These schemes utilize the unused voice capacity to support low-priority, non–real-time data, whereas GPRS and HSCSD support multislot mode for higher rate applications. In case of scarcity of available bandwidth, the transmitted data packets are buffered or suitable flow control techniques are used leading to an increase in the transmission delay. Furthermore, despite the recent auction of the 1850–2000 MHz band by the Federal Communications Commission (FCC) for personal communication services (PCS) users, bandwidth is still the major bottleneck in most real-time multimedia services. Such services can substantially differ in bandwidth requirements, e.g., 9.6 kb/sec for voice service and 76.8 kb/sec for video.

The next generation mobile/wireless networks are envisioned to constitute a variety of infrastructure substrates, including fixed, single-hop, and multihop wireless/mobile networks. A multihop mobile radio network, also called a Mobile Ad hoc NETwork (MANET) [29], is a self-organizing and rapidly deployable network in which neither a wired backbone nor centralized control exist. The network nodes communicate with one another over scarce wireless channels in a multihop fashion. MANETs are a new paradigm of wireless wearable devices enabling immediate, easy, instantaneous person-to-person, person-to-machine, or machine-to-person communications.

The technical challenges to establishing mobile multihop communication are nontrivial. Specifically, multihop mobile networks inherit all the problems and technical challenges that wireless networks present, which fall into three broad categories: First, wireless channels are prone to bursty and location-dependent error. Second, contention for the wireless channel is location dependent. Third, mobile users may move from lightly loaded places (such as clusters, cells) to heavily loaded places. As a result of the first two factors, wireless channel resources are highly dynamic. As a result of the third factor, resource

---

[1] The term QoS is often used both for stringent QoS (e.g., delay, delay jitter, and packet loss) and for requirements to achieve high throughput (see [34] for fundamental discussions on the topic).

negotiations that are made in one place (e.g., cell) may not be valid when the user moves to another place. Besides, user mobility may cause packet flows to be rerouted, also contributing to changes in the resource availability in the backbone and wireless networks. These problems become more pronounced when multihop connectivity is addressed due to the highly dynamic nature of the network.

Future integrated services networks will support multiple classes of service to meet the diverse QoS requirements of applications. To meet these end-to-end QoS requirements, strict resource constraints may have to be imposed on the paths being used. Specifically, when multimedia traffic is concerned, hop-wise routing protocols (that is, routing decisions based on hop distance) may cause network congestion on some particular links; thus, the QoS is degraded. Based on this consideration, it is argued that in order to provide QoS support, it is necessary to effectively control the total traffic that can flow into the network system. And the key to a successful admission control in multimedia networks is QoS routing.

QoS routing presents a great challenge in both wired and wireless networks. The goal is to find a feasible path that provides better end-to-end QoS. The work on QoS routing includes two steps: (1) routing with QoS information; this can help the admission control function in preventing network overload, and (2) finding out QoS satisfactory route(s); this can help load balancing in a low speed wireless network.

The network control mechanism decides whether to accept a new connection by examining whether the route given by the route finding scheme still has sufficient capacity to adapt to this new connection. The network transmission capacity may be limited by the power of the portable radio units as well as by radio signal interference and obstruction; thus a major challenge in multihop, multimedia networks is the ability to account for resources so that bandwidth allocations and reservations can be placed on them. Let us note that in cellular (single-hop) networks, such accountability is made easily by the fact that all stations learn of each other's requirements through a control station, e.g., Home/Foreign Agent in Mobile IP [32], Base Station Sub-system (BSS) in GSM [20] and SGSN in GPRS [19]. However, because of the distributed and dynamic nature of MANETs, this solution cannot be extended to the multihop environment. Hence the resource (i.e., channel) allocation becomes a distributed task.

To this end, in order to support bandwidth (QoS) sensitive multimedia applications, both the calculation of the end-to-end path available bandwidth as well as the distributed channel assignment pose new research challenges. A framework of bandwidth-based routing for QoS support in MANETs [4] is described in this chapter. Specifically, with the goal in mind to select paths for flows with QoS requirements, a new set of slot reservation and distributed channel assignment protocols for mobile ad hoc networks, called MBCA (Minimum Blocking Channel Assignment) and BRCA (Bandwidth Reallocation Channel Assignment), is presented. In addition, a novel mechanism that accounts for the channel usage and seeks alternative channel usage reconfigurations of active channels in order to create or to increase link bandwidth is introduced. The method is referred to as *Bandwidth Reassignment.*

27.2 Overview of Quality-of-Service (QoS) Routing

27.2.1 Current Notion of Quality-of-Service

Best-effort traffic together with other traffic classes that require service, or else quality, guarantees dominate today's Internet traffic. Traditional best-effort traffic, such as electronic mail, telnet, and Remote Procedure Call (RPC), has been mostly small messages with a payload typically less than a few tens of kilobytes of data. Users would like to have their messages arrive at their destination as quickly as possible. For this kind of low-latency traffic, it has been shown [40] that the minimum-hop routing, i.e., packets are sent along the path with the minimum number of hops, may work well when the path is not congested. With best-effort service, all packets are typically treated equally in the network.

Different Quality-of-Service (QoS — in the sense, multimedia) applications require different QoS guarantees: some require stringent end-to-end delay, some require a minimal transmission rate, while others with no strict delay and/or bandwidth requirements may simply require high throughput. In the following, some of these QoS requirements are described in detail [34].

27.2.1.1 Delay Guarantees

A broad class of applications, e.g., interactive multimedia, Internet telephony, and video conferencing, may require stringent delay, delay jitter (or delay variation), and loss guarantees. For example, in real-time playback applications, packets arriving after the playback point will be useless, and the loss of a certain number of packets will seriously degrade the quality of voice and pictures.

The end-to-end delay includes the propagation delay, which is determined by the physical distance between the source and the destination; the transmission delay, which is determined by the capacity of the bottleneck link on the path; and queueing delay, which is determined by the network load, the burstiness of the traffic source, and the service disciplines employed in the network.

27.2.1.2 Bandwidth Guarantees

Transmission of multimedia streams requires a minimum bandwidth to ensure end-to-end QoS guarantees. Bandwidth guarantees can be requested for different time intervals depending on applications. For example, if an application is adaptive and has sufficiently large buffer space at its source and destination, the bandwidth provided by the network can vary over time, as long as the average bandwidth provided is higher than the minimum bandwidth required by the application.

Recent studies [18] suggest that the network should deploy a mechanism to support bandwidth renegotiation, which allows bandwidth reservation to be provided on a finer time scale than per-session bandwidth guarantees allow.

27.2.1.3 High Throughput

Traditional best-effort applications, e.g., RPC, electronic mail, ftp, and telnet, usually send messages as small as a few kilobytes. The main performance index for these applications is the end-to-end per packet delay. As the sophistication of networked applications has grown, so too has the amount of data transmitted. It is now not uncommon to observe applications in which the data payload reaches from several hundred megabytes to several terabytes [30]. In contrast to the transmission of small messages, these new applications can consume as much network bandwidth as is available. It is the end-to-end throughput rather than the per packet end-to-end delay that is the main performance concern. The throughput is determined by the total number of bytes transmitted over the elapsed time, where the elapsed time includes the end-to-end delay experienced by the first packet and the time interval from the arrival of the first packet to the arrival of the last packet.

27.2.2 IETF Proposed Internet QoS Routing Model

Minimum-hop routing, or else *shortest path* routing, computes routes that use the fewest resources (because each route traverses the fewest nodes and links possible). This family of routing algorithms assigns to each hop (i.e., link) a unity cost, thus treating equally a high-speed 5.6 Gb/sec link, for example, and a low-speed 150 kb/sec link, or a poor quality link (due to interference, shadowing etc.) and a link with good quality.

There is a gap, however, between a routing protocol that discovers shortest-path routes with quality unacceptable for large-scale commercial exploitation and one that discovers QoS-based path routes that operate with good and stable quality outputs. If critical flows must be accorded higher priority than other types of flows, a mechanism must be implemented in the network to recognize flow requirements.

The Internet Engineering Task Force (IETF) QoS Routing (QoSR) Working Group [35] has defined new service classes for QoS routing support, including the guaranteed service [38], the controlled load service [41] and, lately, the differentiated service [12] models. Both the guaranteed and controlled load service models involve the establishment of connections through the network with the help of a resource reservation protocol, such as the ReSerVation Protocol (RSVP) [10], and admission control mechanisms

[13] based on measured network state information. The network ensures that sufficient resources are available once a flow (or a session[2]) is admitted.

The guaranteed service model is based on recent studies [13,17,43] of a class of rate-proportional packet scheduling algorithms. Under these service disciplines, packets of different connections sharing the same output link are sent in an order that ensures a weighted fair sharing of link capacity among these connections. As a result, a guaranteed rate is ensured for flows that use the guaranteed service. Packets transmitted on such flows are thus protected from either ill-behaved applications or intentional link sabotage. To invoke the guaranteed service, an application specifies its traffic characteristics and desired performance guarantees. The network, on the other hand, reserves a certain amount of resources at each node (switch or router) on its path. Thus, the traffic specification and the QoS guarantees constitute a "contract" between the network and the application: once a flow is admitted into the network and the traffic source conforms to its traffic specification, the network will provide guaranteed QoS.

The controlled load service is an enhanced best-effort service intended to support applications requiring performance better than what is provided by traditional best-effort service. It limits the amount of traffic entering the network to ensure that once a flow is admitted, it enjoys service equivalent to best-effort service in a lightly loaded network. Even under congestion, network nodes offering controlled load service are expected to provide flows with low delay and low packet loss. To limit the number of flows receiving the service, applications are required to make explicit requests for service. Such requests for service can be made using a reservation setup protocol, such as RSVP, or some other means. Each network node that receives a request for service can either accept or reject the request. Therefore, a flow can receive a guaranteed average rate with no per packet delay guarantees.

27.3 A General Framework for QoS Routing in MANETs

The subsections that follow describe a framework for QoSR support, first proposed in [4,7], tailored for operation in clustered multimedia multihop mobile systems. The goal in mind in the design of a QoSR schema is to deliver a stream of incoming bits with at least the minimum bandwidth fraction of the requested bandwidth, with the minimum number of bandwidth changes, and through preferably the shortest path(s).

Figure 27.1 depicts the structural representation of the proposed QoSR framework. As illustrated, the QoSR architecture consists of five horizontal layers and a virtual vertical layer. The five horizontal layers are the scheduling/MAC layer [21,27], the resource reservation, the resource adaptation layer [26], the dynamic routing protocol [2,3,4,6,25], and the transport layer [15].

As illustrated, in order to provide adaptive service, various resource management algorithms interact in the following sequence of events:

During the call setup period, the higher layers are required to provide the following input parameters to the lower layer modules:

1. Maximum bandwidth desired (HBW)
2. Minimum bandwidth required (LBW)

The bandwidth window [LBW, HBW] is referred to here as *flow bandwidth window* (FBW) or else flow or resource specifications. The HBW and LBW parameters serve as upper and lower bounds during the bandwidth assignment as well as during the resource reservation, adaptation, and negotiation phases.

When a mobile user requests a new connection with QoS bounds, the network will conduct an *admission control* test (and tentatively reserve resources) during the forward pass to the destination

---

[2] Throughout this chapter, the terms "flow," "session," and "connection" are used interchangeably. According to [34], the precise definition of flow captures the concept of a virtual circuit in ATM networks and a flow in IP networks. In other words, a flow can be a hard-state virtual circuit as in an ATM network, a soft-state flow as defined in IPv6, or a stateless connection such as a TCP connection in today's IP networks.

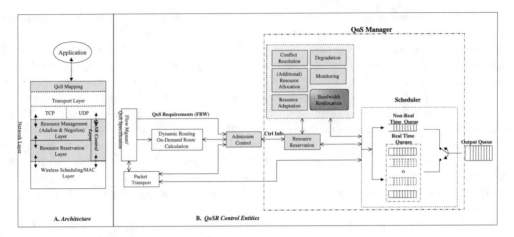

FIGURE 27.1 A framework for QoS routing support in wireless multimedia systems.

via an appropriate route found by the routing algorithm. When performing admission control on a link for a flow, the admission controller tests only whether the link has sufficient available resources to satisfy the minimum requirements of the flow. Once the admission controller at each radio along the multihop path accepts the call, it passes the admission decision and the requirement profile of the user to the Packet Sorter module (this module does not appear in Fig. 27.1 as it is part of the scheduler module). The packet sorter sorts and classifies traffic packets of admitted flows. During the reverse pass, the network will allocate resources for it (i.e., firm reservation). Resources include bandwidth, buffer space, and schedulability. Finally, the network performs *resource adaptation* among the adaptable flows in order to resolve resource conflicts and also distribute additional resources.

In the following subsections a rough description of the individual control building blocks of the proposed QoSR reference model is presented.

27.3.1 MANET Routing

Dynamic routing in MANETs is a multi-objective task that presents stringent requirements, including: high accuracy, low overhead, scalability in a large network, robust path maintenance, etc. In QoS routing, specifically, a major challenge is the ability to account for resources so that bandwidth reservations (in a deterministic or statistical sense) can be placed on them. Similar to cellular (single-hop) networks where such accountability is made easy through a control station (e.g., base station in cellular systems), this solution can be extended to multihop networks by creating clusters of radios in such a way that access can be controlled and bandwidth can be allocated in each cluster.

Due to the requirements of efficient network resource control and multimedia traffic support, clustering [16,25,31,42] provides a convenient framework for the development of important features such as channel access, power control, virtual circuit support, routing, and bandwidth allocation (resource management). Following the clustering approach, the entire population of nodes is grouped into clusters. A cluster is a subset of nodes that can communicate with a clusterhead and (possibly) with each other. The clusterheads are the nodes delegated to act as local coordinators to resolve channel scheduling; perform power measurement/control; maintain time division frame synchronization, channel access, routing, and bandwidth allocation; and enhance the spatial reuse of time slots and codes (among clusters). A node that can hear two or more clusterheads is a gateway.

Within a cluster, time-division scheduling is enforced. Across-cluster spatial reuse of time slots and codes is facilitated.

27.3.2 Channel Access Scheme

With the goal of enhancing the throughput and utilization of the limited channel resources, the problem is how to run the algorithm in real time and how to take advantage of clusterheads to select the appropriate channel access method within each cluster.

In view of the real time traffic component (which requires dedicated bandwidth), time-division multiple access (TDMA) within a cluster has been chosen. Code division multiple access (CDMA) is overlayed on top of the TDMA infrastructure. In this case, the near–far problem and related power control algorithm become critical to the efficiency of the channel access [9]. In addition, separate codes are assigned to different clusters in order to reduce the effect of intercluster interference. Specifically, only a single transmission (per code) within each cluster is permitted, whereas neighboring clusters make use of different sets of codes [22]. Following [22], the transmitter-based code assignment methodology is employed here.

The cluster solution with code separation reduces power interference and maintains spatial frequency reuse. However, note that as the network grows large, the clusters may outnumber the available codes. To overcome this problem, codes can be reused in nonadjacent clusters [11]. The distributed cluster-TDMA approach was also one of the DARPA sponsored WAMIS [8] subnet protocols.

The transmission time scale is organized in frames, each containing a fixed number of time slots. The entire network is synchronized on a frame and slot basis. It should be noted that synchronization is required only within a cluster. This is much easier than maintaining slot synchronization across the entire network as in [25].

A frame is divided into two phases, namely, control phase and info phase:

- *Control phase* — The control phase is used to perform all the control functions, such as slot and frame synchronization, clustering, routing, power measurement, code assignment, QoS path setup, etc. The clustering algorithm elects the clusterheads per cluster. Clusterheads assign the slot(s) to each QoS request within their covering area. The number of slots per frame assigned to a QoS connection (virtual circuit) is determined by the bandwidth requirements of the connection (i.e., FBW).

- *Info (voice/video/data) phase* — The info phase must support both virtual circuit and datagram traffic. Since real time traffic (which is carried on a QoS path) needs guaranteed bandwidth during its active period, bandwidth must be preallocated to the QoS path before actual data transmission. That is, some slots in the info subframe are reserved for QoS paths at call setup time. The remaining slots (free slots) of each cluster can be accessed by datagram traffic, using for example a Slotted-ALOHA scheme.

27.3.3 QoS Manager

The heart of the resource management and control architecture is the QoS Manager. From the functional point of view it is a control plane component primarily responsible for:

1. Creating and maintaining the reservation states of QoS connections
2. Allocating and managing network buffers for these connections
3. Scheduling different classes of packet
4. Call degradation or reducing bandwidth allocation to degradable applications in face of scarcity of available radio channels
5. Bandwidth reassignment to maximize utilization of available channel resources
6. Radio resource usage monitoring

The QoS Manager cooperates with other components in the data plane, such as socket (transport) layer and network interface layer, to coordinate the management of all network related resources.

27.3.3.1 Bandwidth Reservation

As mentioned earlier, in order to provide a stable QoS, it is not sufficient to simply select a route that can provide the correct resources. The resources must be reserved to ensure that they will not be shared or "stolen" by another session [39]. Therefore, guaranteed services require the routers to make a reservation, and for each reservation request routers must make admission control decisions.

Referring to the proposed QoSR model, to create and maintain local reservation states for different flows, a resource reservation signaling protocol is used prior to the flow of data. Several resource reservation protocols have been developed [10,14]. Two protocols, however, are of particular interest and could well be incorporated in the proposed QoSR model: the Resource Reservation Protocol [10] and the Constraint-Routing Label Distribution Protocol [14]. The reader is referred to the respective references for details on these protocols.

27.3.3.2 Call Admission Block

Upon the arrival of a new flow request, the admission control scheme at each hop along the data path determines whether the link has sufficient resources to accommodate this new flow. If so, the resources needed for the flow are set aside and the connection setup packet is sent to the next hop. If insufficient resources are available, either the flow is rejected or other alternative paths are tried again (for example, through the use of a crank back mechanism [33]).

To put this into context for the proposed QoSR model, when a new real-time flow request arrives in a cluster C_a, the cluster's clusterhead node is responsible for checking if the available spectrum in C_a is sufficient to accommodate the minimum requested bandwidth of the new flow. If so, the call admission algorithm proceeds to classify the user, and the call is forwarded to the next cluster along the path. If not, two things may occur: either the call is immediately dropped or a so-called *Bandwidth Reassignment* process (described below) is attempted in both the present and the neighbor clusters, such that common bandwidth can be created to accommodate the call requirements. If this process fails to create available bandwidth at least equal to flow's LBW, the call is dropped. In case the call is dropped, a reservation error message (similar to the RESV ERROR message in RSVP protocol) is returned to the data source.

For non-real-time users, the admission is primarily based on the availability of buffer space in the non-real-time packet queue (NRTQ). This criterion may be set against a queue length threshold, in order to prevent queue overflow once the new call is admitted.

27.3.3.3 Resource Adaptation

Many distributed multimedia applications are adaptive in nature and exhibit flexibility in dealing with fluctuations in network conditions. QoS adaptation algorithms can, for example, trade temporal and spatial quality to available bandwidth or manipulate the playout time of continuous media in response to variations in delay.

Adaptation may be triggered by changes in the measured QoS over the wireless link upon increase or decrease of resource availability on a wireless link, upon portable handoff, or when the application initiates a QoS renegotiation.

The goal of rate adaptation is twofold:

1. Upon resource increase, allocate additional rate to a subset of ongoing flows.
2. Upon resource decrease, reclaim rate from ongoing flows.

In the QoSR model, adaptation is conducted for the following two cases:

1. A real-time connection has been accepted but was not granted its maximum desired share (HBW) during the call setup time.

2. When a real-time call request arrives and finds all channels occupied, it may, under certain circumstances, force one (or more) ongoing non-real-time calls to be temporarily buffered so that the released channel can be used to admit the real-time request.

Note, however, that existing calls can be degraded if and only if these are degradable and adjustable applications. Whether or not an application can be degraded is declared during the connection setup process. At this point, the degradation algorithm is conducted.

27.3.3.4 Degradation Policies

Bandwidth degradation implies that an application occupying multiple channels releases some of them to enter a degraded mode. To put this into context on the proposed QoSR model, when available bandwidth increases within a cluster, for example, due to call termination or due to handover of existing call(s) to neighbor cluster(s), the elected clusterhead node of the cluster could allocate additional bandwidth to QoS flows routed through this cluster that have not been granted their max resource share (HBW). However, if the average rate specified in the requirement profile equals the minimum rate for the application, then the application is not degradable.

Finally, since non–real-time traffic packets can sustain much longer delays, real_time traffic usually has preemptive priority over the former in case of scarcity of available channels in the system.

27.3.3.5 Packet Sorter

The proposed QoSR model differentiates between real-time and non-real-time traffic at the link layer. The packet sorter interfaces with a two-level priority queue in the system: the real-time packet queue (RTQ) and the non-real-time packet queue (NRTQ). Based on the packet type information in the requirement profile, the packet sorter sorts the incoming packets and routes them into the appropriate queue. Note that both RTQ and NRTQ are abstractions for actual physical buffer implementations, of which there can be multiple instances under each type. For example, the system could implement a *multi-level* queue for real-time traffic, based on delay jitter or reliability requirements for further differentiating the real-time traffic packets.

27.3.4 Bandwidth Calculation

As noted above, the transmission time scale is organized in frames, each containing a fixed number of time slots.

The set of the common free slots between two adjacent nodes will be referred to as the *link bandwidth*. A node that knows its link available bandwidth (ABW) to some other node can only determine whether a request can be satisfied through this link or not. Consider the example shown in Fig. 27.2, in which node A in cluster C_1 intends to establish a real-time call to C in cluster C_2 with a bandwidth request (FBW) {2,2}. Let us assume a TDMA frame with six time slots. The admission controller at node A would reject the request, as its ABW to node B, which is one, is smaller than the minimum requested BW, which is two.

The *path bandwidth* (also called end-to-end bandwidth) between two nodes that may not necessarily be adjacent is the minimum link bandwidth of the links that comprise the path. If the destination and source nodes of a flow are adjacent, the path bandwidth is equal to the link bandwidth; the traffic flow

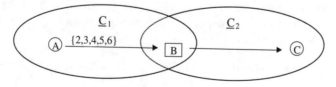

FIGURE 27.2 Example of end-to-end bandwidth calculation.

in this case is called *intra-cluster traffic*. If the destination and source nodes of a flow belong to different clusters, the traffic flow is called *inter-cluster traffic*.

The link bandwidth between nodes A in C_1, which does not function as a gateway, and B, which is GW, can be abstractly illustrated by the following expression:

$$\text{LinkBW(A,B) ITS}_A(C_1) \cap \text{ITS}_B(C_1) \cap \text{ITS}_B(C_2)$$

where $\text{ITS}_i(X)$ denotes the set of time slots where a node i is idle (in the sense of not transmitting or receiving) in cluster X.

Furthermore, to calculate the end-to-end bandwidth (PBW), a careful analysis of the transmission and reception planes of the end nodes of a link is required. Consider the example in the following figure

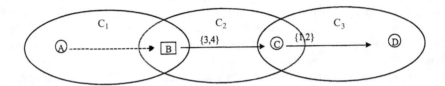

Node A in cluster C_1 intends to send real-time traffic to D in cluster C_3 through cluster C_2. Node B is the gateway node between clusters C_1 and C_2, and node C is the gateway node between clusters C_2 and C_3. Let us assume a TDMA frame with six time slots again and that gateways B and C are busy (either transmitting or receiving) in slots {3,4} and {1,2}, respectively; the set of busy slots of a node i is denoted UBW(i). It is evident that the busy slots of a GW between two clusters C_a and C_b are the union of the busy slots where this GW is transmitting to or receiving from *both* C_a and C_b. It is obvious also that the relationship between ABW and UBW for a node is: ABW = {Slots Per TDMA Frame} – UBW.

As mentioned earlier, failure to satisfy a resource request results in the blocking and/or dropping of the call; this may arise if during the propagation of a resource request a LinkBW value is smaller than the minimum requested bandwidth (LBW). Call dropping manifests not only during periods of high system load but also due to handover of connections from lightly loaded to heavily loaded clusters. Thus, one of the primary objectives of a handover as well as of a channel assignment algorithm is to maintain the blocking/dropping rates at minimum levels.

In the following, two representative cases for call blocking/dropping are presented. These two cases will also constitute the basis of the discussions throughout the following subsections.

The preceding calculation of path bandwidth between nodes A and D, PBW(A,D), involved only the transmission/reception planes of the end nodes and gateways that comprise the path; that is, only the inter-cluster traffic of the *same* path. When, however, end-to-end path bandwidth is calculated and the path length is more than one hop, additional constraints need to be taken into account. As shown below, these additional constraints are imposed by the inter- as well as the intra-cluster traffic requirements of other flows, which share common clusters with path A–D.

Case I

Assume the UBW size of A and B are the same as in Fig. 27.3 (that is, two). Let us further assume that there are internal transmissions in cluster C_1 and C_2 as illustrated in Fig. 27.3. These internal transmissions are placed from nodes M and N to N and K, respectively, in C_1 and X and Y to Y and Z, respectively, in C_2. It turns out that A can transmit to B only in slots {5,6} in C_1, instead of {1,2,5,6}, as illustrated earlier. For B's transmission to gateway C, however, even though the LinkBW(B,C) seems to be {1,2} and thus the size of ABW(B,C) is two, it turns out to be:

$$\text{LinkBW(B,C)} = \text{Slots\_Per\_Frame} - \text{UBW(B)} - \text{IntraCluster\_TXs}(c_1) - \text{UBW}(c_1)$$

$$\Downarrow \qquad\qquad\qquad \Downarrow \qquad\qquad\qquad\qquad \Downarrow$$

$$= \{1,2,3,4,5,6\} \quad - \quad \{3,4,5,6\} \quad - \qquad\qquad \{1,2\} \qquad = 0$$

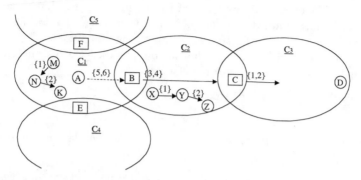

FIGURE 27.3 Example for channel allocation state.

This scenario illustrates the impact of internal (intra-cluster) transmissions on the end-to-end path bandwidth. Even though there is ABW at all nodes along the path (i.e., A, B, C, D), it turns out that BC's link ABW (LinkBW) is zero.

Case II

The same problem occurs with inter-cluster flows. Assume the slot utilization as in Fig. 27.3 again. As illustrated above, node A can transmit to B in slots {5,6} in cluster C_1. This, however, is true only if an additional constraint is satisfied: if there is no inter-cluster traffic that flows through cluster C_1 such that it makes use of the available resource spectrum of C_1. To illustrate this with an example, let us assume that another flow is routed through the gateways E and F of clusters (C_1 and C_4) and (C_1 and C_5), respectively, and also that gateway E transmits to F at slot {5} and F as slot {6}, both in cluster C_1. It is easy then to calculate that LinkBW(A,B) = $\varnothing$. The latter case effectively results in the blocking of not only A's flows but also of any inter- or intra-cluster traffic through cluster C_1, since there is no capacity spectrum left so that new transmissions can effectively be placed on C_1.

Again, although nodes do have ABW on their disposition, still cluster C_1 is effectively blocked as a result of the present channel configuration assignment.

Finally, it is worth mentioning that the bandwidth computation is part of the call admission control and resource reservation mechanisms during the call setup process. Upon computing the size of the path bandwidth, which is fed to the call admission control and resource reservation modules to gauge decisions, the resource guarantee phase effectively ends here.

The slot or channel assignment process, which is responsible for determining which channels from the available spectrum are assigned to a flow, is described in the following subsection.

27.3.5 Slot Assignment Phase

Once the request is successfully admitted along the path, the system needs to allocate slots to the real-time flow hop by hop along the path to the destination. As is argued above, efficient assignment of the available slots during the call setup is vital for minimizing the blocking/dropping rates of the present as well as of future calls.

The figure below is used as an example to describe how the slot assignment algorithm in the proposed QoSR model operates. As illustrated earlier, for a given path, the end nodes (i.e., source/destination), and the intermediate nodes (i.e., gateways) have different levels of resource constraints. End nodes are primarily constrained by their cluster's traffic as well as the *inter-cluster* traffic of their gateways. Gateway nodes are constrained not only by the intra- and inter-cluster traffic of their cluster's flows but also by the *intra-cluster* traffic of their cluster's gateways.

For the correct operation of the proposed resource assignment mechanism, it is assumed that a cluster-head node is aware of the UBW (i.e., transmission/reception scheduling) of its cluster's gateways as well as the intra-cluster UBW of its own cluster and of its gateways' clusters. Clusterhead nodes could be informed about each other's intra-cluster traffic through the exchange of a dedicated signaling that is leveraged by the initiation of new or the termination of existing connections.

The proposed time slot assignment algorithm for the proposed QoSR architecture distinguishes between intra- and inter-cluster traffic flows, as follows:

Rule I: Time Slot Assignment for Intra-Cluster Flows

The rule applied here is:

The scheduling of intra-cluster traffic within a cluster C should favor the time slots that are not used in C's neighbor clusters.

The rationale behind this approach is based on the fact that the used slots of a cluster C cannot be used by inter-cluster traffic that flows through C, thus increasing the *common* available bandwidth spectrum between the two neighbor clusters. The following example illustrates the effectiveness of this approach.

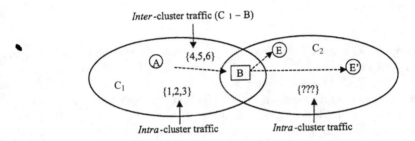

Assume that C_1, C_2 are two neighbor clusters and node B is their elected gateway. A node or a GW in C_1 is to forward a request to B, which in turn is to forward the request, either to an end node (E) or to a GW (E') in C_2. Let us further assume that time slots {1,2,3} are used for *intra*-cluster traffic in C_1 and also that slots {4,5,6} are free slots in C_1. The objective is to devise a flexible mechanism to assign time slots for intra-cluster traffic in C_2 such that the ABW of link B–E is maximized.

It is clear from the figure that any inter-cluster traffic that flows to C_2 through C_1 will be assigned to any time slots but {1,2,3}. One can observe that if time slots {1,2,3} are used for intra-cluster assignment in C_2, the inter-cluster traffic C_1–C_2 (through B) will then make use of the common ABW between C_1 and C_2, which is {4,5,6}. That is, if slot {4} is used from C_1 to B, then B can use any slot from the set of their common ABW, which is {5,6}. Thus the ABW (C_1–B) is 2.

If, however, the proposed rule is applied and slots {4,5,6}, which are the unused slots in C_1, are preferred for intra-cluster traffic in C_2, then the common ABW size is three. This is clear as the link C_1–B can use slots {4,5,6} whereas the link B–E can use slots {1,2,3}. To put this into context, an application with LBW = 3 cannot be accommodated in the first case, whereas in the second case the system can exploit the channel states of current and neighbor clusters to achieve optimum channel allocation.

Rule II: Time Slot Assignment for Inter-Cluster Flows

The following rules capture the time slot assignment to a gateway G for inter-cluster traffic, such that time slots are favored if:

a. The gateways of G's neighbor cluster are busy in these time slots; and/or

b. *Intra-cluster traffic in G's neighbor clusters takes place in either of these time slots; and/or*

c. *The gateways of G's present cluster, except these involved in the forwarding of the flow, use either of these time slots.*

The effectiveness of these rules is illustrated again with an example.

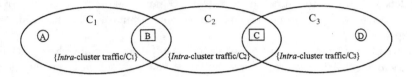

Consider the three clusters C_1, C_2, C_3, as illustrated in this figure, with their associated busy set of slots for intra-cluster traffic. Let us assume that node A from cluster C_1 intends to establish a connection to D from C_3 through gateways B and C. The objective is to efficiently allocate available common slots for the routing of the flow through the three clusters. Let us concentrate on the transmission from gateway B to C and determine according to the rules (a, b, c) the effectiveness of the proposed time slot selection.

According to Rule IIa, gateway B should favor the slots where *all* the gateways of C_1 are busy. That is because these gateways in C_1 cannot use these slots for any future inter-cluster transmission to B (through C_1 of course); thus similar to Rule I, gateway B increases its ABW with C_1.

Rule IIb is complementary to Rule IIa. According to Rule IIb, gateway B should favor the slots that are used for intra-cluster traffic in C_1. The explanation behind this is that inter-cluster traffic from any gateway of C_1 to B could not be placed on these slots; thus by selecting these slots for transmission, B again maximizes its ABW with C_1.

Note here that Rules IIa and IIb contrast with Rule I, which does not favor the used slots from neighbor clusters.

Finally, according to Rule IIc, gateway B should favor the slots where the gateways of its cluster (that is cluster C_2 in the figure), except the next gateway of the present flow (i.e., C) and itself, are busy with inter- or intra-cluster traffic to *other* clusters. The reason is that for future traffic that may be routed through B in cluster C_2, B cannot use these slots to transmit to these gateways and these gateways cannot use these slots to receive from B. As a result, this method minimizes the probability of blocking/dropping future calls through the cluster in C_2.

27.3.6 Bandwidth Reassignment Channel Allocation Policy

Furthermore, in view of rapidly fluctuating wireless link bandwidths, a time slot assignment mechanism should be leveraged to account for the dynamics of the network topology and timeliness of applications. It is easy to see that the termination or handover of ongoing connections within a cluster C results in a change of the intra- as well as inter-cluster traffic in cluster C. This in turn implies that the intra-cluster time slot assignment (Rule I) could be triggered *proactively* in order to account for these changes and reassign to any ongoing connection additional resources (if needed) such that the new channel allocations adapt to the new available channel spectrum of C as well as of C's neighbor clusters.

If the call admission algorithm fails to accommodate a bandwidth request, clearly because the ABW between two clusters is null, a mechanism that accounts for the channel usage between the two clusters is then triggered. The method, referred to as *Bandwidth Reassignment*, seeks alternative channel usage reconfigurations of active channels in order to create or to increase link bandwidth between the two clusters.

Let us assume that a gateway (GW) A attempts to transmit to a GW B through cluster C_1, but there is no common available bandwidth. The *Bandwidth Reassignment* process comprises the following three steps:

Step 1 (Intra-Cluster Time Slot Reassignment): If there is a slot i that is used in C_1 for intra-cluster communication, and i is not used by A and B, then exchange i with a time slot from the pool of free slots of C_i. If the procedure is successful, then communication can be established between A and B as A can use i for transmission and B can use i for reception and the phase end; else continue to Step 2.

Step 2 (Inter-Cluster Time Slot Reassignment at GW A): If there is a slot i where A is busy but B is free, try to free this time slot. To achieve this GW A first finds out the communicating node, say X, at time slot i. Then A determines whether the intersection of its ABW with X's ABW is nonzero. If so, A and X swap the used slot with the free one. This slot is used for communicating with B, since time slot i is free at B as well (see above). If the phase is not successful, then try Step 3.

Step 3 (Inter-Cluster Time Slot Reassignment at GW B): This phase is the same as Step 2 with the only difference that it is gateway B that examines the intersections of its ABW with the ABWs of the communicating nodes at time slots where A is free.

Using this adaptive QoS concept, it is possible to utilize bandwidth more efficiently, thus increasing the network's reward/revenue, while reducing the number of new connection drops/blocks. However the main drawback of the *Bandwidth Reassignment* process is that, under high loads, a large number of bandwidth allocation changes may be triggered, which is an important parameter that influences the cost of ongoing sessions. The reasons frequent bandwidth allocation changes are a problem are that:

1. It takes time to set up the modified bandwidth allocation.
2. In today's routers (and switches), it would normally require the invocation of software in every router on the session path, which would lengthen the response time even more and consume resources at the router.

Thus, it is reasonable to assume that when using the *Bandwidth Reassignment* process an additional goal would be to minimize the number of changes.

27.4 Performance Evaluation

27.4.1 System Model

Each link l has a bandwidth capacity $C(l)$, among which the part reserved by the QoS flows is denoted as $C_{qos}(l)$ and the part available to the best-effort flows is denoted as $C_{best}(l)$. It is clear that the values of $C_{qos}(l)$ and $C_{best}(l)$ are not fixed. When a new QoS flow f is routed through l, $C_{qos}(l)$ is increased by $B(f)$, and $C_{best}(l)$ is decreased by the same amount. In order to prevent the best-effort flows from being starved, $C_{qos}(l)$ is upper-bounded by $\phi C(l)$, where ϕ is a system parameter less than one. For every link l, the condition $C_{qos}(l) \leq \phi C(l)$ must always be satisfied. Hence, the bandwidth for best-effort flows is at least $(1 - \phi)C(l)$.

A new QoS flow f with $B(f)$ requirement can be accepted by a link l only if $\phi C(l) - C_{qos}(l) \geq B(f)$. In the current implementation of the model, in order to experiment with different service classes, link resources are statically partitioned between guaranteed traffic and best-effort traffic. The parameter ϕ therefore is static in a single run experiment; however, different values of ϕ are considered, and the impact on the system's aggregate throughput is studied and discussed. The parameter ϕ is denoted in the following figures as MM-TH (MultiMedia THreshold), which is the multimedia connections threshold per link.

The QoS parameter of interest is rate, thus link bandwidth is the resource of interest. Since, however, future application bit rates are not known in advance, the problem of network QoS representation of application requirements is of an online nature. Hence, the service model used in the proposed QoS routing framework presents a slightly general formulation of the issues related to application requirements. Similar to the work presented in [24], to ensure feasibility it is assumed that the requested bandwidth is expressed in terms of time slots. Given the application bit rate and the number of bits per time slot, the number of time slots per second required to accommodate a specified bit rate can be derived. Using this framework for bandwidth representation, (abstractly) any future application could be mapped upon this.

There are three types of QoS for the offered traffic; QoS_1, QoS_2, and QoS_3 need maximum bandwidths (HBW) of one, two, and three data slots in each frame, respectively. The minimum bandwidth of a QoS connection (LBW) is always assumed to be one slot. An admitted QoS connection maintains its bandwidth requirements within the [LBW, HBW] bounds advertised at connection setup.

The simulated environment consists of 150 mobiles roaming uniformly in a 200×200 m area. Each node has a data rate R and moves randomly at uniform speed V. It is assumed that channel quality has no effect on packet transmissions.

The channel-related values are the same as in [24]: each time frame consists of D data slots in the data phase and each data slot is 5 msec. Since the number of data slots per channel may be less than the number of nodes, nodes need to compete for these data slots. Once a call request is accepted on a link, a transmission window (i.e., data slots) is reserved (on that link) automatically for all the subsequent packets in the connection. The window is released from that link when either the session is finished or is handed over to a neighbor cluster.

Furthermore, the least-ID distributed clustering algorithm [16] is used here for the creation and maintenance of clusters. According to this scheme, the lowest-ID node in a neighborhood is elected as the clusterhead. Comparing to other distributed cluster algorithms such as the higher-connectivity (degree) algorithm also proposed in [31], where the highest degree node in a neighborhood becomes the clusterhead, it is shown [42] that the lowest-ID clustering algorithm yields the fewest changes under different mobility scenarios and communication transmission ranges. That means that the lowest-ID clustering algorithm provides a more stable cluster formation than the highest-connectivity one. This is because in the latter scheme, when the highest-connectivity node drops even one link due to node movement, it may fail to be reelected as a clusterhead. In contrast, the lowest-ID node can still be a clusterhead.

Furthermore, although a constant bandwidth allocation has many advantages from both the user and network perspectives, in applications with traffic of a bursty nature, such as video communication, the required bandwidth may change dramatically over time, usually in an unpredictable manner. In order to accommodate such situations it is reasonable to require the network to accommodate dynamic modification of the bandwidth allocation (see [1] for trends in applying dynamic bandwidth allocation).

Therefore, the basic service model is extended to support adaptive service [28]. In adaptive service, the resource specification for a flow specifies the minimum and maximum bounds for each QoS parameter required by the flow. For example, as mentioned previously, the resource specification for rate is given by a [LBW, HBW] bound. When the network admits a flow, it guarantees that the rate granted for the flow will be at least LBW. The minimum required bandwidth LBW can be considered as the bandwidth required to support the lowest-level quality of the connection the mobile user can "live with." Hence, the minimum required bandwidth, LBW, can be used for admitting/rejecting each new/handed-off connection.

Performance Metrics

The output parameters for the simulation model include the following:

- *Handover Success Probability,* defined as the probability that, while communicating with a particular cluster, an ongoing call requires a handover before the call terminates and the connection is successfully handed over to the new cluster with its minimum requested bandwidth (LBW)

• *System Throughput,* defined as the aggregate traffic (in b/sec) supported by the network

System Performance

Several sets of experiments were carried out in order to evaluate the performance of the QoSR framework, and especially of the two proposed on-demand channel assignment methods, namely the Minimum Blocking Channel Assignment (MBCA) and Bandwidth Reallocation Channel Assignment (BRCA) methods. The MBCA and BRCA methods are compared with the traditional dynamic channel assignment method, which randomly picks a free channel from the pool of available channels given that the interference levels for these free slots are above threshold. The latter method will be referred throughout this study to as Dynamic Channel Allocation (DCA).

The performance of the three methods is evaluated as a function of the average cluster size, channel rate (R), QoS requirements (FBW), and QoS connection threshold, MM-TH (ϕ).

The results, illustrated in Figs. 27.4 o 27.9, to correspond to the following system parameters: 115.2 kb/sec channel rate (R), 10.8 km/h velocity (V), 12.5 Erlang, and eight data slots per frame (D).

An average handover rate of 7.2 is observed throughout these experiments. Rate of handover is defined as the average number of handovers per second.

Figure 27.4 illustrates the aggregate bandwidth supported by the system using the two techniques (BRCA/MBCA and DCA) for different FBWs and MM-TH values. As expected, when the maximum number of multimedia connections per frame (MM-TH) decreases, the aggregate system rate decreases as well. In fact, the observed average decrease is 2.1 kb/sec from MM-TH = 8 to 4.

The system's performance when mixed traffic is involved, for different best-effort traffic probabilities, is depicted in Fig. 27.5.

Overall, significant performance benefits of the proposed BRCA/MBCA scheme against DCA are obtained. The respective handover success probabilities and system throughput for the two schemes are listed in Tables 27.1 and 27.2, respectively, again for different MM-TH values and FBW traffic requirements.

In the next experiment, the effect of increasing the channel rate is considered, such that a larger number of data time slots per TDMA frame is available. In this experiment R is set to 172.8 kb/sec aggregate rate. As expected, both channel assignment techniques improve with higher channel rate.

Figure 27.6 illustrates the aggregate bandwidth supported by the system using the two techniques (BRCA/MBCA and DCA) for different FBWs and MM-TH values.

The system's performance for a mixture of users' traffic flows (QoS-tolerant and QoS-sensitive traffic) under different bandwidth reassignment thresholds is illustrated in Fig. 27.7.

The respective handover success probabilities and system throughput for the two schemes are listed in Tables 27.3 and 27.4, respectively, again for different MM-TH values and FBWs.

Again under all scenarios, the proposed BRCA/MBCA mechanisms outperform the DCA scheme.

The final experiment investigates the impact of cluster size, such that a higher population of mobiles resides within a single cluster.

There are a few tradeoffs with cluster size. On the one hand, the smaller the size, the higher the handover rates and interruption times. On the other hand, the larger the cluster size, the higher the load on a per cluster basis, which translates to a higher number of nodes contending for the (limited) cluster resources. The latter further implies that call dropping/blocking probabilities also increase. A critical parameter that should gauge the cluster size should be the nodal population per cluster instead of transmitting power range.

Figures 27.8 and 27.9 illustrate the aggregate bandwidth supported by the system using the two techniques (BRCA/MBCA and DCA) for different flow types, FBWs, and MM-TH values.

The respective handover success probabilities and system throughput for the two schemes are listed in Tables 27.5 and 27.6, respectively, again for different MM-TH values and FBWs.

Although DCA in this experiment shows a better performance in comparison to the previous experiments, the proposed BRCA/MBCA mechanisms, however, show much better performance compared to the DCA scheme.

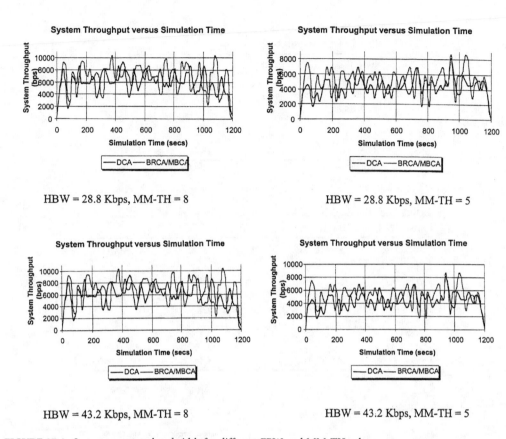

FIGURE 27.4 System aggregate bandwidth for different FBW and MM-TH values.

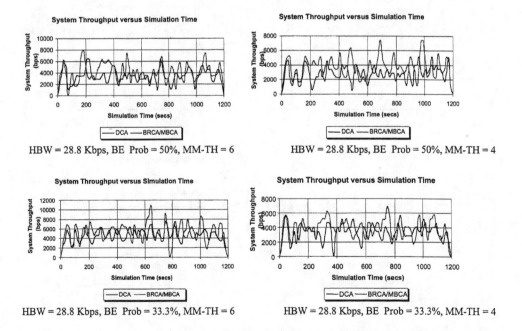

FIGURE 27.5 System aggregate bandwidth for different FBWs, BE traffic probabilities, and MM-TH values.

TABLE 27.1 Handover Success Probabilities for
Different FBW and MM-TH Values

FBW (kb/sec)	MBCA/BRCA	DCA	MM-TH
14.4–28.8	99.5	97.4	8
14.4–43.2	99.5	97.4	8
14.4–28.8	96.8	94.4	5
14.4–28.8	96.2	94.2	4
14.4–43.2	96.8	94.4	5
14.4–43.2	96.2	94.2	4

TABLE 27.2 Handover Success Probabilities for Different FBWs, BE Traffic
Probabilities, and MM-TH Values

FBW (kb/sec)	MBCA/BRCA	DCA	% BE Traffic	MM-TH
14.4–28.8	99.87	98.90	50%	6
14.4–28.8	97.94	96.41	50%	4
14.4–28.8	99.87	98.90	50%	6
14.4–28.8	97.94	96.41	50%	4
14.4–28.8	99.64	98.30	33.3%	6
14.4–28.8	98.09	96.30	33.3%	4
14.4–28.8	99.64	98.30	33.3%	6
14.4–28.8	98.09	96.30	33.3%	4

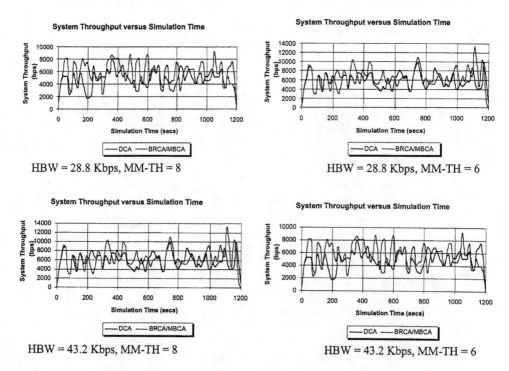

FIGURE 27.6 System aggregate bandwidth for different FBW and MM-TH values.

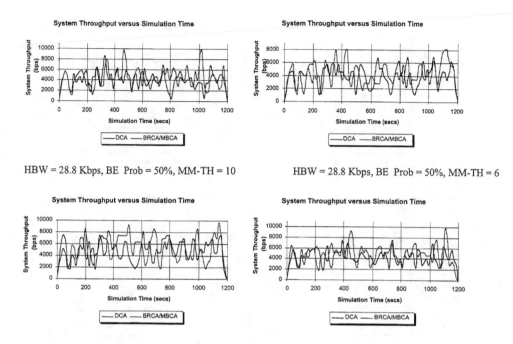

FIGURE 27.7 System aggregate bandwidth for different FBWs, BE traffic probabilities, and MM-TH values.

TABLE 27.3 Handover Success Probabilities for Different FBW and MM-TH Values

FBW (kb/sec)	MBCA/BRCA	DCA	MM-TH
14.4–28.8	99.9	99.7	12
14.4–43.2	99.9	99.7	12
14.4–28.8	99.02	97.3	6
14.4–28.8	99.8	92.2	8
14.4–43.2	99.02	97.3	6
14.4–43.2	99.8	98.2	8

TABLE 27.4 Handover Success Probabilities for Different FBWs, BE Traffic Probabilities, and MM-TH Values

FBW (kb/sec)	MBCA/BRCA	DCA	% BE Traffic	MM-TH
14.4–14.4	99.8	99.04	50%	6
14.4–14.4	100	99.6	50%	8
14.4–28.8	99.8	99.04	50%	6
14.4–28.8	100	100	50%	8
14.4–14.4	99.9	98.2	33.3%	6
14.4–14.4	99.9	99.7	33.3%	8
14.4–28.8	99.5	98.28	33.3%	6
14.4–28.8	99.9	99.7	33.3%	8

27.5 Conclusions

Efficient dynamic routing support for bandwidth-sensitive traffic in cluster-based multihop mobile networks is the main topic of this chapter. A basic QoS routing framework is proposed, described, and evaluated. The protocol incorporates novel techniques on bandwidth calculation and slot reservation for multihop mobile networks. A fundamental feature of the proposed channel assignment methods is that they make use of the congestion states (i.e., ABW and UBW) of clusters. By giving the routing algorithm

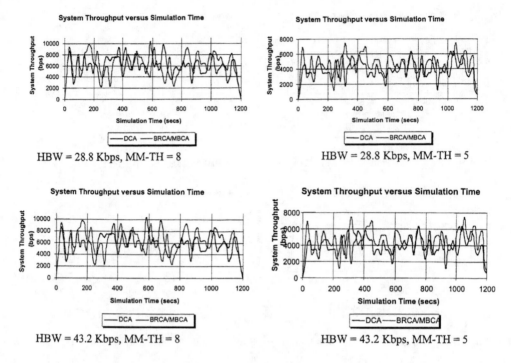

FIGURE 27.8 System aggregate bandwidth for different FBW and MM-TH values.

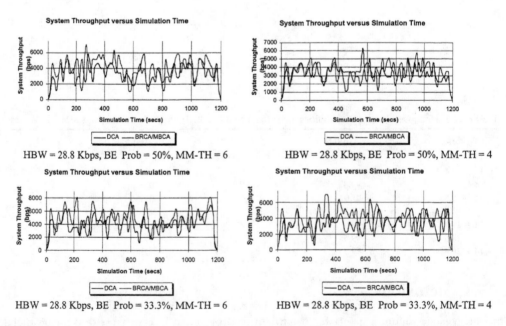

FIGURE 27.9 System aggregate bandwidth for different FBWs, BE traffic probabilities, and MM-TH values.

TABLE 27.5 Handover Success Probabilities for Different FBW and MM-TH Values

FBW (kb/sec)	MBCA/BRCA	DCA	MM-TH
14.4–28.8	99.65	97.65	8
14.4–43.2	99.65	97.65	8
14.4–28.8	96.3	94.8	5
14.4–28.8	95.7	93.37	4
14.4–43.2	96.3	94.8	5
14.4–43.2	95.7	93.37	4

TABLE 27.6 Handover Success Probabilities for Different FBWs, BE Traffic Probabilities, and MM-TH Values

FBW (kb/sec)	MBCA/BRCA	DCA	% BE Traffic	MM-TH
14.4–14.4	99.7	99.6	50%	6
14.4–14.4	98.06	97.54	50%	4
14.4–28.8	99.7	99.6	50%	6
14.4–28.8	98.06	97.45	50%	4
14.4–14.4	99.5	98	33.3%	6
14.4–14.4	97.5	95.47	33.3%	4
14.4–28.8	99.6	98	33.3%	6
14.4–28.8	97.5	95.47	33.3%	4

access to bandwidth information, the coarse grain (routing) and fine grain (congestion control) resource allocation mechanisms are coupled in order to achieve efficient and fair allocation of resources.

Furthermore, a bandwidth reassignment model is introduced. According to this, active channels are swapped with unused channels in order to create or to increase inter-cluster common bandwidth. This effectively translates to efficient handoffs and more effective call admission control decisions.

The performance results assert that a basic channel assignment scheme can be significantly improved using the proposed MBCA and BRCA heuristics. As demonstrated, the proposed bandwidth assignment methods perform consistently well for a number of scenarios for multihop cluster-based mobile networking.

References

[1] Y. Afek, M. Cohen, E. Haalman, and Y. Mansour, Dynamic Bandwidth Allocation, *Proc. IEEE INFOCOM*, San Francisco, CA, 1996, pp. 880–887.

[2] G. Aggélou and R. Tafazolli, RDMAR: A Bandwidth-efficient Routing Protocol for Mobile Ad Hoc Networks, *Proceedings of ACM WoWMoM99*, Seattle, WA, Aug. 1999.

[3] G. Aggélou and R. Tafazolli, Relative Distance Micro-discovery Ad Hoc Routing (RDMAR) Protocol, IETF Internet Draft, draft-ietf-manet-rdmar-00.txt, Nov. 1999.

[4] G. Aggélou, Dynamic IP Routing and Quality-of-Service Support in Mobile Multimedia Ad Hoc Networks, Ph.D. dissertation, University of Surrey, 2001.

[5] G. Aggélou and R. Tafazolli, On the relaying capability of next generation GSM cellular network, *IEEE Personal Communications*, 8, 6–13, 2001.

[6] G. Aggélou and R. Tafazolli, Determining the Optimal Configuration for The Relative Distance Microdiscovery Ad Hoc Routing Protocol, *IEEE Transactions on Vehicular Technology*, Jan. 2002.

[7] G. Aggélou and R. Tafazolli, QoS Support in 4th Generation Mobile Multimedia Ad Hoc Networks, *Proceedings of the Second International Conference on 3G Mobile Communication Technologies*, London, Mar. 26–28, 2001.

[8] D.J. Baker, A. Ephremides, and J.A. Flynn, The Design and Simulation of a Mobile Radio Network with Distributed Control, *IEEE JSAC, SAC-2(1)*, 1984, pp. 226–237.

[9] N. Bambos and G.J. Pottie, On Power Control in High Capacity Cellular Radio Networks, *IEEE GLOBECOM*, 1992, vol. 2, pp. 863–867.

[10] R. Braden, L. Zhang, S. Berson, S. Herzog, and S. Jamin, Resource ReSerVation Protocol (RSVP) — Version 1, Functional Specification, IETF RFC 2205, Sep. 1997.

[11] I. Chlamtac and S.S. Pinter, Distributed nodes organization algorithm for channel access in a multihop dynamic radio network, *IEEE Transactions on Computing*, C_36, 728–737, 1987.

[12] D. Clark and J. Wroclawski, An Approach to Service Allocation in the Internet, IETF Internet Draft, draft_clark_diff_svc_alloc_00.txt, July 1997.

[13] D. Clark, S. Shenker, and L. Zhang, Supporting Real-Time Applications in an Integrated Services Packet Network: Architecture and Mechanism, *ACM SIGCOMM 92*, Aug. 1992.

[14] B. Jamoussi et al., Constrained-Based LSP Setup Using LDP, IETF Internet Draft, Work in progress, Feb. 1999.

[15] D. Dwyer, S. Ha, J. Li, and V. Bharghavan, An Adaptive Transport Protocol for Multimedia Communications, To appear in the *International Conference on Multimedia Computing Systems*, Durham, North Carolina, Aug. 1998.

[16] A. Ephremides, J.E. Wieselthier, and D.J. Baker, A design concept for reliable mobile radio networks with frequency hopping signaling, *Proceedings of IEEE*, 75, 56–73, 1987.

[17] P. Goyal and H.M. Vin, Generalized Guaranteed Rate Scheduling Algorithms: A Framework, Technical Report TR95-30, Department of Computer Science, University of Texas, Austin, 1995.

[18] M. Grossglauser, S. Keshav, and D. Tse, RCBR: A Simple and Efficient Service for Multiple Time-Scale Traffic, *IEEE/ACM Transactions on Networking*, To appear, Dec. 1998.

[19] J. Cai and D. Goodman, General Packet Radio Service in GSM, *IEEE Communications Magazine*, Oct. 1997.

[20] M. Mouly and M. Pautet, *The GSM System for Mobile Communications*, published by the authors, 1992.

[21] S. Ha, K-W. Lee, and V. Bharghavan, Performance Evaluation of Scheduling and Resource Reservation Algorithms in an Integrated Packet Services Network Environment, *Proceedings of ISCC*, Athens, June 1998.

[22] L. Hu, Distributed Code Assignments for CDMA Packet Radio Networks, IEEE/ACM Transactions on Networking, Dec. 1993, pp. 668–677.

[23] L. Kleinrock, Nomadic computing — an opportunity, ACM SIGCOMM, *Computer Communications Review*, 25, 36–40, Jan. 1995.

[24] R. Lin et al., QoS routing in ad hoc wireless networks, *IEEE Journal on Selected Areas on Communications*, 17(8), 1999.

[25] C.R. Lin and M. Gerla, Adaptive clustering for mobile wireless networks, *IEEE Journal on Selected Areas in Communications*, 15, pp. 1265–1275, 1997.

[26] S. Lu and V. Bharghavan, Adaptive Resource Management Algorithms for Indoor Mobile Computing Environments, *Proceedings of ACM SIGCOMM*, Stanford, CA, Aug. 1996.

[27] S. Lu, V. Bharghavan, and R. Srikant, Fair Scheduling in Wireless Packet Networks, *Proceedings of ACM SIGCOMM*, Cannes, Aug. 1997.

[28] S. Lu, K-W. Lee, and V. Bharghavan, Adaptive Service in Mobile Computing Environments, *Proceedings of IFIP IWQoS*, New York, May 1997.

[29] Mobile Ad-hoc Networks (MANET) Working Group, http://www.ietf.org/html.charters/manet-charter.html.

[30] J.M. del Rosario and A. Choudhary, High Performance I/O for Parallel Computers: Problems and Prospects, *IEEE Computer*, March 1994, 59–68.

[31] A.K. Parekh, Selecting Routers in Ad Hoc Wireless Networks, *ITS*, 1994.

[32] C. Perkins, Ed., IPv4 Mobility Support, RFC 2002, Oct. 1996.

[33] PNNI SWG (D. Dykeman, Ed.)., PNNI Draft Specification, ATM Forum 94-0471R10, Oct. 1995.

[34] Q. MA, Routing Traffic with Quality-of-Service Guarantees in Integrated Services Networks, Ph.D. thesis, Carnegie Mellon University, Pittsburgh, PA, 1998.

[35] QoS Routing (QoSR) Working Group, http://www.ietf.org/html.charters/OLD/qosr-charter.html.

[36] A.K. Salkintzis, Packet Data over Cellular Networks: The 36 Approach, *IEEE Communications Magazine*, Jun. 1999, pp. 152–159.

[37] M. Sreetharan and R. Kumar, *Cellular Digital Packet Data*, Artech House Publishers, Norwood, MA, 1996.

[38] S. Shenker, C. Partridge, and R. Guerin, Specification of Guaranteed Quality of Service, IETF Internet Draft, draft_ietf_intserv_guaranteed_svc_07.txt, Feb. 1997.

[39] Z. Wang, *Internet QoS: Architectures and Mechanisms for Quality of Service*, Morgan Kaufmann, 2001.

[40] Z. Wang and J. Crowcroft, Shortest Path First with Emergency Exits, *ACM SIGCOMM 90*, Sep. 1991.

[41] J. Wroclawski, Specification of the Controlled_Load Network Element Service, IETF Internet Draft, draft_ietf_intserv_ctrl_load_svc_04.txt, Nov. 1996.

[42] H.-K. Wu, Multimedia, Mobile, Multihop Networks in Channel Fading Environment, Ph.D. dissertation, University of California, 1999.

[43] H. Zhang, Service disciplines for guaranteed performance service in packet-switching networks, *Proceedings of the IEEE*, 83, 1995.

28

Quality of Service Models for Ad Hoc Wireless Networks

Xiao Hannan
National University of Singapore

Chua Kee Chaing
National University of Singapore

Seah Khoon Guan
Winston
National University of Singapore

Abstract

This chapter discusses Quality of Service (QoS) models for ad hoc wireless networks (also called Mobile Ad hoc Networks [MANETs]). After examining the features of MANETs and QoS models of the Internet, a Flexible QoS Model for MANETs (FQMM) is proposed. Salient features of FQMM include the dynamic roles of nodes and a hybrid provisioning policy that combines the fine-grained QoS of Integrated Service and service differentiation of Differentiated Service. A framework architecture to realize FQMM is discussed, and the functionalities of and the interreactions among its components are elaborated. By simulation, FQMM is evaluated against its performance in service prioritization and service differentiation in MANETs by implementing relevant components in the FQMM framework architecture.

28.1 Introduction

Quality of Service (QoS), as the name suggests, involves studying the level of user satisfaction in the services provided by a communication system. In computer networks, the goal of QoS support is to achieve a more deterministic communication behavior, so that information carried by the network can be better preserved and network resources can be better utilized. Intrinsic to the notion of QoS is an agreement or a guarantee by the network to provide a set of measurable prespecified service attributes to the user in terms of network delay, delay variance (jitter), available bandwidth, probability of packet loss (loss rate), and so on. The QoS forum defines QoS as *the ability of a network element (e.g., an application, a host, or a router) to provide some level of assurance for consistent network data delivery* [13]. The IETF RFC 2386 [10] characterizes QoS as a set of service requirements to be met by the network while transporting a packet stream from source to destination. Different applications require different levels of assurance, or more exactly, QoS requirements, from the network. Real-time applications such as voice and video transmission need a packet by a certain time, otherwise the packet is essentially worthless; non-real-time applications such as file transfer and e-mail emphasize reliability instead.

QoS support has been widely discussed in various networks. The Asynchronous Transfer Mode (ATM) network, for instance, has made a great emphasis on the support of QoS for traffic of different classes. Tremendous efforts have been made by the IETF to enhance the current Internet to support multimedia services. QoS mechanisms are applied to wisely manage network resources (bandwidth, buffer) to meet the various QoS requirements of applications. With the rising popularity of multimedia applications among end users in various networks and the widened scope of applications for ad hoc wireless networks (which are also called Mobile Ad hoc NETworks [MANETs]), QoS support of distributed, non–real-time, and real-time applications in MANETs has emerged as a major area of research. This issue, however, as Corson stated in [8], is really a challenging task faced by researchers.

A network's ability to provide a specific QoS depends upon the inherent properties of the network itself [25], which span over all the elements in the network. For the transmission link, the properties include link delay, throughput, loss rate, and error rate. For the nodes, the properties include hardware capability, such as processing speed and memory size. Above the physical qualities of nodes and transmission links, the QoS control algorithms operating at different layers of the network also contribute to the QoS support in networks. Unfortunately, the inherent features of MANETs show weak support for QoS. The wireless link has low, time-varying raw transmission capacity with relatively high error rate and loss rate. In addition, the possible various wireless physical technologies that nodes may use simultaneously to communicate make MANETs heterogeneous in nature. Each technology will require a MAC layer protocol to support QoS. Therefore, the QoS mechanisms above the MAC layer should be flexible to fit the heterogeneous underlying wireless technologies [9].

In the literature, many proposals have been presented to support QoS in MANETs including QoS MAC protocols [1,17,20,21,27,28,30,36], QoS routing protocols [5,6,17,20,21,25,26], and resource reservation protocols [19,30]. While these proposals are sufficient to meet QoS needs under certain assumptions, none of them proposes a QoS model for MANETs.

It is obvious that a concrete definition of the type of service delivered to the user is needed as we look at the development of a network for commercial application and the diversity of the applications' requirements. This description of the service delivered by the network is called the *service model* and documents the commitments the network makes to the clients that request the provided services. A QoS model describes a set of end-to-end services, and it is up to the network to ensure that the services offered at each link along a path combine meaningfully to support the end-to-end service [14]. More generally, QoS models permit clients to select a number of abstract *guarantees* that apply to a sequence of operations. These guarantees govern properties such as the timing, ordering, and reliability of the operation [29]. The system promises to ensure these properties for operations performed by clients or else inform them that the guarantees cannot be met and perhaps refuse to accept further operations. Different communication networks extend their service models to permit multiple types of services from their own perspectives. The ATM network starts with the assumption of existing Constant Bit Rate (CBR) services, and further services have been derived with relaxations of the bandwidth guarantees and timing constraints. In the IETF, the Integrated Services (IntServ) model starts from the assumption of a best effort service and refines this to add guarantees of various kinds, typically of delay variance, bounds, and then throughput [11]. The Differentiated Service (DiffServ) model divides services into several classes with various requirements and priorities [3].

The purpose of this chapter is to look into the problem of QoS provisioning in MANETs from a systemic view, first by designing a suitable QoS model for MANETs after considering its features; second, by constructing a QoS framework architecture to realize the QoS model. The work on certain aspects of QoS issues in MANETs in the literature can thus fit into the framework architecture.

The rest of this chapter is organized as follows. Section 28.2 discusses how the features of MANETs affect the design of a QoS model for MANETs. Section 28.3 proposes a Flexible QoS Model for MANETs (FQMM), and Section 28.4 presents a framework architecture to realize FQMM. Evaluation of FQMM is introduced in Section 28.5 and finally, conclusions are drawn in Section 28.6.

28.2 MANET Features and QoS Models

A QoS model for MANETs should first consider the features of MANETs, e.g., dynamic topology, time-varying link capacity, and limited power [33]. In addition, as applications of MANETs in the civilian sector become more popular, we assume that MANETs will be seamlessly connected to the Internet in the future. Therefore, the QoS model for MANETs should also consider the existing QoS models for the Internet, viz., IntServ and DiffServ. However, these models are aimed at wired networks. The applicability of IntServ and DiffServ in MANETs therefore needs investigation.

The idea behind IntServ [4] is borrowed from the paradigm of the telephony world, i.e., adopting a virtual circuit connection mechanism. The Resource ReSerVation Protocol (RSVP) [35] is used as a signaling protocol to set up and maintain the virtual connections. Routers apply corresponding resource management schemes, e.g., Class Based Queueing (CBQ) [12], to support the QoS requirements of the connection. Based on these mechanisms, IntServ provides quantitative QoS for every flow.[1] However, such per-flow provisioning results in a huge storage and processing overhead on routers, which is the well-known scalability problem of IntServ.

The tenet of DiffServ [3] is to use a relative-priority scheme to soften the requirements of hard QoS models such as ATM and IntServ, thereby mitigating the scalability problem of the latter. At the boundary of a network, traffic entering the network is classified, conditioned, and assigned to different behavior aggregates by marking the DiffServ field [22] in an IP packet header. Within the core of the network, packets are forwarded according to the Per-Hop Behavior (PHB) [2] associated with the DiffServ field. Implicit reservation is done in the form of a service level agreement, which is agreed upon between users and network providers. DiffServ provides qualitative QoS support for aggregate flows.

We now look into the effects of the salient features of MANETs on the possible QoS model for MANETs.

28.2.1 Multiple Node Functionalities

Each node in MANETs has two functions, i.e., host and router/switch. As a router, a node routes packets for other nodes similar to what the backbone routers do in the Internet. Hence, a MANET is similar to a backbone network in the sense of the functionalities of nodes, although the size of a MANET is not comparable with that of the Internet backbone. DiffServ is designed to overcome the difficulty in implementing and deploying IntServ and RSVP in the Internet backbone. Thus, the DiffServ approach may fit MANETs. In addition, DiffServ is lightweight in interior nodes as it does away with per-flow states and signaling at every hop. In MANETs, keeping the protocol lightweight in interior nodes is important since putting too heavy a load on a temporary forwarding node that is moving is unwise. Therefore, DiffServ and MANETs are also similar in that they are lightweight in interior nodes. Intuitively, these similarities imply a potential usage of the DiffServ approach in MANETs.

28.2.2 Node Mobility and Lack of Infrastructure

Node mobility is the basic cause of the dynamics in MANETs. The MAC layer allocation of bandwidth to each node changes dynamically according to mobility scenarios. Consequently, the aggregate network capacity is also time varying. Furthermore, the absence of infrastructure (without even a base station)

---

[1] A flow is an application session between a sender and a receiver.

makes the control of bandwidth difficult. The roles of nodes as a host or router also change together with node mobility and the dynamic topology. All these dynamics in MANETs require the QoS model and the supporting QoS mechanisms to be adaptive.

28.2.3 Power Constraints

In MANETs, nodes' processing capability is limited because of limited battery power. This feature requires low processing overheads of nodes. Therefore, the control algorithms and QoS mechanisms should use bandwidth and energy efficiently, e.g., try to keep the inter-router information exchange to a minimum. However, in IntServ, all routers must have the four basic components: RSVP, admission control routine, classifier, and packet scheduler. Processing of all these functions is power-consuming and results in high processing overheads. Therefore, the IntServ approach is undesirable for power-constrained MANETs.

28.2.4 Limited Bandwidth and Network Size

At first glance, this feature makes the scalability problem less likely to occur in MANETs. However, as fast radios and efficient low bandwidth compression technology develop rapidly, the emergence of high-speed and large-sized MANETs with plenty of applications is foreseeable. At that time, the pure IntServ approach for MANETs will inevitably meet the scalability problem as in high-speed fixed networks today.

28.2.5 Time-Varying Feature

In MANETs, the link capacity is time varying due to the physical environment of nodes, such as fading and shadowing, the mobility of nodes, and the dynamics of the network topology. This time-varying feature makes the QoS mechanisms in MANETs more difficult than in wired fixed networks. Take the signaling protocol for example. A signaling protocol generally comprises three phases: connection establishment, connection teardown, and connection maintenance. Corson [8] predicts that a larger percentage of link capacity will be allocated to control overhead in a network with smaller and time-varying aggregate network capacity. For MANETs with dynamic topology and link capacities, the overheads of connection maintenance usually outweigh the initial cost of establishing the connection. Therefore, RSVP-like signaling is not practical in MANETs.

The time-varying feature thus makes it hard to provide short timescale QoS by trying to keep up with the time-varying conditions. However, it could be possible to provide QoS over a long timescale for MANETs. DiffServ is aimed at providing service differentiation among traffic aggregates to customers over a long timescale. In particular, Assured Service (AS) [7] is aimed at providing guaranteed, or at least expected, throughput for applications, and it is easy to implement. AS is attractive when throughput is chosen as an important QoS parameter for MANETs. In addition, AS is more qualitative oriented than quantitative oriented [16], and it is not easy, if not impossible, to provide quantitative QoS in MANETs with the physical constraints. Therefore, AS has a potential usage in MANETs.

28.3 FQMM — A Flexible QoS Model for MANETs

Although DiffServ is generally favorable for application in MANETs as discussed in the last section, it is still not straightforward to adopt this approach in MANETs, since DiffServ has been designed for fixed and relatively high-speed networks. On the other hand, it is also desirable to incorporate suitable QoS features provided by IntServ into MANETs. This section describes a Flexible QoS Model for MANETs (FQMM) and its QoS provisioning policy, which takes into consideration the investigation in the previous section.

28.3.1 FQMM

FQMM is designed for small- to medium-sized MANETs, with fewer than 50 nodes and using a flat nonhierarchical topology. It defines three kinds of nodes as in DiffServ:

An ingress node is a mobile node that sends data.

Interior nodes are the nodes that forward data for other nodes.

An egress node is a destination node.

For example, in Fig. 28.1, eight nodes are moving about, and a route is established for communication from node M1 to M6. When data is sent from M1 to M6, M1 behaves as an ingress node — classifying, marking, and policing packets. Nodes M3, M4, and M5 along the route from M1 to M6 behave as interior nodes, forwarding data via certain PHB defined by the DiffServ field. M6 is the egress node, which is the destination of the traffic. In this model, a MANET represents one DiffServ domain where traffic is always generated by applications running on an ingress node and terminating in an egress node.

When nodes move about, another topology may form as shown in Fig. 28.2. In this case, there are two connections: one is from M1 to M6, the other is from M8 to M2, and the roles of the nodes are listed in Table 28.1. As illustrated, the nodes have dynamic roles in FQMM.

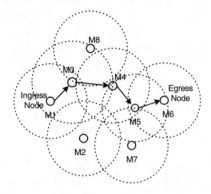

FIGURE 28.1 FQMM example 1.

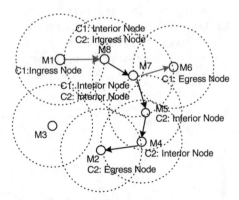

FIGURE 28.2 FQMM example 2.

TABLE 28.1 Roles of Nodes in FQMM (Example 2)

Connections	Ingress Node	Interior Node	Egress Node
C1	M1	M8, M7	M6
C2	M8	M7, M5, M4	M2

28.3.2 Hybrid Provisioning Policy

Provisioning refers to the determination and allocation of resources needed at various points in the network. In the current Internet, provisioning is at a high level, quite coarse, and generally based on the estimation of the scale and utilization of the network. In IntServ, the provisioning policy is to realize ideal per-flow granularity by RSVP signaling along the route and reserving sufficient resources. Provisioning in DiffServ is per-class based with static provisioning performed by human agents on behalf of service providers or users, or dynamic provisioning by signaling or measurement.

We propose a hybrid per-flow and per-class provisioning policy for FQMM. In such a scheme, traffic of the highest priority is given per-flow provisioning while other priority classes are given per-class provisioning. Although like DiffServ, FQMM has service differentiation, we can improve the per-class granularity to per-flow granularity for certain classes of high priority traffic since the traffic load in a MANET is typically less than in the backbone of the Internet. However, it is difficult to provide per-flow granularity to all the traffic types in a MANET due to bandwidth limitation and other constraints. Hence, we try to preserve the per-flow granularity for a small portion of traffic types in a MANET, given that these form a small percentage of the total traffic load. Since the states of per-flow granularity come from only a small fraction of the traffic, the scalability problem as in IntServ becomes insignificant. Therefore, this hybrid scheme combines the per-flow granularity of IntServ and per-class granularity of DiffServ.

28.4 Framework Architecture of FQMM

This section describes a framework architecture to realize FQMM (see Fig. 28.3). FQMM works in the IP layer with the cooperation of the MAC layer. There are two planes: one is the data forwarding plane shown below the dashed line; the other is the control and management plane shown above the dashed line. The components in the data forwarding plane handle the data packet forwarding function while the components in the management plane prepare for the operation of the data forwarding plane according to specific protocols or algorithms. Modules in the two planes communicate either directly or via the information stored in a database. For example, the traffic conditioner configuration module and the adaptive traffic conditioner module operate via the traffic control database. The resource reservation protocol sends or attaches the reservation information to the packet forwarding module directly. We elaborate on each module in the framework architecture below.

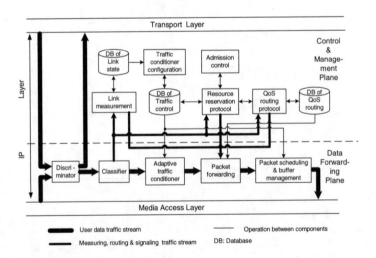

FIGURE 28.3 Framework architecture of FQMM.

28.4.1 Data Forwarding Plane

The *discriminator* module classifies the data from the upper layer (transport layer) and the lower layer (MAC layer). Downstream packets coming from the transport layer are forwarded to the classifier. Upstream packets from the MAC layer are forwarded to the transport layer if the packet's destination is the current node, and to the classifier if the packet's destination is not the current node.

The *classifier* module distributes the control and management packets to corresponding modules in the control and management plane, such as link measurement, resource reservation, and QoS routing. The classifier also maps incoming packets into classes by reading the DiffServ fields in each packet header. For simplicity, one class corresponds to one item in the service profile, which is stored in the database of traffic control. Therefore, a flow is mapped to one class with per-flow provisioning traffic while an aggregate is mapped to one class with the per-aggregate provisioning traffic.

The *adaptive traffic conditioner* module polices and regulates the traffic based on the service profile, which is stored in the traffic control database. Marking, shaping, and dropping can be used alone or together in the traffic conditioner. We call it adaptive because the traffic conditioner configuration module adjusts the service profile according to the link states. Thus, the traffic conditioner can adapt to the dynamics of the network.

The *packet forwarding* module forwards the incoming packets of different classes to a suitable output port. It also accepts control packets from modules in the control and management plane, such as link measurement, resource reservation protocol, and QoS routing protocol. If a proactive routing protocol is used, the packet forwarding module forwards packets after checking the packet's destination address and the routing table. This module forwards the packet according to the routing header in the packet if a reactive routing protocol is used.

The *packet scheduling and buffer management* module manages the queue(s) in the output port. Different scheduling schemes have been proposed in the literature from the simplest FIFO to various complex schemes. Several researchers have also investigated the problem of how to adapt the packet fair queueing algorithms to wireless networks. While the proposed solution is for a cellular wireless network model, it is not clear whether the same solution can be applied directly to MANETs. We will first study the performance of existing scheduling and buffer allocation approaches in FQMM and then find a suitable resource management scheme for the model. The configuration of the packet scheduling and buffer management is set in the database of the traffic control module.

28.4.2 Control and Management Plane

The *link measurement* module measures/monitors the states of the channel and puts the results into the link state database. It may passively monitor link states only or also actively generate measurement packets into the network when necessary. Link measurement could be done on a wide area basis or on a local area basis. The former means that each node gets a picture of a rather wide area around itself by exchanging information with other nodes. The latter means that each node only gets the channel states around itself by applying an estimation algorithm. In existing routing protocols, this task is not explicitly done by a link measurement module, but together with the routing protocol.

The *traffic conditioner configuration* module configures the service profile of applications and puts them into the traffic control database. Thus, the adaptive traffic conditioner module in the data forwarding plane can police the data and control traffic according to the service profile. The traffic conditioner configuration module makes decisions depending on information stored in both the link state database and the traffic control database. The operation timescale of the link measurement and the traffic conditioner configuration modules affect the QoS of the applications. Therefore, the timescale should be set carefully.

The *admission control* module is responsible for comparing the resource requirements arising from the requested QoS against the available resources. The decision whether sufficient resources are available for a new flow or an aggregate of flows at the requested QoS without violating the existing QoS commitments to other applications depends on resource management policies and resource availability. Once admission testing has been successfully completed, local resources are reserved immediately and then committed later if the end-to-end admission control test is successful. Admission control is invoked by the resource reservation protocol before reservation is executed. There are various admission control algorithms such as parameter based and measurement based. INSIGNIA [19], for example, adopts a simple max/min admission control algorithm.

The *resource reservation protocol* module arranges for the allocation of suitable end-system and network resources to satisfy the user QoS specification. In doing so, the resource reservation protocol interacts with the QoS routing protocol to establish a path through the network in the first instance, then, based on admission control at each node, end-to-end resources are allocated. Reservation can be done per-flow or per-class. The protocol maintains per-flow/per-class specific state information of traffic and updates the traffic control database. For in-band signaling protocols for MANET such as INSIGNIA [19], the reservation control message is integrated with the data packet. The reservation scheme also relates to different MAC protocols.

The QoS *routing protocol* is responsible for finding the path for a flow or aggregate of flows and maintaining the path at the required QoS level. A proactive routing protocol, such as DSDV [23], maintains the QoS routing database, while a reactive routing protocol, such as DSR [18], maintains a routing cache. The QoS routing protocol also sends control packets directly to the packet-forwarding module when necessary. The QoS routing module has a close relationship with the network architecture. Self-organized cluster structure is adopted in [15,25,26] since such a structure makes administration easier under certain conditions. On the other hand, Perkins and Royer use a flat network structure for QoS enhancement to the AODV routing protocol [24].

Databases contain state and control information needed for the operation of the modules. There are three databases: the link state database, the traffic control database, and the QoS routing database. The link state database is maintained by the link measurement module and visited by the traffic conditioner configuration module to do traffic conditioner configuration. The traffic control database is maintained by the traffic conditioner configuration module and the resource reservation protocol module. The information stored there includes the service profile, the states of the reservation for flows/aggregates, and other control states that are used to configure the parameters of the adaptive traffic conditioner module and packet scheduling and buffer management module. The QoS routing database is maintained by the QoS routing protocol and used by the packet forwarding module to determine the routes.

The framework architecture that we have described is for a flat architecture assuming that every node has the same position and function. But in other kinds of network architecture (e.g., cluster), some special nodes (e.g., the master node in a cluster) may have more components than other nodes do. It is also possible that the functions of several components may be combined when necessary.

28.5 Evaluation of FQMM

We adopt the "from simple to complex" policy in evaluating FQMM. The framework architecture presented in the previous section is to realize the hybrid QoS provisioning policy of FQMM described in Section 28.3.2. Not all the components in the framework are needed for different QoS provisioning. Coarse granularity QoS provisioning requires fewer components than fine granularity QoS provisioning does.

28.5.1 Service Prioritization in MANET

The first evaluation is done by investigating service prioritization in MANETs. Service prioritization involves giving resource access priority to certain traffic types and has been used in networks for a long time. For example, network administration traffic typically has higher priority over normal user traffic, voice services should have higher priority than data services because of their real-time requirements, or web surfing can have higher priority than the File Transfer Protocol (FTP) and e-mail applications.

Since data services are normally run over the Transport Control Protocol (TCP) and multimedia services over the User Datagram Protocol (UDP), TCP sessions and UDP sessions are used to represent data services and multimedia services respectively. Three kinds of service profiles are considered, viz., "priority within TCP traffic," "priority within UDP traffic," and "priority between TCP and UDP traffic" [31,32]. We only elaborate on the service profile of "priority within TCP traffic" in this chapter (please see [31] for results on the other two service profiles). This service profile is defined as: *there are two levels of priority for FTP services; one is higher than the other*. This service profile is suitable for situations where some users have higher priority than other users. FTP here is used as the representative service running over TCP.

Realization of the above service profile does not require the function of the resource reservation, the QoS routing protocol, and the admission control modules in the FQMM framework architecture (see Fig. 28.3). But the routing protocol is still needed. In addition, since the traffic profile is constant, the link measurement and link state database module can also be omitted. The modules in the control plane needed to realize the service prioritization are thus the traffic conditioner configuration and the database of traffic control modules. However, all the modules in the data forwarding plane are required. Figure 28.4 shows the modules we implement for service prioritization.

In Fig. 28.4, we adopt two schemes in the module of "packet scheduling and buffer management'" to realize service prioritization. The first one is a priority buffer management scheme similar to the well-known Random Early Drop with In/Out (RIO) [7] scheme. Each packet is marked as HIGH (priority) or LOW (priority) first according to the policies in the service profile. In case of network congestion or the sensitivity of network congestion, the scheme drops LOW packets with a higher probability than HIGH packets. The second one is a priority scheduling scheme where packets of higher priority are scheduled more urgently than packets of lower priority. When a higher priority (HIGH) packet arrives, it is inserted before all the lower priority (LOW) packets in the queue. Thus, the higher priority packets are sent before the lower priority packets. In the case of network congestion, the last packet in the queue will be dropped, i.e., drop the tail of the queue.

We simulate ten nodes in a small MANET in a 500 × 500 square meter area. Each node moves according to the *random waypoint* mobility model. Five mobility patterns are generated randomly with a maximum speed of 20 meters/second, and the pause duration time is varied from 0 to 300 seconds. Eight TCP sessions are generated randomly among the nodes, and these are divided into two groups: group 1 has four TCP sessions with lower priority; group 2 has four TCP sessions with higher priority. We evaluate the performance of TCP traffic under three separate cases according to the following priority cases.

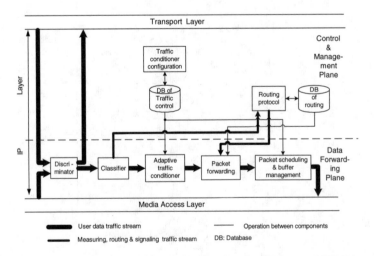

FIGURE 28.4 Components for service prioritization in the framework architecture of FQMM.

Case 1: No priority is given to any packet.

Case 2: Service prioritization is applied using the priority buffer management scheme.

Case 3: Service prioritization is applied using the priority scheduling scheme.

Figure 28.5 shows the average end-to-end delay of TCP packets in group 1 (lower priority) and group 2 (higher priority) in all three cases. The values in the figure are averages of simulations running over the five mobility patterns, and each simulation is run for 900 seconds. Compared with Case 1, Case 2 reduces the average delay of TCP sessions in both group 1 (see Fig. 28.5a) and group 2 (see Fig. 28.5b). However, Case 3 increases the delay of sessions in group 1 (see Fig. 28.5a) and reduces the delay of sessions in group 2 (see Fig. 28.5b). Furthermore, Case 3 achieves the least delay in group 2 among the three cases (see Fig. 28.5b). Thus, combining Figs. 28.5a and 28.5b, we observe that in terms of the average packet end-to-end delay, both Cases 2 and 3 give obvious priority to the higher priority packets, and Case 3 achieves a higher degree of prioritization than Case 2.

This result can be explained by the way that the priority buffer management and priority scheduling schemes arrange packets in the buffer. The priority buffer management scheme does not change the order between the higher priority packets and the lower priority packets. Therefore the higher priority packets and the lower priority packets have the same chance to access the output channel. As a result, the average delays for the higher priority and the lower priority packets are about the same. On the other hand, the priority scheduling scheme keeps the lower priority packets in the buffer longer than the higher priority packets by inserting the higher priority packets before the lower priority packets. Therefore, the average delay of the higher priority packets is lower than that of the lower priority packets.

Extensive simulations have been run [31], which lead to some major conclusions here. First of all, for most of the cases, the priority buffer management scheme and the priority scheduling scheme work well to realize service prioritization according to different service profiles. Therefore, the features of MANET, particularly its mobility and lack of infrastructure, do not affect the execution of services prioritization.

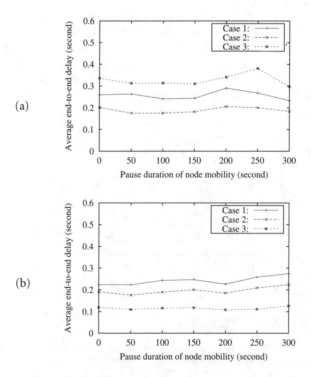

FIGURE 28.5 Average delay of TCP packets under the three cases: (a) Average delay of TCP packets in group 1 (lower priority); (b) Average delay of TCP packets in group 2 (higher priority).

We also compare the performance results of the two priority schemes under the three service profiles. For prioritization within TCP traffic, the priority schemes work well together with the congestion and flow control protocol of TCP. Simulation results show that the priority scheduling scheme outperforms the priority buffer management scheme in terms of the average throughput of TCP sessions and the average end-to-end delay of TCP packets. For prioritization within UDP traffic and prioritization between UDP and TCP traffic, UDP traffic does not respond to traffic congestion and routing failure. When the network congestion is enforced, both the priority buffer management scheme and the priority scheduling scheme work well, and the former scheme outperforms the latter in terms of the average goodput and loss rate of UDP sessions. However, when the network is not congested, service prioritization with UDP traffic is not achieved due to the overwhelming effects of routing failure. To our understanding, this implies that the routing issue is one of the basic issues to be dealt with in MANETs.

28.5.2 Service Differentiation in MANET

The second evaluation is done together with the investigation of service differentiation in MANETs. Service differentiation means that traffic is differentiated into a set of classes, and network nodes provide priority-based treatment based on these classes without explicit resource reservation. Because of its scalability and relative simplicity, service differentiation is one of the major approaches to provide QoS in networks, e.g., DiffServ for the Internet. Service differentiation can be further categorized into absolute service differentiation and relative service differentiation. In absolute service differentiation, an admitted user/class is assured of the requested performance level, such as packet delay, jitter, throughput, and loss rate. The disadvantage is that a user will be rejected if the required resources are not available and the network cannot provide the requested assurances. In relative service differentiation, on the other hand, the assurance from the network is no longer an absolute metric to one class, but the relative relationship between/among classes.

We first study the performance of TCP with AS [7] in MANETs with absolute bandwidth differentiation, where the service profile is defined as an absolute target rate of TCP sessions. Simulation results [34] show that service differentiation in MANETs is possible but the differentiation is not consistent, i.e., the required absolute target rate fails to remain the same over time. This is due to the fact that network conditions in a MANET, e.g., mobility, topology, and the link capacity a node sees when it wants to send packets, change dynamically. The absolute differentiation scheme does not adapt to the dynamic network conditions in MANETs, therefore resulting in the differentiation inconsistency.

A relative differentiation scheme for MANETs is proposed and evaluated in FQMM. The service profile is defined as a relative target rate, which is a percentage of the effective link capacity and ranges between 0 and 1. The effective link capacity is the available bandwidth a node can use to send out packets. It depends on many factors in MANETs, such as power constraints, node mobility, dynamic topology, collision, routing, and traffic load.

To realize the relative service differentiation in the FQMM framework (shown in Fig. 28.3), the admission control module, the QoS routing and database of QoS routing modules, and the resource reservation protocol modules are not required. However, the link measurement and the database of link state modules are required to get the effective link capacity. The components to realize TCP with AS using relative service differentiation are shown in Fig. 28.6.

In Fig. 28.6, the packet scheduling and buffer management module adopts the RIO [7] scheme. The traffic conditioner module is realized by a token bucket profile meter to mark each packet as IN/OUT. The parameters of the token bucket, i.e., token rate ρ, and bucket length, β, are adaptively adjusted by the traffic conditioner configuration module. Parameters of the token bucket profile meter of a certain session i are as follows:

$$\rho_i = \gamma_i \times C_t \times R \tag{28.1}$$

$$\beta_i = \gamma_i \times C_t \times L \tag{28.2}$$

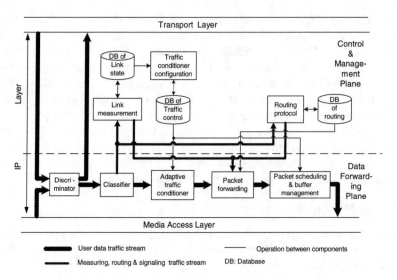

FIGURE 28.6   Components for service differentiation in the framework architecture of FQMM.

where γ_i is the relative target rate of session i, ρ_i is the token generation rate, β_i is the token bucket length, C_t is the effective link capacity, and R and L are constants.

The link measurement module is in charge of the determination of the effective link capacity C_t, which is not an easy task. A simple case study using a *string* topology is presented in [34]. There, since nodes' movement is restricted to along the *string* (see Fig. 28.7), the effects of both mobility and routing can be ignored. It is further assumed that one-way TCP sessions are only generated between the two terminal nodes of the *string*, and power constraint has no effect. TCP reacts to the loss and congestion although it assumes packet loss due to transmission error of the wireless link as congestion. From previous simulation results, it is found that the total TCP throughput changes with the number of hops that TCP sessions cross. The main factor that affects the total throughput is therefore the number of hops that the TCP sessions cross. Therefore, $C_t=N_h$, where N_h is the total TCP throughput when TCP sessions cross h hops.

We simulate using six nodes to form a *string* topology. Four TCP sessions are generated from node 0 to node 5. The four sessions are divided into two groups and each group has two sessions. Sessions in group 1 have a relative target rate of 0.4 and sessions in group 2 have a relative target rate of 0.1. Thus, the four sessions together have a total target rate of 100% of the available bandwidth. All the nodes are static during the simulation except node 0, which moves along the *string*. Simulations are run for 900 seconds. Once the simulation is started, node 0 moves towards node 5 at a constant rate, then turns back towards its original position when it meets node 5. As a result, the number of hops that the TCP sessions travel varies with the simulation time as shown in Fig. 28.8a. Node 0 adapts the token bucket profiler according to the number of hops between itself and node 5. For simplicity, the adaptive token bucket profiler is inserted into the scenario file, which is used in the *NS* simulator. TCP throughput of each session is measured every 10 seconds during the simulation.

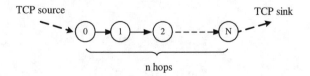

FIGURE 28.7   String topology.

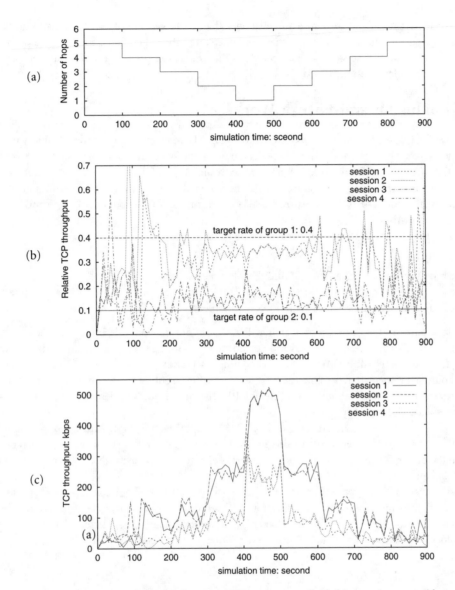

FIGURE 28.8 Relative service differentiation. Four TCP sessions are divided into two groups with target rates of 0.4 and 0.1, respectively. Simulation runs for 900 seconds. (a) Number of hops varies with simulation time; (b) Relative TCP throughput; (c) Absolute TCP throughput.

Figure 28.8b presents the relative TCP throughput of each session. When TCP sessions cross one hop (from 400s to 500s) and two hops (from 300s to 400s and from 500s to 600s), the TCP throughput of sessions in group 1 and group 2 are clearly differentiated. In addition, the achievable throughput of each session is stable and does not vary much. However, when TCP sessions cross three hops (from 300s to 400s and from 600s to 700s) and four hops (from 200s to 300s and from 700s to 800s), the throughput fluctuates much more. Even worse is that the throughputs of the two groups cannot be differentiated when TCP sessions cross five hops (from 0s to 100s and from 800s to 900s). This performance degradation is caused by the fact that the longer the TCP session travels, the higher the probability that the TCP packets and ACKs get dropped and collide with each other along the path. These factors make it difficult for the traffic conditioner and RIO to control the TCP throughput differentiation.

As a whole, Fig. 28.8b shows that the throughput of sessions belonging to the two groups is differentiated, and the differentiation is consistent with the number of hops varying from one up to four. Sessions in group 1 achieve throughput around a value smaller than its target rate of 0.4. Sessions in group 2 achieve throughput around a value larger than its target rate of 0.1. Fig. 28.8c presents the absolute throughput of the four TCP sessions. It again validates that the service profile cannot be kept consistent if absolute service differentiation is used.

28.6 Conclusion and Future Work

We have described FQMM, a Flexible QoS Model for MANETs, after investigating the suitability of the Internet IntServ and DiffServ models for MANETs. To our knowledge, this is the first such QoS model proposed for MANETs. A framework architecture to realize FQMM has also been presented. Each component in the framework architecture is described, and the interaction between components is discussed. Two evaluations of FQMM are presented, i.e., performance of service prioritization and service differentiation in MANETs.

We say that the QoS model is *flexible* because of its features:

1. Nodes have dynamic roles.
2. The provisioning policies are hybrid and flexible.
3. The components of the framework to realize FQMM can be combined in a flexible manner to meet different network situations and QoS requirements.

To realize the hybrid QoS provisioning policy, FQMM should first provide integrated service. The supporting modules such as resource reservation protocol, QoS routing protocol, and admission control will be implemented in the future. We may use the existing reservation protocols such as INSIGNIA [19]. The QoS routing protocol may extend the existing best effort routing protocols that adopt a flat-architecture, such as Dynamic Source Routing (DSR) [18] or Ad-hoc On-demand Distance Vector (AODV) routing [24], to include QoS options.

After implementing the integrated services in FQMM, we can realize the hybrid QoS provisioning policy. We will compare the performance of the hybrid provisioning policy with the pure integrated service and the pure differentiated service policy in terms of system complexity, flexibility, and ability to provide QoS requirements.

One missing piece of QoS support in FQMM is the QoS MAC protocol. The IEEE 802.11 MAC protocol has been used in the simulations. However, this is a MAC protocol without any QoS option. As the MAC protocol is blind to the higher layer QoS mechanisms, the whole system is less efficient than is expected from a protocol that is QoS aware. In the future, we will define a QoS MAC protocol for FQMM.

References

[1] Alwan, A., Bagrodia, R., Bambos, N., Gerla, M., Klenrock, L., Short, J., and Villasenor, J., Adaptive Mobile Multimedia Networks, *IEEE Personal Communications*, Apr. 1996, pp. 34–51.

[2] Black, D., Brim, S., Carpenter, B., and Faucheur, F.L., Per Hop Behavior Identification Codes, Internet IETF RFC3140, June 2001.

[3] Blake, S., An Architecture for Differentiated Services, Internet IETF RFC2475, Dec. 1998.

[4] Braden, R., Clark, D., and Shenker, S., Integrated Services in the Internet Architecture — An Overview, IETF RFC1633, June 1994.

[5] Chen, S. and Nahrstedt, K., Distributed quality-of-service routing in ad hoc networks, *IEEE Journal on Selected Areas in Communications*, 17, 1488–1505, 1999.

[6] Chen, T., Gerla, M., and Tsai, J.T., QoS Routing Performance in a Multi-Hop Wireless Network, *Proceedings of IEEE ICUPC '97*, 1997.

[7] Clark, D.D. and Fang, W., Explicit allocation of best-effort packet delivery service, *IEEE/ACM Transactions on Networking*, 6, 362–373, Aug. 1998.

[8] Corson, M.S., Issues in Supporting Quality of Service in Mobile Ad Hoc Networks, *IFIP 5th Int. Workshop on Quality of Service — (IWQOS '97)*, May 1997.

[9] Corson, M.S. and Campbell, A.T., Towards Supporting Quality of Service in Mobile Ad Hoc Networks, *Proceedings of the First Conference on Open Architecture and Network Programming*, San Francisco, Apr. 1998.

[10] Crawley, E., Nair, R. Rajagopalan, B., and Sandick, H., A Framework for QoS-Based Routing in the Internet, Internet IETF RFC2386, Aug. 1998.

[11] Crowcroft, J. and Wang, Z., A Rough Comparison of the IETF and ATM Service Models, *IEEE Network*, Nov. 1995, pp. 12–16.

[12] Floyd, S. and Jacobson, V., Link-sharing and resource management models for packet networks, *IEEE/ACM Transactions on Networking*, 3, 365–386, Aug. 1995.

[13] QoS Forum, QoS Protocols and Architectures, White Paper of QoS Forum, July 1999, http://www.qosforum.com.

[14] Gevros, P., Crowcroft, J., Kirstein, P., and Bhatti, S., Congestion Control Mechanisms and Best Effort Service Model, *IEEE Networks*, May 2001, 16–26.

[15] Hsu, Y.C., Tsai, T.C., and Lin, Y.D., QoS Routing in Multihop Packet Radio Environment, *Proceedings of IEEE ISCC '98*, 1998, pp. 582–586.

[16] Ibanez, J. and Nichols, K., Preliminary Simulation Evaluation of an Assured Service, IETF Internet Draft, work in progress, Aug. 1998.

[17] Iwata, A., Chiang, C.C., Yu, G., Gerla, M., and Chen, T.W., Scalable routing strategies for ad hoc wireless networks, *IEEE Journal on Selected Areas in Communications*, 17, 1369–1379, 1999.

[18] Johnson, D.B., Routing in Ad Hoc Networks of Mobile Hosts, *IEEE Workshop on Mobile Computing Systems and Applications*, Dec. 1994, pp. 158–163.

[19] Lee, S.B. and Campbell, A.T., INSIGNIA: In-band Signaling Support for QoS in Mobile Ad Hoc Networks, *5th Int. Workshop on Mobile Multimedia Comm. (MoMuc'98)*, Berlin, Oct. 1998.

[20] Lin, C.R. and Gerla, M., Asynchronous Multimedia Multihop Wireless Networks, *Proceedings of IEEE INFOCOM '97*, Kobe, Japan, Apr. 1997, pp. 118–125.

[21] Lin, C.R. and Liu, J.S., QoS routing in ad hoc wireless networks, *IEEE Journal on Selected Areas in Communications*, 17, 1426–1438, 1999.

[22] Nichols, K., Blake, S., Baker, F., and Black, D., Definition of the differentiated services field (DS field) in the IPv4 and IPv6 headers, *Internet IETF RFC2474*, Dec. 1998.

[23] Perkins, C.E. and Bhagwat, P., Highly dynamic destination-sequenced distance-vector routing (DSDV) for mobile computers, *Proceedings of ACM SIGCOMM '94*, London, Aug. 1994, pp. 234–244.

[24] Perkins, C.E. and Royer, E.M., Ad Hoc On-Demand Distance Vector Routing, *IEEE WMCSA '99*, New Orleans, 1999, pp. 90–100.

[25] Ramanathan, R. and Streenstrup, M., Hierarchically organized multihop mobile wireless networks for quality of service support, *Mobile Networks and Applications*, 3, 101–119, 1998.

[26] Sinha, P., Sivakumar, R. and Bharghavan, V., CEDAR: A Core-Extraction Distributed Ad Hoc Routing Algorithm, *Proceedings of IEEE INFOCOM '99*, New York, May 1999, pp. 202–209.

[27] Tang, Z. and Garcia-Luna-Aceves, J.J., Hop-reservation Multiple Access (HRMA) for Ad Hoc Networks, *Proceedings of IEEE INFOCOM '99*, 1999, pp. 194–201.

[28] Tang, Z. and Garcia-Luna-Aceves., J.J., A Protocol for Topology Dependent Transmission Scheduling in Wireless Networks, *Proceedings of WCNC '99*, 1999, pp. 1333–1337.

[29] Terry, D.B., Towards a Quality of Service Model for Replicated Data Access, *The Second International Workshop on Services in Distributed and Networked Environments*, 1995, pp. 118–121.

[30] Tsai, J. and Gerla, M., Multicluster mobile multimedia radio network, *ACM-Baltzer Journal of Wireless Networks*, 1, 255–265, 1995.

[31] Xiao, H., A Flexible Quality of Service Model for Mobile Ad Hoc Networks, Ph.D thesis, National University of Singapore, Mar. 2002.

[32] Xiao, H., Chua, K.C., Seah, K.G., and Lo, A., On Service Prioritization in Mobile Ad Hoc Networks, *IEEE ICC 2001*, Helsinki, June 2001.

[33] Xiao, H., Seah, K.G., Lo, A. and Chua, K.C., A Flexible Quality of Service Model for Mobile Ad-hoc Network, *Proceedings of IEEE VTC2000 — Spring*, Tokyo, May 2000.

[34] Xiao, H., Seah, K.G., Lo, A., and Chua, K.C., On service differentiation in multihop wireless networks, *ITC Specialist Seminar on Mobile Systems and Mobility*, Lillehammer, Norway, Mar. 2000, pp. 1–12.

[35] Zhang, L., Deering, S., Estrin, D, Shenker, S., and Zappala, D., RSVP: A New Resource ReSerVation Protocol, *IEEE Network*, Sep. 1993, pp. 8–18.

[36] Zhu, C. and Corson, M.S., A Five-Phase Reservation Protocol (FPRP) for Mobile Ad Hoc Networks, *Proceedings of IEEE INFOCOM '98*, San Francisco, CA, 1998, pp. 322–331.

29

Scheduling of Broadcasts in Multihop Wireless Networks

Jang-Ping Sheu
National Central University

Pei-Kai Hung
National Central University

Chih-Shun Hsu
National Central University

Abstract

Broadcast is an important operation in wireless networks. However, broadcasting by naïve flooding causes severe contention, collision, and congestion, which is called the broadcast storm problem. Many protocols have been proposed to solve this problem, with some investigations focusing on collision avoidance yet neglecting the reduction of redundant rebroadcasts and broadcasting latency; while other studies have focused on reducing redundant rebroadcasts yet have paid little attention to collision avoidance. Two one-to-all broadcast protocols based on two schemes are proposed herein. The set-covering scheme reduces redundant rebroadcasts, and the independent-transmission-set scheme avoids collisions and reduces latency. Furthermore, an all-to-all broadcast protocol is presented based on the one-to-all protocol. Simulation results show that the novel broadcast protocols are efficient and can achieve high reachability.

29.1 Introduction

Due to advances in wireless communication technology and portable devices, wireless communication systems have recently become increasingly widespread. A wireless network is a collection of hosts that communicate with each other via radio channels. The hosts can be static, such as base stations in packet radio networks, or mobile, such as notebook computers in mobile ad hoc networks (*MANETs*). If all hosts in a wireless network can communicate with each other directly, the network is single hop; otherwise it is multihop. Broadcast is an important operation in all kinds of networks. However, due to the limited transmission range and bandwidth of a radio channel, the broadcast protocol must be designed carefully to avoid severe contention, collision, and congestion, known as the broadcast storm problem [10].

Broadcast problems in wireless networks have been studied extensively in the literature [1–3,5,7–16]. References [1,5,7–9,12,13,15] attempt to design collision-free neighboring broadcast protocols and model the broadcast scheduling problem as a graph-coloring problem. The colors in the graphs represent the channels assigned to hosts (which could be slots, frequencies, or codes). Since no host can be assigned the same color (channel) as any of its neighbors within two hops, collisions can be avoided. The research goal of the protocols presented in [12,13] is to minimize the assigned colors (channels); while the protocols presented in [5,15] aim to increase total throughput. Using a different approach, two collision-free protocols for one-to-all broadcast are proposed in [2], one centralized and the other distributed. In the centralized scheme, the source host schedules the transmission sequence using knowledge of global network topology. Unlike the graph-coloring problem, a host can simultaneously use the same channel as its neighbors within two hops, as long as no collision occurs among the receiving hosts. However, in the distributed scheme, the source host follows the depth first search tree to pass a token to every host in the network. After receiving the token, the host realizes which time slots are collision free and can then decide its transmission sequence based on this information.

The above broadcast protocols aim to alleviate the broadcast storm problem: some works [1,2,5,7–9,12,13,15] focus on avoiding collisions but pay little attention to reducing redundant rebroadcasts and broadcasting latency; other researchers [10,11,16] try to reduce redundant rebroadcasts but cannot guarantee a collision-free broadcast. Here, we propose two efficient one-to-all broadcast protocols, one for low mobility packet radio networks and one for high mobility *MANETs*. Both protocols use the set-covering scheme to reduce redundant rebroadcasts and the independent-transmission-set (IT-set) scheme to avoid collision and reduce broadcasting latency. The broadcast protocol for packet radio networks is efficient and collision free but depends on gathering the global network topology. While the broadcast protocol for *MANETs* is less efficient and cannot guarantee collision-free broadcasts, this protocol requires only information on two-hop neighbors. Additionally, an all-to-all broadcast protocol based on the one-to-all protocol is presented herein. Simulation results show that the broadcast protocols presented herein are more efficient than other broadcast protocols in terms of reachability, rebroadcast ratio, and broadcasting latency.

The rest of this chapter is organized as follow. Section 29.2 describes the system model and introduces two schemes on which the proposed protocols are based. Section 29.3 then proposes two one-to-all broadcast protocols, while Section 29.4 presents an all-to-all broadcast protocol. Section 29.5 shows simulation results. Conclusions are presented in Section 29.6.

29.2 Preliminary Information

The network is modeled as a graph $G = (V, E)$, where V represents the set of nodes (hosts) in the network, and E is the set of links. If $u, v \in V$, then an edge $e = (u,v) \in E$ exists if and only if u is in the transmission range of v and vice versa (assuming that all links in the graph are bidirectional, i.e., if u is in the transmission range of v, v is also in the transmission range of u). The length of the broadcast packet is fixed. While broadcasting, no carrier sense and collision avoidance procedure is executed before transmission. The network is assumed to be connected. If it is partitioned, each component is treated as an independent network.

29.2.1 Set-Covering Scheme

An example of the set-covering scheme is presented before describing it. In Fig. 29.1, broadcast source s has information on its one-hop neighbors (a, b, c, d, e) and two-hop neighbors (f, g, h, i, j, k, l). When node s broadcasts a packet m, the packet will be received by all of its one-hop neighbors. However, it is not necessary for all one-hop neighbors to forward packet m, since if only nodes $a, c,$ and e forward m they can still cover all the two-hop neighbors of s. Set covering is designed to choose the minimum set

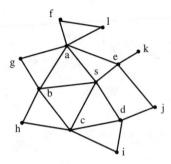

FIGURE 29.1 An example of a set-covering problem.

of one-hop neighbors of source *s* required to cover all the two-hop neighbors of *s*. The nodes in the minimal set serve as the relay nodes of *s*, thus reducing redundant rebroadcasts.

The set-covering problem is defined as follows. Let (*s*, α, β) represent an instance of the set-covering problem, where *s* denotes the source node, α represents a set of *s*'s one-hop neighbors, and β is a set of *s*'s two-hop neighbors. The set β can be represented as follows:

$$\beta = \bigcup_{h \in \alpha} N(h)$$

where node $h \in \alpha$ is *s*'s one-hop neighbor and $N(h) = \{n \mid n \not\subset \alpha \text{ and } e(n,h) \in E\}$.

The set-covering scheme aims to find a minimal set *C* such that the neighbors of the nodes in this set can cover all nodes in β. In Fig. 29.1, the minimal set *C* = {*a, c, e*} because the neighbors of nodes in {*a, c, e*} contain all nodes in β. Since the set-covering problem, which can be reduced to the well-known vertex-cover problem, is NP-hard [4], a greedy algorithm proposed in [4] is used to solve this problem. The greedy algorithm works by picking up a node that covers most of the remaining uncovered elements in β at each iteration step. The greedy algorithm is shown as follows:

Algorithm: Greedy-Set-Cover(α,β)
Input:
    α: a set of *s*'s one-hop neighbors, and
    β: a set of *s*'s two-hop neighbors.
Output: *Set_Cover*: a set of nodes whose neighbors cover set β.
Begin
    Set_Cover = {};
    while β is not *empty*
        Select a node $h \in \alpha$ that maximizes the size of $|N(h) \cap \beta|$;
        Remove nodes $\in N(h)$ from β;
        Add *h* to *Set_Cover*;
    end while
    return *Set_Cover*;
End

29.2.2 Independent-Transmission-Set (IT-Set) Scheme

Consider a three-layer graph as shown in Fig. 29.2, which is derived from Fig. 29.1, where the node in layer 0 is the source node *s*, and the nodes in layers 1 and 2 are the one-hop and two-hop neighbors of the source node, respectively. The set {*a, c, e*} in layer 1 is a minimal set covering all nodes in layer 2. Since nodes *a, c,* and *e* have no common neighbors in layer 2, they can broadcast packets simultaneously without any collision occurring in layer 2. Consider another case in Fig. 29.3. The set {*b, a, c, d*} is a minimal set covering the nodes in layer 2. However, nodes *a* and *b* have a common neighbor *g* in layer

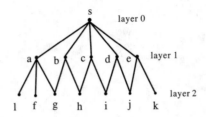

FIGURE 29.2 A three-layer graph derived from Figure 29.1.

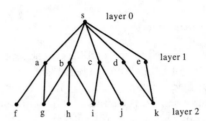

FIGURE 29.3 An example of the IT-set problem.

2, and nodes b and c also have a common neighbor i in layer 2. When nodes a, b, and c transmit a packet simultaneously, collisions will occur at nodes g and i. Consequently, the independent-transmission-set (IT-set) scheme is used to arrange the transmission sequences of the nodes in layer 1 to avoid packet collisions in layer 2.

The IT-set scheme works as follows. Given a three-layer graph, the set-covering scheme is used to find the minimal covering set, *Set_Cover*. The nodes in *Set_Cover* are then taken individually to decide their transmission sequence. When the first node is taken, it forms a new independent-transmission (IT) set, and the subsequent nodes are then individually taken from *Set_Cover* and compared with the nodes in the existing IT sets. If layer 2 contains no common neighbor between the taken node and the nodes in a compared IT set, the node joins the IT set. Otherwise, the node forms a new IT set. Once a new node joins an IT set, the IT sets are sorted in descending order by the number of nodes in layer 2 covered by the nodes in the IT sets. The IT sets with larger coverage numbers can be compared earlier and have a greater chance of including more nodes in layer 1. Since the nodes in the same IT set share no common neighbors in layer 2, they can broadcast simultaneously. Therefore, the IT-set scheme aims to minimize the number of IT sets and reduce broadcasting latency.

For example, executing the IT-set scheme in Fig. 29.3 obtains two IT sets $\{a, c, d\}$ and $\{b\}$. Initially, the first node, b, of *Set_Cover* is taken. Node b forms a new set, $\{b\}$, and node a is then compared to node b. Meanwhile, since nodes a and b share a common neighbor g in layer 2, node a forms a new set, $\{a\}$. Node c is then taken and compared with the two IT sets, $\{b\}$ and $\{a\}$. Nodes c and b have a common neighbor in layer 2, but nodes c and a have no common neighbor in layer 2. Therefore, node c joins set $\{a\}$, and the IT sets are sorted according to their coverage of nodes in layer 2. The two ordered sets are $\{a, c\}$ and $\{b\}$. Next, node d is taken and compared to the nodes in set $\{a, c\}$. Since nodes a, c, and d share no common neighbors in layer 2, node d joins set $\{a, c\}$. Finally, *Set_Cover* is empty, and two ordered IT sets are obtained, $\{a, c, d\}$ and $\{b\}$.

Before presenting our IT-set scheme, we define the following notations. Let *ITS(i)* denote the i-th IT set. A node in *ITS(i)* will delay the forwarding packets for i-1 *time_units* after it receives the broadcast packet, where a *time_unit* is the time spent to forward a broadcast packet. Let *Coverage_of_ITS(i)* represent the number of nodes in layer 2 covered by nodes in *ITS(i)*. *Num_of_ITS* is the number of IT sets.

The scheme is as follows:

Algorithm: Find-ITS(*Set_Cover*)
Input: *Set_Cover*: A set of nodes output from the set-covering scheme.

Output: *ITS()*: An array of IT set.

Initial: *Num_of_ITS* = 0;

Step 1:

    if *Set_Cover* = {} **then** *Return ITS()*;

    else

        Let *f* be the first node listed in *Set_Cover*;

        for *i* = 1 **to** *Num_of_ITS*

            if *f* and every node in *ITS(i)* have no common neighbor in layer 2 **then**

                Add *f* to *ITS(i)*;

                goto Step 3;

            end if

        end for

    end if

Step 2:

    Num_of_ITS = *Num_of_ITS* + 1;

    Create a new set *ITS(Num_of_ITS)*;

    Add *f* to *ITS(Num_of_ITS)*;

Step 3:

    Remove *f* from *Set_Cover*;

    for *i* = 1 **to** *Num_of_ITS*

        Calculate *Coverage_of_ITS(i)*;

    end for

    Sort *ITS()* in descending order according to the value of *Coverage_of_ITS()*;

    goto Step 1;

29.3 Two Broadcast Protocols

This section presents two broadcast protocols designed for different network assumptions. In a packet radio network with low mobility, each node is assumed to know the network topology, and *protocol 1* is proposed to schedule the broadcast by assigning each node a collision-free transmission sequence. In a *MANET* with high mobility, nodes are assumed only to have information about their two-hop neighbors; *protocol 2* is proposed to broadcast packets rapidly and efficiently.

29.3.1 Protocol 1

A collision-free one-to-all broadcast protocol (*protocol 1*) is proposed for a packet radio network in which the network topology is known in advance. Using global information, each node can run *protocol 1* to determine its transmission sequence and calculate waiting time, and thus each node knows when to forward the received packet to avoid collisions. The assignment of the transmission sequence for each node originates from the source node. The source node is first assigned to transmission sequence "0" and its one-hop and two-hop neighbors are then put in sets α and β, respectively. The set-covering scheme is applied in each round to pick the minimal relay nodes in α, and the IT-set scheme is used to divide the minimal relay nodes into different transmission sets. After applying the IT-set scheme, nodes in *ITS(1)*, which has the largest coverage number in the IT sets, *ITS()*, can transmit immediately when they have received the broadcast packet. Accordingly, nodes in *ITS(1)* are scheduled and assigned to transmission sequence "1", while the nodes in the other IT sets are put into set U and considered in the next round. Before executing the next round, the nodes in α and β must be updated as follows. The nodes in *U* and the unscheduled nodes connected to the scheduled nodes in *ITS(1)* are put into α, and the unscheduled nodes connected to nodes in α are put into β. In the next round, the set-covering scheme and IT-set scheme are repeated. Nodes in *ITS(1)* are assigned to the following transmission sequence,

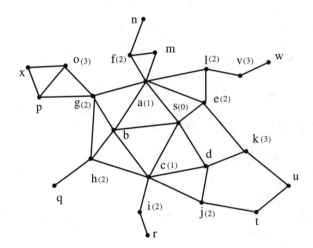

FIGURE 29.4 The results of running *protocol 1*. The number in the parentheses indicates the transmission sequence of a node. Nodes not assigned numbers are end-points.

"2", while nodes in other IT sets are assigned to set *U*. The above procedure is repeated until all the nodes are scheduled and assigned to a dedicated transmission sequence.

When the source node begins to transmit a broadcast packet, its *ID* is recorded in the packet header. Meanwhile, when a node receives a broadcast packet, it performs *protocol 1* to obtain its transmission sequence. Note that a node without a transmission sequence is an end-point and does not need to forward the broadcast packet. For example, in Fig. 29.4, nodes *a*, *b*, *c*, *d* and *e* receive the broadcast packet form node *s*. The transmission sequence of node *s* is "0," while that of node *a* is "1." Consequently, node *a* will forward the broadcast packet as soon as it is received. Meanwhile, node *b* finds it is not assigned a transmission sequence and thus will not forward the packet. The transmission sequence of node *e* is "2," and thus node *e* should wait for a *time_unit* to forward the packet, where a *time_unit* is the time spent to forward a broadcast packet. On receiving the broadcast packet with the source *ID*, each node can perform *protocol 1* to obtain its transmission sequence and set the *Waiting_Timer*. Once the *Waiting_Timer* expires, the node forwards the broadcast packet to its neighbors.

Protocol 1:

Initial:
    S = 1; (*S* is the transmission sequence of a node.)
    α = the set of one-hop neighbors of source node *s*;
    β = the set of two-hop neighbors of source node *s*;
While α and β are not empty
    Step 1: Call Greedy-Set-Cover(α, β) to find the minimal set *Set_Cover*, which covers all nodes in β.
    Step 2: Call Find-ITS(*Set_Cover*) to find *ITS()*. Assign nodes in the maximal covering set *ITS(1)* with transmission sequence *S* and put the nodes in the other sets of *ITS()* into set *U*, *S* = *S* + 1.
    Step 3: The nodes in *U* and the unscheduled nodes connected to nodes in set *ITS(1)* are put into α. Then the unscheduled nodes connected to nodes in α are put into β.
End While

A node *h* will take one of the following actions when it receives a broadcast packet from node *u*:

A1: If node *h* is an end-point, it will not need to forward the packet.
A2: If the packet has already been received, it will be discarded.
A3: If node *h* is receiving the packet for the first time and node *h* is not an end-point, it sets *Waiting_Timer* = (the transmission sequence of *h* – the transmission sequence of *u* – 1) × *time_unit*. Once *Waiting_Timer* expires, node *h* forwards the packet.

29.3.2 Protocol 2

A one-to-all broadcast protocol for a *MANET* (*protocol 2*) whose nodes have information on their two-hop neighbors is proposed here. In *protocol 2*, each sender uses the set-covering scheme to select the minimum number of nodes required for relay from its one-hop neighbors, and uses the IT-set scheme to assign each relay node a proper transmission sequence. Though some collisions may occur among the receiving nodes, simulation results in Section 29.5 show that *protocol 2* has a low rebroadcast ratio and acceptable reachability. The following example illustrates how *protocol 2* works.

In Fig. 29.5, source node *s* assigns a transmission sequence to its one-hop neighbors; nodes *a* and *c* are assigned to transmission sequence "1", while node *e* is assigned to transmission sequence "2." The scheduled results are recorded in the broadcast packet header, and the packet is then forwarded to the one-hop neighbors. Figure 29.6 presents the format of the broadcast packet used in *protocol 2*. In the packet header, the *ID* and *TS_my* fields represent the identity and transmission sequence of the sender, respectively. Meanwhile, the *Forwarding_List* (*FL*) records the *ID*s and transmission sequences of the relay nodes, and if a node is scheduled as an end-point it is not included in the *FL*.

After receiving the broadcast packet from node *s*, nodes *a*, *c*, and *e* in *FL* will act as node *s* to schedule the transmission sequence of their one-hop neighbors. Notably, node *e* will delay a *time_unit* interval before rebroadcasting. When nodes *a*, *c*, and *e* begin to schedule the transmission sequences of their one-hop neighbors, they should update their sets of one-hop and two-hop neighbors to prevent the scheduling of already scheduled nodes. For example, when node *a* receives the broadcast packet from node *s*, it updates its sets of one-hop neighbors α and two-hop neighbors β by applying the following rules:

R1: Let *X* denote a set of nodes including sender *s* and its one-hop neighbors. Removes nodes ∈ *X* from α.

R2: Let *Y* represent a set of nodes including the one-hop neighbors of *a* and ∈ *FL*. For example, in Fig. 29.5, *Y* = {*e*} for node *a*. Remove nodes in *X* and nodes connected to *Y* from β.

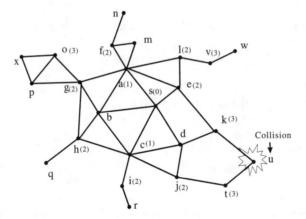

FIGURE 29.5 An example to show how a collision happened in *protocol 2*.

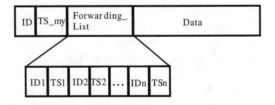

FIGURE 29.6 The format of a broadcast packet in *protocol 2*.

Referring to Fig. 29.5, node a should remove nodes s, b, and e (which are in set X) from α because they have been scheduled. Node a then removes nodes c and d (which are in set X), and k (which is connected to node e in set Y) from β. Node a follows the set-covering and IT-set schemes to assign transmission sequence "2" to nodes f, g, and l. Similarly, node c assigns transmission sequence "2" to nodes h, i, and j, and node e assigns transmission sequence "3" to node k.

In Fig. 29.5, nodes k and t are all assigned to transmission sequence "3." When nodes k and t simultaneously broadcast a packet to next hop, a collision occurs at node u. Since nodes e and j have no way of knowing that nodes k and t have a common neighbor (node u), nodes e and j assign nodes k and t to the same transmission sequence thus causing a collision in node u.

Protocol 2:

Begin
>   When a node h receives a broadcast packet m from node u do
>>   **Case 1:** m is received more than once, discard m;
>>   **Case 2:** receiving m for the first time but h is not in the *FL*, then do nothing;
>>   **Case 3:** receiving m for the first time, and h is in the *FL* do
>>>   **Step 1:**
>>>   Update one-hop neighbors set α and two-hop neighbors set β of h according to rules *R1* and *R2*.
>>>   **Step 2:**
>>>   Call Greedy-Set-Cover(α,β);
>>>   Call Find-ITS(*Set_Cover*);
>>>   **Step 3:**
>>>   **for** $i = 1$ **to** *Num_of_ITS*
>>>>   Assign node a in *ITS(i)* with transmission sequence equals *TS_my* $+ i$;
>>>>   Put the node's *ID* and transmission sequence into *FL*;
>>>   **end for**
>>>   *Waiting_Timer* = (the transmission sequence of h – the transmission sequence of $u - 1$) $\times$ *time_unit*;
>>>   As *Waiting_Timer* is expired, forward m;
End

29.4 All-to-All Broadcast Protocols

An all-to-all broadcast protocol helps all nodes in the network to efficiently broadcast packets to all other nodes. This section presents an efficient all-to-all broadcast protocol in a TDMA-based network. This protocol can also be applied to CDMA- or FDMA-based networks with a slot being replaced by a code or frequency.

The all-to-all broadcast protocol assumes that the broadcast channel consists of frames, with each frame being divided into n slots where n denotes the number of nodes in the network. Figure 29.7 presents the organization of the broadcast channel, where the numbers from one to eight denote the slot numbers in each frame. Meanwhile, the alphabet below each slot number indicates the node that can be transmitted in this slot. Based on the above assumption, two approaches are proposed to solve the all-to-all broadcast problem. A simple approach (termed the *AA1* protocol) assigns each node to a dedicated slot to initialize and forward the broadcast packets. However, numerous packets can be queued in a high degree node because such a node needs to forward the broadcast packets of its neighbors, and the nodes in each frame can transmit only one packet. Even in a well-scheduled network [12,13], where each frame contains fewer slots, the packets are also queued in the high degree nodes. Another approach (termed the *AA2* protocol) is thus proposed, which is source oriented and can efficiently solve the problem. This approach forwards each broadcast packet via the time slot of the source node. Consequently, if a node receives numerous broadcast packets in the previous frame, it can forward them through different time slots in the next

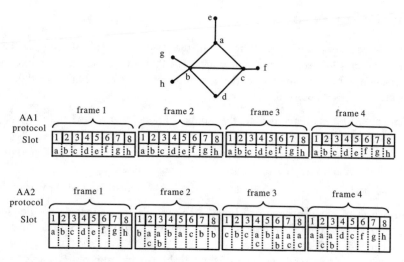

FIGURE 29.7 An example of all-to-all broadcast protocols.

frame. Notably, multiple nodes can use the same time slot for transmission provided that no collision occurs in the receiving nodes. For example, in Fig. 29.7, when running the AA1 protocol, node b receives five packets sent by nodes a, c, d, g, and h in frame 1, and five frames are required to forward the five packets from slot 2. Meanwhile, when using the *AA2* protocol, all five packets received by node b in frame 1 can be forwarded concurrently to slots 1, 3, 4, 7, and 8.

The idea behind the novel protocol is that each node individually performs an independent one-to-all broadcast protocol. Therefore, *protocol 1*, proposed in Section 29.3, can be applied to each node to evaluate the transmission sequences of the received packets. Meanwhile, when a node initiates a broadcast, its *ID* is appended to the broadcast packet and this packet is forwarded via the same slot of each frame. For example, in Fig. 29.7, nodes a, b, c, d, e, f, g, and h all have packets to broadcast to all nodes in the network. First, node a transmits its broadcast packet, m_{a1}, in slot 1 of frame 1, and node a's neighbors (nodes b, c, and e) then receive m_{a1}. Because nodes b and c share a common neighbor d, they cannot forward m_{a1} simultaneously. Applying *protocol 1*, the transmission sequences of nodes b and c for m_{a1} are "1" and "2", respectively. Therefore, node b will transmit m_{a1} in slot 1 of frame 2, and node c will transmit m_{a1} in slot 1 of frame 3. Three frames are required for m_{a1} to be received by all nodes in the network, and node a can then transmit the next broadcast packet, m_{a2}, in frame 4. Node a can transmit a new broadcast packet for every three frames, so the transmission cycle of node a can be defined as 3 (frames) as shown in Fig. 29.8. After node a has transmitted in slot 1, node b successively transmits its broadcast packet, m_{b1}, in slot 2 of frame 1. Nodes a and c can forward m_{b1} in frame 2, and the transmission cycle of node b is 2, as shown in Fig. 29.8. Similarly, nodes c, d, e, f, g, and h can also transmit their packets via their dedicated slots. Each node can continue to initialize a new broadcast packet for every transmission cycle, and the packets will not be queued for long in any node. By applying *protocol 1*, AA2 protocol can not only achieve 100% reachability, but can also efficiently complete the all-to-all broadcast.

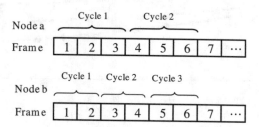

FIGURE 29.8 The transmission cycles of nodes a and b.

29.5 Simulation Results

Simulations are used to evaluate the performance of the proposed broadcast protocols. Each node is assumed to be randomly located in a rectangular area. The following parameters are fixed in the simulations: transmission radius (300 meters), packet size (1 kilobyte), area (1800 m × 1500 m).

The performance metrics observed are:

- *Reachability (RE): r/n*, where *r* denotes the number of nodes that receive the broadcast packet and *n* represents the total number of nodes.
- *Rebroadcast ratio (RR): t/n*, where *t* denotes the number of nodes that actually retransmitted the broadcast packet and *n* represents the total number of nodes.
- *Average latency:* the interval between the initiation and completion of the broadcast packet.

The performances of our two one-to-all broadcast protocols are compared with flooding and the centralized tree-based (*TB*) protocol proposed in [2]. The number of nodes is varied to observe how the protocols perform under different node densities.

Figure 29.9 presents the rebroadcast ratios of the four broadcast protocols. *Protocols 1* and *2*, which use the set-covering scheme to reduce redundant forwarding, have lower rebroadcast ratios than flooding and the *TB* protocols. Furthermore, under high node density, the rebroadcast ratios of the three protocols other than flooding decrease. The rebroadcast ratio of flooding remains close to 100% because each node rebroadcasts the packet it receives one time. However, the rebroadcast ratios of *protocols 1* and *2* decrease with increasing node density because the set-covering scheme can reduce the number of redundant relay nodes.

Figure 29.10 presents the average latency of the four broadcast protocols. *Protocols 1* and *2* finish the broadcast process more quickly than the two other protocols do. On average, *protocols 1* and *2* finish the one-to-all broadcast in just half the time required by flooding. *Protocols 1* and *2* are more efficient than flooding and the *TB* protocol; they use the IT-set scheme to assign each node a proper transmission sequence, avoiding collisions and reducing broadcast latency.

Figure 29.11 presents the reachabilities of the four protocols. *Protocol 1* can achieve 100% reachability because the topology is known in advance and each node is well scheduled to avoid collisions. On the

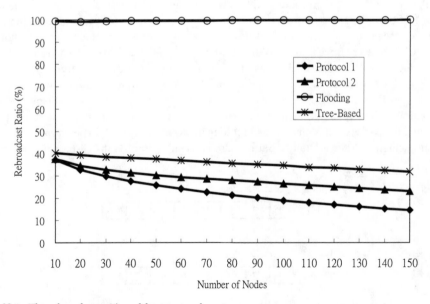

FIGURE 29.9 The rebroadcast ratios of four protocols.

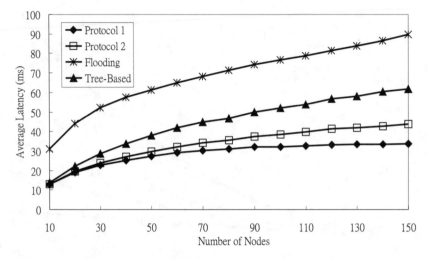

FIGURE 29.10 The average latencies of four protocols.

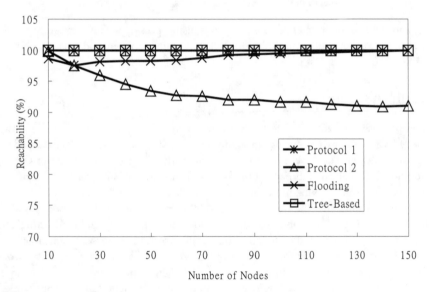

FIGURE 29.11 The reachabilities of four protocols.

other hand, *protocol 2* cannot achieve 100% reachability because hidden terminal problems may cause some collisions. Meanwhile, although high reachability is possible with flooding, the cost is excessive. Finally, the *TB* protocol can achieve 100% reachability because the topology is known in advance, but this protocol is less efficient than the *protocol 1*.

The performance of three all-to-all broadcast protocols is evaluated herein, specifically *AA1* (a simple approach), *AA2* (our approach), and the pure greedy (*PG*) protocols [13]. The *PG* protocol is a collision-free neighboring broadcast protocol in which a vertex is randomly selected and colored in a greedy fashion. This protocol is simple and efficient. The simulation herein only measures the number of time slots needed for all nodes to broadcast their packets to all nodes in the network.

Figure 29.12 shows that the *AA2* protocol performs better than the *AA1* and *PG* protocols. The *AA1* protocol splits each frame into *n* slots, meaning that some slots are wasted when their packet queues become empty. The *PG* protocol, which achieves better scheduling than the *AA1* protocol, has a smaller

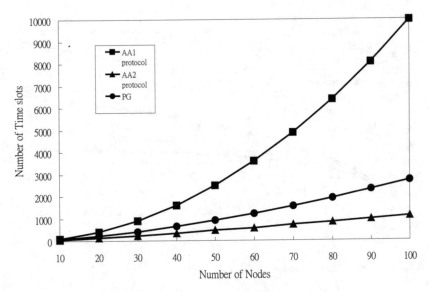

FIGURE 29.12   The average latency of the three all-to-all broadcast protocols.

frame size, and thus each node has more chance to transmit packets within a certain number of time slots. However, the *PG* protocol cannot solve the packet queueing problem, and thus still performs worse than the *AA2* protocol.

29.6 Conclusions

This chapter proposes two efficient one-to-all broadcast protocols, one for packet radio networks and another for *MANETs*. The two protocols can significantly reduce redundant rebroadcasts and thus avoid contention and collision. The novel protocols are particularly effective under conditions of high node density. Additionally, an all-to-all broadcast protocol is proposed based on *protocol 1*; it can prevent the broadcasting packets from queuing in the high degree nodes.

Simulation results show that the two novel one-to-all broadcast protocols can not only significantly reduce the rebroadcast ratio, but can also reduce broadcasting latency. When broadcasting latency and rebroadcast ratio are considered, the two novel one-to-all broadcast protocols perform better than the flooding and *TB* protocols. As for the all-to-all broadcast protocols, the novel *AA2* protocol requires fewer time slots to finish an all-to-all broadcast process than alternative protocols do. The novel one-to-all and all-to-all broadcast protocols developed herein are more efficient than previous works because they focus not only on avoiding collisions but also on reducing redundant rebroadcasts and broadcasting latency.

References

[1]   S. Basagni, D. Bruschi, and I. Chlamtac, A mobility-transparent deterministic broadcast mechanism for ad hoc networks, *IEEE/ACM Transactions on Networking*, 7, 799–807, 1999.

[2]   I. Chlamtac and S. Kutten, Tree-based broadcasting in multihop radio networks, *IEEE Transaction on Computers*, 1209–1225, 1987.

[3]   I. Chlamtac and A. Farago, Making transmission schedules immune to topology changes in multihop packet radio networks, *IEEE/ACM Transactions on Networking*, 2, 23–29, 1994.

[4]   T.H. Cormen, C.E. Leiserson, and R.L. Rivest, *Introduction to Algorithms*, The MIT Press and McGraw-Hill Book Company, New York, 1990.

[5]   A. Ephremides and T.V. Truong, Scheduling broadcasts in multihop radio networks, *IEEE Transactions on Communications*, 38, 456–460, 1990.

 [6] B. Hajek and G. Sasaki, Link scheduling in polynomial time, *IEEE Transactions on Information Theory*, 34, 910–917, 1988.
 [7] M.L. Huson and A. Sen, Broadcast Scheduling Algorithms for Radio Networks, *IEEE Military Communications Conference*, July 1995, pp. 647–651.
 [8] J.H. Ju and V.O.K. Li, An optimal topology-transparent scheduling method in multihop packet radio networks, *IEEE/ACM Transactions on Networking*, 6, 298–306, 1998.
 [9] C.K. Lee, J.E. Burns, and M.H. Ammar, Improved Randomized Broadcast Protocols in Multi-hop Radio Networks, *Proceedings of International Conference on Network Protocols*, Oct. 1993, pp. 234–241.
[10] S.Y. Ni, Y.C. Tseng, Y.S. Chen, and J.P. Sheu, The Broadcast Storm Problem in a Mobile Ad Hoc Network, *Proceedings of the Fifth Annual ACM/IEEE International Conference on Mobile Computing and Networking*, Seattle, WA, Aug. 1999, pp. 151–162.
[11] W. Peng and X.C. Lu, On the Reduction of Broadcast Redundancy in Mobile Ad Hoc Networks, *Proceedings of Mobile and Ad Hoc Networking and Computing*, Boston, MA, 2000, pp. 129–130.
[12] S. Ramanathan and E.L. Lloyd, Scheduling algorithms for multihop radio networks, *IEEE/ACM Transactions on Networking*, 1, 166–177, 1993.
[13] R. Ramaswami and K.K. Parhi, Distributed Scheduling of Broadcasts in a Radio Network, *INFO-COM*, 1989, pp. 497–504.
[14] A. Sen and M.L. Huson, A New Model for Scheduling Packet Radio Networks, *Proceedings of Fifteenth Annual Joint Conference of the IEEE Computer Societies*, Vol. 3, 1996, pp. 1116–1124.
[15] A. Sen and J.M. Capone, Scheduling in Packet Radio Networks — A New Approach, *Proceedings of Global Telecommunications Conference*, 1999, pp. 650–654.
[16] J. Sucec and I. Marsic, An Efficient Distributed Network-Wide Broadcast Algorithm for Mobile Ad Hoc Networks, http://citeseer.nj.nec.com/312658.html.
[17] L. Tassiulas and A. Ephremides, Jointly optimal routing and scheduling in packet ratio networks, *IEEE Transactions on Information Theory*, 8, 165–168, 1992.

IX

Security and Privacy Aspects in Ad Hoc Wireless Networks

30

Security in Wireless Ad Hoc Networks

Amitabh Mishra
*Virginia Polytechnic Institute
and State University*

Ketan M. Nadkarni
*Virginia Polytechnic Institute
and State University*

Abstract

The realm of wireless ad hoc networks is a relatively new one, even with advancement in wireless networks in general. Essentially, these are networks that do not have an underlying fixed infrastructure. Mobile hosts "join" in, on the fly, and create a network on their own. With the network topology changing dynamically and the lack of a centralized network management functionality, these networks tend to be vulnerable to a number of attacks. If such networks are to succeed in the commercial world, the security aspect naturally assumes paramount importance. There is a need to devise security solutions to prevent attacks that jeopardize the secure network operation. This survey chapter aims to provide a state-of-the-art view of security in wireless ad hoc networks. An attempt has been made to discuss the core issues and requirements of security, followed by the current state of research and development in this area, which includes algorithms, architectures, and network management schemes.

30.1 Introduction

Mobile nodes within one another's radio range can communicate through wireless links and thus dynamically form a network. Wireless devices that are not in direct range communicate via intermediate devices (this is multi-hop communication). Thus, an ad hoc network is a collection of autonomous nodes that form a dynamic, purpose-specific, multi-hop radio network in a decentralized fashion. The quintessential nature of such networks is the conspicuous absence of a fixed support infrastructure such as mobile switching centers, base stations, access points, and other centralized machinery traditionally seen in wireless networks. The network topology is constantly changing as a result of nodes joining in and moving out. Packet forwarding, routing, and other network operations are carried out by the individual nodes themselves. An illustration of such a network is shown in Fig. 30.1.

Wireless ad hoc networks find application in military operations in which aircraft, tanks, and moving personnel can communicate. Rescue missions and emergency situations are also suitable applications for such networks. Other examples include virtual classrooms and conferences wherein people can set up a network on the spot through their laptops, personal digital assistants (PDAs), and other mobile devices assuming they share the same physical medium such as Direct Sequence Spread Spectrum (DSSS) or Frequency Hopped Spread Spectrum (FHSS).

The unreliability of wireless links between nodes, constantly changing topology owing to the movement of nodes in and out of the network, and lack of incorporation of security features in statically configured wireless routing protocols not meant for ad hoc environments all lead to increased vulnerability and exposure to attacks. Security in wireless ad hoc networks is particularly difficult to achieve, notably because of the vulnerability of the links, the limited physical protection of each of the nodes, the sporadic nature of connectivity, the dynamically changing topology, the absence of a certification authority (CA[1]; superscript numbers refer to the glossary at the end of the chapter), and the lack of a centralized monitoring or management point. This, in effect, underscores the need for intrusion detection, prevention, and related countermeasures. Security in wireless ad hoc networks is, thus, the focus of this chapter.

The rest of the chapter is organized as follows: In Section 30.2, we present the characteristics of wireless ad hoc networks that make them so vulnerable to attacks. In Section 30.3, we discuss the essence of security needs and requirements. In Section 30.4, we analyze the probable attacks in wireless ad hoc networks. In Section 30.5, we present a state-of-the-art view of the current research and development efforts, including security schemes and algorithms. Finally, in Sections 30.6 and 30.7, we present a summary of the chapter and briefly discuss some of the open issues for further work.

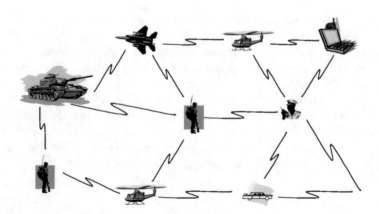

FIGURE 30.1 A generic illustration of a wireless ad hoc network.

30.2 Vulnerable Nature

There are various reasons why wireless ad hoc networks are at risk, from a security point of view. We next discuss the characteristics that make these networks vulnerable to attacks.

The wireless links between nodes are highly susceptible to link attacks, which include passive eavesdropping, active interfering, leakage of secret information, data tampering, impersonation, message replay, message distortion, and denial of service. Eavesdropping might give an adversary access to secret information, violating confidentiality. Active attacks might allow the adversary to delete messages, to inject erroneous messages, to modify messages, and to impersonate a node, thus violating availability, integrity, authentication, and nonrepudiation (these and other security needs are discussed in the next section).

Ad hoc networks do not have a centralized piece of machinery such as a name server, which could lead to a single point of failure and thus, make the network that much more vulnerable. In contrast, however, the lack of support infrastructure and the possibilities of new types of prior context may prevent the application of standard techniques for key agreement. This gives rise to a need for schemes to ensure key agreement.

An additional problem related to the compromised nodes is the potential *byzantine failures* encountered within Mobile Ad hoc NETwork (MANET) routing protocols wherein a set of the nodes could be compromised in such a way that the incorrect and malicious behavior cannot be directly noted at all. The compromised nodes may seemingly operate correctly, but, at the same time, they may make use of the flaws and inconsistencies in the routing protocol to undetectably distort the routing fabric of the network. In addition, such malicious nodes can also create new routing messages and advertise nonexistent links, provide incorrect link state information, and flood other nodes with routing traffic, thus inflicting byzantine failures on the system. Such failures are severe, especially because they may come from seemingly trusted nodes, the malicious intentions of which have not yet been noted. Even if the compromised nodes were noticed and prevented from performing incorrect actions, the erroneous information generated by the byzantine failures might have already been propagated through the network.

No part of the network is dedicated to support individually any specific network functionality, with routing (topology discovery, data forwarding) being the most prominent example. Additional examples of functions that cannot rely on a central service, and that are also of high relevance, are naming services, certification authorities, directory, and other administrative services. Even if such services were assumed, their availability would not be guaranteed, either due to the dynamically changing topology that could easily result in a partitioned network, or due to congested links close to the node acting as a server.

The absence of infrastructure and the consequent absence of authorization facilities impede the usual practice of establishing a line of defense, distinguishing nodes as trusted and nontrusted. Such a distinction would have been based on a security policy, the possession of the necessary credentials, and the ability of nodes to validate them. In the case of wireless ad hoc networks, there may be no ground for an *a priori* classification, since all nodes are required to cooperate in supporting the network operation, while no prior security association (SA[2]) can be assumed for all the network nodes.

Additionally, freely roaming nodes form transient associations with their neighbors; they join and leave subdomains independently and without notice. Thus, it may be difficult, in most cases, to have a clear picture of the ad hoc network membership. Consequently, especially in the case of a large-size network, no form of established trust relationships among the majority of nodes can be assumed.

In such an environment, there is no guarantee that a path between two nodes would be free of malicious nodes that would not comply with the employed protocol and would attempt to harm the network operation. The mechanisms currently incorporated in routing protocols cannot cope with disruptions due to malicious behavior. For example, any node could claim that it is one hop away from the sought destination, causing all routes to the destination to pass through itself. Alternatively, a malicious node could corrupt any in-transit route request (reply) packet and cause data to be misrouted.

The presence of even a small number of adversarial nodes could result in repeatedly compromised routes, and as a result, the network nodes would have to rely on cycles of timeout and new route discoveries to communicate. This would incur arbitrary delays before the establishment of a noncorrupted path, while successive broadcasts of route requests would impose excessive transmission overhead. In particular, intentionally falsified routing messages would result in a denial-of-service (DoS) experienced by the end nodes.

The dynamic and transient nature of an ad hoc network can result in constant changes in trust among nodes. This can create problems, for example, with key management, if cryptography is used in the routing protocol. It must not be trivial, for example, to recover private keys[3] from the device. Evidence that tampering has occurred would be required so as to distinguish a tampered node from the rest. Standard security solutions would not be good enough since they are essentially for statically configured systems. This gives rise to the need for security solutions that adapt to the dynamically changing topology and movement of nodes in and out of the network.

Moreover, the battery-powered operation of ad hoc networks gives attackers ample opportunity to launch a denial-of-service attack by creating additional transmissions or expensive computations to be carried out by a node in an attempt to exhaust its batteries.

In addition, sensor networks (a form of wireless ad hoc network) are made up of devices that tend to have limited computational abilities. For example, the working memory of a sensor node is insufficient to even hold the variables (of sufficient length to ensure security) that are required in asymmetric cryptographic algorithms, let alone perform operations with them. This limitation may exclude techniques such as frequent public key cryptography during normal operation. A particular challenge is broadcasting authenticated data to the entire sensor network. Current proposals for authenticated broadcast rely on asymmetric digital signatures for the authentication and are impractical for multiple reasons (e.g., long signatures with high communication overhead of 50–1000 bytes per packet, very high overhead to create and verify the signature).

Lastly, scalability is another issue that has to be addressed when security solutions are being devised, for the simple reason that an ad hoc network may consist of hundreds or even thousands of nodes. Many ad hoc networking protocols are applied in conditions where the topology must scale up and down efficiently, e.g., due to network partitions or mergers. The scalability requirements also directly affect the scalability requirements targeted to various security services such as key management.

The above discussion makes it clear that ad hoc networks are inherently insecure, more so than their wireline counterparts, and need robust security schemes that take into consideration the inherently susceptible nature of these networks. Coming up with a security scheme, in general, necessitates the discussion of the fundamental factors that make up security. In the next section, we examine the essential security needs of such networks. By this, we mean the factors that ought to be taken into consideration while designing a security scheme.

30.3 Security Needs

Security is a term liberally used in computer network terminology. What exactly is meant by security or, to remain within the scope of this chapter, what constitutes security in ad hoc networks? The basic security needs, more or less, are the same as those for wireline networks. What needs to be examined is their incorporation into security solutions already in place for traditional wireless networks and implementation of certain issues that have not been take care of. In the following, we briefly introduce the standard terms that are used when security aspects of a network are discussed.

30.3.1 Availability

Availability means that services provided by a node continue to be provided irrespective of attacks. Nodes should be available for communication at all times. In other words, it ensures survivability of the network services in the presence of denial-of-service attacks, which can be launched at any layer of an ad hoc network through radio jamming or battery exhaustion.

30.3.2 Authenticity

Authenticity is essentially confirmation that parties in communication with each other are genuine and not impersonators. This would require the nodes to somehow prove that their identities are what they claim to be. Without authentication, an adversary could very well masquerade a node, thus gaining access to sensitive and classified information, and possibly also causing interference with the normal and secure network operation.

30.3.3 Confidentiality

An outsider should not be able to access information in transit between two nodes. This ensures that information is not disclosed to unauthorized entities. For confidentiality, it is necessary to prevent intermediate and nontrusted nodes from understanding the content of the packets being transmitted. If authentication is taken care of properly, then confidentiality is a relatively simple process.

30.3.4 Integrity

Integrity is the guarantee that the message or packet being delivered has not been modified in transit or otherwise, and what has been received is what was originally sent. A message could be corrupted owing to nonmalicious reasons such as radio propagation impairment, but there is always the possibility that an adversary has maliciously modified the content of the message.

30.3.5 Nonrepudiation

Nonrepudiation means that the sender of a message cannot later deny sending the information and the receiver cannot deny the reception. This can be useful while detecting and isolating compromised nodes. Any node that receives an erroneous message can accuse the sender with proof and, thus, convince other nodes about the compromised node. Routers cannot repudiate ownership of routing protocol messages they send. The trust model associated with the propagation of updates that originate from distant nodes forms a major concern.

30.3.6 Ordering

Updates received from routers are in order, the nonoccurrence of which can affect the correctness of routing protocols. Messages may not reflect the true state of the network and may propagate false information.

30.3.7 Timeliness

Routing updates should be delivered in a timely fashion. Update messages that arrive late may not reflect the true state of links or routers on the network. They can cause incorrect forwarding or even propagate false information and weaken the credibility of the update information. If a node that relays information between two large connected components is advertised as "down" by malicious neighbors, large parts of the network become unreachable.

30.3.8 Isolation

Isolation requires that the protocol be able to identify misbehaving nodes and render them unable to interfere with routing. Alternatively, the routing protocol should be designed to be immune to malicious nodes.

30.3.9 Authorization

An authorized user or node is issued an unforgeable credential by the certificate authority. These credentials specify the privileges and permissions associated with the users or the nodes. Currently, credentials are not used in routing protocol packets, and any packet can trigger update propagations and modifications to the routing table.

30.3.10 Lightweight Computations

Many devices connected to an ad hoc network are assumed to be battery powered with limited computational abilities. Such a node cannot be expected to be able to carry out expensive computations. If operations such as public key cryptography or shortest path algorithms for large networks prove necessary, they should be confined to the least possible number of nodes — preferably only the route endpoints at route creation time.

30.3.11 Location Privacy

Often, the information carried in message headers is just as valuable as the message itself. The routing protocol should protect information about the location of nodes in a network and the network structure.

30.3.12 Self-Stabilization

A routing protocol should be able to automatically recover from any problem in a finite amount of time without human intervention. That is, it must not be possible to permanently disable a network by injecting a small number of malformed packets. If the routing protocol is self-stabilizing, an attacker who wishes to inflict continuous damage must remain in the network and continue sending malicious data to the nodes, which makes the attacker easier to locate.

30.3.13 Byzantine Robustness

A routing protocol should be able to function correctly even if some of the nodes participating in routing are intentionally disrupting its operation. Byzantine robustness can be seen as a stricter version of the self-stabilization property; the routing protocol must not only automatically recover from an attack, it should not cease functioning even during the attack. Clearly, if a routing protocol does not have the self-stabilization property, it cannot have byzantine robustness either.

30.3.14 Anonymity

Neither the mobile node nor its system software should default expose any information that allows any conclusions on the owner or current user of the node. In case device or network identifiers are used (e.g., MAC address,[4] Internet Protocol [IP] address), no linking should be possible between the respective identifier and the rover's identity for the communication partner or any outside attacker.

30.3.15 Key Management

The services in key management must provide ways to answer the following questions:

Trust model (How many different elements in the network can trust each other, and trust relationships between network elements.)

Cryptosystems (While public-key cryptography offers more convenience, public-key cryptosystems are significantly slower than their secret-key counterparts when a similar level of security is needed.)

Key creation (Which parties are allowed to generate keys to themselves or other parties and what kind of keys.)

Key storage (Any network element may have to store its own key and possibly keys of other elements as well. In systems with shared keys with parts of keys distributed to several nodes, the compromising of a single node does not yet compromise the secret keys.)

Key distribution (Generated keys have to be securely distributed to their owners. Any key that must be kept secret has to be distributed so that confidentiality, authenticity, and integrity are not violated.)

30.3.16 Access Control

Access control consists of the means to govern the way the users or virtual users such as operating system processes *(subjects)* can have access to data *(objects)*. Only authorized nodes may form, destroy, join, or leave groups. Access control can also mean the way the nodes log into the networking system to be able to communicate with other nodes when initially entering the network. There are various approaches to access control:

Discretionary Access Control (DAC) offers means for defining the access control to the users themselves.
Mandatory Access Control (MAC[5]) involves centralized mechanisms to control access to objects with a formal authorization policy.
Role Based Access Control (RBAC) applies the concept of *roles* within the subjects and objects.

30.3.17 Trust

If the physical security is low and trust relationships are dynamic, then the probability of a security failure may rise rapidly. It is not difficult to see what happens if the suspicion of a secure failure increases. If there is a reason to believe that a portion of the nodes has been compromised, the users will probably become more reluctant to trust the network. Constructing security for the first time may not be so difficult. Maintaining trust and handling dynamic changes seem to need more effort.

30.3.18 Security Needs: Summary

In summary, we can safely say that the mandatory security requirements include confidentiality, authentication, integrity, and nonrepudiation. These would, in turn, require some form of cryptography, certificates, and signatures. Some other ideal characteristics include easy plus strong user authentication, explicit transaction authorization, end-to-end encryption, accepted log-on security (biometrics) instead of separate personal identification numbers (PINs) and passwords, intrusion detection, access control, logging, audit trail, security policy that states the rules for access and plan mechanisms and countermeasures, antivirus scanners (content), firewall, etc. This discussion demarcates the various branches within security *per se*, such as intrusion detection and prevention, key agreement, trust management, data encryption, and access control. Having looked at the essential security needs, we now discuss the various kinds of attacks, practical as well as conceptual. This helps in understanding which security needs are compromised, a direct consequence of which would mean concentrating on a specific security need rather than all, which might really be unnecessary, while countering a particular type of attack. Attacks include external and internal attacks as well as active and passive attacks. These and other types of attacks are discussed in the next section.

30.4 Attacks in Wireless Ad Hoc Networks

One can distinguish between two levels of attack:

Attacks on the basic mechanisms of the ad hoc network such as routing
Attacks on the security mechanisms and notably on the key management mechanisms

Alternatively, attacks against ad hoc networks can again be divided into two groups in a different way:

Passive attacks typically involve only *eavesdropping* of data. Examples of passive attacks include covert channels, traffic analysis, sniffing to compromise keys, etc.

Active attacks involve actions performed by adversaries, for instance, the replication, modification, and deletion of exchanged data. Adversaries actively attempt to change the behavior of the protocol in active attacks whereas adversaries are subtle in their activities in passive attacks. The information inadvertently disclosed to passive attackers by the protocol packets can be used to launch active attacks.

External attacks are typically active attacks that are targeted, e.g., to cause congestion, propagate incorrect routing information, prevent services from working properly, or shut them down completely. External attacks can typically be prevented by using standard security mechanisms such as firewalls, encryption, and so on. *Internal attacks* are typically more severe attacks, since malicious insider nodes already belong to the network as authorized parties and are thus protected by the security mechanisms the network and its services offer. Thus, such malicious insiders, who may even operate in a group, may use the standard security means to actually protect their attacks.

30.4.1 Denial of Service (DoS)

The denial of service threat either produced by an unintentional failure or malicious action constitutes a severe security risk in any distributed system. The classical way to carry out this type of attack is to flood any centralized resource so that it no longer operates correctly or crashes, but in ad hoc networks, this may not be an applicable approach due to the distribution of responsibility. Radio jamming and battery exhaustion are two ways in which service can be denied to other nodes and users. Distributed DoS attack is a more severe threat: if the attackers have enough computing power and bandwidth, smaller ad hoc networks can be crashed or congested rather easily. Compromised nodes may be able to reconfigure the routing protocol or a part of it so that they can send routing information very frequently, thus causing congestion and preventing nodes from gaining new information about the changed topology of the network. In the worst case, the adversary is able to change the routing protocol to operate arbitrarily or perhaps even in the (invalid) way the attacker wants. If the compromised nodes and the changes to the routing protocol are not detected, the consequences are severe, as the network may seem to operate normally to the nodes. This kind of invalid operation of the network initiated by malicious nodes is called a *byzantine failure*.

30.4.2 Impersonation

Impersonation attacks constitute a serious security risk at all levels of ad hoc networking. If proper authentication of parties is not supported, compromised nodes may be able to join the network undetectably, send false routing information, and masquerade as some other trusted node. Within network management, the attacker could gain access to the configuration system as a superuser. At the service level, a malicious party could have its public key[6] certified even without proper credentials. This attack will most likely be noticed very quickly, and the information that is manipulated or accessed is not crucial enough to make the attack worthwhile. A malicious party may be able to masquerade itself as any of the friendly nodes and give false orders or status information to other nodes. Impersonation threats are mitigated by applying strong authentication mechanisms in contexts in which a party has to be able to trust the origin of data it has received or stored. Most often this means, in every layer, the application of digital signature or keyed fingerprints over routing messages, configuration or status information, or exchanged payload data of the services is in use. Digital signatures implemented with public-key cryptography are a problematic issue within ad hoc networks, as they require an efficient and secure key management service and relatively more computation power. Thus, in many cases, lighter solutions such as the use of keyed hash functions or *a priori* negotiated and certified keys and session identifiers are needed. They do not, however, remove the demand for secure key management or proper confidentiality protection mechanisms.

30.4.3 Disclosure

Any communication must be protected from eavesdropping, whenever confidential information is exchanged. Also, critical data that the nodes store must be protected from unauthorized access. In ad hoc networks, such information can include almost anything, e.g., specific status details of a node, the location of nodes, private or secret keys, passwords, and so on. Sometimes the control data include more critical information in respect to security than the actual exchanged data. For instance, the routing directives in packet headers such as the identity or location of the nodes can sometimes be more valuable than the application-level messages. The identities of the observed nodes, compared to the previous traffic patterns of the same nodes, or the detected radio transmissions the nodes generate may be just the information an adversary needs to launch a well-targeted attack.

30.4.4 Trust Attacks

A trust hierarchy is basically an explicit representation of trust levels that reflects organizational privileges. It associates a number with each privilege level, which reflects the security, importance, or capabilities of the mobile node and also of the paths. Attacks on the trust hierarchy can be broadly classified as outsider attacks and insider attacks, based on the trust value associated with the identity or the source of the attack. What is also needed is a binding between the identity of the user with the associated trust level. Without this binding, any user can impersonate anyone else and obtain the privileges associated with higher trust levels. To prevent this, stronger access control mechanisms are required (Authentication, Authorization, and Accounting or AAA). In order to force the nodes and users to respect the trust hierarchy, cryptographic techniques, e.g., encryption, public key certificates, shared secrets, etc., can be employed. Traditionally, strong authentication schemes are used to combat outsider attacks. The identity of a user is certified by a centralized authority and can be verified using a simple challenge–response protocol. Insider attacks are launched by compromised users within a protection domain or trust level. Routing protocol packets in existing ad hoc algorithms do not carry authenticated identities or authorization credentials, and compromised nodes can potentially cause much damage. Insider attacks, in general, are hard to prevent at the protocol level. Some techniques to prevent insider attacks include secure transient associations and tamper proof and tamper resistant nodes.

30.4.5 Attacks on Information in Transit

In addition to exploiting vulnerabilities related to the protection and enforcement of trust levels, compromised or enemy nodes can utilize the information carried in the routing protocol packets to launch attacks. These attacks can lead to corruption of information, disclosure of sensitive information, theft of legitimate service from other protocol entities, or denial of network service to protocol entities. Threats to information in transit include:

Interruption (The flow of routing protocol packets, especially route discovery messages and updates, can be interrupted or blocked by malicious nodes. Attackers can selectively filter control messages and updates, and force the routing protocol to behave incorrectly.)

Interception and subversion (Routing protocol traffic and control messages, e.g., the "keep-alive" and "are-you-up?" messages can be deflected and rerouted.)

Modification (The integrity of the information in routing protocol packets can be compromised by modifying the packets themselves. False routes can be propagated, and legitimate nodes can be bypassed.)

Fabrication (False route and metric information can be inserted into legitimate protocol packets by malicious insider nodes.)

30.4.6 Attacks Against Secure Routing

Attacks against secure routing are basically of two types: internal and external. External attacks can again be classified as active and passive attacks. We briefly discuss the various kinds of routing attacks.

Internal Attacks

An internal attack is a severe threat to ad hoc networks. The attack may broadcast wrong routing information to other nodes within the network. A compromised node is categorized as an internal attack. Detecting such wrong information in routing information is difficult because compromised nodes are able to generate valid signatures using their private keys. Also differentiating between an actual attacker and a change in topology may be problematic because the topology of the ad hoc network dynamically changes.

External Attacks

External attacks on routing can be divided into two categories: passive and active.

Passive attacks involve unauthorized "listening" to the routing packets. The attack might be an attempt to gain routing information from which the attacker could extrapolate data about the positions of each node in relation to the others. For example, an attacker that eavesdrops on all the routing updates transmitted in a certain part of the ad hoc network can begin to piece together which nodes are close together (one or two hops apart) and which nodes are far from each other (many hops apart). In a passive attack, the attacker does not disrupt the operation of a routing protocol but only attempts to discover valuable information by listening to the routed traffic. The attack is usually impossible to detect, which makes defending against such attacks difficult. Furthermore, routing information can reveal relationships between nodes or disclose their IP addresses. If a route to a particular node is requested more often than the route to other nodes, the attacker might expect that the node is important for the functioning of the network and that disabling it could bring the entire network down. Other interesting information that is disclosed by routing data is the location of nodes. Even when it might not be possible to pinpoint the exact location of a node, one may be able to discover information about the network topology. It is worth noting that in an IP network, one cannot defend against these attacks, for example, by only using IPSec.[7] The packets still have most of their IP headers in plaintext, and it may not even be feasible to have symmetric keys distributed to every node in a network.

Active attacks on the network from outside sources are meant to degrade or prevent message flow between the nodes. Active external attacks on the ad hoc routing protocol can collectively be described as denial-of-service attacks, causing a degradation or complete halt in communication between nodes. One type of attack involves insertion of extraneous packets into the network in order to cause congestion. A more subtle method of attack involves intercepting a routing packet, modifying its contents, and sending it back into the network. Alternatively, the attacker can choose not to modify the packet's contents but rather to replay it back to the network at different times, introducing outdated routing information to the nodes. The goal of this form of attack is to confuse the routing nodes with conflicting information, delaying packets or preventing them from reaching their destination. To perform an active attack, the attacker must be able to inject arbitrary packets into the network. The goal may be to attract packets destined to other nodes to the attacker for analysis or just to disable the network. An active attack can sometimes be detected, and this makes active attacks a less inviting option for most attackers. Some types of active attacks that can usually be easily performed against an ad hoc network are:

A. *Black hole:* In this attack, a malicious node uses the routing protocol to advertise itself as having the shortest path to the node whose packets it wants to intercept. In a flooding based protocol, the attacker listens to requests for routes. When the attacker receives a request for a route to the target node, the attacker creates a reply consisting of an extremely short route. If the malicious reply reaches the requesting node before the reply from the actual node does, a forged route has been created. Once the malicious device has been able to insert itself between the communicating nodes, it is able to do anything with the packets passing between them. It can choose to drop the packets to perform a denial-of-service attack or alternatively use its place on the route as the first step in a man-in-the-middle attack.

B. *Routing table overflow:* In this attack, the attacker attempts to create routes to nonexistent nodes. The goal is to create enough routes to prevent new routes from being created or to overwhelm the protocol implementation. Proactive routing algorithms attempt to discover routing information even before it is needed, while a reactive algorithm creates a route only once it is needed. An attacker can disrupt a proactive network simply by sending excessive route advertisements to the routers in the network. Reactive protocols, on the other hand, do not collect routing data in advance.

C. *Sleep deprivation:* Usually, this attack is practical only in ad hoc networks where battery life is a critical parameter. Battery powered devices try to conserve energy by transmitting only when absolutely necessary. An attacker can attempt to consume batteries by requesting routes or by forwarding unnecessary packets to the node using, for example, a black hole attack. This attack is especially suitable against devices that do not offer any services to the network or offer services only to those who have some special credentials. Regardless of the properties of the services, a node must participate in the routing process unless it is willing to risk becoming unreachable to the network.

D. *Location disclosure:* A location disclosure attack can reveal something about the locations of nodes or the structure of the network. The information gained might reveal which other nodes are adjacent to the target or the physical location of a node. Routing messages are sent with inadequate hop-limit values, and the addresses of the devices sending the Internet Control Message Protocol (ICMP[8]) error-messages are recorded. In the end, the attacker knows which nodes are situated on the route to the target node. If the locations of some of the intermediary nodes are known, one can gain information about the location of the target as well.

30.4.7 Military Attacks

A network used by the military needs maximum security. In a military environment, routing attacks can be divided into two types: strategic and tactical attacks.

Strategic routing attacks — Strategic routing attacks include intelligence gathering. They might also cover destruction of enemy networks in preparation for battle. Additionally, because of the attack, the target could gain some information about where the enemy is about to strike next. However, once a routing attack has ended, the network can usually be brought back into use in a short amount of time. Additionally, because of the attack, the target could gain some information about where the enemy is about to strike next. Thus, active attacks are probably best suited to tactical use while passive attacks can be effective in gathering information. One can deduce information about the location of nodes and the roles of each node in the network through passive attacks, with command and control nodes being obvious targets.

Tactical routing attacks — Tactical routing attacks could be used most effectively during battle. The attacks might use information about the network topology or relationships between nodes as well as other information that has been collected earlier using passive attacks. The main goal could be to temporarily disable some important part of a network using denial-of-service attacks.

30.4.8 Types of Attacks: Summary

Ad hoc networks are vulnerable to attacks not only from outside but also from within. Moreover, these attacks can be active as well as passive. We briefly discussed the various attacks possible on wireless ad hoc networks. The significance of the various security needs discussed in an earlier section now comes to the fore, since any attack essentially means disrupting one or more of the security needs. In the next section, we discuss the security schemes proposed to prevent and detect some of these attacks and also to safeguard the security of such networks in general.

30.5 Overview of Security Schemes for Ad Hoc Networks

This section could very well be described as the *pièce de résistance* of this survey chapter. We discuss the pioneering work in the relatively new domain of ad hoc network security, in addition to the security schemes that have been proposed for specific issues. These security schemes are related to key management, intrusion detection, authentication, secure routing, and network management. We begin our discussion with *The Resurrecting Duckling* — a pioneering piece of work in ad hoc network security.

30.5.1 The Resurrecting Duckling

Stajano and Anderson studied ad hoc networks where the nodes are (personal) devices that can communicate with one another over a wireless channel [1]. They present the *resurrecting duckling* security model to solve the secure transient association problem, an example of which would be a person with a remote control not wanting any other person to be able to use another remote control bought at the same shop, to work at his or her place, but wanting it to work for some other person who might buy it from him at the former's place.

In the same way that a duckling considers the first moving object it sees as its mother, a device would recognize the first entity that sends it a secret key as its owner. When necessary, the owner could later clear the imprinting and let the device change its owner. The imprinting — sharing the key — would be done in a physical contact. In the case of several owners with different access rights, the imprinting could be done several times with different keys. In this manner, it could be possible to create a hierarchy among the owners or prioritize the service requests. Tamper resistance or tamper evidence may protect against physical threats against the nodes. The uniqueness of a master is emphasized. A slave has two exclusive states: imprinted and imprintable. The master controls its slave, and they are bound together with a shared secret that is originally transferred from master to slave over a nonwireless, confidential, and integrated channel. The slave is imprinted and made imprintable by its master. The slave becomes imprintable as a consequence of conclusion of a transaction or by an order of its master.

The original *resurrecting duckling* security policy was extended to cover also a peer-to-peer interaction [2]. In the extension, a master can also be a human in addition to devices. In case of a human master, the master can imprint the slave device by typing a Personal Identification Number (PIN). A master can also be a millimeter-sized sensor node (called a "dust mote") of which a system consists. There can be many dust motes in the monitored area, each of them having battery, solar cell, sensors, and some digital computing equipment. A dust mote may also have an active transmitter and receiver. These can be used in military applications. Another feature of this extension is that the master does not have to be unique. Another master that has a credential valid to that slave at the moment can imprint the slave. The slave has a principal master, but it can also receive some kinds of orders from other masters. In the extension of the resurrecting duckling model, the master can upload a new policy to a slave as well.

The next subsection deals with key management, which involves services such as trust model, cryptosystems, key creation, key storage, and key distribution, each of which has been touched upon as part of security needs.

30.5.2 Key Management

We now discuss the various schemes that have been proposed to ensure secure key management within an ad hoc network. To be able to protect nodes, e.g., against eavesdropping by using encryption, the nodes must have made a mutual agreement on a shared secret or exchanged public keys. For very rapidly changing ad hoc networks, the exchange of encryption keys may have to be addressed on demand, thus without assumptions about *a priori* negotiated secrets. In less dynamic environments, the keys may be mutually agreed upon proactively or even configured manually (if encryption is even needed).

A Distributed Asynchronous Key Management Service

A mobile military ad hoc network is both security sensitive and easily exposed to security attacks [3]. The focus is on the goals of secure routing and creating a distributed, asynchronous key management service. Because of the possibility of the nodes being compromised, the network should not have any central entities but instead have a distributed architecture. This protects against denial-of-service attacks and major information disclosure. If public key cryptography is involved, a central Certification Authority is problematic. The authors present a distributed key management where the private key of a trusted service is divided among n servers as shown in Fig. 30.2a. The service, as a whole, has a public/private key pair K/k. The public key K is known to all nodes in the network, whereas the private key k is divided into n shares $S_1, S_2,..., S_n$, one share for each server. Each server i also has a public/private key pair K_i/k_i and knows the public keys of all nodes.

To create a signature with the private key, at least k out of the n servers need to combine their knowledge. Combining the shares would not reveal the actual private key. The correctness of the signature would, as usual, be verifiable with the public key of the service. The method is called *threshold cryptography*: an (n,k) threshold cryptography scheme allows n parties to share the ability to perform a cryptographic operation (e.g., creating a digital signature). This is shown in Fig. 30.2b for three servers, where K/k is the public/private key pair of the service. Using a $(3,2)$ threshold cryptography scheme, each server i gets a share s_i of the private key k. For a message m, server i can generate a partial signature $PS(m,s_i)$ using its share s_i. Correct servers S_1 and S_3 both generate partial signatures and forward the signatures to a combiner c. Even though server S_2 fails to submit a partial signature, c is able to generate the signature $(m)k$ of m signed by service private key k.

Any k parties can perform the operation jointly, whereas it is infeasible for, at most, $k - 1$ parties to do so. If we suppose that, at most, $k - 1$ servers can be compromised at a time, a false signature cannot be created. The key management service also employs share refreshing and is scalable to changes in the number of servers. Periodic share refreshing creates new shares of the private key, so that an adversary cannot collect information about k shares over time. In effect, the proposal applies redundancies in the network topology to provide reliable key management and implements a distributed model owing to the lack of a central authority. The key idea is to use key sharing with the assumption that the ratio of nodes compromised to total nodes can be bounded. If the upper limit on the number of compromised server nodes can be set to $t < 1$, at least $n = (3t + 1)$ nodes are needed to enable the scheme. The architecture does require that the underlying routing protocol manage multiple routes.

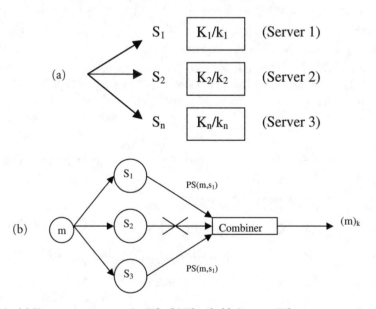

FIGURE 30.2 (a) Key management service K/k. (b) Threshold signature K/k.

A Password Authenticated Key Exchange Protocol

A generic protocol for multi-party password authenticated key exchange has been proposed based on a pioneering piece of work in which two parties M and S share a weak secret P [4]. The underlying protocol's goal is to agree to a strong session key K in spite of weak P. In the first step, a master sends its identifier and an encrypted weak secret to the slave. In the second step, the slave extracts the encrypted key for M, i.e., E_M, and generates a random string R. R is encrypted with E_M and returned to the master. In the third step, R is extracted, and random strings $challenge_M$ and S_M are generated. They are encrypted by R and sent back to the slave. In the fourth step, the slave extracts $challenge_M$ and computes a public function $h(challenge_M)$. Slaves own random strings $challenge_S$ and S_S, which are generated and encrypted with $h(challenge_M)$, using R as a key. These are sent back to that master that, in the fifth step, verifies the given $h(challenge_M)$ to be what it is claimed to be. The master can be convinced that the slave can extract $challenge_M$. This is possible only if the slave knows the used weak secret. The protocol can be extended to a multi-party scenario by electing a leader. In that case, S functions are also used to generate strong session key K. Other parties' knowledge of K has to be confirmed.

Figure 30.3 provides an overview of the underlying scheme where M = master; S = slave; P = weak secret; E_M = encrypted key for M; $Challenge_{M \ or \ S}$ = random string for M or S; $S_{M \ or \ S}$ = random string S for M or S; R = random function used as a key; h() = a public function.

Their password-based authentication protocol is derived from the so-called *encrypted key exchange* (EKE) protocol. In EKE, two participants who share a secret create together a session key. The secret or password can be weak. Nevertheless, anyone who does not know the password cannot successfully participate in the protocol. Finally, EKE provides perfect forward secrecy: even if an attacker later finds out the password, he or she cannot find out the previous session keys. Hence, the messages of the past sessions remain secret. In the group version of EKE, all the participants contribute to the session key. This ensures that the resulting key is not selected from too small a key space even if some participants try to do that. An attacker who tries to participate in the protocol and sends some random messages cannot prevent the construction of the key. As in the original EKE, only the participants who know the original password learn the resulting session key. The secure connections between the participants are created from a manually exchanged password. Hence, no support infrastructure is needed. The key agreement is of significant importance when a secure transient association is to be guaranteed. This will happen best when the key management is done locally. In order to implement location-based key agreement successfully, a set of labels to map the location as well as an identity-based mechanism for key agreement is needed.

NTM — A Progressive Trust Negotiation Scheme

A scheme for Progressive Trust Negotiation in Ad Hoc Networks has been proposed that builds trust, along with a dynamic key agreement scheme to protect the negotiation [5]. NTRG Trust Model (NTM) is subdivided into two main components, namely, the peer-to-peer component and the remote component. The peer-to-peer component deals with securing communication with neighbors. The remote

$$
\begin{array}{lll}
(1) & M \quad S: & M, P \ (E_M) \\
(2) & S \quad M: & P \ (E_M(R)) \\
(3) & M \quad S: & R \ (challenge_M, S_M) \\
(4) & S \quad M: & R \ (h(challenge_M), challenge_S, S_s) \\
(5) & M \quad S: & R \ (h(challenge_S))
\end{array}
$$

FIGURE 30.3 Underlying protocol for password authenticated key exchange.

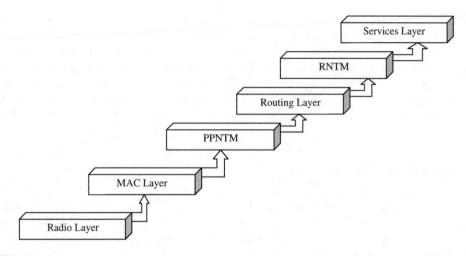

FIGURE 30.4 Layer structure.

component has the dual responsibility of carrying out trust negotiation and establishing secure end-to-end communication. The security system is divided into two distinct layers: the Peer-to-Peer NTM (PPNTM) layer and the Remote NTM (RNTM) layer. The PPNTM layer is situated below the routing layer, as its primary goal is to secure communication between the neighbors (i.e., nodes in the radio range). The RNTM layer provides end-to-end encryption and so, is located just above the routing layer as shown in Fig. 30.4.

The threat of eavesdropping by an external attacker, say X, which is within the listening range of two nodes A and B, is mitigated by the PPNTM layer. Since the symmetric encryption keys are to be negotiated between the neighbors using Station-to-Station (STS[9]), it is impossible for a node to eavesdrop on communication without authenticating itself. The end-to-end key negotiated by the RNTM layer protects against the internal attacker, R. This key formation in the NTM scheme is shown in Fig. 30.5 where K_{1-4} are peer-to-peer keys and K is the end-to-end key between A and D.

Each node has to have at least one *Network Address Certificate* that entitles it to use certain network address(es) and participate in packet relaying. This certificate is used in the PPNTM layer's STS key exchange for authentication. RNTM relies on an *Identity Certificate* so that the user can move between nodes while still maintaining trust. The certificates have to be signed by a third party. The PPNTM layer negotiates a symmetric key with neighbors. The authentication of the remote certificate involved in the STS key formation is done using three models. The RNTM layer has the responsibility of carrying out trust negotiations and negotiation of an end-to-end encryption key. The trust negotiation is carried out by incrementally exchanging certificates. The certificates are asked for by using the attribute name/value pair. Usually a node trying to access some services on the remote node, for which it has not being cleared,

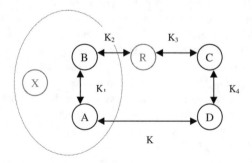

FIGURE 30.5 Key formation in NTM.

triggers the trust negotiation. The trust negotiation can also be explicitly triggered by the RNTM layer. The different models for finding trust in a certificate are necessitated by the absence of an online trusted third party. The simplistic first model assumes that the node is primed for local use and has all the Certificate Revocation Lists (CRL[10]) updated. Any certificate issued by an unknown certificate issuer cannot be verified and will be referred to the user. The second model is a probabilistic model. Each of the trusted certificate issuers has a trust value of one associated with it. There is a distrust value, which is subtracted from the trust value of the CRL if the scheduled update of the CRL is missed. Then the trust negotiation takes place with a default trust value, and it should be exceeded for negotiation to succeed. The third and last model assigns weights to the certificates.

Minimal Public-Key Based Authentication

Sufatrio and Lam have introduced a lightweight and scalable authentication protocol called Minimal Public-Key Based Authentication (Min-PKA) for Mobile IP that does not require any changes to the protocol [6]. Its main purpose is to secure the registration process. It makes use of AAA (Authentication–Authorization–Accounting) server nodes (AAAH for AAA home agent and AAAF for AAA foreign agent). The authors criticize Jacobs' approach [48], which uses only public-key cryptography, since it assumes that the Mobile Nodes (MN) can perform the heavy computations related to the security operations. In contrast, the Min-PKA proposal uses two different approaches, secret-key and public-key based, of which the former requires manual configuration. Since such an approach may offer substantial optimizations in some routing scenarios, Sufatrio and Lam suggest the use of public-key cryptography to be applied in the interdomain authentication. The MN and AAAH can, however, use shared secrets between Home Agent (HA) and MN, since the nodes have a security association.

Their approach introduces three services:

1. Authentication services provide digital signatures and Message Authentication Codes (MAC[11]) between MN and AAAH to protect the routing traffic. The services rely on the correct actions performed by AAAH in the indirect MN-AAAF communication.
2. Integrity services rely on the authentication services to assure integrity when the authenticity is confirmed. Foreign Agent (FA) discoveries form a problem since FA and MN may have no security associations. This problem is, however, solved by putting the advertisements into the Registration Requests from which a MAC can be calculated and which can thus then be authenticated.
3. Antireplay protection services guarantee the freshness and authenticity of the registrations. The mechanism uses nonces, which are basically timestamps, to achieve the goals but has a flaw, since adversaries can fool the AAAFs to sign arbitrary data. Nonces in the message to be signed somewhat reduce the severity of the problem but do not completely remove it.

A Scheme For Learning, Storing, and Distributing Public Keys

The use of public key encryption algorithms requires a scheme for learning, storing, and distributing the public keys of all routing nodes. A certificate guaranteeing that it came from a trusted party must accompany each public key distributed. Distributed key management involves designation of a set of "trustworthy" nodes that share sections of the public key of the management system. Each trusted node keeps a record of all public keys in the network. The number of nodes needed to generate a valid signature is less than the total number of trusted nodes. Therefore, even if an attacker compromises some of the nodes, a complete signature can still be validated. This solution is called *threshold cryptography*. Threshold cryptography is vital in the detection of compromised nodes. Instead of using a signature scheme where the signer is a single entity holding the secret key, the secret key is shared by a group. In order to produce a valid signature on a given message, individual users produce their partial signatures on the message and then combine them to generate a full signature. This protects keys from any single point of failure. However, there still exists a problem because, in the case of sensitive and long-lived keys, the attacker still has a long period of time to gradually break the system. Therefore, a proactive approach is proposed to protect long-lived secrets by periodically refreshing the secret share in such a way that the exposure of a share to an attacker in a given period of time has no value for that attacker after the refreshment of shares is performed [7]. The shares are renewed, but the secret to which they correspond is unchanged. Thus, this proactive approach can be taken to distribute public keys in a mobile ad hoc network because of its resilience to internal attacks from compromised nodes and its lack of a single point of failure.

Non-Disclosure Method

Fasbender et al. have introduced the Non-Disclosure Method (NDM) [8], a solution to the confidentiality of location problem wherein the current location of a mobile node can be easily retrieved by just looking at the address headers of the exchanged packets, and particular registration requests can then be used to generate location profiles. In this approach, every Security Agent (SA) (node) has a public–private key pair. When a sender A wants to send a message M to the receiver B, the message is forwarded to the destination by using a route $(A, SA_1, SA_2, ..., SA_n, B)$ as defined by the intermediate security agents from SA_1 to SA_2. The route is constructed by performing n encryptions E_SA_i with the public keys of the intermediate nodes: Encrypted message $M' = E\_SA_1(SA_2, E\_SA_2(SA_3,...(SA_n, E\_SA_n(B,M))))$. When the sender A sends the encrypted message M', the first security agent SA_1 decrypts the message, thus finding only the location of the next hop in the route (SA_2), and so on. Thus, the security agents see only the location information (addresses) of the next and previous security agents. In addition, the nodes cannot determine where they actually are located in the route and whom the receiver B is. In this approach, the last intermediate node SA_n would know the location and identity of the receiver B, but not M, if it can be assumed that the sender A can encrypt the message with B's public key also. The method can be applied to protect any other vulnerable header information rather than just the location of the nodes. In the NDM approach, the location information as well as the actual message is hidden from the intermediate nodes (SAs). The approach, however, has a problem with respect to MANET networking: the sender must know all the public keys K_SA_i and the identities of the security agents in the route to be able to construct the route. Moreover, the intermediate compromised nodes (or outsiders) can inspect the sizes of the sent packets and try to determine the length of the route. This problem can be mitigated by allowing the SAs to use padding mechanisms with random data to hide the actual length of the payload.

Securing Ad Hoc Jini-Based Services

Jini is an open software architecture that enables developers to create network-centric services that are highly adaptive to change. A security model based on the usage of public keys has been proposed for securing ad hoc Jini-based services [9]. Different from most of the other service discovery methods, Jini uses a distributed model that is built on allowing code to be moved between entities in the ad hoc network. Downloaded code is a security problem in itself. Hence, security is an important part of the system design of Jini ad hoc services, and the authors address it by relying on decentralized authorization. The problem of trusting public keys is separated from the usage of trusted keys.

Given that all of the nodes in the ad hoc network have public key pairs, and that all of the nodes consider the public keys of others good for creating secure connections within the network, any of the public key based authentication schemes can be used. It is possible to use Transport-Layer Security Protocol (TLS[12]) or IPSec protection protocols to secure the actual service. This solution allows the service provider to design a communication security solution that is adapted to the service and uses any cryptographic algorithm.

Unlike other standard approaches, this approach does not assume that the client and the server share a necessarily large set of different symmetric key encryption or MAC algorithms. The authors have suggested a trust distribution protocol that minimizes the number of manual interactions needed when setting up the necessary trust relations as shown in Fig. 30.6. A server that wants to offer a secure communication service has a proxy, which contains the necessary algorithms for authenticated key exchange with the server and also algorithms to encrypt and protect the exchanged data in the client–server interaction. The server digitally signs the proxy with its private key, which ensures that the client can verify the signature. The server packs the signed code together with the signature (and possibly its public key). When a client finds the service, it downloads a proxy corresponding to the service, together with the signature and possibly some included certificates. The client can verify the signature if it has a

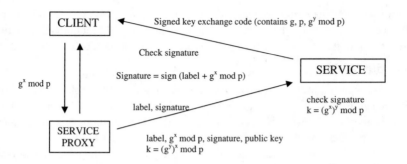

FIGURE 30.6 A basic proxy distribution protocol, with the proxy using Diffie–Hellman (DH[13]) key agreement.

trusted public key that corresponds to the signature or if the client trusts some of the public keys contained in the included certificates. The proxy performs authenticated key exchange with the origin server using a standard authentication and key exchange protocol. On successful authentication, the proxy sets up a secure communication link with the server.

Furthermore, these authors have described a "minimal" preconfiguration solution for securing the communication of an ad hoc service. The technique allows establishment of authenticated connections without allowing undue access to the client's private key. Their example implementation uses Java and Jini, but the same principles can be used with other service discovery techniques.

A Robust Scalable Ubiquitous Security Scheme

A scheme for providing ubiquitous security services for mobile hosts has been proposed; it scales to network size and is robust against break-ins [10]. The scheme distributes the certification authority functions through a threshold secret sharing mechanism, in which each entity holds a secret share and multiple entities in a local neighborhood jointly provide complete services. Localized certification schemes are employed to enable ubiquitous services. Secret shares are updated to further enhance robustness against break-ins. Threshold secret sharing and secret share updates are used to enable intrusion tolerance. No single entity in the network knows or holds the complete system secret (e.g., a certification authority's signing key). Instead, each entity only holds a secret share of the certification authority's signing key. Multiple entities, say K, in a one-hop network locality jointly provide complete security services, as if they were provided by a single and omnipresent certification authority. The system security is not compromised as long as there are fewer than K collaborative intruders in each adversary group. To further resist intrusions on a long-term basis, the secret shares for all entities are periodically updated. A certificate-based approach based on the public key infrastructure (PKI[14]) is employed. Any two communicating entities may establish a temporary trust relationship via unforgeable, renewable, and globally verifiable certificates carried by each of the entities. Security functions such as confidentiality, data integrity, authentication, and nonrepudiation can be readily provided via valid certificates that are usually issued by a globally trusted certification server.

New schemes have been proposed to realize the certificate-related security services to accommodate the unique characteristics of ad hoc wireless networks. They provide ubiquitous services for mobile entities by distributing the CA's functionality to each local neighborhood. A coalition of K neighbors can serve as the CA and jointly provide certification services for a requesting mobile entity. The fully localized and everywhere-available features of this design enable service ubiquity for mobile users. A novel self-initialization protocol is proposed to handle dynamic node membership (i.e., joins and leaves) and secret share updates. Each node can be (re)initialized by K neighbors. Once initialized, a node is qualified to be a coalition member to serve its neighborhood. Security services are effectively provided in the presence of mobility, wireless channel errors, network partitioning, and node failures.

Another Robust Membership Management Scheme

A robust membership management scheme for ad hoc groups based on public key cryptography and on the use of signed certificates has been described in which the members are represented by their public signature keys and each group has a public signature key to represent the group as a whole [11]. Certificates signed by the group key are used to indicate the membership of the nodes. In groups, the owner of the group key is the only member that can let new members join the group. In order to increase the robustness of the membership management, the authority of the leader must be distributed to several members. This is done by letting the original leader (i.e., the group-key owner) delegate the leadership to other members. It can authorize other keys to act as equivalent leaders by issuing *leader certificates*. In order to prove membership in the group, a member that has been certified directly by the group key needs its own private key and the membership certificate. A member that has been certified by another leader needs all the certificates in the path from the group key to its member key. Hence, when a leader certifies other leaders or members, it must pass along all the certificates that prove its own status as a leader in the group. This way, a chain of certificates is formed from the group key to each member key.

Reconstitution of the group is a secure and often a recommendable way to continue with the trusted members only. The members of a group also need some instant mechanisms for canceling the membership of a single key without sacrificing membership of the other, still-trusted members. Unfortunately, canceling a membership that has already been granted is not easy. The membership certificates may be created, stored, and verified concurrently in different parts of the system. There are basically two ways of getting rid of untrusted members: membership expiration (the membership certificates may have a validity period that is decided by the issuer) and membership revocation (membership should be revoked only when there is a reason to suspect that the private key has fallen into the hands of an adversary, and information about the revocation must be propagated to all the parts of the system where relevant certificates may be used).

Increasing the robustness of the scheme, erasing of the group key, and issuing of redundant certificates have been dealt with by the authors as follows:

Increased robustness with erased group keys and redundant certificates: One effect of the expiration or revocation of a leader key is that it not only causes the removal of that leader but also affects every member below the removed leader in the tree structure formed by the certificates. The result is particularly dramatic if the membership of the group key itself is revoked. For if the group key comes under suspicion and needs to be revoked, the whole group perishes. Revocation of the group key may be desired when one wants to replace the group key with a new one and reconstitute the entire group.

Erasing the group key: A perfect way of protecting the private key against a compromise is to erase it. An erased key cannot be recovered or misused in any way. The certificates signed with the erased key continue to be valid, and they can still be verified with the public key. In the group context, the newly generated private group key can be used to certify a few leaders and then erased. Several leaders should be certified with the group key so that if the membership of one of them must be revoked, the remaining leaders can still continue to administer the group. The certificates signed by a key cannot be refreshed after erasure of the key, so when these leader certificates are about to expire, the group needs to be reconstituted. This is shown in Fig. 30.7. In the group context, the newly generated private group key can be used to certify a few leaders and then erased. Several leaders should be certified with the group key so that if the membership of one of them must be revoked, the remaining leaders can still continue to administer the group. The group members should be informed that the group key is protected and cannot be compromised. This way, they know that the group key will never be revoked. For example, Leader 1 in Fig. 30.7 uses the group key to certify its own key and another key as leaders of the group. After signing the certificates, the group key is erased from Leader 1's memory.

Issuing redundant certificates: Erasing the private group key removes a single point of failure, in the sense that there is no single key whose compromise would disable the entire group. However, large

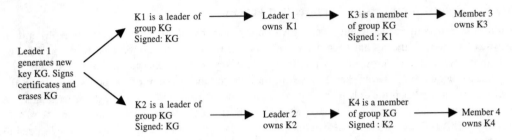

FIGURE 30.7 Erased group key.

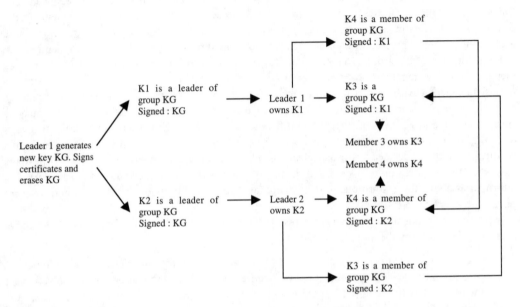

FIGURE 30.8 Redundant certificates.

parts of the certificate tree can still be removed from the group by revoking one of the leaders certified by the group key. A member can alleviate this threat by obtaining multiple independent certificates. The leaders may also issue redundant certificates to each other. This is shown in Fig. 30.8 below. Even if Leader 2 loses its authority (its leader certificate expires or its membership is revoked), Member 4 still remains a member in the group. If the membership of either Member 3 or Member 4 is revoked, redundant certificates do not prevent or complicate the revocation in any way. In fact, by revoking the whole membership of an untrusted member, the leader need not be aware of the certificates that have been issued to the compromised member. This is why the choice has been made to revoke the memberships and not single membership certificates.

Another Robust Scalable Ubiquitous Security Scheme

Yet another design that supports ubiquitous security for mobile nodes, scales to network size, and is robust against adversary break-ins is suggested by Luo and Lu [12]. In their design, they distribute the functionality of conventional security servers, specifically the authentication services, so that each individual node can potentially provide other nodes' certification services. Centralized management is minimized, and the nodes in the network collaboratively self-secure themselves. A suite of fully distributed and localized protocols is proposed that facilitate practical deployment, and these protocols also feature communication efficiency to conserve the wireless channel bandwidth and independence from both the underlying transport layer protocols and the network layer routing protocols. The focus is on the authentication service in ad hoc wireless networks. These authors' work is based on asymmetric cryptographic techniques, specifically the *de facto* standard RSA[15] algorithms. Once the authenticated channels are established with proper access control between communicating parties, confidentiality, integrity, and nonrepudiation can be further realized by following the typical Diffie–Hellman key exchange protocols.

The authors employ several techniques to achieve the design goals:

1. Ubiquitous authentication service availability is achieved by taking a certificate-based approach. Any two communicating nodes establish a temporary trust relationship via globally verifiable certificates. With a scalable threshold sharing of the certificate-signing key, certification services, such as certificate issuing/renewal and revocation, are distributed to each node in the network. No single node holds the complete certificate-signing key. Each node only possesses a share of it. While no single node has the power of providing full certification services, multiple nodes in a network locality can collaboratively provide such services in the same way that an authority with a complete certificate-signing key would.
2. By the distributed certification services, together with the further enhancement of a scalable proactive update mechanism, service robustness is ensured in the presence of short-term computation bounded adversaries.
3. While the certification service distribution and periodic proactive update can be solved in theory using known cryptographic techniques such as threshold secret sharing, threshold multi-signature, and proactive RSA, the approach focuses on scalable and practical solutions in large-scale ad hoc networks with dynamic node membership.

The proposed fully localized (typically within a one-hop neighborhood) and distributed protocols further achieve communication efficiency and load balancing over the network to avoid network congestions [12]. Through the localized design, these communication protocols are immune from the unreliability of the underlying transport layer protocols and routing mechanisms in ad hoc wireless networks. Furthermore, the design has two additional features:

1. *Provable cryptographic security.* The proposed security algorithms are as secure as the underlying cryptographic primitives (e.g., RSA) by the simulatability arguments.
2. *Self-defensive, built-in detection mechanisms.* While the design works with any intrusion detection algorithms and mechanisms that are of each individual node's choice, verifiable techniques are applied as built-in mechanisms to detect adversaries that attack the security protocols.

We have, thus, covered some of the key management proposals in this subsection. We move on to a relatively new field, intrusion detection. Intrusion prevention measures, such as encryption and authentication, can be used to reduce intrusions but cannot eliminate them. For example, encryption and authentication cannot defend against compromised mobile nodes that carry the private keys. Integrity validation using redundant information (from different nodes) also relies on the trustworthiness of other nodes, which could likewise be a weak link for sophisticated attacks. Therefore, a second wall of defense that provides the capability to detect intrusions and to alarm users is needed.

30.5.3 Intrusion Detection

Intrusion detection involves capturing audit data and reasoning about the evidence in the data to determine whether the system is under attack. Based on the type of audit data used, intrusion detection systems (IDS) can be categorized as network-based or host-based. A network-based IDS normally runs at the gateway of a network and "captures" and examines network packets that go through the network hardware interface. A host-based IDS relies on operating system audit data to monitor and analyze the events generated by programs or users on the host. Intrusion detection techniques can be categorized into *misuse detection* and *anomaly detection*. Misuse detection systems use patterns of known attacks or weak spots of the system to match and identify known intrusions. Anomaly detection systems flag observed activities that deviate significantly from the established normal usage profiles as anomalies, i.e., possible (known or new) intrusions. We begin our discussion with a pioneering paper in IDS by Zhang and Lee [13], which forms the basis for most of the current work being done in this area.

Intrusion Detection and Response Architecture

An architecture for intrusion detection and response has been proposed in which every node in the wireless ad hoc network participates in intrusion detection and response [13]. Each node is responsible for detecting signs of intrusion locally and independently, but neighboring nodes can collaboratively investigate over a broader range. In the system's aspect, individual IDS agents are placed on each and every node. Each IDS agent runs independently and monitors local activities (including user and systems activities and communication activities within the radio range). It detects intrusion from local traces and initiates response. If an anomaly is detected in the local data, or if the evidence is inconclusive and a broader search is warranted, neighboring IDS agents will cooperatively participate in global intrusion detection actions. These individual IDS agents collectively form the IDS system to defend the wireless ad hoc network.

The internal of an IDS agent is structured into six pieces as shown in Fig. 30.9. The data collection module is responsible for gathering local audit traces and activity logs. Next, the local detection engine uses these data to detect local anomaly. Detection methods that need broader data sets or that require collaborations among IDS agents use the cooperative detection engine. Both the local response and global response modules provide intrusion response actions. The local response module triggers actions local to this mobile node, for example, an IDS agent alerting the local user, while the global one coordinates actions among neighboring nodes, such as the IDS agents in the network electing a remedy action. Finally, a secure communication module provides a high-confidence communication channel among IDS agents.

The type of intrusion response for wireless ad hoc networks depends on the type of intrusion, the type of network protocols and applications, and the confidence (or certainty) in the evidence. A few likely responses include:

1. Reinitializing communication channels between nodes (e.g., forced rekey)
2. Identifying the compromised nodes and reorganizing the network to preclude the promised nodes. For example, the IDS agent can notify the end-user, who may in turn do an investigation and take appropriate action. It can also send a "reauthentication" request to all nodes in the network to prompt the end-users to authenticate themselves (and hence their wireless nodes), using out-of-bound mechanisms (such as, for example, visual contacts). Only the reauthenticated nodes, which may collectively negotiate a new communication channel, will recognize each other as legitimate. That is, the compromised/malicious nodes can be excluded.

Intrusion detection can complement intrusion prevention techniques (such as encryption, authentication, secure MAC, secure routing, etc.) to improve the network security. The authors recommend that the architecture for better intrusion detection in wireless ad hoc networks should be distributed and

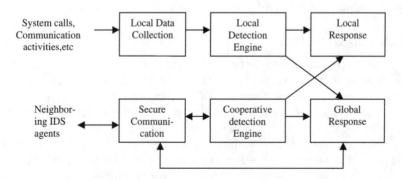

FIGURE 30.9   A conceptual model of an IDS agent.

cooperative. A statistical anomaly detection approach should be used. The trace analysis and anomaly detection should be done locally in each node and possibly through cooperation with all nodes in the network. Further, intrusion detection should take place in all networking layers in an integrated cross-layer manner.

An Intrusion Detection Model for Security Enhancement in AODV Protocol

An Intrusion Detection Model (IDM) has been proposed to enhance security in the Ad hoc On Demand Distance Vector (AODV) protocol [14]. Each node employs IDM that utilizes neighborhood information to detect misbehavior of its neighbors. When the misbehavior count for a node exceeds a predefined threshold, the information is sent out to other nodes as part of global response. The other nodes receive this information and check their local "malcount" for the broadcast malicious node and add their result to the initiator's response. All this leads to secure communication in the AODV protocol.

In the Intrusion Response Model (IRM), a node identifies that another has been compromised when its malcount increases beyond the threshold value for that allegedly compromised node. In such cases, it propagates this information to the entire network by transmitting a special type of packet called a "MAL" packet. If another node also suspects that the node that has been detected is compromised, it reports its suspicion to the network and retransmits another special type of packet called a "REMAL" packet. If two or more nodes report about a particular node, another of the special packets called a "PURGE" packet is transmitted to isolate the malicious node from the network. All nodes that have a route through the compromised node look for newer routes. All packets received from a compromised node are dropped. Figure 30.10 illustrates how the above-mentioned proposed handling of internal attacks is done.

Some of the internal attacks include distributed false route request, DoS, impersonation, and compromise of a destination. The authors have proposed to identify these internal attacks in the following ways:

Distributed false route request: A route request is generated whenever a node has to send data to a particular destination. A malicious node might send frequent, unnecessary route requests. Moreover, if a malicious node generates a false route message from a different range, it will be difficult to identify the malicious node. Route request messages are broadcast messages. When the nodes in the network receive a number of route requests that is greater than a threshold count by a specific source for a destination in a particular time interval, the node is declared malicious, and this information is propagated throughout the network.

Denial-of-service: A malicious node launches a denial-of-service attack by transmitting false control packets and using the entire network resources. DoS can be launched by transmitting false routing messages or data packets. It can be identified if a node is generating control packets that are more than the threshold count in a particular time interval.

Destination is compromised: A destination might not be able to reply if it: (1) is not in the network, (2) is overloaded, (3) did not receive the route request, or (4) is malicious. This attack is identified when the source does not receive the reply from the destination in a particular time interval. The neighbors generate probe/hello packets to determine connectivity. If the node is in the network and does not respond to a route request destined for it, it is identified as malicious.

Impersonation: This attack can be avoided if the sender encrypts the packet with its private key, and other nodes decrypt with the public key of the sender. If the receiver is not able to decrypt the packet, the sender might not be the real source, and hence, the packet is dropped.

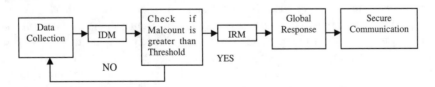

FIGURE 30.10 Handling of internal attacks.

TIARA

Research efforts at Architecture Technology Corporation are aimed at demonstrating a set of innovative design techniques, collectively called TIARA (Techniques for Intrusion-resistant Ad Hoc Routing Algorithms), that strengthen ad hoc networks against denial of service attacks [15]. The TIARA mechanisms limit the damage sustained by ad hoc networks from intrusion attacks and allow for continued network operation at an acceptable level during such attacks. The TIARA mechanisms provide protection against attacks on routing traffic as well as data traffic; routing algorithm independence allows for widespread applicability and supports secure enclaves for dynamic coalitions. These mechanisms are designed to handle attacks on the routing traffic as well as the data traffic in ad hoc networks, thereby providing a comprehensive defense against intruders.

The TIARA approach involves [15]:

Fully distributed lightweight firewalls for ad hoc wireless networks
Distributed traffic policing mechanisms
Intrusion tolerant routing
Distributed intrusion detection mechanisms
Flow monitoring
Reconfiguration mechanisms
Multi-path routing
Source-initiated route switching

The Flow-based Route Access Control (FRAC) rules define admissible flows. Per-flow security association is instituted by a secure session setup signaling protocol and contains information for packet authentication. Also, fast authentication enables low-overhead integrity checks on packet flow-IDs and sequence numbers.

There is referral-based resource allocation, which limits networks' exposure to resource usurpation by spurious sessions, and flows are assigned an initial allowable resource usage. Moreover, additional resources are only granted if the source of the flow can present referrals from a certain number of trusted nodes. Referrals have time-bound validity. Flow-specific sequence numbers limit and contain the impact of traffic replay attacks and sequence number embedded within secret locations within each packet. The destination of flow monitors selects flow parameters to detect intrusion-induced path failures. Multipath-routing and source-initiated route switching diverts flow through available alternate paths to circumvent intruders. Efforts are underway to implement the following:

Dynamic, on-the-fly modifications to FRAC (firewall) policies
Real-time referral based resource allocation
Lightweight implementation of traffic policing
Fast authentication mechanisms that are resistant to traffic analysis
Embedding sequence numbers and path labels in encrypted packets

A summary of countermeasures used in TIARA against intrusion attacks is shown in Table 30.1.

Conceptually, an intrusion detection model, i.e., a misuse detection rule or a normal profile, has these two components: the *features* (or attributes, measures), e.g., "the number of failed login attempts," "the averaged frequency of the *gcc* command," etc., that together describe a logical event, e.g., a user login session; and the *modeling algorithm,* e.g., rule-based pattern matching, that uses the features to identify intrusions. Defining a set of *predictive* features that accurately capture the representative behaviors of intrusive or normal activities is the most important step in building an effective intrusion detection model, and can be independent of the design of modeling algorithms. In the next section, we look at how authentication can be achieved, the most typical case being the use of a password. Authentication, in general, has several security requirements. These include protection against replay attacks, confidentiality, resistance against man-in-the-middle attacks, etc.

TABLE 30.1 Proposed Countermeasures in TIARA Against Intrusion Attacks

Intrusion Attacks → Countermeasures ↓	Spurious Traffic	Packet Replay	Session Flooding	Flow Disruption	Route Hijacking
FRAC	X				
Fast Authentication	X	X	X		
Sequence Numbers		X			
Referrals			X		
Flow Monitoring				X	X
Multi-path Routing				X	X
Source Init Route Switching				X	X

30.5.4 Authentication

Authentication usually means that there is some way to ensure that the entity to which you are talking is what it claims to be. This is called authentication of the channel end point. Usually, you also need to authenticate yourself to the service in order for the service to be sure that you are you, not someone else who is pretending to be you. This is the authentication of the message originator. The use of a password is not really a good choice, because passwords are typically short and easy to break. More secure methods include the use of public key cryptography, challenge–response schemes, symmetric encryption, etc.

MANET Authentication Architecture

The MANET Authentication Architecture (MAA) proposed by Corson and Jacobs [16] has emphasis on building a hierarchy of trust relationship in order to authenticate Internet MANET Encapsulation Protocol (IMEP) messages' security. The proposed scheme details the formats of messages, together with protocols that achieve authentication. The difficulties with proactive schemes are that, first, cryptography is relatively computationally expensive on mobile hosts, where computational capability is comparatively restricted; second, since no central authority can be depended on, the authentication is more difficult to implement, and third, it is only useful to prevent intruders from outside (external attack). If an internal node is compromised (internal attack), such schemes no longer work.

MAA supports multiple authentication options ranging from simple to complex. The IMEP Authentication Object is used to authenticate all IMEP messages between routers; this is accomplished by calculating an Authenticator ("Digital Signature"). Also, MAA identifies the security context between a pair of MANET nodes. The certificate object, though optionally used depending on the security context between corresponding MANET nodes, includes a copy or copies of certificates that bind system "distinguished names" to public keys using a digital signature. The trust hierarchy paths establish a logical chain between two CAs and establish trust relationships through intervening CAs. Certificate validation involves constructing a trust hierarchy path among the sender certificate, the CA that issued the sender certificate, and the CA of the validating system. Establishing a trust hierarchy path must be performed to verify authenticity and usability of certificates within IMEP. The receiver can develop trust in the public key of the sender's CA recursively, if the receiver has a certificate containing the CA's public key signed by a superior CA that it already trusts. Each certificate is processed in turn, starting with that signed using the input trusted public key. There is provision for Certificate Revocation Lists (CRL), and certificates are checked against current CRL from issuing CA. A MANET node caches received certificates along with a value ("staleness value"). The node maintains a maximum degree ("staleness threshold") value of certificate staleness tolerable before the node has to retrieve the appropriate CRL and verify that the certificate has not been revoked.

An End-To-End Data Authentication Scheme

An end-to-end data authentication scheme for ad hoc networks that relies on mutual trust between nodes has been suggested in which the basic strategy is to take advantage of the hierarchical structure that is implemented for routing purposes [17]. The scheme uses TCP at the transport layer and a hierarchical

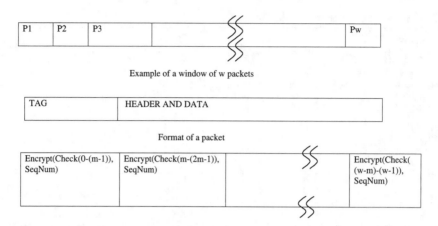

FIGURE 30.11 A packet containing the checks for the tags.

architecture at the IP layer so that the number of encryptions needed is minimized, thus reducing the computational overhead. Also, each node has to maintain keys for fewer nodes. The scheme makes use of a cluster-based network with a Cluster Based Routing Protocol (CBRP)[52]. When a node joins the network, it is given a system public key and a system private key, as well as a cluster key. This cluster key is unique to the cluster. Each node has a table of cluster IDs and the corresponding head's public key. When a node joins a network for the first time, a strong authentication is done by sending a challenge and receiving a response. The system key pair is used for a mutual authentication between the joining node and an existing member of the network. When a node leaves a cluster and joins another cluster, the new cluster head treats this joining node as any other new node joining its cluster. A mutual authentication is performed between the moved node and its new cluster head using the system key pair. The new joining node gets the cluster ID whereas the old cluster head purges the entry for the node that moved out. When a node in one cluster wants to communicate with a node from another cluster, for complete confidentiality, the entire packet is encrypted with a session key. This session key is shared between only the two communicating parties and thus serves as authentication. For replay prevention, strong authentication may be performed for each packet, i.e., a series of challenge and response back and forth.

The proposed algorithm involves exchange of a session key that is valid for just that particular TCP session after mutual authentication with the cluster heads acting as CAs. The heads generate a set of random prime numbers, which are first encrypted with the head's private key and then the cluster key. With each number, a timestamp is also encrypted for limited usage. The head then broadcasts them. These could serve as authentication tags for any of the cluster members (see Fig. 30.11). They are also encrypted with the session key for more protection. The receiver utilizes a check function to verify the origin and authenticity of the tags from the sender. The checksum field of the TCP header is also encrypted with the session key for more security. The check function has encrypted sequence numbers for which it is valid and thus, packets cannot be replayed. The cluster head needs to generate random prime number sets periodically, and a session key needs to be generated for every session.

Authentication, Authorization, and Accounting (AAA)

One way to deal with low physical security and availability constraints is the distribution of trust. Trust can be distributed among a collection of nodes [18]. Public key cryptography can be used to do the *authentication* after having built a key management system. *Authorization* is also needed, to avoid allowing a malicious host to wreak havoc inside the network. This can be prevented by keeping control of what hosts are allowed to do inside the ad hoc network. Authorization also needs some sort of distributed structure, because one cannot rely on one point of failure. In ad hoc networks, individual mobile hosts are providing service to each other, which gives rise to *accounting*. For example, if some mobile node acts as a router in the network, providing connectivity between two nodes that are not within each other's range, then it would be reasonable to charge some money for this service. There exist no protocols to do the actual charging if that is needed. Because connectivity to some central server that takes care of the charging cannot be assumed, there is a clear need for distributed charging protocols as well.

Ad hoc networks and general AAA systems do not fit well together. The biggest problem is related to the varying nature of the network. There are no home domains or foreign domains, because the networks come and go on demand. This affects the AAA systems because some of the basic building blocks of their architecture are missing in ad hoc networks. The basic problem here is that the general AAA model is a centralized trust model, whereas the ad hoc network structure is decentralized. There is a need for some other kinds of methods to achieve the AAA functionality. One approach to provide authentication and authorization functionality in ad hoc networks could be to use trust management-based approaches such as PolicyMaker[16] or Keynote,[17] which are decentralized by nature and can provide the requested functionality in ad hoc networks quite easily. Also, other protocols such as Simple Authentication and Security Layer (SASL[18]) or Internet Security Association and Key Management Protocol/Internet Key Exchange (ISAKMP[19]/IKE[20]) could be used to provide the authentication functionality. Ad hoc networks probably need decentralized models or some other approaches to provide the AAA functionality.

Having discussed some of the schemes to ensure authentication, we take a look at how routing can be secured in the following section. In an ad hoc network, the absence of an underlying infrastructure means that each of the nodes involved has to act as a router by itself, and the likelihood of a node being compromised is high. This necessitates the incorporation of security measures not only in the inter-router communication but also in verifying the nodes *per se*.

30.5.5 Secure Routing

A basic distinction can be made when the issue of secure routing is considered. Attacks on routing can be external as well as internal, and this means that there is a need to come up with schemes to safeguard the routing process. We first briefly discuss how internal and external attacks can be handled before going into routing protocol-specific security features.

Countering External Attacks

To ensure the confidentiality of routing messages, they should be encrypted using DES.[21] This encryption algorithm is chosen for its computational efficiency. To authenticate the messages, the use of digital signatures and keyed one-way functions with windowed sequence numbers is recommended [19]. This provides data origin authentication, content integrity, and antireplay protection for neighbor discovery and transport of subscriber traffic. The RSA algorithm is applied because of its relatively quick signature verification. For external attacks, routing messages should be encrypted with a private key algorithm and authenticated using digitally signed message digests with windowed sequence numbers.

Countering Internal Attacks

In some routing protocols, false routing information is treated as outdated information and the inherent redundancy characteristics are used to circumvent the compromised node [3,19]. If a node is suspected of being compromised, backup routes can be used instead. There is already redundant message transmission in a mobile (wireless) environment because all nodes within radio range of a sending node have the ability to receive all messages that a sending node transmits.

Perlman uses byzantine robustness to protect routing data from compromised nodes [20]. Byzantine robustness is an ability to continue correct operation in the presence of arbitrary nodes with byzantine failures. Perlman's study analyzes the theoretical feasibility of maintaining network connectivity. Basically, detection of inconsistency and isolation of compromised routers are accomplished using the redundant information. Each node maintains a small database, which contains the sequence number packets that it has forwarded from other nodes. Querying the network nodes would allow any inconsistent behavior by a particular node to become apparent. Thus, such compromised nodes can then be isolated. Perlman's Ph.D. thesis, Network-layer Protocol with Byzantine Robustness (NPBR), proposes a theoretical but robust routing protocol applying two modes: flooding and link state modes. In the former NPBR mode the routing robustness is guaranteed by flooding packets, performing public-key encryptions per packet and link and ignoring significant state in routers. The link state NPBR is less robust, but it allows N − 1 compromised routes in case of N redundant routes per two nodes.

Secure Distance-Vector Routing Protocols

There is a method to secure distance-vector routing protocols that could be used in ad hoc networks [21]. This routing protocol performs route computation on a per-destination network basis, and it maintains information about the second-to-last network with distance information from each neighbor to every destination in the network. To ensure the authenticity and integrity of the information, the originating node digitally signs the unchanging fields of each update it generates. An IP address of the originating node is added to each update to allow receiving nodes to validate the signature. The update provides protection against the compromised nodes that have the cryptographic keys as shown in Fig. 30.12.

The digital signature of each update protects the update from fabrication or modification by compromised nodes. Redundant paths, "aging out" of false routing information, and redundancy in routing information at each node are all employed to combat against internal attacks. The following countermeasures effectively implement security services not available from lower level transport or network protocols. Specifically, the routing message digital signature and sequence numbers provide authentication and integrity services of routing messages, which compose the first class of communication. The countermeasures work in these ways:

Routing message sequence number: A sequence number is included in each routing message, which is initialized to zero on the initialization of a newly booted router, and is incremented with each message. On detection of a skipped or repeated sequence number, a reset of the session is forced by the reinitialization of the routing process. The size of this sequence number is made large enough to minimize the chance of its cycling back to zero. However, in the event that it does, the session is reset by the reinitialization of the routing process.

Routing message digital signature: Each routing message is digitally signed by the sender. This provides authenticity and some degree of integrity (protection from message modification but not from replay) of the routing dialog. On detection of corruption, the message is dropped.

The proposal suggests adding sequence number information and predecessor information to updates and digitally sign these updates as shown in Fig. 30.12.

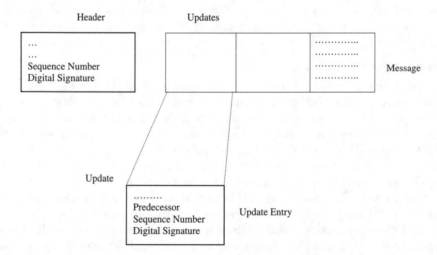

FIGURE 30.12 Proposed routing message changes in securing distance-vector protocols.

Open Shortest Path First Protocol (OSPF)

OSPF is a link state routing protocol used within one autonomous system (AS) or routing domain. Some inherent properties of OSPF as a routing protocol make it very robust to failures and some attacks [22].

OSPF uses flooding for the dissemination of link state advertisements (LSA). This makes sure that within the same *area,* all routers have the identical topological database. Even if a router goes down, other routers can still exchange their link state information provided that an alternate path exists. Furthermore, the link state information propagated in the network is the raw message generated by the original router instead of the summarized information from neighbors as in distance vector routing. This makes it easy to protect the authenticity of the information.

Also, OSPF is a two-level routing protocol with intra-area routing and inter-area routing. Area Border Routers (ABR) connect to the backbone and exchange summarized area information. Since intra-area routing depends only on information from within that area, it is not vulnerable to problems out of the area. Also, problems in one area do not influence the intra-area routing of other areas and inter-area routing among other areas. So hierarchy routing has a security advantage. The latest version of this protocol includes two authentication methods. The first is a simple password scheme wherein the OSPF header carries a plaintext password so that the routers within the routing domain can share a secret for authentication. This is not secure since the password is transmitted in the clear. Another much stronger authentication algorithm is cryptographic message digest, e.g., keyed MD5,[22] with the assumption that routers on a common network share a secret key. This is a symmetric cryptographic scheme. If all the routers share the same secret key, then the security level is low. If each pair of routers shares a secret key, it requires a $O(N^2)$ set of secret keys. So the key distribution process will be very complex.

Murphy and Badger have proposed a digital signature scheme to protect the OSPF routing protocol [23]. Since digital signature is a public key scheme, the number of keys is on the order of $O(N)$. The basic idea of this scheme is to add a digital signature to the OSPF LSA packet and use message digest (such as keyed MD5) to protect all exchanged messages. The originator of the LSA will sign the message, and the signature will stay with the data during the OSPF flooding process, thus protecting the message integrity and providing authentication for LSA data. The key management and distribution also make use of a type of signed LSA. The digital signature scheme can prevent external attackers. Since external attackers cannot generate correct signature for LSAs, if they intercept the LSA and modify it or inject some malicious information into the system, they can be detected. However, some disadvantages still remain, including the following:

1. *MaxAge problem:* The age field is the only element of LSA that is not protected by digital signature. The attacker can modify the age field to the maximum value.
2. *Area Border Routers (ABR):* ABRs run a distance vector routing-like protocol. Even with this protocol, the ABRs can generate false information in the summary LSAs about their attached area and inject into the backbone. They can also inject false information about the backbone into their attached areas.
3. *Autonomous System Boundary Routers (ASBR):* The ASBRs can generate false routing information. It is impossible to double-check the information as the ABRs do.
4. *Internal routers:* Internal routers can generate incorrect routing information because of faulty configuration or bugs. If an internal router is compromised, then the attacker can control the router. This kind of faulty information and attack is difficult to prevent because the digital signature is correctly generated. An internal attacker can also generate bogus information, for example, announcing a nonexistent link.
5. *High cost:* One drawback of the algorithm is that public key cryptography is very expensive, and it will slow performance of the router, which should be fast.

Mitigating Routing Misbehavior

Two schemes have been proposed [24] to alleviate the detrimental effects of packet dropping in secure data forwarding:

1. Detecting misbehaving nodes and reporting such events
2. Maintaining a set of metrics that reflect the past behavior of other nodes

Each node may choose the "best" route, comprised of relatively well-behaved nodes; i.e., nodes that do not have a history of avoiding forwarding packets along established routes. The assumptions include a shared medium, bidirectional links, use of source routing (i.e., packets carry the entire route that becomes known to all intermediate nodes), and no colluding malicious nodes. Nodes operating in the promiscuous mode overhear the transmissions of their successors and may verify whether the packet was forwarded to the downstream node and check the integrity of the forwarded packet. Upon detection of a misbehaving node, a report is generated, and nodes update the rating of the reported misbehaving node. The ratings of nodes along a well-behaved route are periodically incremented, while reception of misbehavior alert dramatically decreases the node rating. When a new route is required, the source node calculates a path metric equal to the average of the ratings of the nodes in each of the route replies, and it selects the route with the highest metric. The detection mechanism exploits two features that frequently appear in MANET: the use of a shared channel and source routing. The possibility of falsely detecting misbehaving nodes could easily create a situation with many nodes falsely suspected for a long period of time. In addition, the metric construction may lead to a route choice that includes a suspected node, if, for example, the number of hops is relatively large, so that a low rating is "averaged out." Finally, the most important vulnerability is the proposed feedback itself; there is no way for the source, or any other node that receives a misbehavior report, to validate its authenticity or correctness. Consequently, the simplest attack would be to generate fake alerts and eventually disable the network operation altogether. The protocol attempts new route discoveries when none of the route replies is free of suspected nodes, with the excessive route request traffic degrading the network performance. At the same time, the adversary can falsely accuse a significant fraction of nodes within the timeout period related to reinstating from a negative rating and, essentially, partition the network.

The Currency Concept

The authors address the issue of service availability and propose a security mechanism to stimulate end users to keep their devices turned on, to refrain from overloading the network, and to thwart tampering aimed at converting the device into a "selfish" node [25]. Their solution is based on the application of a tamper resistant module in each device and cryptographic protection of messages. One approach to ensure that nodes comply with protocol rules, for example, properly relay user data, is to provide incentive to nodes. The concept of fictitious currency is introduced in order to _endogenize_ the behavior of the assumed greedy nodes, which would forward packets in exchange for currency. Each intermediate node purchases from its predecessor the received data packet and sells it to its successor along the path to the destination. Eventually the destination pays for the received packet. This scheme assumes the existence of an overlaid geographic routing infrastructure and a _Public Key Infrastructure (PKI)_. All nodes are preloaded with an amount of currency, have unique identifiers, and are associated with a pair of private/ public keys, and all cryptographic operations related to the currency transfers are performed by a physically tamper-resistant module. The applicability of the scheme, which targets wide-area MANET, is limited by the assumption of an online certification authority in the MANET context. Moreover, nodes could flood the network with packets destined to nonexistent nodes and possibly lead nodes, unable to forward purchased packets, to starvation. The practicality of the scheme is also limited by its assumptions, the high computational overhead (hop-by-hop public key cryptography for each transmitted packet), and the implementation of physically tamper-resistant modules.

Security-Aware Ad Hoc Routing (SAR)

The protection of the route discovery process has been regarded as an additional Quality-of-Service (QoS^{23}) issue [26] because it involves choosing routes that satisfy certain quantifiable security criteria. In particular, nodes in a MANET subnet are classified into different trust and privilege levels. A node initiating a route discovery sets the sought security level for the route; i.e., the required minimal trust level for nodes participating in the query/reply propagation. Nodes at each trust level share symmetric

TABLE 30.2 Security Aware Ad Hoc Routing Properties

Property	Techniques
Timeliness	Timestamp
Ordering	Sequence number
Authenticity	Password; certificate
Authorization	Credential
Integrity	Digest; digital signature
Confidentiality	Encryption
Nonrepudiation	Chaining of digital signatures

encryption and decryption keys. Intermediate nodes of different levels cannot decrypt in-transit routing packets, or determine whether the required *QoS* parameter can be satisfied, and simply drop them. Although this scheme provides protection (e.g., integrity) of the routing protocol traffic, it does not eliminate false routing information provided by malicious nodes. Moreover, the proposed use of symmetric cryptography allows any node to corrupt the routing protocol operation within a level of trust by mounting virtually any attack that would be possible without the presence of the scheme. Finally, the assumed supervising organization and the fixed assignment of trust levels does not pertain to the MANET paradigm. In essence, the proposed solution transcribes the problem of secure routing in a context where nodes of a certain group are assumed to be trustworthy, without actually addressing the global secure routing problem. A generalized SAR protocol has, thus, been proposed for quantifiable secure route discovery and propagation with trust levels and security attributes as metrics.

Table 30.2 illustrates the schemes generally utilized in order to ensure that certain security needs are met; for example, timestamps are used to ensure timeliness, and sequence numbers are used to ensure ordering.

Internet MANET Encapsulation Protocol

A multipurpose network-layer protocol named the Internet MANET Encapsulation Protocol (IMEP) [27] has been designed to support the operation of many routing algorithms, network control protocols, or other Upper Layer Protocols (ULP) (where "upper" denotes any layer above IMEP) intended for use in MANETs as shown in Fig. 30.13. IMEP will run at the network layer and will be an adjunct to whichever network protocol is using it. ULP packets will be encapsulated in IMEP messages, which will be further encapsulated into IP packets.

The protocol incorporates mechanisms for supporting link status and neighbor connectivity sensing, control packet aggregation and encapsulation, one-hop neighbor broadcast (or multicast) reliability,

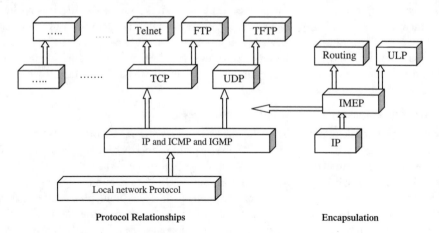

Protocol Relationships   **Encapsulation**

FIGURE 30.13 IMEP in the overall protocol stack.

multipoint relaying and network-layer address resolution, and provides hooks for inter-router authentication procedures. Indirectly, the IMEP also puts forth a framework for MANET router and interface identification and addressing. The primary purpose of the IMEP is to improve overall network performance by reducing the number of network control packet broadcasts through encapsulation and aggregation of multiple MANET control packets (e.g., routing protocol packets, acknowledgments, link status sensing packets, "network-level" address resolution, etc.) into larger IMEP messages. Usage of the IMEP is desirable because per-message, multiple access causes delay in contention-based schemes and, thus, favors the use of fewer, larger messages. It also may be useful in reservation-based, time-slotted access schemes where smaller packets should be aggregated into appropriately sized IP packets for transmission in a given time slot. Its secondary purpose concerns the commonality of certain functionality in many network-level control algorithms. Many algorithms intended for use in a MANET require common functionality such as link status sensing, security authentication with adjacent routers, one-hop neighbor broadcast (or multicast) reliability of control packets, etc. This common functionality can be extracted from these individual protocols and put into a unified, generic protocol useful to all. MANET control algorithms would also benefit from a common approach to router and interface identification and addressing, and this protocol supports a framework for unifying the protocols under a common architecture. Among its many uses, the networking protocol enables inter-router communication for purposes of network control. However, this control function could be a significant vulnerability if IMEP messages are not authenticated.

Secure Ad Hoc On-Demand Distance Vector Routing Protocol

An extension of the *Ad Hoc On-demand Distance Vector* (called *Secure-AODV [S-AODV]*) routing protocol has been proposed [28] to protect the routing protocol messages. The S-AODV scheme assumes that each node has certified public keys of all network nodes, so that intermediate nodes can validate all in-transit routing packets. The basic idea is that the originator of a control message appends a RSA signature and the last element of a *hash chain* (i.e., the result of n consecutive hash calculations on a random number). As the message traverses the network, intermediate nodes cryptographically validate the signature and the hash value, and generate the kth element of the hash chain, with k being the number of traversed hops, and place it in the packet. The route replies are provided either by the destination or intermediate nodes having an active route to the sought destination, with the latter mode of operation enabled by a different type of control packets.

The use of public-key cryptography imposes a high processing overhead on the intermediate nodes and can be considered unrealistic for a wide range of network instances. Furthermore, it is possible for intermediate nodes to corrupt the route discovery by pretending that the destination is their immediate neighbor, advertising arbitrarily high sequence numbers, and altering (either decreasing by one or arbitrarily increasing) the actual route length. Additional vulnerabilities stem from the fact that the IP portion of the S-AODV traffic can be trivially compromised, since it is not protected, unless additional hop-by-hop cryptography and accumulation of signatures is used. Finally, the assumption that certificates are bound with IP addresses is unrealistic; roaming nodes joining MANET subdomains will be assigned IP addresses dynamically or even randomly.

Secure Message Transmission (SMT) Protocol

A different approach is taken by the *Secure Message Transmission (SMT)* protocol [29], which, given a topology view of the network, determines a set of diverse paths connecting the source and the destination nodes. It then introduces limited transmission redundancy across the paths by dispersing the message into N pieces, so that successful reception of any M-out-of-N pieces allows the reconstruction of the original message at the destination. Each piece is equipped with a cryptographic header that provides integrity and replay protection along with origin authentication and is transmitted over one of the paths. Upon reception of a number of pieces, the destination generates an acknowledgment informing the source which pieces, and thus routes, were intact. In order to enhance the robustness of the feedback mechanism, the small-sized acknowledgments are maximally dispersed (i.e., successful reception of at least one piece is sufficient) and are protected by the protocol header as well. If fewer than M pieces were received, the source retransmits the remaining pieces over the intact routes. If too few pieces were acknowledged or too many messages remain outstanding, the protocol adapts its operation by determining a different path set, reencoding undelivered messages, and reallocating pieces over the path set. Otherwise, it proceeds with subsequent message transmissions.

The protocol exploits MANET features such as the topological redundancy, interoperates widely with accepted techniques such as source routing, relies on a security association between the source and the destination, and makes use of highly efficient symmetric-key cryptography. It does not impose processing overhead on intermediate nodes, while the end nodes make the routing decisions, based on the feedback provided by the destination and the underlying topology discovery and route maintenance protocols. It is noteworthy that SMT provides a limited protection against the use of compromised topological information, although its main focus is to safeguard the data forwarding operation. The use of multiple routes compensates for the use of partially incorrect routing information, rendering a compromised route equivalent to a route failure. The usefulness of this protection is also limited, since an attack cannot always be detected by the route endpoints, which would be necessary in order to switch to another route. Nevertheless, the disruption of route discovery can still be the most effective way for adversaries to consistently compromise the communication of one or more pairs of nodes.

Dynamic Destination Multicast Protocol

In the *Dynamic Destination Multicast (DDM)* protocol, the group membership is not restricted in a distributed manner, as only the sender of the data is given the authority to control to whom the information is really delivered. In this way, the DDM nodes become aware of the membership of groups of nodes by inspecting the protocol headers. The DDM approach also prevents outsider nodes from joining the groups arbitrarily. This is not supported in many other protocols directly; if the group membership and the distribution of source data have to be restricted, external means such as the distribution of keys have to be applied. DDM has two modes of operation: the *stateless mode* and the *soft-state mode*. In the stateless mode, the maintenance of multicast associations and restriction of group membership are handled totally by encoding the forwarding information in a special header of the data packets; the nodes do not have to store state information. This kind of reactive approach thus guarantees that there is no unnecessary exchange of control data during idle periods. Thus, in small ad hoc networks that need not scale up substantially, this kind of ultra-reactive approach can be extremely useful. The soft-state mode, on the other hand, requires that the nodes remember the next hops of every destination and thus need not fill up the protocol headers with every destination. In both modes, the nodes must always be able to keep track of the membership of the groups. DDM is best suited for dynamic networks having small multicast groups. Currently the DDM draft [30] does not, however, propose any solutions for securing the DDM networks as such. Moreover, it does not provide any suggestions for a concrete protocol that handles the necessary access control needed in the restriction of group membership.

Optimized Link State Routing Protocol

The *Optimized Link State Routing (OLSR)* protocol [31] is a proactive, table-driven protocol that applies a multi-tiered approach with *multi-point relays (MPRs)*. MPRs allow the network to apply scoped flooding, instead of full node-to-node flooding, with which the amount of exchanged control data can substantially be minimized. This is achieved by propagating the link state information about only the chosen MPR nodes. Since the MPR approach is most suitable for large and dense ad hoc networks, in which the traffic is random and sporadic, the OLSR protocol works best in this kind of an environment. The MPRs are chosen so that only nodes with a one-hop symmetric (bidirectional) link to another node can provide the services. Thus, in very dynamic networks where a substantial number of unidirectional links are constantly present, this approach may not work properly. OLSR works in a totally distributed manner, e.g., the MPR approach does not require the use of centralized resources. The OLSR protocol specification does not include any actual suggestions for the preferred security architecture to be applied with the protocol. The protocol is, however, adaptable to protocols such as IMEP, as it has been designed to work totally independently of other protocols.

On-Demand Multicast Routing Protocol

On-Demand Multicast Routing (ODMRP) protocol is a mesh-based multicast routing protocol for ad hoc networks [32]. It applies the *scoped flooding* approach, in which a subset of nodes — a *forwarding group* — may forward packets. The membership in the forwarding groups is built and maintained dynamically on demand. The protocol does not apply source routing. ODMRP is best suited for MANETs where the topology of the network changes rapidly and resources are constrained. ODMRP assumes bidirectional links, which somewhat restricts the potential area of application for this proposal; ODMRP may not be suitable for use in dynamic networks in which nodes may move rapidly and unpredictably and have varying radio transmission power. Currently, ODMRP does not define or apply any security means. The forwarding group membership is controlled with the protocol itself, though.

Ad Hoc On-Demand Distance-Vector Routing Protocol

Ad Hoc On-Demand Distance-Vector (AODV) routing protocol is a unicast-based reactive routing protocol for mobile nodes in ad hoc networks [33]. It enables multi-hop routing, and the nodes in the network maintain the topology dynamically only when there is traffic. Currently, AODV does not define any security mechanisms. The authors identify the necessity of having proper confidentiality and authentication services within the routing but suggest no solutions for them. IPSec is, however, mentioned as one possible solution. *Multicast Ad Hoc On-Demand Distance-Vector* routing protocol *(MAODV)* [34] extends the AODV protocol with multicast features. The security aspects currently noted in the design of MAODV are similar to those of the AODV protocol.

Topology Broadcast Based on Reverse-Path Forwarding Protocol

Topology Broadcast based on Reverse-Path Forwarding (TBRPF) protocol is a pure proactive, link-state routing protocol for ad hoc networks that can also be applied as the proactive part in hybrid solutions [35]. Each of the nodes of the network in TBRPF carries state information of each link of the network, but the information propagation is optimized by applying *reverse-path forwarding* instead of the costly full flooding or broadcast techniques. TBRPF operates over IPv4 in ad hoc networks and can also be applied within hierarchical network architecture. The proposal, however, does not suggest any specific mechanisms for securing the protocol. Finally, the protocol, just like every other ad hoc network routing protocol, can be protected with IPSec, but this approach is not currently officially in use within TBRPF.

An Authenticated Link-Level Ad Hoc Routing Protocol

Binkley has designed an authenticated link-level ad hoc routing protocol that addresses link security issues [36]. In this protocol, mobile nodes as well as agents broadcast ICMP router discovery packets. The router discovery packets are authenticated and bind the sender's MAC and IP addresses. This is shown in Fig. 30.14, where mobile node B has moved into range of hosts A and D and has transmitted a beacon. When A and D receive the beacon, they check its authenticity and, assuming it is authentic, add the MAC-to-IP address binding contained in the beacon to their tables of known bindings. Node C is shown as already having a prior binding for B and does not need to add to its bindings.

Problems caused by tying IP subnet schemes to routing on radio links are eliminated by this protocol. Security problems associated with ARP spoofing are also reduced. This link-level protocol is integrated with Mobile-IP on links where increased security is needed. The protocol replaces ARP and may be integrated with higher-level multi-hop ad hoc routing protocols.

This protocol solves various reachability problems associated with traditional IP subnetting and ARP. These improvements aid routing survivability: mobile nodes using this protocol are better able to communicate with other mobile nodes in situations where they would be unable to communicate using Mobile-IP or traditional IP subnetting. Also, the protocol improves link-layer routing security, including problems associated with ARP spoofing. All systems on a given link use either ARP or the ad hoc protocol, but not both. Since beacons authenticate the binding of the IP address to the MAC address,

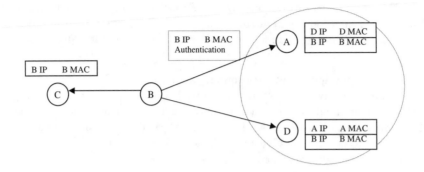

FIGURE 30.14 The protocol being used when mobile node B moves into range of A and D and beacons.

the Mobile-IP ICMP discovery packets are always authenticated for all agents and mobile nodes, unlike standard Mobile-IP. ARP may be easily spoofed, that is, an attacker may send an ARP packet containing its own MAC address and the victim's IP address, thereby usurping the IP-to-MAC address binding of the victim in the other party's ARP cache. This causes the attacker to receive packets intended for the victim. The ad hoc protocol makes spoofing this binding impossible, although MAC address spoofing is still a threat. As another security feature, mobile nodes that lack a link authentication key are unable to gain service from agents on the wired infrastructure. They may send packets to agents, but no packets are sent in return. This makes the perimeter defense against unauthorized mobile nodes more secure. The author presents two key ideas:

1. The conventional IP subnet model that ARP uses is inadequate for mobile systems at the link layer. It may be replaced with a protocol where beacons are used to determine reachability. The protocol is naturally integrated with Mobile-IP as agents already beacon.
2. The beacons binding MAC and IP addresses are authenticated. This gives several security opportunities, including the ability to wall off agents and not let "strangers" access network resources. It minimizes the danger of link-layer spoofing attacks. Authenticated beacons also improve the overall security of Mobile-IP.

Secure Routing Protocol

A route discovery protocol has been presented that mitigates the detrimental effects of malicious behavior, so as to provide correct connectivity information [37]. This protocol guarantees that fabricated, compromised, or replayed route replies would either be rejected or never reach back to the querying node. The sole requirement is the existence of a security association between the node initiating the query and the sought destination. Specifically, no assumption is made regarding the intermediate nodes, which may exhibit arbitrary and malicious behavior. The scheme is robust in the presence of a number of noncolluding nodes and provides accurate routing information in a timely manner. This scheme guarantees that a node initiating a route discovery will be able to identify and discard replies providing false topological information or avoid receiving them. The novelty of the scheme is that false route replies, as a result of malicious node behavior, are discarded partially by benign nodes while in transit towards the querying node or deemed invalid upon reception. As query packets traverse the network, the relaying intermediate nodes append their identifiers (e.g., IP address) in the query packet header. When one or more queries arrive at the sought destination, replies that contain the accumulated routes are returned to the querying node; the source then may use one or more of these routes to forward its data. Reliance on this basic route query broadcasting mechanism allows the proposed *Secure Routing Protocol (SRP)* to be applied as an extension of a multitude of existing routing protocols.

SRP guarantees the acquisition of correct topological information in a timely manner, i.e., the route replies that are validated and accepted by the querying node provide accurate connectivity information, despite the presence of strong adversaries. It combats attacks that disrupt the route discovery process and guarantees the acquisition of correct topological information. It also incorporates mechanisms that safeguard the network functionality from attacks that exploit the protocol itself in order to degrade network performance and possibly lead to denial of service. The protocol introduces a set of features, such as:

The requirement that the query must verifiably arrive at the destination
The explicit binding of network and routing layer functionality
The consequent verifiable return of the query response over the reverse of the query propagation route
Acceptance of route error messages only when generated by nodes on the actual route
Query/reply identification by a dual identifier
Replay protection of the source and destination nodes
Regulation of the query propagation

The protocol is capable of operating without the existence of an online certification authority or the complete knowledge of keys of all network nodes. Its sole requirement is that any two nodes that wish to communicate securely can simply establish *a priori* a shared secret, to be used by their routing protocol modules. Moreover, the correctness of the protocol is retained irrespective of any permanent binding of nodes to IP addresses, a feature of increased importance for the open, dynamic, and cooperative MANET environments.

TABLE 30.3 Security Features in Some of the Routing Protocols in Ad Hoc Networks

Protocol	Security Positives	Security Negatives
OSPF	Flooding and information least dependency; hierarchy routing and information hiding; two authentication methods: a simple password scheme and a cryptographic message digest; a digital signature scheme to protect the OSPF routing protocol	Age field not protected by digital signature; internal routers can generate incorrect routing information; public key cryptography very expensive and will slow performance of the router; Area Border Routers and Autonomous System Boundary Routers can generate false routing information
S-AODV	Public key cryptography used	High overhead; possible route discovery corruption; compromise of IP portion
SMT	Guarantees integrity, replay protection. and origin authentication; interoperability with accepted procedures such as source routing; symmetric key cryptography used	Limited protection against compromised topological information
DDM	Prevents easy joining of outsider nodes; nondistributed group membership; no useless control data exchange in idle periods	No access control (needed for group membership restriction)
OLSR	Minimal control exchange data; can independently work	No guarantee in very dynamic environments
ODMRP		No security means
AODV	Possibility of the use of IPSec	No security means
TBRPF	Possibility of the use of IPSec	No specific mechanisms for security
SRP	Fabricated, compromised, or replay route replies rejected; no online CA; guaranteed acquisition of correct topological information in a timely manner; no complete knowledge of keys by all nodes	Security association as a requirement; possible attack when nodes collude during the two phases of a single route discovery; each SRP query can only discover one route, while diverse routes should be set up to ensure robustness

Dynamics

Dynamics, the hierarchical system for distributing mobility agents proposed by Forsberg et al. [38], mainly focuses on the distribution of nodes and routing issues but includes security features as well. The basic assumption of the proposal is that the Home Agent (HA) and Mobile Node (MN) have a mutual trust relationship and can establish a security association within the registration process. Due to the hierarchical location management system, however, there may be other security associations as well. The associations are protected with *a priori* negotiated shared secrets (session keys), and if no such associations exist, e.g., between HA and Foreign Agents (FAs), RSA public-key encryption is applied to protect the routing traffic. The public keys are not certified, but their MACs are used to associate a public key with its owner. Routing message replay protection and sequence detection are performed using either nonces or timestamps. The use of session keys requires the *a priori* negotiated shared secret between the nodes wanting to use the keys, thus reducing the flexibility and scalability of the system. This is not a difficult problem between the HA and MH, since they have a mutual trust relationship and thus have a fixed security association as such. The use of RSA along with the secret-key cryptography requires the configuration and use of two totally different cryptographic approaches, which set additional requirements on the nodes in the system. Moreover, the public keys are not certified by any trusted authority, decreasing the trust in the keys. The MAC calculated by the FA may not be correct since the FA may be a compromised node. Finally, it is questionable how this kind of hybrid approach can tolerate byzantine failures.

Reducing Overhead in Link State Routing

Hauser et al. have proposed a technique for efficient and secure processing of link state information [39]. This approach is based on Lamport's authentication algorithm using hashing chain. The basic idea of Lamport's algorithm is that there are two parties: A and B. A generates a secret R and computes a hash chain of length *n*: $H_1(R)$, …, $H_i(R)$, …, $H_n(R)$ where $H_i(R) = H[H_{i-1}(R)]$, $0 < i < n$, and the hash function can be MD5. Initially, A sends B the value $H_n(R)$ and n by some means, for example, by mail. (The two values are not secret and can be sent in plaintext.) When A wants to authenticate himself to B, A sends B $H_{n-1}(R)$, and B just checks whether $H_n(R)$ matches $H[H_{n-1}(R)]$. Since only A can generate $H_{n-1}(R)$, B believes that the other party is A. This one-time authentication can be used *n* times.

The most important feature of this algorithm [39] is that the two parties do not need to share any secret before authentication. This allows the nodes to send substantially lighter routing information packets when the link state updates (LSU) would be redundant in respect of the previously exchanged information. The resources are saved by applying a hash chain from the node that sends the routing information to the other nodes with the heavy security mechanisms. The message includes a computation of h(h(......(h(R)))) of some randomly chosen datum R, which is hashed N times. After the first original message, the origin node can then refresh the redundant routing information by sending the nodes the random datum R hashed only N − 1 times. In this way, the receiver nodes can refresh their routing information by computing the hash of the chain and verifying that the result is indeed equal to the original hash chain associated with the routing information that was authenticated within the first routing information exchange. After that, if the same routing information is still valid, the origin node sends a hash chain of N − 2 hashes and so on, until the number of hashes in the chain N is zero, at which point, the routing information must be renewed totally. The proposal does, however, require that the random number generators that produce the random data must be strong. Moreover, the length of the hash chain must be set to be very long.

An advantage of this method is that it does not include any source routing approaches. In addition, the length of the hash chain is always the same, no matter how many hashes are computed. Thus, an adversary cannot determine the length of the chain by just inspecting it. On the other hand, in dynamically changing MANET environments having severe problems with compromised nodes inflicting undetectable byzantine failures, the approach seems to be inadequate. This is because the routing information is sent from individual nodes, thus allowing the nodes to send any kind of malicious routing information they want. If the receiver nodes accept such incorrect information, the origin node may be able to maintain the incorrect states for a long time by setting the hash chain to be very long.

Summary

We have discussed the security features and lack thereof in the routing protocols for wireless ad hoc networks. A summary of the security features in some of the protocols is provided in Table 30.3. The routing protocols must be secured from the viewpoints of authentication, integrity, nonrepudiation, and privacy. These requirements can be at least partially met, for instance, by using strong encryption mechanisms, digital signatures, nonces, and timestamps. Moreover, the protection means can be optimized by analyzing potential redundancies in the routing protocol and applying efficient mechanisms such as secret-key cryptography, hashing functions, and MACs. The use of any keyed method, however, requires a distributed, robust, and secure key management service so that the necessary keys can be generated, distributed, and applied securely. We now turn our attention to current efforts focusing on security architectures for wireless ad hoc networks.

30.5.6 Security Architecture

Several researchers have proposed generic security architectures for wireless ad hoc networks. Such schemes include one or more of the concepts already discussed such as how to ensure secure routing, key management, authentication, etc.

Hierarchical Hybrid Networks

An architecture called Hierarchical Hybrid networks has been proposed for secure wireless networking [40]. In such a network, wireless nodes are organized into groups. A secure communication scheme has been presented to defend against link attacks. Secure mobility support for mobile hosts roaming among groups is also discussed. Mutual authentication is used to protect both foreign groups and mobile hosts. The fault-tolerant authentication scheme proposes to make systems survivable from agent failures. Security schemes are proposed to protect communications among group members from various link attacks and support mobile hosts to roam around the system. The security scheme utilizes encryption/ decryption and public-key based authentication techniques. In the secure mobility support scheme, a mobile host will be detached from the system if its Home Group Agent fails. An algorithm to ensure secure communication has been proposed, as shown in Fig. 30.15. This shows how a mobile node X uses a secret key X to encrypt the header of a packet P being sent to some other node. While receiving a packet, X decrypts the packet header. It then decrypts the body only if it is the destination. If it is not, it forwards the packet to the next hop after making certain changes to the header and encrypting it as well.

A Subscriptionless Service Architecture

A subscriptionless service architecture and a client/provider architecture are proposed that combine a dedicated identity (i.e., pseudonym) management with a security association management that addresses the instant character of ad hoc networks [41]. This architecture is located on the service layer, with additional functional units (personalization components) on the network and MAC layer as shown in

Sending a Packet P	X uses K to encrypt the header and the body of P before sending it
Receiving a Packet P	X decrypts and checks the header IF X is the destination 　　　　THEN it decrypts the body ELSE 　　X makes any necessary modifications to the header and 　　IF X is a group agent AND P is sent from one group to another 　　　　　THEN X encrypts the header with the destination group's key K' 　　　　　　and decrypts the body with K then encrypts it again with K' 　　　ELSE 　　　　　X encrypts the header again with K 　　　X forwards P to the next hop

FIGURE 30.15　Algorithm for secure communication.

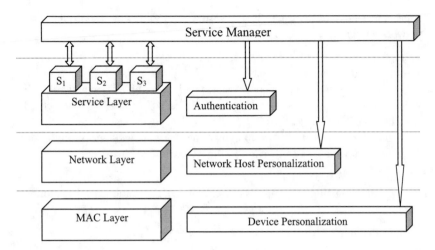

FIGURE 30.16 A subscriptionless service architecture.

Fig. 30.16. In this architecture, services that are located on a client node communicate with services that are located on another client or provider node. A service manager that starts, stops, and personalizes them according to a policy defined by the node owner controls these services. Furthermore, the service manager controls authentication of remote services that seek communication with a local service. A dedicated identity management and a security association management are key components of the architecture (hidden in the service manager).

Identity management — The various business relationships between clients on the one side and clients or providers on the other side require a flexible identity system, in which various pseudonyms with different lifespans and relationship characters are used for different types of relationships. Identity management is accomplished using three different types of pseudonyms, namely, person pseudonym, relationship pseudonym, and transaction pseudonym. Thus, service personalization (i.e., association with the selected identity) is achieved within the subscriptionless mobile networking project.

Security association management — A security association is established as soon as two communication partners initiate a communication session for the first time. The association contains the user identities (pseudonyms) and security-relevant information (e.g., minimum key length requirements for confidentiality and integrity protection), as well as personal profile data of the session. A valid security association is a precondition for any communication session establishment. It may span more than one communication session. For security associations between communication partners, three modes differing in their lifetime are possible, namely, long-term association, event association, and transaction association.

Group Collaboration

Research is being done on security models and mechanisms to support group collaboration on ad hoc wireless networks [42]. The emphasis is on establishing ad hoc groups in such a network to enable short-term or long-term collaboration and sharing of resources without extensive manual involvement or reliance on technical staff and system administrators. This requires a new security model and the use of context-sensitive and implicit security paradigms. In order to enable this type of ad hoc grouping infrastructure, it is argued that it is essential to develop an effective security model that allows grouping to work. Typically, security for collaboration emphasizes traditional network contexts such as an editing model of collaboration, in which two users collaboratively edit a document, and focuses on data structures for collaborative tasks and collaborative activity by organized teams. Although context-sensitive security is likely to be a key component of any security model supporting ad hoc grouping, according to the authors, it appears that none of the extant literature addresses some important questions that arise in the ad hoc network and group environment, including:

What constitutes a context in an ad hoc network?

How can multiple contexts be handled?

How can contexts be used to implicitly apply security policies in an ad hoc network?

Can one context and security model extend beyond the local ad hoc network to the broader local area network (LAN) and WAN?

Can a context-based security model work in a disconnected setting where a group of devices creates an ad hoc network without a connection to the back-end fixed network?

The overall goal is to develop an interaction model for collaboration on ad hoc wireless networks that fits with users' ways of interacting in groups, is object centered, and minimizes the amount of administrative overhead.

The Archipelago Project

The Archipelago Project investigates efficient ways to form a secure extended ad hoc network of laptops, handhelds, and other wireless capable devices, and bridge it to the Internet [43]. Archipelago constructs a multi-hop dynamic network using the wireless devices of participating users. Encryption and authentication provide protection against eavesdropping, snooping by nonauthorized participants, and impersonation.

Internet bridge architecture — Archipelago allows users to transparently tunnel connections to the Internet through the ad hoc peer-to-peer wireless network. This is accomplished by intercepting the networking calls made by the users' applications. Archipelago then establishes a peer-to-peer connection to an Archipelago node that has access to the Internet (see Fig. 30.17). This node acts as a proxy and opens a connection to the intended Internet destination. The user's node and intermediate nodes remain mobile. The tunnel is maintained until the connection is complete as long as any path exists between the user's node and the proxy node.

System architecture — Archipelago uses an extensible modular architecture (see Fig. 30.18). User level applications initiate connections to one of the Archipelago session layer modules to gain access to the system. For example, Internet applications use the proxy session module to tunnel. Each session module can provide nonsecure service as well as secure service by using the key manager module to set up secure channels. Transport layer modules provide services such as ordering, reliability, flow control, and congestion control. The network layer provides routing, topology maintenance, and best effort delivery of data packets. The network layer is also responsible for using the encryption engine module to encrypt/decrypt data flowing over established secure channels. Data link modules act as low-level hardware abstractions. This allows Archipelago to support multiple wireless standards simultaneously.

Security architecture — Wireless networks operate using a broadcast medium. This allows everyone in range to hear what is being sent. By using the Wired Equivalent Privacy (WEP[24]) algorithm (which is used to protect wireless communication from eavesdropping and prevent unauthorized access to a wireless network) 802.11 cards attempt to mitigate this risk. Unfortunately, WEP encrypts using only a static shared key, which is vulnerable to attack. In addition to the problem of eavesdropping off the air, in Archipelago, packets are explicitly passed through intermediate nodes giving them even easier access to the data. Archipelago uses end-to-end key exchange and encryption to combat these two challenges. Other security challenges include man-in-the-middle, impersonation, and topology disruption attacks [43].

To summarize, security architecture for wireless ad hoc networks needs to go beyond concentrating on a specific security need such as authentication or confidentiality. What needs to be addressed is a range of issues that threaten the overall security of such a network. An ideal architecture should, at least, take care of the security issues in routing, have a key management scheme in place, have an intrusion detection and response scheme, and have an authentication scheme. In our last section, we briefly discuss

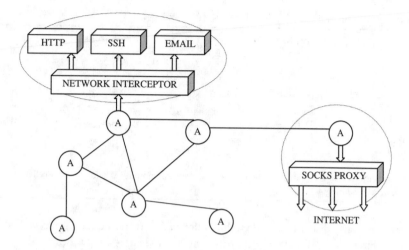

FIGURE 30.17 Internet bridge architecture where A represents Archipelago nodes.

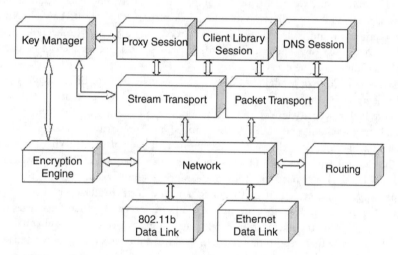

FIGURE 30.18 System architecture.

network management in ad hoc networks, an approach that does have some of the security features needed.

30.5.7 Network Management

Ad hoc Network Management Protocol (ANMP) implements the unicast security of SNMPv3 [55] and supports secure multicasts [44]. ANMP also implements the military security model where security clearances are assigned to nodes. Likewise, security classification levels are assigned to data. Cluster heads are unable to read MIB[25] data that are at a higher classification level. They can, however, read data at classification levels equal to their own or lower. In order to enforce this mechanism, a problem to be solved is ensuring that an agent knows the security clearance of the cluster head it uses to determine how to encrypt its data. In ANMP, the manager assigns security levels to nodes, and this information is distributed as an unforgeable ticket to cluster heads. Cluster heads transmit this information to all their cluster nodes when the clusters are formed.

Table 30.4 illustrates the intricacies of information transfer between entities A and B. If A and B are trustworthy, then information is exchanged as a secure digest and both parties believe .everything. If on

TABLE 30.4 Information Transfer from A to B in ANMP

	A trusts B B trusts A	A trusts B B distrusts A	A distrusts B B trusts A	A distrusts B B distrusts A
A and B are trustworthy	(SD, E)	(SD, N)	(UD, E)	(UD, N)
Only A is trustworthy	(SD, E) Breach	(SD, N) Breach	(UD, E)	(UD, N)
Only B is trustworthy	(UD, E)	(UD, N)	(UD, E)	(UD, N)
Neither A nor B is trustworthy	(UD, E)	(UD, N)	(UD, E)	(UD, N)

Note: What is sent — SD, secure digest; UD, unclassified digest or nothing is sent; What is believed — E, believe everything; N, believe nothing.

the other hand, any party distrusts the other, then information is transferred as unclassified digest. It is possible that nothing is sent. If neither party trusts the other, then nothing is believed

Secure multicasting — Secure multicast is needed to allow the manager to configure networking elements safely and to ensure secret distribution of information between cluster heads and the manager. The same message is multicast to all group members. The plain text message to be multicast is encrypted using an encryption key and is decrypted by all receivers using a decryption key. All members of the multicast group have their own encryption and decryption keys. The message can be decrypted by all group members if they have the decryption key.

Level-Based Access Control Model (LACM) — Each node is thought of as having a clearance level and each data item as having a security level [44]. If a data item has a security level of three (small numbers denote higher security levels), then only nodes with clearance of zero, one, two, or three can read this data item. Data items and nodes can be grouped into compartments (read individual projects), where a node may have different clearance levels in different compartments. When an agent is queried for data by the cluster head or by the manager, the agent first determines the security clearance level of the cluster head. If the cluster head has a lower clearance level, the agent encrypts the data item such that only the manager can decrypt it. If, on the other hand, the cluster head has a higher or equal clearance level, the data item is encrypted so that the cluster head can decrypt it. When the cluster head prepares a summary of the data for the manager, it assigns the summary a clearance level equal to the highest clearance level of the data that were summarized. In the LACM model, a user is allowed to assign security classification levels to managers in the form of certificates. Managers then respond to information requests from other managers by appropriately editing and/or summarizing the data.

30.6 Summary

This survey chapter is meant to provide a comprehensive state-of-the-art view of security in wireless ad hoc networks. We discussed the characteristics of ad hoc wireless networks that make them so vulnerable to attacks and make security of such networks a difficult task. We then looked at security needs and the types of attacks that can take place. This was followed by a comprehensive look at the security schemes that have been proposed as well as the current work that is going on in this area. To summarize, we present the following:

30.6.1 Network-Layer Security

The security of ad hoc networks can be based on protection in the link or network layer. In some ad hoc solutions, the link layer offers strong security services for protecting confidentiality and authenticity, in which case all of the security requirements need not be addressed in the network or upper layers. However, in most cases, the security services are implemented in higher layers, for instance, in the network layer, since many ad hoc networks apply IP-based routing and recommend or suggest the use of IPSec. Authenticity and integrity of routing information are often handled in parallel, if public-key cryptosystems are in use, since digital signatures are applied for confirming both the origin of the data and their integrity.

30.6.2 Service Aspects

The necessary services such as the routing of packets and key management have to be distributed so that all nodes have responsibility in providing the service. As there are no dedicated server nodes, any node may be able to provide a necessary service to another. Moreover, if a tolerable number of nodes in the ad hoc network crash or leave the network, this does not break the availability of the services. Finally, the protection of services against denial of service is, in theory, impossible. In ad hoc networks, redundancies in the communication channels can increase the possibility that each node can receive proper routing information. The redundancies in the communication paths may reduce the denial of service threat and allow the system to detect malicious nodes and prevent them from performing malicious actions more easily than in service provisioning approaches that rely on single paths between the source and destination.

30.6.3 Security of Key Management

In ad hoc networks, an environment-specific and efficient key management system is needed. To be able to protect nodes against threats such as eavesdropping by using encryption, the nodes must have made a mutual agreement on a shared secret key or exchanged public keys. For very rapidly changing ad hoc networks, the exchange of encryption keys may have to be addressed on demand, without assumptions about *a priori* negotiated secrets. In less dynamic environments, the keys may be mutually agreed upon proactively or even configured manually (if encryption is even needed). If public-key cryptography is applied, the whole protection mechanism relies on the security of the private key. Consequently, as the physical security of nodes may be poor, private keys have to be stored in the nodes confidentially, for instance, encrypted with a system key. For dynamic ad hoc networks, this is not a desired feature and thus the security of the private key must be guaranteed with proper hardware protection (smart cards) or by distributing the key in parts to several nodes. Hardware protection is, however, never an adequate solution by itself for preventing attacks as such. The mechanical replication of the private keys or other information is an inadequate protection approach, since the private keys of the nodes then simply have multiple opportunities to be compromised. Thus a distributed approach in key management — for any cryptosystem in use — is needed.

30.6.4 Access Control

There is a need to control access to the network and to the services it provides. Moreover, as the networking approach may allow or require the forming of groups in, for instance, the network layer, several access control mechanisms working in parallel may be needed. In the network layer, the routing protocol must guarantee that no unauthorized nodes are allowed to join the network or a *packet forwarding group* such as the clusters in the hierarchical routing approach. At the application level, the access control mechanism must guarantee that unauthorized parties cannot have accesses to services, for instance, the vital key management service. Access control is often related to the identification and authentication wherein the main issue is that the parties can be confirmed to be authorized to gain the access. In some systems, however, identification or authentication of nodes is not required; for example, nodes may be given delegate certificates with which the nodes can gain access to services. In this case, actual authentication mechanisms are not needed if the nodes are able to present adequate credentials to the access control system. If a centralized ad hoc networking approach with low security requirements is applied, the access control can be managed by the server party with simple means such as a user ID/password scheme. In ad hoc networks that operate in more difficult conditions without any centralized resources, the implementation of access control is much more difficult. The access to the network, its groups and resources, must be defined when the network is formed, which is very inflexible. The other possibility is to define and use a very complex, scalable, and dynamic access control protocol, which brings flexibility but is prone to various kinds of attacks and may even be impossible to apply properly and efficiently.

30.6.5 Integrity

As certificates and identities are not secret, it is possible to clone a device. By securing the public key and letting only the device know its private key, it is possible to use the certificate in a sensible way. The device has to be tamper proof, so that its private key cannot be read out and used in a fake device. The best way to avoid attacks is to try to make the node tamper proof, which is not simple at all. The systems have to be tested also against hostile attackers. Physical tamper-evidence mechanisms are often more suitable. The tampering itself can be more than changing a code or key; it can also be a direct attack to a device. The software has to be able to be uploaded into nodes. This has to be possible also after deployment. If a part of the core code of the node is tamper proof, uploading an infiltrating or malicious code is possible.

30.6.6 Trust

If an *a priori* trust relationship exists between the nodes of an ad hoc network, entity authentication can be sufficient to ensure the correct execution of critical network functions. *A priori* trust can only exist in a few special scenarios, such as military networks, and requires tamper-proof hardware for the implementation of critical functions. Entity authentication in a large network, on the other hand, raises key management requirements. If tamper-proof hardware and a strong authentication infrastructure are not available, the reliability of basic functions such as routing can be endangered by any node of an ad hoc network. No classical security mechanism can help counter a misbehaving node in this context. The correct operation of the network requires not only the correct execution of critical network functions by each participating node but also that each node perform a fair share of the functions. The latter requirement seems to be a strong limitation for wireless mobile nodes whereby power saving is a major concern. With lack of *a priori* trust, cooperative security schemes seem to offer the only reasonable solution. In a cooperative security scheme, malicious behavior can be detected through collaboration among a number of nodes, assuming that a majority of nodes do not behave maliciously. The threats considered in such a scenario are not limited to maliciousness, and a new type of misbehavior called selfishness should also be taken into account to prevent failure of nodes to cooperate.

30.6.7 Routing Security

Two types of security schemes are proposed for ad hoc routing networks. One type consists of proactive schemes; generally, standard schemes such as key generation and management service are used in a distributed manner among individual hosts participating in authentication, so as to ensure routing information's authenticity and integrity. Another approach is to detect intrusions from existing information and respond promptly. Intrusion detection schemes can detect both external and internal attacks. The requirements of authentication, integrity, nonrepudiation, and privacy can be at least partially met, for instance, by using strong encryption mechanisms, digital signatures, nonces, and timestamps. Moreover, the protection means can be optimized by analyzing potential redundancies in the routing protocol and applying efficient mechanisms such as secret-key cryptography, hashing functions, and MACs. The use of any keyed method will, however, require a distributed, robust, and secure key management service so that the necessary keys can be generated, distributed, and applied securely. The choice of the security mechanisms involves also the choice between secret and public-key cryptography. While shared secret keys offer significant efficiency, such an approach will require static and manual key negotiations or complex secure dynamic key negotiation services between the nodes. Moreover, the use of secret keys is highly dependent on the assumed security associations and trust relationships among the network nodes, which may produce unpredictable vulnerabilities to the system. On the other hand, the mere use of public-key cryptography requires extensive key management. Such services may produce too much overhead especially for energy- and computational power-constrained MANET nodes and thus the concurrent use of both public- and shared secret key techniques may have to be supported.

IPSec can be applied within the protocol to achieve the necessary security goals. This kind of approach is not totally adequate, due to problems with, for example, configuration requirements and the retrofitting of security mechanisms. Methods such as NDM are also probably too expensive to be used in ad hoc networks, and their use also necessitates the use of security agents, which need to be present in the network at all times. Hierarchy appears to be a desirable property in routing protocols because it can sometimes limit failures to smaller areas in a network. As it also limits the number of routing messages in comparison with flat routing, it may also limit vulnerability against denial-of-service attacks based on excessive route requests. Moreover, the traditional security mechanisms such as link-level encryption or bidirectional tunnels are not adequate, due to the dynamic and unpredictable nature of MANET networks. The possibility of byzantine failures and other malicious actions performed by compromised nodes requires a distributed approach, both at the abstract level (trust model) and in the security mechanism level, such as the necessary key management. Byzantine robustness can indeed be achieved adequately, but at the expense of huge computational and network traffic overheads. Therefore such an approach cannot directly be applied within MANET networking.

Since security in wireless ad hoc networks is a relatively unexplored territory, much effort is underway to develop algorithms and mechanisms not only to ensure security but also to work around loopholes in the proposed solutions. There is much scope for further research and development in this area. We briefly touch upon some of these open issues for future work in the final section of this chapter.

30.6.8 Open Issues

As far as multi-party key agreement goes, the implementation needs to address issues of synchronization and resilience in the face of benign faults. Also, with dynamic groups, the session key needs to be updated every time the group formation changes. Several existing Diffie-Hellman–based protocols for group key establishment have been proposed. None of these protocols has been found suitable for all types of ad hoc networks. This is mainly because they demand that the network topology follow a predetermined structure. The protocol to be used should be selected specifically for the implementation at hand. It should be chosen so that it is always possible to arrange the nodes according to the required topology. An open problem is to define a group key establishment protocol for a network with an arbitrary topology. Some open issues in key management are an optimal best-effort revocation procedure and the relation of the groups with routing and network management. Some attacks, especially those that are not deterministic, have not been handled. Intrusion detection schemes that analyze traffic profiles to detect intruders would be another challenging area to explore. Some questions that need to be answered are:

What information should a routing protocol include to make intrusion detection effective?
What is the best anomaly detection model for a given routing protocol?

MANET and Mobile IP routing protocol proposals seem to meet the basic requirements for protocols, such as dynamically changing network topologies, rather well. The security issues have, however, been left for small notice, especially within the MANET protocols; Mobile IP extension proposals do have various approaches for enhancing security. The common concerns in ad hoc networks include access control: there needs to exist a method for restricting the access of foreign nodes to the network, which requires the use of a proper authentication mechanism. Moreover, communication between insider nodes in the network must be protected from attacks on confidentiality. This is especially important in military applications.

Glossary

CA[1] — A certificate authority is an authority in a network that issues and manages security credentials and public keys for message encryption and decryption. As part of a public key infrastructure (PKI), a CA checks with a registration authority (RA) to verify information provided by the requestor of a digital certificate. If the RA verifies the requestor's information, the CA can then issue a certificate.

SA[2] — A Security Association (SA) is a relationship between two or more entities that describes how the entities will utilize security services to communicate securely. This relationship is represented by a set of information that can be considered a contract between the entities. The information must be agreed upon and shared among all the entities.

Private key[3] — In cryptography, a private or secret key is an encryption/decryption key known only to the party or parties that exchange secret messages. In traditional secret key cryptography, a key would be shared by the communicators so that each could encrypt and decrypt messages. The risk in this system is that if either party loses the key or it is stolen, the system is broken. A more recent alternative is to use a combination of public and private keys. In this system, a public key is used together with a private key.

MAC address[4] — A MAC address is used by the Data Link Layer to deliver a frame to the destination node. MAC addresses are also called hardware addresses or NIC addresses, because this address is hard-coded into each NIC. Each type of network hardware has its own MAC addressing scheme. For example, Ethernet uses 48-bit hardware addresses assigned by the vendor.

Mandatory Access Control (MAC)[5] — Mandatory Access Control is an added security constraint enforced by a trusted operating system. MAC governs access to data, devices, or networks based on their sensitivity levels. MAC, as the name implies, is mandated by the trusted operation system and cannot be changed or removed by users.

Public key[6] — A public key is a value provided by some designated authority as a key that, combined with a private key derived from the public key, can be used to effectively encrypt and decrypt messages and digital signatures. The use of combined public and private keys is known as asymmetric encryption. A system for using public keys is called a public key infrastructure (PKI).

IPSec[7] — A developing standard for security at the network or packet processing layer of network communication. IPSec provides two choices of security service: Authentication Header (AH), which essentially allows authentication of the sender of data, and Encapsulating Security Payload (ESP), which supports both authentication of the sender and encryption of data.

ICMP[8] — Short for Internet Control Message Protocol, an extension to the Internet Protocol (IP). ICMP supports packets containing error, control, and informational messages. The PING command, for example, uses ICMP to test an Internet connection.

STS[9] — Station-to-Station protocol (STS) is the three-pass variation of the basic Diffie–Hellman protocol. It allows the establishment of a shared secret key between two parties with mutual entity authentication and mutual explicit key authentication. The protocol also facilitates anonymity; the identities of A and B may be protected from eavesdroppers. The method employs digital signatures.

CRL[10] — A document maintained and published by a certification authority (CA) that lists certificates issued by the CA that are no longer valid.

MAC[11] — A bit string that is a function of both data (either plaintext or ciphertext) and a secret key, and that is attached to the data in order to allow data authentication. The function used to generate the message authentication code must be a one-way function.

TLS[12] — Transport Layer Security (TLS) protocol. The TLS protocol provides communications privacy over the Internet. The protocol allows client/server applications to communicate in a way that is designed to prevent eavesdropping, tampering, or message forgery. The primary goal of the TLS protocol is to provide privacy and data integrity between two communicating applications.

The protocol is composed of two layers: the TLS Record Protocol and the TLS Handshake Protocol.

DH[13] — The Diffie–Hellman Method For Key Agreement allows two hosts to create and share a secret key. VPNs operating on the IPSec standard use the Diffie–Hellman method for key management. Key management in IPSec begins with the overall framework called the Internet Security Association and Key Management Protocol (ISAKMP). Within that framework is the Internet Key Exchange (IKE) protocol. IKE relies on yet another protocol known as OAKLEY, and it uses Diffie–Hellman.

PKI[14] — A policy that establishes a secure method for exchanging information within an organization, industry, or country. PKI includes cryptography, the use of digital certificates and certificate authorities, and the system for managing the process. A PKI (public key infrastructure) enables users of a basically nonsecure public network such as the Internet to securely and privately exchange data and money through the use of a public and a private cryptographic key pair that is obtained and shared through a trusted authority.

RSA[15] — One of the fundamental encryption algorithms or series of mathematical actions developed in 1977 by Ron Rivest, Adi Shamir, and Leonard Adleman. RSA relies on the relative ease of finding large primes and the comparative difficulty of factoring integers for its security. RSA is a public-key cryptosystem for both encryption and authentication. It works as follows: take two large primes, p and q, and find their product $n = pq$; n is called the modulus. Choose a number, e, less than n and relatively prime to $(p - 1)(q - 1)$, which means that e and $(p - 1)(q - 1)$ have no common factors except 1. Find another number d such that $(ed - 1)$ is divisible by $(p - 1)(q - 1)$. The values e and d are called the public and private exponents, respectively. The public key is the pair (n,e); the private key is (n,d). The factors p and q may be kept with the private key or destroyed.

PolicyMaker[16] — PolicyMaker is a trust management system. It is concerned with defining policies, credentials, and trust relationships. It uses a "safe" programming language to define the trust relationships, credentials, and policies. It is designed to be flexible enough to be used in large network applications and to integrate easily with the existing protocols [56].

Keynote[17] — Keynote is a simple and flexible trust management system that is designed for small and large Internet-based applications. It is fast enough to be used even in real time applications. It uses one easily human readable language to specify the policies and credentials [57].

SASL[18] — Simple Authentication and Security Layer (SASL) is a way to add authentication support to connection-based protocols. In order to use SASL, the protocol must include a command for identifying and authenticating the user to a server. The protocol may also include optional negotiation of a security layer for the subsequent protocol interactions. The command contains an argument that identifies the SASL mechanism to be used. If the server supports this mechanism, it initiates the authentication protocol exchange. This typically consists of changing challenge response pairs between the client and the server. These are specific to each protocol used. During the authentication protocol exchange, the mechanism performs authentication, transmits an authentication identity, and negotiates the use of a mechanism specific security layer. If a security layer is to be used, it is put into use immediately, and all the subsequent data exchanges are encrypted [58].

ISAKMP[19] — Internet Security Association and Key Management Protocol (ISAKMP) is a key exchange independent framework for authentication, SA (Security Association) management, and establishment. It does not define the actual protocols to be used. ISAKMP uses a two-phase approach in establishing SAs. In the first phase, the ISAKMP SA is established between the entities to protect further negotiation traffic. In the second phase, the ISAKMP SA is used to establish other security protocol SAs such as IPSec [59].

IKE[20] — *Internet Key Exchange* is one implementation of ISAKMP. It is used to negotiate and exchange keying material between entities in the Internet. For example, IKE can be used to establish the IPsec security association. In IKE, the Diffie–Hellman algorithm is used for the key exchange.

IKE uses the same kind of two-phase SA establishing as ISAKMP. In the first phase IKE SA is created, and in the second phase keying information is changed and non-IKE SAs are established. The first phase may take place in one of the two modes. One of these protects the identity and the other does not [60].

DES[21] — A widely-used method of data encryption using a private (secret) key that was judged so difficult to break by the U.S. government that it was restricted for exportation to other countries. There are 72,000,000,000,000,000 (72 quadrillion) or more possible encryption keys that can be used. For each given message, the key is chosen at random from among this enormous number of keys. As with other private key cryptographic methods, both the sender and the receiver must know and use the same private key.

MD5[22] — MD5 was developed by Professor Ronald L. Rivest. The MD5 algorithm takes as input a message of arbitrary length and produces as output a 128-bit "fingerprint" or "message digest" of the input. It has been conjectured that it is computationally infeasible to produce two messages having the same message digest or to produce any message having a given prespecified target message digest. The MD5 algorithm is intended for digital signature applications, where a large file must be "compressed" in a secure manner before being encrypted with a private (secret) key under a public-key cryptosystem such as RSA. In essence, MD5 is a way to verify data integrity and is much more reliable than checksum and many other commonly used methods.

QoS[23] — On the Internet and in other networks, QoS is the idea that transmission rates, error rates, and other characteristics can be measured, improved, and, to some extent, guaranteed in advance. QoS is of particular concern for the continuous transmission of high-bandwidth video and multimedia information.

WEP[24] — The 802.11 standard describes the communication that occurs in wireless local area networks (LANs). The Wired Equivalent Privacy (WEP) algorithm is used to protect wireless communication from eavesdropping. A secondary function of WEP is to prevent unauthorized access to a wireless network; this function is not an explicit goal in the 802.11 standard, but it is frequently considered to be a feature of WEP. WEP relies on a secret key that is shared between a mobile station (e.g., a laptop with a wireless ethernet card) and an access point (i.e., a base station). The secret key is used to encrypt packets before they are transmitted, and an integrity check is used to ensure that packets are not modified in transit. The standard does not discuss how the shared key is established.

MIB[25] — The set of variables or database that a gateway running network management protocol maintains. It defines variables needed by the SNMP protocol to monitor and control components in a network. Managers fetch or store into these variables.

References

[1] F. Stajano and R. Anderson, The Resurrecting Duckling: Security Issues for Ad Hoc Wireless Networks, *Proceedings of the 7th International Workshop on Security Protocols,* Lecture Notes in Computer Science, Springler-Verlag, Berlin, Apr. 1999. Available from http://www.cl.ac.uk/~fms27/duckling/duckling.htm.

[2] F. Stajano, The Resurrecting Duckling – What Next? *Proceedings of the 8th International Workshop on Security Protocols,* Lecture Notes in Computer Science, Springler-Verlag, Berlin, Apr. 2000. Available from http://www.cl.ac.uk/~fms27/duckling/ duckling-what-next.htm.

[3] L. Zhou and Z.J. Haas, Securing Ad Hoc Networks, *IEEE Network Magazine,* Nov/Dec 1999.

[4] N. Asokan and P. Ginzboorg, Key agreement in ad hoc networks, *Computer Communications,* 23, 1627–1637, 2000.

[5] R.R.S. Verma, D. O'Mahony, and H. Tewari, NTM — Progressive Trust Negotiation in Ad Hoc Networks, http://www.cs.tcd.ie/omahony/iei-ntm.pdf.

[6] Sufatrio and K.-Y. Lam, Scalable Authentication Framework for Mobile-IP (SAFe-MIP), Internet draft, IETF, Nov. 1999.

[7]   A. Herzberg, M. Jakobsson, S. Jarecki, H. Krawczyk, and M. Yung, Proactive Public Key and Signature Systems, *ACM Security '97*, 1997.

[8]   A. Fasbender et al., Variable and Scalable Security: Protection of Location Information in Mobile IP, *Mobile Technology for the Human Race, IEEE 46th Vehicular Technology Conference*, 1996.

[9]   P. Eronen, C. Gehrmann, and P. Nikander, Securing Ad Hoc Jini Services, http://www.cs.hut.fi/~peronen/publications/nordsec_2000.pdf.

[10] J. Kong, P. Zerfos, H. Luo, S. Lu, and L. Zhang, Providing Robust and Ubiquitous Security Support for Mobile Ad hoc Networks, *IEEE ICNP (International Conference on Network Protocols) 2001*, Riverside, CA, Nov. 2001.

[11] S. Mäki, M. Hietalahti, and T. Aura, A Survey of Ad Hoc Network Security, Interim report of project 007 — Security of Mobile Agents and Ad Hoc Societies, Helsinki University of Technology, Sep. 2000.

[12] H. Luo and S. Lu, Ubiquitous and Robust Authentication Services for Ad Hoc Wireless Networks, Technical Report TR-200030, Dept. of Computer Science, UCLA, 2000.

[13] Y. Zhang and W. Lee, Intrusion Detection in Wireless Ad Hoc Networks, *Sixth International Conference on Mobile Computing and Networking (MobiCom '00)*, Boston, MA, 2000, pp. 275–283.

[14] S. Bhargava and D.P. Agrawal, Security Enhancements in AODV Protocol for Wireless Ad Hoc Networks, *Vehicular Technology Conference*, Atlantic City, NJ, October 7–11, 2001.

[15] TIARA, http://www.atcorp.com/research/cur_projects/tiara/index.html.

[16] M.S. Corson and S. Jacobs, Manet authentication architecture, http://www.ietf.org/internet-drafts/draft-jacobs-imep-auth-arch-00.txt, Aug. 1998.

[17] L. Venkatraman and D.P. Agrawal, A Novel Authentication Scheme for Ad hoc Networks, *Second IEEE Wireless Communications and Networking Conference*, Chicago, IL, September 2000.

[18] S. Levijoki, Authentication, Authorization and Accounting in Ad Hoc Networks, http://www.tml.hut.fi/Opinnot/Tik-110.551/2000/papers/authentication/aaa.htm#chap1.

[19] D. Nguyen, L. Zhao, P. Uisawang, and J. Platt, Security Routing Analysis for Mobile Ad Hoc Networks, http://198.11.21.25/capstoneTest/Students/Papers/docs/Final_revision37165.PDF.

[20] R. Perlman, Network Layer Protocols with Byzantine Robustness, Ph.D. thesis, Massachusetts Institute of Technology, Cambridge, MA, 1988.

[21] B.R. Smith, S. Murthy, and J.J. Garcia-Luna-Aceves, Securing Distance-Vector Routing Protocols, http://www.isoc.org/isoc/conferences/ndss/97/smith_sl.pdf.

[22] F. Wang, On the Vulnerability and Protection of OSPF Routing Protocol, *IEEE Seventh International Conference on Computer Communications and Networks*, Lafayette, LA, Oct. 12–15, 1998.

[23] S. Murphy and M. Badger, Digital signature protection of the OSPF routing protocol, *Proceedings of the Symposium on Network and Distributed System Security (SNDSS '96)*, Feb. 1996, pp. 93–102.

[24] S. Marti, T.J. Giuli, K. Lai, and M. Baker, Mitigating Routing Misbehavior in Mobile Ad Hoc Networks, *Sixth International Conference on Mobile Computing and Networking (MobiCom '00)*, Boston, MA, Aug. 2000, pp. 255–265.

[25] L. Buttyan and J.P. Hubaux, Enforcing Service Availability in Mobile Ad Hoc WANs, *1st MobiHoc*, Boston, MA, August 2000.

[26] S. Yi, P. Naldurg, and R. Kravets, Security-Aware Ad Hoc Routing for Wireless Networks, UIUCDCS-R-2001–2241 Technical Report, Aug. 2001.

[27] M.S. Corson, S. Papademetriou, P. Papadopoulos, V. Park, and A. Qayyum, An Internet MANET Encapsulation Protocol (IMEP) Specification, <draft-ietf-manet-imep-spec-02.txt>.

[28] M.G. Zapata, Secure Ad Hoc On-Demand Distance Vector (SAODV) Routing, <draft-guerrero-manet-saodv-00.txt>.

[29] P. Papadimitratos and Z.J. Haas, Secure Message Transmission in Mobile Ad Hoc Networks, http://wnl.ece.cornell.edu/Publications/cnds02.pdf.

[30] L. Ji and M.S. Corson, Differential Destination Multicast Specification (DDM), http://www.ietf.org/proceedings/00jul/I-D/manet-ddm-00.txt.

[31] P. Jacquet, et al., Optimized Link-State Routing Protocol (OLSR). IETF draft, http://www.ietf.org/internet-drafts/draft-ietf-manet-olsr-02.txt>.

[32] Lee, S.-J. et al., On-Demand Multicast Routing Protocol (ODMRP), IETF draft, http://www.ietf.org/internet-drafts/draft-ietf-manet-odmrp-02.txt.

[33] C. Perkins et al., Ad Hoc On-Demand Distance-Vector Routing (AODV), IETF draft, http://www.ietf.org/internet-drafts/draft-ietf-manet-aodv-06.txt.

[34] C. Perkins and E. Royer, Multicast Ad Hoc On-Demand Distance-Vector Routing (MAODV), IETF draft, http://www.ietf.org/internet-drafts/draft-ietf-manet-maodv-00.txt.

[35] B. Bellur et al., Topology Broadcast Based on Reverse-Path Forwarding (TBRPF), IETF draft, http://www.ietf.org/internet-drafts/draft-ietf-manet-tbrpf-00.txt.

[36] J. Binkley, Authenticated Ad Hoc Routing at the Link Layer for Mobile Systems, http://citeseer.nj.nec.com/cachedpage/121413/1.

[37] P. Papadimitratos and Z.J. Haas, Secure Routing for Mobile Ad Hoc Networks, *SCS Communication Networks and Distributed Systems Modeling and Simulation Conference (CNDS 2002)*, San Antonio, TX, January 27–31, 2002.

[38] D. Forsberg et al., Distributing Mobility Agents Hierarchically Under Frequent Location Updates, *Sixth IEEE International Workshop on Mobile Multimedia Communications*, 1999.

[39] R. Hauser et al., Lowering Security Overhead In Link State Routing, *Proceedings of the Symposium on Network and Distributed System Security (SNDSS '97)*, Feb. 1997, pp. 93–99.

[40] Yi Lu Bharat Bhargava Mohamed Hefeeda, An Architecture for Secure Wireless Networking, http://www.cs.purdue.edu/homes/yilu/papers/wireless-sec.pdf.

[41] M. Schmidt, Subscriptionless Mobile Networking: Anonymity and Privacy Aspects Within Personal Area Networks, research paper to be presented at IEEE WCNC 2002.

[42] A. Danesh and K. Inkpen, Collaborating on Ad Hoc Wireless Networks, http://www.parc.xerox.com/csl/projects/ubicomp-workshop/positionpapers/danesh.pdf.

[43] The Archipelago Project at Johns Hopkins University, Baltimore, http://www.cnds.jhu.edu/research/networks/archipelago.

[44] W. Chen, N. Jain, and S. Singh, ANMP: ad hoc network management protocol, *IEEE Journal On Selected Areas In Communications*, 17(8), 1999.

[45] Z. Haas, J. Deng, B. Liang, P. Papadimitratos, and S. Sajama, Wireless Ad Hoc Networks. http://wnl.ece.cornell.edu/Publications/ency01.pdf.

[46] Internet Engineering Task Force, IP Routing for Wireless/Mobile Hosts (MobileIP) Working Group, http://www.ietf.org/html.charters/mobileip-charter.html.

[47] Internet Engineering Task Force, Mobile Ad Hoc Networks (MANET) Working Group, http://www.ietf.org/html.charters/manet-charter.html.

[48] S. Jacobs, Mobile IP Public Key Based Authentication, Internet draft, IETF, Mar. 1999 [referred 24.3.2000], http://search.ietf.org/internet-drafts/draft-jacobs-mobileip-pki-auth-02.txt.

[49] C. Perkins, Mobile Ad Hoc Networking Terminology Internet draft (expired), IETF, 1998 [referred 25.9.2000], <http://www.ctron.com/support/internet/Internet-Drafts/draft-ietf-manet-term-01.txt> [in ASCII format].

[50] J. Lundberg, Routing Security in Ad hoc Networks, http://citeseer.nj.nec.com.

[51] A. Vanhala, Security in Ad Hoc Networks, http://citeseer.nj.nec.com.

[52] V. Karpijoki, Signaling and Routing Security in Mobile and Ad Hoc Networks, http://www.hut.fi/vkarpijo/iwork00/, May 2000.

[53] Y. Huang, Anomaly Detection for Wireless Ad Hoc Routing Protocols, Master's thesis, North Carolina State University, Raleigh, NC.

[54] M. Jiang, J. Li, and Y.C. Tay, Cluster Based Routing Protocol, http://cram.comp.nus.edu.sg/cbrp/draft-ietf-manet-cbrp-spec-01.txt, Aug 1999.

[55] W. Stallings, SNMPv3: A Security Enhancement for SNMP, http://www.comsoc.org/livepubs/surveys/public/4q98issue/stallings.html.

[56] M. Blaze, J. Feigenbaum, and J. Lacy, Decentralized Trust Management, May 1996.

[57] M. Blaze, J. Feigenbaum, J. Ioannidis, and A. Keromytis, The KeyNote Trust-Management System, Version 2, RFC2704, Sep. 1999, < http://www.ietf.org/rfc/rfc2704.txt >.

[58] J. Myers, Simple Authentication and Security Layer (SASL), RFC 2222, Oct. 1997, < http://www.ietf.org/rfc/rfc2222.txt >.

[59] D. Maughan, M. Schertler, M. Schneider, and J. Turner, Internet Security Association and Key Management Protocol (ISAKMP), RFC2048, < http://www.ietf.org/rfc/rfc2408.txt >.

[60] D. Harkins and D. Carrel, The Internet Key Exchange (IKE), RFC 2409, Nov. 1998, < http://www.ietf.org/rfc/rfc2409.txt >.

31

Securing Mobile Ad Hoc Networks

Panagiotis Papadimitratos
Cornell University

Zygmunt J. Haas
Cornell University

Abstract[1]

The vision of nomadic computing with its ubiquitous access has stimulated much interest in the Mobile Ad Hoc Networking (MANET) technology. These infrastructureless, self-organized networks, which either operate autonomously or as an extension to the wired networking infrastructure, are expected to support new MANET-based applications. However, the proliferation of this networking paradigm strongly depends on the availability of security provisions, among other factors. The absence of infrastructure, the nature of the envisioned applications, and the resource-constrained environment pose some new challenges in securing the protocols in ad hoc networking environments. The security requirements can differ significantly from those for infrastructure-based networks, while the provision of security enhancements may take completely different directions as well. In this chapter, we study the schemes proposed to secure mobile ad hoc networks. We explain the primary goals of security enhancements, shed light on the commensurate challenges, survey the current literature on this topic, and finally introduce our approach to this multifaceted and intriguing topic.

31.1 Introduction

Mobile ad hoc networks comprise freely roaming wireless nodes that cooperatively make up for the absence of fixed infrastructure; that is, the nodes themselves support the network functionality. Nodes transiently associate with peers that are within the radio connectivity range of their transceivers and

---
[1] This work has been supported in part by the DoD Multidisciplinary University Research Initiative (MURI) program administered by the Office of Naval Research under the Grant number N00014-00-1-0564 and by the National Science Foundation grant number ANI-9980521.

implicitly agree to assist in provision of the basic network services. These associations are dynamically created and torn down, often without prior notice or the consent of the communicating parties. MANET technology targets networks that can be rapidly deployed or formed in an arbitrary environment to enable communications or to serve a common objective dictated by the supported application. Such networks can be highly heterogeneous, with various types of equipment, usage, transmission, and mobility patterns.

Secure communication, an important aspect of any networking environment, becomes an especially significant challenge in ad hoc networks. This is due to the particular characteristics of this new networking paradigm and due to the fact that traditional security mechanisms may be inapplicable.

The absence of a central authority deprives the network of the administrative and management services that would otherwise greatly facilitate its operation. MANET has to rely on continuous self-configuration, especially because of the highly dynamic nature of the network. Problems such as scheduling, address assignment, provision of naming services, and formation of network hierarchy cannot be solved by traditional centralized protocols. Instead, distributed operation is necessary in all aspects of network control, including basic security-related operations, such as the validation of node credentials. In the fully distributed and open environment of ad hoc networking, the provision of such services may not only incur a high overhead, but also provide additional opportunities for misbehaving nodes to harm the network operation.

In general, nodes participate in a protocol execution as peers, which implies that potentially any network node can abuse the protocol operation. As a result, it is fairly difficult to identify trustworthy and supportive nodes based on the network interaction. Additionally, determining the protocol or network components that have to be safeguarded is far from straightforward, something that makes the design of adequate security countermeasures even more difficult.

Meanwhile, the practically invisible (or nonexistent) administrative or domain boundaries make the enforcement of any security measures an even more complex problem. Migrating nodes may face varying "rules" even when they run the same application, as they move through different network areas and become associated with different groups of nodes. Or, they may lack the ground for the establishment of trust associations, that is, the establishment of some type of a secret, so that cryptographic mechanisms can be employed.

Below, we discuss in further detail the vulnerability of mobile ad hoc networks, clarify how security goals may have to be modified, and explain which types of solutions are plausible for different network instances. Although the discussion throughout a great part of the chapter applies to all types of ad hoc networks, it is important to realize that strictly not all solutions can be applied in all ad hoc networking environments. Moreover, it is necessary to emphasize the relative importance of addressing certain security issues, which can be considered, to some extent, as prerequisites for solutions to other security problems. In the following sections, we present the challenges posed by the MANET environment, survey the relevant literature, identify the limitations of the proposed approaches, and suggest directions for future solutions.

31.2 Security Goals

The overall problem of securing a distributed system comprises the security of the networked environment and the security of each individual network node. The latter issue is important due to the pervasive nature of MANET, which does not allow us to assume that networked devices will always be under the continuous control of their owners. As a result, the physical security of the node becomes an important issue, leading to the requirement of tamper-resistant nodes [24], if comprehensive security is to be provided. However, security problems manifest themselves in a more emphatic manner in a networked environment, and especially in mobile ad hoc networks. This is why in this work we focus on the network-related security issues.

Security encompasses a number of attributes that have to be addressed: availability, integrity, authentication, confidentiality, nonrepudiation, and authorization. These goals, which are not MANET-specific only, call for approaches that have to be adapted to the particular features of MANETs.

Availability ensures the survivability of network services despite misbehavior of network nodes; for instance, when nodes exhibit selfish behavior or when denial-of-service (DoS) attacks are mounted. DoS attacks can be launched at any layer of an ad hoc network. For example, an adversary could use jamming to interfere with communication at the physical layer, or, at the network layer, it could disable the routing protocol operation, by disrupting the route discovery procedure. Moreover, an adversary could bring down high-level services. One such target is the key management service, an essential service for implementation of any security framework.

Integrity guarantees that an in-transit message is not altered. A message could be altered because of benign failures, such as radio propagation impairments, or because of malicious attacks on the network. Integrity viewed in the context of a specific connection, that is, the communication of two or more nodes, can provide the assurance that no messages are removed, replayed, reordered (if reordering would cause loss of information), or unlawfully inserted.

Authentication enables a node to ensure the identity of the peer node that it is communicating with. Without authentication, an adversary could masquerade a node, possibly gain unauthorized access to resources and sensitive information, and interfere with the operation of other nodes.

Confidentiality ensures that certain information is never disclosed to unauthorized entities. Confidentiality is required for the protection of sensitive information, such as strategic or tactical military information. However, confidentiality is not restricted to user information only; routing information may also need to remain confidential in certain cases. For example, routing information might be valuable for an enemy to identify and to locate targets on a battlefield.

Nonrepudiation ensures that the originator of a message cannot deny having sent the message. Nonrepudiation is useful for detection and isolation of compromised nodes. When node A receives an erroneous message from node B, A can use this message to accuse B and convince other nodes that B is compromised.

Finally, *authorization* establishes rules that define what each network node is or is not allowed to do. In many cases, it is required to determine which resources or information across the network a node can access. This requirement can be the result of the network organization or the supported application when, for instance, a group of nodes or a service provider wishes to regulate the interaction with the rest of the network. Another example could be when specific roles are attributed to nodes in order to facilitate network operation.

The security of mobile ad hoc networks has additional dimensions, such as privacy, correctness, reliability, and fault tolerance. In particular, the resilience to failures, which in our context can be the result of malicious acts, and the protection of the correct operation of the employed protocols are of critical importance and should be considered in conjunction with the security of the mobile ad hoc network.

31.3 Threats and Challenges

Mobile ad hoc networks are vulnerable to a wide range of active and passive attacks that can be launched relatively easily, since all communications take place over the wireless medium. In particular, wireless communication facilitates eavesdropping, especially because continuous monitoring of the shared medium, referred to as promiscuous mode, is required by many MANET protocols. Impersonation is another attack that becomes more feasible in the wireless environment. Physical access to the network is gained simply by transmitting with adequate power to reach one or more nodes in proximity, which may have no means to distinguish the transmission of an adversary from that of a legitimate source. Finally, wireless transmissions can be intercepted, and an adversary with sufficient transmission power and knowledge of the physical and medium access control layer mechanisms can obstruct its neighbors from gaining access to the wireless medium.

Assisted by these "opportunities" that wireless communication offers, malicious nodes can meaningfully alter, discard, forge, inject, and replay control and data traffic, generate floods of spurious messages, and, in general, avoid complying with the employed protocols. The impact of such malicious behavior can be severe, especially because the cooperation of all network nodes provides for the functionality of the absent fixed infrastructure. In particular, as part of the normal operation of the network, nodes are transiently associated with a dynamically changing, over time, subset of their peers; that is, the nodes within the range of their transceivers, or the ones that provide routing information and implicitly agree to relay their data packets. As a result, a malicious node can obstruct the communications of potentially any node in the network, exactly because it is entitled or even expected to assist in the network operation.

In addition, freely roaming nodes join and leave MANET subdomains independently, possibly frequently, and without notice, making it difficult in most cases to have a clear picture of the ad hoc network membership. In other words, there may be no ground for an *a priori* classification of a subset of nodes as trusted to support the network functionality. Trust may only be developed over time, while trust relationships among nodes may also change, when, for example, nodes in an ad hoc network dynamically become affiliated with administrative domains. This is in contrast to other mobile networking paradigms, such as Mobile IP or cellular telephony, where nodes continue to belong to their administrative domain in spite of mobility. Consequently, security solutions with static configuration would not suffice, and the assumption that all nodes can be bootstrapped with the credentials of all other nodes would be unrealistic for a wide range of MANET instances.

From a slightly different point of view, it becomes apparent that nodes cannot be easily classified as "internal" or "external," that is, nodes that belong to the network or not; i.e., nodes that are expected to participate and be dedicated to supporting a certain network operation and those that are not. In other words, the absence of an infrastructure impedes the usual practice of establishing a line of defense, separating nodes into trusted and nontrusted. As a result, attacks cannot be classified as internal or external either, especially at the network layer. Of course, such a distinction could be made at the application layer, where access to a service or participation to its collaborative support may be allowed only to authorized nodes. In the latter example, an attack from a compromised node within the group, that is, a group node under the control of an adversary, would be considered an internal one.

The absence of a central entity makes the detection of attacks a very difficult problem, since highly dynamic large networks cannot be easily monitored. Benign failures, such as transmission impairments, path breakages, and dropped packets, are naturally a fairly common occurrence in mobile ad hoc networks, and, consequently, malicious failures will be more difficult to distinguish. This will be especially true for adversaries that vary their attack pattern and misbehave intermittently against a set of their peers that also changes over time. As a result, short-lived observations will not allow detection of adversaries. Moreover, abnormal situations may occur frequently because nodes behave in a selfish manner and do not always assist the network functionality. It is noteworthy that such behavior may not be malicious, but only necessary when, for example, a node shuts its transceiver down in order to preserve its battery.

Most of the currently considered MANET protocols were not originally designed to deal with malicious behavior or other security threats. Thus, they are easy to abuse. Incorrect routing information can be injected by malicious nodes that respond with or advertise nonexistent or stale routes and links. In addition, compromised routes, i.e., routes that are not free of malicious nodes, may be repeatedly chosen with the "encouragement" provided by the malicious nodes themselves.[2] The result is that the pair of communicating end-nodes will experience DoS, and they may have to rely on cycles of timeout and new route discovery to find operational routes, with successive query broadcasts imposing additional overhead. Or even worse, the end nodes may be easily deceived for some period of time that the data flow is undisrupted, while no actual communication takes place. For example, the adversary may drop a route error message, "hiding" a route breakage, or forge network and transport layer acknowledgments.

---

[2] For instance, the malicious nodes may claim that they possess an inexpensive (short) route to the destination.

Finally, mobile or nomadic hosts have limited computational capabilities, due to constraints stemming from the nature of the envisioned MANET applications. Expensive cryptographic operations, especially if they have to be performed for each packet and over each link of the traversed path, make such schemes implausible for the vast majority of mobile devices. Cryptographic algorithms may require significant computation delays, which in some cases would range from one to several seconds for low-end devices [5,11]. These delays, imposed for example by the generation or verification of a single digital signature, affect the data rate of secure communication. But, more importantly, mobile devices could become ideal targets of DoS attacks due to their limited computational resources. An adversary could generate bogus packets, forcing the device to consume a substantial portion of its resources. Even worse, a malicious node with valid credentials could frequently generate control traffic, such as route queries, at a high rate not only to consume bandwidth, but also to impose cumbersome cryptographic operations on a sizable portion of the network nodes.

31.4 Trust Management

The use of cryptographic techniques is necessary for the provision of any type of security services, and mobile ad hoc networks are not an exception to this rule. The definition and the mechanisms for security policies, credentials, and trust relationships, i.e., the components of what is collectively identified as trust management, are a prerequisite for any security scheme. A large number of solutions have been presented in the literature for distributed systems, but they cannot be readily transplanted into the MANET context, since they rely on the existence of a network hierarchy and a central entity. Envisioned applications for the ad hoc networking environment may require a completely different notion of establishing a trust relationship, while the network operation may impose additional obstacles to the effective implementation of such solutions.

For small-scale networks, of the size of a personal or home network, trust can be established in a truly ad hoc manner, since relationships can be static and sporadically reconfigured manually. In such an environment, the owner of a number of devices or appliances can imprint them, that is, distribute their credentials along with a set of rules that determine the allowed interaction with and between devices [24]. The proposed security policy follows a master-slave model, with the master device being responsible for reconfiguring slave devices, issuing commands, or retrieving data. The return to the initial state can be performed only by the master device or by some trusted key escrow service.

This model naturally lends itself to represent personal area networking, in particular network instances such as Bluetooth [4], in the sense that within a piconet the interactions between nodes can be determined by the security policy. The model can be extended by allowing partial control or access rights to be delegated, so that the secure interaction of devices becomes more flexible [25]. However, if the control over a node can be delegated, the new master should be prevented from eradicating prior associations and assuming full control of the node.

A more flexible configuration, independent of initial bindings, can be useful when a group of people wish to form a collaborative computing environment [9]. In such a scenario, the problem of establishing a trust relationship can be solved by a secure key agreement, so that any two or more devices are able to communicate securely. The mutual trust among users allows them to share or establish a password using an offline secure channel or perform a "pre-authentication" step through a localized channel [1]. Then, they can execute a password-based authenticated key exchange over the nonsecure wireless medium. Schemes that derive a shared symmetric key could use a two- or a multi-party version of the password authenticated Diffie–Hellman key-exchange algorithm [3].

Human judgment and intervention can greatly facilitate the establishment of spontaneous connectivity among devices. Users can select a shared password or manually configure the security bindings between devices, as seen above. Furthermore, they could assess subjectively the "security" of their physical and networking environment and then proceed accordingly. However, human assistance may be impossible for the envisioned MANET environment with nodes acting as mobile routers, even though the distinction between an end device and a router may be only logical, with nodes assuming both roles. Frequently, the sole requirement for two transiently associated devices will be to mutually assist each other in the provision of basic networking services, such as route discovery and data forwarding. This could be so since mobile nodes do not necessarily pursue collectively a common goal. As a result, the users of the devices may have no means to establish a trust relationship in the absence of a prior context.

However, there is no reason to believe that a more general trust model would not be required in the MANET context. For instance, a node joining a domain may have to present its credentials in order to access an available service and, at the same time, authenticate the service itself. Similarly, two network nodes may wish to employ a secure mode of multi-hop communication and verify each other's identity. Clearly, support for such types of secure interaction, either at the network or at the application layer, will be needed.

A public key cryptosystem can be a solution, with each node bound to a pair of keys, one publicly known and one private. However, the deployment of a public key infrastructure (PKI) requires the existence of a certification authority (CA), a trusted third party responsible for certifying the binding between nodes and public keys. The use of a single point of service for key management can be a problem in the MANET context, especially because such a service should always remain available. It is possible that network partitions or congested links close to the CA server, although they may be transient, could cause significant delays in getting a response. Moreover, in the presence of adversaries, access to the CA may be obstructed, or the resources of the CA node may be exhausted by a DoS attack. One approach is not to rely on a CA and thus abolish all the advantages of such a facility. Another approach is to institute the CA in a way that answers the particular challenges of the MANET environment.

The former approach can be based on the bootstrapping of all network nodes with the credentials of every other node. However, such an assumption would dramatically narrow the scope of ad hoc networking, since it can be applied only to short-lived mission-oriented and thus closed networks. An additional limitation may stem from the need to ensure a sufficient level of security, which implies that certificates should be refreshed from time to time, requiring, again, the presence of a CA.

Alternatively, it has been suggested that users certify the public keys of other users. One such scheme proposes that any group of K nodes may provide a certificate to a requesting node. Such a node broadcasts the request to its one-hop neighborhood, each neighbor provides a partial certificate, and if sufficient K such certificates are collected, the node acquires the complete certificate [14, 29]. Another scheme proposes that each node select a number of certificates to store, so that, when a node wants the public key of one of its peers, the two certificate repositories are merged, and if a chain of certificates is discovered, the public key is obtained [13].

The solution of a key management facility that meets the requirement of the MANET environment has been proposed in [29]. To do so, the proposed instantiation of the public key infrastructure provides increased availability and fault tolerance. The distributed instantiation of the certification authority (CA) is equipped with a private/public key pair. All network nodes know the public key of the CA and trust all certificates signed by the CA's private key. Nodes that wish to establish secure communication with a destination query the CA and retrieve the required certificate, thus being able to authenticate the other end, and establish a secret shared key for improved efficiency. Similarly, nodes can request an update from the CA, that is, change their own public key and acquire a certificate for the new key.

The distributed CA is instantiated by a set of nodes (servers), as shown in Fig. 31.1, for enhanced availability. However, this is not done through naïve replication, which would increase the vulnerability of the system, since the compromise of a single replica would be sufficient for the adversary to control the CA. Instead, the trust is distributed among a set of nodes that share the key management responsibility. In particular, each of the n servers has its own pair of public/private keys and they collectively share the ability to sign certificates. This is achieved with the use of threshold cryptography, which allows any t + 1 out of n parties to perform a cryptographic operation, while t parties cannot do so. To accomplish this, the private key of the service, as a whole, is divided into n shares, with each of the servers holding one share. When a signature has to be computed, each server uses its share and generates a partial signature. All partial signatures are submitted to a combiner, a server with the special role to generate

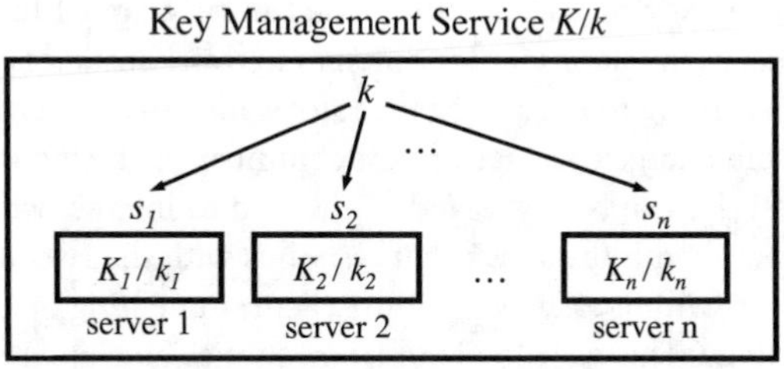

Key Management Service *K/k*

K_1/k_1 server 1 / K_2/k_2 server 2 / K_n/k_n server n

FIGURE 31.1 The configuration of a key management service comprising n servers. The service, as a whole, has a public/private key pair K/k. The public key K is known to all nodes in the network, whereas the private key k is divided into n shares $s_1, s_2,...,s_n$, with one share for each server. Moreover, each server has a public/private key pair K_i/k_i and knows the public keys of all nodes. (Reprinted with permission from Zhou, L. and Haas, Z.J., Securing ad hoc networks, *IEEE Network Magazine*, 13(6), Nov./Dec. 1999. © 1999 IEEE.)

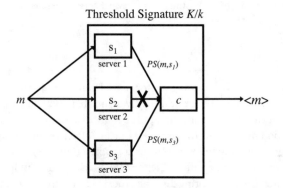

Threshold Signature *K/k*

FIGURE 31.2 The calculation of a threshold signature. As an example, the service consists of three servers and uses a (3,2) threshold cryptography scheme. K/k is the public/private key pair of the service and each server has a share s_i of the private key. To calculate the threshold signature on a message m, each server generates a partial signature $PS(m, s_i)$ and correct servers 1 and 3 forward their partial signatures to a combiner c. Even though server 2 fails to submit a partial signature, c is able to generate the signature $<m>_k$ (m signed by the service private key k). (Reprinted with permission from Zhou, L. and Haas, Z.J., Securing ad hoc networks, *IEEE Network Magazine*, 13(6), Nov./Dec. 1999. © 1999 IEEE.)

the certificate signature out of the collected partial signatures, as shown in the example of Fig. 31.2. This is possible only with at least t + 1 valid partial signatures.

The application of threshold cryptography provides protection from compromised servers, since more than t servers have to be compromised before it assumes control of the service. If fewer than t + 1 servers are under the control of an adversary, the operation of the CA can continue, since purposefully invalid partial signatures, "contributed" by rogue servers, will be detected. Moreover, the service provides the assurance that the adversary will not be able to compromise enough servers over a long period of time. This is done with the help of share refreshing, a technique that allows the servers to calculate new shares from the old ones without disclosing the private key of the service. The new shares are independent from the older ones and cannot be combined with the old shares in an attempt to recover the private key of the CA. As a result, to compromise the system, all t + 1 shares have to be compromised within one refresh period, which can be chosen appropriately short in order to decrease vulnerability. The vulnerability can be decreased even further, when a quorum of correct servers detects compromised or unavailable servers and reconfigures the service, that is, generates and distributes a new set of n' shares, t' + 1 of which need be combined now to calculate a valid signature. It is noteworthy that the public/private key pair of the service is not affected by share refreshing and reconfiguration operations, which are transparent to all clients.

The threshold cryptography key management scheme can be adapted further by selecting different configurations of the key management service for different network instances. For example, the numbers of servers can be selected according to the size or the rate of membership changes of the network; for a large number of nodes within a large coverage area, the number of servers should also be large, so that the responsiveness of the service can be high. Nodes will tend to interact with the closest server, which can be only a few hops away, or with the server that responds with the least delay. Another possibility is to alternate among the servers within easy reach of the client, something that can happen naturally in a dynamically changing topology. This way, the load from queries and updates will be balanced among different servers, and the chances of congestion near one of the servers will be reduced. At the same time, the storage requirements can be traded off for interserver communication, by storing at each server a fraction of the entire database.

Additionally, the efficient operation of the CA can be enhanced when it is combined with secure route discovery and data forwarding protocols. Such protocols could, in fact, approximate the assumption of reliable links between servers in [29] even in the presence of adversaries. In particular, two of the protocols that will be discussed below, SRP and SMT, lend themselves naturally to this model. Any two servers[3] can discover and maintain routes to each other and forward service-related traffic, regardless of whether or not intermediate nodes are trusted.

31.5 Secure Routing

The secure operation of the MANET routing protocol is of central importance because of the absence of a fixed infrastructure. Instead, nodes are transiently associated and will cooperate with virtually any node, including those that could potentially disrupt the route discovery and data forwarding operations. In particular, the disruption of the route discovery may be an "effective" means to systematically obstruct the flow of data. Adversaries can respond with stale or corrupted route replies or broadcast forged control packets in order to obstruct the propagation of legitimate queries and routing updates.

However, the usual practice for securing Internet routing protocols [19] cannot be applied in the MANET context. The schemes proposed to secure Internet routing rely mainly on the existence of a line of defense, separating the fixed routing infrastructure from all other network entities. This is achieved by distributing a set of public keys/certificates, which signify the authority of the router to act within the limits of the employed protocol (e.g., advertise certain routes), and allow all routing data exchanges to be authenticated, not repudiated, and protected from tampering. However, such approaches cannot combat a malicious router disseminating incorrect topological information. More importantly, they are not applicable in the MANET context because of impediments such as the absence of a fixed infrastructure and a central entity.

Although the appropriate design could provide increased assurances of the availability of an online certification authority (CA), the use of digital signatures and the hop-by-hop validation of control traffic may not be practical. First, mobile nodes lack sufficient computational power, as discussed above, and second, the interaction with the CA could become a limiting factor. In order to verify the correctness of the discovered routes, a node will have to acquire and validate the credentials of the responding nodes. Clearly, at least one route to the server has to be discovered before the node can contact the node instituting the CA server. But the problem is that, in the presence of adversaries, forged replies would still require the server's response to be validated. Another important limitation arises from the frequently changing topology and network membership, which would incur frequent queries addressed to the CA. In addition, congested links close to the server, although they may be transient or intermittent, could result in significant delays or even total failure to provide the certification services. Even relatively small delays may render the validation process obsolete.

---

[3] Any two servers of the key management service have a mutual security binding.

The protection of the route discovery process has been regarded as an additional Quality-of-Service (QoS) issue [28], by choosing routes that satisfy certain quantifiable security criteria. In particular, nodes in a MANET subnet are classified into different trust and privilege levels. A node initiating a route discovery sets the sought "security" for the route, that is, the required minimum trust level for nodes participating in the query/reply propagation. Nodes at each trust level share symmetric encryption and decryption keys. Intermediate nodes of different levels that cannot determine whether the required QoS parameter can be satisfied or decrypt in-transit routing packets drop them. This scheme provides protection (e.g., integrity) of the routing protocol traffic against adversaries outside a specific trust level.

An extension of the Ad Hoc On-demand Distance Vector (AODV) [20] routing protocol has been proposed [10] in order to protect the routing protocol messages. The Secure-AODV scheme assumes that each node has the certified public keys of all network nodes, since intermediate nodes validate all in-transit routing packets. The basic idea is that the originator of a control message appends an RSA signature [23] and the last element of a hash chain [15], i.e., the result of n consecutive hash calculations of a random number. As the message traverses the network, intermediate nodes cryptographically validate the signature and the hash value, generate the k-th element of the hash chain, with k being the number of traversed hops, and place it in the packet. Route replies are provided either by the destination or by intermediate nodes that have an active route to the sought destination.

A second proposal to secure AODV makes use of public key cryptography as well and operates in two stages, an end-to-end authentication, and an optional secure shortest path discovery [7]. First, a signed route request propagates to the sought destination, which returns a signed response to the querying node. At each hop, for either direction, the receiving node validates the received control packet and forwards it after signing it. At the second stage, a "shortest path confirmation" packet is sent towards the destination, while now intermediate nodes sign the message in an onion-like manner in order to disallow changes of the path length.

31.5.1 The Secure Routing Protocol

The Secure Routing Protocol (SRP) [17] for mobile ad hoc networks provides correct end-to-end routing information over an unknown, frequently changing network, in the presence of malicious nodes. It is assumed that any two nodes that wish to employ SRP have a Security Association (SA), such as a symmetric shared secret key. Communication takes place over a broadcast medium, and it is assumed that malicious nodes, which may concurrently corrupt the route discovery, cannot collude during a single route discovery. Moreover, we assume that nodes have a single data link interface, with a one-to-one correspondence between data link and IP addresses. Under these assumptions, the protocol is proven robust.

SRP provides one or more route replies, the correctness of which is verified by the route "geometry" itself, while compromised and invalid routing information is discarded. The route request packets verifiably propagate to the destination, and route replies are returned to the querying node strictly over the reversed route, as accumulated in the route request packet. In order to guarantee this crucially important functionality, SRP employs explicit interaction with the network layer; i.e., the IP-related functionality. Moreover, a number of novel features allow SRP to safeguard the route discovery operation, as explained below.

The Neighbor Lookup Protocol

The Neighbor Lookup Protocol (NLP) is an integral part of SRP responsible for the following tasks: (i) It maintains a mapping of Medium Access Control and IP layer addresses of the node's neighbors, (ii) it identifies potential discrepancies, such as the use of multiple IP addresses by a single data-link interface, and (iii) it measures the rates at which control packets are received from each neighbor, by differentiating the traffic primarily based on Medium Access Control addresses. The measured rates of incoming control packets are provided to the routing protocol as well. This way control traffic originating from nodes that selfishly or maliciously attempt to overload the network can be discarded.

Basically, NLP extracts and retains the 48-bit hardware source address for each received (overheard) frame along with the encapsulated IP address. This requires a simple modification of the device driver [27], so that the data link address is "passed up" to the routing protocol with each packet. With nodes operating in promiscuous mode, the extraction of such pairs of addresses from all overheard packets leads to a reduction in the use of the neighbor discovery and query/reply mechanisms for medium access control address resolution. Each node updates its neighbor table by retaining both addresses. The mappings between data-link and network interface addresses are retained in a table as long as transmissions from the corresponding neighboring nodes are overheard; a timeout period is associated with each entry removed from the table upon expiration.

NLP issues a notification to SRP in the event that according to the content of a received packet: (i) a neighbor used an IP address different from the address currently recorded in the neighbor table, (ii) two neighbors used the same IP address (that is, a packet appears to originate from a node that may have "spoofed" an IP address), (iii) a node uses the same medium access control address as the detecting node (in that case, the data link address may be "spoofed"). Upon reception of the notification, the routing protocol discards the packet bearing the address that violated the aforementioned policies.

Even though NLP does not rely on cryptographic validation, it thwarts adversaries from presenting themselves at the routing layer as more than one node. This would have been possible if different IP addresses were inserted in or used as the source address of the control traffic the adversary relays or originates. However, the effectiveness of NLP relies on the fact that medium access control addresses are either hardwired or may be changed only with substantial latency. In the former case, NLP can provide very strong assurances; in the latter one, it will be a significant line of defense, deterring, for example a malicious node from flooding the network with spurious traffic. In any case, we should note that it is not of interest for SRP whether a relay node indeed presented itself with its "actual" IP address, but whether the node participated in the discovery of the route.[4]

The Basic Secure Route Discovery Procedure

The querying node maintains a Query Sequence number, Q_{seq}, for each destination it securely communicates with. The monotonically increasing sequence number allows the destination to detect outdated route requests. At the same time, route requests are assigned a pseudorandom Query Identifier, Q_{ID}, which is used by intermediate nodes. Q_{ID} is statistically indistinguishable from a random number and thus unpredictable by an adversary with limited computational power. As a result, *broadcasted* fabricated requests will fail to cause subsequent legitimate queries to be dropped as previously seen, if, for example, the forged packets carry a higher sequence number.

Both Q_{ID} and Q_{seq} are placed in the SRP header, along with a Message Authentication Code (MAC) that covers the shared key, $K_{S,T}$, and the protocol header. Fields that are updated as the packet propagates towards the destination, such as the accumulated addresses, are excluded from the MAC calculation.

Nodes compare the last entry in the accumulated route to the IP datagram source address, which belongs to the neighboring node that relayed the request. If there is a mismatch, or NLP provides a notification that the relaying neighbor violated one of the enforced policies, the query is dropped. Otherwise the Q_{ID} and the source and destination addresses are placed in the query table, so that previously seen queries are discarded. "Fresh" route requests are rebroadcasted, with intermediate nodes inserting their IP address in the request packet.

The destination validates the integrity and freshness of queries originating from nodes it is securely associated with. It generates a number of replies that does not exceed the number of its neighbors, so that a malicious neighbor does not control more than one route. The reversed accumulated route serves as the source route of the reply packet, which is identified by Q_{ID} and Q_{seq}. The appended MAC covers the SRP header, including the source route. This way the source can be provided with evidence that the

---

[4] The special case of using the address of a node already on the path is equivalent to any other malicious alteration of the control traffic, which the adversary could do in the first place. Of course, such a duplicate address will be perceived as a loop and the route will be readily discarded.

request had reached the destination and, in conjunction with the source route, that the reply was indeed returned along the reverse of the discovered route.

As the reply propagates along the reverse route, each intermediate node simply checks whether the source address of the route reply datagram is the same as the one of its downstream node, as determined by the route reply; if not, the reply is discarded. Ultimately, the source validates the reply by first checking whether it corresponds to a pending query. Then, it is sufficient to validate the MAC, since the IP source-route already provides the (reversed) route itself.

Priority-Based Query Handling

In order to guarantee the responsiveness of the routing protocol, nodes maintain a priority ranking of their neighbors according to the rate of queries observed by NLP. The highest priority is assigned to the nodes generating (or relaying) requests with the lowest rate and vice versa. Quanta are allocated proportionally to the priorities, and low-priority queries that are not serviced are eventually discarded. Within each class, queries are serviced in a round robin manner. Selfish or malicious nodes that broadcast requests at a very high rate are throttled back, first by their immediate neighbors and then by nodes farther from the source of potential misbehavior. Nonmalicious queries, that is, queries originating from benign nodes that regulate in a nonselfish manner the rate of query generation, will be affected only for a period equal to the time it takes to update the priority (weight) assigned to a misbehaving neighbor. In the mean time, the round robin servicing of requests provides the assurance that benign requests will be relayed even amid a "storm" of malicious or extraneous requests.

The Route Maintenance Procedure

The route-error packets are source-routed to the end node along the prefix of the route that is being reported as broken. The intermediate upstream nodes, with respect to the point of breakage, check whether the source address of the route error datagram is the same as the one of their downstream node as reported in the broken route. Then, if there is no notification from NLP that the relaying neighbor violated one of the enforced policies, they relay the packet towards the source. In this case, NLP prevents an adversary that does not belong to the route, but lies at a one-hop distance from it, from generating an error message, since an inconsistency with the addresses already used (during the route discovery) by the actual downstream neighbor will be detected.

The notified source compares the source-route of the error message to the prefix of the corresponding active route. This way, it verifies that the provided route error message refers to the actual route and that it is not generated by a node that is not part of the route. The correctness of the feedback (i.e., whether it reports an actual failure to forward a packet) cannot be verified, though. As a result, a malicious node lying on a route can mislead the source by corrupting error messages generated by another node or by masking a dropped packet as a link failure. However, it can harm only the route it belongs to, something that was possible in the first place if it simply dropped or corrupted the data packets.

The SRP Extension

The basic operation of SRP can be extended in order to allow for nodes other than the destination to provide route replies. This would be possible only under additional trust assumptions, when, for example, nodes sharing a common objective belong to the same group and mutually trust all the group members. In particular, this could be the case when all group members share a secret key.

Under this assumption, a querying node appends to each query an additional MAC calculated with the group key, which we call Intermediate Node Reply Token (INRT). The functionality of SRP remains as described above, with the following addition: each group member maintains the latest query identifier seen from each of its peers and can thus validate both the freshness and origin authenticity of queries generated from other group nodes.

If a node other than the sought destination receives such a valid query, it can respond to the request if it has knowledge of a route to the destination in question. However, the correctness of such a route is conditional upon the correctness of the second portion of the route, which is provided by the intermediate node.

This functionality can be provided independently from and in parallel with the one relying solely on the end-to-end security associations. For example, it could be useful for frequent intragroup communication; any two members can benefit from the assistance of their trusted peers, which may already have useful routes.

31.6 Secure Data Forwarding

The frequent interaction with a CA and the frequent use of computationally expensive cryptographic tools are restrictive assumptions, especially for secure data-forwarding schemes. Such protocols must also take into account the inherent limitations of the MANET paradigm, exploit its features, and incorporate widely accepted and evaluated techniques in order to be efficient and effective. Moreover, a secure routing protocol is a prerequisite for an effective secure data-forwarding scheme. The above Secure Routing Protocol (SRP) for mobile ad hoc networks satisfies the above-stated goals.

However, SRP or any other underlying routing protocol cannot guarantee that the nodes along a correctly discovered route will, indeed, relay the data as expected. An adversary may misbehave in an intermittent manner, that is, provide correct routing information during the route discovery stage, and later forge or corrupt data packets during the data forwarding stage. This is exactly the function that is required by any secure data forwarding protocol; to secure the flow of data traffic in the presence of malicious nodes, after the routes between the source and the destination have been discovered.

One of the solutions targeting the MANET environment proposes two mechanisms that (1) detect misbehaving nodes and report such events and (2) maintain a set of metrics reflecting the past behavior of other nodes [16]. To alleviate the detrimental effects of packet dropping, nodes choose the "best" route, comprised of relatively well-behaved nodes; i.e., nodes that do not have a history of avoiding forwarding packets. Among the assumptions in the above-mentioned work are a shared medium, bidirectional links, use of source routing (i.e., packets carry the entire route that becomes known to all intermediate nodes), and no colluding malicious nodes. Nodes operating in promiscuous mode overhear the transmissions of their successors and may verify whether the packet was forwarded intact to the downstream node. Upon detection of a misbehaving node, a report is generated, and nodes update the rating of the reported misbehaving node. The ratings of nodes along a well-behaved route are periodically incremented, while reception of a misbehavior alert dramatically decreases the nodes rating. When a new route is required, the source node calculates a path metric equal to the average of the ratings of the nodes in each of the route replies and selects the route with the highest metric.

A different approach is to provide incentive to nodes so that they comply with protocol rules, i.e., properly relay user data. The concept of fictitious currency is introduced in [6], in order to endogenize the behavior of the assumed greedy nodes, which would forward packets in exchange for currency. Each intermediate node purchases from its predecessor the received data packet and sells it to its successor along the path to the destination. Eventually, the destination pays for the received packet.[5] This scheme assumes the existence of an overlaid geographic routing infrastructure and a Public Key Infrastructure (PKI). All nodes are preloaded with an amount of currency, have unique identifiers, and are associated with a pair of private/public keys. Finally, the cryptographic operations related to the currency transfers are performed by a physically tamper-resistant module.

Another approach appropriate for MANET, which departs significantly from the two above-mentioned schemes, is presented below. Low-cost cryptography is used to protect the integrity and origin authenticity

---

[5] An alternative implementation, with each packet carrying a purse of fictitious currency from which nodes remove their reward, has been proposed as well.

of exchanged data, without placing any overhead at intermediate nodes. Moreover, the feedback that determines the "security" of the chosen paths originates only from trusted destinations, thus allowing "safe" inferences on the quality of the paths. Finally, the reliability and fault tolerance of data transmissions is enhanced significantly.

31.6.1 Secure Message Transmission Protocol

The Secure Message Transmission (SMT) protocol [18] is a network-layer secure and fault-tolerant data-forwarding scheme, tailored to the MANET characteristics. In short, SMT is provided with routing information by a protocol such as SRP. This allows SMT to determine a set of diverse paths connecting the source and the destination, as shown in the example of Fig. 31.3. Then, it introduces limited transmission redundancy across the paths, by dispersing a message into N pieces, so that successful reception of any M-out-of-N pieces allows the reconstruction of the original message at the destination. Each piece, transmitted over one path, is equipped with a cryptographic header that provides origin authentication, integrity, and replay protection. Upon reception of a number of pieces, the destination informs the source of which pieces, and thus routes, were intact. In order to enhance the robustness of the feedback mechanism, the small-sized acknowledgments, also protected by a cryptographic header, are maximally dispersed, so that successful reception of one piece is sufficient. If less than M pieces of the message were received, the source retransmits the remaining pieces over the intact routes, or in general the ones deemed as more "secure." If too few pieces were acknowledged or too many messages remain outstanding, the protocol adapts its operation by determining a different path set, reencoding undelivered messages, and reallocating pieces over the path set. Otherwise, it proceeds with subsequent message transmissions.

SMT exploits MANET features such as the topological redundancies, interoperates widely with accepted techniques such as on-demand route discovery and source routing, relies on a security association only between the source and the destination, and makes use of highly efficient symmetric-key cryptography. Moreover, the routing decisions are made by the querying node, based on the feedback that the destination and the underlying secure routing protocol provide. At the same time, no additional processing overhead is imposed on intermediate nodes, which do not perform any cryptographic operation but simply relay the message pieces. However, the use of multiple paths and the resultant greater number of nodes involved in the forwarding of a single message can be admittedly considered as the price to pay in order to achieve the sought robustness.

On the one hand, SMT's robustness can be enhanced by the adaptation of parameters such as the number of paths, and the ratio of the numbers of transmitted to required pieces, termed as the redundancy or dispersion factor. On the other hand, in a low-risk environment with limited malicious failures, the same parameters can be adjusted, so that the imposed transmission overhead is reduced to a level close to that of a single-path scheme. An additional element that contributes to the flexibility of SMT is that different algorithms can be implemented for the selection of the path set, based on different metrics and interpretations of the network feedback. SMT can yield 100% successful message reception even in a highly adverse environment, when, for example, 20% of the network nodes are malicious, while keeping the message and computation overhead low.

The two communicating end nodes make use of the Active Path Set (APS), comprising diverse paths that are not deemed failed. The sender invokes the underlying route discovery protocol, updates its network topology view, and then determines the APS for a specific destination. This model can be extended to multiple destinations, with one APS per destination. At the receiver's side, the APS is used for the transmission of the feedback, but if links are not bidirectional, the destination will have to determine its own "reverse" APS.

The dispersion of messages, which is performed by the information dispersal algorithm (IDA) [22], is coupled to the APS characteristics through an appropriate selection of the dispersion algorithm parameters. For example, in low connectivity conditions (small number of disjoint paths), the sender

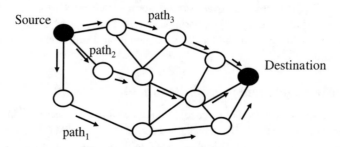

FIGURE 31.3   The Secure Message Transmission Protocol makes use of multiple diverse paths connecting the source and the destination. In particular, the Active Path Set (APS) contains paths that have not been detected as failed, either due to path breakage or because of the presence of an adversary on the path.

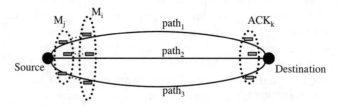

FIGURE 31.4   For an APS with three paths, the source can disperse each message into three pieces and transmit them across APS. The destination responds to each message M_k with an acknowledgment ACK_k, notifying explicitly which pieces were received. This feedback allows the source to quickly update the rating of the APS paths and retransmit lost pieces across the operational paths if the message cannot be reconstructed at the destination.

may increase the redundancy factor in order to provide increased assurance and possibly low transmission delay. The adaptation of the protocol is the result of the interplay among the following parameters:

1. K, the (sought) cardinality of APS
2. k, the S,T-connectivity, i.e., the maximum number of S→T node-disjoint paths from the source (S) to the destination (T)
3. r, the redundancy factor of the IDA encoding
4. x, the maximum number of malicious nodes

Clearly, the condition for successful reception is $x \leq \lceil K \times (1 - r^{-1}) \rceil$, which demonstrates the coupling among choices of parameters.

In particular, K can be determined as a function of r, so that the probability of successful transmission is maximized. (Note that K is equal to N when one message piece is allocated per path.) In order to do so, the source starts by determining an APS of the k shortest, in number of hops, node disjoint paths. Then, let P_{GOAL} be the sought probability of successful reconstruction of a dispersed message. P_{GOAL} can be provided from the application layer and may correspond to the features of the supported application, for example. Given P_{GOAL} and k, the node calculates the required redundancy factor, r_{GOAL}, for a given or estimated fraction of present adversaries. The source disperses outgoing messages with the redundancy value closest to r_{GOAL}, with M and N selected to minimize the transmission overhead.

Once dispersed, the message pieces are transmitted across APS. If N < k, the node selects the N paths of the APS with the highest rating. If the receiver cannot reconstruct the message, the source retransmits the pieces that were not received, according to the feedback provided by the destination. Message pieces are retransmitted by SMT a maximum number of times, $Retry_{MAX}$, which is a protocol selectable parameter. If all retransmissions fail, the message is discarded. This way, limited retransmissions enhance the efficiency of SMT by alleviating the overhead from retransmitting the entire amount of data. On the other hand, SMT does not assume the role of a transport or application layer protocol; its goal is to promptly detect and tolerate failures and thus adapt its operation to remain effective and efficient (see Fig. 31.4).

The transmission of data is continuous over the APS, with retransmissions placed at the head of the queue, upon reception of the feedback. The continuous usage of the APS allows SMT to update quickly its assessment on the quality of the paths. For each successful or failed piece, the rating of the corresponding path is increased or decreased, respectively. When the rating drops below a threshold, the path is discarded. The path rating is also decreased slowly as time goes by in order to reduce the chance of using a stale path. Moreover, the simultaneous routing over a number of paths, if not the entire APS, provides the opportunity for low-cost probing of the paths. In particular, the source can easily tolerate the loss of a piece that was transmitted over a low-rated path.

31.7 Discussion

The fast development of mobile ad hoc networking technology over the last few years, with satisfactory solutions to a number of technical problems, supports the vision of widely deployed mobile ad hoc networks with self-organizing features and without the necessity of a preexisting infrastructure. In this context, the secure operation of such infrastructureless networks becomes a primary concern. Nevertheless, the provision of security services is dependent on the characteristics of the supported application and the networked environment, which may vary significantly. At one extreme, we can think of a library or an Internet café, which provides short-range wireless connectivity to patrons, without any access constraint other than the location of the mobile device. At the other extreme, a military or law enforcement unit can make use of powerful mobile devices capable of performing expensive cryptographic operations. Such devices would communicate only with other trusted devices.

Between these two ends of the spectrum, a multitude of MANET instances will provide different services, assume different modes of interaction and trust models, and admit solutions such as the ones surveyed above. Moreover, it is probable that instead of a clear-cut distinction among network instances, devices and users with various security requirements will coexist in a large, open, frequently changing ubiquitous network.

In this context, an important related issue is the IP addressing scheme employed in the MANET environment. The common assumption that node credentials, e.g., certificates, are bound to IP addresses may need to be revisited, since one can imagine that roaming nodes will join MANET subdomains, and IP addresses will be assigned dynamically (e.g., DHCP [8] or IPv6 auto-configuration [26]) or even randomly (e.g., Zero-Configuration [12]).

A type of ad hoc network with particular requirements is a sensor network, which requires multi-hop communication throughout a network of hundreds or even thousands of nodes, with relatively infrequent topological changes. It is expected that a single organization will undertake the deployment and administration of these networks. Moreover, sensing devices have very limited computational capabilities, network transmission rates are relatively low, and communications are mostly data driven. These requirements may affect in different ways the design of security measures for sensor networks, as demonstrated by the schemes proposed in the literature.

One of the proposals to secure sensor networks provides a protocol for data authentication, integrity, and freshness and a lightweight implementation of an authenticated broadcast protocol [21]. The scheme targets a restricted, infrastructure-oriented environment, with a trusted central entity instituted by a set of base stations. Sensor nodes communicate only with a base station, which broadcasts messages towards the sensors. The base station and all nodes initially possess a symmetric encryption and authentication key, which secures the exchanged traffic, while later, the base station periodically broadcasts the key that was used to authenticate transmissions during the last period.

An approach that has similarities but targets a more general setting proposes a key management scheme for sensor networks [2]. The focus is on resource-constrained large sensor networks, comprising nodes that are assumed tamper resistant and equipped with a secret group key. Similarly to the previous scheme, the use of symmetric key cryptography is proposed as the only feasible, low-cost solution. However, frequent rekeying, that is, periodic regeneration of the single key that is used to encrypt all data trans-mitted by sensors, is proposed to protect it from possible compromise. In order to make this reconfig-uration operation efficient, the sensors are organized into clusters with a two-hop diameter, while clusterheads are elected and form a backbone. Then, from a subset of the backbone, a randomly elected node generates the new key.

The simplified trust models of sensor networks, which, nonetheless, lead to efficient solutions, may not necessarily be usable in other ad hoc networking instances. The circumstantial coexistence of disparate nodes, or the requirement of fine-grained trust relationships, call for solutions that can adapt to specific contexts and support the corresponding application. However, although the requirements of the appli-cation are expected to dictate the characteristics of the required security mechanisms, some aspects of security, such as confidentiality, may not be different at all in the MANET context. Instead, the greatest challenge is to safeguard the basic network operation.

In particular, the securing of the network topology discovery and data forwarding is a prerequisite for the secure operation of mobile ad hoc networks in any adverse environment. Furthermore, the protection of the functionality of the networking protocols will be in many cases orthogonal to the security require-ments and the security services provided at the application layer. For example, a transaction can be secured when the two communicating end nodes execute a cryptographic protocol based on established mutual trust, with the adversary being practically unable to attack the protocol. But this does not imply that the nodes are secure against denial of service attacks; the adversary can still abuse the network protocols, and in fact, do it with little effort compared to the effort needed to compromise the crypto-graphic protocol.

The self-organizing networking infrastructure has to be protected against misbehaving nodes, with the use of low-cost cryptographic tools, under the least restrictive trust assumptions. Moreover, the overhead stemming from such security measures should be imposed mostly, if not entirely, on nodes that communicate in a secure manner and that directly benefit from these security measures. Further-more, we believe that the salient MANET features and the unique operational requirements of these networks call for security mechanisms that are primarily present at, and closely interwoven with, the network-layer operation, in order to realize the full potential of this promising new technology.

References

[1] D. Balfanz, D.K. Smetters, P. Stuart, H. C. Wang, Talking to Strangers: Authentication in Ad Hoc Networks, *Network and Distributed System Security Symposium,* San Diego, CA, Feb. 2002.

[2] S. Basagni, K. Herrin, E. Rosti, and D. Bruschi, Secure Pebblenets, *2nd MobiHoc,* Long Beach, CA, Oct. 2001.

[3] S.M. Bellovin and M. Merritt, Encrypted Key Exchange: Password-Based Protocols Secure Against Dictionary Attacks, *Proceedings of the IEEE Symposium on Security and Privacy,* May 1992.

[4] Bluetooth Special Interest Group, Specifications of the Bluetooth System, http://www.blue-tooth.com.

[5] M. Brown, D. Cheung, D. Hankerson, J.L. Hernadez, M. Kirkup, and A. Menezes, PGP in Con-strained Wireless Devices, *Proceedings of 9th USENIX Symposium,* Denver, CO, Aug. 2000.

[6] L. Buttyan and J.P. Hubaux, Enforcing Service Availability in Mobile Ad Hoc WANs, *1st MobiHoc,* Boston, MA, Aug. 2000.

[7] B. Dahill, B.N. Levine, E. Royer, and C. Shields, A Secure Routing Protocol for Ad Hoc Networks, Technical Report UM-CS-2001–037, EE&CS, University of Michigan, Ann Arbor, Aug. 2001.

[8] R. Droms, Dynamic Host Configuration Protocol, IETF RFC 2131, Mar. 1997.

[9] L.M. Feeney, B. Ahlgren, and A. Westerlund, Spontaneous Networking: An Application-Oriented Approach to Ad Hoc Networking, *IEEE Communications Magazine,* Jun. 2001, pp. 176–181.

[10] M. Guerrero, Secure AODV, Draft sent to the manet@itd.nrl.navy.mil mailing list.

[11] V. Gupta and S. Gupta, Securing the Wireless Internet, *IEEE Communications Magazine,* Dec. 2001, pp. 68–74.

[12] M. Hattig, Ed., Zero-conf IP Host Requirements, Draft-ietf-zeroconf-reqts-09.txt, IETF MANET Working Group, Aug. 2001.

[13] J.P. Hubaux, L. Buttyan, and S. Capkun, The Quest for Security in Mobile Ad Hoc Networks, *2nd MobiHoc*, Long Beach, CA, Oct. 2001.

[14] J. Kong, P. Zerfos, H. Luo, S. Lu, and L. Zhang, Providing Robust and Ubiquitous Security Support for Mobile Ad-Hoc Networks, *IEEE ICNP (International Conference on Network Protocols) 2001*, Riverside, CA, Nov. 2001.

[15] L. Lamport, Password authentication with insecure communication, *Comm. of ACM*, 24, 770–772, 1981.

[16] S. Marti, T.J. Giuli, K. Lai, and M. Baker, Mitigating Routing Misbehavior in Mobile Ad Hoc Networks, *6th MobiCom*, Boston, MA, Aug. 2000.

[17] P. Papadimitratos and Z.J. Haas, Secure Routing for Mobile Ad Hoc Networks, *SCS Communication Networks and Distributed Systems Modeling and Simulation Conference (CNDS 2002)*, San Antonio, TX, Jan. 27–31, 2002.

[18] P. Papadimitratos and Z.J. Haas, Secure Message Transmission in Mobile Ad Hoc Networks, submitted for publication.

[19] P. Papadimitratos and Z.J. Haas, Securing the Internet Routing Infrastructure, *IEEE Communications Magazine*, 40(10), Oct. 2002.

[20] C.E. Perkins, E.M. Royer, and S.R. Das, Ad hoc On-Demand Distance Vector Routing, Draft-ietf-manet-aodv-08.txt, IETF MANET Working Group, June, 2001.

[21] A. Perrig, R. Szewczyk, V. Wen, D. Culler, and J.D. Tygar, SPINS: Security Protocols for Sensor Networks, *Proc. 7th Ann. Intl. Conf. Mobile Computing and Networks (MobiCom 2001)*, Rome, Italy, 2001, pp. 189–199.

[22] M.O. Rabin, Efficient dispersal of information for security, load balancing, and fault tolerance, *Journal of ACM*, 36, 335–348, 1989.

[23] R. Rivest, A. Shamir, and L. Adleman, A method for obtaining digital signatures and public key cryptosystems, *Comm. of ACM*, 21, 120–126, 1978.

[24] F. Stajano and R. Anderson, The Resurrecting Duckling: Security Issues for Ad Hoc Wireless Networks, *Security Protocols, 7th International Workshop, LNCS*, 1999.

[25] F. Stajano, The Resurrecting Duckling – What Next? *Security Protocols, 8th International Workshop, LNCS*, 2000.

[26] S. Thomson and T. Narten, IPv6 Stateless Address Autoconfiguration, IETF RFC 1971, www.ietf.org.

[27] G.R. Wright and W. Stevens, *TCP/IP Illustrated, Vol. 2, The Implementation*, Addison-Wesley, Reading, MA, 1997.

[28] S. Yi, P. Naldurg, and R. Kravets, Security-Aware Ad-Hoc Routing for Wireless Networks, UIUCDCS-R-2001–2241 Technical Report, Aug. 2001.

[29] L. Zhou and Z.J. Haas, Securing Ad Hoc Networks, *IEEE Network Magazine*, Nov./Dec. 1999.

32

Security Issues in Ad Hoc Networks

Dan Zhou
Florida Atlantic University

Abstract

In this chapter, we discuss issues and survey current solutions in securing ad hoc wireless networks. The characteristics of ad hoc networks render the trust a host could place on other hosts and a network more precarious than in a conventional network. Any viable security approach ought to address trust concerns for specific applications. As examples, we look at three specific issues and their current proposed solutions: a transient association as host access control policy for mobile appliances, link-by-link and end-to-end authentication for securing routing in open networks, and split control for a centralized service or distributed services for survivable services in a rapidly changing network.

32.1 Introduction

Numerous task forces are formed on a need basis, such as a search committee for the president of a university, or a military deployment in a foreign country. A logical communication vehicle for these task forces is a mobile ad hoc wireless network because it does not require a prior physical infrastructure (i.e., wired network) [1]. Task forces have valuable assets such as transcripts of a search committee meeting or a program that controls the movement of tanks. These resources could come under attack, from both within and without, with malicious intention or through mere carelessness.

To protect networks from adversaries, we investigate security issues in Ad Hoc Networks (*AHNs*), based on our knowledge in securing wired networks. AHNs are prone to the same types of attacks as wired networks. Furthermore, the openness of wireless communication media and node mobility make AHNs more vulnerable than traditional networks to attacks. Anyone with a scanner can monitor traffic from the comfort of his or her home or the ease of a street corner. With a powerful jamming machine, an attacker can reduce the channel availability or even shut down communication channels [24].

Wired networks are built over time. They reflect security policies of organizations. Trust between entities, an essential element of a security policy, is also built over time. System administrators support network operations such as implementing security policies. In comparison, AHNs are built quickly and as needed. Trust and policies may be put together in a hurry. Mobility and some physical features (e.g., small size) of nodes make them more easily compromised and lost than those in wired networks.

Different AHNs have different initial contexts and requirements for security depending on applications. However, they all share one characteristic: no fixed infrastructure. The lack of infrastructure support leads to the absence of dedicated machines providing naming and routing service. Every node in an AHN becomes a router. Thus network operations have higher dependence on individual nodes than in wired networks. The mobility of nodes brings constant change in network topology and membership, making it impractical to provide traditional, centralized services [1,24].

In this chapter, we look at security challenges presented in ad hoc mobile wireless networks and how they are addressed currently. In Section 32.2 Introduction to Security, we introduce security requirements and traditional security mechanisms. We describe security requirements specific to AHNs and the particular challenges in implementing security mechanisms in AHNs in Section 32.3. We then present some of the current work in the research community in attempting to address these challenges in the rest of the chapter. In Section 32.4, we sketch an access control model that defines what access nodes can have to each other. We then describe routing security issues and some proposed solutions in Section 32.5 Routing Security. The state of the art in implementing traditional security mechanisms is explained in Section 32.6 Key Distribution. We conclude with a discussion of future work in Section 32.7 Future Directions.

32.2 Introduction to Security

"Security is the possibility of a system withstanding an attack."[1] During the 20th century, we refined our requirements for security and mechanisms to satisfy them. There are two types of security mechanisms: preventative and detective [17]. The majority of the preventive mechanisms have cryptography as a building component.

32.2.1 Security Requirements

The goal of system security is to have controlled access to resources. The key requirements for networks are confidentiality, authentication, integrity, nonrepudiation, and availability [10,17]. We define them as follows:

- *Availability:* no interruption of services
- *Confidentiality:* no unauthorized divulge of information
- *Authentication:* knowing the identity of a communicating party or the source of a piece of information
- *Integrity:* no unauthorized modification of resources
- *Nonrepudiation:* nondeniability of committed actions

---

[1] Discussion with Shaoying, Liu.

Traditionally, we categorize the attacks that computer and network systems experience in four broad categories: interruption, interception, modification, and fabrication [19]. Interruption renders a resource unavailable, interception discloses classified information, modification changes the attributes of a resource, and fabrication creates a false resource.

Security controls are put in place to deter attacks, therefore providing system-desired security services. A security mechanism follows three steps — identification, authentication, and authorization — to control access to resources. Identification names entities. Authentication checks that an entity is who or what it claims to be. Authorization either grants or refuses access rights based on some security policies, which are a part of an organization policy. Policies define access control rules and translate the trust that we place on entities into access control decisions.

32.2.2 Cryptography Basis

Preventive security controls are often protocols that utilize cryptography. Cryptographic algorithms are functions that transform information to conceal it [12]. There are three types of cryptographic algorithms: hash, secret-key cryptography, and public-key cryptography. Hash algorithms do not use keys. Secret-key cryptography uses one key. Public-key cryptography uses two keys.

A hash algorithm is a one-way function that maps a message of any size into a fixed size digest (see Fig. 32.1). Message digests are fingerprints of messages. A hash function is considered secure if it is computationally infeasible to find a corresponding message given a fingerprint, or to find one message that has the same fingerprint as a given message, or to find two arbitrary messages that have the same fingerprint [12].

Secret-key cryptography makes use of a pair of functions: encryption and decryption (see Fig. 32.2). The encryption function uses a key to mangle a message. The message before encryption is called plaintext. The encrypted message is called ciphertext. The decryption function uses the same key to unmangle the ciphertext. The key is a shared secret between communicating entities. Secret-key encryption provides confidentiality, as only those entities knowing the secret can uncover the plaintext messages.

Public-key cryptography uses a pair of keys, a public key and a private key, which are uniquely associated with each other (see Fig. 32.3). Each entity has a key pair, $<K_E, K_E^{-1}>$, where K_E is the public key of entity E, and K_E^{-1} is E's private key. The private key is only known to the owner, while the public key is widely publicized. Public key encryption uses a public key for encryption and a private key for

FIGURE 32.1 Hash function uses no key.

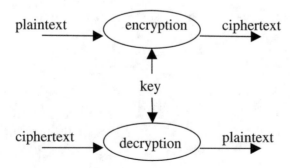

FIGURE 32.2 Secret-key cryptography uses one key.

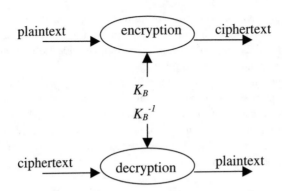

FIGURE 32.3 Public-key encryption uses a key pair: K_B and K_B^{-1}.

decryption. To send Bob a message that only Bob can read, Alice uses Bob's public key, K_B, to encrypt the message. Bob uses his private key, K_B^{-1}, to decrypt the ciphertext.

A voice mail service provides an analogy to public key cryptography. Your phone number is your public key. Anyone can leave you a message by dialing that number. The personal identification number (PIN) to your mailbox is your private key. Only you can listen to the messages left in your mailbox.

Public-key cryptography can also generate digital signatures that can be verified by an arbitrator (see Fig. 32.4). A digital signature binds a signature with an entity and a message. Alice signs a message using her private key K_A^{-1}. An arbitrator can verify the signature using Alice's public key K_A. We can again make a crude analogy to the voice mail service. You can make an announcement on your voice mail using your PIN. Anyone can dial your phone number and verify that you made that statement.

Public-key encryption provides confidentiality because a private key is only known to the key owner. Public-key signature provides authentication, integrity, and nonrepudiation because of the binding of a message, a signature, and the private key that was used to generate the signature. In practice, we sign a digest of a message instead of the message itself, which takes less processing time because of the reduced size.

A comprehensive solution to communication security includes protocols, algorithms, and key management [19]. The breakdown of any of these components compromises security. We now turn to the key management aspect of a security solution.

32.2.3 Key Management

Keys are an essential component of security because they allow us to read otherwise unintelligible messages and to sign documents, among other things. Cryptographic protocols use keys to authenticate

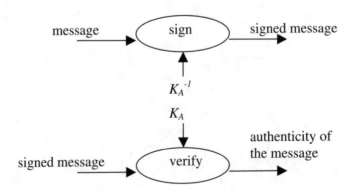

FIGURE 32.4 Public-key digital signature.

entities and grant access to guarded information to those who exhibit their knowledge of the keys [12]. Therefore, it is imperative that keys be securely generated and distributed to appropriate entities.

Secret keys are shared between communicating entities. A secret key can be generated by one party and distributed to another entity, either through direct physical contact or a secure channel. The key can also be negotiated among entities, in which case key generation and distribution are accomplished simultaneously.

In public-key cryptography, a public key is made public, while the corresponding private key is kept secret. A public-key certificate certifies the binding between a public key and an entity. Certificates are signed bindings by a trusted party whose public key is known beforehand.

Public-key certificates can be generated and distributed through a central server (similar to publishing phone numbers in a phone book) or a network of nodes that provides such services (similar to distributing cell phone numbers by the word of mouth), or a combination of the two. Public-key cryptography is often used to distribute secret keys.

32.3 Security Issues in Ad Hoc Networks

Security requirements in AHNs do not differ dramatically from their wired network counterparts [24]. Traditional security mechanisms still play a role in achieving AHN security. However, the context to achieve security goals is different. Changes in network topology and membership occur rapidly in this new context [15]. Consequently, some issues that are only of concern to high-assurance applications in wired networks are now essential to general AHN applications.

In wired networks we assume the following are in place:

1. Availability of routing service, which implies knowledge of network topology and membership
2. Availability of supporting services, such as naming and key distribution, through central, static system control
3. Security policy for networks and systems

Security policies (i.e., access control policies) are embedded in the networked nodes and protocols as prevention and detection mechanisms. Prevention mechanisms include identification, authentication, authorization, and firewall.

32.3.1 Access Control Policy

The underlying characteristic of the key issues in AHN security is the ad hoc, mobile, and wireless nature of the network. In AHNs, the physical boundary between internal and external networks disappears. Collaborations of nodes can no longer be taken for granted. Each node makes decisions regarding access to the network in addition to controlling access to itself. The roles each node takes on are more critical in AHNs. Hence it is critical that security policies be clearly defined before they are embedded in the network protocols and applications [20]. A good policy should encompass access rules to the network and individual nodes.

Policy decisions are based on a trust relationship [5]. The ad hoc nature of trust in a dynamic network raises some issues to the forefront:

How do individual nodes establish trust among themselves?
How does a network (a collection of nodes) establish trust with individual nodes?
How do trust relationships evolve over time?
How much risk is there in trusting a node or a network?

In Section 32.4 Recurrent Duckling Transient Association, we will examine one access control model that provides a framework to address some of these concerns.

32.3.2 Routing Security

Routing in AHNs is a collective work of nodes in the network [15,16]. Accordingly, the availability of routing service depends on the good behavior of nodes within transmission range of one another.

Nodes in AHNs have less physical security than in wired networks because they are not within a physical protection boundary. They are more easily compromised as a result. Malicious nodes can fabricate routing information and modify routing packets that pass through them. Subsequently, networks can be fragmented by the wrong routing information advertised by these nodes. Cryptography is a commonly used preventive measure to counter fabrication and modification attacks [21].

Nodes could behave more subtly to affect the effectiveness of AHNs. A chatty node could occupy valuable bandwidth. A passive–aggressive node could either drop packets that pass through or not respond to routing requests. Detective mechanisms help to curb those behaviors. Tactics such as auditing, quota-and-reward, and trading induce collaboration from these nodes [4,11].

An ideal secure routing algorithm for AHNs withstands the behavior of both malicious and selfish nodes. In Section 32.5 Routing Security, we relate several routing algorithms that provide some level of security.

32.3.3 Service Survivability

Mobility and the increased vulnerability of nodes in AHNs necessitate decentralization for a viable security solution [24]. Networks are partitioned and combined as nodes move around. A centralized service provider is a single point of failure and attack: services would be rendered unavailable if the server is partitioned into a different network and when it fails. A decentralized service would lessen the severity of these problems and increase the survivability of the service. In Section 32.6 Key Distribution, we investigate one attempt to adapt security mechanisms from wired networks to AHNs.

32.4 Recurrent Duckling Transient Association

Home appliances form an AHN in which there are clearly defined roles for each node: controllers (e.g., remote control) and controlled devices (e.g., TV and oven). However, the association between a controller and the controlled is not permanent. Stajano and Anderson developed an access control model, *resurrecting duckling transition association*, to describe this transient master-slave relationship among appliances [18].

In this model, a device is initially in a prebirth stage where it is free but latent. It is born when a controller comes into contact with it. The controller becomes its master, and it becomes a slave. This process is called *imprint*. The master controls the fate of the slave, from when it should die (i.e., be deactivated) to what services it can provide to other appliances. When a device is deactivated, it goes back to the prebirth stage. It can be reborn through another imprint (*resurrection*).

Take as an example the appliances at the home of Alice and Bob (see Fig. 32.5). Alice purchases *TV-small* and *VCR-cool* that are in prebirth stage and imprints them with her remote control *RT-Alice*, which gives her full control of both devices. She also instructs *TV-small* to receive control signals from the VCR for tape recording. When Bob adds *TV-large* later, Alice deactivates *TV-small* and imprints *TV-large* with *RT-Alice*. Now *RT-Alice* controls both *TV-large* and *VCR-cool*.

She then imprints *TV-small* with Bob's remote control *RT-Bob*. Through *RT-Alice*, she also instructs *VCR-cool* to accept control signals from *RT-Bob*. In the end Alice has *TV-large* all to herself; Bob has full control of *TV-small*; Alice controls the fate of *VCR-cool* and shares with Bob general access rights to it.

32.5 Routing Security

Every node is a router in an AHN. In a wired network, routers are a part of the network infrastructure that is oblivious from regular nodes. In an AHN, routing is a shared responsibility among all the nodes

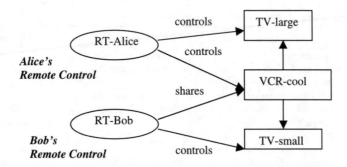

FIGURE 32.5 Security association among home appliances.

that are a part of the network [15,16]. Routers play two roles: that of a relay, which forwards packets, and that of a pathfinder, which discovers routes in collaboration with other nodes. In its capacity as a pathfinder, a router shares its knowledge of network topology, seeks information from other nodes, and calculates routes between end nodes in the network.

32.5.1 Threats to Routing Security

In a friendly environment, we expect a node to relay packets passing through it, share information truthfully, and generate packets only when needed. However, not every node is cooperative in a network [3,11]. Nodes could be noncooperative or even malicious. A noncooperative node could simply and quietly drop packets that pass through it and might not respond to solicitation from other nodes. A malicious node might be chatty to take up limited bandwidth. Or it might spread rumors about the network topology, that is, it could either fabricate routing information or distort routing information that passes through it.

As a concrete example of attacks, consider the topology shown in Fig. 32.6, where there is a route S-A-B-C-D [7]. As an example of passive-aggressive behavior, node B, when compromised, could silently drop routing requests from S, thereby rendering the route unavailable. As an example of malicious node behavior, node E as an adversary broadcasts a distorted message stating that it has a shorter route to D. A routing protocol that selects paths based on distance would select S-A-E-D instead. By doing so, E successfully directs communication from S to D to itself, and it can drop packets silently.

Cryptography is a powerful defense against many types of attacks. Message authentication codes (MACs) based on cryptography could identify and authenticate nodes that participate in the routing, thereby detecting the fabricated and distorted information and preventing nodes from impersonation [7,14,22]. Encryption could protect routing messages from disclosure. Auditing combined with authentication could detect noncooperative behaviors from nodes, such as dropping packets [3,11]. Table 32.1 lists threats against routing and security controls to counter these attacks. Attacks to routing of an AHN could come from inside or outside, if we have a notion of a network boundary. To defend against outsiders, we could use distributed firewall and intrusion detection tools.

Every node that comes into the transmission range of an AHN physically becomes a part of the AHN. Logically, however, the AHN may be only partially open or even closed to visitors. Furthermore, nodes that are a part of a network can have different classification levels; accordingly, an AHN could have routes with different levels of sensitivity and security. A multilevel communication model could address this issue [22].

32.5.2 End-to-End Routing Authentication

Papadimitratos and Haas developed a routing protocol that provides end-to-end authentication based on shared secrets [14]. It assumes a security association (SA) between a source S and a destination D. An SA between two nodes establishes security parameters that they could use to achieve end-to-end

TABLE 32.1   Threats, Attacks, Defense and Reaction in AHN Routing

Threats	Attacks	Prevention	Detection/Reaction
Interception	Sniff traffic Sniff traffic pattern Probe network topology	Use cryptography for traffic confidentiality	
Interruption	Jam communication channel Do not respond to routing requests Drop packets Overflow traffic	Spread spectrum and frequency hopping	Audit nodes and revoke membership of offending nodes
Modification	Change routing data	Use MAC	
Fabrication	Send wrong routing data as another node Send wrong routing data as itself Replay old routing data from the network	Use MAC Timestamp	Use MAC for nonrepudiation

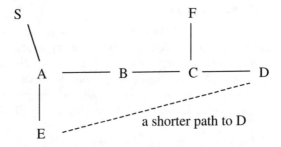

FIGURE 32.6   Fabrication results in denial-of-service attack.

security [13]. In this protocol, routing reply (RREP) packets are MAC protected. Only those RREP packets from trusted nodes are accepted by source S. Message origin authentication of RREP is achieved through a shared secret between S and D, which is a part of their SA. Alternatively, if a node T, which S trusts, has a valid path to D, it can generate an RREP with a MAC using a shared secret between S and T (see Fig.32.7).

All nodes in the network participate in the route discovery and can be a part of the final route. However, there is no accountability of intermediate nodes. The next protocol addresses this issue.

32.5.3 Link-Based End-to-End Route Authentication

Dahill and associates proposed a routing protocol that provides both end-to-end and link-by-link authentication [7]. A routing request (RREQ) packet is a message signed by the source S. Each intermediate node verifies the integrity of the received RREQ, signs it, and passes it along. Routing reply (RREP) is a message signed by D. Each intermediate node processes RREP the same way it processes RREQ (see Fig. 32.8). Only RREP originated from D is accepted. Every node in the network contributes to routing security, as in neighborhood watch. Public-key infrastructure is needed for the deployment of this protocol.

32.5.4 Security Metrics for Routing Path

Existing routing protocols use distance [15,16] as a metric in selecting optimal routing. Sueng proposed using a security metric that is based on the classification level of nodes on the path from a source to a destination. Routing discovery packets are encrypted using a key of desired sensitivity level [22]. Only

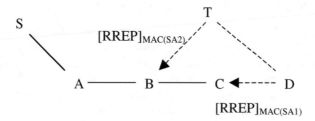

FIGURE 32.7 End-to-end security. SA$_1$ is a security association between S and D. SA$_2$ is a security association between S and T.

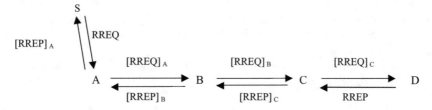

FIGURE 32.8 Link-based end-to-end authentication. RREQ and RREP are messages signed by their originators.

those nodes that have access to classified routing information participate in the route discovery. Alternatively, all the nodes on a path could attach their highest clearance level to RREP. The source can then select a path with a clearance level acceptable for the data to be transmitted.

32.5.5 Abnormal Flow Detection

There have been several attempts to curb passive-aggressive behavior of nodes [3,11]. One approach models socialism, while another models capitalism. The main idea behind the socialist approach is to have every node be vigilant. Nodes watch their neighbors' behavior. The group has an accepted norm. Any deviation from norm would trigger an alarm. When the warning signals exceed a predefined capacity, the ill-behaved nodes are marked by their neighbors as outlaws to be avoided. The sentence could be decided by a single judge or a jury, depending on the severity of the suspect's vicious behavior [23].

A capitalist approach uses a quota-reward system to induce good behavior of citizens [6]. Every node is initially assigned a certain amount of tokens. Tokens are currency. Routing is a commodity to be traded. Nodes provide services to other nodes to accumulate wealth. They can buy routing services later. Chatty nodes deplete their currency and slip into poverty. Cooperative nodes sleep with money under their pillows.

32.6 Key Distribution

Distribution of keys is at the center of protocols that employ cryptography. Secret keys are shared by multiple entities. Public keys are a public knowledge. There are two ways to distribute secret keys: through a preestablished secure channel or an open channel [12].

Public keys are distributed through certificates. A certificate binds a public key with an entity. Certificates are certified, stored, and distributed by one or more trusted parties. In a centralized approach there is only one trusted third party, which is called *Certificate Authority* (CA). There are two approaches to decentralized public-key distribution [10]:

1. Through a decentralized key distribution center
2. Through individual nodes that comprise the network

In this section, we describe two examples of decentralized public-key distribution and one example of a secret-key establishment.

32.6.1 Decentralized Key-Distribution Center

Zhou and Haas proposed a decentralized Key Distribution Center (KDC) that splits responsibilities of key certification and distribution among a group of servers [24]. Any subset of the group with a size greater than a threshold can issue a certificate. No other subset can issue certificates. The decentralized KDC is based on homomorphic secret-sharing schemes, which can be achieved through proactive threshold cryptography [12]. This scheme provides survivability to the service. The service tolerates failure and compromise of some servers as long as there are still no less than t nodes functioning. The scheme allows for changes in configuration. Consequently, we can add nodes and remove failed or compromised nodes without interrupting the service. The scheme also allows for refreshing of pieces of the secret for each node; this increases the difficulty of compromising the service.

A (n, t) threshold scheme shares a secret s among n entities by dividing it into n shares, with each entity holding one share. Any t $(< n)$ entities can pool their shares to reproduce s. Any set of fewer entities cannot. We can refresh shares of each member and add or delete members if we use a special kind of threshold scheme [12,24].

Let us illustrate a (4, 3) secret-sharing scheme using a plane in three-dimensional space [12]. We use $<a, b, c>$ to represent a plane, where

$$ax + by + cz = 1$$

The secret is the plane $<a, b, c>$. For four nodes sharing $<a, b, c>$, we select any four points $p1$, $p2$, $p3$, and $p4$ on the plane and securely distribute a different point to each different node.

Any three of the group can poll their shares (points) together to find the value of $<a, b, c>$, as three points uniquely define a plane. Two members polling their shares together will not reveal the secret because a plane is undefined with only two points [12].

In Zhou and Haas's scheme for public key distribution, a KDC has a public key, K_{CA}, and a private key, K_{CA}^{-1}. Each service provider (or server) has a share of K_{CA}^{-1}. Let us assume a $<4,3>$ threshold scheme (Fig. 32.9). Four nodes collectively act as certificate authority to certify and distribute public keys. Every server knows the public keys of all the servers and the service. Each server maintains a repository of public keys of all the nodes in the network.

Alice retrieves Bob's public key by contacting all servers (see Fig. 32.10). With its share of K_{CA}^{-1}, a server CA_i generates a partial signature s_i to bind Bob's name, his public key, and its validity period, $B_{Bob} = <Bob, K_{Bob}, t_{valid}>$. A combiner receives partial signatures from servers and generates Bob's certificate, the binding B_{Bob} signed with CA's private key K_{CA}^{-1}. The combiner is a trusted entity that stores neither keys nor certificates. Collectively, servers can refresh the shares of K_{CA}^{-1} and change configuration through secure channels among them (e.g., using public key cryptography).

32.6.2 Democratic Key Distribution

In a democratic society, every citizen participates in the political process. Hubaux, Buttyan, and Capkun proposed a self-organized public-key infrastructure for AHN, in which every node participate in the key

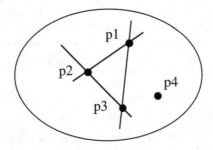

FIGURE 32.9 A $<4,3>$ threshold scheme: any three points define a plane.

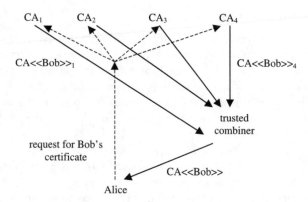

FIGURE 32.10 Alice retrieves Bob's certificate.

distribution process [8]. Certificates are issued by individual nodes in the network. Nodes are assumed to be honest, that is, they do not issue fake certificates. Each node maintains its own repository of certificates, which are issued by itself and other nodes, hence avoiding a single points of failure.

For Alice to have a secure communication with Bob, she determines Bob's public key through a certificate chain that runs from her to Bob by combining their private repositories. Let us use Bob<<David>> to represent a certificate for David issued by Bob. Assume in Alice's repository there are three certificates:

$$\text{Alice}<<G>>, \; G<<F>>, \; F<<E>>$$

In Bob's repository there are three certificates:

$$E<<D>>, \; D<<C>>, \; C<<Bob>>$$

There is a certificate chain from Alice to Bob as follows:

$$\text{Alice} \rightarrow G \rightarrow F \rightarrow E \rightarrow D \rightarrow C \rightarrow \text{Bob}$$

where Alice issued a certificate for G, G issued a certificate for F, … …, and C issued a certificate for Bob (see Fig. 32.11).

With Alice<<G>>, a certificate issued by her, Alice verifies the certificate and learns K_G, G's public key. With K_G, she verifies G<<F>> and learns K_F, F's public key. Eventually she learns K_B, Bob's public key, through C<<Bob>> and K_C, C's public key.

One difficulty in democratic key distribution is the complexity of trust. In centralized key distribution, we have a certain level of trust on certificates as we place our trust in the KDC. In a DKD, the trust we place on a certificate is a function of the trusts we place in each individual nodes along the chain that we use.

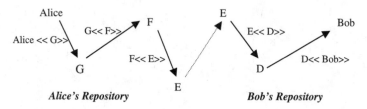

FIGURE 32.11 Finding a certificate chain from Alice to Bob by combing their repositories.

32.6.3 Conference Key Establishment

In some situations, there is neither a central certificate authority that everyone trusts, nor is there a certificate chain running from one node to another. A group of conference participants gathering in a meeting room is one such example [1,15]. A shared secret key among them is needed to protect their wireless communication. Asokan and Ginzboorg proposed a password-based authenticated-key exchange protocol that establishes a strong shared secret for conference participants, hence achieving strong secrecy for their communication for that particular session [1]. We now illustrate the protocol for four participants: Alice, Bob, Catherine, and David.

They first agree on three values: a password P, a prime q, and a number g (which is a generator of the multiplicative group Z_q^*), and a public function H. Password P is their shared secret, while q and g can be public information. Each of them then selects two random secrets: S_a and R_a for Alice, S_b and R_b for Bob, S_c and R_c for Catherine, and S_d and R_d for David. They then communicate over a public channel as follows (see Fig. 32.12). Let G_{ABC} denote Alice, Bob, and Catherine.

1. *(1.1):* Alice $\rightarrow$ Bob: g^{Sa}
   (1.2): Bob $\rightarrow$ Catherine: g^{SaSb}
2. Catherine $\rightarrow$ {Alice, Bob, David}: $\pi = g^{SaSbSc}$
3. Each member of the group G_{ABC} carries out this step: calculates c_i and then sends it securely to David. For instance, Alice calculates c_A and sends it to David.
   (3.A): Alice $\rightarrow$ David: $E_P[c_A = \pi^{(Ra/Sa)}]$
4. David sends a different message to each member of G_{ABC}. Again, we use Alice as an example.
   (4.A): David $\rightarrow$ Alice: $c_A{}^{Sd}$
   Everyone then calculates $K = g^{SaSbScSd}$, which is their shared secret.
5. One of G_{ABC}, say, Bob, carries out the last step.
   Bob $\rightarrow$ {Alice, Catherine, David}: *Bob, E_K[Bob, H(Alice, Bob, Catherine, David)]*

This way they can verify that they arrived at the same secret.

32.7 Future Directions

Though ad hoc mobile network security only recently started to gain attention, experiences from securing other types of systems shed light on the issues presented here. Notably among them are network security and secure group communication in mobile computation.

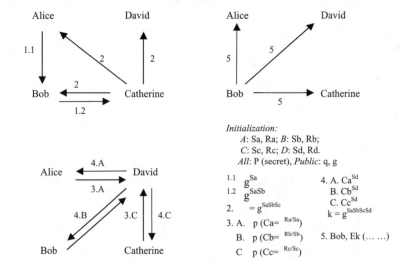

FIGURE 32.12 Conference key establishment with four participants.

In network security, cryptographic protocols protect private communication over a public network [10]. Standard cryptographic techniques are encryption, digital signature, message authentication code, and distribution of secret keys and public-key certificates [12]. Firewalls set up boundaries between external and internal networks. Intrusion detection techniques monitor internal networks for suspicious activities. Distributed firewalls maintain the increasingly blurred network boundaries as employees take work home and corporate visitors carry their laptops with them [2,9].

In an ad hoc mobile network, the physical boundary between internal and external network is nonexistent. This feature is desirable for some applications; for other applications, a strict logical boundary is required and should not be crossed. Still other applications prefer to have some control over their boundaries while still permitting visitors from outside and travelers aboard [7].

The major concern in open networks is the availability of network services where nodes move around. Traditionally centralized services such as naming and key distribution are adapted to mobile networks through decentralization, such as the emulated KDC proposed by Zhou and Haas [24] and the distributed service proposed by Hubaux, Buttyan, and Capkun [8]. Traditionally distributed services such as routing and packet forwarding are now a collective effort of the whole community [16]. Nodes that need reliable services act with extra vigilance to monitor their fellow citizens. These nodes either avoid needing their troublesome neighbors or stop trading with them [11,23]. Current work in securing routing is in its infancy. Solutions addressing subsets of threats are emerging. More elaborated solutions that address specific applications will surge as the needs of applications become known.

In the real world, we have public groups formed by concerned citizens acting as watchdogs to monitor some well-known service providers (such as government agencies). We have neighborhood watch groups to monitor suspicious activities in a neighborhood. We also have groups who monitor their own behavior. Group-specific characteristics are critical in deciding the level of vigilance needed and the actions performed, as well as what is considered abnormal behavior.

For applications with a strict logical boundary requirement, key management is a major concern. Key management issues in AHN security are similar to those in secure group communication. Solutions in secure-group key management and secure multicast can be borrowed and adapted to AHNs. One major research area is the interaction of mobility and secure multicasting.

Open network communication concerns itself with a physical group while a logic group layer is added for closed and managed networks. A managed network is then a multilevel group in which trust building plays an eminent role. Clearly specified security policies are essential for both managed and closed networks. In AHNs, policies are embedded in and enforced by individual nodes. They are much more dynamic than in wired networks, and trust is much more fluid. When we move beyond applications born out of a research lab to real world, a user-friendly, precise, and concise language is a major challenge to describing trust and policies and a management framework for their evolution.

References

[1] N. Asokan and P. Ginzboorg, Key-Agreement in Ad-hoc Networks, *Proceedings of the Fourth Nordic Workshop on Secure IT Systems (Nordsec '99)*, 1999.

[2] S.M. Bellovin, Distributed Firewalls, ;login:, Nov. 1999, pp. 39–47.

[3] S. Bhargava and D.P. Agrawal, Security Enhancements in AODV Protocol for Wireless Ad Hoc Networks, *Vehicular Technology Conference, 2001*, 2001, vol. 4, pp. 2143–2147.

[4] L. Blazevic, L. Buttyan, S. Capkun, S. Giordano, J.-P. Hubaux, and J.-Y. Le Boudec, Self-Organization in Mobile Ad-Hoc Networks: the Approach of Terminodes, *IEEE Communications Magazine*, June 2001.

[5] E. Brickell, J. Feigenbaum, and D. Maher, *DIMACS Workshop on Trust Management in Networks*, South Plainfield, NJ, Sep. 1996.

[6] L. Buttyan and J.-P. Hubaux, Enforcing Service Availability in Mobile Ad-Hoc WANs, *Proceedings of the First IEEE/ACM Workshop on Mobile Ad Hoc Networking and Computing (MobiHoc)*, Boston, MA, Aug. 2000.

[7] B. Dahill, B.N. Levine, E. Royer, and C. Shields, A Secure Routing Protocol for Ad Hoc Networks, Technical Report UM-CS-2001–037, University of Massachusetts, Amherst, Aug. 2001.

[8] J.-P. Hubaux, L. Buttyan, and S. Capkun, The Quest for Security in Mobile Ad Hoc Networks, *Proceedings of the ACM Symposium on Mobile Ad Hoc Networking & Computing (MobiHoc 2001)*, Long Beach, CA, Oct. 2001.

[9] S. Ioannidis, A.D. Keromytis, S.M. Bellovin, and J.M. Smith, Implementing a Distributed Firewall, *Proceedings of Computer and Communications Security (CCS) 2000*, Nov. 2000.

[10] C. Kaufman, R. Perlman, and M. Speciner, *Network Security: Private Communication in a Public World*, Prentice Hall, Englewood Cliffs, NJ, 1995.

[11] S. Marti, T.J. Giuli, K. Lai, and M. Baker, Mitigating Routing Misbehavior in Mobile Ad Hoc Networks, *Proceedings of the Sixth Annual ACM/IEEE International Conference on Mobile Computing and Networking*, Boston, MA, 2000, pp. 255–265.

[12] A.J. Menzes, P.C. van Oorschot, and S.A. Vanstone, *Handbook of Applied Cryptography*, CRC Press, Boca Raton, FL, 1997.

[13] R. Oppliger, Internet and Intranet Security, Artech House Publishers, Norwood, MA, 1998.

[14] P. Papadimitratos and Z.J. Haas, Secure Routing for Mobile Ad Hoc Networks, *SCS Communication Networks and Distributed Systems Modeling and Simulation Conference (CNDS 2002)*, San Antonio, TX, Jan. 27–31, 2002.

[15] C. Perkins, Ed., *Ad Hoc Networking*, Addison-Wesley Publishers, Reading, MA, 2000.

[16] E.M. Royer and C.-K. Toh, A Review of Current Routing Protocols for Ad-Hoc Mobile Wireless Networks, *IEEE Personal Communications Magazine*, Apr. 1999, pp. 46–55.

[17] B. Schnerer, *Secrets and Lies: Digital Security in a Networked World*, John Wiley & Sons, Inc, New York, 2000.

[18] F. Stajano and R. Anderson, The Resurrecting Duckling: Security Issues for Ad-Hoc Wireless Networks, *Seventh International Workshop on Security Protocols*, 1999.

[19] W. Stallings, *Cryptography and Network Security*, 2nd Ed., Prentice Hall, Englewood Cliffs, NJ, 1999.

[20] R.C. Summers, *Secure Computing: Threats and Safeguards*, McGraw-Hill, New York, 1996.

[21] F. Wang, B. Vetter, and S. Wu, Secure Routing Protocols: Theory and Practice, North Carolina State University, Raleigh, May 1997.

[22] S. Yi, P. Naldurg, and R. Kravets, Security-Aware Ad Hoc Routing for Wireless Networks, Technical Report UIUCDCS-R-2001–2241, Aug. 2001.

[23] Y. Zhang and W. Lee, Intrusion Detection in Wireless Ad-Hoc Networks, *Proceedings of the Sixth Annual International Conference on Mobile Computing and Networking (MobiCom '2000)*, Boston, MA, Aug. 6–11, 2000.

[24] L. Zhou and Z.J. Haas, Securing Ad Hoc Networks, *IEEE Network Magazine*, Nov./Dec. 1999.

Index